A Vector Space Approach to Models and Optimization

Wiley Series on Systems Engineering and Analysis
HAROLD CHESTNUT, Editor

Chestnut

Systems Engineering Tools

Wilson and Wilson

Information, Computers, and System Design

Hahn and Shapiro

Statistical Models in Engineering

Chestnut

Systems Engineering Methods

Rudwick

Systems Analysis for Effective Planning: Principles and Cases

Wilson and Wilson

From Idea to Working Model

Rodgers

Introduction to System Safety Engineering

Reitman

Computer Simulation Applications

Miles

Systems Concepts

Chase

Management of Systems Engineering

Weinberg

An Introduction to General Systems Thinking

Dorny

A Vector Space Approach to Models and Optimization

A Vector Space Approach to Models and Optimization

C. Nelson Dorny
Moore School of Electrical Engineering
University of Pennsylvania

A WILEY-INTERSCIENCE PUBLICATION
JOHN WILEY & SONS, New York • London • Sydney • Toronto

Library of Congress Cataloging in Publication Data:

Dorny, C. Nelson, 1937–
A vector space approach to models and optimization.

(Wiley series on systems engineering and analysis)
"A Wiley-Interscience publication."
Includes bibliographies and index.
1. Mathematical optimization. 2. Mathematical models. 3. Vector spaces. I. Title.
QA402.5.D67 511′.8 75-20395
ISBN 0–471–21920–7

Printed in the United States of America

10 9 8 7 6 5 4 3 2 1

To

Patricia
Scott
Brett
Jonathan
Jennifer
Christopher

Every important idea is simple.

War and Peace
Count Leo Tolstoy

SYSTEMS ENGINEERING AND ANALYSIS SERIES

In a society which is producing more people, more materials, more things, and more information than ever before, systems engineering is indispensable in meeting the challenge of complexity. This series of books is an attempt to bring together in a complementary as well as unified fashion the many specialties of the subject, such as modeling and simulation, computing, control, probability and statistics, optimization, reliability, and economics, and to emphasize the interrelationship between them.

The aim is to make the series as comprehensive as possible without dwelling on the myriad details of each specialty and at the same time to provide a broad basic framework on which to build these details. The design of these books will be fundamental in nature to meet the needs of students and engineers and to insure they remain of lasting interest and importance.

Preface

Models and optimization are fundamental to the design and operation of complex systems. This book is intended as an intuitive, probing, unified treatment of the mathematics of model analysis and optimization. It explores in a unifying framework the structure of deterministic linear system models and the optimization of both linear and nonlinear models. The unification is accomplished by means of the vector space language and a relatively small number of vector space concepts. The mathematical concepts and techniques, although not new, become more accessible when treated in an intuitive, unified manner.

This book is broader in coverage than most books on the subject, and is relatively low in its level of mathematical sophistication. I have de-emphasized mathematical proofs; I have attempted instead to develop concepts by means of geometrical intuition and analogies to ideas familiar to engineering graduates. All concepts are illustrated with specific detailed examples. In addition, I have tried to relate the mathematical concepts to the real world by presenting practical applications and by discussing practical computer implementations of techniques for model analysis and optimization. Thus the development is less sterile than the treatments found in mathematics books.

I have attempted to build up the mathematical machinery in a way that demonstrates what can and cannot be accomplished with each tool. This methodical buildup helps to develop a fundamental feel for the mathematical concepts. For example, I withhold the definition of the inner product until late in the development in order that it be clear that perpendicular coordinate systems are not fundamental to the modeling process.

The background required of the reader is a familiarity with elementary matrix manipulations and elementary differential equation concepts. The selection of topics and the order of presentation reflect seven years of experience in presenting the material to full-time graduate students and to practicing engineers at the Moore School of Electrical Engineering, University of Pennsylvania. The book is designed for use as a text in a two-semester course sequence for first-year graduate students in engineering, operations research, and other disciplines which deal with systems. As a consequence of the extensive cross-referencing and the numerous detailed examples, the book is also suitable for self-study. At the end of each chapter references that are good general references for much of the material in that chapter are indicated with asterisks (*). Answers to selected problems are included.

The symbols P & C appear frequently throughout the text in reference to the Problems and Comments sections at the end of each chapter. These problems and comments form an important part of the book. Those problems that are in the form of statements are intended to be proved or verified by simple examples. The problems marked with asterisks (*) present concepts which are used later in the book, or which are significant extensions of the text material; these problems should at least be read and understood.

The reader will find that abstract symbols can be understood more easily if they are thought of in terms of simple examples. If possible, concepts should be illustrated geometrically with two- or three-dimensional arrow vectors.

In a two-semester course sequence, it would be appropriate to treat Sections 1.1–5.3 in the first semester (a vector space approach to models) and Sections 5.4–8.5 in the second (a vector space approach to optimization). There is not sufficient time in two semesters to include all the applications if all the mathematical concepts are covered. (I have usually omitted some of Section 4.4 and some of the applications.) By deleting Chapter 3 and by de-emphasizing differential systems and nondiagonalizable matrices in the remaining chapters, the two semesters can be reduced to two quarters. The accompanying diagram shows how the chapters depend on each other.

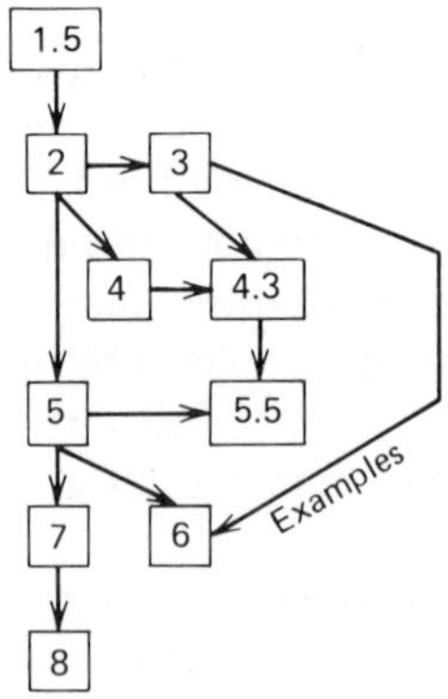

Because the concepts treated in this book find application in many fields, it is difficult to avoid conflicts between different standards in notation. I have tried to be as consistent as possible with previous standards. Because instructors cannot use bold print at the blackboard, I have avoided the use of boldfaced type as a *primary* means of distinguishing vectors and transformations. However, I do use boldfaced type redundantly to *emphasize* the interpretation of an object as a vector or transformation of vectors.

I wish to express my appreciation to H. R. Howland, W. A. Gruver, and C. N. Campopiano, who read the full manuscript and suggested helpful improvements. Thanks are also due to Renate Schulz for her help in proofreading and drawing, to Pam Dorny and Nancy Maguire who did most of the typing, and to the Moore School of Electrical Engineering of the University of Pennsylvania which provided support for much of the effort. Most of all, I wish to express my gratitude to my wife and children who waited patiently for the long nights, weekends, and summers to end.

C. Nelson Dorny

May 1975
Philadelphia, Pennsylvania

Contents

Symbols

Scalars

$a, b, c, d, s, t,$ $\alpha, \beta, \gamma, \epsilon, \sigma, \tau$	general scalars	36
$\bar{a}$	complex conjugate of a	239
i, j, k, r	integer subscripts	37
l, m, n, p, q	positive integers	184
ξ_i, η_i	elements of vectors	37
λ_i	eigenvalues	151
λ_i, μ_i, ν_i	Lagrange multipliers	413

Vectors

$\mathbf{x}, \mathbf{y}, \mathbf{z}, \mathbf{w}$	general vectors	36
$\mathbf{f}, \mathbf{g}, \mathbf{h}, \mathbf{u}, \mathbf{v}, \mathbf{w}, \boldsymbol{\phi}$	general functions	40
$\boldsymbol{\theta}$	zero vector or zero function	36
$\mathbf{1}$	unit function	455
$\boldsymbol{\varepsilon}_i$	standard basis vectors for $\mathcal{R}^n$ or $\mathcal{M}^{n\times 1}$	48
$\boldsymbol{\lambda}, \boldsymbol{\mu}, \boldsymbol{\nu}$	Lagrange multiplier vectors	415

Vector Spaces

$\mathcal{V}, \mathcal{W}, \mathcal{U}$	general spaces or subspaces	36
$\mathcal{H}$	Hilbert space	276
$\mathcal{R}^n$	real n-tuple space	39
$\mathcal{P}^n$	polynomial functions of order less than n	40
$\mathcal{C}(a,b)$	functions continuous on $[a,b]$	42
$\mathcal{C}^n(a,b)$	functions with continuous nth derivatives	130, 95
ℓ_2	square-summable infinite sequences	38, 276
$\mathcal{L}_2(a,b)$	functions square-integrable on $[a,b]$	42
$\mathcal{M}^{n\times 1}$	$n\times 1$ matrices	39

Ordered Bases

$\mathscr{X} = \{\mathbf{x}_i\}$, $\mathscr{Y} = \{\mathbf{y}_i\}$, $\mathscr{Z} = \{\mathbf{z}_i\}$	general bases	48
$\mathscr{F} = \{\mathbf{f}_i\}$, $\mathscr{G} = \{\mathbf{g}_i\}$, $\mathscr{H} = \{\mathbf{h}_i\}$	function space bases	50
$\mathscr{E} = \{\boldsymbol{\varepsilon}_i\}$	standard basis for $\mathscr{R}^n$ or $\mathscr{M}^{n\times 1}$	48
$\mathscr{N}$	natural basis	49

Transformations

S, T, U, F, G, H	general transformations	56
F, G	usually nonlinear transformations	400
F	usually a functional	403
B	bounded linear functional	286
$\boldsymbol{\Theta}$	zero transformation	59
I	identity transformation	58
D	differentiation operator	64
L	general differential operator	103, 64
$\boldsymbol{\nabla}^2$	Laplacian operator	85
$\mathscr{L}$	Lagrangian functional	420
E	expected value operator	189
$\boldsymbol{\Phi}_i$	penalty function	538
$\boldsymbol{\psi}_c$	penalized objective function	538
$\boldsymbol{\beta}_i$	boundary condition	104
$\mathfrak{L}$	Laplace transform	65

Matrices

A, B, C	general matrices	559, 62
Q, R	positive definite matrices	319
S	change-of-coordinates matrix	77
$\mathbf{E}_{ij}$	constituent matrices	212
$\boldsymbol{\Theta}$	zero matrix	559
I	identity matrix	561
$\boldsymbol{\Lambda}$	diagonal or Jordan form	156, 188

Miscellaneous Symbols

$\mathscr{S}$	general set	56
$\mathscr{S}^{\perp}$	orthogonal complement of $\mathscr{S}$	250
$\mathscr{N}_g$, $\mathscr{R}_g$	generalized nullspace and range	182
$\stackrel{\Delta}{=}$	"defined as"	18, 37
$\Rightarrow$	"implies"	47

$[a,b]$	real numbers between and including the end points a and b	42
(a,b)	real numbers between but excluding the end points a and b	115
$\langle\cdot,\cdot\rangle$	inner product	239
$\cdot\rangle\langle\cdot$	outer product	326
$\|\cdot\|$	norm	240, 464, 325
$\times$	cartesian product	39
$\{\mathbf{x}_i\}$	a set of vectors denoted $\mathbf{x}_1$, $\mathbf{x}_2$, and so on	47
$\mathbf{A}^{\mathrm{T}}$	transpose of matrix **A**	560
$\mathbf{A}^{-1}$, $\mathbf{T}^{-1}$	inverse of **A** or **T**	58, 564, 22
$\overline{\mathbf{A}}$	complex conjugate of **A**	245
$\mathbf{A}^{\dagger}$, $\mathbf{T}^{\dagger}$	pseudoinverse of **A** or **T**	368
$\mathbf{T}^*$	adjoint of **T**	288
$(\mathbf{A} \vdots \mathbf{B})$	**A** augmented with **B**	20
$\det(\mathbf{A})$	determinant of **A**	561
$\dim(\mathcal{V})$	dimension of $\mathcal{V}$	53
$\oplus$	direct sum	145
$\overset{\perp}{\oplus}$	orthogonal direct sum	294
$\nabla\mathbf{F}$	gradient of **F**	404
$\mathbf{dG}(\mathbf{x},\mathbf{h})$	Fréchet differential of **G** at **x**	400
$\mathbf{G}'(\mathbf{x})$	Fréchet derivative of **G** at **x**	400
$\mathbf{F}''$	Second Fréchet derivative of **F**	465
$\dfrac{\partial \mathbf{G}}{\partial \mathbf{x}}$	Jacobian matrix of **G**	405
$[\mathbf{x}]_{\mathcal{X}}$	coordinate matrix of **x** relative to $\mathcal{X}$	49
$[\mathbf{T}]_{\mathcal{X}\mathcal{Y}}$	matrix of **T** relative to $\mathcal{X}$ and $\mathcal{Y}$	72
$\mathbf{Q}_{\mathcal{X}}$	matrix of an inner product relative to $\mathcal{X}$	245
$k(t,s)$	Green's function	110
$\rho_j(t)$	boundary kernel	110
δ	delta function	568, 575

A Vector Space Approach to Models and Optimization

1

Introduction

1.1 System Models

This book analyzes mathematical models for systems and explores techniques for optimizing systems described by these models. We use the term system in its broad sense; by a **system** we mean a collection of things which are related in such a way that it makes sense to think of them as a whole. Examples of systems are an electric motor, an automobile, a transportation system, and a city. Each of these systems is part of a larger system. Small systems are usually well understood; large, complex systems are not.

Rational decision making concerning the design and operation of a system is always based upon a model of that system. A **model** of a system is a simpler system that behaves sufficiently like the system of interest to be of use in predicting the behavior of the system. The choice of appropriate model depends upon the complexity of the system, the available resources, and the questions that need to be answered by the model. Many decisions are based upon nothing more than the conceptual model which the decision maker develops by observing the operation of other systems. In this book we concern ourselves with a more quantitative class of models, mathematical models.

Most systems can be thought of (or modeled) as an operation on the system inputs (or independent variables) which produces the system outputs (or dependent variables); we state this input–output relationship symbolically by means of the following mathematical equation:

$$\mathbf{T}\mathbf{x} = \mathbf{y} \tag{1.1}$$

In this equation $\mathbf{x}$ represents the set of inputs to the system and $\mathbf{y}$ the set of outputs of the system.* The symbol $\mathbf{T}$ represents the operation which the system performs on the inputs; thus $\mathbf{T}$ is a mathematical model of the system.

*See Section 2.3 for a more complete discussion of inputs and outputs.

In order for a model of a system to be conceptually simple, it must be abstract. The more details we include explicitly in the model, the more complicated it becomes. The more details we make implicit, the more abstract it becomes. Thus if we seek conceptual simplicity, we cannot avoid abstraction. The model **T** of (1.1) epitomizes this simplicity and abstraction.

The generality of the model given in (1.1) allows it to be applied to many different systems. In the simplest of situations **T** might represent a simple economic transaction: let p be the unit price of a particular commodity; then (1.1) means $\mathbf{y}=p\mathbf{x}$, where $\mathbf{x}$ is the quantity purchased and $\mathbf{y}$ is the total cost of the purchase. At the other extreme, **T** might represent a large city. Figure 1.1 shows the system output $\mathbf{y}$ that might result from a given input $\mathbf{x}$; obviously, many pertinent variables are not explicit in Figure 1.1.

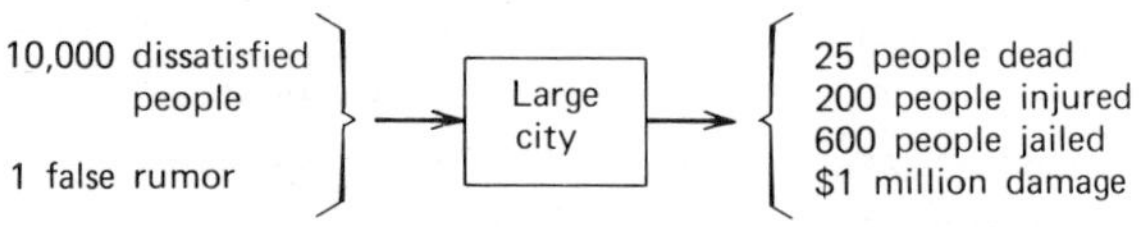

Figure 1.1. A conceptual model of a large city.

Equation (1.1) is the focus of this book. The first five chapters are devoted to a detailed analysis of (1.1) for models **T** which are linear.* By decomposing linear models into smaller, simpler pieces we develop an intuitive feel for their properties and determine the practical computational difficulties which can arise in using linear models. Chapter 6 treats the least-square optimization of systems that can be represented by linear models. The analysis and optimization of systems that are described by nonlinear models are considered in Chapters 7–8.

We emphasize linear models because most known analytical results pertain only to linear models. Furthermore, most of the successful techniques for analyzing and optimizing nonlinear systems consist in the repetitive application of linear techniques (Chapter 7–8). We dwell extensively on the two most frequently used linear models—linear algebraic equations and linear differential equations. These models are the most frequently used because they are well understood and relatively easy to deal with. In addition, they are satisfactory models for a large number of practical systems.

Throughout the text we explore the computational implications of the analytical techniques which we develop, but we do not develop computer

*See Section 2.4 for the definition of a linear system.

algorithms. We do not discuss stochastic systems; we treat systems with stochastic inputs only by means of examples.

System Questions

Questions concerning a system usually fall into one of the following categories:

1. *System operation*: in terms of (1.1), given the model $\mathbf{T}$ and the input $\mathbf{x}$, find the output $\mathbf{y}$.
2. *System inversion*: given the model $\mathbf{T}$ and output $\mathbf{y}$, find the input $\mathbf{x}$.
3. *System synthesis* or *identification*: given several different choices of input $\mathbf{x}$ and the corresponding output $\mathbf{y}$ for each input, determine a suitable system model $\mathbf{T}$. (If the system is to be identified, the inputs and outputs are measurements from a real system. If the system is to be synthesized, $\mathbf{T}$ would be chosen to provide some desired input–output relationship.)
4. *System optimization*: pick the input $\mathbf{x}$, the output $\mathbf{y}$, or the system $\mathbf{T}$ so that some criterion is optimized.

Note that we have expressed these questions in terms of the system model rather than in terms of the system itself. Although experimentation with actual systems may be appropriate in certain circumstances, these questions are usually explored by means of a model. We discuss the modeling process briefly in Section 1.4. We also examine in Chapter 6 some techniques for making an optimum choice of model parameters once a model structure has been established. However, we do not dwell extensively on techniques for obtaining good models. Rather, we work with the models themselves, assuming that they are good models for the systems they represent. Questions 1 and 2 are treated in Chapters 1, 2, 4, and 5 for linear algebraic equation models and in Chapters 2–5 for linear differential equation models. Question 4 is treated in Chapters 6–8. We do not consider question 3.*

The concepts explored in this book apply directly to any field which uses equations to represent systems or portions of systems. Although we focus on linear algebraic equations and linear differential equations, we also demonstrate the applicability of the concepts to partial differential equations and difference equations; we include equations which are probabilistic, "time-varying," and nonlinear. Our examples pertain to models and optimization in such fields as automatic control, electric power, circuits, statistical communications, coding, heat flow, economics, operations research, etc.

*See Sage [1.10] for a discussion of identification.

1.2 Approach

All students of science and engineering have noticed occasional similarities between the physical laws of different fields. For instance, gravitational attraction, electrostatic attraction, and magnetic attraction all obey an inverse-square law. Electrical resistance to the flow of current has its analogue in the resistance of materials to the conduction of heat. Not only does the physical world tend to repeat itself; it also tends toward simplicity and economy. Most natural phenomena can be explained by simple differential relationships: the net force on a rigid object is proportional to its acceleration; the rate of flow of heat is proportional to the gradient of the temperature distribution.

If we put a number of simple relationships together to describe the motion of a nonrigid object (fuel in a rocket) or the heat flow in an irregular nonhomogeneous object (a nuclear reactor), then nature appears complicated. The human mind is not good at thinking of several things at once. The development of large-scale digital computers has provided the capability for solving complex sets of equations; it has made system study a reality. However, the engineer, the designer of a system, still must conceive of the variables and interactions in the system to such an extent that he can describe for a computer what it is he wants to know. He needs simple conceptual models for systems.

We can simplify models for complex systems by stretching our imagination in a search for analogies. For instance, the multiplication of an electrical current by a resistance to determine a voltage has an analogue in the differentiation of a current and then multiplication by an inductance; both actions are operations on a current to yield a voltage. This analogy suggests that we think of differentiation as analogous to multiplication by a number. By reducing the number of "different concepts" necessary to understand the parts of a system, such analogies help the system designer to achieve greater economy of thought; he can conceive of the system in simpler terms, hopefully gaining insight in the process. William K. Linvill [1.7] has coined the term "portable concept" to describe a concept that is transferable from one setting to another. This book is concerned with *portable mathematical concepts*. The purpose of exploring such concepts is to enhance the ability of the reader to model systems, understand them, synthesize them, and optimize them. Our basic premise is that this ability is enhanced by an intuitive understanding of the models and optimization techniques that have proved useful in many settings in the past. By an intuitive understanding, we mean the type of "intuitive feel" that an engineer obtains by applying and reapplying a concept to many different situations.

It would seem, then, that we must fully absorb most of mathematics. However, much of the mathematical literature is directed toward the modeling and optimization of pathological cases, those cases for which "standard" models or techniques are insufficient. Because techniques for handling these cases are new, it is appropriate that they be the focus of the current literature. Yet this emphasis on exceptional cases can distort our perspective. In maximizing a function, we should not become so concerned about nondifferentiability of functions that we forget to try setting the derivative equal to zero. Rather than try to explore *all* cases, we focus on well-behaved systems. By making analogies, we organize the most common models and optimization techniques into a framework which contains only a relatively few fundamental concepts. The exceptional cases can be more clearly understood in comparison to this basic framework.

The importance of learning the *structure* of a subject is stressed by Bruner [1.1]: "Grasping the structure of a subject is understanding it in a way that permits many other things to be related to it meaningfully... the transfer of principles is dependent upon mastery of the structure of the subject matter.... Perhaps the most basic thing that can be said about human memory, after a century of intensive research, is that unless detail is placed into a structured pattern, it is rapidly forgotten." In order to simplify and unify the concepts used in model analysis and optimization, we organize fundamental mathematical principles into a mnemonic structure—a structure which draws extensively on geometrical analogies as an aid to the memory. We also develop a mathematical language suitable for communicating these structural concepts.

The first half of this book is concerned with models and their analysis. Mathematically speaking, this is the subject of algebra—the use of symbols to express quantitative concepts and their relations. In the latter half of the book we turn to geometry—the measurement and comparison of quantitative concepts—in order to further analyze models and to optimize their parameters and inputs. Because the bulk of known analytical results are concerned with linear models, these models necessarily dominate our discussions. Our emphasis is on geometrical insight rather than mathematical theorems. We reach deep into the mathematical literature for concepts. We try to be rigorously correct. Yet we develop concepts by means of analogies and simple examples rather than proofs, in order to nurture the intuition of the reader. We concern ourselves with the practical aspects of computation. To engineers the material seems like mathematics; to mathematicians it seems like engineering.

1.3 Portable Concepts

To illustrate the portability of the mathematical model (1.1) we compare the two most common mathematical models: (*a*) a set of linear algebraic equations; and (*b*) a linear differential equation. The following algebraic equations might represent the relationship between the voltages and the currents in a resistive circuit:

$$\begin{aligned} 2\xi_1+3\xi_2&=\eta_1\\ \xi_1+\xi_2&=\eta_2 \end{aligned} \tag{1.2}$$

Such a set of equations is often expressed in the matrix form:

$$\begin{pmatrix} 2 & 3\\ 1 & 1 \end{pmatrix}\begin{pmatrix} \xi_1\\ \xi_2 \end{pmatrix}=\begin{pmatrix} \eta_1\\ \eta_2 \end{pmatrix} \tag{1.3}$$

In the form (1.3), we can interpret the set of equations as an operation (matrix multiplication) on the pair of variables ξ_1 and ξ_2 to obtain the pair of quantities η_1 and η_2. The relationship (1.2) between the pairs of variables can also be expressed in terms of the "inverse equations":

$$\begin{aligned} \xi_1&=-\eta_1+3\eta_2\\ \xi_2&=\eta_1-2\eta_2 \end{aligned} \tag{1.4}$$

Equations (1.4) can be verified by substitution into (1.2). The coefficients in (1.4) indicate what must be done to the "right-hand side" variables in order to determine the solution to (1.2). Equations (1.4) can be expressed in the "inverse matrix" form:

$$\begin{pmatrix} \xi_1\\ \xi_2 \end{pmatrix}=\begin{pmatrix} -1 & 3\\ 1 & -2 \end{pmatrix}\begin{pmatrix} \eta_1\\ \eta_2 \end{pmatrix} \tag{1.5}$$

In Section 1.5 we explore in detail the process of solving or inverting equations such as (1.2). In Chapter 2 we begin the discussion of algebraic equation models in a manner which is consistent with the notation of (1.1). Chapters 4 and 5 are, to a great extent, devoted to analyzing these models.

The angular velocity $\omega(t)$ of a particular loaded dc motor, initially at rest, can be expressed in terms of its armature voltage $u(t)$ as

$$\frac{d\omega(t)}{dt}+\omega(t)=u(t), \qquad \omega(0)=0 \tag{1.6}$$

We can think of the differential equation and boundary condition as an abstract operation on ω to obtain u. Equation (1.6) also can be expressed in the inverse form:

$$\omega(t)=\int_0^t e^{-(t-s)}u(s)\,ds \tag{1.7}$$

That the integral equation (1.7) is, in fact, the solution to (1.6) is easily verified for a *particular* armature voltage, say, $u(t)=e^{2t}$, by evaluating $\omega(t)$, then substituting it into (1.6). We can think of (1.7) as an abstract "integral" operation on u to determine ω; this is the "inverse" of the "differential" operation in (1.6). These two abstract operations and techniques for determining the inverse operation are the subject of Chapter 3. The analysis of these abstract operations carries into Chapters 4 and 5.

The algebraic equations (1.2) and the differential equation with its boundary condition (1.6) have much in common. We must not let details cloud the issue; in each case, an "input" is affecting an "output" according to certain (linear) principles. We can think of the pair of variables ξ_1 and ξ_2 and the function ω as each constituting a single "vector" variable. The analogy between these entities is carried further in the comparison of Figure 1.2, wherein the pair of variables ξ_1, ξ_2 is treated as a "discrete" function. This analogy is discussed further in Section 2.1. It seems evident that concepts are more clearly portable if they are abstracted—stripped of their details.

A Portable Optimization Concept

We again employ the analogy between a "discrete vector" variable and a "continuous vector" variable to discuss the portability of an optimization

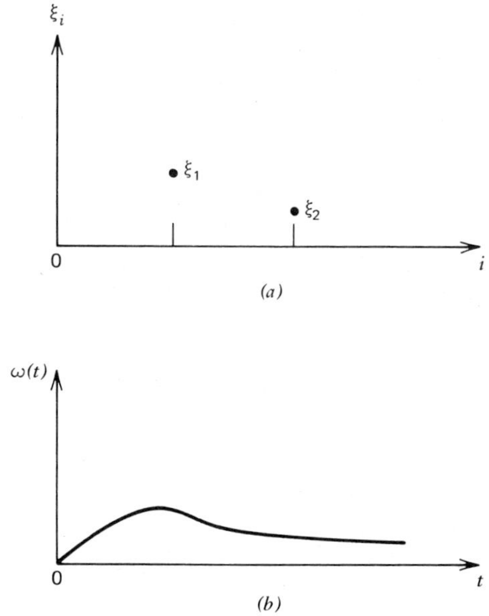

Figure 1.2. Vector variables plotted as functions: (*a*) discrete variables of (1.2); (*b*) continuous variable of (1.6).

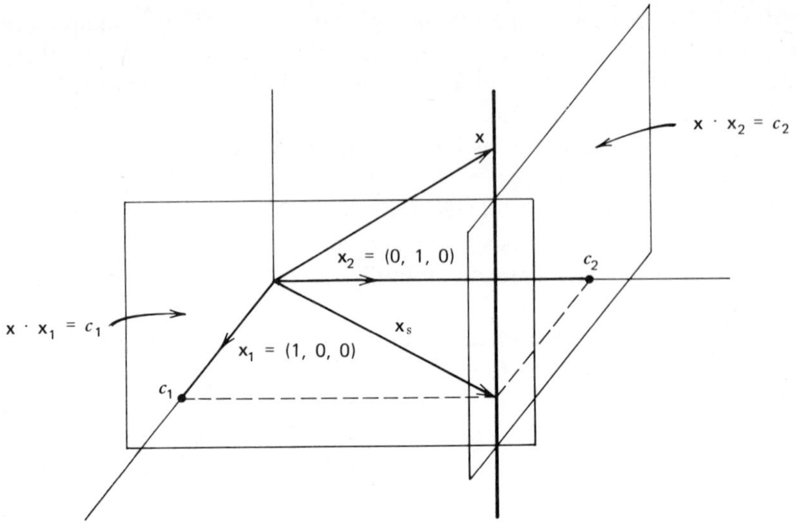

Figure 1.3. A vector of minimum length.

concept. Figure 1.3 shows the locus of all vectors $\mathbf{x}$ in a three-dimensional space which lie in the intersection of two planes. We seek that vector $\mathbf{x}$ which is of minimum length. The solution vector $\mathbf{x}_s$ is perpendicular to the line which constitutes the locus of the candidate vectors $\mathbf{x}$.

Using the standard notation of analytic geometry, we think of the vector $\mathbf{x}$ as $\mathbf{x} = (\xi_1, \xi_2, \xi_3)$. The plane that is perpendicular to the vector $\mathbf{x}_1$ can be expressed mathematically in terms of the dot product of vectors as $\mathbf{x} \cdot \mathbf{x}_1 = \xi_1 = c_1$. Similarly, the second plane consists in vectors $\mathbf{x}$ which satisfy $\mathbf{x} \cdot \mathbf{x}_2 = c_2$. Since $\mathbf{x}_s$ must be perpendicular to the intersection of the planes, it must be some combination of the vectors $\mathbf{x}_1$ and $\mathbf{x}_2$ that determine the planes; that is, $\mathbf{x}_s = d_1\mathbf{x}_1 + d_2\mathbf{x}_2$ for some constants d_1 and d_2. Substituting $\mathbf{x}_s$ into the equations that determine the planes, we obtain a pair of algebraic equations in d_1 and d_2:

$$\begin{aligned} \mathbf{x}_s \cdot \mathbf{x}_1 &= d_1\mathbf{x}_1 \cdot \mathbf{x}_1 + d_2\mathbf{x}_2 \cdot \mathbf{x}_1 = c_1 \\ \mathbf{x}_s \cdot \mathbf{x}_2 &= d_1\mathbf{x}_1 \cdot \mathbf{x}_2 + d_2\mathbf{x}_2 \cdot \mathbf{x}_2 = c_2 \end{aligned} \tag{1.8}$$

Since the vectors $\mathbf{x}_1$ and $\mathbf{x}_2$ are perpendicular and of unit length, then

$$\mathbf{x}_1 \cdot \mathbf{x}_1 = \mathbf{x}_2 \cdot \mathbf{x}_2 = 1, \qquad \mathbf{x}_1 \cdot \mathbf{x}_2 = \mathbf{x}_2 \cdot \mathbf{x}_1 = 0, \qquad d_i = c_i,$$

and

$$\mathbf{x}_s = c_1\mathbf{x}_1 + c_2\mathbf{x}_2 = (c_1, c_2, 0)$$

The geometric minimization problem described above is simple. By using geometric notions, we have found the vector **x** which satisfies two linear equations and for which the quantity $\xi_1^2+\xi_2^2+\xi_3^2$ (the length of **x** squared) is minimum. The same geometric principles can be used to solve other, more complicated, problems wherein linear equations must be satisfied and a quadratic quantity minimized. For instance, the angular position $\phi(t)$ of the shaft of the dc motor of (1.6) satisfies

$$\frac{d^2\phi(t)}{dt^2}+\frac{d\phi(t)}{dt}=u(t) \tag{1.9}$$

Suppose we seek that armature voltage function $u(t)$ that will drive the motor shaft from one position to another in a fixed time, while consuming a minimum amount of energy; that is, let $\phi(0)=\dot{\phi}(0)=0$, $\phi(1)=1$, $\dot{\phi}(1)=0$, and pick u to minimize $\int_0^1 u^2(t)\,dt$. In our search for a technique for solving this problem we should not cloud the issue by thinking about techniques for solving differential equations. Equation (1.9) is linear; the quantity to be minimized is quadratic. Chapter 6 is devoted to solving such problems by using analogues of the planes and perpendicular vectors of Figure 1.3.

1.4 System Modeling

The rationale for modeling a system is a desire to determine how to design and/or operate a system without experimenting with actual systems. If a system is large, experimenting is usually very time consuming, extremely expensive, and often socially unacceptable. A designer uses models to predict the performance characteristics of a system or to aid in modifying the design of the system so that it meets a desired set of specifications. He will probably be interested in the degree of stability of the system, its accuracy, and its speed of response to commands. The designer also uses models to predict the nature of the interaction of the system with other systems. For example, he may wish to predict the effect of the system or of a particular system operating policy on the environment or on a related energy distribution system. Or he may wish to predict the performance of the system in the presence of extraneous inputs (noise) or sudden changes in load. The reliability of the system and the sensitivity of the system performance to changes in the environment are also important.

Types of Models

A single system has many models. One or more models of the system pertain to its electrical behavior, others to its thermal behavior, still others to its mechanical behavior. An investigation of the social or economic characteristics of the system requires additional models.

Physical models are appropriate in many situations. One example of such a model is a scale model of a building or bridge. The conceptual representation of a rocket by a solid cylinder is another example. In most system studies, a *mathematical model* for the system (or part of the system) facilitates analysis. An appropriate mathematical model usually can be derived more easily from a simplified physical model than from the original system. The resulting mathematical model usually consists of a set of algebraic and/or differential equations. Often these equations can be solved (for given system inputs) on a digital, analogue, or hybrid computer.* In some instances, the distributed nature of the system requires a mathematical model consisting of partial differential equations, and computer solutions are difficult to obtain even if the equations are linear.

The behavior of some systems fluctuates randomly with time. For such systems (or portions of systems) it is common to build a discrete-event *simulation model*.† Rather than predicting the precise behavior of the system, such a model simulates the behavior numerically in a manner that is statistically correct. For instance, we might be interested in the flow of customers through a set of checkout counters. A simple physical model of such a customer service system consists of a single checkout counter, where customers arrive, wait for service, are served, then leave; arrival times and service times are random with known statistics. By means of a digital computer, we would generate a random sequence of arrivals (with correct statistical properties). We would also determine a service time for each customer by an appropriate random number generation process. Then we would observe the simulated flow of customers over time. The simulation would predict not only the average flow through the system, but also the frequency of occurrence of various queue lengths and waiting times. Thus the dynamic performance of certain types of systems can be predicted by digital simulation.

As a practical matter, a model should contain no more detail than is necessary to accomplish the purposes of the model. One is seldom *sure* of the accuracy of a model. Yet if a model is accurate enough to improve one's decision-making capability, it serves a useful purpose. Generally speaking, the more complex the model is, the more expensive will be the process of developing and using the model. In the extreme, the most accurate model is a copy of the system itself.

*Special computer programs have been developed to facilitate the solving of certain classes of equations. For example, the program CSMP is designed to solve linear differential equations associated with problems of automatic control.

†Specialized computer languages have been developed to facilitate digital simulation. See Meier, Newell, and Pazer [1.8] and Forrester [1.3].

Unfortunately, it is probable that some complex systems will never be represented in sufficient detail by manageable mathematical models. Yet a *conceptual model* can be applied in situations where it is difficult to obtain meaningful quantitative models; for example, the principle of negative feedback (with its beneficial effects on stability and sensitivity) often is applied successfully without the use of a mathematical model. The system concepts that are associated with mathematical models serve as a guide to the exploration of complex systems. By the use of specific models for small subsystems, by computer analysis of the combined subsystem models, and by the application of model concepts (such as feedback) to the whole system, we can better understand large systems.

The Modeling Process

The process of modeling can be divided into two closely related steps: (1) establishing the model structure and (2) supplying the data. We focus primarily on the first step. However, we cannot ignore the second; it is seldom useful to establish a model structure for which we cannot obtain data.

We begin the modeling process by examining the system of interest. In many complex systems, even the boundaries of the system are not clear. The motivation for modeling such a system is usually a desire to solve a problem, to improve an unsatisfactory situation, or to satisfy a felt need. We must describe the system and the manner in which it performs in a simple fashion, omitting unnecessary detail. As we begin to understand better the relationship between the system and the problem which motivates study of the system, we will be able to establish suitable boundaries for the system.

Suppose a housing official of a large city is concerned because the number of vacant apartments in his city cycles badly, some times being so high as to seriously depress rental rates, other times being so low as to make it difficult for people to find or afford housing.* What is the reason for the cycling? To answer this question, we need to explore the "housing system." Should we include in "the system" the financial institutions which provide capital? The construction industry and labor unions which affect new construction? The welfare system which supports a significant fraction of low-income housing? Initially, we would be likely to concern ourselves only with the direct mechanisms by which vacant apartments are generated (new construction, people moving out, etc.) and eliminated (new renters).

Should the model account for different sizes of apartments? Different styles? Different locations? Seasonal variations in the number of vacan-

*The idea for this example was obtained from Truxal [1.11, Chapter 2].

cies? A model that accounts for all these factors would require detailed data (as a function of time) for each factor. These data are not likely to be readily available. Rather, obtaining the data would require the cooperation of many apartment managers and an extensive data-taking operation over at least a 1-year period. A more likely approach, at least initially, would be to develop a simple model which predicts the average number of vacancies (of any type) in the city in a 1-year period. Data concerning this quantity are probably available for at least a large fraction of the large apartment complexes in the city.

Once the approximate extent of the system and the approximate degree of detail of the model have been determined, the course of model development usually progresses through the following steps:

1. Development of a simple physical model.
2. Derivation of a mathematical model of the physical model.
3. Obtaining of data from which model parameters are determined.
4. Validation of the model.

In deriving a model for a system it usually helps to visualize the behavior of the unfamiliar system in terms of the behavior of familiar systems which are similar. It is for this reason that we start with a simple physical model. The physical model of the system is likely to be conceptual rather than actual. It is a simple abstraction which retains only the essential characteristics of the original system. In the case of the apartment vacancy model introduced above, a simple physical model might consist of a set of identical empty boxes (vacant apartments). At 1-year intervals some number of boxes is added by construction or renters moving out; another number of boxes is removed by new renters. See Figure 1.4.

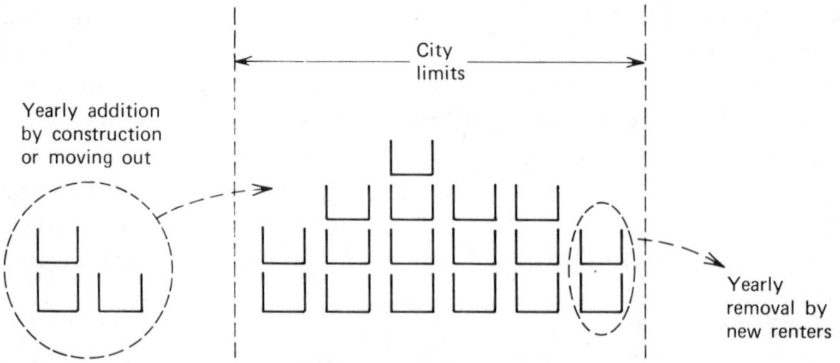

Figure 1.4. Simple physical model of apartment vacancies.

A mathematical model of a system is usually easier to derive from a simple physical model than from the system itself. In most instances the mathematical model consists of algebraic and/or differential equations. The mathematical model must be kept simple in order that it be solvable analytically or by means of practical computer techniques. Generally, the model simplifications that reduce data requirements also reduce the complexity of the mathematical model. For example, in the housing system described above, the aggregation of the various types of apartments into a single type greatly reduces the number of variables in the mathematical model. Other simplifying approximations which may be appropriate in some situations are (1) ignoring interaction between the system and its environment; (2) neglecting uncertainty and noise; (3) lumping distributed characteristics; and (4) assuming linearity and time invariance. Sage [1.10] describes some techniques that are useful in identifying the structure and parameter values of those systems that act in a linear fashion.

Mathematical Model of Apartment Vacancies

In order to demonstrate the logical thought process entailed in the derivation of a mathematical model, we derive a mathematical model of the physical model of apartment vacancies illustrated in Figure 1.4.

We expect that the number of "apartment construction starts" in a given year is approximately equal to the apparent need for new apartments. We formalize this statement by postulating the following relationship:

$$\begin{aligned} S(n) &= \alpha(V_d - V(n)), \qquad V(n) \leqslant V_d \\ &= 0, \qquad\qquad\qquad\quad\; V(n) \geqslant V_d \end{aligned} \tag{1.10}$$

where $S(n)$ = number of apartment construction starts in year n;

$V(n)$ = average number of vacant apartments during the 1-year period centered at the beginning of year n.

Underlying (1.10) is the assumption that the people who build apartments feel that the city should have approximately V_d vacancies. The proportionality factor α and the number of vacancies V_d should be selected in such a manner that (1.10) most nearly describes recent historical data for the city.

Of course, actual apartment completions lag behind the starts by an appreciable time. We formalize this statement by the equation

$$C(n) = S(n-l) \tag{1.11}$$

where $C(n)$ is the number of completions in year n, and l is the average construction time. A suitable value for the lag l should be determined from historical data.

Let $R(n)$ denote the number of new apartments rented during year n. We can include in $R(n)$ the families who move out of apartments during the year [$R(n)$ can be negative]. From Figure 1.4, it is apparent that

$$\Delta V(n) = C(n) - R(n) \tag{1.12}$$

where $\Delta V(n) = V(n+1) - V(n)$, the increase in vacant apartments during the 1-year period.

The empirical relations (1.10)–(1.11) and the logical statement (1.12) can be related pictorially by means of a **block diagram**. A block diagram is a conceptual tool which is useful for clarifying the structure of a model or for portraying sequences of events. It dramatizes cause and effect relationships. A block diagram of the mathematical model (1.10)–(1.12) is shown in Figure 1.5. Each block in the diagram displays one of the relationships in the mathematical model.*

Figure 1.5 establishes the model structure. In order to determine the values of the model parameters and to validate the model, we need historical data for each variable in the model. The data that we need in order to pick appropriate values for the parameters α, V_d, and l are historical values of yearly starts $S(n)$, yearly completions $C(n)$, and yearly average vacancies $V(n)$. We would probably pick the values of α, V_d, and l by the least-square data-fitting process known as *linear regression* (see Section 6.1).

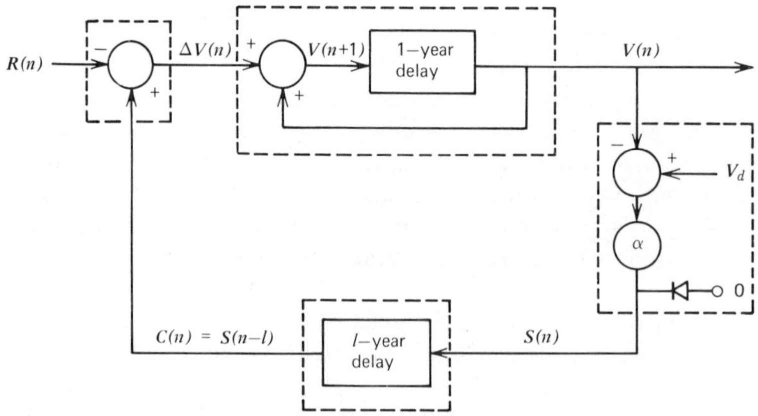

Figure 1.5. Block diagram model of apartment vacancies.

*See Cannon [1.2] for a detailed discussion of block diagrams and their use.

After parameter values have been determined, we need to verify that the mathematical model is a sufficiently good representation of the actual apartment vacancy system. In order to validate the model, we need historical values of the model input $R(n)$ and output $V(n)$. Since we required data for $V(n)$ previously, the only additional data needed are a corresponding set of yearly rentals $R(n)$ (new rentals minus renters moving out). We use the input data $R(n)$ for a sequence of years together with the mathematical model to obtain a predicted sequence of values of $V(n)$. The model is validated if the predicted values of $V(n)$ agree sufficiently with the corresponding historical values of $V(n)$. If the model were verified to be accurate to a certain precision for historical data, we would feel confident that it would exhibit approximately the same accuracy in predicting future apartment vacancies. A housing official would probably be satisfied if the predicted vacancies were within 10% of the actual average vacancies. Of course, predictions of future values of $V(n)$ have to be based on assumed future values of $R(n)$. If future values of $R(n)$ cannot be predicted with reasonable confidence, then another model must be developed to relate the demand for apartments $R(n)$ to those variables which affect demand.

If the data do not validate the model to a sufficient degree, then the model structure must be modified; additional factors must be accounted for. Specifically, the number of apartment construction starts $S(n)$ is likely to depend not only on the demand for housing $R(n)$, but also on the number of uncompleted housing starts (starts from the previous $l-1$ years). The number of starts $S(n)$ is also likely to depend on the availability of capital at a favorable interest rate. Thus an improved apartment vacancy model would probably have more than one input variable.

Once a validated model has been obtained, it can be used to aid city officials in determining an appropriate housing policy. City officials can affect the number of apartment vacancies by modifying the variables which are inputs to the model. Demand for apartments $R(n)$ can be affected by adjusting tax rates, rent subsidies, urban renewal plans, etc. If the final model includes interest rate as an input, this interest rate can be affected by means of interest rate subsidies.

Suppose that low interest capital has been plentiful, and there has been an overabundance of housing. Specifically, suppose recent historical data indicate that the best values for the parameters of the model in Figure 1.5 are $V_d = 1000$ apartments, $\alpha = 0.5$, and $l = 2$ years, and that reasonable initial conditions are $V(0) = 1500$ vacancies, and $S(-2) = S(-1) = 0$ apartments. Suppose that as a result of a new rent subsidy program we expect the future demand to be $R(n) = 500$ apartments, $n = 0, 1, 2, \ldots$. According to the mathematical model of (1.10)–(1.12) and Figure 1.5, the new rent subsidy program will cause the apartment vacancies in the city to exhibit the behavior shown in Table 1.1 and Figure 1.6.

Table 1.1 Apartment Vacancies Predicted by Figure 1.5

n	$V(n)$	$S(n)$	$C(n)$	$R(n)$	$\Delta V(n)$	$V(n+1)$
0	1500	0	0	500	−500	1000
1	1000	0	0	500	−500	500
2	500	250	0	500	−500	0
3	0	500	0	500	−500	−500
4	−500	750	250	500	−250	−750
5	−750	875	500	500	0	−750
6	−750	875	750	500	250	−500
7	−500	750	875	500	375	−125
8	−125	563	875	500	375	250
9	250	375	750	500	250	500
10	500	250	563	500	63	563
11	563	219	375	500	−125	438
12	438	281	250	500	−250	188

According to Figure 1.6, the model predicts that severe housing shortages will result from the new housing policy. If the model is correct, and if social pressures make the rent subsidy program mandatory, then the city officials must compensate for the policy by encouraging builders to expand the available housing. (Perhaps this expansion could be encouraged by publicizing the predicted housing shortage, or by having the city assume some of the risk of investment in new construction.)

If the model has not been carefully validated, however, the predictions that result from the model should be used with caution. The fact that builders themselves might predict future housing shortages is ignored in

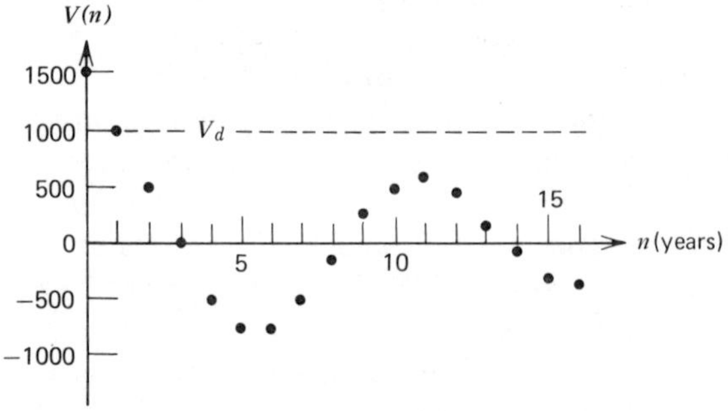

Figure 1.6. Apartment vacancies predicted by Figure 1.5.

(1.10). Thus this simple model of the relationship between vacancies and construction starts should probably be modified to more accurately describe the process by which builders decide to start new construction. Then the determination of model parameter values and the validation of the model should be repeated for the new model before it is used to predict the effect of housing policies.

The modeling process we have described has been used extensively to describe such situations as the flow of electric power in large transmission line networks and the growth of competing species in ecosystems. It is apparent that the same modeling process can be used to describe the relationships among the variables in many other types of systems. For example, it is suitable for describing the response of an eye pupil to variations in light intensity, the response of a banking system to market fluctuations, or the response of the people of a given country to variations in the world price of oil. It is in the social, economic, and biological fields that system modeling is likely to have its greatest impact in the future.

1.5 Solution of Linear Algebraic Equations

To this point our discussion has been of an introductory nature. The development of vector space concepts and the vector space language begins in Chapter 2. We now explore briefly, in a matrix format, the process of solving sets of linear algebraic equations, in order that we be able to use such sets of equations in the examples of Chapter 2 and later chapters. In this discussion we emphasize practical techniques for computing solutions to sets of linear algebraic equations and for computing the inverses of square matrices.

Models of most systems eventually lead to the formation and solution of sets of linear algebraic equations. For example, it is common practice to replace the derivatives in a differential equation by finite differences, thereby producing a set of linear algebraic equations which can be solved by a digital computer. The solution of nonlinear equations almost always requires linearization and, again, involves solution of linear algebraic equations (Chapter 8). Thus simultaneous algebraic equations are fundamental to practical analysis.

There is a wide variety of methods for solving a set of linear algebraic equations.* The design of *practical* computer algorithms which will obtain accurate solutions in an efficient manner calls upon most of the concepts of this book: spectral analysis, least-square optimization, orthogonalization, iteration, etc. Frequently, the sets of equations that arise in practice

*See Forsythe [1.6].

are nearly degenerate; that is, they border on being unsolvable by computers which have finite accuracy. Furthermore, the number of equations can be large; finite-difference approximations for partial differential equations sometimes involve more than 100,000 equations (P&C 2.17). Thus the solution of linear algebraic equations constitutes one of the easiest, and yet one of the most difficult problems.

Any set of linear algebraic equations can be written in the form

$$\begin{aligned} a_{11}\xi_1 + a_{12}\xi_2 + \cdots + a_{1n}\xi_n &= \eta_1 \\ \vdots \qquad\qquad\qquad & \quad \vdots \\ a_{m1}\xi_1 + a_{m2}\xi_2 + \cdots + a_{mn}\xi_n &= \eta_m \end{aligned} \tag{1.13}$$

Equation (1.13) easily fits the symbolic structure of the basic system model (1.1). Suppose we define $\mathbf{x} \triangleq \{\xi_1, \xi_2, \ldots, \xi_n\}$ and $\mathbf{y} \triangleq \{\eta_1, \eta_2, \ldots, \eta_m\}$ as the unknown outputs and known inputs, respectively, of the model, **T**. Our immediate goal is to clarify the manner in which **T**, by way of the coefficients a_{ij}, relates **x** to **y**. Associated with (1.13) are three basic questions:

1. Do the equations possess a solution **x** for each given **y**; that is, are the equations consistent?
2. Is the solution unique; that is, are there enough independent equations to determine **x**?
3. What is the solution (or solutions)?

It is appropriate to ask the same questions concerning (1.1). Although the third question may appear to be the most pertinent for a specific problem, the answers to the other two give valuable insight into the structure of the model and its applicability to the situation it is supposed to represent. Such insight is generally the real reason for solving the equations, and certainly the prime purpose of our present analysis.

We rephrase the problem in matrix notation in order to separate the information about the system $\{a_{ij}\}$ from the information about the "state" or "condition" of the system (the variables $\{\xi_i\}$, $\{\eta_j\}$).

$$\begin{pmatrix} a_{11} & \cdots & a_{1n} \\ \vdots & & \vdots \\ a_{m1} & \cdots & a_{mn} \end{pmatrix} \begin{pmatrix} \xi_1 \\ \vdots \\ \xi_n \end{pmatrix} = \begin{pmatrix} \eta_1 \\ \vdots \\ \eta_m \end{pmatrix} \tag{1.14}$$

Matrix multiplication is defined in such a way that (1.13) and (1.14) are equivalent.* The notation of (1.14) is close to the abstract symbolism of (1.1). In order to be more direct concerning the meaning of **T**, we redefine **x** and **y** as the column matrices:

$$\mathbf{x} \triangleq \begin{pmatrix} \xi_1 \\ \vdots \\ \xi_n \end{pmatrix} \qquad \mathbf{y} \triangleq \begin{pmatrix} \eta_1 \\ \vdots \\ \eta_m \end{pmatrix}$$

Then (1.14) states

$$\mathbf{A}\mathbf{x} = \mathbf{y} \tag{1.15}$$

where **A** is the $m \times n$ matrix of equation coefficients. The system **T** can be defined explicitly by $\mathbf{T}\mathbf{x} \triangleq \mathbf{A}\mathbf{x}$; that is, the abstract operation of the system model **T** on the "vector" **x** is multiplication of **x** by the matrix **A**.

Typical of the classical methods of solution of (1.15) is Cramer's formula (Appendix 1):

$$\xi_i = \frac{\det(\mathbf{A}(i))}{\det(\mathbf{A})}$$

where $\mathbf{A}(i)$ is the matrix **A** with its ith column replaced by **y**. The formula applies only when **A** is square ($m = n$) and $\det(\mathbf{A}) \neq 0$. The method indicates that for square **A**, $\det(\mathbf{A}) \neq 0$ is a necessary and sufficient condition to guarantee a unique solution **x** to (1.15).

The most efficient scheme for evaluating a determinant requires approximately $n^3/3$ multiplications (Appendix 1 and P&C 1.3). Thus solution for **x** using Cramer's formula requires $(n+1)n^3/3$ multiplications. Compared with other techniques, Cramer's formula is not a practical tool for analyzing linear equations.

Row Reduction

Ordinary elimination of variables forms the basis for an efficient method of solution to (1.15). In point of fact, it is the basis for most computer algorithms for solving sets of linear algebraic equations. In essence, the method consists in successively adding some multiple of one equation to another until only one variable remains in each equation; then we obtain

*See Appendix 1 for a brief introduction to matrices and determinants.

the unknowns by inspection. For example:

$$\begin{aligned}\xi_1+2\xi_2&=2\\3\xi_1+4\xi_2&=6\end{aligned}\;\rightarrow\;\begin{aligned}\xi_1+2\xi_2&=2\\-2\xi_2&=0\end{aligned}\;\rightarrow$$

$$\begin{aligned}\xi_1+2\xi_2&=2\\\xi_2&=0\end{aligned}\;\rightarrow\;\begin{aligned}\xi_1\quad&=2\\\xi_2&=0\end{aligned}$$

The elimination method reduces to an automatable procedure (or algorithm) which requires no creative decision making by the user. Since the unknowns are unaffected by the procedure, they need not be written down; the above elimination process is expressed in matrix notation by

$$\begin{pmatrix}1&2&\vdots&2\\3&4&\vdots&6\end{pmatrix}\rightarrow\begin{pmatrix}1&2&\vdots&2\\0&-2&\vdots&0\end{pmatrix}\rightarrow$$
$$\begin{pmatrix}1&2&\vdots&2\\0&1&\vdots&0\end{pmatrix}\rightarrow\begin{pmatrix}1&0&\vdots&2\\0&1&\vdots&0\end{pmatrix}$$

The first matrix in this elimination process is $(\mathbf{A}\,\vdots\,\mathbf{y})$; we call it the **augmented matrix** (we augmented **A** with **y**). We refer to the matrix version of this elimination process as row reduction of the matrix $(\mathbf{A}\,\vdots\,\mathbf{y})$. Specifically, **row reduction of a matrix B** consists in systematically operating on the rows of **B** as if they were equations until (*a*) the first nonzero element in each row is 1; (*b*) each column which contains the leading 1 for some row has all its other entries 0; and (*c*) the leading 1's are in an order which descends from the left, with all zero rows at the bottom. We need the last requirement only to make the row-reduced matrix unique. We call the row-reduced matrix the **echelon form** (or Hermite normal form) of **B**.

There are two basic techniques for row reducing a matrix. In **Gauss–Jordan elimination** we complete the operations on each column, obtaining a single 1 with all other elements 0, before concerning ourselves with succeeding columns (Example 1). In **Gaussian elimination** we first eliminate all elements below the main diagonal, one column at a time, thereby making the matrix "upper triangular." We then eliminate elements above the diagonal by a process commonly called "back substitution." In Example 2 the first three steps demonstrate the triangularization, the last two the back substitution. Although the two methods are similar, Gaussian elimination is 33% more efficient than Gauss–Jordan elimination for large sets of equations (say, $n>5$); Gaussian elimination requires about $n^3/3$ multiplications to row reduce $(\mathbf{A}\,\vdots\,\mathbf{y})$ for an $n\times n$ matrix **A**. Gauss–Jordan elimination requires about $n^3/2$ multiplications. Both methods are far superior to Cramer's formula for solving linear algebraic equations (P&C 1.3).

Example 1. Gauss–Jordan Elimination

$$\begin{pmatrix} 1 & 2 & 2 & 1 \\ 2 & 3 & 5 & 1 \\ 3 & 2 & 5 & 1 \end{pmatrix} \rightarrow \begin{pmatrix} 1 & 2 & 2 & 1 \\ 0 & -1 & 1 & -1 \\ 0 & -4 & -1 & -2 \end{pmatrix}$$

$$\rightarrow \begin{pmatrix} 1 & 0 & 4 & -1 \\ 0 & 1 & -1 & 1 \\ 0 & 0 & -5 & 2 \end{pmatrix} \rightarrow \begin{pmatrix} 1 & 0 & 0 & \frac{3}{5} \\ 0 & 1 & 0 & \frac{3}{5} \\ 0 & 0 & 1 & -\frac{2}{5} \end{pmatrix}$$

Example 2. Gaussian Elimination

$$\begin{pmatrix} 1 & 2 & 2 & 1 \\ 2 & 3 & 5 & 1 \\ 3 & 2 & 5 & 1 \end{pmatrix} \rightarrow \begin{pmatrix} 1 & 2 & 2 & 1 \\ 0 & -1 & 1 & -1 \\ 0 & -4 & -1 & -2 \end{pmatrix} \rightarrow \begin{pmatrix} 1 & 2 & 2 & 1 \\ 0 & 1 & -1 & 1 \\ 0 & 0 & -5 & 2 \end{pmatrix}$$

$$\rightarrow \begin{pmatrix} 1 & 2 & 2 & 1 \\ 0 & 1 & -1 & 1 \\ 0 & 0 & 1 & -\frac{2}{5} \end{pmatrix} \rightarrow \begin{pmatrix} 1 & 2 & 0 & \frac{9}{5} \\ 0 & 1 & 0 & \frac{3}{5} \\ 0 & 0 & 1 & -\frac{2}{5} \end{pmatrix} \rightarrow \begin{pmatrix} 1 & 0 & 0 & \frac{3}{5} \\ 0 & 1 & 0 & \frac{3}{5} \\ 0 & 0 & 1 & -\frac{2}{5} \end{pmatrix}$$

In the row reduction of small matrices by hand, the number of multiplications is of less concern than is accuracy. To guard against errors during row reduction of a matrix **B**, we can add a "check" column whose ith element is the sum of the elements in the ith row of **B**. Throughout the row-reduction process the ith element in the check column should remain equal to the sum of all other elements in the ith row; wherever it is not equal to that sum, one of the elements in that row is in error. Because adding fractions by hand is complicated, we can avoid fractions by not forcing nonzero elements to be 1 until the last step in the row-reduction process.

Example 3. Row Reduction by Hand

$$(\mathbf{B} \vdots \text{check column}) \stackrel{\Delta}{=} \begin{pmatrix} 3 & 1 & 2 & \vdots & 6 \\ 4 & 2 & 1 & \vdots & 7 \end{pmatrix} \rightarrow \begin{pmatrix} 12 & 4 & 8 & \vdots & 24 \\ 12 & 6 & 3 & \vdots & 21 \end{pmatrix}$$

$$\rightarrow \begin{pmatrix} 12 & 4 & 8 & \vdots & 24 \\ 0 & 2 & -5 & \vdots & -3 \end{pmatrix} \rightarrow \begin{pmatrix} 6 & 2 & 4 & \vdots & 12 \\ 0 & 2 & -5 & \vdots & -3 \end{pmatrix} \rightarrow \begin{pmatrix} 6 & 0 & 9 & \vdots & 15 \\ 0 & 2 & -5 & \vdots & -3 \end{pmatrix}$$

$$\rightarrow \begin{pmatrix} 1 & 0 & \frac{3}{2} & \vdots & \frac{5}{2} \\ 0 & 1 & -\frac{5}{2} & \vdots & -\frac{3}{2} \end{pmatrix}$$

If we are interested in the solution to a set of equations $\mathbf{Ax}=\mathbf{y}$ as a function of $\mathbf{y}$, we can carry an unspecified $\mathbf{y}$ through the row-reduction process.

Example 4. Row Reduction with an Unspecified Column

$$(\mathbf{A}\,\vdots\,\mathbf{y}) \triangleq \begin{pmatrix} 1 & 2 & 2 & \vdots & \eta_1 \\ 2 & 3 & 5 & \vdots & \eta_2 \\ 3 & 2 & 5 & \vdots & \eta_3 \end{pmatrix} \rightarrow \begin{pmatrix} 1 & 2 & 2 & \vdots & \eta_1 \\ 0 & -1 & 1 & \vdots & \eta_2-2\eta_1 \\ 0 & -4 & -1 & \vdots & \eta_3-3\eta_1 \end{pmatrix}$$

$$\rightarrow \begin{pmatrix} 1 & 0 & 4 & \vdots & -3\eta_1+2\eta_2 \\ 0 & 1 & -1 & \vdots & 2\eta_1-\eta_2 \\ 0 & 0 & -5 & \vdots & 5\eta_1-4\eta_2+\eta_3 \end{pmatrix} \rightarrow \begin{pmatrix} 1 & 0 & 0 & \vdots & \eta_1-\frac{6}{5}\eta_2+\frac{4}{5}\eta_3 \\ 0 & 1 & 0 & \vdots & \eta_1-\frac{1}{5}\eta_2-\frac{1}{5}\eta_3 \\ 0 & 0 & 1 & \vdots & -\eta_1+\frac{4}{5}\eta_2-\frac{1}{5}\eta_3 \end{pmatrix}$$

The solution to the equations represented by the matrix $(\mathbf{A}\,\vdots\,\mathbf{y})$ of Example 4 can be expressed

$$\mathbf{x} = \begin{pmatrix} \eta_1-\frac{6}{5}\eta_2+\frac{4}{5}\eta_3 \\ \eta_1-\frac{1}{5}\eta_2-\frac{1}{5}\eta_3 \\ -\eta_1+\frac{4}{5}\eta_2-\frac{1}{5}\eta_3 \end{pmatrix} = \begin{pmatrix} 1 & -\frac{6}{5} & \frac{4}{5} \\ 1 & -\frac{1}{5} & -\frac{1}{5} \\ -1 & \frac{4}{5} & -\frac{1}{5} \end{pmatrix} \begin{pmatrix} \eta_1 \\ \eta_2 \\ \eta_3 \end{pmatrix}$$

Clearly, the final coefficients on the variables $\{\eta_i\}$ constitute the **inverse matrix** $\mathbf{A}^{-1}$. The coefficients which multiply these variables during the row reduction keep a record of the elimination operations on the rows of $\mathbf{A}$. The variables $\{\eta_i\}$ merely serve to keep the coefficients separated. The row reduction of Example 4 was, in effect, performed on $(\mathbf{A}\,\vdots\,\mathbf{I})$ to obtain $(\mathbf{I}\,\vdots\,\mathbf{A}^{-1})$, where $\mathbf{I}$ is the identity matrix; that is,*

$$\begin{pmatrix} 1 & 2 & 2 & \vdots & 1 & 0 & 0 \\ 2 & 3 & 5 & \vdots & 0 & 1 & 0 \\ 3 & 2 & 5 & \vdots & 0 & 0 & 1 \end{pmatrix} \rightarrow \begin{pmatrix} 1 & 0 & 0 & \vdots & 1 & -\frac{6}{5} & \frac{4}{5} \\ 0 & 1 & 0 & \vdots & 1 & -\frac{1}{5} & -\frac{1}{5} \\ 0 & 0 & 1 & \vdots & -1 & \frac{4}{5} & -\frac{1}{5} \end{pmatrix}$$

Row reduction is an efficient method for computing $\mathbf{A}^{-1}$. Yet in most instances, computation of $\mathbf{A}^{-1}$ is, in itself, inefficient. Computing $\mathbf{A}^{-1}$ by using Gaussian elimination on $(\mathbf{A}\,\vdots\,\mathbf{I})$ requires $\frac{4}{3}n^3$ multiplications for an $n\times n$ matrix $\mathbf{A}$ (P&C 1.3). Since this is four times the number of multiplications needed to find the solution $\mathbf{x}$ for a given $\mathbf{y}$, we find the inverse only when we actually need it—when we are interested in the properties of the system model (the set of equations) and of the matrix $\mathbf{A}$ which represents it.

*In Appendix 1, $\mathbf{A}^{-1}$ is defined as a matrix which satisfies $\mathbf{A}^{-1}\mathbf{A}=\mathbf{AA}^{-1}=\mathbf{I}$. In P&C 1.4 we find that if such a matrix exists, the row reduction of $(\mathbf{A}\,\vdots\,\mathbf{I})$ will produce it.

Many system models lead to matrices which are not square; there can be more equations than unknowns; there can be fewer. Even if the matrix is square, its inverse need not exist. Yet for any $m \times n$ matrix $\mathbf{A}$, row reduction of $(\mathbf{A} \vdots \mathbf{I})$ yields complete information about the equation $\mathbf{Ax} = \mathbf{y}$, including answers to the questions of existence and uniqueness of the solutions (P&C 1.1, 1.2).

Example 5. Solution by Row Reduction—a Nonsquare Matrix. Suppose we obtain the following equations from three independent measurements of some quantity

$$\xi_1 + \xi_2 = 1.2$$

$$\xi_1 + \xi_2 = 1.3$$

$$\xi_1 + \xi_2 = 1.2$$

Then

$$(\mathbf{A} \vdots \mathbf{I}) = \begin{pmatrix} 1 & 1 & \vdots & 1 & 0 & 0 \\ 1 & 1 & \vdots & 0 & 1 & 0 \\ 1 & 1 & \vdots & 0 & 0 & 1 \end{pmatrix}$$

which we row reduce to

$$\begin{pmatrix} 1 & 1 & \vdots & 1 & 0 & 0 \\ 0 & 0 & \vdots & -1 & 1 & 0 \\ 0 & 0 & \vdots & -1 & 0 & 1 \end{pmatrix}$$

We interpret the row reduced matrix to mean

$$\xi_1 + \xi_2 = \eta_1$$

$$0 = \eta_2 - \eta_1$$

$$0 = \eta_3 - \eta_1$$

Unless $\eta_1 = \eta_2 = \eta_3$, the equations allow no solution. In our example the equations are not consistent; $\eta_1 = \eta_3 = 1.2$, but $\eta_2 = 1.3$. If the equations were consistent, the row-reduced equations indicate that the solution would not be unique; for example, if η_2 were 1.2, the solution would be

$$\xi_1 + \xi_2 = \eta_1$$

Row and Column Interpretations

We have, to this point, viewed the matrix multiplication in (1.14) as the operation of the system on $\mathbf{x}$ to produce $\mathbf{y}$. This interpretation is expressed in (1.15). We now suggest two more interpretations that will be useful throughout our discussions of modeling. It is apparent from (1.14) and (1.15) that the columns of the matrix $\mathbf{A}$ are in some sense similar to $\mathbf{y}$; they both contain the same number (m) of elements. We call them **column vectors of A**, and denote the jth column vector by $\mathbf{A}_{(j)}$. Again, the rows of $\mathbf{A}$ are similar to $\mathbf{x}$, both containing n elements; we denote the ith **row vector of A** by $\mathbf{A}^{(i)}$. If we focus on the column vectors of $\mathbf{A}$, (1.14) becomes

$$\xi_1\mathbf{A}_{(1)}+\xi_2\mathbf{A}_{(2)}+\cdots+\xi_n\mathbf{A}_{(n)}=\mathbf{y} \tag{1.16}$$

That is, $\mathbf{y}$ is a simple combination of the column vectors of $\mathbf{A}$; the elements of $\mathbf{x}$ specify the combination. We will make use of this column vector interpretation in Section 2.2 and thereafter.

Changing our focus to the row vectors of $\mathbf{A}$, (1.14) becomes

$$\begin{aligned} \mathbf{A}^{(1)}\mathbf{x}&=\eta_1\\ \mathbf{A}^{(2)}\mathbf{x}&=\eta_2\\ \vdots\quad&\quad\vdots\\ \mathbf{A}^{(m)}\mathbf{x}&=\eta_m \end{aligned} \tag{1.17}$$

Each element of $\mathbf{y}$ is determined by the corresponding row vector of $\mathbf{A}$. By this interpretation, we are merely focusing separately on each of the equations of (1.13). We can use the geometrical pictures of analytic geometry to help develop a physical feel for the individual algebraic equations of (1.17). Suppose

$$\mathbf{A}\mathbf{x}=\begin{pmatrix}2 & 1\\ 2 & 1+\epsilon\end{pmatrix}\begin{pmatrix}\xi_1\\ \xi_2\end{pmatrix}=\begin{pmatrix}2\\ 3\end{pmatrix} \tag{1.18}$$

where ϵ is some constant. The 2×1 matrix $\mathbf{x}$ and the 1×2 matrices $\mathbf{A}^{(i)}$ are each equivalent to a vector (or arrow) in a plane. We simply pick coordinate axes and associate with each element of $\mathbf{x}$ or $\mathbf{A}^{(i)}$ a component along one of the axes. Thus we can represent (1.18) geometrically as in Figure 1.7. The vectors $\mathbf{x}$ such that

$$\mathbf{A}^{(1)}\mathbf{x}=\text{a constant}$$

terminate on a line perpendicular to the vector $\mathbf{A}^{(1)}$. The solution $\mathbf{x}$ to the pair of equations lies at the intersection of the lines $\mathbf{A}^{(1)}\mathbf{x}=2$ and $\mathbf{A}^{(2)}\mathbf{x}=3$. Since the lines in Figure 1.7 have a well-defined intersection, the equations of (1.18) possess a well-defined (unique) solution. However, if $\epsilon\to 0$, $\mathbf{A}^{(2)}\to\mathbf{A}^{(1)}$ and the system becomes degenerate; the lines become parallel, the equations become inconsistent, and there is no solution (intersection). If

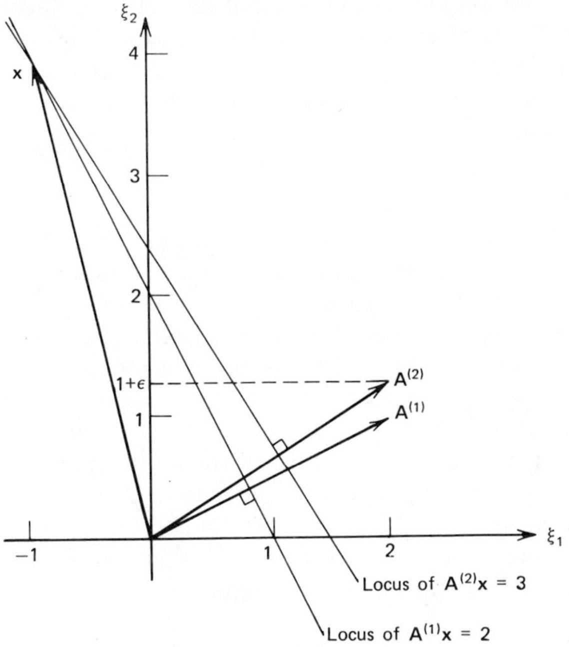

Figure 1.7. Row vector interpretation of (1.18) for $\epsilon = 0.25$.

the numbers on the right side of (1.18) were equal, the lines would overlap, the equations would be consistent, but the solution would not be unique—any **x** terminating on the common line would satisfy both equations.

The geometrical example of (1.18) and Figure 1.7 introduces a significant computational difficulty which exists in nearly degenerate systems of equations. Slight changes in the numbers on the right side of (1.18) result in slight shifts in the positions of the lines in Figure 1.7. Slight changes in the equation coefficients cause slight tilts in these lines. If ϵ is nearly zero, the lines are nearly parallel, and slight perturbations in the line positions or angles cause large swings in the intersection (or solution) **x**. A solution to a matrix equation which is very sensitive to small changes (or errors) in the data is called an **unstable solution**. A matrix (or the corresponding set of equations) which leads to an unstable solution is said to be **ill-conditioned**. Assume the matrix is normalized so that the magnitude of its largest element is approximately one. Then the magnitudes of the elements of the inverse matrix indicate the degree of sensitivity of the solution **x** of (1.14) to errors in the data, $\{a_{ij}\}$ or $\{\eta_i\}$. In Section 6.6 we define a condition number which indicates the size of the largest elements of the inverse. A very large condition number implies that the matrix is ill-conditioned. The size of det(**A**) is another indication of the ill-conditioning of the equations; as the equations become more degenerate, det(**A**) must approach zero (P&C 1.6). However, det(**A**) is not an absolute measure of ill-conditioning as is the condition number.

Numerical Error

There are two fundamental sources of error in the solution to a set of linear algebraic equations, measurement error and computer roundoff. When the data that are used to make up a set of equations come from physical measurements, these data usually contain empirical error. Even if the data are exact, however, the numbers are rounded by the computer; the data can be represented only to a finite number of significant digits. Thus inaccuracies in the equation data are the rule, not the exception. As computations are carried out, further rounding occurs. Although individual inaccuracies are slight, their cumulative effect can be disastrous if handled carelessly.

The following example demonstrates that slight errors in the data can be vastly magnified by straightforward use of row-reduction techniques. Let

$$(\mathbf{A} \,\vdots\, \mathbf{y}) = \begin{pmatrix} 2 & 1 & 3 & \vdots & 1 \\ 2 & 1.01 & 1 & \vdots & 2 \\ 2 & 3 & 2 & \vdots & 3 \end{pmatrix} \tag{1.19}$$

Suppose the element a_{22} is in error by 0.5%; that is, $a_{22} = 1.01 \pm 0.005$. Elimination operations on the first column reduces (1.19) to

$$\begin{pmatrix} 2 & 1 & 3 & \vdots & 1 \\ 0 & 0.01 & -2 & \vdots & 1 \\ 0 & 2 & -1 & \vdots & 2 \end{pmatrix} \tag{1.20}$$

where the subtraction of two nearly equal numbers has magnified the error at the element in question to about 50%, that is, the new element in row 2, column 2, is 0.01 ± 0.005. Were we to use this element to eliminate the other elements in column 2, we would propagate this 50% error throughout the matrix; that is, we would obtain

$$\begin{pmatrix} 2 & 0 \mp 0.5 & 203 \pm 100 & \vdots & -\ 99 \mp 50 \\ 0 & 1 \pm 0.5 & -200 \mp 100 & \vdots & 100 \pm 50 \\ 0 & 0 \mp 1 & 399 \pm 200 & \vdots & -198 \mp 100 \end{pmatrix} \tag{1.21}$$

Further computations would be meaningless. Fortunately, we do not need to divide by the inaccurate element. We merely interchange rows 2 and 3 in (1.20) to obtain

$$\begin{pmatrix} 2 & 1 & 3 & \vdots & 1 \\ 0 & 2 & -1 & \vdots & 2 \\ 0 & 0.01 & -2 & \vdots & 1 \end{pmatrix} \tag{1.22}$$

This interchange is equivalent to writing the equations in a different order. We now use the larger and more accurate element "2" of row 2, column 2 to eliminate the other elements in column 2:

$$\begin{pmatrix} 4 & 0 & 7 & \vdots & 0 \\ 0 & 2 & -1 & \vdots & 2 \\ 0 & 0 \pm 0.005 & -1.995 & \vdots & 0.99 \end{pmatrix} \tag{1.23}$$

The element moved into position for elimination of other elements in its column is called a pivot. The process of interchanging rows to avoid division by relatively small (and therefore inaccurate) numbers is called **pivoting** or **positioning for size**. We also can move the inaccurate element from row 2, column 2 of (1.20) by interchanging *columns* 2 and 3 if we change the order of the variables ξ_2 and ξ_3 which multiply these columns. This column interchange is also used in pivoting. All good computer algorithms for solving sets of linear algebraic equations or for inverting square matrices use some form of pivoting to minimize the magnification and propagation of errors in the data. Scaling of the equations is also an important part of these algorithms.

The matrix of (1.19) is not ill-conditioned. It is apparent, therefore, that we must compute solutions carefully, regardless of the conditioning of the equations, if we are to avoid magnification of errors. If the equations are ill-conditioned, however, careful computing (scaling and pivoting) and the use of double precision arithmetic (additional significant digits) are crucial. Furthermore, division by small numbers is inevitable at some point in the process of solving ill-conditioned equations, and errors *will* be magnified. An iterative technique for improving the computed solution to a set of ill-conditioned equations is described in P&C 1.5.

If a set of equations is very ill-conditioned, it may be that the underlying system is degenerate. Perhaps the matrix would be singular, were it not for empirical error in the data. (That is, perhaps ϵ should be zero in (1.18) and Figure 1.7.) Then in order to completely solve the set of equations, we not only need to compute a particular solution $\mathbf{x}$ as described above, but we also need to estimate the full set of "near solutions" (the locus of the "nearly-overlapping" lines of Figure 1.7 for $\epsilon = 0$). We describe a technique for computing this set of "near solutions" in Section 2.4. Further information on the solution of linear algebraic equations is contained in Forsythe and Moler [1.4] and Forsythe [1.5].

1.6 Problems and Comments

*1.1 *Exploring matrix equations by row reduction*: let **A** be an $m \times n$ matrix. Row reduction of $(\mathbf{A} \vdots \mathbf{y})$ for an unspecified column vector $\mathbf{y} = (\eta_1 \cdots \eta_m)^T$, or the equivalent row reduction of $(\mathbf{A} \vdots \mathbf{I})$ for an $m \times m$ matrix **I**, determines the conditions which must be satisfied by **y** in order for the equation $\mathbf{Ax} = \mathbf{y}$ to have a solution; the set of vectors **y** for which a solution **x** exists is called the **range of A**. The same row reduction determines the set of solutions **x** for $\mathbf{y} = (0 \cdots 0)^T$; this set of solutions is referred to as the **nullspace of A**. If the nullspace of **A** contains nonzero vectors, the solutions to $\mathbf{Ax} = \mathbf{y}$ cannot be unique. Let the matrix equation be

$$\begin{pmatrix} 1 & 2 & 1 & 3 \\ 2 & 1 & 1 & 3 \\ 4 & 5 & 3 & 9 \end{pmatrix} \begin{pmatrix} \xi_1 \\ \xi_2 \\ \xi_3 \\ \xi_4 \end{pmatrix} = \begin{pmatrix} 2 \\ 1 \\ 5 \end{pmatrix}$$

(*a*) Row reduce $(\mathbf{A} \vdots \mathbf{I})$.
(*b*) Determine the range of **A**; that is, determine the relationships that must exist among the elements $\{\eta_i\}$ of **y** in order for the matrix equation $\mathbf{Ax} = \mathbf{y}$ to have a solution.
(*c*) Determine the nullspace of **A**.
(*d*) Determine the solutions **x** for the specified right-hand side **y**.
(*e*) Give an example of a matrix equation that is both inconsistent and underdetermined; that is, an equation for which **y** is not in the range of **A** and for which the nullspace of **A** is nonzero.

1.2 Use the row-reduction technique to determine the solutions to the following sets of equations:

$$(a) \qquad \begin{aligned} \xi_1 + 6\xi_2 - 18\xi_3 &= 0 \\ -4\xi_1 \qquad + 5\xi_3 &= 0 \\ -3\xi_1 + 6\xi_2 - 13\xi_3 &= 0 \\ -7\xi_1 + 6\xi_2 - 8\xi_3 &= 0 \end{aligned}$$

(*b*)
$$\begin{aligned} 2\xi_1+3\xi_2+4\xi_3&=9\\ \xi_1\ +\xi_2\ +\xi_3&=3\\ 3\xi_1+2\xi_2+2\xi_3&=7 \end{aligned}$$

(*c*)
$$\begin{aligned} 2\xi_1+3\xi_2+4\xi_3&=9\\ 3\xi_1+4\xi_2+5\xi_3&=12\\ 4\xi_1+3\xi_2+3\xi_3&=10\\ 5\xi_1+5\xi_2+6\xi_3&=10 \end{aligned}$$

(*d*)
$$\begin{aligned} \xi_1\qquad\ -2\xi_3&=\eta_1\\ 2\xi_1+2\xi_2\qquad\ &=\eta_2\\ 2\xi_1\qquad\ -4\xi_3&=\eta_3\\ \xi_1\ +\xi_2+3\xi_3&=\eta_4 \end{aligned}$$

1.3 *Efficiency of computations*: the number of multiplications performed during a computation is a measure of the efficiency of a computational technique. Let **A** be an invertible $n\times n$ matrix. Determine the number of multiplications required:

(*a*) To compute $\mathbf{A}^{-1}$ by Gaussian elimination;

(*b*) To compute $\mathbf{A}^{-1}$ by Gauss–Jordan elimination;

(*c*) To compute det(**A**), using Gaussian elimination to triangularize **A** (Example 2, Appendix 1).

Determine the number of multiplications required to solve $\mathbf{Ax}=\mathbf{y}$ for a specific vector **y** by:

(*d*) Cramer's rule [Hint: use the answer to (*c*)].

(*e*) The computation in (*a*) and the multiplication $\mathbf{A}^{-1}\mathbf{y}$;

(*f*) Direct row reduction of $(\mathbf{A}\,\vdots\,\mathbf{y})$.

1.4 *Elementary matrices*: the row reduction of an $m\times n$ matrix **A** consists in performing elementary operations on the rows of **A**. Each such operation is equivalent to the multiplication of **A** by a simple $m\times m$ matrix which we refer to as an **elementary matrix.**

(*a*) For $m=5$, find the elementary matrices corresponding to the following:

(1) the multiplication of row 3 by a constant c;

(2) the addition of row 4 to row 1;

(3) the interchange of row 3 with row 5.

(*b*) Every elementary matrix is invertible. Find the inverses of the elementary matrices determined in (*a*).

(*c*) The row reduction of $(\mathbf{A} \vdots \mathbf{I})$ is equivalent to multiplication of $(\mathbf{A} \vdots \mathbf{I})$ by an invertible matrix $\mathbf{B}$ (a product of elementary matrices). Show that if $\mathbf{A}$ is square and $(\mathbf{A} \vdots \mathbf{I})$ can be row reduced to the form $(\mathbf{I} \vdots \mathbf{B})$, then $\mathbf{AB}=\mathbf{BA}=\mathbf{I}$, and therefore $\mathbf{B}=\mathbf{A}^{-1}$.

1.5 *Iterative improvement of solutions*: the solution to the matrix equation $\mathbf{Ax}=\mathbf{y}$ can be obtained by Gaussian elimination. As a result of roundoff, the computed solution $\mathbf{x}_1$ is usually in error. Denote the error by $\mathbf{x}-\mathbf{x}_1$, where $\mathbf{x}$ is the exact solution. A computable measure of the error is the residual $r_1 \triangleq \mathbf{y}-\mathbf{Ax}_1$. If we could solve exactly for $(\mathbf{x}-\mathbf{x}_1)$ in the equation $\mathbf{A}(\mathbf{x}-\mathbf{x}_1)=\mathbf{y}-\mathbf{Ax}_1=r_1$, we could obtain the exact solution. We solve $\mathbf{Az}_1=r_1$ by Gaussian elimination to obtain a correction $\mathbf{z}_1$; $\mathbf{x}_2 \triangleq \mathbf{x}_1+\mathbf{z}_1$ is an improved solution. By repeating the improvement process iteratively, we obtain an approximate solution which is accurate to the number of significant digits used in the computation. However, the residuals $\mathbf{r}_k=\mathbf{y}-\mathbf{Ax}_k$ must be computed to double precision; otherwise the corrections, $\mathbf{z}_k$, will not be improvements. See Forsythe and Moler [1.4, p. 49]. Let

$$\mathbf{A}=\begin{pmatrix} 2.1 & 1.9 \\ 1.9 & 2.0 \end{pmatrix} \quad \text{and} \quad \mathbf{y}=\begin{pmatrix} 1.2 \\ 1.3 \end{pmatrix}$$

To five figures, the solution to $\mathbf{Ax}=\mathbf{y}$ is $\mathbf{x}=(-0.11864 \;\; 0.76271)^{\mathrm{T}}$.

(*a*) Compute an approximate solution $\mathbf{x}_1$ by Gaussian elimination, rounding all computations to three significant digits (slide rule accuracy).

(*b*) Find the residual r_1 by hand computation to *full* accuracy.

(*c*) Round r_1 to three significant digits, if necessary, and compute the correction $\mathbf{z}_1$. Find $\mathbf{x}_2=\mathbf{x}_1+\mathbf{z}_1$.

1.6 *Determinants and volumes*: using a natural correspondence between row vectors and arrows in a plane, we associate a parallelogram with the rows of every real 2×2 matrix $\mathbf{A}$. For example,

$$\mathbf{A}=\begin{pmatrix} 4 & 1 \\ 2 & 2 \end{pmatrix} \Leftrightarrow$$

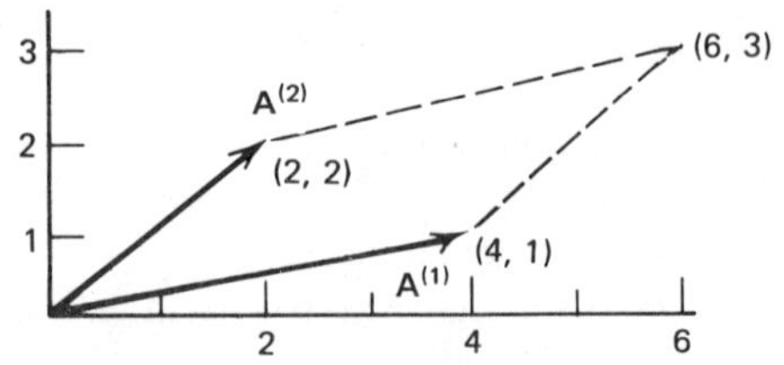

(*a*) Show that the area of the above parallelogram is equal to the determinant of the matrix **A** which is associated with it.

(*b*) For the right-hand coordinate system shown above, we define the *sign of the area* to be positive if $\mathbf{A}^{(1)}$ turns counterclockwise inside the parallogram in order to reach $\mathbf{A}^{(2)}$; if $\mathbf{A}^{(1)}$ turns clockwise, the area is negative. Show graphically that the area of the above parallelogram obeys the following properties of determinants:

(1) The value of det(**A**) is not changed if we add to one row of **A** a multiple of another row of **A**;

(2) The sign of det(**A**) is reversed if we interchange two rows of **A**;

(3) If we multiply one row of **A** by c, then det(**A**) is multiplied by c;

(4) If the rows of **A** are dependent (i.e., one is a multiple of the other), then $\det(\mathbf{A})=0$.

(*c*) The geometrical interpretation of det(**A**) can be extended to $n\times n$ matrices by defining n-dimensional spaces, n-dimensional parallelopipeds, and signed volumes. See Martin and Mizel [1.9]. Since $\det(\mathbf{A}^{\mathrm{T}})=\det(\mathbf{A})$, the volume of the parallelopiped described by the columns of **A** equals the volume described by the rows of **A**. Verify graphically that the geometrical interpretation of determinants extends to 3×3 matrices.

1.7 *Partitioned matrices*: it is sometimes useful to partition a matrix into an array of submatrices. If **P** and **Q** are conformable, we can form the partitions

$$\mathbf{P}=\begin{pmatrix}\mathbf{P}_{11} & \mathbf{P}_{12}\\ \mathbf{P}_{21} & \mathbf{P}_{22}\end{pmatrix} \qquad \mathbf{Q}=\begin{pmatrix}\mathbf{Q}_{11} & \mathbf{Q}_{12}\\ \mathbf{Q}_{21} & \mathbf{Q}_{22}\end{pmatrix}$$

in a manner which allows us to express **PQ** as

$$\mathbf{PQ}=\begin{pmatrix}\mathbf{P}_{11}\mathbf{Q}_{11}+\mathbf{P}_{12}\mathbf{Q}_{21} & \mathbf{P}_{11}\mathbf{Q}_{12}+\mathbf{P}_{12}\mathbf{Q}_{22}\\ \mathbf{P}_{21}\mathbf{Q}_{11}+\mathbf{P}_{22}\mathbf{Q}_{21} & \mathbf{P}_{21}\mathbf{Q}_{12}+\mathbf{P}_{22}\mathbf{Q}_{22}\end{pmatrix}$$

(*a*) Assume that **A** is an invertible matrix. The following factorization can be verified by the block multiplication described above:

$$\begin{pmatrix}\mathbf{A} & \mathbf{B}\\ \mathbf{C} & \mathbf{D}\end{pmatrix}=\begin{pmatrix}\mathbf{I} & \mathbf{O}\\ \mathbf{CA}^{-1} & \mathbf{I}\end{pmatrix}\begin{pmatrix}\mathbf{A} & \mathbf{O}\\ \mathbf{O} & \mathbf{D}-\mathbf{CA}^{-1}\mathbf{B}\end{pmatrix}\begin{pmatrix}\mathbf{I} & \mathbf{A}^{-1}\mathbf{B}\\ \mathbf{O} & \mathbf{I}\end{pmatrix}$$

(*b*) Show that for any submatrix **P** of appropriate dimensions,

$$\begin{vmatrix} \mathbf{I} & \mathbf{O} \\ \mathbf{P} & \mathbf{I} \end{vmatrix} = 1$$

Use this result with (*a*) to show that

$$\begin{vmatrix} \mathbf{A} & \mathbf{B} \\ \mathbf{C} & \mathbf{D} \end{vmatrix} = |\mathbf{A}|\,|\mathbf{D} - \mathbf{C}\mathbf{A}^{-1}\mathbf{B}|$$

(*c*) Use (*a*) to show that

$$\begin{pmatrix} \mathbf{A} & \mathbf{B} \\ \mathbf{C} & \mathbf{D} \end{pmatrix}^{-1} = \begin{pmatrix} \mathbf{I} & -\mathbf{A}^{-1}\mathbf{B} \\ \mathbf{O} & \mathbf{I} \end{pmatrix} \begin{pmatrix} \mathbf{A}^{-1} & \mathbf{O} \\ \mathbf{O} & (\mathbf{D} - \mathbf{C}\mathbf{A}^{-1}\mathbf{B})^{-1} \end{pmatrix} \begin{pmatrix} \mathbf{I} & \mathbf{O} \\ -\mathbf{C}\mathbf{A}^{-1} & \mathbf{I} \end{pmatrix}$$

The number of multiplications required to compute the determinant or the inverse of an $n \times n$ matrix can be reduced by a factor of eight (if n is large) by use of the partitioning schemes in (*b*) or (*c*), respectively.

1.7 References

[1.1] Bruner, Jerome S., *The Process of Education*, Harvard University Press, Cambridge, Mass., 1960.

[1.2] Cannon, Robert H., Jr., *Dynamics of Physical Systems*, McGraw-Hill, New York, 1969.

[1.3] Forrester, Jay W., *Urban Dynamics*, M.I.T. Press, Cambridge, Mass., 1969.

*[1.4] Forsythe, George E. and Cleve B. Moler, *Computer Solution of Linear Algebraic Systems*, Prentice-Hall, Englewood Cliffs, N.J., 1967.

[1.5] Forsythe, George E., "Today's Computational Methods of Linear Algebra," *SIAM Rev.*, **9**, 3 (July 1967), 489–515.

[1.6] Forsythe, George E., "Solving Linear Algebraic Equations Can be Interesting," *Bull. Am. Math. Soc.*, **59** (1953), 299–329.

[1.7] Linvill, William K., "Models and Model Construction," *IRE Trans. Educ.*, **E-5**, 2 (June 1962), 64–67.

[1.8] Meier, Robert C., William T. Newell, and Harold L. Pazer, *Simulation in Business and Economics*, Prentice-Hall, Englewood Cliffs, N.J., 1969.

[1.9] Martin, Allan D. and Victor J. Mizel, *Introduction to Linear Algebra*, McGraw-Hill, New York, 1966.

[1.10] Sage, Andrew P. and James L. Melsa, *System Identification*, Academic Press, New York, 1971.

[1.11] Truxal, John G., *Introduction to Systems Engineering*, McGraw-Hill, New York, 1972.

2

System Models: Transformations on Vector Spaces

The fundamental purpose in modeling a system is to develop a mechanism for predicting the condition or change in condition of the system. In the abstract model $\mathbf{Tx}=\mathbf{y}$ of (1.1), $\mathbf{T}$ represents (or is a model of) the system, whereas $\mathbf{x}$ and $\mathbf{y}$ have to do with the condition of the system. We explore first some familiar models for the condition or changes in condition of systems. These examples lead us to use a generalization of the usual notion of a vector as a model for the condition of a system. We then develop the concept of a transformation of vectors as a model of the system itself. The rest of the chapter is devoted to examination of the most commonly used models—linear models—and their matrix representations.

2.1 The Condition of a System

The physical condition (or change in condition) of many simple systems has been found to possess a magnitude and a direction in our physical three-dimensional space. It is natural, therefore, that a mathematical concept of condition (or change in condition) has developed over time which has these two properties; this concept is the vector. Probably the most obvious example of the use of this concept is the use of arrows in a two-dimensional plane to represent changes in the position of an object on the two-dimensional surface of the earth (see Figure 2.1). Using the usual techniques of analytic geometry, we can represent each such arrow by a pair of numbers that indicates the components of that arrow along each of a pair of coordinate axes. Thus pairs of numbers serve as an equivalent model for changes in position.

An ordinary road map is another model for the two-dimensional surface of the earth. It is equivalent to the arrow diagram; points on the map are

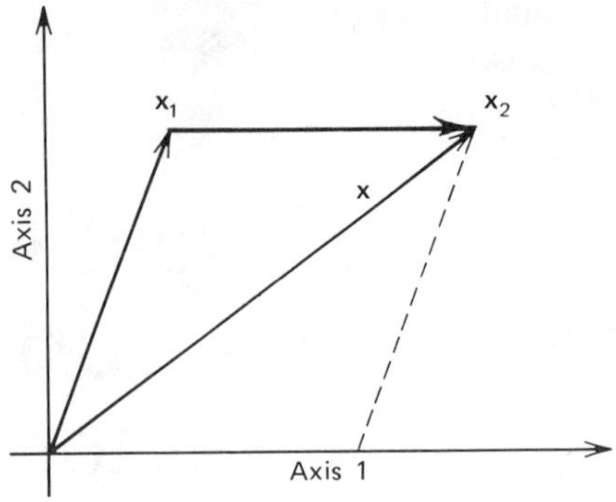

Figure 2.1. An "arrow vector" diagram.

equivalent to the arrow tips of Figure 2.1. The only significant difference between these two models is that the map emphasizes the position (or condition) of an object on the earth, whereas the arrow diagram stresses the changes in position and the manner in which intermediate changes in position add to yield a total change in position. We can also interpret a position on the map as a change from some reference position. The manner in which we combine arrows or changes in position (the parallelogram rule) is the most significant characteristic of either model. Consequently we focus on the arrow model which emphasizes the combination process.

Reference arrows (coordinate axes) are used to tie the arrow model to the physical world. By means of a reference position and a pair of reference "position changes" on the surface of the earth, we relate the positions and changes in position on the earth to positions and arrows in the arrow diagram. However, there are no inherent reference axes on either the physical earth or the two-dimensional plane of arrows.

The same vector model that we use to represent changes in position can be used to represent the forces acting at a point on a physical object. The reason we can use the same model is that the magnitudes and directions of forces also combine according to the parallelogram rule. The physical natures of the reference vectors are different in these three situations: in one case they are changes in position on the earth, in another they are arrows, in the third, forces. Yet once reference vectors are chosen in each, all three situations become in some sense equivalent; corresponding to each vector in one situation is a vector in the other two; corresponding to each sum of vectors in one is a corresponding sum in the other two. We use the set of arrows as a model for the other two situations because it is the most convenient of the three to work with.

The set of complex numbers is one more example of a set of objects which is equivalent to the set of arrows. We usually choose as references in the set of complex numbers the two numbers 1 and i. Based on these reference numbers and two reference arrows, we interpret every arrow as a complex number. Here we have one set of mathematical (or geometrical) objects serving as a model for another set of mathematical objects.

Consider now a physical system which is more complicated than the two physical systems discussed above. Imagine a flat metal sheet exposed to the sun and partly submerged in a stream. (The sheet is representative of any object subject to heat sources and coolants.) The thermal condition of the sheet is described by the temperature distribution over the surface of the sheet. A change in the cloud cover in the sky will change the pattern in which the sun falls on the sheet. As a result, the temperature distribution will change. Assuming the temperature distribution reaches a new steady state, the new distribution equals the old distribution plus the change in the distribution. We model this situation as follows. Let (s,t) denote a position in some two-dimensional coordinate system on the surface of the sheet. Let $\mathbf{f}(s,t)$ be the temperature at the point (s,t), measured in degrees centigrade, for all points (s,t) on the sheet. We model a change in the thermal condition of the sheet by

$$\mathbf{f}_{\text{new}}(s,t)=\mathbf{f}_{\text{old}}(s,t)+\mathbf{f}_{\text{change}}(s,t) \tag{2.1}$$

for all (s,t) on the sheet. In effect, (2.1) *defines* $\mathbf{f}_{\text{change}}$. However, we hope to use a model of the system to *predict* $\mathbf{f}_{\text{change}}$. Then (2.1) will determine $\mathbf{f}_{\text{new}}$. Equation (2.1) is a "distributed" equivalent of the arrow diagram in Figure 2.1; each of these models illustrates the manner in which changes in condition combine to yield a net condition of the system in question. Once again, references have been chosen in both the physical system and the model (mathematical system) in order to equate the two systems; choosing physical units of measurement (degrees centigrade) amounts to fixing the relationship between the physical and mathematical systems.

The most significant difference between a system modeled by Figure 2.1 and a system modeled by (2.1) consists in the nature of the conditions in each system. In one case we have a quantity with magnitude and direction (e.g., force); in the other, a quantity without magnitude and direction—a quantity that is distributed over a two-dimensional region. Yet there are important similarities between the two systems. The changes in condition of the system are under scrutiny; also, several changes in condition combine by simple rules to yield a total or net condition.

Vector Spaces

By expressing various types of problems in a common framework, we learn to use concepts derived from one type of problem in understanding other types of problems. In particular, we are able to draw useful analogies between algebraic equations and differential equations by expressing both types of equations as "vector" equations. Therefore, we now generalize the common notion of a vector to include all the examples discussed in the previous section.

Definition. A **linear space** (or **vector space**) $\mathcal{V}$ is a set of elements $\mathbf{x}$, $\mathbf{y}$, $\mathbf{z}, \ldots$, called vectors, together with definitions of *vector addition* and *scalar multiplication*.

a. The definition of vector addition is such that:
   1. To every pair, $\mathbf{x}$ and $\mathbf{y}$, of vectors in $\mathcal{V}$ there corresponds a unique vector $\mathbf{x}+\mathbf{y}$ in $\mathcal{V}$, called the **sum** of $\mathbf{x}$ and $\mathbf{y}$.
   2. $\mathbf{x}+\mathbf{y}=\mathbf{y}+\mathbf{x}$.
   3. $(\mathbf{x}+\mathbf{y})+\mathbf{z}=\mathbf{x}+(\mathbf{y}+\mathbf{z})$.
   4. There is a unique vector $\boldsymbol{\theta}$ in $\mathcal{V}$, called the **zero vector** (or origin), such that $\mathbf{x}+\boldsymbol{\theta}=\mathbf{x}$ for all $\mathbf{x}$ in $\mathcal{V}$.
   5. Corresponding to each $\mathbf{x}$ in $\mathcal{V}$ there is a unique vector "$-\mathbf{x}$" in $\mathcal{V}$ such that $\mathbf{x}+(-\mathbf{x})=\boldsymbol{\theta}$.

b. The definition of scalar multiplication is such that:
   1. To every vector $\mathbf{x}$ in $\mathcal{V}$ and every scalar a there corresponds a unique vector $a\mathbf{x}$ in $\mathcal{V}$, called the **scalar multiple** of $\mathbf{x}$.*
   2. $a(b\mathbf{x})=(ab)\mathbf{x}$.
   3. $1(\mathbf{x})=\mathbf{x}$ (where 1 is the unit scalar).
   4. $a(\mathbf{x}+\mathbf{y})=a\mathbf{x}+a\mathbf{y}$.
   5. $(a+b)\mathbf{x}=a\mathbf{x}+b\mathbf{x}$.

Notice that a vector space includes not only a set of elements (vectors) but also "valid" definitions of vector addition and scalar multiplication. Also inherent in the definition is the fact that the vector space $\mathcal{V}$ contains all "combinations" of its own vectors: if $\mathbf{x}$ and $\mathbf{y}$ are in $\mathcal{V}$, then $a\mathbf{x}+b\mathbf{y}$ is also in $\mathcal{V}$. The rules of algebra are so much a part of us that some of the requirements may at first appear above definition; however, they are necessary. A few more vector space properties which may be deduced from the above definition are as follows:

1. $0\mathbf{x}=\boldsymbol{\theta}$ (where "0" is the zero scalar).
2. $a\boldsymbol{\theta}=\boldsymbol{\theta}$.
3. $(-1)\mathbf{x}=-\mathbf{x}$.

Example 1. The Real 3-tuple Space $\mathcal{R}^3$. The space $\mathcal{R}^3$ consists in the set of all

*The scalars are any set of elements which obey the usual rules of algebra. A set of elements which obeys these rules constitutes a field (see Hoffman and Kunze [2.6]). We usually use as scalars either the real numbers or the complex numbers. There are *other* useful fields, however (P&C 2.4).

real 3-tuples (all ordered sequences of three real numbers), $\mathbf{x}=(\xi_1,\xi_2,\xi_3)$, $\mathbf{y}=(\eta_1,\eta_2,\eta_3)$, with the following definitions of addition and scalar multiplication:

$$\begin{aligned} \mathbf{x}+\mathbf{y} &\triangleq (\xi_1+\eta_1,\xi_2+\eta_2,\xi_3+\eta_3) \\ a\mathbf{x} &\triangleq (a\xi_1,a\xi_2,a\xi_3) \end{aligned} \tag{2.2}$$

It is clear that the zero vector for this 3-tuple space, $\boldsymbol{\theta}=(0,0,0)$, satisfies $\mathbf{x}+\boldsymbol{\theta}=\mathbf{x}$. We show that $\boldsymbol{\theta}$ is unique by assuming another vector $\mathbf{y}$ also satisfies $\mathbf{x}+\mathbf{y}=\mathbf{x}$; that is,

$$(\xi_1+\eta_1,\xi_2+\eta_2,\xi_3+\eta_3)=(\xi_1,\xi_2,\xi_3)$$

or $\xi_i+\eta_i=\xi_i$. The properties of scalars then require $\eta_i=0$ (or $\mathbf{y}=\boldsymbol{\theta}$). It is easy to prove that $\mathcal{R}^3$, as defined above, satisfies the other requirements for a linear space. In each instance, questions about vectors are reduced to questions about scalars.

We emphasize that the definition of $\mathcal{R}^3$ says nothing about coordinates. Coordinates are multipliers for reference vectors (reference arrows, for instance). The 3-tuples are vectors in their own right. However, there is a commonly used correspondence between $\mathcal{R}^3$ and the set of vectors (arrows) in the usual three-dimensional space which makes it difficult not to think of the 3-tuples as coordinates. The two sets of vectors are certainly equivalent. We will, in fact, use this natural correspondence to help illustrate vector concepts graphically.

Example 2. The Two-Dimensional Space of Points (or Arrows). This space consists in the set of all points in a plane. Addition is defined by the parallelogram rule using a fixed reference point (see Figure 2.2). Scalar multiplication is defined as "length" multiplication using the reference point. The zero vector is obviously the

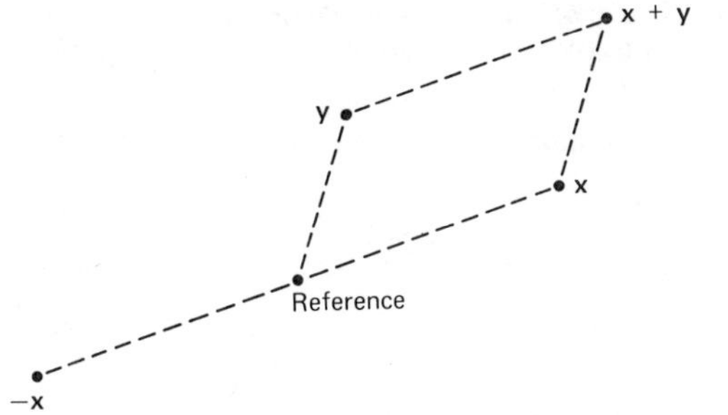

Figure 2.2. The two-dimensional space of points.

reference point. Each of the requirements can be verified by geometrical arguments.

An equivalent (but not identical) space is one where the vectors are not the points, but rather, arrows to the points from the reference point. We distinguish only the magnitude and direction of each arrow; *two parallel arrows of the same length are considered identical.*

Both the arrow space and the point space are easily visualized: we often use the arrow space in two or three dimensions to demonstrate concepts graphically. Although the arrow space contains no *inherent* reference arrows, we sometimes *specify* reference arrows in order to equate the arrows to vectors in $\mathcal{R}^3$. Because of the equivalence between vectors in $\mathcal{R}^3$ and vectors in the three-dimensional space of points, we occasionally refer to vectors in $\mathcal{R}^3$ and in other spaces as *points*.

Example 3. The Space of Column Vectors $\mathfrak{M}^{3\times 1}$. The space $\mathfrak{M}^{3\times 1}$ consists in the set of all real 3×1 column matrices (or column vectors), denoted by

$$\mathbf{x}=\begin{pmatrix}\xi_1\\ \xi_2\\ \xi_3\end{pmatrix} \qquad \mathbf{y}=\begin{pmatrix}\eta_1\\ \eta_2\\ \eta_3\end{pmatrix}$$

with the following definitions of addition and scalar multiplication:

$$\mathbf{x}+\mathbf{y}\overset{\Delta}{=}\begin{pmatrix}\xi_1+\eta_1\\ \xi_2+\eta_2\\ \xi_3+\eta_3\end{pmatrix} \qquad a\mathbf{x}\overset{\Delta}{=}\begin{pmatrix}a\xi_1\\ a\xi_2\\ a\xi_3\end{pmatrix} \tag{2.3}$$

In order to save space in writing, we occasionally write vectors from $\mathfrak{M}^{3\times 1}$ in the transposed form $\mathbf{x}=(\xi_1\ \ \xi_2\ \ \xi_3)^{\mathrm{T}}$. The equivalence between $\mathfrak{M}^{3\times 1}$ and $\mathcal{R}^3$ is obvious. The only difference between the two vector spaces is in the nature of their vectors. Vectors in $\mathfrak{M}^{3\times 1}$ can be multiplied by $m\times 3$ matrices (as in Section 1.5), whereas vectors in $\mathcal{R}^3$ cannot.

Example 4. The Space of Real Square-Summable Sequences, l_2. The space l_2 consists in the set of all infinite sequences of real numbers, $\mathbf{x}=(\xi_1,\xi_2,\xi_3,\ldots)$, $\mathbf{y}=(\eta_1,\eta_2,\eta_3,\ldots)$ which are square summable; that is, for which $\sum_{i=1}^{\infty}\xi_i^2<\infty$. Addition and scalar multiplication in l_2 are defined by

$$\begin{aligned}\mathbf{x}+\mathbf{y}&\overset{\Delta}{=}(\xi_1+\eta_1,\xi_2+\eta_2,\xi_3+\eta_3,\ldots)\\ a\mathbf{x}&\overset{\Delta}{=}(a\xi_1,a\xi_2,a\xi_3,\ldots)\end{aligned} \tag{2.4}$$

Most of the properties required by the definition of a linear space are easily verified for l_2; for instance, the zero vector is obviously $\boldsymbol{\theta}=(0,0,0,\ldots)$. However, there is one subtle difference between l_2 and the space $\mathcal{R}^3$ of Example 1. Because the sequences in l_2 are infinite, it is not obvious that if **x** and **y** are in l_2, $\mathbf{x}+\mathbf{y}$ is also in l_2. It can be shown that

$$\sqrt{\Sigma_{i=1}^{\infty}(\xi_i+\eta_i)^2} \leq \sqrt{\Sigma_{i=1}^{\infty}\xi_i^2} + \sqrt{\Sigma_{i=1}^{\infty}\eta_i^2}$$

[This fact is known as the triangle inequality (P&C 5.4)]. Therefore,

$$\sum_{i=1}^{\infty}(\xi_i+\eta_i)^2 < \infty$$

and $\mathbf{x}+\mathbf{y}$ is square-summable. The requirement of square summability is a definite restriction on the elements of l_2; the simple sequence $(1,1,1,\ldots)$, for instance, is not in l_2.

The definition of $\mathcal{R}^3$ extends easily to $\mathcal{R}^n$, the space of n-tuples of real numbers (where n is a positive integer). The space $\mathfrak{M}^{n\times 1}$ is a similar extension of $\mathfrak{M}^{3\times 1}$. Mathematically these "n-dimensional" spaces are no more complicated than their three-dimensional counterparts. Yet we are not able to draw arrow-space equivalents because our physical world is three-dimensional. Visualization of an abstract vector space is most easily accomplished by thinking in terms of its three-dimensional counterpart.

The spaces $\mathcal{R}^n$, $\mathfrak{M}^{n\times 1}$, and l_2 can also be redefined using complex numbers, rather than real numbers, for scalars. We denote by $\mathcal{R}_c^n$ the complex n-tuple space. We use the symbol $\mathfrak{M}_c^{n\times 1}$ for the space of complex $n\times 1$ column vectors. Let l_2^c represent the space of complex square-summable sequences. (We need a slightly different definition of square summability for the space l_2^c:$\Sigma_{i=1}^{\infty}|\xi_i|^2<\infty$). In most vector space definitions, either set of scalars can be used. A notable exception to interchangeability of scalars is the arrow space in two or three dimensions. The primary value of the arrow space is in graphical illustration. We have already discussed the equivalence of the set of complex scalars to the two-dimensional space of arrows. Therefore, substituting complex scalars in the real two-dimensional arrow space would require four-dimensional graphical illustration.

We eventually find it useful to combine simple vector spaces to form more complicated spaces.

Definition. Suppose $\mathcal{V}$ and $\mathcal{W}$ are vector spaces. We define the **cartesian product** $\mathcal{V}\times\mathcal{W}$ of the spaces $\mathcal{V}$ and $\mathcal{W}$ to be the set of pairs of vectors $\mathbf{z} \triangleq (\mathbf{x},\mathbf{y})$, with $\mathbf{x}$ in $\mathcal{V}$ and $\mathbf{y}$ in $\mathcal{W}$. We define addition and scalar multiplication of vectors in $\mathcal{V}\times\mathcal{W}$ in terms of the corresponding operations in $\mathcal{V}$ and in $\mathcal{W}$: if $\mathbf{z}_1=(\mathbf{x}_1,\mathbf{y}_1)$ and $\mathbf{z}_2=(\mathbf{x}_2,\mathbf{y}_2)$, then

$$\mathbf{z}_1+\mathbf{z}_2 \triangleq (\mathbf{x}_1+\mathbf{x}_2,\mathbf{y}_1+\mathbf{y}_2)$$

$$a\mathbf{z}_1 \triangleq (a\mathbf{x}_1,a\mathbf{y}_1)$$

Example 5. A Cartesian Product. Let $\mathbf{x}=(\xi_1,\xi_2)$, a vector in $\mathfrak{R}^2$. Let $\mathbf{y}=(\eta_1)$, a vector in $\mathfrak{R}^1$. Then $\mathbf{z} \stackrel{\Delta}{=} ((\xi_1,\xi_2),(\eta_1))$ is a typical vector in $\mathfrak{R}^2\times\mathfrak{R}^1$. This cartesian product space is clearly equivalent to $\mathfrak{R}^3$. Strictly speaking, however, $\mathbf{z}$ is not in $\mathfrak{R}^3$. It is not a 3-tuple, but rather a 2-tuple followed by a 1-tuple. Yet we have no need to distinguish between $\mathfrak{R}^3$ and $\mathfrak{R}^2\times\mathfrak{R}^1$.

Function Spaces

Each vector in the above examples has discrete elements. It is a small conceptual step from the notion of an infinite sequence of discrete numbers (a vector in l_2) to the usual notion of a function—a "continuum" of numbers. Yet vectors and functions are seldom related in the thinking of engineers. We will find that vectors and functions can be viewed as essentially equivalent objects; functions can be treated as vectors, and vectors can be treated as functions. A **function space** is a linear space whose elements are functions. We usually think of a function as a rule or graph which associates with each scalar in its domain a single scalar value. We do not confuse the graph with particular values of the function. Our notation should also keep this distinction. Let $\mathbf{f}$ denote a **function**; that is, the symbol $\mathbf{f}$ recalls to mind a particular rule or graph. Let $\mathbf{f}(t)$ denote the **value of the function at t**. By $\mathbf{f}=\mathbf{g}$, we mean that the scalars $\mathbf{f}(t)$ and $\mathbf{g}(t)$ are equal for each t of interest.

Example 6. $\mathfrak{P}^n$, The Polynomials of Degree Less Than n. The space $\mathfrak{P}^n$ consists in all real-valued polynomial functions of degree *less* than n: $\mathbf{f}(t)=\xi_1+\xi_2 t+\cdots+\xi_n t^{n-1}$ for all real t. Addition and scalar multiplication of vectors (functions) in $\mathfrak{P}^n$ are defined by

$$
\begin{aligned}
(\mathbf{f}+\mathbf{g})(t) &\stackrel{\Delta}{=} \mathbf{f}(t)+\mathbf{g}(t)\\
(a\mathbf{f})(t) &\stackrel{\Delta}{=} a(\mathbf{f}(t))
\end{aligned}
\tag{2.5}
$$

for all t. The zero function is $\boldsymbol{\theta}(t)=0$ for all t. This zero function is unique; if the function $\mathbf{g}$ also satisfied $\mathbf{f}+\mathbf{g}=\mathbf{f}$, then the values of $\mathbf{f}$ and $\mathbf{g}$ would satisfy

$$(\mathbf{f}+\mathbf{g})(t)=\mathbf{f}(t)+\mathbf{g}(t)=\mathbf{f}(t)$$

It would follow that $\mathbf{g}(t)=0$ for all t, or $\mathbf{g}=\boldsymbol{\theta}$. The other requirements for a vector space are easily verified for $\mathfrak{P}^n$.

We emphasize that the vector $\mathbf{f}$ in Example 6 is the entire portrait of the function $\mathbf{f}$. The scalar variable t is a "dummy" variable. The only purpose of this variable is to order the values of the function in precisely the same way that the subscript i orders the elements in the following vector from l_2:

$$\mathbf{x}=(\xi_1,\xi_2,\ldots,\xi_i,\ldots)$$

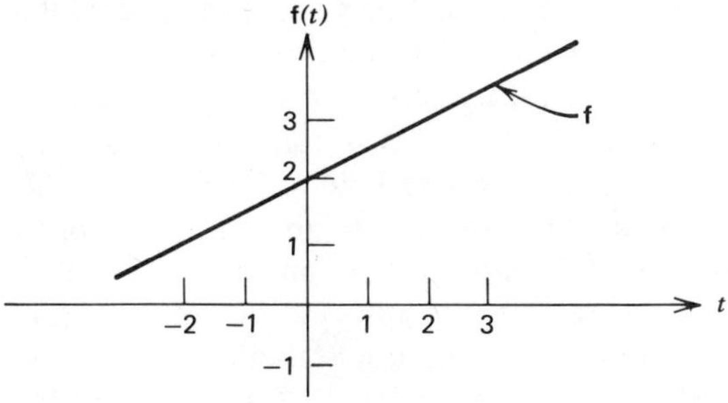

Figure 2.3. A function **f** and its values **f**(t).

Figure 2.3 distinguishes graphically between the vector **f** and its value at t for the specific function **f** defined by $\mathbf{f}(t)=2+0.5t$. Figure 2.4 distinguishes in a similar manner between an infinite sequence **x** and its ith element.

It is evident that the vector **x** from l_2 is just as much a function as is the polynomial **f** from $\mathcal{P}^n$. In the space of polynomials, the index t is continuous; in the space of infinite sequences the index i is discrete—it takes on only positive integral values. In the latter case, we could as well refer to the ith element ξ_i as the value of **x** at i. In point of fact, most vector spaces can be interpreted as spaces of functions; the terms vector space and function space are somewhat interchangeable. However, it is common practice to use the term function space only for a space in which the index t varies continuously over an interval.

It is unfortunate that the symbol **f**(t) is commonly used to represent both a function and the value of that function at t. This blurring of the meaning of symbols is particularly true of the sinusoidal and exponential functions. We will try to be explicit in our distinction between the two concepts. As discussed in the preface, boldface type is used to *emphasize* the interpretation of a function as a vector. However, to avoid overuse of boldface type, it is not used where emphasis on the vector interpretation appears un-

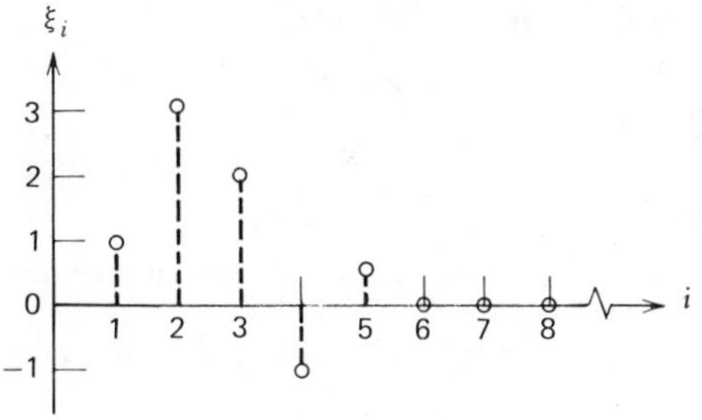

Figure 2.4. The elements ξ_i of an infinite sequence **x**.

necessary; thus the value of a function $\mathbf{f}$ at t may appear either as $\mathbf{f}(t)$ or as $f(t)$. Furthermore, where confusion is unlikely, we sometimes use standard mathematical shorthand; for example, we use $\int_a^b \mathbf{fg}\,dt$ to mean $\int_a^b \mathbf{f}(t)\mathbf{g}(t)\,dt$.

It is difficult to describe or discuss functions in any detail except in terms of their scalar values. In Example 6, for instance, the definitions of addition and scalar multiplication were given in terms of function values. Furthermore, we resorted again to function values to verify that the vector space requirements were met. We will find ourselves continually reducing questions about functions to questions about the scalar values of those functions. Why then do we emphasize the function $\mathbf{f}$ rather than the value $\mathbf{f}(t)$? Because system models act on the whole vector $\mathbf{f}$ rather than on its individual values. As an example, we turn to the one system model we have explored thus far—the matrix equation $\mathbf{Ax}=\mathbf{y}$ which was introduced in Section 1.5. If $\mathbf{A}$ is an $m\times n$ matrix, the vector $\mathbf{x}$ is a column matrix in $\mathfrak{M}^{n\times 1}$; $\mathbf{y}$ is in $\mathfrak{M}^{m\times 1}$. Even though the matrix multiplication requires manipulation of the individual elements (or values) of $\mathbf{x}$, it is impossible to determine *any* element of $\mathbf{y}$ without operating on *all* elements of $\mathbf{x}$. Thus it is natural to think in terms of $\mathbf{A}$ operating on the whole vector $\mathbf{x}$. Similarly, equations involving functions require operations on the whole function (e.g., integration), as we shall see in Section 2.3.

Example 7. The Space $\mathcal{C}(a,b)$ of Continuous Functions. The vectors in $\mathcal{C}(a,b)$ are those real functions which are defined and continuous on the interval $[a,b]$. Addition and scalar multiplication of functions in $\mathcal{C}(a,b)$ are defined by the standard function space definitions (2.5) for all t in $[a,b]$. It is clear that the sums and scalar multiples of continuous functions are also continuous functions.

Example 8. $\mathcal{L}_2(a,b)$ The Real Square-Integrable Functions. The space $\mathcal{L}_2(a,b)$ consists in all real functions which are defined and square integrable on the interval $[a,b]$; that is, functions $\mathbf{f}$ for which*

$$\int_a^b \mathbf{f}^2(t)\,dt < \infty$$

Addition and scalar multiplication of functions in $\mathcal{L}_2(a,b)$ are defined by (2.5) for all t in $[a,b]$. The space $\mathcal{L}_2(a,b)$ is analogous to l_2. It is not clear that the sum of two square-integrable functions is itself square integrable. As in Example 4, we must rely on P&C 5.4 and the concepts of Chapter 5 to find that

$$\sqrt{\int_a^b [\mathbf{f}(t)+\mathbf{g}(t)]^2\,dt} \leqslant \sqrt{\int_a^b \mathbf{f}^2(t)\,dt} + \sqrt{\int_a^b \mathbf{g}^2(t)\,dt}$$

*The integral used in the definition of $\mathcal{L}_2(a,b)$ is the Lebesgue integral. For all practical purposes, Lebesgue integration can be considered the same as the usual Riemann integration. Whenever the Riemann integral exists, it yields the same result as the Lebesgue integral. (See Royden [2.10].)

It follows that if $\mathbf{f}$ and $\mathbf{g}$ are square integrable, then $\mathbf{f}+\mathbf{g}$ is square integrable.

Example 9. A Set of Functions. The set of positive real functions [together with the definitions of addition and scalar multiplication in (2.5)] does *not* form a vector space. This set contains a positive valued function $\mathbf{f}$, but not the negative valued function $-\mathbf{f}$; therefore, this set does not include all sums and multiples of its members.

Example 10. Functions of a Complex Variable. Let $\mathcal{V}$ be the space of all complex functions $\mathbf{w}$ of the complex variable z which are defined and analytic on some region Ω of the complex z plane.* For instance, Ω might be the circle $|z| \leqslant 1$. We define addition and scalar multiplication of functions in $\mathcal{V}$ by

$$\begin{aligned}(\mathbf{w}_1+\mathbf{w}_2)(z) &\triangleq \mathbf{w}_1(z)+\mathbf{w}_2(z)\\ (a\mathbf{w})(z) &\triangleq a(\mathbf{w}(z))\end{aligned} \tag{2.6}$$

for all z in Ω. In this example, the zero vector $\boldsymbol{\theta}$ is defined by $\boldsymbol{\theta}(z)=0$ for all z in Ω. (We do not care about the values of the functions $\boldsymbol{\theta}$ and $\mathbf{w}$ outside of Ω.)

Exercise 1. Show that if $\mathbf{w}_1$ and $\mathbf{w}_2$ are in the space $\mathcal{V}$ of Example 10, then $\mathbf{w}_1+\mathbf{w}_2$ is also in $\mathcal{V}$.

Example 11. A Vector Space of Random Variables.† A **random variable x** is a numerical-valued function whose domain consists in the possible outcomes of an experiment or phenomenon. Associated with the experiment is a probability distribution. Therefore, there is a probability distribution associated with the values of the random variable. For example, the throwing of a single die is an experiment. We define the random variable $\mathbf{x}$ in terms of the possible outcomes σ by

$$\begin{aligned}\mathbf{x}(\sigma) &\triangleq 0 \quad \text{for } \sigma=2,4,6 \text{ (the die is even)}\\ &\triangleq 1 \quad \text{for } \sigma=1,3,5 \text{ (the die is odd)}\end{aligned}$$

The probability mass function ω associated with the outcome σ of the experiment is given by

$$\omega(\sigma)=\tfrac{1}{6} \quad \text{for } \sigma=1,2,3,4,5,6$$

*Express the complex variable z in the form $s+it$, where s and t are real. Let the complex function $\mathbf{w}$ be written as $\mathbf{u}+i\mathbf{v}$, where $\mathbf{u}(z)$ and $\mathbf{v}(z)$ are real. Then $\mathbf{w}$ is **analytic** in Ω if and only if the partial derivatives of $\mathbf{u}$ and $\mathbf{v}$ are continuous and satisfy the Cauchy–Riemann conditions in Ω:

$$\frac{\partial \mathbf{u}(z)}{\partial s}=\frac{\partial \mathbf{v}(z)}{\partial t}, \quad \frac{\partial \mathbf{v}(z)}{\partial s}=-\frac{\partial \mathbf{u}(z)}{\partial t}$$

For instance, $\mathbf{w}(z) \triangleq z^2$ is analytic in the whole z plane. See Wylie [2.11].

†See Papoulis [2.7], or Cramér and Leadbetter [2.2].

Then the probability mass function $\omega_{\mathbf{x}}$ associated with the values of the random variable $\mathbf{x}$ is

$$\omega_{\mathbf{x}}(x) = \tfrac{1}{2} \quad \text{for } x = 0, 1$$

We can define many other random variables (functions) for the same die-throwing experiment. One other random variable is

$$\begin{aligned} \mathbf{y}(\sigma) &\triangleq 1 \quad \text{for } \sigma = 1 \text{ (the die is 1)} \\ &\triangleq 0 \quad \text{for } \sigma = 2, 3, 4, 5, 6 \text{ (the die is not 1)} \end{aligned}$$

where

$$\begin{aligned} \omega_{\mathbf{y}}(y) &= \tfrac{5}{6} \quad \text{for } y = 0 \\ &= \tfrac{1}{6} \quad \text{for } y = 1 \end{aligned}$$

Two random variables $\mathbf{x}_1$ and $\mathbf{x}_2$ are equal if and only if their values $\mathbf{x}_1(\sigma)$ and $\mathbf{x}_2(\sigma)$ are identical for all possible outcomes σ of the experiment.

A **vector space of random variables** defined on a given experiment consists in a set of functions defined on the possible outcomes of the experiment, together with the following definitions of addition and scalar multiplication*:

$$(a\mathbf{x})(\sigma) \triangleq a(\mathbf{x}(\sigma)) \qquad (\mathbf{x}+\mathbf{y})(\sigma) \triangleq \mathbf{x}(\sigma) + \mathbf{y}(\sigma)$$

for all possible outcomes σ of the experiment. Let $\mathcal{V}$ be the space of all possible random variables defined on the above die-throwing experiment. If $\mathbf{x}$ and $\mathbf{y}$ are the particular vectors described above, then $\mathbf{x}+\mathbf{y}$ is given by

$$\begin{aligned} (\mathbf{x}+\mathbf{y})(\sigma) &\triangleq 2 \quad \text{for } \sigma = 1 \\ &\triangleq 1 \quad \text{for } \sigma = 3, 5 \\ &\triangleq 0 \quad \text{for } \sigma = 2, 4, 6 \end{aligned}$$

and

$$\begin{aligned} \omega_{\mathbf{x}+\mathbf{y}}(z) &= \tfrac{1}{2} \quad \text{for } z = 0 \\ &= \tfrac{1}{3} \quad \text{for } z = 1 \\ &= \tfrac{1}{6} \quad \text{for } z = 2 \end{aligned}$$

What is the zero random variable for the vector space $\mathcal{V}$?

*We note that the set of functions must be such that it includes all sums and scalar multiples of its members.

2.2 Relations Among Vectors

Combining Vectors

Assuming a vector represents the condition or change in condition of a system, we can use the definitions of addition and scalar multiplication of vectors to find the net result of several successive changes in condition of the system.

Definition. A vector $\mathbf{x}$ is said to be a **linear combination** of the vectors $\mathbf{x}_1, \mathbf{x}_2, \ldots, \mathbf{x}_n$ if it can be expressed as

$$\mathbf{x} = c_1\mathbf{x}_1 + c_2\mathbf{x}_2 + \cdots + c_n\mathbf{x}_n \tag{2.7}$$

for some set of scalars $c_1, \ldots, c_n$. This concept is illustrated in Figure 2.5 where $\mathbf{x} = \frac{1}{2}\mathbf{x}_1 + \mathbf{x}_2 - \mathbf{x}_3$.

A vector space $\mathcal{V}$ is simply a set of elements and a definition of linear combination (addition and scalar multiplication); the space $\mathcal{V}$ includes all linear combinations of its own elements. If $\mathcal{S}$ is a subset of $\mathcal{V}$, the set of all linear combinations of vectors from $\mathcal{S}$, using the same definition of linear combination, is also a vector space. We call it a subspace of $\mathcal{V}$. A line or plane through the origin of the three-dimensional arrow space is an example of a subspace.

Definition. A subset $\mathcal{W}$ of a linear space $\mathcal{V}$ is a **linear subspace** (or **linear manifold**) of $\mathcal{V}$ if along with every pair, $\mathbf{x}_1$ and $\mathbf{x}_2$, of vectors in $\mathcal{W}$, every linear combination $c_1\mathbf{x}_1 + c_2\mathbf{x}_2$ is also in $\mathcal{W}$.* We call $\mathcal{W}$ a *proper subspace* if it is smaller than $\mathcal{V}$; that is if $\mathcal{W}$ is not $\mathcal{V}$ itself.

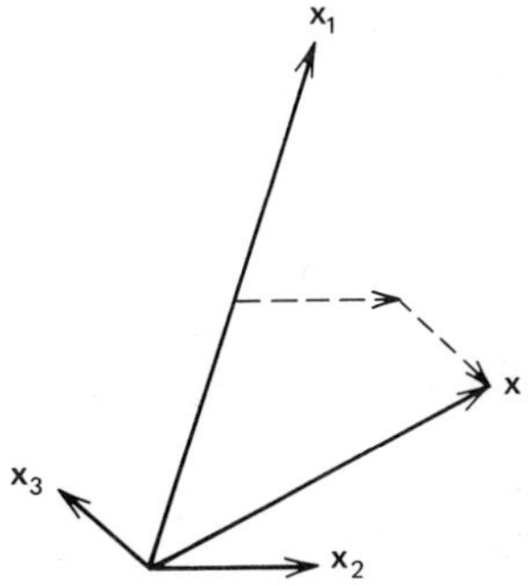

Figure 2.5. A linear combination of arrows.

*In the discussion of infinite-dimensional Hilbert spaces (Section 5.3), we distinguish between a linear subspace and a linear manifold. Linear manifold is the correct term to use in this definition. Yet because a finite-dimensional linear manifold is a linear subspace as well, we emphasize the physically motivated term subspace.

Example 1. A Linear Subspace. The set of vectors from $\mathcal{R}^3$ which are of the form (c_1, c_2, c_1+c_2) forms a subspace of $\mathcal{R}^3$. It is, in fact, the set of all linear combinations of the two vectors (1, 0, 1) and (0, 1, 1).

Example 2. A Solution Space. The set $\mathcal{W}$ of all solutions to the matrix equation

$$\begin{pmatrix} 1 & 1 & 1 \\ 1 & 2 & 2 \\ 2 & 3 & 3 \end{pmatrix}\begin{pmatrix} \xi_1 \\ \xi_2 \\ \xi_3 \end{pmatrix}=\begin{pmatrix} 0 \\ 0 \\ 0 \end{pmatrix}$$

is a subspace of $\mathfrak{M}^{3\times 1}$. By elimination (Section 1.5), we find that $\mathcal{W}$ contains all vectors of the form $(0 \;\; \xi_2 \;\; -\xi_2)^{\mathrm{T}}$. Clearly, $\mathcal{W}$ consists in all linear combinations of the single vector $(0 \;\; 1 \;\; -1)^{\mathrm{T}}$. This example extends to general matrices. Let $\mathbf{A}$ be an $m\times n$ matrix. Let $\mathbf{x}$ be in $\mathfrak{M}^{n\times 1}$. Using the rules of matrix multiplication (Appendix 1) it can be shown that if $\mathbf{x}_1$ and $\mathbf{x}_2$ are solutions to $\mathbf{Ax}=\boldsymbol{\theta}$, then an arbitrary linear combination $c_1\mathbf{x}_1+c_2\mathbf{x}_2$ is also a solution. Thus the space of solutions is a subspace of $\mathfrak{M}^{n\times 1}$.

Example 3. Subspaces (Linear Manifolds) of Functions. Let $\mathcal{C}^2(\Omega)$ be the space of all real-valued functions which are defined and have continuous second partial derivatives in the two-dimensional region Ω. (This region could be the square $0\leqslant s\leqslant 1$, $0\leqslant t\leqslant 1$, for instance.) Let Γ denote the boundary of the region Ω. Linear combination in $\mathcal{C}^2(\Omega)$ is defined by

$$\begin{aligned}(\mathbf{f}+\mathbf{g})(s,t) &\triangleq \mathbf{f}(s,t)+\mathbf{g}(s,t)\\ (a\mathbf{f})(s,t) &\triangleq a(\mathbf{f}(s,t))\end{aligned} \tag{2.8}$$

for all (s,t) in Ω. The functions $\mathbf{f}$ in $\mathcal{C}^2(\Omega)$ which satisfy the homogeneous boundary condition $\mathbf{f}(s,t)=0$ for (s,t) on Γ constitute a linear manifold of $\mathcal{C}^2(\Omega)$. For if $\mathbf{f}_1$ and $\mathbf{f}_2$ satisfy the boundary condition, then $(c_1\mathbf{f}_1+c_2\mathbf{f}_2)(s,t)=c_1\mathbf{f}_1(s,t)+c_2\mathbf{f}_2(s,t)=0$, and the arbitrary linear combination $c_1\mathbf{f}_1+c_2\mathbf{f}_2$ also satisfies the boundary condition.

The set of solutions to Laplace's equation,

$$\frac{\partial^2\mathbf{f}(s,t)}{\partial s^2}+\frac{\partial^2\mathbf{f}(s,t)}{\partial t^2}=0 \tag{2.9}$$

for all (s,t) in Ω, also forms a linear manifold of $\mathcal{C}^2(\Omega)$. For if $\mathbf{f}_1$ and $\mathbf{f}_2$ both satisfy (2.9), then

$$\frac{\partial^2[c_1\mathbf{f}_1(s,t)+c_2\mathbf{f}_2(s,t)]}{\partial s^2}+\frac{\partial^2[c_1\mathbf{f}_1(s,t)+c_2\mathbf{f}_2(s,t)]}{\partial t^2}=0$$

and the arbitrary linear combination $c_1\mathbf{f}_1+c_2\mathbf{f}_2$ also satisfies (2.9). Equation (2.9) is phrased in terms of the values of $\mathbf{f}$. Laplace's equation can also be expressed in the vector notation

$$\nabla^2\mathbf{f}=\boldsymbol{\theta} \tag{2.10}$$

The domain of definition Ω is implicit in (2.10). The vector $\boldsymbol{\theta}$ is defined by $\boldsymbol{\theta}(s,t)=0$ for all (s,t) in Ω.

In using vector diagrams to analyze physical problems, we often resolve a vector into a linear combination of component vectors. We usually do this in a unique manner. In Figure 2.5, $\mathbf{x}$ is not a unique linear combination of $\mathbf{x}_1$, $\mathbf{x}_2$, and $\mathbf{x}_3$; $\mathbf{x}=0\mathbf{x}_1+3\mathbf{x}_2+2\mathbf{x}_3$ is a second resolution of $\mathbf{x}$; the number of possible resolutions is infinite. In point of fact, $\mathbf{x}$ can be represented as a linear combination of any two of the other vectors; the three vectors $\mathbf{x}_1$, $\mathbf{x}_2$, and $\mathbf{x}_3$ are redundant as far as representation of $\mathbf{x}$ is concerned.

Definition. The vectors $\mathbf{x}_1, \mathbf{x}_2, \ldots, \mathbf{x}_n$ are **linearly dependent** (or coplanar) if at least one of them can be written as a linear combination of the others. Otherwise they are **linearly independent**. (We often refer to sets of vectors as simply "dependent" or "independent.")

In Figure 2.5 the set $\{\mathbf{x}_1, \mathbf{x}_2, \mathbf{x}_3\}$ is dependent. Any two of the vectors form an independent set. In any vector space, a set which contains the $\boldsymbol{\theta}$ vector is dependent, for $\boldsymbol{\theta}$ can be written as zero times any other vector in the set. We define the $\boldsymbol{\theta}$ vector by itself as a dependent set.

The following statement is equivalent to the above definition of independence: the vectors $\mathbf{x}_1, \mathbf{x}_2, \ldots, \mathbf{x}_n$ are linearly independent if and only if

$$c_1\mathbf{x}_1+c_2\mathbf{x}_2+\cdots+c_n\mathbf{x}_n=\boldsymbol{\theta} \quad\Rightarrow\quad c_1=\cdots=c_n=0 \tag{2.11}$$

Equation (2.11) says the "zero combination" is the only combination that equals $\boldsymbol{\theta}$. For if c_i were not 0, we could simply divide by c_i to find $\mathbf{x}_i$ as a linear combination of the other vectors, and the set $\{\mathbf{x}_j\}$ would be dependent. If $c_i=0$, $\mathbf{x}_i$ cannot be a linear combination of the other vectors. Equation (2.11) is a practical tool for determining independence of vectors.

Exercise 1. Explore graphically and by means of (2.11) the following set of vectors from $\mathfrak{R}^3$: $\{\mathbf{x}_1=(1, 0, 0), \mathbf{x}_2=(0, 1, 0), \mathbf{x}_3=(1, 1, 0), \mathbf{x}_4=(0, 0, 1)\}$.

Example 4. Determining Independence. In the space $\mathfrak{R}^3$ let $\mathbf{x}_1=(1,2,1)$, $\mathbf{x}_2=(2, 3, 1)$, and $\mathbf{x}_3=(4, 7, 3)$. Equation (2.11) becomes

$$\begin{aligned}c_1(1,2,1)&+c_2(2,3,1)+c_3(4,7,3)\\&=(c_1+2c_2+4c_3,\quad 2c_1+3c_2+7c_3,\quad c_1+c_2+3c_3)\\&=(0,0,0)\end{aligned}$$

Each component of this vector equation is a scalar-valued linear algebraic equation. We write the three equations in the matrix form:

$$\begin{pmatrix} 1 & 2 & 4 \\ 2 & 3 & 7 \\ 1 & 1 & 3 \end{pmatrix}\begin{pmatrix} c_1 \\ c_2 \\ c_3 \end{pmatrix}=\begin{pmatrix} 0 \\ 0 \\ 0 \end{pmatrix}$$

We solve this equation by elimination (Section 1.5) to find $c_1 = -2c_3$ and $c_2 = -c_3$. Any choice for c_3 will yield a particular nonzero linear combination of the vectors $\mathbf{x}_1$, $\mathbf{x}_2$, $\mathbf{x}_3$ which equals $\boldsymbol{\theta}$. The set is linearly dependent.

Definition. Let $\mathcal{S} \triangleq \{\mathbf{x}_1, \mathbf{x}_2, \ldots, \mathbf{x}_n\}$ be a set of vectors from a linear space $\mathcal{V}$. The set of all linear combinations of vectors from $\mathcal{S}$ is called the subspace of $\mathcal{V}$ **spanned** (or generated) by $\mathcal{S}$.* We often refer to this subspace as span($\mathcal{S}$) or span$\{\mathbf{x}_1, \ldots, \mathbf{x}_n\}$.

Bases and Coordinates

We have introduced the vector space concept in order to provide a common mathematical framework for different types of systems. We can make the similarities between systems more apparent by converting their vector space representations to a standard form. We perform this standardization by introducing coordinate systems. In the example of Figure 2.5, the vectors $\{\mathbf{x}, \mathbf{x}_1, \mathbf{x}_2, \mathbf{x}_3\}$ span a plane; yet any two of them will span the same plane. Two of them are redundant as far as generation of the plane is concerned.

Definition. A **basis** (or coordinate system) for a linear space $\mathcal{V}$ is a linearly independent set of vectors from $\mathcal{V}$ which spans $\mathcal{V}$.

Example 5. The Standard Bases for $\mathcal{R}^n$, $\mathcal{M}^{n\times 1}$, and $\mathcal{P}^n$. It is evident that any three linearly independent vectors in $\mathcal{R}^3$ form a basis for $\mathcal{R}^3$. The n-tuples

$$\begin{aligned} \boldsymbol{\varepsilon}_1 &= (1,0,\ldots,0) \\ \boldsymbol{\varepsilon}_2 &= (0,1,0,\ldots,0) \\ &\vdots \\ \boldsymbol{\varepsilon}_n &= (0,\ldots,0,1) \end{aligned} \tag{2.12}$$

form a basis for $\mathcal{R}^n$. The set $\mathcal{E} \triangleq \{\boldsymbol{\varepsilon}_1, \ldots, \boldsymbol{\varepsilon}_n\}$ is called the **standard basis for** $\mathcal{R}^n$. We use the same notation to represent the **standard basis for** $\mathcal{M}^{n\times 1}$: $\mathcal{E} \triangleq \{\boldsymbol{\varepsilon}_i\}$, where $\boldsymbol{\varepsilon}_i$ is a column vector of zeros except for a 1 in the ith place. The set $\mathcal{N} \triangleq \{\mathbf{f}_1, \mathbf{f}_2, \ldots, \mathbf{f}_n\}$ defined by $\mathbf{f}_k(t) = t^{k-1}$ forms a basis for $\mathcal{P}^n$; it is analogous to the standard bases for $\mathcal{R}^n$ and $\mathcal{M}^{n\times 1}$.

*The definition of the space spanned by an infinite set of vectors depends on limiting concepts. We delay the definition until Section 5.3.

Example 6. The Zero Vector Space. The set $\{\boldsymbol{\theta}\}$ together with the obvious definitions of addition and scalar multiplication forms a vector space which we denote $\mathcal{O}$. However, the vector $\boldsymbol{\theta}$, by itself, is a dependent set. Therefore $\mathcal{O}$ has no basis.

If $\mathcal{X} \triangleq \{\mathbf{x}_1, \mathbf{x}_2, \ldots, \mathbf{x}_n\}$ is a basis for the space $\mathcal{V}$, any vector $\mathbf{x}$ in $\mathcal{V}$ can be written uniquely as some linear combination

$$\mathbf{x} = c_1\mathbf{x}_1 + c_2\mathbf{x}_2 + \cdots + c_n\mathbf{x}_n \tag{2.13}$$

of vectors in $\mathcal{X}$. The multipliers c_i are called the **coordinates of x relative to the ordered basis** $\mathcal{X}$. It is easy to show that the coordinates relative to a particular ordered basis are unique: just expand $\mathbf{x}$ as in (2.13) for a second set $\{d_i\}$ of coordinates; then independence of the basis vectors implies $d_i = c_i$.

It is common to write the coordinates of a vector relative to a particular basis as a column matrix. We will denote by $[\mathbf{x}]_{\mathcal{X}}$ the **coordinate matrix** of the vector $\mathbf{x}$ relative to the (ordered) basis $\mathcal{X}$; thus corresponding to (2.13) we have

$$[\mathbf{x}]_{\mathcal{X}} \triangleq \begin{pmatrix} c_1 \\ c_2 \\ \vdots \\ c_n \end{pmatrix} \tag{2.14}$$

Some bases are more natural or convenient than others. We use the term **natural basis** to mean a basis relative to which we can find coordinates by inspection. The bases of Example 5 are natural bases for $\mathcal{R}^n$, $\mathfrak{M}^{n\times 1}$, and $\mathcal{P}^n$. Thus if $\mathbf{f}(t) = \xi_1 + \xi_2 t + \cdots + \xi_n t^{n-1}$, then $[\mathbf{f}]_{\mathfrak{N}} = (\xi_1\ \xi_2 \cdots \xi_n)^{\mathrm{T}}$.

Example 7. Coordinates for Vectors in $\mathcal{R}^3$. Let $\mathcal{X} \triangleq \{\mathbf{x}_1, \mathbf{x}_2, \mathbf{x}_3\}$ be an ordered basis for $\mathcal{R}^3$, where $\mathbf{x}_1 = (1, 2, 3)$, $\mathbf{x}_2 = (2, 3, 2)$, and $\mathbf{x}_3 = (2, 5, 5)$. Let $\mathbf{x} = (1, 1, 1)$. To find $[\mathbf{x}]_{\mathcal{X}}$, we must solve (2.13):

$$\begin{aligned}(1,1,1) &= c_1(1,2,3) + c_2(2,3,2) + c_3(2,5,5).\\ &= (c_1 + 2c_2 + 2c_3,\quad 2c_1 + 3c_2 + 5c_3,\quad 3c_1 + 2c_2 + 5c_3)\end{aligned}$$

We rewrite the vector (3-tuple) equation in the matrix notation:

$$\begin{pmatrix} 1 & 2 & 2 \\ 2 & 3 & 5 \\ 3 & 2 & 5 \end{pmatrix}\begin{pmatrix} c_1 \\ c_2 \\ c_3 \end{pmatrix} = \begin{pmatrix} 1 \\ 1 \\ 1 \end{pmatrix} \tag{2.15}$$

We solved this equation in Example 1 of Section 1.5. The result is

$$[\mathbf{x}]_{\mathfrak{X}} \stackrel{\Delta}{=} [(1,1,1)]_{\mathfrak{X}} = \begin{pmatrix} c_1 \\ c_2 \\ c_3 \end{pmatrix} = \begin{pmatrix} \frac{3}{5} \\ \frac{3}{5} \\ -\frac{2}{5} \end{pmatrix}$$

The coordinate matrix of Example 7 is merely a simple way of stating that $\mathbf{x} = \frac{3}{5}\mathbf{x}_1 + \frac{3}{5}\mathbf{x}_2 - \frac{2}{5}\mathbf{x}_3$. We choose to write the coordinates of a vector $\mathbf{x}$ as a column matrix because it allows us to carry out in a standard matrix format all manipulations involving the coordinates of $\mathbf{x}$.

In Example 4 of Section 1.5 we solved (2.15) with a general right-hand side; that is, for $\mathbf{x} = (\eta_1, \eta_2, \eta_3)$. That solution allows us to determine quickly the coordinate matrix, relative to the basis $\mathfrak{X}$ of Example 7, for *any* vector $\mathbf{x}$ in $\mathfrak{R}^3$, including the case $\mathbf{x} = (0, 0, 0)$. In general, (2.13) includes (2.11); inherent in the process of finding coordinates for an arbitrary vector $\mathbf{x}$ is the process of determining whether $\mathfrak{X}$ is a basis. If $\mathfrak{X}$ is not independent, there will exist nonzero coordinates for $\mathbf{x} = \boldsymbol{\theta}$. If $\mathfrak{X}$ does not span the space, there will be some vector $\mathbf{x}$ for which no coordinates exist (P&C 2.7).

Example 8. Coordinates for Vectors in $\mathcal{P}^3$. Let $\mathcal{F} \stackrel{\Delta}{=} \{\mathbf{f}_1, \mathbf{f}_2, \mathbf{f}_3\}$ be an ordered basis for $\mathcal{P}^3$, where $\mathbf{f}_1(t) = 1 + 2t + 3t^2$, $\mathbf{f}_2(t) = 2 + 3t + 2t^2$, and $\mathbf{f}_3(t) = 2 + 5t + 5t^2$. Let $\mathbf{f}$ be defined by $\mathbf{f}(t) = 1 + t + t^2$. To find $[\mathbf{f}]_{\mathcal{F}}$, we solve (2.13), $\mathbf{f} = c_1\mathbf{f}_1 + c_2\mathbf{f}_2 + c_3\mathbf{f}_3$. To solve this equation, we evaluate both sides at t:

$$\begin{aligned} \mathbf{f}(t) &= (c_1\mathbf{f}_1 + c_2\mathbf{f}_2 + c_3\mathbf{f}_3)(t) \\ &= c_1\mathbf{f}_1(t) + c_2\mathbf{f}_2(t) + c_3\mathbf{f}_3(t) \end{aligned} \tag{2.16}$$

or

$$\begin{aligned} 1 + t + t^2 &= c_1(1 + 2t + 3t^2) + c_2(2 + 3t + 2t^2) + c_3(2 + 5t + 5t^2) \\ &= (c_1 + 2c_2 + 2c_3) + (2c_1 + 3c_2 + 5c_3)t + (3c_1 + 2c_2 + 5c_3)t^2 \end{aligned}$$

Equating coefficients on like powers of t we again obtain (2.15). The coordinate matrix of $\mathbf{f}$ is

$$[\mathbf{f}]_{\mathcal{F}} = \begin{pmatrix} c_1 \\ c_2 \\ c_3 \end{pmatrix} = \begin{pmatrix} \frac{3}{5} \\ \frac{3}{5} \\ -\frac{2}{5} \end{pmatrix}$$

In order to solve the vector (function) equation (2.16) we converted it to a set of scalar equations expressed in matrix form. A second method for converting (2.16) to a matrix equation in the unknowns $\{c_i\}$ is to evaluate the equation at three different values of t. Each such evaluation yields an algebraic equation in $\{c_i\}$. The resulting matrix equation is different from (2.15), but the solution is the same. We now describe a general method, built around a natural basis, for converting (2.13) to a matrix equation.The coordinate matrix of a vector $\mathbf{x}$ relative to the basis $\mathcal{X}=\{\mathbf{x}_1,\ldots,\mathbf{x}_n\}$ is $[\mathbf{x}]_{\mathcal{X}}=(c_1\cdots c_n)^{\mathrm{T}}$, where the coordinates c_i are obtained by solving the vector equation

$$\mathbf{x}=c_1\mathbf{x}_1+\cdots+c_n\mathbf{x}_n$$

A general method for obtaining an equivalent matrix equation consists in taking coordinates of the vector equation relative to a natural basis $\mathcal{N}$—a basis relative to which coordinates can be obtained by inspection. The vector equation becomes

$$\begin{aligned}[\mathbf{x}]_{\mathcal{N}} &= \left[\sum_{i=1}^{n} c_i\mathbf{x}_i\right]_{\mathcal{N}}\\ &= \sum_{i=1}^{n} c_i[\mathbf{x}_i]_{\mathcal{N}}\\ &= \left([\mathbf{x}_1]_{\mathcal{N}} \vdots \cdots \vdots [\mathbf{x}_n]_{\mathcal{N}}\right)[\mathbf{x}]_{\mathcal{X}} \qquad (2.17)\end{aligned}$$

We determine $[\mathbf{x}]_{\mathcal{N}}$, $[\mathbf{x}_1]_{\mathcal{N}},\cdots,[\mathbf{x}_n]_{\mathcal{N}}$ by inspection. Then we solve (2.17) routinely for $[\mathbf{x}]_{\mathcal{X}}$.

Example 9. Finding Coordinates via a Natural Basis. Let the set $\mathcal{F}\triangleq\{\mathbf{f}_1,\mathbf{f}_2,\mathbf{f}_3\}$ be a basis for $\mathcal{P}^3$, where $\mathbf{f}_1(t)=1+2t+3t^2$, $\mathbf{f}_2(t)=2+3t+2t^2$, and $\mathbf{f}_3(t)=2+5t+5t^2$. We seek $[\mathbf{f}]_{\mathcal{F}}$ for the vector $\mathbf{f}(t)=1+t+t^2$. To convert the defining equation for coordinates into a matrix equation, we use the natural basis $\mathcal{N}\triangleq\{\mathbf{g}_1,\mathbf{g}_2,\mathbf{g}_3\}$, where $\mathbf{g}_k(t)=t^{k-1}$. For this problem, (2.17) becomes

$$[\mathbf{f}]_{\mathcal{N}}=\left([\mathbf{f}_1]_{\mathcal{N}} \vdots [\mathbf{f}_2]_{\mathcal{N}} \vdots [\mathbf{f}_3]_{\mathcal{N}}\right)[\mathbf{f}]_{\mathcal{F}}$$

or

$$\begin{pmatrix}1\\1\\1\end{pmatrix}=\begin{pmatrix}1 & \vdots & 2 & \vdots & 2\\ 2 & \vdots & 3 & \vdots & 5\\ 3 & \vdots & 2 & \vdots & 5\end{pmatrix}\begin{pmatrix}c_1\\c_2\\c_3\end{pmatrix}$$

The solution to this equation is $[\mathbf{f}]_{\mathcal{F}}=(\frac{3}{5}\ \ \frac{3}{5}\ \ -\frac{2}{5})^{\mathrm{T}}$. (Compare with Example 8.)

Typically, the solution of (2.17) requires the elimination procedure

$$([\mathbf{x}_1]_{\mathfrak{N}} \vdots \cdots \vdots [\mathbf{x}_n]_{\mathfrak{N}} \vdots [\mathbf{x}]_{\mathfrak{N}}) \rightarrow (\mathbf{I} \vdots [\mathbf{x}]_{\mathfrak{X}}) \tag{2.18}$$

If we wish to solve for the coordinates of more than one vector, we still perform the elimination indicated in (2.18), but augment the matrix with all the vectors whose coordinates we desire. Thus if we wish the coordinates for $\mathbf{z}_1$, $\mathbf{z}_2$, and $\mathbf{z}_3$, we perform elimination on

$$([\mathbf{x}_1]_{\mathfrak{N}} \vdots \cdots \vdots [\mathbf{x}_n]_{\mathfrak{N}} \vdots [\mathbf{z}_1]_{\mathfrak{N}} \vdots [\mathbf{z}_2]_{\mathfrak{N}} \vdots [\mathbf{z}_3]_{\mathfrak{N}})$$

This elimination requires less computation than does the process which goes through inversion of the matrix $([\mathbf{x}_1]_{\mathfrak{N}} \vdots \cdots \vdots [\mathbf{x}_n]_{\mathfrak{N}})$, regardless of the number of vectors whose coordinates we desire (P&C 1.3).

Example 10. A Basis and Coordinates for a Subspace. Let $\mathcal{W}$ be the subspace of $\mathcal{P}^3$ consisting in all functions $\mathbf{f}$ defined by the rule $\mathbf{f}(t)=\xi_1+\xi_2 t+(\xi_1+\xi_2)t^2$ for some ξ_1 and ξ_2. Note that the standard basis functions for $\mathcal{P}^3$ are not contained in $\mathcal{W}$. The functions defined by $\mathbf{g}_1(t)=1+t^2$ and $\mathbf{g}_2(t)=t+t^2$ are clearly independent vectors in $\mathcal{W}$. Because there are two "degrees of freedom" in $\mathcal{W}$ (i.e., two parameters ξ_1 and ξ_2 must be given to specify a particular function in $\mathcal{W}$) we expect the set $\mathcal{G} \triangleq \{\mathbf{g}_1, \mathbf{g}_2\}$ to span $\mathcal{W}$ and thus be a basis. We seek the coordinate matrix $[\mathbf{f}]_{\mathcal{G}}$ of an arbitrary vector $\mathbf{f}$ in $\mathcal{W}$. That is, we seek c_1 and c_2 such that

$$\mathbf{f}(t)=c_1\mathbf{g}_1(t)+c_2\mathbf{g}_2(t)$$

The matrix equation (2.17) can be written by inspection using the natural basis $\mathfrak{N}$ of Example 9:

$$[\mathbf{f}]_{\mathfrak{N}}=\left([\mathbf{g}_1]_{\mathfrak{N}} \vdots [\mathbf{g}_2]_{\mathfrak{N}}\right)[\mathbf{f}]_{\mathcal{G}}$$

or

$$\begin{pmatrix} \xi_1 \\ \xi_2 \\ \xi_1+\xi_2 \end{pmatrix}=\begin{pmatrix} 1 & 0 \\ 0 & 1 \\ 1 & 1 \end{pmatrix}\begin{pmatrix} c_1 \\ c_2 \end{pmatrix}$$

Then $c_i=\xi_i$ and

$$[\mathbf{f}]_{\mathcal{G}}=\begin{pmatrix} \xi_1 \\ \xi_2 \end{pmatrix}$$

Because we were able to solve uniquely for the coordinates, we know that $\mathcal{G}$ is indeed a basis for $\mathcal{W}$. The subspace $\mathcal{W}$ is equivalent to the subspace of Example 1. Note that the elimination procedure does not agree precisely with (2.18) because there are only two degrees of freedom among the three coefficients of the arbitrary vector **f** in $\mathcal{W}$.

Dimension

The equivalence between the three vector spaces $\mathcal{R}^3$, $\mathcal{P}^3$, and $\mathfrak{M}^{3\times 1}$ is apparent from Examples 7 and 8. The subspace $\mathcal{W}$ of Example 10, however, is equivalent to $\mathfrak{M}^{2\times 1}$ rather than $\mathfrak{M}^{3\times 1}$, even though the elements of $\mathcal{W}$ are polynomials in $\mathcal{P}^3$. The key to the equivalence lies not in the nature of the elements, but rather in the number of "degrees of freedom" in each space (the number of scalars which must be specified in order to specify a vector); more to the point, the key lies in the number of vectors in a basis for each space.

Definition. A vector space is **finite dimensional** if it is spanned by a finite number of vectors. It is intuitively clear that all bases for a finite-dimensional space contain the same number of vectors. The number of vectors in a basis for a finite-dimensional space $\mathcal{V}$ is called the **dimension** of $\mathcal{V}$ and is denoted by $\dim(\mathcal{V})$.

Thus $\mathcal{R}^3$ and $\mathcal{P}^3$ are both three-dimensional spaces. The subspace $\mathcal{W}$ of Example 10 has dimension 2. Knowledge of the dimension of a space (or a subspace) is obtained in the course of determining a basis for the space (subspace). Since the space $\mathcal{O} \triangleq \{\boldsymbol{\theta}\}$ has no basis, we assign it dimension zero.

Example 11. A Basis for a Space of Random Variables. A vector space $\mathcal{V}$ of random variables, defined on the possible outcomes of a single die-throwing experiment, is described in Example 11 of Section 2.1. A natural basis for $\mathcal{V}$ is the set of random variables $\mathcal{X} \triangleq \{\mathbf{x}_i,\ i=1,\ldots,6\}$, where

$$\begin{aligned} \mathbf{x}_i(\sigma) &\triangleq 1 \text{ for } \sigma = i \text{ (the die equals } i) \\ &\triangleq 0 \text{ for } \sigma \neq i \text{ (the die does not equal } i) \end{aligned}$$

That $\mathcal{X}$ is a basis for $\mathcal{V}$ can be seen from an attempt to determine the coordinates with respect to $\mathcal{X}$ of an arbitrary random variable **z** defined on the experiment. If

$$\begin{aligned} \mathbf{z}(\sigma) &\triangleq \quad c_1 \quad \text{for} \quad \sigma = 1 \\ &\qquad \vdots \\ &\triangleq \quad c_6 \quad \text{for} \quad \sigma = 6 \end{aligned}$$

then $[\mathbf{z}]_{\mathcal{X}} = (c_1 \cdots c_6)^{\mathrm{T}}$; a unique representation exists.

The random variables $\{\mathbf{x}_1,\ldots,\mathbf{x}_6\}$ are linearly independent. However, they are not *statistically* independent. **Statistical independence** of two random variables $\mathbf{x}$ and $\mathbf{y}$ means that knowledge of the *value* of one variable, say, $\mathbf{x}$, does not tell us anything about the outcome of the experiment which determines the value of the other variable $\mathbf{y}$, and therefore it tells us nothing about the value of $\mathbf{y}$. The random variables $\{\mathbf{x}_i\}$ are related by the underlying die-throwing experiment. If we know $\mathbf{x}_1 = 0$, for instance, then we know $\sigma \neq 1$ (the die is not equal to 1); the probability mass functions for $\mathbf{x}_2,\ldots,\mathbf{x}_6$ and for all other vectors in $\mathcal{V}$ are modified by the information concerning the value of $\mathbf{x}_1$. The new probability mass functions for $\mathbf{x}$ and $\mathbf{y}$ of Example 11, Section 2.1, given that $\mathbf{x}_1 = 0$, are

$$\begin{aligned} \omega_{\mathbf{x}}(x;\mathbf{x}_1=0) &= \tfrac{3}{5} \quad \text{for } x=0 \qquad & \omega_{\mathbf{y}}(y;\mathbf{x}_1=0) &= 1 \quad \text{for } y=0 \\ &= \tfrac{2}{5} \quad \text{for } x=1 & &= 0 \quad \text{for } y=1 \end{aligned}$$

The space l_2 of square-summable sequences described in Example 4 of Section 2.1 is obviously *infinite dimensional*. A direct extension of the standard basis for $\mathcal{R}^n$ seems likely to be a basis for l_2. It is common knowledge that functions $\mathbf{f}$ in $\mathcal{C}(0, 2\pi)$, the space of functions continuous on $[0, 2\pi]$, can be expanded uniquely in a Fourier series of the form $\mathbf{f}(t) = b_0 + \sum_{k=1}^{\infty} (a_k \,\mathbf{sin}\, kt + b_k \,\mathbf{cos}\, kt)$. This fact leads us to suspect that the set of functions

$$\mathcal{F} \triangleq \{\mathbf{1}, \mathbf{sin}\, t, \mathbf{cos}\, t, \mathbf{sin}\, 2t, \mathbf{cos}\, 2t, \ldots\} \tag{2.19}$$

forms a basis for $\mathcal{C}(0, 2\pi)$, and that the coordinates of $\mathbf{f}$ relative to this basis are

$$(b_0, a_1, b_1, a_2, b_2, \ldots)$$

This suspicion is correct. The coordinates (or Fourier coefficients) actually constitute a vector in l_2. We show in Example 11 of Section 5.3 that l_2 serves as a convenient standard space of coordinate vectors for infinite-dimensional spaces; in that sense, it plays the same role that $\mathcal{M}^{n \times 1}$ does for n-dimensional spaces. Unfortunately, the concepts of independence, spanning sets, and bases do not extend easily to infinite-dimensional vector spaces. The concept of linear combination applies only to the combination of a finite number of vectors. We cannot add an infinite number of vectors without the concept of a limit; this concept is introduced in Chapter 5. Hence detailed examination of infinite-dimensional function spaces is left for that chapter.

Summary

There is no inherent basis in any space—one basis is as good as another. Yet a space may have one basis which appears more convenient than others. The standard basis for $\mathcal{R}^n$ is an example. By picking units of measurement in a physical system (e.g., volts, feet, degrees centigrade) we tie together the system and the model; our choice of units may automatically determine convenient or standard basis vectors for the vector space of the model (based on, say, 1 V, 1 ft, or 1° C).

By choosing a basis for a space, we remove the most distinguishing feature of that space, the nature of its elements, and thus tie each vector in the space to a unique coordinate matrix. Because of this unique connection which a basis establishes between the elements of a particular vector space and the elements of the corresponding space of coordinate matrices, we are able to carry out most vector manipulations in terms of coordinate matrices which represent the vectors. We have selected $\mathfrak{M}^{n\times 1}$, rather than $\mathcal{R}^n$, as our standard n-dimensional space because matrix operations are closely tied to computer algorithms for solving linear algebraic equations (Section 1.5). Most vector space manipulations lead eventually to such equations.

Because coordinate matrices are themselves vectors in a vector space ($\mathfrak{M}^{n\times 1}$), we must be careful to distinguish vectors from their coordinates. The confusion is typified by the problem of finding the coordinate matrix of a vector $\mathbf{x}$ from $\mathfrak{M}^{n\times 1}$ relative to the standard basis for $\mathfrak{M}^{n\times 1}$. In this instance $[\mathbf{x}]_{\mathcal{E}} = \mathbf{x}$; the difference between the vector and its coordinate matrix is only conceptual. A vector is simply one of a set of elements, although we may use it to represent the physical condition of some system. The coordinate matrix of the vector, on the other hand, is the unique set of multipliers which specifies the vector as a linear combination of arbitrarily chosen basis vectors.

2.3 System Models

The concept of a vector as a model for the condition or change in condition of a system is explored in Sections 2.1 and 2.2. We usually separate the variables which pertain to the condition of the system into two broad sets: the independent (or input) variables, the values of which are determined outside of the system, and the dependent (or output) variables, whose values are determined by the system together with the independent variables. A model for the system itself consists in expressions of relations among the variables. In this section we identify properties of system models.

Example 1. An Economic System. Let **x** represent a set of inputs to the U. S. national economy (tax rates, interest rates, reinvestment policies, etc.); let **y** represent a set of economic indicators (cost of living, unemployment rate, growth rate, etc.). The system model **T** must describe the economic laws which relate **y** to **x**.

Example 2. A Baking Process. Suppose **x** is the weight of a sample of clay before a baking process and **y** is the weight after baking. Then the system model **T** must describe the chemical and thermodynamic laws insofar as they relate **x** and **y**.

Example 3. A Positioning System. Suppose the system of interest is an armature-controlled motor which is used to position a piece of equipment. Let **x** represent the armature voltage, a function of time; let **y** be the shaft position, another function of time. The system model **T** should describe the manner in which the dynamic system relates the function **y** to the function **x**.

The variables in the economic system of Example 1 clearly separate into input (or independent) variables and output (or system condition) variables. In Example 2, both the independent and dependent variables describe the condition of the system. Yet we can view the condition before baking as the input to the system and view the condition after baking as the output. The dynamic system of Example 3 is reciprocal; **x** and **y** are mutually related by **T**. Since the system is used as a motor, we view the armature voltage **x** as the input to the system and the shaft position **y** as the output. We could, as well, use the machine as a dc generator; then we would view the shaft position as the input and the armature voltage as the output.

The notation $\mathbf{Tx}=\mathbf{y}$ that we introduced in (1.1) implies that the model **T** does something to the vector **x** to yield the vector **y**. As a result, we may feel inclined to call **x** the input and **y** the output. Yet in Section 1.3 we note that equations are sometimes expressed in an inverse form. The positions of the variables in an equation do not determine whether they are independent or dependent variables. Furthermore, we can see from Example 3 that the input and output of a system in some instances may be determined arbitrarily. In general, we treat one of the vectors in the equation $\mathbf{Tx}=\mathbf{y}$ as the input and the other as the output. However, unless we are exploring a problem for which the input is clearly defined, we use the terms input and output loosely in reference to the known and unknown variables, respectively.

Transformations on Vector Spaces

Our present purpose is to make more precise the vaguely defined model **T** introduced in (1.1) and illustrated above.

Definition. A **transformation** or **function T**: $\mathcal{S}_1 \rightarrow \mathcal{S}_2$ is a rule that associates with each element of the set $\mathcal{S}_1$ a unique element from the set $\mathcal{S}_2$*. The set $\mathcal{S}_1$ is called the **domain** of **T**; $\mathcal{S}_2$ is the **range of definition** of **T**.

Our attention is directed primarily toward transformations where $\mathcal{S}_1$ and $\mathcal{S}_2$ are linear spaces. We speak of **T**: $\mathcal{V}\to\mathcal{W}$ as a transformation from the vector space $\mathcal{V}$ into the vector space $\mathcal{W}$. An **operator** is another term for a transformation between vector spaces. We use this term primarily when the domain and range of definition are identical; we speak of **T**: $\mathcal{V}\to\mathcal{V}$ as an operator **on** $\mathcal{V}$. If $\mathcal{S}_{\mathcal{V}}$ is a subset of $\mathcal{V}$, we denote by $\mathbf{T}(\mathcal{S}_{\mathcal{V}})$ the set of all vectors **Tx** in $\mathcal{W}$ for which **x** is in $\mathcal{S}_{\mathcal{V}}$; we refer to $\mathbf{T}(\mathcal{S}_{\mathcal{V}})$ as the **image of** $\mathcal{S}_{\mathcal{V}}$ **under T**. The **range of T** is $\mathbf{T}(\mathcal{V})$, the image of $\mathcal{V}$ under **T**. The **nullspace of T** is the set of all vectors **x** in $\mathcal{V}$ such that $\mathbf{Tx}=\boldsymbol{\theta}_{\mathcal{W}}$ ($\boldsymbol{\theta}_{\mathcal{W}}$ is the zero vector in the space $\mathcal{W}$). If $\mathcal{S}_{\mathcal{W}}$ is a subset of $\mathcal{W}$, we call the set of vectors **x** in $\mathcal{V}$ for which **Tx** is in $\mathcal{S}_{\mathcal{W}}$ the **inverse image** of $\mathcal{S}_{\mathcal{W}}$. Thus the nullspace of **T** is the inverse image of the set $\{\boldsymbol{\theta}_{\mathcal{W}}\}$. See Figure 2.6.

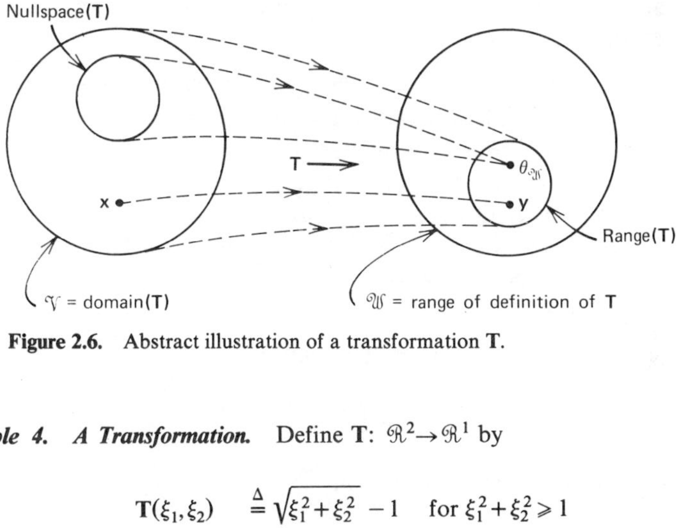

Figure 2.6. Abstract illustration of a transformation **T**.

Example 4. A Transformation. Define **T**: $\mathcal{R}^2\to\mathcal{R}^1$ by

$$\begin{aligned}\mathbf{T}(\xi_1,\xi_2) &\triangleq \sqrt{\xi_1^2+\xi_2^2}-1 && \text{for } \xi_1^2+\xi_2^2\geqslant 1\\ &\triangleq 0 && \text{for } \xi_1^2+\xi_2^2<1\end{aligned}\tag{2.20}$$

Physically, the vector **Tx** can be interpreted as the distance between **x** and the unit circle in the two-dimensional arrow space. The variables ξ_1 and ξ_2 are "dummy" variables; they merely assist us in cataloguing the "values" of **T** in the defining

*In the modeling process we use the function concept twice: once as a vector—a model for the condition of a system—and once as a relation between input and output vectors—a model for the system itself. In order to avoid confusion, we use the term function in referring to vectors in a vector space, but the term transformation in referring to the relation between vectors.

equation; we can use any other symbols in their place without changing the definition of **T**. The range of **T** is the set of positive numbers in $\mathcal{R}^1$. The nullspace of **T** is the set consisting of all vectors in the domain $\mathcal{R}^2$ which satisfy $\xi_1^2+\xi_2^2 \leqslant 1$.

Suppose we wish to solve the equation $\mathbf{Tx}=1$ for the transformation of Example 4. In effect, we ask which points in the arrow space are a unit distance from the unit circle—all points on the circle of radius 2. The solution is not unique because **T** assigns to the single number 1 in $\mathcal{R}^1$ more than one vector in $\mathcal{R}^2$. The equation $\mathbf{Tx}=-1$, on the other hand, has no solution because **T** does not assign the number -1 in $\mathcal{R}^1$ to any vector in $\mathcal{R}^2$. We now proceed to specify the properties of a transformation which are necessary in order that the transformation be uniquely reversible.

Definition. Let $\mathbf{T}\colon \mathcal{V}\rightarrow\mathcal{W}$. Then **T** is **one-to-one** if

$$\mathbf{x}_1 \neq \mathbf{x}_2 \quad \Rightarrow \quad \mathbf{Tx}_1 \neq \mathbf{Tx}_2 \tag{2.21}$$

for all $\mathbf{x}_1$ and $\mathbf{x}_2$ in $\mathcal{V}$; that is, if **T** does not assign more than one **x** in $\mathcal{V}$ to a single **y** in $\mathcal{W}$.

If **T** is one-to-one, any solution to $\mathbf{Tx}=\mathbf{y}$ is unique. It might appear that the effect of **T** is reversible if **T** is one-to-one. The nonreversibility of **T** in Example 4, however, arises only in part because **T** is not one-to-one. In general, there may be vectors in the range of definition $\mathcal{W}$ which are not associated in any way with vectors in $\mathcal{V}$. In point of fact, range(**T**) consists precisely of those vectors **y** in $\mathcal{W}$ for which the equation $\mathbf{Tx}=\mathbf{y}$ is solvable. Unless we know which vectors are in range(**T**), we cannot reverse the transformation.

Definition. Let $\mathbf{T}\colon \mathcal{V}\rightarrow\mathcal{W}$. Then **T** is **onto** if

$$\text{range}(\mathbf{T}) = \mathcal{W} \tag{2.22}$$

That is, **T** is onto if every vector **y** in $\mathcal{W}$ is associated with at least one vector **x** in $\mathcal{V}$.

Definition. If a transformation is one-to-one and onto, then it is **invertible** —it can be reversed uniquely. If $\mathbf{T}\colon \mathcal{V}\rightarrow\mathcal{W}$ is invertible, we define the **inverse of T** to be the transformation $\mathbf{T}^{-1}\colon \mathcal{W}\rightarrow\mathcal{V}$ which associates with each **y** in $\mathcal{W}$ the unique vector **x** in $\mathcal{V}$ for which $\mathbf{Tx}=\mathbf{y}$. See (2.29) for another characterization of $\mathbf{T}^{-1}$.

***Example 5. The Identity Operator,* I.** Let $\mathcal{V}$ be a vector space. Define the operator **I** on $\mathcal{V}$ by

$$\mathbf{Ix} \triangleq \mathbf{x} \tag{2.23}$$

for all $\mathbf{x}$ in $\mathcal{V}$. The nullspace of $\mathbf{I}$ is $\boldsymbol{\theta}_{\mathcal{V}}$. Range $(\mathbf{I}) = \mathcal{V}$; thus $\mathbf{I}$ is onto. Furthermore, $\mathbf{I}$ is one-to-one. Therefore, the identity operator is invertible.

Example 6. The Zero Transformation, $\boldsymbol{\Theta}$. Let $\mathcal{V}$ and $\mathcal{W}$ be vector spaces. Define $\boldsymbol{\Theta}: \mathcal{V} \to \mathcal{W}$ by

$$\boldsymbol{\Theta}\mathbf{x} \triangleq \boldsymbol{\theta}_{\mathcal{W}} \tag{2.24}$$

for all $\mathbf{x}$ in $\mathcal{V}$. The nullspace of $\boldsymbol{\Theta}$ is $\mathcal{V}$. The range of $\boldsymbol{\Theta}$ is $\boldsymbol{\theta}_{\mathcal{W}}$. The zero transformation is neither one-to-one nor onto. It is clearly not invertible.

Example 7. A Transformation on a Function Space. Define $\mathbf{T}: \mathcal{C}(a,b) \to \mathcal{R}^1$ by

$$\mathbf{Tf} \triangleq \int_a^b \mathbf{f}^2(t)\,dt \tag{2.25}$$

for all $\mathbf{f}$ in $\mathcal{C}(a,b)$. This transformation specifies an integral-square measure of the size of the function $\mathbf{f}$; this measure is used often in judging the performance of a control system. The function $\mathbf{f}$ is a dummy variable used to define $\mathbf{T}$; the scalar t is a dummy variable used to define $\mathbf{f}$. In order to avoid confusion, we must carefully distinguish between the concept of the function $\mathbf{f}$ in the vector space $\mathcal{C}(a,b)$ and the concept of the transformation $\mathbf{T}$ which relates each function $\mathbf{f}$ in $\mathcal{C}(a,b)$ to a vector in $\mathcal{R}^1$. The transformation acts on the whole function $\mathbf{f}$—we must use all values of $\mathbf{f}$ to find $\mathbf{Tf}$. The range of $\mathbf{T}$ is the set of positive numbers in $\mathcal{R}^1$; thus $\mathbf{T}$ is not onto the range of definition $\mathcal{R}^1$. The nullspace of $\mathbf{T}$ is the single vector $\boldsymbol{\theta}_{\mathcal{V}}$. If we define $\mathbf{f}_1$ and $\mathbf{f}_2$ by $\mathbf{f}_1(t) = 1$ and $\mathbf{f}_2(t) = -1$, then $\mathbf{Tf}_1 = \mathbf{Tf}_2$; therefore $\mathbf{T}$ is not one-to-one.

The transformations of Examples 4 and 7 are scalar valued; that is, the range of definition in each case is the space of scalars. We call a scalar-valued transformation a **functional**. Most functionals are not one-to-one.

Example 8. A Transformation for a Dynamic System. Let $\mathcal{C}^2(a,b)$ be the space of functions which have continuous second derivatives on $[a,b]$. Define $\mathbf{L}: \mathcal{C}^2(a,b) \to \mathcal{C}(a,b)$ by

$$(\mathbf{Lf})(t) \triangleq \mathbf{f}''(t) + \alpha\big(\mathbf{f}(t) + 0.01\mathbf{f}^3(t)\big) \tag{2.26}$$

for all $\mathbf{f}$ in $\mathcal{C}^2(a,b)$ and all t in $[a,b]$. This transformation is a model for a particular mass–spring system in which the spring is nonlinear. The comments under Example 7 concerning the dummy variables $\mathbf{f}$ and t apply here as well. As usual, the definition is given in terms of scalars, functions evaluated at t. Again, $\mathbf{L}$ acts on the whole function $\mathbf{f}$. Even in this example we cannot determine any value of the function $\mathbf{Lf}$ without using an "interval" of values of $\mathbf{f}$, because the derivative function $\mathbf{f}'$ is defined in terms of a limit of values of $\mathbf{f}$ in the neighborhood of t:

$$\mathbf{f}'(t) \stackrel{\Delta}{=} \lim_{\Delta t \to 0} \frac{\mathbf{f}(t+\Delta t)-\mathbf{f}(t)}{\Delta t}$$

The nullspace of $\mathbf{L}$ consists in all solutions of the nonlinear differential equation, $\mathbf{Lf} = \boldsymbol{\theta}_{\mathcal{W}}$; restated in terms of the values of $\mathbf{Lf}$, this equation is

$$\mathbf{f}''(t) + \alpha\big(\mathbf{f}(t) + 0.01\mathbf{f}^3(t)\big) = 0 \qquad a \leqslant t \leqslant b$$

To determine these solutions is not a simple task. By selecting $\mathcal{C}(a,b)$ as the range of definition, we ask that the function $\mathbf{Lf}$ be continuous; since $\mathbf{Lf}$ represents a force in the mass–spring system described by (2.26) continuity seems a practical assumption. By choosing $\mathcal{C}^2(a,b)$ as the domain, we guarantee that $\mathbf{Lf}$ is continuous. Yet the range of $\mathbf{L}$ is not clear. It is in the range of definition, but is it equal to the range of definition? In other words, can we solve the nonlinear differential equation $\mathbf{Lf} = \mathbf{u}$ for *any* continuous $\mathbf{u}$? The function $\mathbf{f}$ represents the displacement versus time in the physical mass-spring system. The function $\mathbf{u}$ represents the force applied to the system as a function of time. Physical intuition leads us to believe that for given initial conditions there is a unique displacement pattern $\mathbf{f}$ associated with each continuous forcing pattern $\mathbf{u}$. Therefore, $\mathbf{L}$ should be onto. On the other hand, since no initial conditions are specified, we expect two degrees of freedom in the solution to $\mathbf{Lf} = \mathbf{u}$ for each continuous $\mathbf{u}$. Thus the dimension of nullspace ($\mathbf{L}$) is two, and $\mathbf{L}$ is not one-to-one.

Combining Transformations

The transformation introduced in Example 8 is actually a composite of several simpler transformations. In developing a model for a system, we usually start with simple models for portions of the system, and then combine the parts into the total system model. Suppose $\mathbf{T}$ and $\mathbf{U}$ are both transformations from $\mathcal{V}$ into $\mathcal{W}$. We define the transformation $a\mathbf{T} + b\mathbf{U}$: $\mathcal{V} \to \mathcal{W}$ by

$$(a\mathbf{T} + b\mathbf{U})\mathbf{x} \stackrel{\Delta}{=} a\mathbf{T}\mathbf{x} + b\mathbf{U}\mathbf{x} \tag{2.27}$$

for all $\mathbf{x}$ in $\mathcal{V}$. If $\mathbf{G}$: $\mathcal{W} \to \mathcal{U}$, we define the transformation $\mathbf{GT}$: $\mathcal{V} \to \mathcal{U}$ by

$$(\mathbf{GT})\mathbf{x} \stackrel{\Delta}{=} \mathbf{G}(\mathbf{T}\mathbf{x}) \tag{2.28}$$

for all $\mathbf{x}$ in $\mathcal{V}$. Equations (2.27) and (2.28) define **linear combination** and **composition** of transformations, respectively.

Example 9. Composition of Matrix Multiplications. Define **G**: $\mathfrak{R}^3 \to \mathfrak{R}^2$ by

$$\mathbf{G}\begin{pmatrix}\xi_1\\ \xi_2\\ \xi_3\end{pmatrix} \triangleq \begin{pmatrix}1 & 0 & 1\\ 2 & 1 & 3\end{pmatrix}\begin{pmatrix}\xi_1\\ \xi_2\\ \xi_3\end{pmatrix}$$

and **T**: $\mathfrak{R}^2 \to \mathfrak{R}^3$ by

$$\mathbf{T}\begin{pmatrix}\eta_1\\ \eta_2\end{pmatrix} \triangleq \begin{pmatrix}2 & 2\\ 1 & 2\\ 3 & 1\end{pmatrix}\begin{pmatrix}\eta_1\\ \eta_2\end{pmatrix}$$

Then **GT**: $\mathfrak{R}^2 \to \mathfrak{R}^2$ is described by

$$\begin{aligned}\mathbf{GT}\begin{pmatrix}\eta_1\\ \eta_2\end{pmatrix} &= \mathbf{G}\begin{pmatrix}2 & 2\\ 1 & 2\\ 3 & 1\end{pmatrix}\begin{pmatrix}\eta_1\\ \eta_2\end{pmatrix}\\ &= \begin{pmatrix}1 & 0 & 1\\ 2 & 1 & 3\end{pmatrix}\begin{pmatrix}2 & 2\\ 1 & 2\\ 3 & 1\end{pmatrix}\begin{pmatrix}\eta_1\\ \eta_2\end{pmatrix}\\ &= \begin{pmatrix}5 & 3\\ 14 & 9\end{pmatrix}\begin{pmatrix}\eta_1\\ \eta_2\end{pmatrix}\end{aligned}$$

Exercise 1. Let **T**: $\mathcal{V} \to \mathcal{W}$. Show that **T** is invertible if and only if $\mathcal{V} = \mathcal{W}$ and there is a transformation $\mathbf{T}^{-1}$: $\mathcal{W} \to \mathcal{V}$ such that

$$\mathbf{T}^{-1}\mathbf{T} = \mathbf{T}\mathbf{T}^{-1} = \mathbf{I} \tag{2.29}$$

Exercise 2. Suppose **G** and **T** of (2.26) are invertible. Show that

$$(\mathbf{GT})^{-1} = \mathbf{T}^{-1}\mathbf{G}^{-1} \tag{2.30}$$

The composition (or product) of two transformations has two nasty characteristics. First, unlike scalars, transformations usually **do not commute**; that is, $\mathbf{GT} \neq \mathbf{TG}$. As illustrated in Example 9, **G** and **T** generally do not even act on the same vector space, and **TG** has no meaning. Even if **G** and **T** both act on the same space, we must not expect commutability, as demonstrated by the following matrix multiplications:

$$\begin{pmatrix}1 & 0\\ 0 & 0\end{pmatrix}\begin{pmatrix}0 & 1\\ 0 & 0\end{pmatrix} = \begin{pmatrix}0 & 1\\ 0 & 0\end{pmatrix}$$

$$\begin{pmatrix}0 & 1\\ 0 & 0\end{pmatrix}\begin{pmatrix}1 & 0\\ 0 & 0\end{pmatrix} = \begin{pmatrix}0 & 0\\ 0 & 0\end{pmatrix}$$

Commutable operators do exist. In fact, since any operator commutes with itself, we can write $\mathbf{G}^2$, as we do in Example 10 below, without being ambiguous. Operators which commute act much like scalars in their behavior toward each other (see P&C 4.29).

If two scalars satisfy $ab=0$, then either $a=0$, $b=0$, or both. The second matrix multiplication above demonstrates that this property does not extend even to simple transformations. This second difficulty with the composition of transformations is sometimes called the existence of **divisors of zero**. If $\mathbf{GT}=\boldsymbol{\Theta}$ and $\mathbf{G}\neq\boldsymbol{\Theta}$, we cannnot conclude that $\mathbf{T}=\boldsymbol{\Theta}$; the cancellation laws of algebra do not apply to transformations. The difficulty lies in the fact that for transformations there is a "gray" region between being invertible and being zero. The range of **T** can lie in the nullspace of **G**.

Example 10. Linear Combination and Composition of Transformations. The space $\mathcal{C}^n(a,b)$ consists in all functions with continuous nth derivatives on $[a,b]$. Define $\mathbf{G}$: $\mathcal{C}^n(a,b)\to\mathcal{C}^{n-1}(a,b)$ by $\mathbf{Gf}\triangleq\mathbf{f}'$ for all $\mathbf{f}$ in $\mathcal{C}^n(a,b)$. Then $\mathbf{G}^2$: $\mathcal{C}^2(a,b)\to\mathcal{C}(a,b)$ is well defined. Let $\mathbf{U}$: $\mathcal{C}^2(a,b)\to\mathcal{C}(a,b)$ be defined by $(\mathbf{Uf})(t)\triangleq\mathbf{f}(t)+0.01\mathbf{f}^3(t)$ for all $\mathbf{f}$ in $\mathcal{C}^2(a,b)$ and all t in $[a,b]$. The transformation **L** of Example 8 can be described by $\mathbf{L}\triangleq\mathbf{G}^2+\alpha\mathbf{U}$.

As demonstrated by the above examples, the domain and range of definition are essential parts of the definition of a transformation. This importance is emphasized by the notation $\mathbf{T}$: $\mathcal{V}\to\mathcal{W}$. The spaces $\mathcal{V}$ and $\mathcal{W}$ are selected to fit the structure of the situation we wish to model. If we pick a domain that is too large, the operator will not be one-to-one. If we pick a range of definition that is too large, the operator will not be onto. Thus both $\mathcal{V}$ and $\mathcal{W}$ affect the invertibility of **T**. We apply loosely the term *finite (infinite) dimensional transformation* to those transformations that act on a finite (infinite) dimensional domain.

2.4 Linear Transformations

One of the most common and useful transformations is the matrix multiplication introduced in Chapter 1. It is well suited for automatic computation using a digital computer. Let **A** be an $m\times n$ matrix. We define $\mathbf{T}$: $\mathfrak{M}^{n\times 1}\to\mathfrak{M}^{m\times 1}$ by

$$\mathbf{Tx}\triangleq\mathbf{Ax} \tag{2.31}$$

for all $\mathbf{x}$ in $\mathfrak{M}^{n\times 1}$. We distinguish carefully between **T** and **A**. **T** is not **A**, but rather *multiplication* by **A**. The nullspace of **T** is the set of solutions to the matrix equation $\mathbf{Ax}=\boldsymbol{\theta}$. Even though $\mathbf{T}$ and $\mathbf{A}$ are conceptually different, we sometimes refer to the nullspace of $\mathbf{T}$ as the nullspace of $\mathbf{A}$. Similarly, we define range$(\mathbf{A}) \triangleq$ range$(\mathbf{T})$.

Suppose $\mathbf{A}$ is square ($m=n$) and invertible; then the equation $\mathbf{Tx}=\mathbf{Ax}=\mathbf{y}$ has a unique solution $\mathbf{x}=\mathbf{A}^{-1}\mathbf{y}$ for each $\mathbf{y}$ in $\mathfrak{M}^{n\times 1}$. But $\mathbf{T}^{-1}$ is defined as precisely that transformation which associates with each $\mathbf{y}$ in $\mathfrak{M}^{n\times 1}$ the unique solution to the equation $\mathbf{Tx}=\mathbf{y}$. Therefore, $\mathbf{T}$ is invertible, and $\mathbf{T}^{-1}$: $\mathfrak{M}^{n\times 1}\to\mathfrak{M}^{n\times 1}$ is given by $\mathbf{T}^{-1}\mathbf{y} \triangleq \mathbf{A}^{-1}\mathbf{y}$.

The properties of matrix multiplication (Appendix 1) are such that $\mathbf{A}(a\mathbf{x}_1+b\mathbf{x}_2)=a\mathbf{Ax}_1+b\mathbf{Ax}_2$. That is, matrix multiplication preserves linear combinations. This property of matrix multiplication allows **superposition** of solutions to a matrix equation: if $\mathbf{x}_1$ solves $\mathbf{Ax}=\mathbf{y}_1$ and $\mathbf{x}_2$ solves $\mathbf{Ax}=\mathbf{y}_2$, then the solution to $\mathbf{Ax}=\mathbf{y}_1+\mathbf{y}_2$ is $\mathbf{x}_1+\mathbf{x}_2$. From one or two input–output relationships we can infer others. Many other familiar transformations preserve linear combinations and allow superposition of solutions.

Definition. The transformation $\mathbf{T}$: $\mathcal{V}\to\mathcal{W}$ is **linear** if

$$\mathbf{T}(a\mathbf{x}_1+b\mathbf{x}_2)=a\mathbf{Tx}_1+b\mathbf{Tx}_2 \tag{2.32}$$

for all vectors $\mathbf{x}_1$ and $\mathbf{x}_2$ in $\mathcal{V}$ and all scalars a and b.

Example 1. Integration. Define $\mathbf{T}$: $\mathcal{C}(0,1)\to\mathcal{C}(0,1)$ by

$$(\mathbf{Tf})(t) \triangleq \int_0^t \mathbf{f}(s)\,ds \tag{2.33}$$

for all $\mathbf{f}$ in $\mathcal{C}(0,1)$ and all t in $[0,1]$. The linearity of this indefinite integration operation is a fundamental fact of integral calculus; that is,

$$\int_0^t [a\mathbf{f}_1(s)+b\mathbf{f}_2(s)]\,ds = a\int_0^t \mathbf{f}_1(s)\,ds + b\int_0^t \mathbf{f}_2(s)\,ds$$

The operator (2.33) is a special case of the linear integral operator $\mathbf{T}$: $\mathcal{C}(a,b)\to\mathcal{C}(c,d)$ defined by

$$(\mathbf{Tf})(t) \triangleq \int_a^b k(t,s)\mathbf{f}(s)\,ds \tag{2.34}$$

for all $\mathbf{f}$ in $\mathcal{C}(a,b)$ and all t in $[c,d]$. We can substitute for the domain $\mathcal{C}(a,b)$ any other space of functions for which the integral exists. We can use any range of definition which includes the integrals (2.34) of all functions in the domain. The function k is called the **kernel** of the integral transformation. Another special case of (2.34) is $\mathbf{T}$: $\mathcal{L}_2(-\infty,\infty)\to\mathcal{L}_2(-\infty,\infty)$ defined by

$$(\mathbf{Tf})(t) \triangleq \int_{-\infty}^{\infty} \mathbf{g}(t-s)\mathbf{f}(s)\,ds$$

for some $\mathbf{g}$ in $\mathcal{L}_2(-\infty,\infty)$, all $\mathbf{f}$ in $\mathcal{L}_2(-\infty,\infty)$, and all t in $(-\infty,\infty)$. This $\mathbf{T}$ is known as the convolution of $\mathbf{f}$ with the function $\mathbf{g}$. It arises in connection with the solution of linear constant-coefficient differential equations (Appendix 2).

The integral transformation (2.34) is the analogue for function spaces of the matrix multiplication (2.31). That matrix transformation can be expressed

$$(\mathbf{Tx})_i \triangleq \sum_{j=1}^{n} \mathbf{A}_{ij}\xi_j \qquad i=1,\ldots,m \tag{2.35}$$

for all vectors $\mathbf{x}$ in $\mathfrak{M}^{n\times 1}$. The symbol ξ_j represents the jth element of $\mathbf{x}$; the symbol $(\mathbf{Tx})_i$ means the ith element of $\mathbf{Tx}$. In (2.35) the matrix is treated as a function of two discrete variables, the row variable i and the column variable j. In analogy with the integral transformation, we call the matrix multiplication [as viewed in the form of (2.35)] a **summation transformation**; we refer to the function $\mathbf{A}$ (with values $\mathbf{A}_{ij}$) as the **kernel** of the summation transformation.

Example 2. Differentiation. Define $\mathbf{D}$: $\mathcal{C}^1(a,b)\rightarrow\mathcal{C}(a,b)$ by

$$(\mathbf{Df})(t) \triangleq \mathbf{f}'(t) \triangleq \lim_{\Delta t\rightarrow 0} \frac{\mathbf{f}(t+\Delta t)-\mathbf{f}(t)}{\Delta t} \tag{2.36}$$

for all $\mathbf{f}$ in $\mathcal{C}^1(a,b)$ and all t in $[a,b]$; $\mathbf{f}'(t)$ is the slope of the graph of $\mathbf{f}$ at t; $\mathbf{f}'$ (or $\mathbf{Df}$) is the whole "slope" function. We also use the symbols $\dot{\mathbf{f}}$ and $\mathbf{f}^{(1)}$ in place of $\mathbf{Df}$. We can substitute for the above domain and range of definition any pair of function spaces for which the derivatives of all functions in the domain lie in the range of definition. Thus we could define $\mathbf{D}$ on $\mathcal{C}(a,b)$ if we picked a range of definition which contains the appropriate discontinuous functions. The nullspace of $\mathbf{D}$ is span$\{\mathbf{1}\}$, where $\mathbf{1}$ is the function defined by $\mathbf{1}(t)=1$ for all t in $[a,b]$. It is well known that differentiation is linear; $\mathbf{D}(c_1\mathbf{f}_1+c_2\mathbf{f}_2)=c_1\mathbf{Df}_1+c_2\mathbf{Df}_2$.

We can define more general differential operators in terms of (2.36). The general linear constant-coefficient differential operator $\mathbf{L}$: $\mathcal{C}^n(a,b)\rightarrow\mathcal{C}(a,b)$ is defined, for real scalars $\{a_i\}$, by

$$\mathbf{L} \triangleq \mathbf{D}^n + a_1\mathbf{D}^{n-1} + \cdots + a_n\mathbf{I} \tag{2.37}$$

where we have used (2.27) and (2.28) to combine transformations. A variable-coefficient (or "time-varying") extension of (2.37) is the operator $\mathbf{L}$: $\mathcal{C}^n(a,b)\rightarrow\mathcal{C}(a,b)$ defined by*

$$(\mathbf{Lf})(t) \triangleq g_0(t)\mathbf{f}^{(n)}(t) + g_1(t)\mathbf{f}^{(n-1)}(t) + \cdots + g_n(t)\mathbf{f}(t) \tag{2.38}$$

*Note that we use boldface print for some of the functions in (2.38) but not for others. As indicated in the Preface, we use boldface print only to emphasize the vector or transformation interpretation of an object. We sometimes describe the same function both ways, **f** and *f*.

for all $\mathbf{f}$ in $\mathcal{C}^n(a,b)$ and all t in $[a,b]$. (We have denoted the kth derivative $\mathbf{D}^k\mathbf{f}$ by $\mathbf{f}^{(k)}$.) If the interval $[a,b]$ is finite, if the functions g_i are continuous, and if $g_0(t)\neq 0$ on $[a,b]$, we refer to (2.38) as a **regular** nth-order differential operator. [With $g_0(t)\neq 0$, we would lose no generality by letting $g_0(t)=1$ in (2.38).] We can apply the differential operators (2.37) and (2.38) to other function spaces than $\mathcal{C}^n(a,b)$.

Example 3. Evaluation of a Function. Define $\mathbf{T}$: $\mathcal{C}(a,b)\to \mathcal{R}^1$ by

$$\mathbf{Tf} \triangleq \mathbf{f}(t_1) \tag{2.39}$$

for all $\mathbf{f}$ in the function space $\mathcal{C}(a,b)$. In this example, $\mathbf{f}$ is a dummy variable, but t_1 is not. The transformation is a *linear functional* called "evaluation at t_1." The range of $\mathbf{T}$ is $\mathcal{R}^1$; $\mathbf{T}$ is onto. The nullspace of $\mathbf{T}$ is the set of continuous functions which pass through zero at t_1. Because many functions have the same value at t_1, $\mathbf{T}$ is not one-to-one. This functional can also be defined using some other function space for its domain.

Example 4. The One-Sided Laplace Transform, $\mathfrak{L}$. Suppose $\mathcal{W}$ is the space of complex-valued functions defined on the positive-real half of the complex plane. (See Example 10, Section 2.1.) Let $\mathcal{V}$ be the space of functions which are defined and continuous on $[0,\infty]$ and for which $e^{-ct}|f(t)|$ is bounded for some constant c and all values of t greater than some finite number. We define the one-sided Laplace transform $\mathfrak{L}$: $\mathcal{V}\to\mathcal{W}$ by

$$(\mathfrak{L}\mathbf{f})(s) \triangleq \int_0^{\infty} e^{-st}\mathbf{f}(t)\,dt \tag{2.40}$$

for all complex s with real$(s)>0$. The functions in $\mathcal{V}$ are such that (2.40) converges for real$(s)>0$. We sometimes denote the transformed function $\mathfrak{L}\mathbf{f}$ by $\mathbf{F}$. This integral transform, like that of (2.34), is linear. The Laplace transform is used to convert linear constant-coefficient differential equations into linear algebraic equations.*

Exercise 1. Suppose the transformations $\mathbf{T}$, $\mathbf{U}$, and $\mathbf{G}$ of (2.27) and (2.28) are linear and $\mathbf{T}$ is invertible. Show that the transformations $a\mathbf{T}+b\mathbf{U}$, $\mathbf{GT}$, and $\mathbf{T}^{-1}$ are also linear.

Exercise 2. Let $\mathcal{V}$ be an n-dimensional linear space with basis $\mathcal{X}$. Define $\mathbf{T}$: $\mathcal{V}\to\mathfrak{M}^{n\times 1}$ by

$$\mathbf{Tx} \triangleq [\mathbf{x}]_{\mathcal{X}} \tag{2.41}$$

Show that $\mathbf{T}$, the process of taking coordinates, is a linear, invertible transformation.

*It can be shown that $[\mathfrak{L}(\mathbf{Df})](s)=s(\mathfrak{L}\mathbf{f})(s)-\mathbf{f}(0^+)$, where $\mathbf{f}(0^+)$ is the limit of $\mathbf{f}(t)$ as $t\to 0$ from the positive side of 0.

The vector space $\mathcal{V}$ of Exercise 2 is equivalent to $\mathfrak{M}^{n\times 1}$ in every sense we might wish. The linear, invertible transformation is the key. We say two vector spaces $\mathcal{V}$ and $\mathcal{W}$ are **isomorphic** (or equivalent) if there exists an invertible linear transformation from $\mathcal{V}$ into $\mathcal{W}$. Each real n-dimensional vector space is isomorphic to each other real n-dimensional space and, in particular, to the real space $\mathfrak{M}^{n\times 1}$. A similar statement can be made using complex scalars for each space. Infinite-dimensional spaces also exhibit isomorphism. In Section 5.3 we show that all well behaved infinite-dimensional spaces are isomorphic to l_2.

Nullspace and Range—Keys to Invertibility

Even *linear* transformations may have troublesome properties. In point of fact, the example in which we demonstrate *noncommutability* and *noncancellation* of products of transformations uses linear transformations (matrix multiplications). Most difficulties with a linear transformation can be understood through investigation of the range and nullspace of the transformation.*

Let $\mathbf{T}: \mathcal{V}\rightarrow\mathcal{W}$ be linear. Suppose $\mathbf{x}_h$ is a vector in the nullspace of $\mathbf{T}$ (any solution to $\mathbf{Tx}=\boldsymbol{\theta}$); we call $\mathbf{x}_h$ a homogeneous solution for the transformation $\mathbf{T}$. Denote by $\mathbf{x}_p$ a particular solution to the equation $\mathbf{Tx}=\mathbf{y}$. (An $\mathbf{x}_p$ exists if and only if $\mathbf{y}$ is in range($\mathbf{T}$).) Then $\mathbf{x}_p+\alpha\mathbf{x}_h$ is also a solution to $\mathbf{Tx}=\mathbf{y}$ for any scalar α. One of the most familiar uses of the principle of superposition is in obtaining the general solution to a linear differential equation by combining particular and homogeneous solutions. The general solution to any linear operator equation can be obtained in this manner.

Example 5. The General Solution to a Matrix Equation. Define the linear operator $\mathbf{T}: \mathfrak{M}^{2\times 1}\rightarrow\mathfrak{M}^{2\times 1}$ by

$$\mathbf{T}\begin{pmatrix}\xi_1\\ \xi_2\end{pmatrix}\triangleq\begin{pmatrix}2 & 1\\ 2 & 1\end{pmatrix}\begin{pmatrix}\xi_1\\ \xi_2\end{pmatrix}$$

Then the equation

$$\mathbf{Tx}=\begin{pmatrix}2 & 1\\ 2 & 1\end{pmatrix}\begin{pmatrix}\xi_1\\ \xi_2\end{pmatrix}=\begin{pmatrix}2\\ 2\end{pmatrix}\triangleq\mathbf{y} \tag{2.42}$$

has as its general solution $\mathbf{x}=(\xi_1\ \ 2-2\xi_1)^{\mathrm{T}}$. A particular solution is $\mathbf{x}_p=(1\ \ 0)^{\mathrm{T}}$. The nullspace of $\mathbf{T}$ consists in the vector $\mathbf{x}_h=(-1\ \ 2)^{\mathrm{T}}$ and all its multiples. The general solution can be expressed as $\mathbf{x}=\mathbf{x}_p+\alpha\mathbf{x}_h$ where α is arbitrary. Figure 2.7 shows an

*See Sections 4.4 and 4.6 for further insight into noncancellation and noncommutability of linear operators.

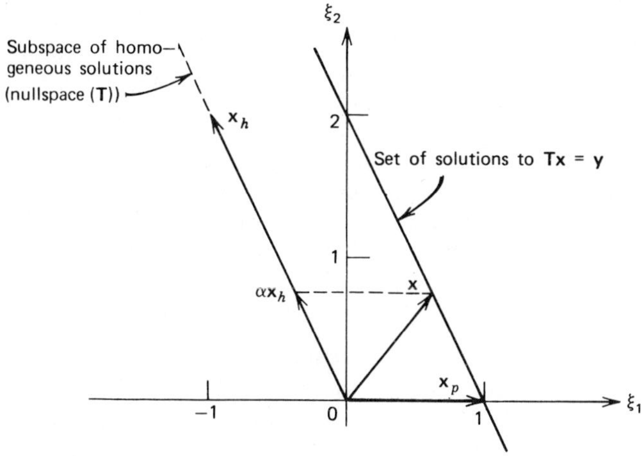

Figure 2.7. Solutions to the linear equation of Example 5.

arrow-space equivalent of these vectors. The nullspace of $\mathbf{T}$ is a subspace of $\mathfrak{M}^{2\times 1}$. The general solution (the set of all solutions to $\mathbf{Tx}=\mathbf{y}$) consists of a line in $\mathfrak{M}^{2\times 1}$; specifically, it is the nullspace of $\mathbf{T}$ shifted by the addition of any particular solution.

The nullspace of a linear transformation is always a subspace of the domain $\mathcal{V}$. The freedom in the general solution to $\mathbf{Tx}=\mathbf{y}$ lies only in nullspace($\mathbf{T}$), the subspace of homogeneous solutions. For if $\hat{\mathbf{x}}_p$ is another particular solution to $\mathbf{Tx}=\mathbf{y}$, then

$$\mathbf{T}(\mathbf{x}_p-\hat{\mathbf{x}}_p)=\mathbf{Tx}_p-\mathbf{T}\hat{\mathbf{x}}_p=\mathbf{y}-\mathbf{y}=\boldsymbol{\theta}$$

The difference between $\mathbf{x}_p$ and $\hat{\mathbf{x}}_p$ is a vector in nullspace($\mathbf{T}$). If nullspace($\mathbf{T}$)$=\boldsymbol{\theta}$, there is no freedom in the solution to $\mathbf{Tx}=\mathbf{y}$; it is unique.

Definition. A transformation $\mathbf{G}$: $\mathcal{V}\rightarrow\mathcal{W}$ is **nonsingular** if nullspace($\mathbf{G}$)$=$ $\boldsymbol{\theta}$.

Exercise 3. Show that a *linear* transformation is one-to-one if and only if it is nonsingular.

Because a linear transformation $\mathbf{T}$: $\mathcal{V}\rightarrow\mathcal{W}$ preserves linear combinations, it necessarily transforms $\boldsymbol{\theta}_{\mathcal{V}}$ into $\boldsymbol{\theta}_{\mathcal{W}}$. Furthermore, $\mathbf{T}$ acts on the vectors in $\mathcal{V}$ by subspaces—whatever $\mathbf{T}$ does to $\mathbf{x}$ it does also to $c\mathbf{x}$, where c is any scalar. The set of vectors in $\mathcal{V}$ which are taken to zero, for example, is the subspace which we call nullspace($\mathbf{T}$). Other subspaces of $\mathcal{V}$ are "rotated" or "stretched" by $\mathbf{T}$. This fact becomes more clear during our discussion of spectral decomposition in Chapter 4.

Example 6. The Action of a Linear Transformation on Subspaces. Define $\mathbf{T}$: $\mathcal{R}^3 \to \mathcal{R}^2$ by $\mathbf{T}(\xi_1,\xi_2,\xi_3) \triangleq (\xi_3,0)$. The set $\{\mathbf{x}_1=(1,0,0),\ \mathbf{x}_2=(0,1,0)\}$ forms a basis for nullspace($\mathbf{T}$). By adding a third independent vector, say, $\mathbf{x}_3=(1,1,1)$, we obtain a basis for the domain $\mathcal{R}^3$. The subspace spanned by $\{\mathbf{x}_1,\mathbf{x}_2\}$ is annihilated by $\mathbf{T}$. The subspace spanned by $\{\mathbf{x}_3\}$ is transformed by $\mathbf{T}$ into a subspace of $\mathcal{R}^2$—the range of $\mathbf{T}$. The vector $\mathbf{x}_3$ itself is transformed into a basis for range($\mathbf{T}$). Because $\mathbf{T}$ acts on the vectors in $\mathcal{R}^3$ by subspaces, the dimension of nullspace($\mathbf{T}$) is a measure of the degree to which $\mathbf{T}$ acts like zero; the dimension of range($\mathbf{T}$) indicates the degree to which $\mathbf{T}$ acts invertible. Specifically, of the three dimensions in $\mathcal{R}^3$, $\mathbf{T}$ takes two to zero. The third dimension of $\mathcal{R}^3$ is taken into the one-dimensional range($\mathbf{T}$).

The characteristics exhibited by Example 6 extend to any linear transformation on a finite-dimensional space. Let $\mathbf{T}$: $\mathcal{V} \to \mathcal{W}$ be linear with $\dim(\mathcal{V})=n$. We call the dimension of nullspace($\mathbf{T}$) the **nullity of T**. The **rank of T** is the dimension of range($\mathbf{T}$). Let $\{\mathbf{x}_1,\ldots,\mathbf{x}_k\}$ be a basis for nullspace($\mathbf{T}$). Pick vectors $\{\mathbf{x}_{k+1},\ldots,\mathbf{x}_n\}$ which extend the basis for nullspace($\mathbf{T}$) to a basis for $\mathcal{V}$ (P&C 2.9). We show that $\mathbf{T}$ takes $\{\mathbf{x}_{k+1},\ldots,\mathbf{x}_n\}$ into a basis for range($\mathbf{T}$). Suppose $\mathbf{x}=c_1\mathbf{x}_1+\cdots+c_n\mathbf{x}_n$ is an arbitrary vector in $\mathcal{V}$. The linear transformation $\mathbf{T}$ annihilates the first **k** components of $\mathbf{x}$. Only the remaining $n-k$ components are taken into range($\mathbf{T}$). Thus the vectors $\{\mathbf{T}\mathbf{x}_{k+1},\ldots,\mathbf{T}\mathbf{x}_n\}$ must span range($\mathbf{T}$). To show that these vectors are independent, we use the test (2.11):

$$\xi_{k+1}(\mathbf{T}\mathbf{x}_{k+1})+\cdots+\xi_n(\mathbf{T}\mathbf{x}_n)=\boldsymbol{\theta}_{\mathcal{W}}$$

Since $\mathbf{T}$ is linear,

$$\mathbf{T}(\xi_{k+1}\mathbf{x}_{k+1}+\cdots+\xi_n\mathbf{x}_n)=\boldsymbol{\theta}_{\mathcal{W}}$$

Then $\xi_{k+1}\mathbf{x}_{k+1}+\cdots+\xi_n\mathbf{x}_n$ is in nullspace($\mathbf{T}$), and

$$\xi_{k+1}\mathbf{x}_{k+1}+\cdots+\xi_n\mathbf{x}_n=d_1\mathbf{x}_1+\cdots+d_k\mathbf{x}_k$$

for some $\{d_i\}$. The independence of $\{\mathbf{x}_1,\ldots,\mathbf{x}_n\}$ implies $d_1=\cdots=d_k=\xi_{k+1}=\cdots=\xi_n=0$; thus $\{\mathbf{T}\mathbf{x}_{k+1},\ldots,\mathbf{T}\mathbf{x}_n\}$ is an independent set and is a basis for range($\mathbf{T}$).

We have shown that a linear transformation $\mathbf{T}$ acting on a finite-dimensional space $\mathcal{V}$ obeys a "conservation of dimension" law:

$$\dim(\mathcal{V})=\text{rank}(\mathbf{T})+\text{nullity}(\mathbf{T}) \tag{2.43}$$

Nullity($\mathbf{T}$) is the "dimension" annihilated by $\mathbf{T}$. Rank($\mathbf{T}$) is the "dimension" $\mathbf{T}$ retains. If nullspace($\mathbf{T}$)$=\{\boldsymbol{\theta}\}$, then nullity($\mathbf{T}$)$=0$ and rank($\mathbf{T}$) $=\dim(\mathcal{V})$. If, in addition, $\dim(\mathcal{W})=\dim(\mathcal{V})$, then rank($\mathbf{T}$)$=\dim(\mathcal{W})$ ($\mathbf{T}$ is onto), and $\mathbf{T}$ is invertible. A linear $\mathbf{T}$: $\mathcal{V} \rightarrow \mathcal{W}$ cannot be invertible unless $\dim(\mathcal{W}) = \dim(\mathcal{V})$.

We sometimes refer to the vectors $\mathbf{x}_{k+1}, \ldots, \mathbf{x}_n$ as **progenitors of the range of T**. Although the nullspace and range of $\mathbf{T}$ are unique, the space spanned by the progenitors is not; we can add any vector in nullspace($\mathbf{T}$) to any progenitor without changing the basis for the range (see Example 6).

The Near Nullspace

In contrast to mathematical analysis, mathematical *computation* is not clear-cut. For example, a set of equations which is mathematically invertible can be so "nearly singular" that the inverse cannot be computed to an acceptable degree of precision. On the other hand, because of the finite number of significant digits used in the computer, a mathematically singular system will be indistinguishable from a "nearly singular" system. The phenomenon merits serious consideration.

The matrix operator of Example 5 is singular. Suppose we modify the matrix slightly to obtain the nonsingular, but "nearly singular" matrix equation

$$\begin{pmatrix} 2 & 1 \\ 2 & 1+\epsilon \end{pmatrix} \begin{pmatrix} \xi_1 \\ \xi_2 \end{pmatrix} = \begin{pmatrix} 2 \\ 2 \end{pmatrix} \tag{2.44}$$

where ϵ is small. Then the arrow space diagram of Figure 2.7 must also be modified to show a pair of almost parallel lines. (Figure 1.7 of Section 1.5 is the arrow space diagram of essentially this pair of equations.) Although the solution (the intersection of the nearly parallel lines) is unique, it is difficult to compute accurately; the nearly singular equations are very ill conditioned. Slight errors in the data and roundoff during computing lead to significant uncertainty in the computed solution, even if the computation is handled carefully (Section 1.5). The uncertain component of the solution lies essentially in the nullspace of the operator; that is, it is almost parallel to the nearly parallel lines in the arrow-space diagram. The above pair of nearly singular algebraic equations might represent a nearly singular system. On the other hand, the underlying system might be precisely singular; the equations in the model of a singular system may be only *nearly singular* because of inaccuracies in the data. Regardless of which of these interpretations is correct, determining the "near nullspace" of the matrix is an important part of the analysis of the system. If the underlying system is singular, a description of the near nullspace is a description of the *freedom* in the solutions for the system. If the underlying system is just nearly singular, a description of the near nullspace is a description of the *uncertainty* in the solution.

Definition. Suppose **T** is a *nearly singular* linear operator on a vector space $\mathcal{V}$. We use the term **near nullspace of T** to mean those vectors that are taken *nearly* to zero by **T**; that is, those vectors which **T** drastically reduces in "size."*

In the two-dimensional example described above, the near nullspace consists in vectors which are *nearly* parallel to the vector $\mathbf{x}=(-1\ \ 2)^{\mathrm{T}}$. The near nullspace of **T** is *not a subspace* of $\mathcal{V}$. Rather, it consists in a set of vectors which are *nearly* in a subspace of $\mathcal{V}$. We can think of the near nullspace as a "fuzzy" subspace of $\mathcal{V}$.

We now present a method, referred to as **inverse iteration**, for describing the near nullspace of a nearly singular operator **T** acting on a vector space $\mathcal{V}$. Let $\mathbf{x}_0$ be an arbitrary vector in $\mathcal{V}$. Assume $\mathbf{x}_0$ contains a component which is in the near nullspace of **T**. (If it does not, such a component will be introduced by roundoff during the ensuing computation.) Since **T** reduces such components drastically, compared to its effect on the other components of $\mathbf{x}_0$, $\mathbf{T}^{-1}$ must drastically emphasize such components. Therefore, if we solve $\mathbf{T}\mathbf{x}_1=\mathbf{x}_0$ (in effect determining $\mathbf{x}_1=\mathbf{T}^{-1}\mathbf{x}_0$), the computed solution $\mathbf{x}_1$ contains a *significant* component in the near nullspace of **T**. (This component is the error vector which appears during the solution of the nearly singular equation.) The inverse iteration method consists in iteratively solving $\mathbf{T}\mathbf{x}_{k+1}=\mathbf{x}_k$. After a few iterations, $\mathbf{x}_k$ is dominated by its near-nullspace component; we use $\mathbf{x}_k$ as a partial basis for the near nullspace of **T**. (The number of iterations required is at the discretion of the analyst. We are not looking for a precisely defined subspace, but rather, a subspace that is fuzzy.) By repeating the above process for several different starting vectors $\mathbf{x}_0$, we usually obtain a set of vectors which spans the near nullspace of **T**.

Example 7. Describing a Near Nullspace. Define a linear operator **T** on $\mathcal{M}^{2\times 1}$ by means of the nearly singular matrix multiplication described above:

$$\mathbf{T}\mathbf{x}\triangleq\begin{pmatrix}2 & 1\\ 2 & 1+\epsilon\end{pmatrix}\mathbf{x}$$

For this simple example we can invert **T** explicitly

$$\mathbf{T}^{-1}\mathbf{x}=\frac{1}{2\epsilon}\begin{pmatrix}1+\epsilon & -1\\ -2 & 2\end{pmatrix}\mathbf{x}$$

We apply the inverse iteration method to the vector $\mathbf{x}_0=(1\ \ 1)^{\mathrm{T}}$; of course, we have no roundoff in our computations:

$$\mathbf{x}_1=\begin{pmatrix}\frac{1}{2}\\ 0\end{pmatrix},\quad \mathbf{x}_2=\frac{1}{2\epsilon}\begin{pmatrix}(1+\epsilon)/2\\ -1\end{pmatrix},\quad \mathbf{x}_3=\frac{1}{(2\epsilon)^2}\begin{pmatrix}(\epsilon^2+2\epsilon+3)/2\\ -(\epsilon+3)\end{pmatrix},\ldots$$

*In Section 4.2 we describe the near nullspace more precisely as the eigenspace for the smallest eigenvalue of **T**.

If ϵ is small, say $\epsilon=0.01$, then

$$\mathbf{x}_2=50\begin{pmatrix}0.505\\-1\end{pmatrix} \quad \text{and} \quad \mathbf{x}_3=(50)^2\begin{pmatrix}1.51\\-3.01\end{pmatrix}$$

After only three iterations, the sequence $\mathbf{x}_k$ has settled; the vector $\mathbf{x}_3$ provides a good description of the near nullspace of $\mathbf{T}$. If $\epsilon=0$, $\mathbf{T}$ is singular; $\mathbf{x}_3$ lies almost in the nullspace of this singular operator (Figure 2.7). Were we to try other starting vectors $\mathbf{x}_0$, we would obtain other vectors $\mathbf{x}_k$ nearly parallel to $(-1\ \ 2)^{\mathrm{T}}$. This near nullspace of $\mathbf{T}$ should be considered one-dimensional.

We note from Example 7 that the vector $\mathbf{x}_k$ in the inverse iteration grows drastically in size. Practical computer implementations of inverse iteration include normalization of $\mathbf{x}_k$ at each step in order to avoid numbers too large for the computer. A description for a two-dimensional near nullspace is sought in P&C 2.26. In Section 4.2 we analyze the inverse iteration more precisely in terms of eigenvalues and eigenvectors. Forsythe [2.3] gives some interesting examples of the treatment of nearly singular operators.

The Role of Linear Transformations

The purpose of modeling a system is to develop insight concerning the system, to develop an intuitive feel for the input–output relationship. In order to decide whether or not a particular model, linear or nonlinear, is a good model, we must compare the input–output relationship of the model with the corresponding, but measurable, input–output relationship of the system being modeled. If the model and the system are sufficiently in agreement for our purposes, we need not distinguish between the system and the model.

Almost all physical systems are to some degree nonlinear. Yet most systems act in a nearly linear manner if the range of variation of the variables is restricted. For example, the current through a resistor is essentially proportional to the applied voltage if the current is not large enough to heat the resistor significantly. We are able to develop adequate models for a wide variety of static and dynamic physical systems using only linear transformations. For linear models there is available a vast array of mathematical results; most mathematical analysis is linear analysis. Furthermore, the analysis or optimization of a *nonlinear* system is usually based on linearization (Chapters 7 and 8). Even in solving a nonlinear equation for a given input, we typically must resort to repetitive linearization.

The examples and exercises of this section have demonstrated the variety of familiar transformations which are linear: matrix multiplication, differentiation, integration, etc. We introduce other linear transformations as we need them. The next few chapters pertain only to linear transformations. In Chapter 3 we focus on the peculiarities of linear differential systems. In Chapter 4 we develop the concepts of spectral decomposition of linear systems. The discussion of infinite-dimensional systems in Chapter 5 is also directed toward linear systems. Because we use the symbols **T** and **U** so much in reference to linear transformations, hereinafter we employ the symbols **F** and **G** to emphasize concepts which apply as well to nonlinear transformations. We begin to examine nonlinear concepts in Chapter 6. We do not return fully to the subject of nonlinear systems, however, until we introduce the concepts of linearization and repetitive linearization in Chapters 7 and 8.

2.5 Matrices of Linear Transformations

By the process of picking an ordered basis for an n-dimensional vector space $\mathcal{V}$, we associate with each vector in $\mathcal{V}$ a unique $n\times 1$ column matrix. In effect, we convert the vectors in $\mathcal{V}$ into an equivalent set of vectors which are suitable for matrix manipulation and, therefore, automatic computation by computer. By taking coordinates, we can also convert a linear equation, $\mathbf{Tx}=\mathbf{y}$, into a matrix equation. Suppose $\mathbf{T}$: $\mathcal{V}\rightarrow\mathcal{W}$ is a linear transformation, $\dim(\mathcal{V})=n$, and $\dim(\mathcal{W})=m$. Pick as bases for $\mathcal{V}$ and $\mathcal{W}$ the sets $\mathcal{X} \triangleq \{\mathbf{x}_1,\ldots,\mathbf{x}_n\}$ and $\mathcal{Y} \triangleq \{\mathbf{y}_1,\ldots,\mathbf{y}_m\}$, respectively. The vectors $\mathbf{x}$ in $\mathcal{V}$ and $\mathbf{Tx}$ in $\mathcal{W}$ can be represented by their coordinate matrices $[\mathbf{x}]_{\mathcal{X}}$ and $[\mathbf{Tx}]_{\mathcal{Y}}$. The vectors $\mathbf{x}$ and $\mathbf{Tx}$ are linearly related (by the linear transformation $\mathbf{T}$). By (2.41), we know that a vector and its coordinates are also linearly related. Therefore, we expect $[\mathbf{x}]_{\mathcal{X}}$ and $[\mathbf{Tx}]_{\mathcal{Y}}$ to be linearly related as well. Furthermore, we intuitively expect the linear relation between the $n\times 1$ matrix $[\mathbf{x}]_{\mathcal{X}}$ and the $m\times 1$ matrix $[\mathbf{Tx}]_{\mathcal{Y}}$ to be multiplication by an $m\times n$ matrix. We denote this matrix by $[\mathbf{T}]_{\mathcal{X}\mathcal{Y}}$ and refer to it as the **matrix of T relative to the ordered bases $\mathcal{X}$ and $\mathcal{Y}$**; it must satisfy

$$[\mathbf{T}]_{\mathcal{X}\mathcal{Y}}[\mathbf{x}]_{\mathcal{X}} \triangleq [\mathbf{Tx}]_{\mathcal{Y}} \tag{2.45}$$

for all $\mathbf{x}$ in $\mathcal{V}$. Assume we can find such a matrix. Then by taking coordinates (with respect to $\mathcal{Y}$) of each side of the linear equation $\mathbf{Tx}=\mathbf{y}$, we convert the equation to the equivalent matrix equation.

$$[\mathbf{T}]_{\mathcal{X}\mathcal{Y}}[\mathbf{x}]_{\mathcal{X}} = [\mathbf{y}]_{\mathcal{Y}} \tag{2.46}$$

We will show that we can represent any linear transformation of $\mathcal{V}$ into $\mathcal{W}$ by a matrix multiplication by selecting bases for $\mathcal{V}$ and $\mathcal{W}$—we can convert any linear equation involving finite-dimensional vector spaces into a matrix equation. We first show how to determine the matrix of **T**, then we show that it satisfies the defining equation (2.45) for all vectors **x** in $\mathcal{V}$.

Example 1. Determining the Matrix of a Linear Transformation. Let $\mathbf{x}=(\xi_1, \xi_2, \xi_3)$, an arbitrary vector in $\mathcal{R}^3$. Define $\mathbf{T}$: $\mathcal{R}^3 \to \mathcal{R}^2$ by

$$\mathbf{T}(\xi_1,\xi_2,\xi_3) \triangleq (2\xi_2-\xi_1, \xi_1+\xi_2+\xi_3)$$

We now find $[\mathbf{T}]_{\mathcal{E}_3\mathcal{E}_2}$, where $\mathcal{E}_3$ and $\mathcal{E}_2$ are the standard bases for $\mathcal{R}^3$ and $\mathcal{R}^2$, respectively. By (2.45), we have

$$[\mathbf{T}]_{\mathcal{E}_3\mathcal{E}_2}[(\xi_1,\xi_2,\xi_3)]_{\mathcal{E}_3}=[(2\xi_2-\xi_1,\xi_1+\xi_2+\xi_3)]_{\mathcal{E}_2}$$

for all vectors (ξ_1, ξ_2, ξ_3), or

$$\begin{pmatrix} a_{11} & a_{12} & a_{13} \\ a_{21} & a_{22} & a_{23} \end{pmatrix}\begin{pmatrix} \xi_1 \\ \xi_2 \\ \xi_3 \end{pmatrix}=\begin{pmatrix} 2\xi_2-\xi_1 \\ \xi_1+\xi_2+\xi_3 \end{pmatrix} \tag{2.47}$$

where we have used $\{a_{ij}\}$ to represent the elements of $[\mathbf{T}]_{\mathcal{E}_3\mathcal{E}_2}$. By making three independent choices of the scalars ξ_1, ξ_2, and ξ_3, we could convert this matrix equation into six equations in the six unknowns $\{a_{ij}\}$. However, by using a little ingenuity, we reduce this effort. Think of the matrix multiplication in terms of the columns of $[\mathbf{T}]_{\mathcal{E}_3\mathcal{E}_2}$. The ith element of $[\mathbf{x}]_{\mathcal{E}_3}$ multiplies the ith column of $[\mathbf{T}]_{\mathcal{E}_3\mathcal{E}_2}$. If we choose $\mathbf{x}=(1, 0, 0)$, then $[(1, 0, 0)]_{\mathcal{E}_3}=\begin{pmatrix} 1 \\ 0 \\ 0 \end{pmatrix}$, and (2.47) becomes

$$\begin{pmatrix} a_{11} & a_{12} & a_{13} \\ a_{21} & a_{22} & a_{23} \end{pmatrix}\begin{pmatrix} 1 \\ 0 \\ 0 \end{pmatrix}=\begin{pmatrix} a_{11} \\ a_{21} \end{pmatrix}=\begin{pmatrix} -1 \\ 1 \end{pmatrix}$$

We have found the first column of $[\mathbf{T}]_{\mathcal{E}_3\mathcal{E}_2}$ directly. We obtain the other two columns of $[\mathbf{T}]_{\mathcal{E}_3\mathcal{E}_2}$ from (2.47) by successive substitution of $\mathbf{x}=(0, 1, 0)$ and $\mathbf{x}=(0, 0, 1)$. The result is

$$[\mathbf{T}]_{\mathcal{E}_3\mathcal{E}_2}=\begin{pmatrix} -1 & 2 & 0 \\ 1 & 1 & 1 \end{pmatrix}$$

In Example 1 we avoided the need for simultaneous equations by substituting the basis vectors $\boldsymbol{\varepsilon}_1$, $\boldsymbol{\varepsilon}_2$, and $\boldsymbol{\varepsilon}_3$ into (2.47) to pick out the columns of $[\mathbf{T}]_{\mathcal{E}_3\mathcal{E}_2}$. This same technique can be used to find the matrix of any linear transformation acting on a finite-dimensional space. We refer again to $\mathbf{T}$: $\mathcal{V} \to \mathcal{W}$, with $\dim(\mathcal{V})=n$, $\dim(\mathcal{W})=m$, $\mathcal{X}$ a basis for $\mathcal{V}$, and $\mathcal{Y}$ a basis for $\mathcal{W}$. If we substitute into (2.45) the vector $\mathbf{x}_i$, the ith vector of the basis $\mathcal{X}$, we pick out the ith column of $[\mathbf{T}]_{\mathcal{X}\mathcal{Y}}$:

$$[\mathbf{T}]_{\mathcal{X}\mathcal{Y}}[\mathbf{x}_i]_{\mathcal{X}}=[\mathbf{T}]_{\mathcal{X}\mathcal{Y}}\begin{pmatrix}0\\ \vdots\\ 0\\ 1_i\\ 0\\ \vdots\\ 0\end{pmatrix}=i\text{th column of }[\mathbf{T}]_{\mathcal{X}\mathcal{Y}}=[\mathbf{T}\mathbf{x}_i]_{\mathcal{Y}}$$

We can find each column of $[\mathbf{T}]_{\mathcal{X}\mathcal{Y}}$ independently. The only computational effort is that in determining the coordinate matrices $[\mathbf{T}\mathbf{x}_i]_{\mathcal{Y}}$. Therefore,

$$[\mathbf{T}]_{\mathcal{X}\mathcal{Y}}=([\mathbf{T}\mathbf{x}_1]_{\mathcal{Y}}\ \vdots\ [\mathbf{T}\mathbf{x}_2]_{\mathcal{Y}}\ \vdots\ \cdots\ \vdots\ [\mathbf{T}\mathbf{x}_n]_{\mathcal{Y}}) \tag{2.48}$$

Example 2. The Matrix of a Linear Operator. Define the differential operator $\mathbf{D}$: $\mathcal{P}^3\rightarrow\mathcal{P}^3$ as in (2.36). The set $\mathcal{N}\triangleq\{\mathbf{f}_1, \mathbf{f}_2, \mathbf{f}_3\}$, where $\mathbf{f}_1(t)=1$, $\mathbf{f}_2(t)=t$, $\mathbf{f}_3(t)=t^2$, is a natural basis for $\mathcal{P}^3$. We use (2.48) to find

$$\begin{aligned}[\mathbf{D}]_{\mathcal{N}\mathcal{N}}&=([\mathbf{D}\mathbf{f}_1]_{\mathcal{N}}\ \vdots\ [\mathbf{D}\mathbf{f}_2]_{\mathcal{N}}\ \vdots\ [\mathbf{D}\mathbf{f}_3]_{\mathcal{N}})\\ &=([\boldsymbol{\theta}]_{\mathcal{N}}\ \vdots\ [\mathbf{f}_1]_{\mathcal{N}}\ \vdots\ [2\mathbf{f}_2]_{\mathcal{N}})\\ &=\begin{pmatrix}0 & 1 & 0\\ 0 & 0 & 2\\ 0 & 0 & 0\end{pmatrix}\end{aligned}$$

From the method used to determine $[\mathbf{T}]_{\mathcal{X}\mathcal{Y}}$ in (2.48), we know that this matrix correctly represents the action of $\mathbf{T}$ on the basis vectors $\{\mathbf{x}_i\}$. We now show that the matrix (2.48) also represents correctly the action of $\mathbf{T}$ on all other vectors in $\mathcal{V}$. An arbitrary vector $\mathbf{x}$ in $\mathcal{V}$ may be written in terms of the basis vectors for $\mathcal{V}$:

$$\mathbf{x}=\sum_{i=1}^{n}\xi_i\mathbf{x}_i$$

Since the transformation $\mathbf{T}$ is linear,

$$\mathbf{T}\mathbf{x}=\sum_{i=1}^{n}\xi_i\mathbf{T}\mathbf{x}_i$$

Because the process of taking coordinates is linear [see (2.41)],

$$\begin{aligned}[\mathbf{Tx}]_{\mathcal{Y}} &= \sum_{i=1}^{n} \xi_i [\mathbf{Tx}_i]_{\mathcal{Y}} \\ &= ([\mathbf{Tx}_1]_{\mathcal{Y}} \vdots \cdots \vdots [\mathbf{Tx}_n]_{\mathcal{Y}}) \begin{pmatrix} \xi_1 \\ \vdots \\ \xi_n \end{pmatrix} \\ &= [\mathbf{T}]_{\mathcal{X}\mathcal{Y}} [\mathbf{x}]_{\mathcal{X}}\end{aligned}$$

Thus, continuing Example 2 above, if **f** is the arbitrary vector defined by $\mathbf{f}(t) \stackrel{\Delta}{=} \xi_1 + \xi_2 t + \xi_3 t^2$, then

$$(\mathbf{Df})(t) = \xi_2 + 2\xi_3 t, \; [\mathbf{f}]_{\mathcal{N}} = \begin{pmatrix} \xi_1 \\ \xi_2 \\ \xi_3 \end{pmatrix}, \; [\mathbf{Df}]_{\mathcal{N}} = \begin{pmatrix} \xi_2 \\ 2\xi_3 \\ 0 \end{pmatrix}, \text{ and } [\mathbf{D}]_{\mathcal{N}\mathcal{N}} [\mathbf{f}]_{\mathcal{N}} = [\mathbf{Df}]_{\mathcal{N}}$$

When the domain and range space of **T** are identical, and the same basis is used for both spaces (as it is in Example 2), we sometimes refer to the matrix $[\mathbf{T}]_{\mathcal{X}\mathcal{X}}$ as the **matrix of the operator T relative to the basis** $\mathcal{X}$.

We expect the matrix of a linear transformation to possess the basic characteristics of that transformation. The only basic characteristics of a linear transformation that we have discussed thus far are its rank and nullity. The picking of coordinate systems $\mathcal{X}$ and $\mathcal{Y}$ converts the transformation equation $\mathbf{Tx} = \mathbf{y}$ to a precisely equivalent matrix equation, $[\mathbf{Tx}]_{\mathcal{Y}} = [\mathbf{T}]_{\mathcal{X}\mathcal{Y}} [\mathbf{x}]_{\mathcal{X}} = [\mathbf{y}]_{\mathcal{Y}}$; for every **x** and **y** in the one equation, there is a unique $[\mathbf{x}]_{\mathcal{X}}$ and $[\mathbf{y}]_{\mathcal{Y}}$ in the other. The dimensions of the nullspace and range of the transformation "multiplication by $[\mathbf{T}]_{\mathcal{X}\mathcal{Y}}$" must be the same, therefore, as the dimensions of the nullspace and range of **T**. We speak loosely of the rank and nullity of $[\mathbf{T}]_{\mathcal{X}\mathcal{Y}}$ when we actually mean the rank and nullity of the transformation "multiplication by $[\mathbf{T}]_{\mathcal{X}\mathcal{Y}}$." We refer to the nullity and rank of a matrix as if it were the matrix of a linear transformation. The nullspace and range of matrix multiplications are explored in P&C 2.19; the problem demonstrates that for an $m \times n$ matrix **A**,

$$\begin{aligned}\text{rank}(\mathbf{A}) &= \text{the number of independent columns of } \mathbf{A} \\ &= \text{the number of independent rows of } \mathbf{A} \\ \text{nullity}(\mathbf{A}) &= n - \text{rank}(\mathbf{A}) \\ \text{nullity}(\mathbf{A}^{\mathrm{T}}) &= m - \text{rank}(\mathbf{A})\end{aligned}$$

Once again referring to Example 2, we see that the nullity of **D** is 1 [the vector $\mathbf{f}_1$ is a basis for nullspace(**D**)]. The nullity of $[\mathbf{D}]_{\mathfrak{N}\mathfrak{N}}$ is also 1 ($[\mathbf{D}]_{\mathfrak{N}\mathfrak{N}}$ contains one dependent column). The matrix $[\mathbf{D}]_{\mathfrak{N}\mathfrak{N}}$ does possess the same nullity and rank as the operator **D**.

It is apparent that determination of the matrix of a transformation reduces to the determination of coordinate matrices for the set of vectors $\{\mathbf{Tx}_i\}$ of (2.48). We found in Section 2.2 that determination of the coordinate matrix of a vector $\mathbf{x}$ with respect to a basis $\mathfrak{X}=\{\mathbf{x}_i\}$ can be reduced to performing elimination on the matrix equation (2.17):

$$[\mathbf{x}]_{\mathfrak{N}}=([\mathbf{x}_1]_{\mathfrak{N}}\ \vdots\ \cdots\ \vdots\ [\mathbf{x}_n]_{\mathfrak{N}})[\mathbf{x}]_{\mathfrak{X}}$$

where $\mathfrak{N}$ is a natural basis for the space $\mathcal{V}$ of which $\mathbf{x}$ is a member (i.e., a basis with respect to which coordinates can be determined by inspection).

Exercise 1. Show that $[\mathbf{T}]_{\mathfrak{X}\mathfrak{Y}}$ of (2.48) can be obtained by the row reduction

$$\left([\mathbf{y}_1]_{\mathfrak{N}}\ \vdots\ \cdots\ \vdots\ [\mathbf{y}_n]_{\mathfrak{N}}\ \vdots\ [\mathbf{Tx}_1]_{\mathfrak{N}}\ \vdots\ \cdots\ \vdots\ [\mathbf{Tx}_n]_{\mathfrak{N}}\right)\rightarrow\left(\mathbf{I}\ \vdots\ [\mathbf{T}]_{\mathfrak{X}\mathfrak{Y}}\right) \tag{2.49}$$

where $\mathfrak{N}$ is a natural basis for the range of definition $\mathcal{W}$. (Hint: if the elements of $[\mathbf{Tx}_i]_{\mathfrak{Y}}$ are denoted by $[\mathbf{Tx}_i]_{\mathfrak{Y}}=(c_{1i}\cdots c_{ni})^{\mathrm{T}}$, then $\mathbf{Tx}_i=\Sigma_j c_{ji}\mathbf{y}_i$, and $[\mathbf{Tx}_i]_{\mathfrak{N}}=\Sigma_j c_{ji}[\mathbf{y}_i]_{\mathfrak{N}}$.) Use this approach to find $[\mathbf{T}]_{\mathcal{E}_3\mathcal{E}_2}$ of Example 1.

Example 3. The Matrix of a Matrix Transformation. Let $\mathbf{T}$: $\mathfrak{M}^{n\times 1}\rightarrow\mathfrak{M}^{m\times 1}$ be defined by $\mathbf{Tx}\triangleq\mathbf{Ax}$, where $\mathbf{A}$ is an $m\times n$ matrix. Denoting the standard bases for $\mathfrak{M}^{n\times 1}$ and $\mathfrak{M}^{m\times 1}$ by $\mathcal{E}_n$ and $\mathcal{E}_m$, respectively, we find $[\mathbf{T}]_{\mathcal{E}_n\mathcal{E}_m}=\mathbf{A}$. Although $[\mathbf{x}]_{\mathfrak{X}}$ and $\mathbf{x}$ are identical in this example, we should distinguish between them, for it is certainly incorrect to equate the matrix $[\mathbf{T}]_{\mathcal{E}_n\mathcal{E}_m}$ to the transformation $\mathbf{T}$.

Suppose $\mathbf{T}$: $\mathcal{V}\rightarrow\mathcal{W}$ is invertible and linear; $\mathcal{V}$ and $\mathcal{W}$ are finite-dimensional with bases $\mathfrak{X}$ and $\mathfrak{Y}$, respectively. It follows from (2.45) that

$$[\mathbf{T}^{-1}]_{\mathfrak{Y}\mathfrak{X}}[\mathbf{y}]_{\mathfrak{Y}}=[\mathbf{T}^{-1}\mathbf{y}]_{\mathfrak{X}} \tag{2.50}$$

for all $\mathbf{y}$ in $\mathcal{W}$. Then, for each $\mathbf{x}$ in $\mathcal{V}$,

$$[\mathbf{x}]_{\mathfrak{X}}=[\mathbf{T}^{-1}\mathbf{Tx}]_{\mathfrak{X}}=[\mathbf{T}^{-1}]_{\mathfrak{Y}\mathfrak{X}}[\mathbf{Tx}]_{\mathfrak{Y}}=[\mathbf{T}^{-1}]_{\mathfrak{Y}\mathfrak{X}}[\mathbf{T}]_{\mathfrak{X}\mathfrak{Y}}[\mathbf{x}]_{\mathfrak{X}}$$

A similar relationship can be established with $\mathbf{T}$ and $\mathbf{T}^{-1}$ reversed. Then as a consequence of (2.29),

$$[\mathbf{T}^{-1}]_{\mathfrak{Y}\mathfrak{X}}=[\mathbf{T}]_{\mathfrak{X}\mathfrak{Y}}^{-1} \tag{2.51}$$

Exercise 2. Suppose $\mathcal{V}$, $\mathcal{W}$, and $\mathcal{U}$ are finite-dimensional vector spaces with bases $\mathcal{X}$, $\mathcal{Y}$, and $\mathcal{Z}$, respectively. Show that

a. If $\mathbf{T}$: $\mathcal{V}\to\mathcal{W}$ and $\mathbf{U}$: $\mathcal{V}\to\mathcal{W}$ are linear, then

$$[a\mathbf{T}+b\mathbf{U}]_{\mathcal{X}\mathcal{Y}}=a[\mathbf{T}]_{\mathcal{X}\mathcal{Y}}+b[\mathbf{U}]_{\mathcal{X}\mathcal{Y}} \tag{2.52}$$

b. If $\mathbf{T}$: $\mathcal{V}\to\mathcal{W}$ and $\mathbf{U}$: $\mathcal{W}\to\mathcal{U}$ are linear, then

$$[\mathbf{UT}]_{\mathcal{X}\mathcal{Z}}=[\mathbf{U}]_{\mathcal{Y}\mathcal{Z}}[\mathbf{T}]_{\mathcal{X}\mathcal{Y}} \tag{2.53}$$

Changes in Coordinate System

In Chapter 4 we discuss coordinate systems which are particularly suitable for analysis of a given linear transformation—coordinate systems for which the matrix of the transformation is diagonal. In preparation for that discussion we now explore the effect of a change of coordinate system on a coordinate matrix $[\mathbf{x}]$ and on the matrix of a transformation $[\mathbf{T}]$.

Suppose $\mathcal{X}$ and $\mathcal{Z}$ are two different bases for an n-dimensional vector space $\mathcal{V}$. We know by (2.41) that the transformations

$$\mathbf{x}\to[\mathbf{x}]_{\mathcal{X}} \quad\text{and}\quad \mathbf{x}\to[\mathbf{x}]_{\mathcal{Z}}$$

are linear and invertible. Thus we expect $[\mathbf{x}]_{\mathcal{X}}$ and $[\mathbf{x}]_{\mathcal{Z}}$ to be related by

$$\mathbf{S}[\mathbf{x}]_{\mathcal{X}}=[\mathbf{x}]_{\mathcal{Z}} \tag{2.54}$$

where $\mathbf{S}$ is an $n\times n$ invertible matrix. In fact, multiplication of $[\mathbf{x}]_{\mathcal{X}}$ by *any* invertible matrix represents a change from the coordinate system $\mathcal{X}$ to some new coordinate system. We sometimes denote the matrix $\mathbf{S}$ of (2.54) by the symbol $\mathbf{S}_{\mathcal{X}\mathcal{Z}}$, thereby making explicit the fact that $\mathbf{S}$ converts coordinates relative to $\mathcal{X}$ into coordinates relative to $\mathcal{Z}$. Then $(\mathbf{S}_{\mathcal{X}\mathcal{Z}})^{-1}$ $=S_{\mathcal{Z}\mathcal{X}}$.

Determination of the specific change-of-coordinates matrix $\mathbf{S}$ defined in (2.54) follows the same line of thought as that used to determine $[\mathbf{T}]$ in (2.48). By successively substituting into (2.54) the vectors $\mathbf{x}_1, \mathbf{x}_2,\ldots,\mathbf{x}_n$ from the basis $\mathcal{X}$, we isolate the columns of $\mathbf{S}$: the ith column of $\mathbf{S}$ is $[\mathbf{x}_i]_{\mathcal{Z}}$. Thus the unique invertible matrix $\mathbf{S}$ which transforms coordinate matrices relative to $\mathcal{X}$ into coordinate matrices relative to $\mathcal{Z}$ is

$$\mathbf{S}=\mathbf{S}_{\mathcal{X}\mathcal{Z}}=([\mathbf{x}_1]_{\mathcal{Z}} \vdots \cdots \vdots [\mathbf{x}_n]_{\mathcal{Z}}) \tag{2.55}$$

where the $\mathbf{x}_i$ are the vectors in the basis $\mathcal{X}$.

Since a change-of-coordinates matrix is always invertible, we determine from (2.54) that

$$\mathbf{S}^{-1}[\mathbf{x}]_{\mathcal{Z}} = [\mathbf{x}]_{\mathcal{X}}$$

and

$$\mathbf{S}^{-1} = \mathbf{S}_{\mathcal{X}\mathcal{Z}}^{-1} = \mathbf{S}_{\mathcal{Z}\mathcal{X}} = ([\mathbf{z}_1]_{\mathcal{X}} \vdots \cdots \vdots [\mathbf{z}_n]_{\mathcal{X}}) \tag{2.56}$$

where the $\mathbf{z}_i$ are the vectors in the basis $\mathcal{Z}$. If $\mathcal{Z}$ is a natural basis for the space, then $\mathbf{S}$ can be found by inspection. On the other hand, if $\mathcal{X}$ is a natural basis, we find $\mathbf{S}^{-1}$ by inspection. It is appropriate to use either (2.55) or (2.56) in determining $\mathbf{S}$. We need both $\mathbf{S}$ and $\mathbf{S}^{-1}$ to allow conversion back and forth between the two coordinate systems. Besides, the placing of $\mathbf{S}$ on the left side of (2.54) was arbitrary.

Example 4. A Change-of-Coordinates Matrix. Let $\mathcal{E}$ be the standard basis for $\mathcal{R}^3$. Another basis for $\mathcal{R}^3$ is $\mathcal{Z} = \{\mathbf{z}_1, \mathbf{z}_2, \mathbf{z}_3\}$, where $\mathbf{z}_1 = (1, 1, 1)$, $\mathbf{z}_2 = (1, 1, 0)$, and $\mathbf{z}_3 = (1, 0, 0)$. Since $\mathcal{E}$ is a natural basis for $\mathcal{R}^3$, we use (2.56) to find

$$\mathbf{S}^{-1} = ([\mathbf{z}_1]_{\mathcal{E}} \vdots [\mathbf{z}_2]_{\mathcal{E}} \vdots [\mathbf{z}_3]_{\mathcal{E}}) = \begin{pmatrix} 1 & 1 & 1 \\ 1 & 1 & 0 \\ 1 & 0 & 0 \end{pmatrix} \tag{2.57}$$

A straightforward elimination (Section 1.5) yields

$$\mathbf{S} = \begin{pmatrix} 0 & 0 & 1 \\ 0 & 1 & -1 \\ 1 & -1 & 0 \end{pmatrix} \tag{2.58}$$

We note that for an arbitrary vector $\mathbf{x} = (\xi_1, \xi_2, \xi_3)$ in $\mathcal{R}^3$, $[\mathbf{x}]_{\mathcal{E}} = (\xi_1 \ \ \xi_2 \ \ \xi_3)^{\mathrm{T}}$. By (2.54),

$$[\mathbf{x}]_{\mathcal{Z}} = \mathbf{S}[\mathbf{x}]_{\mathcal{E}} = (\xi_3 \quad \xi_2 - \xi_3 \quad \xi_1 - \xi_2)^{\mathrm{T}} \tag{2.59}$$

But then,

$$\begin{aligned} \mathbf{x} &= (\xi_3)\mathbf{z}_1 + (\xi_2 - \xi_3)\mathbf{z}_2 + (\xi_1 - \xi_2)\mathbf{z}_3 \\ &= (\xi_3)(1, 1, 1) + (\xi_2 - \xi_3)(1, 1, 0) + (\xi_1 - \xi_2)(1, 0, 0) \\ &= (\xi_1, \xi_2, \xi_3) \end{aligned} \tag{2.60}$$

and the validity of the change of coordinates matrix $\mathbf{S}$ is verified.

If neither $\mathcal{X}$ nor $\mathcal{Z}$ is a natural basis, the determination of $\mathbf{S}$ can still be systematized by the introduction of an intermediate step which does involve a natural basis.

Exercise 3. Suppose we need the change-of-coordinates matrix $\mathbf{S}$ such that $\mathbf{S}[\mathbf{x}]_{\mathcal{X}} = [\mathbf{x}]_{\mathcal{Z}}$, where neither $\mathcal{X}$ nor $\mathcal{Z}$ is a natural basis for $\mathcal{V}$. Suppose $\mathcal{N}$ is a natural basis. Show, by introducing an intermediate change to the coordinates $[\mathbf{x}]_{\mathcal{N}}$, that

$$\mathbf{S} = \big([\mathbf{z}_1]_{\mathcal{N}} \vdots \cdots \vdots [\mathbf{z}_n]_{\mathcal{N}}\big)^{-1}\big([\mathbf{x}_1]_{\mathcal{N}} \vdots \cdots \vdots [\mathbf{x}_n]_{\mathcal{N}}\big) \tag{2.61}$$

Example 5. Change of Coordinates via an Intermediate Natural Basis. Two bases for $\mathcal{P}^3$ are $\mathcal{F} \triangleq \{\mathbf{f}_1, \mathbf{f}_2, \mathbf{f}_3\}$ and $\mathcal{G} \triangleq \{\mathbf{g}_1, \mathbf{g}_2, \mathbf{g}_3\}$, where

$$\mathbf{f}_1(t) = 1, \quad \mathbf{f}_2(t) = 1 + t, \quad \mathbf{f}_3(t) = 1 + t^2$$

$$\mathbf{g}_1(t) = 1 + t, \quad \mathbf{g}_2(t) = t, \quad \mathbf{g}_3(t) = t + t^2$$

To find $\mathbf{S}$ such that $\mathbf{S}[\mathbf{f}]_{\mathcal{F}} = [\mathbf{f}]_{\mathcal{G}}$, we introduce the natural basis $\mathcal{N} \triangleq \{\mathbf{h}_1, \mathbf{h}_2, \mathbf{h}_3\}$, where $\mathbf{h}_i(t) = t^{i-1}$. Then, by (2.61),

$$\mathbf{S} = ([\mathbf{g}_1]_{\mathcal{N}} \vdots [\mathbf{g}_2]_{\mathcal{N}} \vdots [\mathbf{g}_3]_{\mathcal{N}})^{-1}([\mathbf{f}_1]_{\mathcal{N}} \vdots [\mathbf{f}_2]_{\mathcal{N}} \vdots [\mathbf{f}_3]_{\mathcal{N}})$$

$$= \begin{pmatrix} 1 & 0 & 0 \\ 1 & 1 & 1 \\ 0 & 0 & 1 \end{pmatrix}^{-1} \begin{pmatrix} 1 & 1 & 1 \\ 0 & 1 & 0 \\ 0 & 0 & 1 \end{pmatrix}$$

$$= \begin{pmatrix} 1 & 0 & 0 \\ -1 & 1 & -1 \\ 0 & 0 & 1 \end{pmatrix} \begin{pmatrix} 1 & 1 & 1 \\ 0 & 1 & 0 \\ 0 & 0 & 1 \end{pmatrix} = \begin{pmatrix} 1 & 1 & 1 \\ -1 & 0 & -2 \\ 0 & 0 & 1 \end{pmatrix}$$

Similarity and Equivalence Transformations

Now that we have a process for changing coordinate systems, we explore the effect of such a change on the matrix of a transformation. Suppose $\mathbf{T}$ is a linear operator on $\mathcal{V}$, and that $\mathcal{X}$ and $\mathcal{Z}$ are two different bases for $\mathcal{V}$. Then $[\mathbf{T}]_{\mathcal{X}\mathcal{X}}$ is defined by

$$[\mathbf{T}]_{\mathcal{X}\mathcal{X}}[\mathbf{x}]_{\mathcal{X}} = [\mathbf{T}\mathbf{x}]_{\mathcal{X}}$$

The change from the $\mathcal{X}$ to the $\mathcal{Z}$ coordinate system is described by

$$\mathbf{S}[\mathbf{x}]_{\mathcal{X}} = [\mathbf{x}]_{\mathcal{Z}}$$

The change-of-coordinates matrix $\mathbf{S}$ also applies to the vector $\mathbf{T}\mathbf{x}$ in $\mathcal{V}$:

$$\mathbf{S}[\mathbf{T}\mathbf{x}]_{\mathcal{X}} = [\mathbf{T}\mathbf{x}]_{\mathcal{Z}}$$

By substituting $[\mathbf{x}]_{\mathcal{X}}$ and $[\mathbf{Tx}]_{\mathcal{X}}$ from these last two equations into the defining equation for $[\mathbf{Tx}]_{\mathcal{X}\mathcal{X}}$, we find

$$[\mathbf{T}]_{\mathcal{X}\mathcal{X}}\mathbf{S}^{-1}[\mathbf{x}]_{\mathcal{Z}}=\mathbf{S}^{-1}[\mathbf{Tx}]_{\mathcal{Z}}$$

or

$$(\mathbf{S}[\mathbf{T}]_{\mathcal{X}\mathcal{X}}\mathbf{S}^{-1})[\mathbf{x}]_{\mathcal{Z}}=[\mathbf{Tx}]_{\mathcal{Z}}$$

But this is the defining equation for $[\mathbf{T}]_{\mathcal{Z}\mathcal{Z}}$. It is apparent that

$$[\mathbf{T}]_{\mathcal{Z}\mathcal{Z}}=\mathbf{S}[\mathbf{T}]_{\mathcal{X}\mathcal{X}}\mathbf{S}^{-1} \tag{2.62}$$

where $\mathbf{S}$ converts from the $\mathcal{X}$ coordinate system to the $\mathcal{Z}$ coordinate system. Equation (2.62) describes an invertible linear transformation on $[\mathbf{T}]_{\mathcal{X}\mathcal{X}}$ known as a **similarity transformation.** In Section 4.2, we find that a similarity transformation preserves the basic spectral properties of the matrix. It is comforting to know that any two matrix representations of a linear system have the same properties—these properties are inherent in the model, $\mathbf{T}$, and should not be affected by the coordinate system we select.

Example 6. A Similarity Transformation. In Example 2 we found the matrix of the differential operator on $\mathcal{P}^3$ relative to the natural basis for $\mathcal{P}^3$:

$$[\mathbf{D}]_{\mathcal{N}\mathcal{N}}=\begin{pmatrix}0&1&0\\0&0&2\\0&0&0\end{pmatrix}$$

Another basis for $\mathcal{P}^3$ is $\mathcal{G}=\{\mathbf{g}_1,\mathbf{g}_2,\mathbf{g}_3\}$, where $\mathbf{g}_1(t)=1+t$, $\mathbf{g}_2(t)=t$, and $\mathbf{g}_3(t)=t+t^2$. The change-of-coordinates matrix which relates the two bases $\mathcal{N}$ and $\mathcal{G}$ is defined by $\mathbf{S}[\mathbf{f}]_{\mathcal{N}}=[\mathbf{f}]_{\mathcal{G}}$; we find it using (2.56):

$$\begin{aligned}\mathbf{S}^{-1}&=([\mathbf{g}_1]_{\mathcal{N}}\ \vdots\ [\mathbf{g}_2]_{\mathcal{N}}\ \vdots\ [\mathbf{g}_3]_{\mathcal{N}})\\&=\begin{pmatrix}1&0&0\\1&1&1\\0&0&1\end{pmatrix}\end{aligned}$$

The inverse matrix is

$$\mathbf{S}=\begin{pmatrix}1&0&0\\-1&1&-1\\0&0&1\end{pmatrix}$$

Then, by (2.62),

$$[\mathbf{D}]_{\mathcal{G}\mathcal{G}} = \mathbf{S}[\mathbf{D}]_{\mathcal{N}\mathcal{N}}\mathbf{S}^{-1}$$

$$= \begin{pmatrix} 1 & 0 & 0 \\ -1 & 1 & -1 \\ 0 & 0 & 1 \end{pmatrix} \begin{pmatrix} 0 & 1 & 0 \\ 0 & 0 & 2 \\ 0 & 0 & 0 \end{pmatrix} \begin{pmatrix} 1 & 0 & 0 \\ 1 & 1 & 1 \\ 0 & 0 & 1 \end{pmatrix}$$

$$= \begin{pmatrix} 1 & 1 & 1 \\ -1 & -1 & 1 \\ 0 & 0 & 0 \end{pmatrix}$$

Exercise 4. Let $\mathbf{T}$: $\mathcal{V} \to \mathcal{W}$ be a linear transformation. Assume $\mathcal{V}$ and $\mathcal{W}$ are finite dimensional. Let the invertible matrix $\mathbf{S}_{\mathcal{X}\mathcal{F}}$ convert from the basis $\mathcal{X}$ to the basis $\mathcal{F}$ in $\mathcal{V}$. Let the invertible matrix $\mathbf{S}_{\mathcal{Y}\mathcal{G}}$ convert from the basis $\mathcal{Y}$ to the basis $\mathcal{G}$ in $\mathcal{W}$. Show that

$$[\mathbf{T}]_{\mathcal{F}\mathcal{G}} = \mathbf{S}_{\mathcal{Y}\mathcal{G}}[\mathbf{T}]_{\mathcal{X}\mathcal{Y}}\mathbf{S}_{\mathcal{X}\mathcal{F}}^{-1} \tag{2.63}$$

This transformation of the matrix $[\mathbf{T}]_{\mathcal{X}\mathcal{Y}}$ is called an **equivalence transformation.** The similarity transformation (2.62) is a special case. The term "equivalence" is motivated by the fact that $[\mathbf{T}]_{\mathcal{X}\mathcal{Y}}$ and $[\mathbf{T}]_{\mathcal{F}\mathcal{G}}$ are equivalent models of the system. The system equation $\mathbf{Tx} = \mathbf{y}$ is equally well represented by the matrix equations which result from the introduction of *any* coordinate systems for $\mathcal{V}$ and $\mathcal{W}$.

The discussion of matrices of transformations has been limited to transformations on finite-dimensional vector spaces. The primary reason for avoiding the infinite-dimensional counterparts is our inability to speak meaningfully about bases for infinite-dimensional spaces before discussing convergence of an infinite sequence of vectors (Section 5.3). However, matrices of infinite dimension are more difficult to work with (to invert, etc.) than are finite-dimensional matrices.

2.6 Problems and Comments

*2.1 Let $\mathcal{S}_1$ and $\mathcal{S}_2$ be subsets of a vector space $\mathcal{V}$. Let $\mathcal{W}_1$ and $\mathcal{W}_2$ be subspaces of $\mathcal{V}$.

(*a*) The **intersection** $\mathcal{S}_1 \cap \mathcal{S}_2$ of the sets $\mathcal{S}_1$ and $\mathcal{S}_2$ is the set of vectors which belong to both $\mathcal{S}_1$ and $\mathcal{S}_2$; if $\mathcal{S}_1 \cap \mathcal{S}_2$ is empty or if $\mathcal{S}_1 \cap \mathcal{S}_2 = \boldsymbol{\theta}$, we say $\mathcal{S}_1$ and $\mathcal{S}_2$ are **disjoint**.

(*b*) The **union** $\mathcal{S}_1 \cup \mathcal{S}_2$ of the sets $\mathcal{S}_1$ and $\mathcal{S}_2$ is the set of vectors which belong either to $\mathcal{S}_1$ or to $\mathcal{S}_2$ or to both.

(*c*) The **sum** $\mathcal{S}_1+\mathcal{S}_2$ of the sets $\mathcal{S}_1$ and $\mathcal{S}_2$ is the set of vectors of the form $\mathbf{x}_1+\mathbf{x}_2$, where $\mathbf{x}_1$ is in $\mathcal{S}_1$ and $\mathbf{x}_2$ is in $\mathcal{S}_2$.
(*d*) $\mathcal{W}_1 \cap \mathcal{W}_2$ is a subspace.
(*e*) $\mathcal{W}_1 \cup \mathcal{W}_2$ is usually not a subspace.
(*f*) $\mathcal{W}_1+\mathcal{W}_2$ is the subspace spanned by $\mathcal{W}_1 \cup \mathcal{W}_2$.
(*g*) $\dim(\mathcal{W}_1)+\dim(\mathcal{W}_2)=\dim(\mathcal{W}_1+\mathcal{W}_2)+\dim(\mathcal{W}_1 \cap \mathcal{W}_2)$.

2.2 Prove that the real 3-tuple space $\mathcal{R}^3$ introduced in Equation (2.2) is a vector space.

2.3 Determine whether or not the following sets of vectors are linearly independent:
(*a*) The column vectors $(2\ 1\ 0\ 1)^{\mathrm{T}}$, $(1\ 2\ -1\ 1)^{\mathrm{T}}$, and $(3\ 0\ 1\ 1)^{\mathrm{T}}$ in $\mathcal{M}^{4\times 1}$.
(*b*) The functions $\mathbf{f}_1(t)=1+2t-t^2$, $\mathbf{f}_2(t)=2+2t+t^2$, and $\mathbf{f}_3(t)=-1+3t+t^2$ in $\mathcal{P}^3$.
(*c*) The functions $\mathbf{g}_1(t)=1+2t+t^2-t^3$, $\mathbf{g}_2(t)=1+t-t^2+t^3$, and $\mathbf{g}_3(t)=1+3t+3t^2-3t^3$ in $\mathcal{P}^4$.

*2.4 *Modulo-2 scalars*: data transmitted by radio or telephone usually consist in strings of binary numbers (ones and zeros). A character or number to be transmitted is represented by a binary code word of length n. It is a sequence of these code words which makes up the transmitted string. We can think of the set of all possible code words of length n as vectors in a vector space. We call the space a binary linear code (see [2.8]). The scalars used in vector space manipulations can be restricted to binary numbers if ordinary addition of scalars is replaced by **modulo-2 addition**:

$$0+0=0 \qquad 0+1=1$$
$$1+0=1 \qquad 1+1=0$$

The rules for multiplication of scalars need not be changed. One way to check for errors in data transmission is to let the nth element of each code word equal the sum (mod-2) of the other elements in the word. If a single error appears in the transmitted word, the nth element will fail to give the proper sum.
(*a*) Let $\mathcal{V}$ be the set of 5×1 matrices with the mod-2 scalars as elements. Show that $\mathcal{V}$ is a vector space. (Assume that addition and scalar multiplication of the matrices is based on the mod-2 scalars.)
(*b*) Let $\mathcal{W}$ be the subset of $\mathcal{V}$ consisting in vectors for which the fifth element equals the sum of the other four elements. Show that $\mathcal{W}$ is a subspace of $\mathcal{V}$.
(*c*) Find a basis $\mathcal{X}$ for $\mathcal{W}$. Determine $[\mathbf{x}]_{\mathcal{X}}$, where $\mathbf{x}=(1\ 1\ 0\ 1\ 1)^{\mathrm{T}}$.

(*d*) The subspace $\mathcal{W}$ is a binary linear code. A code can also be described by a "parity check" matrix $\mathbf{P}$ for which the code is the nullspace. Find the parity check matrix for the code $\mathcal{W}$.

2.5 The set of all real $m\times n$ matrices, together with the usual definitions of addition and scalar multiplication of matrices, forms a vector space which we denote by $\mathfrak{M}^{m\times n}$. Determine the dimension of this linear space by exhibiting a basis for the space.

*2.6 Let $\mathcal{V}$ and $\mathcal{W}$ be vector spaces. With the definition of linear combination of transformations given in (2.27),

(*a*) The set of all transformations from $\mathcal{V}$ into $\mathcal{W}$ forms a vector space.

(*b*) The set $\mathcal{L}(\mathcal{V},\mathcal{W})$ of all linear transformations from $\mathcal{V}$ into $\mathcal{W}$ forms a subspace of the vector space in (*a*).

(*c*) The set of all linear transformations which take a particular subspace of $\mathcal{V}$ into $\boldsymbol{\theta}_{\mathcal{W}}$ constitutes a subspace of $\mathcal{L}(\mathcal{V},\mathcal{W})$.

(*d*) If $\dim(\mathcal{V})=n$ and $\dim(\mathcal{W})=m$, then $\dim(\mathcal{L}(\mathcal{V},\mathcal{W}))=mn$.

*2.7 *Exploring linear combinations by row reduction.* let $\mathcal{Y}\triangleq\{\mathbf{y}_1,\ldots,\mathbf{y}_n\}$ be a set of $m\times 1$ column vectors. The linear combination $\mathbf{y}=c_1\mathbf{y}_1+\cdots+c_n\mathbf{y}_n$ can be expressed as $\mathbf{y}=\mathbf{A}\mathbf{x}$ by defining $\mathbf{A}\triangleq(\mathbf{y}_1 \vdots \mathbf{y}_2 \vdots \cdots \vdots \mathbf{y}_n)$ and $\mathbf{x}\triangleq(c_1\cdots c_n)^{\mathrm{T}}$. Row reduction of the matrix $(\mathbf{A}\vdots\mathbf{y})$ for an unspecified vector $\mathbf{y}\triangleq(\eta_1\cdots\eta_m)^{\mathrm{T}}$, or the equivalent row reduction of $(\mathbf{A}\vdots\mathbf{I})$ for an $m\times m$ matrix $\mathbf{I}$, determines the form of the vectors in span($\mathcal{Y}$) and pinpoints any linear dependency in the set $\mathcal{Y}$. If $\mathcal{Y}$ is linearly independent, the row reduction also determines the coordinates with respect to $\mathcal{Y}$ of each vector $\mathbf{y}$ in span($\mathcal{Y}$). Let

$$\mathbf{y}=\begin{pmatrix}2\\1\\5\end{pmatrix},\quad \mathbf{y}_1=\begin{pmatrix}1\\2\\4\end{pmatrix},\quad \mathbf{y}_2=\begin{pmatrix}2\\1\\5\end{pmatrix},\quad \mathbf{y}_3=\begin{pmatrix}1\\1\\3\end{pmatrix},\quad \mathbf{y}_4=\begin{pmatrix}3\\3\\9\end{pmatrix}$$

(*a*) Row reduce $(\mathbf{A}\vdots\mathbf{I})$.

(*b*) Determine the space spanned by $\mathcal{Y}$; that is, determine the relationships that must exist among the elements $\{\eta_i\}$ of $\mathbf{y}$ in order that $\mathbf{y}$ be some linear combination of the vectors in $\mathcal{Y}$. Determine a basis for span($\mathcal{Y}$).

(*c*) Determine which linear combinations of the vectors in $\mathcal{Y}$ equal the specific vector $\mathbf{y}$ given above.

(*d*) The form of span($\mathcal{Y}$) can also be determined by row reduction of $\mathbf{A}^{\mathrm{T}}$. The nonzero rows of the row-reduced matrix constitute a basis for span($\mathcal{Y}$). Any zero rows which appear indicate the linear dependence of the set $\mathcal{Y}$.

2.8 For the following sets of vectors, determine if $\mathbf{y}$ is in span$\{\mathbf{y}_i\}$. If so, express $\mathbf{y}$ as a linear combination of the vectors $\{\mathbf{y}_i\}$.

$$(a)\quad \mathbf{y}=\begin{pmatrix}9\\3\\7\end{pmatrix},\quad \mathbf{y}_1=\begin{pmatrix}2\\1\\3\end{pmatrix},\quad \mathbf{y}_2=\begin{pmatrix}3\\1\\2\end{pmatrix},\quad \mathbf{y}_3=\begin{pmatrix}4\\1\\2\end{pmatrix}$$

$$(b)\quad \mathbf{y}=\begin{pmatrix}9\\12\\10\\10\end{pmatrix},\quad \mathbf{y}_1=\begin{pmatrix}2\\3\\4\\5\end{pmatrix},\quad \mathbf{y}_2=\begin{pmatrix}3\\4\\3\\5\end{pmatrix},\quad \mathbf{y}_3=\begin{pmatrix}4\\5\\3\\6\end{pmatrix}$$

$$(c)\quad \mathbf{y}=\begin{pmatrix}\eta_1\\\eta_2\\\eta_3\\\eta_4\end{pmatrix},\quad \mathbf{y}_1=\begin{pmatrix}1\\2\\2\\1\end{pmatrix},\quad \mathbf{y}_2=\begin{pmatrix}0\\2\\0\\1\end{pmatrix},\quad \mathbf{y}_3=\begin{pmatrix}-2\\1\\-4\\3\end{pmatrix}$$

2.9 Find a basis for the subspace of $\mathcal{P}^4$ spanned by the functions $\mathbf{f}_1(t)=1+t+2t^2$, $\mathbf{f}_2(t)=2t+t^2+t^3$, and $\mathbf{f}_3(t)=2+3t^2-t^3$. Extend the basis for the subspace to a basis for $\mathcal{P}^4$ by adding appropriate vectors to the basis.

2.10 Find the coordinate matrix of the vector $\mathbf{x}=(1, 1, 1)$ in $\mathcal{R}^3$:
(a) Relative to the basis $\mathcal{X}=\{(1, 0, 0), (1, -1, 0), (0, 1, -1)\}$.
(b) Relative to the basis $\mathcal{Y}=\{(1, 1, -1), (1, -1, 1), (-1, 1, 1)\}$.

2.11 Find the coordinate matrix of the function $\mathbf{f}$ relative to the basis $\mathcal{G}=\{\mathbf{g}_1, \mathbf{g}_2, \mathbf{g}_3\}$, where $\mathbf{f}(t)\triangleq t$, $\mathbf{g}_1(t)\triangleq 1+t$, $\mathbf{g}_2(t)\triangleq 1+t^2$, and $\mathbf{g}_3(t)=1-t^2$.

2.12 Find the coordinate matrix of the function $\mathbf{g}$ relative to the basis $\mathcal{F}=\{\mathbf{f}_1, \mathbf{f}_2, \mathbf{f}_3\}$, where $\mathbf{f}_1(t)\triangleq 1-t$, $\mathbf{f}_2(t)\triangleq 1-t^2$, $\mathbf{f}_3(t)\triangleq 1+t-t^2$, and $\mathbf{g}(t)\triangleq \xi_1+\xi_2 t+\xi_3 t^2$.

2.13 Find the coordinates of the vector $\mathbf{x}$ in $\mathfrak{M}^{2\times 2}$ relative to the basis $\mathcal{X}=\{\mathbf{x}_1, \mathbf{x}_2, \mathbf{x}_3, \mathbf{x}_4\}$, where

$$\mathbf{x}_1\triangleq\begin{pmatrix}1&1\\0&0\end{pmatrix},\quad \mathbf{x}_2\triangleq\begin{pmatrix}1&-1\\0&0\end{pmatrix},$$

$$\mathbf{x}_3\triangleq\begin{pmatrix}0&0\\1&1\end{pmatrix},\quad \mathbf{x}_4\triangleq\begin{pmatrix}0&0\\1&-1\end{pmatrix},\quad\text{and}\quad \mathbf{x}\triangleq\begin{pmatrix}3&-1\\-3&1\end{pmatrix}$$

2.14 Let $\mathcal{P}^{2\times 2}$ denote the space of polynomial functions of the form $\mathbf{f}(s,t)=a_{11}+a_{12}s+a_{21}t+a_{22}st$. Find a basis for $\mathcal{P}^{2\times 2}$ which includes the function $\mathbf{f}_1(s,t)=2s-t-1$. Find the coordinate matrix of the general vector $\mathbf{f}$ in $\mathcal{P}^{2\times 2}$ relative to that basis.

2.15 Let $\mathcal{V}$ be the space of continuous functions. Define the forward difference operator $\boldsymbol{\Delta}_\delta$: $\mathcal{V}\to\mathcal{V}$ by $(\boldsymbol{\Delta}_\delta \mathbf{f})(t) \triangleq [\mathbf{f}(t+\delta)-\mathbf{f}(t)]/\delta$ for all $\mathbf{f}$ in $\mathcal{V}$ and for all t, where $\delta>0$ is a fixed real number. Show that $\boldsymbol{\Delta}_\delta$ is linear.

2.16 *Financial planning*: the financial condition of a family unit at time t can be described by $\mathbf{f}(t)=\mathbf{f}(t-\delta)+a\mathbf{f}(t-\delta)+\mathbf{g}(t)$ where $\mathbf{f}(t)$ is the family savings at time t, $\mathbf{f}(t-\delta)$ is the savings at a previous time $t-\delta$, a is the interest rate per time interval δ, and $\mathbf{g}(t)$ is the deposit at time t. (No deposits occur between $t-\delta$ and t.)

(*a*) Let the time interval δ be 1 month. If we consider t only at monthly intervals, the above financial model can be expressed as the difference equation, $\mathbf{f}(k)=(1+a)\mathbf{f}(k-1)+\mathbf{g}(k)$. Given $\mathbf{f}(0)=\$100$, $a=0.005$ (i.e., 6% compounded monthly), and $\mathbf{g}(k)=\$10$ for $k=1, 2,\ldots,$ determine the savings versus time over 1 year by computing $\mathbf{f}(1)$ from $\mathbf{f}(0)$, $\mathbf{f}(2)$ from $\mathbf{f}(1)$, etc. (This computation is known as "marching.")

(*b*) The above financial model can be rewritten as

$$\frac{\mathbf{f}(t)-\mathbf{f}(t-\delta)}{\delta}=\frac{a}{\delta}\mathbf{f}(t-\delta)+\frac{\mathbf{g}(t)}{\delta}$$

The quantity $b\triangleq a/\delta$ is the interest rate per unit time; $\mathbf{u}(t)\triangleq \mathbf{g}(t)/\delta$ is the deposit rate for the interval. If we let $\delta\to 0$, the model becomes a differential equation, $\dot{\mathbf{f}}(t)=b\mathbf{f}(t)+\mathbf{u}(t)$. Let $\mathbf{f}(0)=\$100$, $b=0.005$ per month, and $\mathbf{u}(t)=\$10$ per month for $t>0$; find the savings versus time over 1 year by solving the differential equation. Compare the result with (*a*).

(*c*) An arbitrary nonlinear time-varying differential equation with initial conditions can be approximated by a difference equation in order to obtain an approximate solution via the simple marching technique of (*a*). Approximate the differential equation of (*b*) by using the *forward-difference approximation* $\dot{\mathbf{f}}(t) \approx (1/\epsilon)(\mathbf{f}(t+\epsilon)-\mathbf{f}(t)$, $\epsilon=1$ month, and considering t only at monthly intervals. Solve the difference equation for a 1 year period using $\mathbf{f}(0)$, b, and $\mathbf{u}(t)$ as given in (*b*). Compare the result with (*b*). How can the difference approximation be improved?

2.17 The electrostatic potential distribution within a two-dimensional chargefree region satisfies Laplace's equation:

$$(\nabla^2\mathbf{f})(s,t)\triangleq\frac{\partial^2\mathbf{f}(s,t)}{\partial s^2}+\frac{\partial^2\mathbf{f}(s,t)}{\partial t^2}=0$$

For the potential distribution between two parallel plates of spacing d, the model reduces to $\mathbf{f}''(s)=0$ with $\mathbf{f}(0)$ and $\mathbf{f}(d)$ given.

(*a*) Assume the differential operator $\mathbf{D}^2$ acts on $\mathcal{C}^2(0,d)$, a space of twice-differentiable functions. Find the nullspace of $\mathbf{D}^2$, a subspace of $\mathcal{C}^2(0,d)$. The nullspace is the solution space for the above differential equation. Express the solution space in terms of the known boundary values $\mathbf{f}(0)$ and $\mathbf{f}(d)$. What is the dimension of the nullspace of $\mathbf{D}^2$?

(*b*) Define the *central-difference operator* $\mathbf{\Delta}$ on $\mathcal{C}^2(0,d)$ by

$$(\mathbf{\Delta f})(s) \triangleq \mathbf{f}\left(s+\frac{\delta}{2}\right)-\mathbf{f}\left(s-\frac{\delta}{2}\right)$$

The derivative of $\mathbf{f}$ can be expressed as the limit of the central-difference approximation, $\mathbf{f}'(s)\approx(\mathbf{\Delta f})(s)/\delta$. Verify that $\mathbf{D}^2$, as it acts on $\mathcal{C}^2(0,d)$, can be approximated arbitrarily closely by the second-central-difference approximation, $\mathbf{D}^2 \approx\mathbf{\Delta}^2/\delta^2$.

(*c*) Suppose the plate spacing is $d=5$. Let $\delta=1$, and evaluate the finite-difference approximation $\mathbf{\Delta}^2\mathbf{f}=\boldsymbol{\theta}$ at $s=1$, 2, 3, and 4 to obtain four algebraic equations in the variables $\mathbf{f}(0)$, $\mathbf{f}(1),\ldots,\mathbf{f}(5)$. Formulate these algebraic equations as a 4×6 matrix equation $\mathbf{Ax}=\boldsymbol{\theta}$. Compare this matrix equation with the differential equation $\mathbf{D}^2\mathbf{f}=\boldsymbol{\theta}$; that is, compare the spaces on which the operators act; also compare the dimensions of their solution spaces. Solve the matrix equation in terms of the boundary values $\mathbf{f}(0)$ and $\mathbf{f}(5)$. Compare the discrete solution with the continuous solution found in (*a*).

This problem can also be carried out for the two-dimensional case, where $\mathbf{f}(s,t)$ is given on a closed boundary. The finite-difference approach in (*b*) and (*c*) is widely used in the solution of practical problems of this type. The equations, sometimes numbering as many as 100,000, are solved by iterative computer techniques. See Forsythe and Wasow [2.4].

2.18 According to the trapezoidal rule for approximate integration, if we subdivide the interval $[a,b]$ into n segments of length δ, and denote $\mathbf{g}(a+j\delta)$ by g_j, $j=0, 1,\ldots,n$, then for a continuous $\mathbf{g}$,

$$\int_a^b \mathbf{g}(s)\,ds\approx\frac{\delta}{2}(g_0+g_1)+\frac{\delta}{2}(g_1+g_2)+\cdots+\frac{\delta}{2}(g_{n-1}+g_n)$$

$$=\delta\left(\frac{g_0}{2}+g_1+\cdots+g_{n-1}+\frac{g_n}{2}\right)$$

We can view the trapezoidal rule as an approximation of a function space integral operation by a matrix multiplication **Ax**, where **A** is $1 \times n$ and $\mathbf{x} \triangleq (g_0 \cdots g_n)^{\mathrm{T}}$.

(*a*) Find the matrix **A** which expresses the trapezoidal rule for $\delta = 1$ and $n = 5$. Apply the trapezoidal rule to accurately represent the integral of the discontinuous function $\mathbf{g}(s) \triangleq 1$ for $0 \leqslant s < 2$, $\mathbf{g}(s) \triangleq 0$ for $2 < s \leqslant 5$. Hint: at the discontinuity use the midpoint value, $(g_2^- + g_2^+)/2$.

(*b*) We can also approximate a *general* integral operator by a matrix multiplication. Suppose $(\mathbf{Tf})(t) \triangleq \int_a^b k(t,s)\mathbf{f}(s)\, ds$ for t in $[a,b]$. We can treat the function $k(t,s)\mathbf{f}(s)$ as we did $\mathbf{g}(s)$ in (*a*). Subdivide both the s and t intervals into n segments of length δ, and use the same subscript notation for function values as above. Then if $k(t,s)\mathbf{f}(s)$ is continuous,

$$(\mathbf{Tf})_j = \delta\left(\frac{k_{j,0}f_0}{2} + k_{j,1}f_1 + \cdots + k_{j,n-1}f_{n-1} + \frac{k_{j,n}f_n}{2}\right)$$

for $j = 0, 1, \ldots, n$. We can approximate the integral operation by a matrix multiplication, $\mathbf{y} = \mathbf{Ax}$, where $\mathbf{x} = (f_0 \cdots f_n)^{\mathrm{T}}$ and $\mathbf{y} = ((\mathbf{Tf})_0 \cdots (\mathbf{Tf})_n)^{\mathrm{T}}$. Find **A** for $\delta = 1$, $n = 5$, $a = 0$, $b = 5$, and

$$\begin{aligned} k(t,s) &= 1 \quad \text{for } 0 \leqslant s < t \\ &= 0 \quad \text{for } t < s \leqslant 5 \end{aligned}$$

Hint: use midpoint values as in (*a*). Note that the operator is ordinary indefinite integration.

(*c*) Apply the matrix multiplication found in (*b*) to obtain the approximate integral of $\mathbf{f}(s) = 3s^2$. Compare the approximation to the actual integral at the points $t = 0, 1, \ldots, 5$.

*2.19 *Exploring the nullspace and range by row reduction*: Let

$$\mathbf{A} = \begin{pmatrix} 1 & 2 & 0 & 4 \\ 2 & 1 & 3 & 0 \\ 4 & 5 & 3 & 8 \end{pmatrix}$$

Multiplication by **A** is a linear transformation from $\mathfrak{M}^{4\times 1}$ into $\mathfrak{M}^{3\times 1}$. Multiplication by $\mathbf{A}^{\mathrm{T}}$ is a linear transformation from $\mathfrak{M}^{3\times 1}$ into $\mathfrak{M}^{4\times 1}$. In Section 5.4 we find that if **y** is in range(**A**) and **x** is in nullspace($\mathbf{A}^{\mathrm{T}}$), then $\mathbf{x} \cdot \mathbf{y} = 0$ where $\mathbf{x} \cdot \mathbf{y}$ is the dot product of analytic geometry. Furthermore, if **z** is in range($\mathbf{A}^{\mathrm{T}}$) and **w** is in nullspace(**A**), then $\mathbf{z}\cdot\mathbf{w}=0$. By means of these dot product equations, we can use bases for nullspace($\mathbf{A}^{\mathrm{T}}$) and range($\mathbf{A}^{\mathrm{T}}$) to find bases for range(**A**) and nullspace(**A**), and vice versa. We can also show that rank(**A**)=rank($\mathbf{A}^{\mathrm{T}}$). In this problem we obtain nullspace(**A**) and range(**A**) directly from $\mathbf{A}^{\mathrm{T}}$.

(*a*) Row reduce $(\mathbf{A} \vdots \mathbf{I})$. Use the results of the row reduction to determine bases for nullspace(**A**) and range(**A**).

(*b*) Row reduce $(\mathbf{A}^{\mathrm{T}} \vdots \mathbf{I})$. Show that the nonzero rows in the left-hand block of the row-reduced matrix constitute a basis for range(**A**). Show that the rows of the right-hand block which correspond to zero rows of the left-hand block of the row-reduced matrix constitute a basis for nullspace(**A**).

2.20 Define **T**: $\mathcal{P}^3 \to \mathcal{C}(0, 1)$ by $(\mathbf{Tf})(t) \triangleq \int_0^1 k(t,s)\mathbf{f}(s)\,ds$ for all **f** in $\mathcal{P}^3$, where

$$k(t,s)=t(1-s) \quad \text{for} \quad t\leqslant s$$
$$=(1-t)s \quad \text{for} \quad t\geqslant s$$

Find a basis for range(**T**). Describe nullspace(**T**).

2.21 Let $\mathcal{W}$ be the space of polynomial functions **f** of the form $\mathbf{f}(s,t) \triangleq c_1+c_2s+c_3t+c_4st$ for all s and t. Define **T**: $\mathcal{W}\to\mathcal{W}$ by $(\mathbf{Tf})(s,t) \triangleq (\partial/\partial s)\mathbf{f}(s,t)$ for all **f** in $\mathcal{W}$.

(*a*) Find a basis for the range of **T**.

(*b*) Determine the rank and nullity of **T**.

2.22 Define **T**: $\mathfrak{M}^{2\times2}\to\mathfrak{M}^{2\times2}$ by

$$\mathbf{T}\begin{pmatrix} c_1 & c_2 \\ c_3 & c_4 \end{pmatrix} \triangleq \begin{pmatrix} c_1-c_2 & c_1 \\ c_2 & c_4-c_3 \end{pmatrix}$$

for all choices of the scalars c_1, c_2, c_3, and c_4. Find nullspace(**T**) and range(**T**) by exhibiting a basis for each.

2.23 *Expected value*: the throws of a single die constitute an experiment. Let $\mathcal{V}$ be the space of random variables defined on this experiment. We can think of the probability mass function $\omega(\sigma)$ as the relative frequency with which the outcome σ occurs: $\omega(\sigma)=\frac{1}{6}$ for $\sigma=1, 2,\ldots,6$.

(*a*) A random variable **x** in $\mathcal{V}$ associates a value $\mathbf{x}(\sigma)$ with each possible outcome of the experiment. The value which **x** associates with an actual trial of the experiment is called a sample value of **x**. The probability mass function $\omega_{\mathbf{x}}(x)$ specifies the relative frequency with which the sample value x occurs during repeated trials. Find $\omega_{\mathbf{y}}(y)$ for the random variable $\mathbf{y}$ defined by $\mathbf{y}(\sigma) \stackrel{\Delta}{=} 2$ for $\sigma = 1$ or 2 and by $\mathbf{y}(\sigma) \stackrel{\Delta}{=} 0$ for $\sigma = 3, 4, 5$, or 6.

(*b*) The *expected value* of $\mathbf{x}$ is the average, over many trials, of the sample values of $\mathbf{x}$. Thus

$$\mathbf{E}(\mathbf{x}) = \sum_{x} x\,\omega_{\mathbf{x}}(x) = \sum_{\sigma} \mathbf{x}(\sigma)\omega(\sigma)$$

Find $\mathbf{E}(\mathbf{y})$ for the random variable $\mathbf{y}$ given in (*a*).

(*c*) Show that the functional $\mathbf{E}: \mathcal{V} \to \mathcal{R}$ is linear. Pick a basis $\mathcal{X}$ for $\mathcal{V}$. Let $\mathcal{E} \stackrel{\Delta}{=} \{(1)\}$ be a basis for $\mathcal{R}$. Find $[\mathbf{y}]_{\mathcal{X}}$ and $[\mathbf{E}]_{\mathcal{X}\mathcal{E}}$, where $\mathbf{y}$ is the random variable in (*a*).

(*d*) If $\mathbf{f}: \mathcal{V} \to \mathcal{V}$ then $\mathbf{f}(\mathbf{x})$ is a random variable. Express $\mathbf{E}(\mathbf{f}(\mathbf{x}))$ in terms of $\omega(\sigma)$. Find $\mathbf{E}(\mathbf{y}^2)$ for the random variable $\mathbf{y}$ given in (*a*). If $\mathbf{g}: \mathcal{V} \times \mathcal{V} \to \mathcal{V}$, can $\mathbf{E}$ be applied to $\mathbf{g}(\mathbf{x}, \mathbf{y})$?

2.24 *Hadamard matrices*: let $\mathbf{f}(s)$ represent the light intensity versus position in one line of a television picture. Let the $n \times 1$ column vector $\mathbf{x}$ be a discrete approximation to $\mathbf{f}$. Then $\mathbf{x}$ can be viewed as a one-dimensional photograph. Suppose the data $\mathbf{x}$ must be transmitted for remote viewing. One way to reduce the effect of transmission errors and to reduce the amount of data transmitted is to transmit, instead, a transformed version of $\mathbf{x}$. A computationally simple transformation is the Hadamard transform—multiplication by a Hadamard matrix. A symmetric Hadamard matrix $\mathbf{H}$ consists in plus and minus ones, and satisfies $\mathbf{H}^{-1} = \mathbf{H}$ (see [2.9]). Denote the transformed vector by $\mathbf{X} = \mathbf{H}\mathbf{x}$. Let $n = 8$ and

$$\mathbf{H} = \frac{1}{8}\begin{pmatrix} 1 & 1 & 1 & 1 & 1 & 1 & 1 & 1 \\ 1 & -1 & 1 & -1 & 1 & -1 & 1 & -1 \\ 1 & 1 & -1 & -1 & 1 & 1 & -1 & -1 \\ 1 & -1 & -1 & 1 & 1 & -1 & -1 & 1 \\ 1 & 1 & 1 & 1 & -1 & -1 & -1 & -1 \\ 1 & -1 & 1 & -1 & -1 & 1 & -1 & 1 \\ 1 & 1 & -1 & -1 & -1 & -1 & 1 & 1 \\ 1 & -1 & -1 & 1 & -1 & 1 & 1 & -1 \end{pmatrix}$$

The Hadamard transform spreads throughout the elements of $\mathbf{X}$ the information which is concentrated in a single element of $\mathbf{x}$; it concentrates information which is spread out.

(*a*) Determine the effect of $\mathbf{H}$ on the photographs $\mathbf{x} = (1\,1\,1\,1\,1\,1\,1\,1)^{\mathrm{T}}$ and $\mathbf{x} = \boldsymbol{\varepsilon}_i$, where $\boldsymbol{\varepsilon}_i$ is the ith standard basis vector for $\mathcal{M}^{8 \times 1}$.

(*b*) Find the transform of the photograph $\mathbf{x}=(22232222)^{\mathrm{T}}$. Assume that an error during transmission of $\mathbf{X}$ reduces the third element of $\mathbf{X}$ to zero. Determine the effect of the error on the reconstructed photograph.

(*c*) The inverse transform, $\mathbf{x}=\mathbf{HX}$, can be interpreted as an expansion of $\mathbf{x}$ in terms of the columns of $\mathbf{H}$. The columns of $\mathbf{H}$ are analogous to sinusoidal functions; the number of zero crossings corresponds to frequency. Let $\mathbf{x}$ be the photograph in (*b*). Determine the effect on the reconstructed photograph of not transmitting the highest frequency component of $\mathbf{X}$ (i.e., the effect of making the second element of $\mathbf{X}$ zero). Determine the effect on the reconstructed photograph of eliminating the zero frequency component (i.e., the effect of making the first component of $\mathbf{X}$ zero).

2.25 The space $\mathcal{C}^1(0,\infty)$ consists in the continuously differentiable functions on $[0,\infty]$. Define the cartesian product space $\mathcal{V}$ by $\mathcal{V} \stackrel{\Delta}{=} \mathcal{C}^1(0,\infty) \underset{1}{\times} \cdots \underset{n-1}{\times} \mathcal{C}^1(0,\infty)$. Denote the vector-valued functions in $\mathcal{V}$ by $\mathbf{x}$. We can treat the values of $\mathbf{x}$ as vectors in $\mathfrak{M}^{n\times 1}$; that is, $\mathbf{x}(t)=(\mathbf{f}_1(t)\cdots \mathbf{f}_n(t))^{\mathrm{T}}$, where $\mathbf{f}_i$ is in $\mathcal{C}^1(0,\infty)$. Let $\mathbf{A}$ be a real $n\times n$ matrix. Define the linear transformation $\mathbf{T}\colon \mathcal{V}\to\mathcal{W}$ by

$$(\mathbf{Tx})(t)=\begin{pmatrix}\dot{\mathbf{f}}_1(t)\\ \vdots\\ \dot{\mathbf{f}}_n(t)\end{pmatrix}-\mathbf{A}\begin{pmatrix}\mathbf{f}_1(t)\\ \vdots\\ \mathbf{f}_n(t)\end{pmatrix}=\dot{\mathbf{x}}(t)-\mathbf{Ax}(t)$$

This transformation is central to the state-space analysis of dynamic systems.

(*a*) Determine an appropriate range of definition $\mathcal{W}$ for $\mathbf{T}$.

(*b*) Find a basis for nullspace($\mathbf{T}$) if $n=2$ and

$$\mathbf{A}=\begin{pmatrix}0 & 1\\ 0 & -1\end{pmatrix}$$

2.26 Assume $\epsilon<\delta\ll 1$. Then the following matrix is nearly singular:

$$\mathbf{A}=\begin{pmatrix}1 & 0 & 0\\ 1 & \epsilon & 0\\ 1 & 0 & \delta\end{pmatrix}$$

Use inverse iteration to find a basis for the near nullspace of $\mathbf{A}$.

2.27 Define $\mathbf{T}\colon \mathcal{R}^2\to\mathcal{R}^2$ by $\mathbf{T}(\xi_1,\xi_2)\stackrel{\Delta}{=}(\xi_1+2\xi_2,\xi_1-2\xi_2)$ for all (ξ_1,ξ_2) in $\mathcal{R}^2$. Let $\mathcal{X}=\{(1,1),(1,-1)\}$. Find $[\mathbf{T}]_{\mathcal{X}\mathcal{X}}$.

2.28 Define **T**: $\mathcal{R}^2 \to \mathcal{R}^2$ by

$$\mathbf{T}(\xi_1, \xi_2, \xi_3) \triangleq (\xi_1 + \xi_2, 2\xi_3 - \xi_1)$$

(*a*) Determine $[\mathbf{T}]_{\mathcal{E}_3 \mathcal{E}_2}$, the matrix of **T** relative to the standard bases for $\mathcal{R}^3$ and $\mathcal{R}^2$.

(*b*) Determine $[\mathbf{T}]_{\mathcal{X}\mathcal{Y}}$, where $\mathcal{X} = \{(1,0,-1), (1,1,1,), (1,0,0)\}$ and $\mathcal{Y} = \{(1,0), (1,1)\}$.

2.29 Define **T**: $\mathcal{R}^2 \to \mathcal{R}^3$ by $\mathbf{T}(\xi_1, \xi_2) \triangleq (\xi_1 + \xi_2, \xi_1 - \xi_2, 2\xi_2)$ for all (ξ_1, ξ_2) in $\mathcal{R}^2$. Let $\mathcal{X} = \{(1,1), (1,-1)\}$ and $\mathcal{Y} = \{(1,1,-1), (1,-1,1), (-1,1,1)\}$. Find $[\mathbf{T}]_{\mathcal{X}\mathcal{Y}}$.

2.30 Let $\mathcal{P}^{2\times 2}$ denote the space of polynomial functions of the form $\mathbf{f}(s,t) = a_{11} + a_{12}s + a_{21}t + a_{22}st$. Define **T**: $\mathcal{P}^{2\times 2} \to \mathcal{W}$ by

$$(\mathbf{Tf})(s,t) = \int_0^s \mathbf{f}(\sigma, t)\, d\sigma$$

where $\mathcal{W} = \text{range}(\mathbf{T})$.

(*a*) Find bases, $\mathcal{F}$ for $\mathcal{P}^{2\times 2}$ and $\mathcal{G}$ for $\mathcal{W}$.

(*b*) Find $[\mathbf{T}]_{\mathcal{F}\mathcal{G}}$.

(*c*) Determine $\mathbf{T}^{-1}$ and $[\mathbf{T}^{-1}]_{\mathcal{G}\mathcal{F}}$. How else might $[\mathbf{T}^{-1}]_{\mathcal{G}\mathcal{F}}$ be obtained?

2.31 The sets $\mathcal{X} = \{(1,-1,0), (1,0,1), (1,1,1)\}$ and $\mathcal{Y} = \{(1,1,0), (0,1,1), (1,-1,1)\}$ are bases for $\mathcal{R}^3$. Find the change of coordinates matrix $\mathbf{S}_{\mathcal{X}\mathcal{Y}}$ which converts coordinates relative to $\mathcal{X}$ into coordinates relative to $\mathcal{Y}$.

2.32 Let $\mathbf{g}_1(t) = 1 - t$, $\mathbf{g}_2(t) = 1 - t^2$, and $\mathbf{g}_3(t) = 1 + t - t^2$. The set $\mathcal{G} = \{\mathbf{g}_1, \mathbf{g}_2, \mathbf{g}_3\}$ is a basis for $\mathcal{P}^3$. Another basis is $\mathcal{F} = \{\mathbf{f}_1, \mathbf{f}_2, \mathbf{f}_3\}$ where $\mathbf{f}_k(t) = t^{k-1}$.

(*a*) Find $[\mathbf{f}]_{\mathcal{G}}$ for the arbitrary vector $\mathbf{f}(t) = \xi_1 + \xi_2 t + \xi_3 t^2$.

(*b*) Find the coordinate-transformation matrix **S** such that $[\mathbf{f}]_{\mathcal{G}} = \mathbf{S}[\mathbf{f}]_{\mathcal{F}}$.

2.33 Define **T**: $\mathcal{R}^2 \to \mathcal{R}^3$ by $\mathbf{T}(\xi_1, \xi_2) \triangleq (\xi_2 - \xi_1, \xi_1, 2\xi_1 - \xi_2)$ for all (ξ_1, ξ_2) in $\mathcal{R}^2$. The sets $\mathcal{X} = \{(1,1), (1,-1)\}$ and $\mathcal{Z} = \{(1,2), (2,1)\}$ are bases for $\mathcal{R}^2$. The sets $\mathcal{Y} = \{(1,1,-1), (1,-1,1), (-1,1,1)\}$ and $\mathcal{K} = \{(1,1,1), (0,1,1), (0,0,1)\}$ are bases for $\mathcal{R}^3$.

(*a*) Find $[\mathbf{T}]_{\mathcal{X}\mathcal{Y}}$.

(*b*) Find the coordinate transformations $\mathbf{S}_{\mathcal{X}\mathcal{Z}}$ and $\mathbf{S}_{\mathcal{Y}\mathcal{K}}$.

(*c*) Use the answers to (*a*) and (*b*) to compute $[\mathbf{T}]_{\mathcal{Z}\mathcal{K}}$ by means of an equivalence transformation.

2.34 Define **T**: $\mathcal{R}^2 \to \mathcal{R}^2$ by $\mathbf{T}(\xi_1, \xi_2) \triangleq (\xi_1 + 2\xi_2, \xi_1 - 2\xi_2)$ for all (ξ_1, ξ_2) in $\mathcal{R}^2$. The sets $\mathcal{X} = \{(1,2), (2,1)\}$ and $\mathcal{Y} = \{(1,1), (1,-1)\}$ are bases for $\mathcal{R}^2$.

(*a*) Find $[\mathbf{T}]_{\mathfrak{X}\mathfrak{X}}$.
(*b*) Find the coordinate transformation $\mathbf{S}_{\mathfrak{X}\mathfrak{Y}}$.
(*c*) Use the answers to (*a*) and (*b*) to compute $[\mathbf{T}]_{\mathfrak{Y}\mathfrak{Y}}$ by means of a similarity transformation.

2.35 Multiplication by an invertible matrix can be interpreted either as a linear transformation or as a change of coordinates. Let $\mathfrak{X} = \{\mathbf{x}_1, \mathbf{x}_2\}$ be a basis for a two-dimensional space $\mathcal{V}$ and $\mathbf{x}$ a vector in $\mathcal{V}$. Then $[\mathbf{x}_1]_{\mathfrak{X}} = \begin{pmatrix}1\\0\end{pmatrix}$ and $[\mathbf{x}_2]_{\mathfrak{X}} = \begin{pmatrix}0\\1\end{pmatrix}$. Let

$$[\mathbf{x}]_{\mathfrak{X}} = \begin{pmatrix}2\\1\end{pmatrix}, \qquad \mathbf{A} = \begin{pmatrix}1 & -1\\1 & 0\end{pmatrix}$$

(*a*) Alias interpretation: assume $\mathbf{A}[\mathbf{x}]_{\mathfrak{X}} = [\mathbf{x}]_{\mathfrak{Y}}$, where $\mathfrak{Y} = \{\mathbf{y}_1, \mathbf{y}_2\}$ is a second basis for $\mathcal{V}$. Find $[\mathbf{y}_1]_{\mathfrak{X}}$ and $[\mathbf{y}_2]_{\mathfrak{X}}$. Sketch $[\mathbf{x}_1]_{\mathfrak{X}}$, $[\mathbf{x}_2]_{\mathfrak{X}}$, $[\mathbf{x}]_{\mathfrak{X}}$, $[\mathbf{y}_1]_{\mathfrak{X}}$, and $[\mathbf{y}_2]_{\mathfrak{X}}$ as arrows in a plane. What is the relationship between $[\mathbf{x}]_{\mathfrak{X}}$ and the basis $\{[\mathbf{y}_1]_{\mathfrak{X}}, [\mathbf{y}_2]_{\mathfrak{X}}\}$; that is, what is meant by the notation $[\mathbf{x}]_{\mathfrak{Y}}$?
(*b*) Alibi interpretation: assume $\mathbf{A}[\mathbf{x}]_{\mathfrak{X}} = [\mathbf{Tx}]_{\mathfrak{X}}$. Sketch $[\mathbf{x}_1]_{\mathfrak{X}}$, $[\mathbf{x}_2]_{\mathfrak{X}}$, $[\mathbf{x}]_{\mathfrak{X}}$, and $[\mathbf{Tx}]_{\mathfrak{X}}$ as arrows in a plane. What is the relationship between $[\mathbf{Tx}]_{\mathfrak{X}}$ and the basis $\{[\mathbf{x}_1]_{\mathfrak{X}}, [\mathbf{x}_2]_{\mathfrak{X}}\}$; that is, what is meant by the notation $[\mathbf{Tx}]_{\mathfrak{X}}$?

2.7 References

[2.1] Churchill, R. V., *Fourier Series and Boundary Value Problems*, McGraw-Hill, New York, 1941.

[2.2] Cramér, Harald and M. R. Leadbetter, *Stationary and Related Stochastic Processes*, Wiley, New York, 1967.

[2.3] Forsythe, George E., "Singularity and Near Singularity in Numerical Analysis," *Am. Math. Mon.*, **65** (1958), 229–40.

[2.4] Forsythe, George E. and Wolfgang R. Wasow, *Finite Difference Methods for Partial Differential Equations*, Wiley, 1960.

*[2.5] Halmos, P. R., *Finite-Dimensional Vector Spaces*, Van Nostrand, Princeton, N. J., 1958.

*[2.6] Hoffman, Kenneth and Ray Kunze, *Linear Algebra*, Prentice-Hall, Englewood Cliffs, N. J., 1961.

[2.7] Papoulis, Athanasios, *Probability, Random Variables, and Stochastic Processes*, McGraw-Hill, New York, 1965.

[2.8] Peterson, W. Wesley, *Error-Correcting Codes*, M.I.T. Press and Wiley, New York, 1961.

[2.9] Pratt, William K., Julius Kane, and Harry C. Andrews, "Hadamard Transform Image Coding," *Proc. IEEE*, **57**, 1 (January 1969), 58–68.

[2.10] Royden, H. L., *Real Analysis*, 2nd ed., Macmillan, New York, 1968.

[2.11] Wylie, C. R., Jr., *Advanced Engineering Mathematics*, 3rd ed., McGraw-Hill, New York, 1966.

3

Linear Differential Operators

Differential equations seem to be well suited as models for systems. Thus an understanding of differential equations is at least as important as an understanding of matrix equations. In Section 1.5 we inverted matrices and solved matrix equations. In this chapter we explore the analogous inversion and solution process for linear differential equations.

Because of the presence of boundary conditions, the process of inverting a differential operator is somewhat more complex than the analogous matrix inversion. The notation ordinarily used for the study of differential equations is designed for easy handling of boundary conditions rather than for understanding of differential operators. As a consequence, the concept of the inverse of a differential operator is not widely understood among engineers. The approach we use in this chapter is one that draws a strong analogy between linear differential equations and matrix equations, thereby placing both these types of models in the same conceptual framework. The key concept is the Green's function. It plays the same role for a linear differential equation as does the inverse matrix for a matrix equation.

There are both practical and theoretical reasons for examining the process of inverting differential operators. The inverse (or integral form) of a differential equation displays explicitly the input–output relationship of the system. Furthermore, integral operators are computationally and theoretically less troublesome than differential operators; for example, differentiation emphasizes data errors, whereas integration averages them. Consequently, the theoretical justification for applying many of the computational procedures of later chapters to differential systems is based on the inverse (or integral) description of the system. Finally, the application of the optimization techniques of Chapters 6–8 to differential systems often depends upon the prior determination of the integral forms of the systems.

One of the reasons that matrix equations are widely used is that we have a practical, automatable scheme, Gaussian elimination, for inverting a matrix or solving a matrix equation. It is also possible to invert certain types of differential equations by computer automation. The greatest progress in understanding and automation has been made for linear, constant-coefficient differential equations with initial conditions. These equations are good models for many dynamic systems (systems which evolve with time). In Section 3.4 we examine these linear constant-coefficient models in state-space form and also in the form of nth-order differential equations. The inversion concept can be extended to partial differential equations.

3.1 A Differential Operator and Its Inverse

Within the process of inverting a differential operator there is an analogue of the elimination technique for matrix inversion. However, the analogy between the matrix equation and the differential equation is clouded by the presence of the boundary conditions. As an example of a linear differential equation and its associated boundary conditions, we use

$$-\mathbf{f}''=\mathbf{u} \qquad \text{with} \quad \mathbf{f}(0)=\alpha_1 \quad \text{and} \quad \mathbf{f}(b)=\alpha_2 \tag{3.1}$$

Equation (3.1) can be viewed as a description of the relationship between the steady-state temperature distribution and the sources of heat in an insulated bar of length b. The temperature distribution $\mathbf{f}$ varies only as a function of position t along the bar. The temperature distribution is controlled partly by $\mathbf{u}$, the heat generated (say, by induction heating) throughout the bar, and partly by constant temperature baths (of temperatures α_1 and α_2, respectively) at the two ends $t=0$ and $t=b$. Thus both the distributed input $\mathbf{u}$ and the boundary inputs $\{\alpha_i\}$ have practical significance. The concepts of *distributed and boundary inputs* extend to other ordinary and partial differential equations.

A Discrete Approximation of the Differential System

In order to obtain a more transparent analogy to matrix equations and thereby clarify the role of the boundary conditions, we temporarily approximate the differential equation by a set of difference equations.* Let $b=4$, substitute into (3.1) the finite-difference approximation

$$-\frac{d^2\mathbf{f}(t)}{dt^2}\approx-\frac{[\mathbf{f}(t+1)-\mathbf{f}(t)]/1-[\mathbf{f}(t)-\mathbf{f}(t-1)]/1}{1}$$

$$=-\mathbf{f}(t-1)+2\mathbf{f}(t)-\mathbf{f}(t+1)$$

*The approximation of derivatives by finite differences is a practical numerical approach to the solution of ordinary and partial differential equations. The error owing to the finite-difference approximation can be made as small as desired by using a sufficiently fine approximation to the derivatives. (See Forsythe and Wasow [3.3].) Special techniques are usually used to solve the resulting algebraic equations. See P&C 3.3 and Varga [3.12].

and evaluate the equation at $t=1$, 2, and 3:

$$\begin{array}{llll} -\mathbf{f}(0)+2\mathbf{f}(1) & -\mathbf{f}(2) & & =\mathbf{u}(1) \\ & -\mathbf{f}(1)+2\mathbf{f}(2) & -\mathbf{f}(3) & =\mathbf{u}(2) \\ & & -\mathbf{f}(2)+2\mathbf{f}(3)-\mathbf{f}(4) & =\mathbf{u}(3) \\ \mathbf{f}(0) & & & =\alpha_1 \\ & & \mathbf{f}(4) & =\alpha_2 \end{array} \tag{3.2}$$

It is obvious that this set of algebraic equations would not be invertible without the boundary conditions. We can view the boundary conditions either as an increase in the number of equations or as a decrease in the number of unknowns. The left side of (3.2), including the boundary conditions, is a matrix multiplication of the general vector $(\mathbf{f}(0)\ \mathbf{f}(1)\cdots\mathbf{f}(4))^{\mathrm{T}}$ in the space $\mathfrak{M}^{5\times 1}$. The corresponding right-hand side of (3.2) is $(\mathbf{u}(1)\ \mathbf{u}(2)\ \mathbf{u}(3)\ \alpha_1\,\alpha_2)^{\mathrm{T}}$; the boundary values increase the dimension of the range of definition by two. On the other hand, if we use the boundary conditions to eliminate two variables, we reduce the dimension of the domain of the matrix operator by two, $\mathbf{f}(0)$ and $\mathbf{f}(4)$ become part of the right-hand side, and the reduced matrix operates on the general vector $(\mathbf{f}(1)\ \mathbf{f}(2)\ \mathbf{f}(3))^{\mathrm{T}}$ in $\mathfrak{M}^{3\times 1}$. By either the "expanded" or the "reduced" view, the transformation with its boundary conditions is invertible. In the next section we explore the differential equation and its boundary conditions along the same lines as we have used for this discrete approximation.

The Role of the Boundary Conditions

A differential operator without boundary conditions is like a matrix with fewer rows than columns: it leads to an underdetermined differential equation. In the same manner as in the discrete approximation (3.2), appropriate boundary conditions make a linear differential operator invertible. In order that we be able to denote the inverse of (3.1) in a simple manner as we do for matrix equations, we must combine the differential operator $-\mathbf{D}^2$ and the two boundary conditions into a single operator on a vector space. We can do so using the "increased equations" view of the boundary conditions. Let $\mathbf{f}$ be a function in the space $\mathcal{C}^2(0,b)$ of twice continuously differentiable functions; then $-\mathbf{f}''$ will be in $\mathcal{C}(0,b)$, the space of continuous functions. Define the differential system operator $\mathbf{T}$: $\mathcal{C}^2(0,b)\rightarrow\mathcal{C}(0,b)\times\mathfrak{R}^2$ by

$$\mathbf{Tf}\triangleq(-\mathbf{f}'',\mathbf{f}(0),\mathbf{f}(b)) \tag{3.3}$$

The system equations become

$$\mathbf{Tf}=(\mathbf{u},\alpha_1,\alpha_2) \tag{3.4}$$

We are seeking an explicit expression of $\mathbf{T}^{-1}$ such that $\mathbf{f}=\mathbf{T}^{-1}(\mathbf{u},\alpha_1,\alpha_2)$. Because of the abstractness of $\mathbf{T}$, an operation which produces a mixture of a distributed quantity $\mathbf{u}$ and discrete quantities $\{\alpha_i\}$, it is not clear how to proceed to determine $\mathbf{T}^{-1}$.

Standard techniques for solution of differential equations are more consistent with the "decreased unknowns" interpretation of the boundary conditions. Ordinarily, we solve the differential equation, $-\mathbf{f}''=\mathbf{u}$, ignoring the boundary conditions. Then we apply the boundary conditions to eliminate the arbitrary constants in the solution. If we think of the operator $-\mathbf{D}^2$ as being restricted through the whole solution process to act only on functions which satisfy the boundary conditions, then the "arbitrary" constants in the solution to the differential equation are not arbitrary; rather, they are specific (but unknown) functions of the boundary values, $\{\alpha_i\}$. We develop this interpretation of the inversion process into an explicit expression for the inverse of the operator $\mathbf{T}$ of (3.3).

How do we express the "restriction" of $-\mathbf{D}^2$ in terms of an operator on a *vector space*? The set of functions which satisfy the boundary conditions is not a subspace of $\mathcal{C}^2(0,b)$; it does not include the zero function (unless $\alpha_1=\alpha_2=0$). The analogue of this set of functions in the three-dimensional arrow space is a plane which does not pass through the origin. We frame the problem in terms of vector space concepts by separating the effects of the distributed and boundary inputs. In point of fact, it is the difference in the nature of these two types of inputs that has prevented the differential equation and the boundary conditions from being expressed as a single equation.* Decompose the differential system (3.1) into two parts, one involving only the distributed input, the other only the boundary inputs:

$$-\mathbf{f}_d''=\mathbf{u} \quad \text{with} \quad \mathbf{f}_d(0)=\mathbf{f}_d(b)=0 \tag{3.5}$$

$$-\mathbf{f}_b''=\boldsymbol{\theta} \quad \text{with} \quad \mathbf{f}_b(0)=\alpha_1, \quad \mathbf{f}_b(b)=\alpha_2 \tag{3.6}$$

Equations (3.5) and (3.6) possess unique solutions. By superposition, these solutions combine to yield the unique solution $\mathbf{f}$ to (3.1); that is, $\mathbf{f}=\mathbf{f}_d+\mathbf{f}_b$. Each of these differential systems can be expressed as a single operator on a vector space. We invert the two systems separately.

We work first with (3.5). The operator $-\mathbf{D}^2$ is *onto* $\mathcal{C}(0,b)$; that is, we can obtain any continuous function by twice differentiating some function in $\mathcal{C}^2(0,b)$. However, $-\mathbf{D}^2$ is singular; the general vector in nullspace $(-\mathbf{D}^2)$ is of the form $\mathbf{f}(t)=c_1+c_2t$. We modify the definition of the operator $-\mathbf{D}^2$ by reducing its domain. Let $\mathcal{V}$ be the subspace of functions in $\mathcal{C}^2(0,b)$ which satisfy the homogeneous boundary conditions of (3.5),

*Friedman [3.4] does include the boundary conditions in the differential equation by treating the boundary conditions as delta functions superimposed on the distributed input.

$\mathbf{f}(0)=\mathbf{f}(b)=0$. Define the modified differential operator $\mathbf{T}_d\colon \mathcal{V}\to\mathcal{C}(0,b)$ by $\mathbf{T}_d\mathbf{f} \triangleq -\mathbf{D}^2\mathbf{f}$ for all $\mathbf{f}$ in $\mathcal{V}$. The "distributed input" differential system (3.5) becomes

$$\mathbf{T}_d\mathbf{f}_d=\mathbf{u} \tag{3.7}$$

The boundary conditions are now included in the definition of the operator; in effect, we have "reduced" the operator $-\mathbf{D}^2$ to the operator $\mathbf{T}_d$ by using the two boundary conditions to eliminate two "variables" or two degrees of freedom from the domain of the operator $-\mathbf{D}^2$. The operator $\mathbf{T}_d$ is nonsingular; the equation $-\mathbf{f}''(t)=0$ has no nonzero solutions in $\mathcal{V}$. Furthermore, $\mathbf{T}_d$ is onto; eliminating from the domain of $-\mathbf{D}^2$ those functions which do not satisfy the zero boundary conditions of (3.5) does not eliminate any functions from the range of $-\mathbf{D}^2$. Suppose $\mathbf{g}$ is in $\mathcal{C}^2(0,b)$, and that $\mathbf{g}(0)$ and $\mathbf{g}(b)$ are not zero. Define the related function $\mathbf{f}$ in $\mathcal{V}$ by $\mathbf{f}(t) \triangleq \mathbf{g}(t)-[\mathbf{g}(0)+t(\mathbf{g}(b)-\mathbf{g}(0))/b]$. We have simply subtracted a "straight line" to remove the nonzero end points from $\mathbf{g}$; as a result, $\mathbf{f}(0)=\mathbf{f}(b)=0$. But $-\mathbf{D}^2\mathbf{f}=-\mathbf{D}^2\mathbf{g}$. Both $\mathbf{f}$ and $\mathbf{g}$ lead to the same function in $\mathcal{C}(0,b)$. Every vector in $\mathcal{C}(0,b)$ comes (via $-\mathbf{D}^2$) from some function in $\mathcal{V}$. Thus $\mathbf{T}_d$ is onto and invertible.

The differential system (3.6) can also be expressed as a single invertible operator. The nonzero boundary conditions of (3.6) describe a transformation $\mathbf{U}\colon \mathcal{C}^2(0,b)\to\mathcal{R}^2$, where

$$\mathbf{U}\mathbf{f} \triangleq (\mathbf{f}(0),\mathbf{f}(b))$$

Since $\mathcal{C}^2(0,b)$ is infinite dimensional but $\mathcal{R}^2$ is not, $\mathbf{U}$ must be singular. We modify the operator $\mathbf{U}$ by reducing its domain. Let $\mathcal{W}$ be the subspace of functions in $\mathcal{C}^2(0,b)$ which satisfy the homogeneous differential equation of (3.6), $-\mathbf{f}_b''(t)=0$; $\mathcal{W}$ is the two-dimensional space $\mathcal{P}^2$ consisting in functions of the form $\mathbf{f}(t)=c_1+c_2t$. We define the modified operator $\mathbf{T}_b$: $\mathcal{P}^2\to\mathcal{R}^2$ by $\mathbf{T}_b\mathbf{f} \triangleq (\mathbf{f}(0),\mathbf{f}(b))$ for all $\mathbf{f}$ in $\mathcal{P}^2$. The "boundary input" differential system (3.6) can be expressed as the two-dimensional equation

$$\mathbf{T}_b\mathbf{f}_b=(\alpha_1,\alpha_2) \tag{3.8}$$

The differential equation and boundary conditions of (3.6) have been combined into the single operator, $\mathbf{T}_b$. It is apparent that $\mathbf{T}_b$ is invertible—the operator equation is easily solved for its unique solution.

The Inverse Operator

We have rephrased (3.5) and (3.6) in terms of the invertible operators $\mathbf{T}_d$ and $\mathbf{T}_b$, respectively. Because (3.5) and (3.6) constitute a restructuring of

(3.1), we can express the solution to (3.1) as

$$
\begin{aligned}
\mathbf{f} &= \mathbf{f}_d + \mathbf{f}_b \\
&= \mathbf{T}_d^{-1}\mathbf{u} + \mathbf{T}_b^{-1}(\alpha_1, \alpha_2) \\
&= \mathbf{T}^{-1}(\mathbf{u}, \alpha_1, \alpha_2)
\end{aligned} \tag{3.9}
$$

where $\mathbf{T}$ is the operator of (3.3).

Since $\mathbf{T}_d$ is a differential operator, we expect $\mathbf{T}_d^{-1}: \mathcal{C}(0,b) \to \mathcal{V}$ to be an integral operator. We express it explicitly in the general form (2.34):

$$
\mathbf{f}_d(t) = (\mathbf{T}_d^{-1}\mathbf{u})(t) = \int_0^b k(t,s)\mathbf{u}(s)\,ds \tag{3.10}
$$

The kernel function k is commonly referred to as the **Green's function** for the differential system (3.1). In order that (3.10) correctly express the inverse of $\mathbf{T}_d$, $\mathbf{f}_d(t)$ must satisfy the differential system (3.5) from which $\mathbf{T}_d$ is derived. Substituting (3.10) into (3.5) yields

$$
\begin{aligned}
-\mathbf{f}_d''(t) &= -\frac{d^2}{dt^2}\int_0^b k(t,s)\mathbf{u}(s)\,ds \\
&= \int_0^b -\frac{d^2k(t,s)}{dt^2}\mathbf{u}(s)\,ds = \mathbf{u}(t)
\end{aligned}
$$

with

$$
\mathbf{f}_d(0) = \int_0^b k(0,s)\mathbf{u}(s)\,ds = 0
$$

$$
\mathbf{f}_d(b) = \int_0^b k(b,s)\mathbf{u}(s)\,ds = 0
$$

for all $\mathbf{u}$ in $\mathcal{C}(0,b)$. These equations are satisfied for all continuous $\mathbf{u}$ if and only if

$$
-\frac{d^2k(t,s)}{dt^2} = \delta(t-s) \qquad \text{with} \quad k(0,s) = k(b,s) = 0 \tag{3.11}
$$

That is, the Green's function k, *as a function of its first variable* t, must satisfy the differential equation and boundary conditions (3.5) for $\mathbf{u}(t) = \delta(t-s)$, where $\delta(t-s)$ is a unit impulse (or Dirac delta function) applied at the point $t = s$.* We can use (3.11) to determine the Green's function.

*See Appendix 2 for a discussion of delta functions. We use some license in interchanging the order of differentiation and integration when delta functions are present. The interchange can be justified, however, through the theory of distributions (Schwartz [3.10]).

For practical purposes we can think of $\delta(t-s)$ as a narrow continuous pulse of unit area, centered at $t=s$. [In terms of the steady-state heat-flow problem (3.1), the function $\delta(t-s)$ in (3.11) represents the generation of a unit quantity of heat per unit time in the cross section of the bar at $t=s$.] However, $\delta(t-s)$ is not a function in the usual sense; its value is not defined at $t=s$. It is not in $\mathcal{C}^2(0,b)$. Therefore, the solution k to (3.11) cannot be in $\mathcal{C}^2(0,b)$. We simply note that the domain $\mathcal{C}^2(0,b)$ and range of definition $\mathcal{C}(0,b)$ of the operator $-\mathbf{D}^2$ were defined somewhat arbitrarily. We can allow a "few" discontinuities or delta functions in $-\mathbf{D}^2\mathbf{f}$ if we also add to $\mathcal{C}^2(0,b)$ those functions whose second derivatives contain a "few" discontinuities or delta functions.

The operator $\mathbf{T}_b^{-1}: \mathcal{R}^2 \to \mathcal{P}^2$ can also be expressed explicitly. Since $\mathbf{T}_b^{-1}$ acts linearly on the vector (α_1, α_2) in $\mathcal{R}^2$ to yield a polynomial in $\mathcal{P}^2$, we express $\mathbf{T}_b^{-1}$ as

$$\mathbf{f}_b = \mathbf{T}_b^{-1}(\alpha_1, \alpha_2) = \alpha_1\rho_1 + \alpha_2\rho_2 \tag{3.12}$$

where ρ_1 and ρ_2 are functions in $\mathcal{P}^2$. We refer to the function $\rho_j(t)$ as the **boundary kernel** for the differential system (3.1). Just as the Green's function is a function of two variables, t and s, so the boundary kernel is a function of both the continuous variable t and the discrete variable j. Because of the simplicity of the differential operator of this example, the introduction of the boundary kernel seems unnecessary and artificial. For more complicated differential operators, however, the boundary kernel provides a straightforward approach to determination of the full inverse operator. In order that (3.12) correctly describe $\mathbf{T}_b^{-1}$, $\mathbf{f}_b$ must satisfy the differential system (3.6):

$$-\mathbf{f}_b'' = -\alpha_1\rho_1'' - \alpha_2\rho_2'' = 0$$

$$\mathbf{f}_b(0) = \alpha_1\rho_1(0) + \alpha_2\rho_2(0) = \alpha_1$$

$$\mathbf{f}_b(b) = \alpha_1\rho_1(b) + \alpha_2\rho_2(b) = \alpha_2$$

for all α_1 and α_2. Thus the boundary kernel ρ must obey

$$\begin{aligned} -\rho_1''(t) &= 0 \quad \text{with} \quad \rho_1(0) = 1, \quad \rho_1(b) = 0 \\ -\rho_2''(t) &= 0 \quad \text{with} \quad \rho_2(0) = 0, \quad \rho_2(b) = 1 \end{aligned} \tag{3.13}$$

We can use (3.13) to determine the boundary kernel.

We have defined carefully the differential system operator $\mathbf{T}$, the "distributed input" system operator $\mathbf{T}_d$, and the "boundary input" system operator $\mathbf{T}_b$ in order to be precise about the vector space concepts involved with inversion of differential equations. However, to continue use of this precise notation would require an awkward transition back and forth between the vector space notation and the notation standard to the field of differential equations. We rely primarily on the standard notation. We use the term **differential system** to refer to the differential operator with its boundary conditions (denoted $\{-\mathbf{D}^2, \mathbf{f}(0), \mathbf{f}(b)\}$ in this example) and also to the differential equation with its boundary conditions [denoted as in (3.1)]. We refer to both the inverse of the operator and the inverse of the equation as the *inverse of the differential system.* Where we refer to the purely differential part of the system separately, we usually denote it explicitly, for example, as $-\mathbf{D}^2$ or as $-\mathbf{f}''=\mathbf{u}$.

A Green's Function and Boundary Kernel

We solve for the Green's function k of the system (3.1) by direct integration of (3.11). The successive integration steps are depicted graphically in Figure 3.1. It is clear from the figure that the integral of $-d^2k/dt^2$ is constant for $t<s$ and $t>s$, and contains a jump of size 1 at $t=s$. We permit the value of the constant c to depend upon the point s at which the unit impulse is applied.

$$\begin{aligned} -\frac{dk(t,s)}{dt} &= c(s), \qquad && t<s \\ &= c(s)+1, && t>s \end{aligned}$$

Integration of $-dk/dt$ yields continuity of $-k$ at s:

$$\begin{aligned} -k(t,s) &= c(s)t+d(s), && t\leqslant s \\ &= c(s)s+d(s)+(c(s)+1)(t-s), \qquad && t\geqslant s \end{aligned}$$

Applying the boundary conditions we find

$$\begin{aligned} -k(0,s) &= c(s)(0)+d(s)=0 && \Rightarrow d(s)=0 \\ -k(b,s) &= c(s)s+(c(s)+1)(b-s)=0 \quad && \Rightarrow c(s)=\frac{s-b}{b} \end{aligned}$$

Thus

$$\begin{aligned} k(t,s) &= \frac{(b-s)t}{b}, \qquad && t\leqslant s \\ &= \frac{(b-t)s}{b}, && t\geqslant s \end{aligned} \tag{3.14}$$

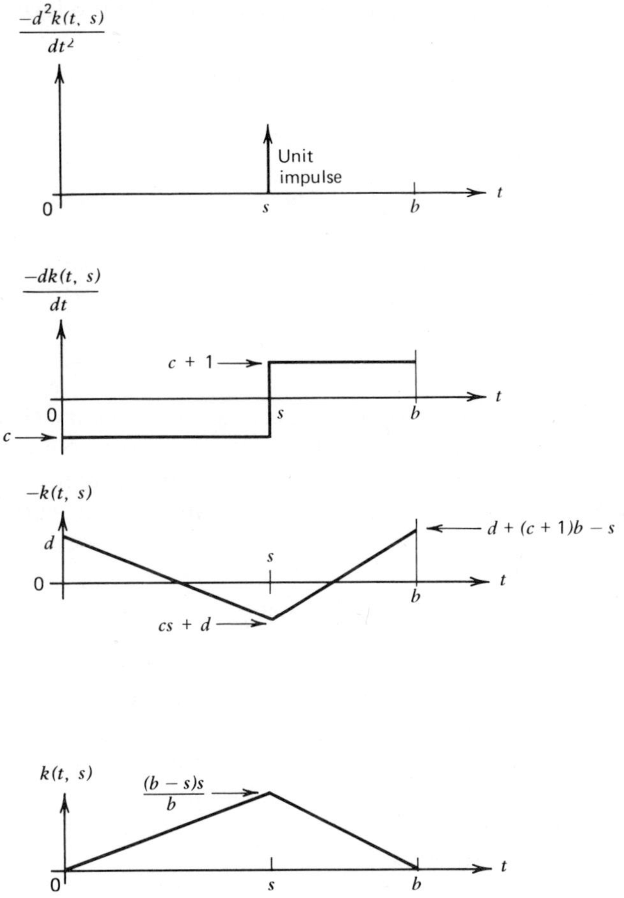

Figure 3.1. Graphical integration of (3.11).

where both t and s lie in the interval $[0,b]$.

By integration of (3.13) we determine the boundary kernel ρ associated with (3.1). The general solution to the jth differential equation is $\rho_j(t) = c_{j1} + c_{j2}t$. Using the boundary conditions we find

$$\rho_1(t) = \frac{b-t}{b}$$
$$\rho_2(t) = \frac{t}{b} \tag{3.15}$$

Having found k and ρ, we insert them into (3.10) and (3.12) to obtain

$\mathbf{T}_d^{-1}$ and $\mathbf{T}_b^{-1}$. Combining the two inverses as in (3.9) produces

$$\mathbf{f}(t) = \int_0^b k(t,s)\mathbf{u}(s)\,ds + \alpha_1\rho_1(t) + \alpha_2\rho_2(t)$$

$$= \int_0^t \frac{(b-t)s}{b}\mathbf{u}(s)\,ds + \int_t^b \frac{(b-s)t}{b}\mathbf{u}(s)\,ds + \alpha_1\frac{b-t}{b} + \alpha_2\frac{t}{b} \qquad (3.16)$$

Equation (3.16) is an explicit description of the inverse of the linear differential system (3.1).

A Matrix Analogy

A differential equation with an appropriate set of boundary conditions is analogous to a square matrix equation. We explore this analogy in order to remove some of the abstractness and mystery from differential operators and their inverses. An example of a matrix equation and its corresponding inverse is

$$\begin{pmatrix} 1 & 2 \\ 1 & 3 \end{pmatrix}\begin{pmatrix} \xi_1 \\ \xi_2 \end{pmatrix} = \begin{pmatrix} \eta_1 \\ \eta_2 \end{pmatrix} \quad \text{and} \quad \begin{pmatrix} \xi_1 \\ \xi_2 \end{pmatrix} = \begin{pmatrix} 3 & -2 \\ -1 & 1 \end{pmatrix}\begin{pmatrix} \eta_1 \\ \eta_2 \end{pmatrix}$$

Any such pair of equations can be expressed as $\mathbf{Ax}=\mathbf{y}$ and $\mathbf{x}=\mathbf{A}^{-1}\mathbf{y}$, respectively, for some square matrix $\mathbf{A}$. The inverse matrix equation is more clearly analogous to an inverse differential equation (or integral equation) if we express the matrix multiplication in the form of a summation. Denote the elements of $\mathbf{x}$ and $\mathbf{y}$ by ξ_i and η_i, respectively. Then the equation $\mathbf{x}=\mathbf{A}^{-1}\mathbf{y}$ becomes*

$$\xi_i = \sum_{j=1}^{n} (\mathbf{A}^{-1})_{ij}\eta_j, \qquad i=1,\ldots,n \qquad (3.17)$$

The symbol $(\mathbf{A}^{-1})_{ij}$ represents the element in row i and column j of the $n\times n$ matrix $\mathbf{A}^{-1}$. Thus the inverse matrix, a function of the two integer variables i and j, is the kernel of a summation operator. In the form (3.17), the inverse matrix equation $\mathbf{x}=\mathbf{A}^{-1}\mathbf{y}$ is obviously a discrete analogue of the integral equation $\mathbf{f}_d(t)=\int_0^b k(t,s)\mathbf{u}(s)\,ds$ of (3.10). The Green's function $k(t,s)$ is the analogue of the inverse matrix $\mathbf{A}^{-1}$. If we compare the inverse matrix equation (3.17) to (3.16), the full inverse of the differential system (3.1), the analogy is clouded somewhat by the presence of the boundary terms. The true analogue of $\mathbf{A}^{-1}$ is the *pair* of kernel functions, k and ρ.

*See (2.35).

Because $k(t,s)$ and $\rho_j(t)$ appear as "weights" in an integral or summation, the inverse form of the differential system is somewhat more useful to the intuition than is the differential system itself.

We can also draw an analogy between the process of inverting the matrix $\mathbf{A}$ and the process of solving for k and ρ. The solution to the equation $\mathbf{Ax}=\boldsymbol{\varepsilon}_i$, where $\boldsymbol{\varepsilon}_i$ is the ith standard basis vector for $\mathfrak{M}^{n\times 1}$, is the ith column of $\mathbf{A}^{-1}$. The solution process is analogous to solving (3.11) for $k(t,s)$ with s fixed; it is also analogous to solving (3.13) for ρ_j with j fixed. The row reduction $(\mathbf{A} \vdots \mathbf{I}) \rightarrow (\mathbf{I} \vdots \mathbf{A}^{-1})$ produces all columns of the inverse matrix simultaneously. Thus the inversion of $\mathbf{A}$ by row reduction is analogous to the determination of k and ρ by solving (3.11) and (3.13), respectively. In general, the process of computing k and ρ requires more effort than does the direct solution of (3.1) for specific inputs $\mathbf{u}$ and $\{\alpha_i\}$. However, the resulting inverse equation (3.16) contains information about the solution for any set of inputs.

3.2 Properties of *n*th-Order Systems and Green's Functions

In Section 3.1 we introduced the concepts of a differential operator and its inverse by means of a simple second-order example, (3.1). We now explore these concepts in detail for more general linear differential systems. Included in this section is an examination of noninvertible differential systems and a development of conditions for invertibility. Techniques for explicit determination of the Green's function and boundary kernel are treated in Section 3.3.

We define a **regular** nth-order linear differential operator $\mathbf{L}\colon \mathcal{C}^n(a,b) \rightarrow \mathcal{C}(a,b)$ by

$$(\mathbf{Lf})(t) \triangleq g_0(t)\mathbf{f}^{(n)}(t) + g_1(t)\mathbf{f}^{(n-1)}(t) + \cdots + g_n(t)\mathbf{f}(t) \qquad (3.18)$$

where the coefficients $\{g_i\}$ are continuous and $g_0(t) \neq 0$ on $[a,b]$.* The corresponding nth-order differential equation is $\mathbf{Lf}=\mathbf{u}$, where the distributed input function $\mathbf{u}$ is continuous on $[a,b]$. It is well known that $\mathbf{L}$ is onto $\mathcal{C}(a,b)$; the nth-order differential equation without boundary conditions always has solutions (Ince [3.6]). The **homogeneous differential equation** is defined as the equation $\mathbf{Lf}=\boldsymbol{\theta}$, without boundary conditions (the input $\mathbf{u}$ is zero). The homogeneous differential equation for the operator

*If the interval $[a,b]$ were infinite, if g_0 were zero at some point, or if one of the coefficient functions were discontinuous, we would refer to (3.18) as a *singular* differential operator. In Section 5.5 we refer to the regular second-order linear differential operator as a regular Sturm–Liouville operator.

(3.18) always has n linearly independent solutions[†]; we call a set $\{\mathbf{v}_1,\ldots,\mathbf{v}_n\}$ of independent solutions a **fundamental set of solutions** for **L**. We sometimes express such a set as the **complementary function** for **L**:

$$\mathbf{f}_c \stackrel{\Delta}{=} c_1\mathbf{v}_1 + \cdots + c_n\mathbf{v}_n \tag{3.19}$$

where $c_1,\ldots,c_n$ are unspecified constants. Both the complementary function $\mathbf{f}_c$ and the fundamental set of solutions $\{\mathbf{v}_i\}$ are, in reality, descriptions of the n-dimensional nullspace of **L**.

In order that **L** of (3.18) be invertible, we must add n appropriate boundary conditions to eliminate the n arbitrary constants in the complementary function. We denote the ith boundary condition for (3.18) by $\boldsymbol{\beta}_i(\mathbf{f}) = \alpha_i$, where α_i is a scalar and $\boldsymbol{\beta}_i$ is a linear functional on $\mathcal{C}^n(a,b)$.[‡] A typical boundary condition is some linear combination of **f** and its first $n-1$ derivatives evaluated at the end points of the interval of definition. For example,

$$\boldsymbol{\beta}_1(\mathbf{f}) \stackrel{\Delta}{=} \gamma_1\mathbf{f}(a) + \gamma_2\mathbf{f}'(a) + \gamma_3\mathbf{f}(b) + \gamma_4\mathbf{f}'(b) = \alpha_1 \tag{3.20}$$

(where the $\{\gamma_i\}$ are scalars) is as general a boundary condition as we would normally expect to encounter for a second-order differential operator acting on functions defined over $[a,b]$. The second boundary condition for the second-order differential equation, $\boldsymbol{\beta}_2(\mathbf{f}) = \alpha_2$, would be of the same form, although the particular linear combination of derivatives which constitutes $\boldsymbol{\beta}_2$ would have to be linearly independent of that specified by the coefficients $(\gamma_1,\gamma_2,\gamma_3,\gamma_4)$ in $\boldsymbol{\beta}_1$. There is, of course, no reason why the boundary conditions could not involve evaluations of **f** and its derivatives at interior points of the interval of definition. We refer to the boundary condition $\boldsymbol{\beta}_i(\mathbf{f}) = 0$, where the boundary input α_i is zero, as a **homogeneous boundary condition.**

Consider the following nth-order differential system:

$$\begin{aligned} \mathbf{Lf} &= \mathbf{u} \\ \boldsymbol{\beta}_i(\mathbf{f}) &= \alpha_i, \qquad i = 1,\ldots,m \end{aligned} \tag{3.21}$$

where **L** is defined in (3.18) and $\boldsymbol{\beta}_i$ is an nth-order version of (3.20); m is typically but not necessarily equal to n. We call a solution $\mathbf{f}_p$ to (3.21) a

[†]See P&C 3.4.

[‡]Of course, it is possible for the boundary conditions associated with a physical system to be nonlinear functions of **f**. We consider here only linear differential equations and linear boundary conditions.

particular solution for the differential system. A **completely homogeneous solution** $\mathbf{f}_h$ for the differential system is a solution to the homogeneous differential equation with homogeneous boundary conditions (the homogeneous differential system):

$$\begin{aligned} \mathbf{Lf} &= \boldsymbol{\theta} \\ \beta_i(\mathbf{f}) &= 0, \qquad i=1,\ldots,m \end{aligned} \tag{3.22}$$

Thus a completely homogeneous solution for the differential system is a solution with all inputs zero. Any solution $\mathbf{f}$ to (3.21) can be written as $\mathbf{f}=\mathbf{f}_p+\mathbf{f}_h$, where $\mathbf{f}_p$ is any particular solution and $\mathbf{f}_h$ is some homogeneous solution. The set of completely homogeneous solutions constitutes the nullspace of the differential system (or the nullspace of the underlying differential operator).* A system with a nonzero nullspace is not invertible.

Exercise 1. Suppose

$$(\mathbf{Lf})(t) \triangleq \mathbf{f}''(t)=\mathbf{u}(t) \tag{3.23}$$

with the boundary conditions

$$\beta_1(\mathbf{f}) \triangleq \mathbf{f}'(0)=\alpha_1 \qquad \beta_2(\mathbf{f}) \triangleq \mathbf{f}'(1)=\alpha_2 \tag{3.24}$$

What is the completely homogeneous solution to (3.23)–(3.24)? Show that the general solution to (3.23)–(3.24) is

$$\mathbf{f}(t)=\int_0^t\int_0^\sigma \mathbf{u}(\tau)\,d\tau\,d\sigma+\alpha_1 t+\mathbf{f}(0) \tag{3.25}$$

where

$$\int_0^1 \mathbf{u}(\tau)\,d\tau=\alpha_1-\alpha_2 \tag{3.26}$$

Note that the differential system (3.23)–(3.24) is not invertible. No solution exists unless the inputs $\mathbf{u}$ and $\{\alpha_i\}$ satisfy (3.26).

The Role of the Homogeneous Differential System

The matrix analogue of the *n*th-order differential system (3.21) is the matrix equation $\mathbf{Ax}=\mathbf{y}$ (where $\mathbf{A}$ is not necessarily square). Row reduction of $\mathbf{A}$ determines the nullspace of $\mathbf{A}$ (the solution to $\mathbf{Ax}=\boldsymbol{\theta}$); it also shows

*See (3.4) and (3.9).

the dependencies in the rows of **A** and the degree of degeneracy of the equation—the degree to which the range of the matrix transformation fails to fill the range of definition. To actually find the range of the matrix transformation (specific conditions on **y** for which the equation is solvable), we can follow either of two approaches: (*a*) row reduce $\mathbf{A}^{\mathrm{T}}$ (the rows of $\mathbf{A}^{\mathrm{T}}$ span the range of **A**)*; or (*b*) row reduce $(\mathbf{A} \vdots \mathbf{I})$. If **A** is square and invertible, approach (*b*) amounts to inversion of **A**.

For the differential system (3.21), the analogue of row reduction of **A** is the analysis of the completely homogeneous system (3.22). We focus first on this analysis, thereby determining the extent to which (3.21) is underdetermined or overdetermined. Then assuming the system (3.21) is invertible, we perform the analogue of row reduction of $(\mathbf{A} \vdots \mathbf{I})$—inversion of the differential operator.

The solutions to the homogeneous differential equation, $\mathbf{Lf}=\boldsymbol{\theta}$, are expressed as the complementary function $\mathbf{f}_c$ of (3.19). We apply the m homogeneous boundary conditions to $\mathbf{f}_c$, thereby eliminating some of the arbitrary constants in $\mathbf{f}_c$:

$$\begin{aligned}
\beta_1(\mathbf{f}_c) &= c_1\beta_1(\mathbf{v}_1) + \cdots + c_n\beta_1(\mathbf{v}_n) = 0\\
&\vdots\\
\beta_m(\mathbf{f}_c) &= c_1\beta_m(\mathbf{v}_1) + \cdots + c_n\beta_m(\mathbf{v}_n) = 0
\end{aligned}$$

or

$$\mathbf{B}\begin{pmatrix} c_1\\ \vdots\\ c_n\end{pmatrix} \triangleq \begin{pmatrix} \beta_1(\mathbf{v}_1)\cdots\beta_1(\mathbf{v}_n)\\ \vdots \qquad\qquad \vdots\\ \beta_m(\mathbf{v}_1)\cdots\beta_m(\mathbf{v}_n)\end{pmatrix}\begin{pmatrix} c_1\\ \vdots\\ c_n\end{pmatrix} = \begin{pmatrix} 0\\ \vdots\\ 0\end{pmatrix} \tag{3.27}$$

The nullspace of the differential system (3.21) consists in the functions $\mathbf{f}_c = c_1\mathbf{v}_1 + \cdots + c_n\mathbf{v}_n$, where some of the arbitrary constants $\{c_i\}$ are eliminated by (3.27).

The key to the differential system (3.21) lies in **B**, a **boundary condition matrix** (or **compatibility matrix**) for the system. In point of fact, **B** completely characterizes (3.22). It describes not just the boundary conditions, but rather the effect of the boundary conditions on a set of fundamental solutions for **L**. In general, the m boundary conditions, in concert with the

*See P&C 2.19. In Section 5.4 we introduce the adjoint operator, the analogue of $\mathbf{A}^{\mathrm{T}}$. The orthogonal decomposition theorem (5.67) is the basis of a method for determining the range of an operator from the nullspace of its adjoint; this method is the analogue of row reduction of $\mathbf{A}^{\mathrm{T}}$.

nth-order differential equation $\mathbf{Lf}=\mathbf{u}$, can specify either an underdetermined or overdetermined set of equations. Exercise 1 exhibits symptoms of both the underdetermined and overdetermined cases. Of course, $\mathbf{B}$ is not unique since it can be based on any fundamental set of solutions. Yet the rank of $\mathbf{B}$ is unique; rank($\mathbf{B}$) tells much about the solutions to (3.21)*:

1. If $\text{rank}(\mathbf{B})=m=n$, then (3.27) precisely eliminates the completely homogeneous solution, and (3.21) is then the analogue of an invertible square matrix equation; the system is invertible.
2. If $\text{rank}(\mathbf{B})=p<n$, then $(n-p)$ of the constants $\{c_i\}$ in the characteristic function remain arbitrary and the nullspace of the system has dimension $(n-p)$. There are $(n-p)$ degrees of freedom in the solutions to (3.21); the system is singular.
3. If $\text{rank}(\mathbf{B})=p<m$, then $(m-p)$ rows of $\mathbf{B}$ are dependent on the rest. As demonstrated by Exercise 1, these dependencies in the rows of $\mathbf{B}$ must be matched by $(m-p)$ scalar-valued relations among the boundary values $\{\alpha_i\}$ and the distributed input $\mathbf{u}$, or there can be no solutions to (3.21) (P&C 3.5). The system is not onto $\mathcal{C}(a,b)$.

The following example demonstrates the relationship between rank($\mathbf{B}$) and the properties of the differential system.

Example 1. The Rank of the Boundary Condition Matrix. Let

$$(\mathbf{Lf})(t) \triangleq \mathbf{f}''(t)=\mathbf{u}(t)$$

for t in $[0,1]$. The set $\{\mathbf{v}_1,\mathbf{v}_2\}$, where $\mathbf{v}_1(t)=1$ and $\mathbf{v}_2(t)=t$, is a fundamental set of solutions for t. We apply several different sets of boundary conditions, demonstrating the three cases mentioned above.

1. $\boldsymbol{\beta}_1(\mathbf{f}) \triangleq \mathbf{f}(0)=\alpha_1$, $\boldsymbol{\beta}_2(\mathbf{f}) \triangleq \mathbf{f}(1)=\alpha_2$. In this case,

$$\mathbf{B}=\begin{pmatrix} \mathbf{v}_1(0) & \mathbf{v}_2(0) \\ \mathbf{v}_1(1) & \mathbf{v}_2(1) \end{pmatrix}=\begin{pmatrix} 1 & 0 \\ 1 & 1 \end{pmatrix}$$

Since $\text{rank}(\mathbf{B})=2=m=n$, the system is invertible. We find the unique solution by direct integration:

$$\mathbf{f}(t)=\int_0^t\int_0^s \mathbf{u}(\tau)\,d\tau\,ds+\left[\alpha_2-\alpha_1-\int_0^1\int_0^s \mathbf{u}(\tau)\,d\tau\,ds\right]t+\alpha_1$$

2. $\boldsymbol{\beta}_1(\mathbf{f}) \triangleq \mathbf{f}(0)=\alpha_1$. For this single boundary condition,

$$\mathbf{B}=\begin{pmatrix} \mathbf{v}_1(0) & \mathbf{v}_2(0) \end{pmatrix}=\begin{pmatrix} 1 & 0 \end{pmatrix}$$

*Ince [3.6].

and rank$(\mathbf{B})=1$. Since $n=2$, we should expect one degree of freedom in the solution. Since $m=\text{rank}(\mathbf{B})$, we should expect a solution to exist for all scalars α_1 and all continuous functions $\mathbf{u}$. By direct integration, the solution is

$$\mathbf{f}(t)=\int_0^t\int_0^s \mathbf{u}(\tau)\,d\tau\,ds+d_1t+\alpha_1$$

where d_1 is an arbitrary constant.

3. $\beta_1(\mathbf{f})\stackrel{\Delta}{=}\mathbf{f}(0)=\alpha_1$, $\beta_2(\mathbf{f})\stackrel{\Delta}{=}\mathbf{f}(1)=\alpha_2$, $\beta_3(\mathbf{f})\stackrel{\Delta}{=}\mathbf{f}'(0)=\alpha_3$. Then,

$$\mathbf{B}=\begin{pmatrix}\mathbf{v}_1(0) & \mathbf{v}_2(0)\\ \mathbf{v}_1(1) & \mathbf{v}_2(1)\\ \mathbf{v}_1'(0) & \mathbf{v}_2'(0)\end{pmatrix}=\begin{pmatrix}1 & 0\\ 1 & 1\\ 0 & 1\end{pmatrix}$$

Because rank$(\mathbf{B})=2$ and $m=3$, one scalar-valued function of $\mathbf{u}$, α_1, α_2, and α_3 must be satisfied in order that a solution exist. Since rank$(\mathbf{B})=n$, if a solution exists for a given set of inputs $(\mathbf{u},\alpha_1,\alpha_2)$, that solution is unique. We find the solution by direct integration and application of the three boundary conditions:

$$\mathbf{f}(t)=\int_0^t\int_0^s \mathbf{u}(\tau)\,d\tau\,ds+\alpha_3t+\alpha_1$$

where $\mathbf{u}$, α_1, α_2, and α_3 must satisfy

$$\alpha_2-\alpha_1-\alpha_3-\int_0^1\int_0^s \mathbf{u}(\tau)\,d\tau\,ds=0$$

4. $\beta_1(\mathbf{f})\stackrel{\Delta}{=}\mathbf{f}'(0)=\alpha_1$, $\beta_2(\mathbf{f})\stackrel{\Delta}{=}\mathbf{f}'(1)=\alpha_2$. This case is presented in Exercise 1.

$$\mathbf{B}=\begin{pmatrix}\mathbf{v}_1'(0) & \mathbf{v}_2'(0)\\ \mathbf{v}_1'(1) & \mathbf{v}_2'(1)\end{pmatrix}=\begin{pmatrix}0 & 1\\ 0 & 1\end{pmatrix}$$

Rank$(\mathbf{B})=1$, but $m=n=2$. We expect one scalar-valued condition on the inputs, and one degree of freedom in the solutions. The general solution and the restriction on the inputs are given in (3.25) and (3.26), respectively.

It is apparent from Example 1 that if $m<n$, the system is underdetermined; there are at least $(n-m)$ degrees of freedom in the solutions. On the other hand, if $m>n$, the system is usually overdetermined; since rank$(\mathbf{B})\leqslant n$ for an nth-order differential system, the input data must satisfy at least $(m-n)$ different scalar-valued restrictions in order that the differential equation and boundary conditions be solvable.

Ordinarily, $m=n$; that is, the differential equation which represents a physical system usually has associated with it n independent boundary conditions $\{\beta_i\}$. These n boundary conditions are independent in the sense that they represent independent linear combinations of $\mathbf{f}$, $\mathbf{f}^{(1)},\ldots,\mathbf{f}^{(n-1)}$ evaluated at one or more points of $[a,b]$. However, we see from the fourth case of Example 1 that a boundary condition matrix $\mathbf{B}$ can be degenerate even if the boundary conditions are independent. Thus for a "square" differential operator, the condition for invertibility (or compatibility) is

$$\det(\mathbf{B}) \neq 0 \tag{3.28}$$

where $\mathbf{B}$ is a boundary condition matrix as defined in (3.27).

It can be shown that (3.28) is satisfied for any differential operator for which $m=n$ and for which the boundary conditions are linearly independent and are all at one point (P&C 3.4). Only for multipoint boundary value problems can the test (3.28) fail. Exercise 1 is such a case.

For the rest of this chapter we assume (3.28) is satisfied, and proceed to determine the inverse of the differential system (3.21). In Section 4.3, where we determine eigenvalues and eigenfunctions of differential operators, we seek conditions under which (3.28) is *not* satisfied. These conditions occur, of course, only with multipoint boundary value problems.

The Green's Function and the Boundary Kernel

Our procedure for inverting the system (3.21) parallels the procedure used with the second-order example (3.1). Of course, the compatibility condition (3.28) must be satisfied. Assume $m=n$. We begin by splitting (3.21) into two parts, one involving only the distributed input, the other only the boundary inputs:

$$\begin{aligned} &\mathbf{Lf}=\mathbf{u} \\ &\beta_i(\mathbf{f})=0 \quad i=1,\ldots,n \end{aligned} \tag{3.29}$$

$$\begin{aligned} &\mathbf{Lf}=\boldsymbol{\theta} \\ &\beta_i(\mathbf{f})=\alpha_i \quad i=1,\ldots,n \end{aligned} \tag{3.30}$$

where $\mathbf{L}$ is given in (3.18). The completely homogeneous equation (3.22) is a special case of both (3.29) and (3.30). Thus both are characterized by any boundary condition matrix $\mathbf{B}$ derived from (3.22). If (3.28) is satisfied, both (3.29) and (3.30) are invertible. The inverse of (3.29) is an integral operator with a distributed kernel. The inverse of (3.30) is a summation operator involving a boundary kernel. These two kernels describe explicitly the dependence of $\mathbf{f}(t)$ on the input data $\mathbf{u}(t)$ and $\{\alpha_j\}$.

Assume the inverse of (3.29) is representable in the integral form

$$\mathbf{f}(t)=\int_a^b k(t,s)\mathbf{u}(s)\,ds \tag{3.31}$$

for all t in $\{a,b\}$. The kernel k is known as the **Green's function** for the system (3.21). If (3.31) is the correct inverse for (3.29), $\mathbf{f}$ must satisfy (3.29):

$$(\mathbf{Lf})(t)=\mathbf{L}\int_a^b k(t,s)\mathbf{u}(s)\,ds$$

$$=\int_a^b \mathbf{L}k(t,s)\mathbf{u}(s)\,ds=\mathbf{u}(t)$$

$$(\boldsymbol{\beta}_i\mathbf{f})(t)=\boldsymbol{\beta}_i\int_a^b k(t,s)\mathbf{u}(s)\,ds$$

$$=\int_a^b \boldsymbol{\beta}_i k(t,s)\mathbf{u}(s)\,ds=0 \qquad i=1,\ldots,n$$

for all $\mathbf{u}$ in $\mathcal{C}(a,b)$. Both $\mathbf{L}$ and $\boldsymbol{\beta}_i$ treat the variable s as a constant, *acting on* $k(t,s)$ *only as a function of* t. Each operator acts on the whole "t" function $k(\cdot,s)$. It is evident that $\mathbf{L}k(t,s)$ exhibits the "sifting" property of a delta function (see Appendix 2). On the other hand, $\boldsymbol{\beta}_i k(t,s)$ acts like the zero function. Consequently, the Green's function k must satisfy

$$\begin{aligned}\mathbf{L}k(t,s) &\triangleq g_0(t)\frac{d^nk(t,s)}{dt^n}+\cdots+g_n(t)k(t,s)=\delta(t-s)\\ \boldsymbol{\beta}_i k(t,s)&=0 \qquad i=1,\ldots,n\end{aligned} \tag{3.32}$$

for all t and s in $[a,b]$. Because the delta function appears in (3.32), we cannot rigorously interchange the order of the differential operator $\mathbf{L}$ and the integration without resorting to the theory of generalized functions (Appendix 2). However, we can justify the formal interchange for each specific problem by showing that the Green's function k derived from (3.32) does indeed lead to the solution of (3.29) for every continuous function $\mathbf{u}$.

Assume the inverse of (3.30) is representable as a summation operator of the form:

$$\mathbf{f}(t)=\sum_{j=1}^{n}\rho_j(t)\alpha_j \tag{3.33}$$

We can think of ρ as a kernel function of the two variables j and t. We call ρ the **boundary kernel** for (3.21). To find the equations which determine ρ, we substitute (3.33) into (3.30):

$$\mathbf{Lf} = \mathbf{L}\sum_{j=1}^{n} \rho_j \alpha_j$$

$$= \sum_{j=1}^{n} \alpha_j (\mathbf{L}\rho_j) = \boldsymbol{\theta}$$

$$\boldsymbol{\beta}_i(\mathbf{f}) = \boldsymbol{\beta}_i \sum_{j=1}^{n} \rho_j \alpha_j$$

$$= \sum_{j=1}^{n} \alpha_j \boldsymbol{\beta}_i(\rho_j) = \alpha_i \qquad i = 1, \ldots, n$$

for all $\{\alpha_i\}$. Suppose we let $\alpha_k = 1$ and $\alpha_j = 0$ for $j \neq k$. It follows that for $k = 1, \ldots, n$, $\mathbf{L}\rho_k = \boldsymbol{\theta}$, $\boldsymbol{\beta}_i \rho_k = 1$ for $k = i$, and $\boldsymbol{\beta}_i \rho_k = 0$ for $k \neq i$. Thus the boundary kernel ρ must satisfy

$$\begin{aligned} (\mathbf{L}\rho_j)(t) &\triangleq g_0(t)\frac{d^n \rho_j(t)}{dt^n} + \cdots + g_n(t)\rho_j(t) = 0 \qquad j = 1, \ldots, n \\ \boldsymbol{\beta}_i(\rho_j) &= \delta_{ij} \qquad\qquad i = 1, \ldots, n; \quad j = 1, \ldots, n \end{aligned} \tag{3.34}$$

for all t in $[a,b]$, where δ_{ij} is the Kronecker delta (see A2.11 of Appendix 2). According to (3.34), the n components $\{\rho_j\}$ of the boundary kernel constitute a fundamental set of solutions for the operator $\mathbf{L}$; furthermore, $\{\rho_j\}$ is a fundamental set for which the boundary condition matrix $\mathbf{B}$ of (3.27) is the $n \times n$ identity matrix.

By solving (3.32) and (3.34), we can invert any regular nth-order differential system which has a nonsingular boundary condition matrix. The inverse of the differential system (3.21) (with $m = n$) consists in the sum of the inverses of (3.29) and (3.30), namely,

$$\mathbf{f}(t) = \int_a^b k(t,s)\mathbf{u}(s)\,ds + \sum_{j=1}^{n} \alpha_j \rho_j(t) \tag{3.35}$$

where k and ρ are determined by (3.32) and (3.34), respectively.

Theoretically, we can invert any linear differential operator, ordinary or partial, which has appropriate boundary conditions. That is, we can convert any invertible linear differential equation to an integral equation analogous to (3.35). As a model for a system, the integral equation is more desirable than the differential equation from two standpoints. First, the boundary conditions are included automatically. Second, integral operators tend to "smooth" functions whereas differential operators introduce discontinuities and delta functions.* It is well known that numerical differentiation amplifies errors in empirical data, but numerical integration does not (Ralston [3.9, p. 79]). The rest of this chapter is devoted to techniques for determining the inverse (or integral) model for various types of ordinary differential operators. Techniques and examples which apply to partial differential operators can be found in Friedman [3.4], Stakgold [3.11], Morse and Feshbach [3.8], and Bergman and Schiffer [3.1].

3.3 Inversion of *n*th-Order Differential Systems

In Section 3.1 we determined the Green's function and boundary kernel for a simple second-order system, (3.1). The Green's function and boundary kernel for the general *n*th-order differential systems of Section 3.2 cannot be determined by the direct integration technique used for that simple system. In this section we describe general procedures for solving (3.32) and (3.34) to obtain k and ρ for the *n*th-order differential system (3.21) with n independent boundary conditions. The procedures are demonstrated in detail for regular second-order variable-coefficient differential systems.

Obtaining a Complementary Function

Most techniques for determining particular solutions to differential systems are based on the complementary function (3.19). Techniques for determining the Green's function k and the boundary kernel ρ also depend heavily on the complementary function (or the equivalent, a fundamental set of solutions). In point of fact, the individual segments or components of k and ρ are of the form of the complementary function.

It is well known that the complementary function for a **constant-coefficient** differential operator consists in sums of exponentials. Let $\mathbf{L}$ of (3.18) be the constant-coefficient operator

$$\mathbf{L} \triangleq \mathbf{D}^n + a_1\mathbf{D}^{n-1} + \cdots + a_n\mathbf{I} \tag{3.36}$$

To find which exponentials are contained in the complementary function for $\mathbf{L}$, we insert a particular exponential $\mathbf{v}(t) = e^{\mu t}$ into the equation $\mathbf{Lf} = \boldsymbol{\theta}$

*Integral operators are continuous, whereas differential operators are not. See the discussion of continuous operators in Section 5.4.

and solve for μ. The result is

$$\mu^n + a_1 \mu^{n-1} + \cdots + a_n = 0 \tag{3.37}$$

This equation, known as the **characteristic equation for L**, has n roots $\mu_1, \mu_2, \ldots, \mu_n$. If the n roots are distinct, the complementary function is

$$\mathbf{f}_c(t) = c_1 \exp(\mu_1 t) + \cdots + c_n \exp(\mu_n t) \tag{3.38}$$

Equation (3.38) can be verified by substituting $\mathbf{f}_c$ into $\mathbf{Lf} = \boldsymbol{\theta}$. If two roots are equal, say, $\mu_1 = \mu_2$, then the corresponding fundamental solutions in (3.38) must be replaced by $c_1 \exp(\mu_1 t) + c_2 t \exp(\mu_1 t)$. This equal root case is discussed further in Section 4.4.

We are unable to deal with the variable-coefficient operator (3.18) with much generality. An approach that can be used to *seek* the complementary function for the variable-coefficient operator is the **power series method** (the method of Frobenius). The method consists in assuming a power series form for the complementary function, substituting the series into the homogeneous differential equation, equating the coefficient on each power of t to zero, and solving for the coefficients of the power series. The sum of the series, where it converges, is at least part of the complementary function. The sum will not, in general, consist of elementary functions. For example, Bessel functions arise as fundamental solutions to Bessel's equation (a second-order variable-coefficient differential equation); the power series method provides an expression for one of the two fundamental solutions to Bessel's equation. In the event that the power series method does not provide a full set of fundamental solutions for the differential equation, other methods must be used to complete the complementary function. See Ince [3.6] or Wiley [3.13, p. 255].

Example 1. Power Series Method—Variable Coefficients. Suppose

$$(\mathbf{Lf})(t) \triangleq \mathbf{f}'(t) + t\mathbf{f}(t) \tag{3.39}$$

We find the complementary function for (3.39) by assuming a power series of the general form

$$\mathbf{f}_c(t) = t^a (c_0 + c_1 t + c_2 t^2 + \cdots)$$

where the constant a allows for noninteger powers of t. We first insert $\mathbf{f}_c$ into the homogeneous equation and regroup terms:

$$\begin{aligned}\mathbf{f}'(t) + t\mathbf{f}(t) &= ac_0 t^{a-1} + (a+1)c_1 t^a + [(a+2)c_2 + c_0]t^{a+1} \\ &\quad + [(a+3)c_3 + c_1]t^{a+2} + [(a+4)c_4 + c_2]t^{a+3} + \cdots \\ &= 0\end{aligned}$$

Equating each coefficient to zero, we obtain

$$\begin{aligned} ac_0&=0\\ (a+1)c_1&=0\\ (a+2)c_2+c_0&=0\\ (a+3)c_3+c_1&=0\\ (a+4)c_4+c_2&=0\\ &\vdots \end{aligned}$$

We assume, without loss of generality, that $c_0 \neq 0$. It follows that $a=0$ and c_0 is arbitrary; then

$$c_1=c_3=c_5=\cdots=0,$$

$$c_2=-\frac{c_0}{2},\qquad c_4=\frac{c_0}{4(2)},\qquad c_6=-\frac{c_0}{6(4)(2)},\qquad \cdots$$

and

$$\begin{aligned} \mathbf{f}_c(t)&=c_0\left[1-\frac{t^2}{2}+\frac{1}{2!}\left(\frac{t^2}{2}\right)^2-\frac{1}{3!}\left(\frac{t^2}{2}\right)^3+\cdots\right]\\ &=c_0\exp\left(\frac{-t^2}{2}\right) \end{aligned}$$

Determination of the Green's Function and Boundary Kernel—An Example

We solved for the kernel functions k and ρ associated with (3.1) by direct integration of the differential equation. Unfortunately, that simple approach does not apply to most differential equations. In the following example we introduce a general technique for finding k and ρ.

The model for a particular armature-controlled dc motor and load is the differential equation

$$\ddot{\phi}(t)+\dot{\phi}(t)=\mathbf{u}(t) \tag{3.40}$$

where $\mathbf{u}(t)$ is the armature voltage at time t and $\phi(t)$ is the angular position of the motor shaft relative to some reference position. Let the boundary conditions be

$$\phi(0)=\alpha_1 \quad \text{and} \quad \phi(b)=\alpha_2 \tag{3.41}$$

That is, we seek the "trajectory" (or angular position versus time), of the shaft in order that it be in position α_1 at time 0 and pass through position α_2 at time b. Comparing this problem to that of (3.21), we note that $\mathbf{L}=\mathbf{D}^2+\mathbf{D}$, $\boldsymbol{\beta}_1(\boldsymbol{\phi})=\boldsymbol{\phi}(0)$, and $\boldsymbol{\beta}_2(\boldsymbol{\phi})=\boldsymbol{\phi}(b)$. The symbol $\boldsymbol{\phi}$ replaces the symbol $\mathbf{f}$ used earlier.

Finding the Green's function for the differential system (3.40)–(3.41) is equivalent to exploring the trajectory $\boldsymbol{\phi}$ of the motor shaft for all possible applied voltages $\mathbf{u}(t)$, but for $\alpha_1=\alpha_2=0$. The Green's function must satisfy (3.32):

$$\frac{d^2k(t,s)}{dt^2}+\frac{dk(t,s)}{dt}=\delta(t-s)$$

$$k(0,s)=k(b,s)=0$$

Clearly $k(t,s)$ satisfies the homogeneous differential equation in each of the regions $[0,s)$ and $(s,b]$; that is, in the regions where $\delta(t-s)$ is zero. We let $k(t,s)=\mathbf{f}_c(t)$ for each of the two regions $[0,s)$ and $(s,b]$:

$$\begin{aligned} k(t,s)&=c_1+c_2e^{-t}, \qquad && t \text{ in } [0,s) \\ &=d_1+d_2e^{-t}, && t \text{ in } (s,b] \end{aligned}$$

Since $k(t,s)$ is a function of s, the arbitrary constants must depend on s. We eliminate half of the arbitrary constants by applying the boundary conditions

$$k(0,s)=c_1+c_2=0 \qquad \Rightarrow c_2=-c_1$$

$$k(b,s)=d_1+d_2e^{-b}=0 \Rightarrow d_2=-e^bd_1$$

It is the second (or highest) derivative of k that introduces the delta function in (3.32); for if the first derivative included a delta function, the second derivative would introduce the derivative of the delta function.* Since d^2k/dt^2 includes a unit impulse at $t=s$, dk/dt must include a unit step at $t=s$, and k itself must be continuous at $t=s$. We express these facts by the two "discontinuity" conditions:

$$k(s^+,s)=k(s^-,s) \qquad \text{(continuity of } k \text{ at } t=s)$$

$$\frac{dk(s^+,s)}{dt}-\frac{dk(s^-,s)}{dt}=1 \qquad \left(\text{unit step in } \frac{dk}{dt} \text{ at } t=s\right)$$

*See Appendix 2 for a discussion of unit steps, delta functions, and derivatives of delta functions.

Applying these conditions to $k(t,s)$, we find

$$d_1 + d_2 e^{-s} = c_1 + c_2 e^{-s}$$

$$-d_2 e^{-s} - (-c_2 e^{-s}) = 1$$

A messy elimination procedure among the boundary condition equations and discontinuity condition equations yields

$$c_1(s) = \frac{e^s - e^b}{e^b - 1} \quad \text{and} \quad d_1(s) = \frac{e^s - 1}{e^b - 1}$$

It follows that

$$\begin{aligned} k(t,s) &= \frac{(1-e^{-t})(e^s - e^b)}{e^b - 1} & t \leqslant s \\ &= \frac{(1-e^b e^{-t})(e^s - 1)}{e^b - 1} & t \geqslant s \end{aligned} \tag{3.42}$$

To get a feel for the nature of this system (for which $\phi(0) = \phi(b) = 0$),we use k to determine the shaft trajectory ϕ and velocity profile $\dot{\phi}$ for a specific input $\mathbf{u}(t) = 1$:

$$\begin{aligned} \phi(t) &= \int_0^b k(t,s)\mathbf{u}(s)\,ds \\ &= \frac{1-e^b e^{-t}}{e^b - 1}\int_0^t (e^s - 1)\,ds + \frac{1-e^{-t}}{e^b - 1}\int_t^b (e^s - e^b)\,ds \\ &= t - \left(\frac{be^b}{e^b - 1}\right)(1 - e^{-t}) \end{aligned}$$

$$\dot{\phi}(t) = 1 - \left(\frac{be^b}{e^b - 1}\right)e^{-t}$$

The trajectory ϕ and the velocity profile $\dot{\phi}$ are plotted in Figure 3.2 for $b = 1$. Observe that, in general, the motor shaft cannot be at rest at $t = 0$ and at $t = b$ if the shaft positions are specified; it is precisely the freedom in the initial and terminal velocities which allows us to choose both the end points, $\phi(0)$ and $\phi(b)$, and an arbitrary continuous input voltage $\mathbf{u}$.

The boundary kernel ρ for the system (3.40)–(3.41) describes the trajectory $\phi(t)$ as a function of the boundary conditions $\phi(0) = \alpha_1$ and

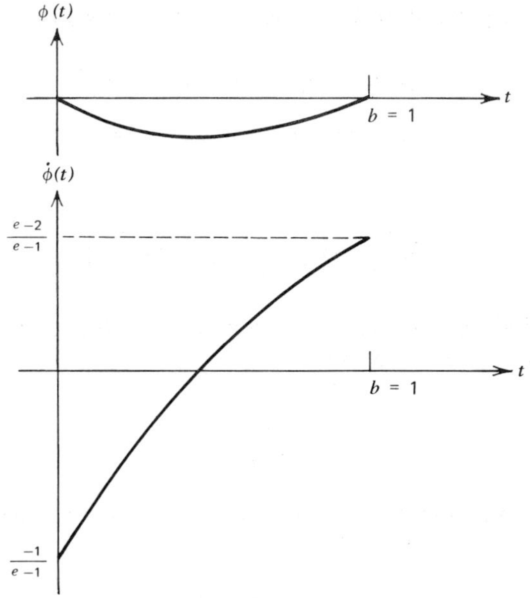

Figure 3.2. Shaft position and velocity for $\boldsymbol{\phi}(0)=\boldsymbol{\phi}(1)=0$ and $\mathbf{u}(t)=1$.

$\boldsymbol{\phi}(b)=\alpha_2$ with no voltage applied to the motor; that is,

$$\boldsymbol{\phi}(t)=\rho_1(t)\alpha_1+\rho_2(t)\alpha_2$$

Perhaps the most direct approach to the determination of $\rho_j(t)$ is to let $\boldsymbol{\phi}(t)=c_1\mathbf{v}_1(t)+c_2\mathbf{v}_2(t)$, a linear combination of the fundamental solutions for (3.40), then apply the boundary conditions (3.41) to obtain the coefficients c_i as a function of α_1 and α_2. Rather than use this approach, we attack the defining equations for $\rho_j(t)$ in a more formal manner which parallels the determination of the Green's function. The two approaches are equivalent in the amount of computation they require. The boundary kernel satisfies (3.34):

$$\begin{aligned} &\ddot{\rho}_1(t)+\dot{\rho}_1(t)=0 && \ddot{\rho}_2(t)+\dot{\rho}_2(t)=0\\ &\boldsymbol{\beta}_1(\rho_1)=\rho_1(0)=1 && \boldsymbol{\beta}_1(\rho_2)=\rho_2(0)=0\\ &\boldsymbol{\beta}_2(\rho_1)=\rho_1(b)=0 && \boldsymbol{\beta}_2(\rho_2)=\rho_2(b)=1 \end{aligned}$$

The boundary condition statements are reminiscent of the boundary condition matrix (3.27). In point of fact, ρ_1 and ρ_2 each consist in a linear combination of the fundamental solutions $\mathbf{v}_1(t)=1$ and $\mathbf{v}_2(t)=e^{-t}$. Applying the boundary conditions to $\rho_1(t)=c_1+c_2e^{-t}$, we get

$$\beta_1(\rho_1)=\rho_1(0)=c_1+c_2e^{-0}=1$$

$$\beta_2(\rho_1)=\rho_1(b)=c_1+c_2e^{-b}=0$$

or

$$\mathbf{B}\begin{pmatrix} c_1 \\ c_2 \end{pmatrix}=\begin{pmatrix} 1 & 1 \\ 1 & e^{-b} \end{pmatrix}\begin{pmatrix} c_1 \\ c_2 \end{pmatrix}=\begin{pmatrix} 1 \\ 0 \end{pmatrix}$$

where **B** is, indeed, the boundary condition matrix of (3.27). Similarly, using $\rho_2(t)=d_1+d_2e^{-t}$, we find

$$\mathbf{B}\begin{pmatrix} d_1 \\ d_2 \end{pmatrix}=\begin{pmatrix} 0 \\ 1 \end{pmatrix}$$

We can combine the two coefficient equations into the single matrix equation

$$\mathbf{B}\begin{pmatrix} c_1 & d_1 \\ c_2 & d_2 \end{pmatrix}=\begin{pmatrix} 1 & 0 \\ 0 & 1 \end{pmatrix}$$

which has the solution

$$\begin{pmatrix} c_1 & d_1 \\ c_2 & d_2 \end{pmatrix}=\mathbf{B}^{-1}=\left(\frac{1}{e^b-1}\right)\begin{pmatrix} -1 & e^b \\ e^b & -e^b \end{pmatrix}$$

The function ρ_j is a specific linear combination of the two fundamental solutions specified above; the jth column of $\mathbf{B}^{-1}$ specifies the linear combination. Thus

$$\begin{aligned} \rho_1(t)&=\frac{-1}{e^b-1}+\frac{e^b}{e^b-1}e^{-t} \\ \rho_2(t)&=\frac{e^b}{e^b-1}+\frac{-e^b}{e^b-1}e^{-t} \end{aligned} \tag{3.43}$$

The shaft position and velocity, as functions of the boundary conditions, are

$$\phi(t)=\frac{e^be^{-t}-1}{e^b-1}\alpha_1+\frac{e^b(1-e^{-t})}{e^b-1}\alpha_2$$

$$\dot{\phi}(t)=\frac{e^be^{-t}}{e^b-1}(\alpha_2-\alpha_1)$$

Figure 3.3 shows the position and velocity of the motor shaft for $\alpha_1=0$ and $\alpha_2=1$. The shaft is already in motion at $t=0$, and exhibits an "undriven" decay in velocity until it reaches the position $\phi(b)=1$ rad. If the boundary conditions were $\alpha_1=\alpha_2=1$, the shaft would sit at rest in the position $\phi(t)=1$ rad; again an undriven trajectory.

The inverse of the system (3.40)–(3.41) is the sum of the separate solutions for the distributed and boundary inputs. That is,

$$\mathbf{f}(t)=\int_0^b k(t,s)\mathbf{u}(s)\,ds+\rho_1(t)\alpha_1+\rho_2(t)\alpha_2$$

where k and ρ are given in (3.42) and (3.43), respectively. The nature of the system (3.40)–(3.41) does not seem in keeping with the nature of dynamic (real-time) systems. The motor must anticipate the input $\mathbf{u}(t)$ (or the impulse $\delta(t-s)$) and appropriately select its velocity at $t=0$ in order to be able to meet the requirement on its position at $t=b$. We are more likely to meet such a two-point boundary value problem when the independent variable t represents not time, but rather a space variable. Yet a two-point

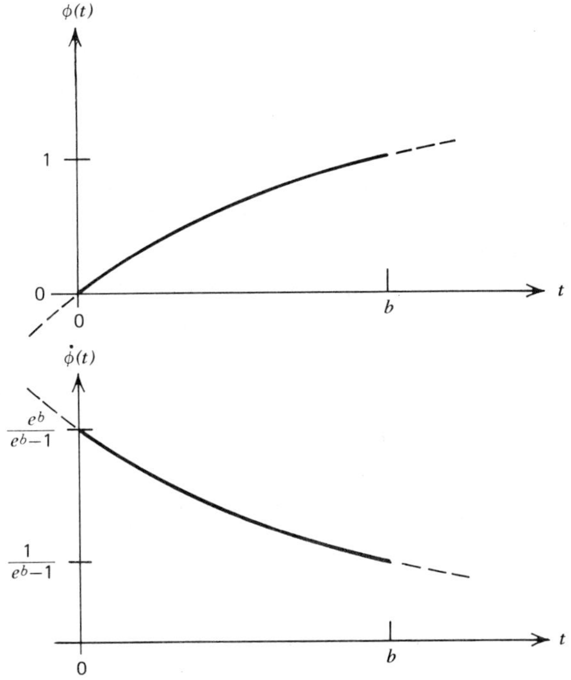

Figure 3.3. Undriven shaft position and velocity for $\alpha_1=0$ and $\alpha_2=1$.

boundary value problem can arise in a dynamic system if we impose requirements on the future behavior of the system as we did in (3.41).

Summary of the Technique

The technique demonstrated above for determining the Green's function and the boundary kernel depends upon knowledge of the complementary function. We can apply the technique to the regular nth-order system (3.21) if the corresponding complementary function can be determined. Assume $\mathbf{L}$ of (3.21) has the complementary function $\mathbf{f}=c_1\mathbf{v}_1+\cdots+c_n\mathbf{v}_n$. Further assume that the system is invertible (i.e., we have n independent boundary conditions for which (3.28) is satisfied). We obtain the Green's function k and the boundary kernel ρ for the system (3.21) by following the technique used for the system (3.40)–(3.41).

Equation (3.32) determines the Green's function k. The unit impulse $\delta(t-s)$ is zero for all $t\neq s$. Therefore, $k(t,s)$ satisfies the homogeneous differential equation for $t\neq s$; $k(t,s)$ is equal to the complementary function (3.19) in each of the two regions $[a,s)$ and $(s,b]$. Because the complementary function $\mathbf{f}_c$ is used in two separate regions, we must determine two sets of n arbitrary constants:

$$\begin{aligned} k(t,s) &= b_1\mathbf{v}_1(t)+\cdots+b_n\mathbf{v}_n(t), \quad t \text{ in } [a,s) \\ &= d_1\mathbf{v}_1(t)+\cdots+d_n\mathbf{v}_n(t), \quad t \text{ in } (s,b] \end{aligned} \tag{3.44}$$

Half of the $2n$ constants can be eliminated by the *homogeneous* boundary conditions of (3.21): $\beta_i k(t,s)=0,\ i=1,\ldots,n$. The rest are determined by appropriate "discontinuity" conditions at $t=s$. Only the highest derivative term, $g_0(t)d^nk(t,s)/dt^n$, can introduce the delta function into (3.32) (otherwise derivatives of delta functions would appear); therefore, we match the two halves of $k(t,s)$ at $t=s$ in such a way that we satisfy the following n conditions:

$$\begin{gathered} k, \frac{dk}{dt}, \ldots, \frac{d^{n-2}k}{dt^{n-2}} \quad \text{are continuous at } t=s \\ \frac{d^{n-1}k(s^+,s)}{dt^{n-1}} - \frac{d^{n-1}k(s^-,s)}{dt^{n-1}} = \frac{1}{g_0(s)} \end{gathered} \tag{3.45}$$

That is, $d^{n-1}k(t,s)/dt^{n-1}$ must contain a step of size $1/g_0(s)$ at $t=s$. Then $g_0(t)d^nk(t,s)/dt^n$ will include the term $\delta(t-s)$.*

*See Appendix 2 for a discussion of steps, delta functions, and derivatives of delta functions.

The boundary kernel ρ is specified by (3.34). Each component of ρ is a linear combination of the fundamental solutions for **L**:

$$\rho_j = c_{1j}\mathbf{v}_1 + \cdots + c_{nj}\mathbf{v}_n \qquad j=1,\ldots,n \tag{3.46}$$

Applying the n boundary conditions of (3.21) as required by (3.34), we find

$$\begin{pmatrix} \beta_1(\rho_j) \\ \vdots \\ \beta_n(\rho_j) \end{pmatrix} = \begin{pmatrix} \beta_1(\mathbf{v}_1)\cdots\beta_1(\mathbf{v}_n) \\ \vdots \qquad\quad \vdots \\ \beta_n(\mathbf{v}_1)\cdots\beta_n(\mathbf{v}_n) \end{pmatrix} \begin{pmatrix} c_{1j} \\ \vdots \\ c_{nj} \end{pmatrix} = \begin{pmatrix} 0 \\ \vdots \\ 1_j \\ \vdots \\ 0 \end{pmatrix}, \qquad j=1,\ldots,n$$

These n sets of equations can be expressed as

$$\begin{pmatrix} \beta_1(\mathbf{v}_1)\cdots\beta_1(\mathbf{v}_n) \\ \vdots \qquad\quad \vdots \\ \beta_n(\mathbf{v}_1)\cdots\beta_n(\mathbf{v}_n) \end{pmatrix} \begin{pmatrix} c_{11}\cdots c_{1n} \\ \vdots \qquad \vdots \\ c_{n1}\cdots c_{nn} \end{pmatrix} = \begin{pmatrix} 1 & 0 & \cdots & 0 \\ 0 & 1 & \cdots & 0 \\ \vdots & \vdots & & \vdots \\ 0 & 0 & \cdots & 1 \end{pmatrix} \tag{3.47}$$

It follows that the coefficients for ρ_j in (3.46) are the elements in the jth column of $\mathbf{B}^{-1}$, where $\mathbf{B}$ is the boundary condition matrix defined in (3.27). Specifically, c_{ij} is the element in row i and column j of $\mathbf{B}^{-1}$.

Exercise 1. Let $\mathbf{f}'(t) + t\mathbf{f}(t) = \mathbf{u}(t)$ with $\mathbf{f}(0) = \alpha_1$. (The complementary function for this differential equation was determined in Example 1.) Show that the inverse of this differential system is

$$\mathbf{f}(t) = \exp(-t^2/2)\int_0^t \exp(s^2/2)\mathbf{u}(s)\,ds + \alpha_1 \exp(-t^2/2) \tag{3.48}$$

Second-Order Differential Systems

Many of the ordinary and partial differential equations that arise in the modeling of physical systems are second order. Some of the second-order partial differential equations can be reduced, by a substitution of variables or by integral transforms, to second-order ordinary differential equations.* Furthermore, use of the "separation of variables" technique in solving second-order partial differential equations produces sets of second-order

*See Kaplan [3.7].

ordinary differential equations. Thus the general second-order ordinary differential equation with variable coefficients is of considerable practical importance. We present explicit expressions for the Green's function and the boundary kernel for an arbitrary regular second-order differential system; these expressions are obtained in terms of a fundamental set of solutions for the differential operator.

The regular second-order differential system is†

$$\begin{gathered} g_0(t)\mathbf{f}''(t)+g_1(t)\mathbf{f}'(t)+g_2(t)\mathbf{f}(t)=\mathbf{u}(t) \\ \boldsymbol{\beta}_1(\mathbf{f})=\alpha_1 \quad \text{and} \quad \boldsymbol{\beta}_2(\mathbf{f})=\alpha_2 \end{gathered} \tag{3.49}$$

where g_i is continuous and $g_0(t)\neq 0$ in the region of interest. Assume $\mathbf{v}_1$ and $\mathbf{v}_2$ are independent solutions to the homogeneous differential equation. By (3.44), the Green's function is of the form

$$\begin{aligned} k(t,s)&=b_1\mathbf{v}_1(t)+b_2\mathbf{v}_2(t), \quad t<s \\ &=d_1\mathbf{v}_1(t)+d_2\mathbf{v}_2(t), \quad t>s \end{aligned}$$

The discontinuity conditions (3.45) become

$$d_1\mathbf{v}_1(s)+d_2\mathbf{v}_2(s)=b_1\mathbf{v}_1(s)+b_2\mathbf{v}_2(s) \text{ (continuity of } k\text{)}$$

$$d_1\mathbf{v}_1'(s)+d_2\mathbf{v}_2'(s)-b_1\mathbf{v}_1'(s)-b_2\mathbf{v}_2'(s)=\frac{1}{g_0(s)} \quad \left(\text{step of size } \frac{1}{g_0(s)} \text{ in } \frac{dk}{dt}\right)$$

Since dk/dt has a step of size $1/g_0(s)$, then $g_0(t)d^2k/dt^2$ includes a *unit* impulse. These two discontinuity equations can be put in the matrix form

$$\begin{pmatrix} \mathbf{v}_1(s) & \mathbf{v}_2(s) \\ \mathbf{v}_1'(s) & \mathbf{v}_2'(s) \end{pmatrix}\begin{pmatrix} d_1-b_1 \\ d_2-b_2 \end{pmatrix}=\begin{pmatrix} 0 \\ 1/g_0(s) \end{pmatrix}$$

The solution is

$$d_1-b_1=-\frac{\mathbf{v}_2(s)}{w(s)\,g_0(s)}, \qquad d_2-b_2=\frac{\mathbf{v}_1(s)}{w(s)\,g_0(s)}$$

where $w(s)$ is the *Wronskian determinant**:

$$w(s)\triangleq\begin{vmatrix} \mathbf{v}_1(s) & \mathbf{v}_2(s) \\ \mathbf{v}_1'(s) & \mathbf{v}_2'(s) \end{vmatrix} \tag{3.50}$$

†In Section 5.5 we refer to the differential operator of (3.49) as a regular Sturm–Liouville operator.

*Note that the solution is undefined for $w(s)=0$. It can be shown that if $\mathbf{v}_1$ and $\mathbf{v}_2$ are independent solutions to the homogeneous differential equation, then $w(s)\neq 0$ for all s in the interval of interest. See P&C 3.7.

The boundary conditions $\boldsymbol{\beta}_1 k(t,s)=\boldsymbol{\beta}_2 k(t,s)=0$ provide two more linear algebraic equations which, together with the above pair of equations, determine the constants b_1, b_2, d_1, and d_2, and therefore, $k(t,s)$. However, without specific information about the nature of the boundary conditions, we can carry the solution no further. The solution for a dynamic system (initial conditions) is given in Exercise 2. Two-point boundary conditions are treated in Exercise 3.

Exercise 2. Let the boundary conditions of (3.49) be

$$\boldsymbol{\beta}_1(\mathbf{f}) \triangleq \mathbf{f}(a) \quad \text{and} \quad \boldsymbol{\beta}_2(\mathbf{f}) \triangleq \mathbf{f}'(a) \tag{3.51}$$

Show that the corresponding Green's function is

$$\begin{aligned} k(t,s) &= 0, & t \text{ in } [a,s) \\ &= \frac{\Delta(s,t)}{g_0(s)w(s)}, & t \text{ in } (s,\infty) \end{aligned} \tag{3.52}$$

where w is given by (3.50), and

$$\Delta(s,t) \triangleq \begin{vmatrix} \mathbf{v}_1(s) & \mathbf{v}_2(s) \\ \mathbf{v}_1(t) & \mathbf{v}_2(t) \end{vmatrix}$$

Show also that the corresponding boundary kernel is

$$\begin{aligned} \rho_1(t) &= \frac{\mathbf{v}_2'(a)\mathbf{v}_1(t)-\mathbf{v}_1'(a)\mathbf{v}_2(t)}{w(a)} \\ \rho_2(t) &= \frac{\mathbf{v}_1(a)\mathbf{v}_2(t)-\mathbf{v}_2(a)\mathbf{v}_1(t)}{w(a)} \end{aligned} \tag{3.53}$$

Exercise 3. Let the boundary conditions of (3.49) be $\boldsymbol{\beta}_1(\mathbf{f}) \triangleq \mathbf{f}(a)$ and $\boldsymbol{\beta}_2(\mathbf{f}) \triangleq \mathbf{f}(b)$. Show that for this two-point boundary value problem

$$\begin{aligned} k(t,s) &= \frac{1}{g_0(s)w(s)} \begin{cases} \dfrac{\Delta(b,s)\Delta(a,t)}{\Delta(a,b)}, & a \leqslant t \leqslant s \\ \dfrac{\Delta(b,s)\Delta(a,t)}{\Delta(a,b)} + \Delta(s,t), & s \leqslant t \leqslant b \end{cases} \\ \rho_1(t) &= -\frac{\Delta(b,t)}{\Delta(a,b)} \quad \text{and} \quad \rho_2(t) = \frac{\Delta(a,t)}{\Delta(a,b)} \end{aligned} \tag{3.54}$$

where $\Delta(s,t)$ is given beneath (3.52) and $w(s)$ is defined in (3.50).

Exercise 4. Use (3.54) to find k and ρ for the dc motor system (3.40)–(3.41). Compare the result with (3.42) and (3.43).

It is apparent that we could derive an explicit expression for the inverse of a regular nth-order linear differential system [assuming the boundary conditions satisfy the invertibility condition (3.28)]. The inverse would involve n independent fundamental solutions and the nth-order Wronskian determinant of these n solutions. Of course, as indicated by Exercise 3, the manipulation can be complicated. The determination of the Green's function for an nth-order two-point boundary value problem requires the solution of $2n$ simultaneous algebraic equations with coefficients which are functions of s. In contrast, the Green's function for the initial condition problem (or one-point boundary value problem) requires the solution of only n simultaneous equations because $k(t,s)=0$ for $t<s$. Of particular interest is the constant-coefficient initial condition problem, for which determination of the Green's function reduces to inversion of an $n\times n$ matrix of constants.

3.4 Time-Invariant Dynamic Systems

The initial value problem is at the heart of dynamic systems—systems for which the variable t represents time. The linear time-invariant (or constant-coefficient) dynamic system merits special attention if only because its inversion is easily automated using standard computer programs for solving matrix equations. Furthermore, many dynamic systems are adequately represented as linear time-invariant systems. We examine these systems in detail in this section.

The Inverse of the nth-Order System

The general nth-order constant-coefficient differential equation with initial conditions is

$$\begin{aligned} &\mathbf{f}^{(n)}(t)+a_1\mathbf{f}^{(n-1)}(t)+\cdots+a_n\mathbf{f}(t)=\mathbf{u}(t) \\ &\beta_i(\mathbf{f}) \triangleq \mathbf{f}^{(i-1)}(0)=\alpha_i \qquad i=1,\ldots,n \end{aligned} \tag{3.55}$$

for real scalars $\{a_i\}$ and $t\geqslant 0$. The characteristic equation for (3.55) is (3.37); assume it has n distinct roots $\mu_1,\ldots,\mu_n$ (the multiple root case is considered in Section 4.4). Then the fundamental solutions for (3.55) are $\mathbf{v}_i(t)\triangleq \exp(\mu_i t)$, $i=1,\ldots,n$.

The Green's function, as given by (3.44), is

$$\begin{aligned} k(t,s) &= b_1 \exp(\mu_1 t) + \cdots + b_n \exp(\mu_n t), \quad t \text{ in } [0,s) \\ &= d_1 \exp(\mu_1 t) + \cdots + d_n \exp(\mu_n t), \quad t \text{ in } (s,\infty) \end{aligned}$$

All n boundary conditions apply to the first half of $k(t,s)$, the half involving the unknowns $b_1,\ldots,b_n$. As a result,

$$\begin{pmatrix} k(0,s) \\ \dfrac{dk}{dt}(0,s) \\ \vdots \\ \dfrac{d^{n-1}k(0,s)}{dt^{n-1}} \end{pmatrix} = \begin{pmatrix} 1 & \cdots & 1 \\ \mu_1 & \cdots & \mu_n \\ \vdots & & \vdots \\ \mu_1^{n-1} & \cdots & \mu_n^{n-1} \end{pmatrix} \begin{pmatrix} b_1 \\ \\ \vdots \\ b_n \end{pmatrix} = \begin{pmatrix} 0 \\ \vdots \\ 0 \end{pmatrix}$$

This boundary condition matrix is the **Wronskian matrix** of the functions $\{\exp(\mu_i t)\}$ at $t=0$. The matrix is also known as the **Vandermond matrix** for the system (3.55). It is invertible if and only if the roots $\mu_1,\ldots,\mu_n$ are distinct as assumed.* Therefore, $b_1 = \cdots = b_n = 0$, and $k(t,s)=0$ for t in $[0,s)$. The discontinuity conditions (3.45) at $t=s$ are

$$\begin{aligned} d_1 \exp(\mu_1 s) + \cdots + d_n \exp(\mu_n s) &= 0 \qquad (k \text{ continuous}) \\ d_1 \mu_1 \exp(\mu_1 s) + \cdots + d_n \mu_n \exp(\mu_n s) &= 0 \qquad (dk/dt \text{ continuous}) \\ &\vdots \\ d_1 \mu_1^{n-2} \exp(\mu_1 s) + \cdots + d_n \mu_n^{n-2} \exp(\mu_n s) &= 0 \qquad (d^{n-2}k/dt^{n-2} \text{ continuous}) \\ d_1 \mu_1^{n-1} \exp(\mu_1 s) + \cdots + d_n \mu_n^{n-1} \exp(\mu_n s) &= 1 \qquad (\text{unit step in } d^{n-1}k/dt^{n-1}) \end{aligned}$$

We substitute the new variables $\hat{d}_i \triangleq d_i \exp(\mu_i s)$, $i=1,\ldots,n$ into the discontinuity equations to obtain

$$\begin{pmatrix} 1 & \cdots & 1 \\ \mu_1 & \cdots & \mu_n \\ \vdots & & \vdots \\ \mu_1^{n-1} & \cdots & \mu_n^{n-1} \end{pmatrix} \begin{pmatrix} \hat{d}_1 \\ \\ \vdots \\ \hat{d}_n \end{pmatrix} = \begin{pmatrix} 0 \\ \vdots \\ 0 \\ 1 \end{pmatrix} \tag{3.56}$$

*If the roots were not distinct, we would use a different set of fundamental solutions $\{\mathbf{v}_i\}$, and obtain a different boundary condition matrix. The Wronskian matrix is explored in P&C 3.7. The Vandermond matrix is examined in P&C 4.16.

Because the roots $\{\mu_i\}$ are distinct, the Vandermond matrix is invertible, and (3.56) can be solved by means of a standard computer program to obtain $\{\hat{d}_i\}$. Notice that the new variables $\{\hat{d}_i\}$ are independent of s. The s dependence of the variables $\{d_i\}$ has been removed by the substitution. In terms of the new variables, the Green's function becomes

$$\begin{aligned} k(t,s) &= 0 && \text{for } 0 \leqslant t \leqslant s \\ &= \hat{d}_1 \exp[\mu_1(t-s)] + \cdots + \hat{d}_n \exp[\mu_n(t-s)] && \text{for } t \geqslant s \end{aligned} \tag{3.57}$$

The boundary kernel for the system (3.55) is found from (3.46) and (3.47). Equation (3.47) is

$$\begin{pmatrix} 1 & \cdots & 1 \\ \mu_1 & \cdots & \mu_n \\ \vdots & & \vdots \\ \mu_1^{n-1} & \cdots & \mu_n^{n-1} \end{pmatrix} \begin{pmatrix} c_{11} & \cdots & c_{1n} \\ \vdots & & \vdots \\ c_{n1} & \cdots & c_{nn} \end{pmatrix} = \mathbf{I}$$

Then, by (3.46),

$$\rho_j(t) = c_{1j} \exp(\mu_1 t) + \cdots + c_{nj} \exp(\mu_n t), \qquad j = 1, \ldots, n \tag{3.58}$$

where the coefficients for ρ_j are obtained from the jth column of the inverse Vandermond matrix:

$$\begin{pmatrix} c_{11} & \cdots & c_{1n} \\ \vdots & & \vdots \\ c_{n1} & \cdots & c_{nn} \end{pmatrix} = \begin{pmatrix} 1 & \cdots & 1 \\ \mu_1 & \cdots & \mu_n \\ \vdots & & \vdots \\ \mu_1^{n-1} & \cdots & \mu_n^{n-1} \end{pmatrix}^{-1} \tag{3.59}$$

The inverse of the differential equation and boundary conditions of (3.55) is

$$\begin{aligned} \mathbf{f}(t) &= \int_0^\infty k(t,s)\mathbf{u}(s)\,ds + \sum_{j=1}^n \rho_j(t)\alpha_j \\ &= \int_0^t \{\hat{d}_1 \exp[\mu_1(t-s)] + \cdots + \hat{d}_n \exp[\mu_n(t-s)]\}\mathbf{u}(s)\,ds \\ &\quad + \sum_{j=1}^n [c_{1j} \exp(\mu_1 t) + \cdots + c_{nj} \exp(\mu_n t)]\mathbf{f}^{(j-1)}(0) \end{aligned} \tag{3.60}$$

where $\{\hat{d}_i\}$ and $\{c_{ij}\}$ are specified by (3.56) and (3.59), respectively. The computer program which produces (3.59) will simultaneously solve (3.56). Section 4.4 explores the computational difficulties which arise when the characteristic equation of the system has nearly equal roots.

The shape of the Green's function for a time-invariant (i.e., constant-coefficient) dynamic system depends only on $t-s$, the delay between the time s that an impulse is applied at the system input and the time t that the output $k(t,s)$ is observed. That is, $k(t,s)=k(t-s,0)$. Therefore, actual measurement of the response of the physical system to an approximate impulse is a suitable method for determining the Green's function. The response of such a system, initially at rest, to an impulse input $\mathbf{u}(t)=\delta(t)$ is commonly referred to as the **impulse response** of the system. We denote the impulse response by g, where $g(t) \stackrel{\Delta}{=} k(t,0)$. Then the integral term in (3.60) can be rewritten as a convolution of $\mathbf{u}$ and g.*

$$\int_0^t k(t,s)\mathbf{u}(s)\,ds = \int_0^t g(t-s)\mathbf{u}(s)\,ds$$

The components of the boundary kernel also can be measured physically; $\rho_j(t)$ is the response of the system with no distributed input $\mathbf{u}$, and with the initial conditions $\alpha_j=1$, $\alpha_i=0$, for $i\neq j$. Furthermore, we see from (3.56)–(3.59) that $\hat{d}_i=c_{in}$ for $i=1,\ldots,n$. Therefore, the impulse response is equal to one of the initial condition responses; specifically,

$$\rho_n(t)=k(t,0)=g(t) \tag{3.61}$$

Applying a unit impulse $\delta(t)$ is equivalent to instantaneously applying to the system (at rest) the unit initial condition $\mathbf{f}^{(n-1)}(0)=1$ (all other initial conditions remaining zero); if we can apply this initial condition some other way, we do not need an approximate impulse in order to measure the impulse response of the system.

Exercise 1. The differential equation (3.40) for an armature-controlled dc motor is

$$\ddot{\phi}(t)+\dot{\phi}(t)=\mathbf{u}(t)$$

Show that for given initial conditions, $\phi(0)$ and $\dot{\phi}(0)$, the Green's function,

*See Appendix 2 for a discussion of convolution.

boundary kernel, and inverse equation are

$$
\begin{aligned}
k(t,s) &= 0, \qquad && 0 \leqslant t \leqslant s \\
&= 1 - e^{-(t-s)}, && t \geqslant s \\
\rho_1(t) &= 1 \\
\rho_2(t) &= 1 - e^{-t} \\
\phi(t) &= \int_0^t [1 - e^{-(t-s)}]\mathbf{u}(s)\,ds + \phi(0) + (1 - e^{-t})\dot{\phi}(0)
\end{aligned}
\qquad (3.62)
$$

Compare (3.62) with (3.52) and (3.53).

The State-Space Model

The nth-order constant-coefficient differential equation with initial conditions, (3.55), can be expressed as a first-order vector differential equation by redefining the variables. If $\mathbf{u}(t)=0$, the quantities $\mathbf{f}(0), \mathbf{f}^{(1)}(0), \ldots, \mathbf{f}^{(n-1)}(0)$ determine the trajectory $\mathbf{f}(t)$ for all t; these n quantities together form a more complete description of the state (or condition) of the system at $t=0$ than does $\mathbf{f}(0)$ alone. Let $\mathbf{f}_1 \triangleq \mathbf{f}, \mathbf{f}_2 \triangleq \mathbf{f}^{(1)}, \ldots, \mathbf{f}_n \triangleq \mathbf{f}^{(n-1)}$. Then (3.55) can be expressed as the following set of n first-order differential equations.

$$
\begin{aligned}
\dot{\mathbf{f}}_1(t) &= \mathbf{f}_2(t) \\
\dot{\mathbf{f}}_2(t) &= \mathbf{f}_3(t) \\
&\;\vdots \\
\dot{\mathbf{f}}_{n-1}(t) &= \mathbf{f}_n(t) \\
\dot{\mathbf{f}}_n(t) &= -a_n\mathbf{f}_1(t) - \cdots - a_1\mathbf{f}_n(t) + \mathbf{u}(t)
\end{aligned}
$$

By defining $\mathbf{x} \triangleq (\mathbf{f}_1 \cdots \mathbf{f}_n)^{\mathrm{T}}$ and $\dot{\mathbf{x}} \triangleq (\dot{\mathbf{f}}_1 \cdots \dot{\mathbf{f}}_n)^{\mathrm{T}}$, we write the n individual equations as

$$
\dot{\mathbf{x}}(t) = \begin{pmatrix} 0 & 1 & 0 & \cdots & 0 \\ 0 & 0 & 1 & \cdots & 0 \\ \vdots & \vdots & \vdots & \vdots & \vdots \\ 0 & 0 & \cdots & 0 & 1 \\ -a_n & -a_{n-1} & \cdots & -a_2 & -a_1 \end{pmatrix} \mathbf{x}(t) + \begin{pmatrix} 0 \\ \vdots \\ \\ 0 \\ 1 \end{pmatrix} \mathbf{u}(t) \qquad (3.63)
$$

The square matrix of (3.63) is known as the **companion matrix** for the nth order differential operator (3.55). The initial conditions of (3.55) become

$$\mathbf{x}(0)=\begin{pmatrix}\mathbf{f}(0)\\ \mathbf{f}^{(1)}(0)\\ \vdots\\ \mathbf{f}^{(n-1)}(0)\end{pmatrix}=\begin{pmatrix}\alpha_1\\ \\ \vdots\\ \alpha_n\end{pmatrix} \tag{3.64}$$

We call $\mathbf{x}(t)$ the *state vector* of the system at time t. Since the differential system (3.55) has a unique solution, the state at time t can be determined from the state at any time previous to t. The state provides precisely enough information concerning the condition of the system to determine the future behavior of the system for a given input. The vector $\mathbf{x}(t)$ is in $\mathfrak{M}^{n\times 1}$. Therefore, we call $\mathfrak{M}^{n\times 1}$ the *state space* of the system. The variables $\{\mathbf{f}_i(t)\}$ are known as *state variables*.

Equations (3.63) and (3.64) are of the general form

$$\dot{\mathbf{x}}(t)=\mathbf{A}\mathbf{x}(t)+\mathbf{B}\mathbf{u}(t), \qquad \mathbf{x}(0) \text{ given} \tag{3.65}$$

However, the notation of (3.65) is more general than that of (3.63) and (3.64). The input $\mathbf{u}$ can include more than one function. A meaningful equation is defined by any $n\times n$ matrix $\mathbf{A}$ and $n\times m$ matrix $\mathbf{B}$; the resulting vector equation describes the evolution in time of a system with m inputs and n outputs. A general set of coupled linear time-invariant differential equations can be expressed in this state-space form (P&C 3.18). We refer to (3.65) as a **state equation**. We call $\mathbf{x}(t)$ the **state vector** and its elements the **state variables**; $\mathbf{A}$ and $\mathbf{B}$ are the **system matrix** and the **input matrix**, respectively.*

We should note that the description of a dynamic system by a state-space model is not unique. If we multiply both sides of (3.65) by an arbitrary invertible $n\times n$ matrix $\mathbf{S}$, we obtain

$$\mathbf{S}\dot{\mathbf{x}}(t)=\mathbf{S}\mathbf{A}\mathbf{x}(t)+\mathbf{S}\mathbf{B}\mathbf{u}(t)$$

Defining $\mathbf{y}=\mathbf{S}\mathbf{x}$, we find

$$\begin{aligned}\dot{\mathbf{y}}(t)&=\mathbf{S}\mathbf{A}\mathbf{S}^{-1}\mathbf{y}(t)+\mathbf{S}\mathbf{B}\mathbf{u}(t)\\ &=\hat{\mathbf{A}}\mathbf{y}(t)+\hat{\mathbf{B}}\mathbf{u}(t)\end{aligned}$$

*See Zadeh and Desoer [3.14] or DeRusso, Roy, and Close [3.2] for a more complete discussion of state-space models.

with $\mathbf{y}(0)=\mathbf{Sx}(0)$ given. This second state-space differential equation is equivalent to (3.65) as a representative of the system. The state vector $\mathbf{y}(t)$ is a representation of $\mathbf{x}(t)$ in new coordinates. Thus the state variables and system matrix which describe a given system are not unique. In Section 4.2 we explore the essential characteristics of a matrix, its eigenvalues. We find that the similarity transformation $\mathbf{SAS}^{-1}$ does not affect the eigenvalues. Consequently, all system matrices which represent the same system have the same essential characteristics. State space models of dynamic systems are analyzed in terms of their eigenvalues in Sections 4.3 and 4.5.

Example 1. A State Equation. The differential equation for the armature-controlled dc motor of Exercise 1 is

$$\ddot{\phi}(t)+\dot{\phi}(t)=\mathbf{u}(t), \qquad \phi(0)=\alpha_1, \qquad \dot{\phi}(0)=\alpha_2$$

Defining the state variables $\mathbf{f}_1(t) \triangleq \phi(t)$ and $\mathbf{f}_2(t) \triangleq \dot{\phi}(t)$, we obtain the following state equation

$$\dot{\mathbf{x}}(t)=\begin{pmatrix}0 & 1\\ 0 & -1\end{pmatrix}\mathbf{x}(t)+\begin{pmatrix}0\\ 1\end{pmatrix}\mathbf{u}(t), \qquad \mathbf{x}(0)=\begin{pmatrix}\alpha_1\\ \alpha_2\end{pmatrix}$$

The system matrix is the companion matrix for the second-order differential equation.

Let us find an integral equation which is the inverse (or explicit solution) of the first-order vector-valued differential system (3.65). Although we work directly with the system in the specific form (3.65), we note that the equation can be expressed in terms of a general differential operator $\mathbf{L}$ acting on a vector-valued function space. Let $\mathbf{f}$ be in $\mathcal{C}^n(0,\infty)$; then $\mathbf{f}^{(k)}$ is in $\mathcal{C}^{n-k}(0,\infty)$ and $\mathbf{x}$ is in the cartesian product space:

$$\mathcal{V}=\mathcal{C}^n(0,\infty)\times\mathcal{C}^{n-1}(0,\infty)\times\cdots\times\mathcal{C}^1(0,\infty)$$

The system (3.65) is equivalent to the following operator equation on $\mathcal{V}$:

$$\mathbf{Lx} \triangleq \dot{\mathbf{x}}-\mathbf{Ax}=\mathbf{Bu} \quad \text{with } \mathbf{x}(0) \text{ given} \tag{3.66}$$

We express $\mathbf{x}$ as an integral operation on the whole vector-valued function $\mathbf{Bu}$.

Inversion of the State Equation

The state equation for an nth-order time-invariant dynamic system is

$$\dot{\mathbf{x}}(t)=\mathbf{Ax}(t)+\mathbf{Bu}(t), \qquad \mathbf{x}(0)=\mathbf{x}_0 \tag{3.67}$$

where **A** and **B** are arbitrary $n \times n$ and $n \times m$ matrices, respectively. The state vector $\mathbf{x}(t)$ and the input vector $\mathbf{u}(t)$ are in $\mathfrak{M}^{n\times 1}$ and $\mathfrak{M}^{m\times 1}$, respectively. We invert (3.67) by the same approach we used for the nth-order differential system; we invert separately the two component equations

$$\dot{\mathbf{x}}(t) - \mathbf{A}\mathbf{x}(t) = \mathbf{B}\mathbf{u}(t), \qquad \mathbf{x}(0) = \boldsymbol{\theta} \tag{3.68}$$

$$\dot{\mathbf{x}}(t) - \mathbf{A}\mathbf{x}(t) = \boldsymbol{\theta}, \qquad \mathbf{x}(0) = \mathbf{x}_0 \tag{3.69}$$

Assume the inverse of the "boundary input" system (3.69) is of the form

$$\mathbf{x}(t) = \mathbf{\Phi}(t)\mathbf{x}(0) \tag{3.70}$$

where the boundary kernel $\mathbf{\Phi}(t)$ is a $n \times n$ matrix commonly referred to as the **state transition matrix**. (It describes the "undriven" transition from the state at "0" to the state at t.) In order that (3.70) be the correct inverse, $\mathbf{x}(t)$ must satisfy (3.69),

$$\dot{\mathbf{x}}(t) - \mathbf{A}\mathbf{x}(t) = \frac{d\mathbf{\Phi}(t)}{dt}\mathbf{x}(0) - \mathbf{A}\mathbf{\Phi}(t)\mathbf{x}(0) = \boldsymbol{\theta}, \qquad \mathbf{x}(0) = \mathbf{\Phi}(0)\mathbf{x}(0)$$

for any initial condition vector $\mathbf{x}(0)$. Therefore, the state transition matrix must satisfy

$$\frac{d\mathbf{\Phi}(t)}{dt} - \mathbf{A}\mathbf{\Phi}(t) = \mathbf{\Theta}, \qquad \mathbf{\Phi}(0) = \mathbf{I} \tag{3.71}$$

Rather than treat the system (3.71) one element at a time, we work with the whole $n \times n$ matrix-valued system. We use the power series method to find the complementary function for the system. Assume

$$\mathbf{\Phi}(t) = \mathbf{C}_0 + \mathbf{C}_1 t + \mathbf{C}_2 t^2 + \cdots$$

where each $\mathbf{C}_i$ is a constant $n \times n$ matrix. We substitute $\mathbf{\Phi}(t)$ into the differential equation of (3.71) and equate the coefficient on each power of t to the zero matrix $\mathbf{\Theta}$ to find

$$\mathbf{C}_1 = \mathbf{A}\mathbf{C}_0, \qquad \mathbf{C}_2 = \left(\frac{1}{2!}\right)\mathbf{A}^2\mathbf{C}_0, \qquad \mathbf{C}_3 = \left(\frac{1}{3!}\right)\mathbf{A}^3\mathbf{C}_0, \qquad \cdots$$

It follows that $\mathbf{C}_0$ is arbitrary and

$$\mathbf{\Phi}(t) = \left(\mathbf{I} + \mathbf{A}t + \frac{\mathbf{A}^2 t^2}{2!} + \cdots\right)\mathbf{C}_0 \triangleq e^{\mathbf{A}t}\mathbf{C}_0 \tag{3.72}$$

We have used the symbol $e^{\mathbf{A}t}$ to represent the sum of the "exponential-looking" matrix series of (3.72):

$$e^{\mathbf{A}t} \triangleq \mathbf{I}+\mathbf{A}t+\frac{\mathbf{A}^2t^2}{2!}\cdots$$

We call $e^{\mathbf{A}t}$ a **fundamental matrix** for the state equation of (3.67); the matrix is analogous to a fundamental set of solutions for an nth-order differential equation. Applying the boundary conditions of (3.71) to (3.72), we find $\boldsymbol{\Phi}(0)=e^{\mathbf{A}0}\mathbf{C}_0=\mathbf{I}$. It is clear from the definition of $e^{\mathbf{A}t}$ that $e^{\mathbf{A}0}=\mathbf{I}$; therefore, $\mathbf{C}_0=\mathbf{I}$ and the state transition matrix (or boundary kernel) for the state-space system (3.67) is

$$\boldsymbol{\Phi}(t)=e^{\mathbf{A}t} \tag{3.73}$$

Example 2. A State Transition Matrix. In Example 1 we found the system matrix $\mathbf{A}$ for the differential equation $\ddot{\phi}(\mathrm{t})+\dot{\phi}(\mathrm{t})=\mathbf{u}(\mathrm{t})$:

$$\mathbf{A}=\begin{pmatrix}0 & 1\\ 0 & -1\end{pmatrix}$$

To find the fundamental matrix for this system, we sum the defining infinite series:

$$\boldsymbol{\Phi}(t)=e^{\mathbf{A}t}=\mathbf{I}+\mathbf{A}t+\frac{\mathbf{A}^2t^2}{2!}+\cdots$$

$$=\begin{pmatrix}1 & 0\\ 0 & 1\end{pmatrix}+\begin{pmatrix}0 & 1\\ 0 & -1\end{pmatrix}t+\begin{pmatrix}0 & -1\\ 0 & 1\end{pmatrix}\frac{t^2}{2!}+\cdots$$

$$=\begin{pmatrix}1 & \left(t-\frac{t^2}{2!}+\frac{t^3}{3!}-\cdots\right)\\ 0 & \left(1-t+\frac{t^2}{2!}-\cdots\right)\end{pmatrix}=\begin{pmatrix}1 & 1-e^{-t}\\ 0 & e^{-t}\end{pmatrix}$$

If the matrix $\mathbf{A}$ of Example 2 were not simple, it would be difficult to sum the infinite series for $e^{\mathbf{A}t}$ by the method of that example. It would not be easy to recognize the function to which each scalar series converges. Arbitrary functions of matrices are examined in detail in Section 4.6, and practical techniques for computing functions of matrices are developed. These techniques can be used to compute $e^{\mathbf{A}t}$ for an arbitrary square matrix $\mathbf{A}$.

Exercise 2. Show that for the fundamental matrix $e^{\mathbf{A}t}$ of Example 2,

$$\begin{aligned} e^{\mathbf{A}t}e^{\mathbf{A}\tau} &= e^{\mathbf{A}(t+\tau)} \\ \left(e^{\mathbf{A}t}\right)^{-1} &= e^{-\mathbf{A}t} \end{aligned} \tag{3.74}$$

Properties (3.74) apply to all time-invariant systems, that is, all systems which have a constant system matrix (P&C 3.19).

We can view $\mathbf{Bu}$ as a vector-valued distributed input to (3.67). Therefore, we assume the inverse of the distributed-input state equation (3.68) is an integral equation of the form

$$\mathbf{x}(t) = \int_0^\infty \mathbf{K}(t,s)\mathbf{B}\mathbf{u}(s)\,ds \tag{3.75}$$

where the $n \times n$ matrix $\mathbf{K}(t,s)$ is called the **matrix Green's function** for the system (3.67). (By the integral of a matrix we mean the matrix of integrals.) We substitute (3.75) into (3.68) to determine the equations which describe $\mathbf{K}$:

$$\begin{aligned} \dot{\mathbf{x}}(t) - \mathbf{A}\mathbf{x}(t) &= \int_0^\infty \left[\frac{d}{dt}\mathbf{K}(t,s) - \mathbf{A}\mathbf{K}(t,s)\right]\mathbf{B}\mathbf{u}(s)\,ds = \mathbf{B}\mathbf{u}(t) \\ \mathbf{x}(0) &= \int_0^\infty \mathbf{K}(0,s)\mathbf{B}\mathbf{u}(s)\,ds = \boldsymbol{\theta} \end{aligned} \tag{3.76}$$

for all vectors $\mathbf{u}$ with elements which are continuous functions. To see more clearly the conditions on $\mathbf{K}(t,s)$ which follow from (3.76), note that

$$\int_0^\infty \delta(t-s)\mathbf{I}\begin{pmatrix} g_{11}(s) & \cdots & g_{1m}(s) \\ \vdots & & \vdots \\ g_{n1}(s) & \cdots & g_{nm}(s) \end{pmatrix} ds = \begin{pmatrix} g_{11}(t) & \cdots & g_{1m}(t) \\ \vdots & & \vdots \\ g_{n1}(t) & \cdots & g_{nm}(t) \end{pmatrix}$$

In other words, if we let $\mathbf{G}(s)$ denote the matrix with elements $g_{ij}(s)$, then the equation $\int_0^\infty \mathbf{K}(t,s)\mathbf{G}(s)\,ds = \mathbf{G}(t)$ is satisfied by $\mathbf{K}(t,s) = \delta(t-s)\mathbf{I}$. Thus in order to satisfy (3.76), it is sufficient that $\mathbf{K}$ meet the following requirements:

$$\begin{aligned} \frac{d\mathbf{K}(t,s)}{dt} - \mathbf{A}\mathbf{K}(t,s) &= \delta(t-s)\mathbf{I} \\ \mathbf{K}(0,s) &= \boldsymbol{\Theta} \end{aligned} \tag{3.77}$$

The approach we use to solve (3.77) for $\mathbf{K}$ is essentially the same as that used for the nth-order scalar system (3.55). For $t \neq s$, $\mathbf{K}(t,s)$ satisfies the same $n \times n$ differential equation, (3.71), as does the state transition matrix. Thus, using the general solution to (3.71) found earlier,

$$\begin{aligned}\mathbf{K}(t,s) &= e^{\mathbf{A}t}\mathbf{B}_0, \quad t \text{ in } [0,s)\\ &= e^{\mathbf{A}t}\mathbf{D}_0, \quad t \text{ in } (s,\infty)\end{aligned}$$

where $\mathbf{B}_0$ and $\mathbf{D}_0$ are $n \times n$ constant matrices. The boundary conditions of (3.76) require $\mathbf{K}(0,s) = e^{\mathbf{A}0}\mathbf{B}_0 = \boldsymbol{\Theta}$; since $e^{\mathbf{A}0} = \mathbf{I}$, $\mathbf{B}_0 = \boldsymbol{\Theta}$. From (3.77), we also note that $\mathbf{K}$ must satisfy a discontinuity condition at $t = s$. The delta functions on the right-hand side of (3.77) must be introduced by the highest derivative, $d\mathbf{K}/dt$; otherwise derivatives of delta functions would appear. Consequently, the diagonal elements of $\mathbf{K}$ contain a unit step at $t = s$, whereas off-diagonal elements are continuous:

$$\mathbf{K}(s^+,s) - \mathbf{K}(s^-,s) = e^{\mathbf{A}s}\mathbf{D}_0 - \boldsymbol{\Theta} = \mathbf{I}$$

Then, using (3.74), $\mathbf{D}_0 = (e^{\mathbf{A}s})^{-1} = e^{-\mathbf{A}s}$, and

$$\begin{aligned}\mathbf{K}(t,s) &= \boldsymbol{\Theta}, \quad t<s\\ &= e^{\mathbf{A}(t-s)}, \quad t>s\end{aligned} \tag{3.78}$$

The inverse of the state-space system (3.67) is the sum of (3.75) and (3.70); $\boldsymbol{\Phi}$ and $\mathbf{K}$ are given by (3.73) and (3.78), respectively:

$$\mathbf{x}(t) = \int_0^t e^{\mathbf{A}(t-s)}\mathbf{B}\mathbf{u}(s)\,ds + e^{\mathbf{A}t}\mathbf{x}(0) \tag{3.79}$$

The inverse system is fully determined by the state transition matrix $e^{\mathbf{A}t}$ and the input matrix $\mathbf{B}$. In Section 4.6 we determine how to evaluate $e^{\mathbf{A}t}$ by methods other than summing of the series (3.72).

At the heart of the solution (3.60) for the nth-order dynamic system (3.55) is the Vandermond matrix for the system. If the state equation is derived from the nth-order differential equation as in (3.63), we would expect the Vandermond matrix to be involved in the solution (3.79) of the state equation. We find in P&C 4.16 and (4.98) that if the system matrix $\mathbf{A}$ is the companion matrix for an nth-order dynamic system, the Vandermond matrix is intimately related to both $\mathbf{A}$ and $e^{\mathbf{A}t}$.

Exercise 3. Show that for the system of Examples 1 and 2,

$$\mathbf{x}(t) = \begin{pmatrix}\phi(t)\\ \dot{\phi}(t)\end{pmatrix} = \int_0^t \begin{pmatrix}1-e^{-(t-s)}\\ e^{-(t-s)}\end{pmatrix}\mathbf{u}(s)\,ds + \begin{pmatrix}\phi(0)+\dot{\phi}(0)-\dot{\phi}(0)e^{-t}\\ \dot{\phi}(0)e^{-t}\end{pmatrix} \tag{3.80}$$

Equation (3.80) should be compared with its second-order scalar equivalent (3.62). The state-space solution usually contains more information than its scalar counterpart—information about derivatives of the solution is stated explicitly.

Exercise 4. Use the solution (3.79) at $t=a$ to determine the form of the solution to the state-space system (3.67) if the initial conditions are given at $t=a$ instead of $t=0$; that is, show that

$$\mathbf{f}(t)=\int_a^t e^{\mathbf{A}(t-s)}\mathbf{B}\mathbf{u}(s)\,ds+e^{\mathbf{A}(t-a)}\mathbf{x}(a)$$

The discussion beneath the nth-order scalar solution (3.60) extends to the more general state-space solution (3.79). We can interpret $\mathbf{K}(t,0)=e^{\mathbf{A}t}$ as the matrix impulse response of the state-space system. Since the matrix $\mathbf{A}$ is constant, it is appropriate to measure physically the state transition matrix $e^{\mathbf{A}t}$. By (3.70), the jth column of $\mathbf{\Phi}(t)$ (or $e^{\mathbf{A}t}$) consists in the "undriven" decay of $\mathbf{x}(t)$ from the initial condition $\mathbf{x}(0)=\boldsymbol{\varepsilon}_j$, the jth standard basis vector for $\mathfrak{M}^{n\times 1}$. From measurements of the n columns of $e^{\mathbf{A}t}$ we can determine the full inverse equation (3.79) without explicit determination of the system matrix $\mathbf{A}$ (P&C 3.20).

The techniques used to invert the first-order state-space system (3.67) are applied to a *second-order* vector differential system in P&C 4.32. As with the state-space system, the Green's function for this system can be obtained from the boundary kernel; the latter can be measured physically. The inverse for this second-order vector system involves several functions of matrices. We discuss methods for evaluating general functions of matrices in Section 4.6.

3.5 Problems and Comments

3.1 *Forward integration*: the differential system $\mathbf{f}''(t)+\frac{1}{4}\mathbf{f}(t)+(1/400)\mathbf{f}^3(t)=0$, $\mathbf{f}(0)=10$, $\mathbf{f}'(0)=0$ describes the unforced oscillations of a mass hanging on a spring. The spring has a nonlinear force-elongation characteristic; $\mathbf{f}(t)$ denotes the position of the mass at time t. There are many numerical integration techniques for obtaining an approximate solution to such a nonlinear differential equation with initial conditions (see [3.9]). The following technique is one of the simplest. We concern ourselves only with integer values of t, and replace the derivatives by the finite-difference approximations $\mathbf{f}'(n)\approx\mathbf{f}(n+1)-\mathbf{f}(n)$ and $\mathbf{f}''(n)\approx\mathbf{f}(n+1)-2\mathbf{f}(n)+\mathbf{f}(n-1)$. Use these finite-difference approximations and the differential system to express $\mathbf{f}(n+1)$ in terms of $\mathbf{f}(n)$ and $\mathbf{f}(n-1)$. Compute $\mathbf{f}(1), \mathbf{f}(2), \ldots, \mathbf{f}(8)$. How might the above finite-difference approximation be modified to obtain a more accurate solution to the differential equation?

3.2 *Backward integration*: a (nonlinear) differential equation with final end-point conditions (rather than initial conditions) can be solved by backward numerical integration. Backward integration can be carried out be means of any forward integration routine. Suppose the differential system is of the form $\mathbf{f}^{(n)}(t)+\mathbf{F}(\mathbf{f}(t), \mathbf{f}'(t), \ldots, \mathbf{f}^{(n-1)}(t), t)=0$ with $\mathbf{f}(t_f), \mathbf{f}'(t_f), \ldots, \mathbf{f}^{(k-1)}(t_f)$ specified. Show that the change of variables $\mathbf{f}(t)=\mathbf{f}(t_f-s)=\mathbf{g}(s)$ converts the final conditions on $\mathbf{f}$ to initial conditions on $\mathbf{g}$ and produces a differential equation in $\mathbf{g}$ which differs from the differential equation in $\mathbf{f}$ in the sign on the odd-order derivations.

3.3 *Relaxation*: the finite-difference approximation to a two-point boundary value problem can be solved by a simple iterative technique known as relaxation [3.3]. Suppose $\mathbf{f}''(s)=1$ with $\mathbf{f}(0)=\mathbf{f}(5)=0$. Consider the values of $\mathbf{f}$ only at integer values of s. Replace the second derivative by the approximation $\mathbf{f}''(n) \approx \mathbf{f}(n+1)-2\mathbf{f}(n)+\mathbf{f}(n-1)$, and express $\mathbf{f}(n)$ in terms of $\mathbf{f}(n-1)$ and $\mathbf{f}(n+1)$. Let the initial values of $\mathbf{f}(1), \ldots, \mathbf{f}(4)$ be zero. A single step in the iteration consists in solving successively for each of the values $\mathbf{f}(1), \ldots, \mathbf{f}(4)$ in terms of current values of $\mathbf{f}$ at the two neighboring points. Repetitive improvement of the set of values $\{\mathbf{f}(k)\}$ results in convergence of this set of values to the solution of the set of difference equations, regardless of numerical errors, and regardless of the order in which the values are improved during each iteration.

(*a*) Carry out six iterations for the above problem.

(*b*) Find the exact solution to the set of difference equations by solving the equations simultaneously. Compare the results of the iteration of (*a*) with the exact solution for the differential system.

*3.4 An intuitive understanding of the following properties of differential systems can be gained by examining a finite-difference approximation to the second-order case. See [3.6] for a rigorous discussion of these statements.

(*a*) A regular nth-order linear differential equation has n independent solutions.

(*b*) A boundary condition consisting in a linear combination of values of $\mathbf{f}, \mathbf{f}', \ldots, \mathbf{f}^{(n-1)}$ need not be independent of the regular nth-order differential equation; consider, for example, $\mathbf{f}''(s)=0$ with $\mathbf{f}'(0)-\mathbf{f}'(1)=0$.

(*c*) If the boundary conditions associated with a regular nth-order differential equation consist in n independent linear combinations of the values $\mathbf{f}(a), \mathbf{f}'(a), \ldots, \mathbf{f}^{(n-1)}(a)$, at a single point a in the domain of $\mathbf{f}$, then the differential system has a unique solution.

3.5 The following differential system is degenerate:

$$\phi'' + \phi' = \mathbf{u} \quad \text{with} \begin{cases} \phi(0) - \phi(1) = \alpha_1 \\ 2\phi(1) - 2\phi(0) + \phi'(0) - \phi'(1) = \alpha_2 \\ \phi'(1) - \phi'(0) = \alpha_3 \end{cases}$$

Find the solutions to the differential system in terms of the inputs $\mathbf{u}$, α_1, α_2, and α_3. Also find the relations among the inputs that must be satisfied in order that solutions exist. (Hint: the solution to the differential equation is expressed in terms of $\phi(0)$ and $\phi'(0)$ in (3.80).) What relationship exists between the number of dependent rows in a boundary condition matrix for a system and the number of different relations which must be satisfied by the inputs to that system?

3.6 Let $\mathbf{L}$ be a regular nth-order differential operator and $\{\beta_i(\mathbf{f}) = 0, i = 1, \ldots, m\}$ a set of homogeneous boundary conditions. Let $\mathcal{V}$ be the space of functions in $\mathcal{C}^n(a,b)$ which satisfy the homogeneous differential equation $\mathbf{Lf} = \boldsymbol{\theta}$. Let $\mathcal{F} \triangleq \{\mathbf{v}_1, \mathbf{v}_2, \ldots, \mathbf{v}_n\}$ be a fundamental set of solutions for $\mathbf{L}$; $\mathcal{F}$ is a basis for $\mathcal{V}$. Define $\mathbf{T}$: $\mathcal{V} \to \mathcal{R}^n$ by $\mathbf{Tf} \triangleq (\beta_1(\mathbf{f}), \ldots, \beta_m(\mathbf{f}))$ for all $\mathbf{f}$ in $\mathcal{V}$. Let $\mathcal{E}$ be the standard basis for $\mathcal{R}^n$. Show that the matrix $[\mathbf{T}]_{\mathcal{F}\mathcal{E}}$ is a boundary condition matrix for the differential system $\{\mathbf{L}, \beta_1, \ldots, \beta_m\}$.

*3.7 *The Wronskian*: let $\mathbf{f}_1, \ldots, \mathbf{f}_n$ be in $\mathcal{C}^n(a,b)$. The Wronskian matrix of $\mathbf{f}_1, \ldots, \mathbf{f}_n$ at t is defined by

$$\mathbf{W}(t) \triangleq \begin{pmatrix} \mathbf{f}_1(t) & \cdots & \mathbf{f}_n(t) \\ \mathbf{f}_1^{(1)}(t) & \cdots & \mathbf{f}_n^{(1)}(t) \\ \vdots & & \vdots \\ \mathbf{f}_1^{(n-1)}(t) & \cdots & \mathbf{f}_n^{(n-1)}(t) \end{pmatrix}$$

The Wronskian determinant is $w(t) \triangleq \det(\mathbf{W}(t))$.

(*a*) Show that $\{\mathbf{f}_1, \ldots, \mathbf{f}_n\}$ cannot be linearly dependent unless $w(t) = 0$ for all t in $[a,b]$.

(*b*) The fact that $w(t)=0$ for *some* t does not ordinarily imply that the set $\{\mathbf{f}_1,\ldots,\mathbf{f}_n\}$ is dependent; try, for example, $\mathbf{f}_1(t)=t^2$ and $\mathbf{f}_2(t)=t^3$ at $t=0$. Suppose, however, that $\mathbf{f}_1,\ldots,\mathbf{f}_n$ are solutions to an nth-order homogeneous differential equation defined on $[a,b]$. Then if $w(t)=0$ for *any* t in $[a,b]$, $\{\mathbf{f}_1,\ldots,\mathbf{f}_n\}$ is a linearly dependent set.

3.8 *Difference equations*: an arbitrary linear constant-coefficient difference equation can be expressed in the form

$$a_0(\mathbf{E}^n\mathbf{f})(k)+a_1(\mathbf{E}^{n-1}\mathbf{f})(k)+\cdots+a_n\mathbf{f}(k)=\mathbf{u}(k), \qquad k=0,1,2,\ldots$$

where $\mathbf{E}$ is the shift operator defined by $(\mathbf{E}\mathbf{f})(k)\triangleq\mathbf{f}(k+1)$; we concern ourselves only with integer values of the argument of $\mathbf{f}$. The order of the difference equation is the number of boundary conditions needed to specify a unique solution to the equation; that is, the order is $n-p$, where p is the lowest power of $\mathbf{E}$ to appear in the equation. (See [3.2].)

(*a*) The solutions to the homogeneous difference equation (the equation with $\mathbf{u}(k)=0$) usually consist of combinations of geometric sequences. Substitution of the sequence $\mathbf{f}(k)=r^k$, $k=0,1,2,\ldots$, into the homogeneous equation shows that nontrivial sequences must satisfy the following characteristic equation: $a_0r^n+a_1r^{n-1}+\cdots+a_n=0$. Find a basis for the nullspace of the difference operator $\mathbf{T}$ defined by

$$\begin{aligned}(\mathbf{T}\mathbf{f})(k)&\triangleq 2(\mathbf{E}^2\mathbf{f})(k)-3(\mathbf{E}\mathbf{f})(k)+\mathbf{f}(k)\\&=2\mathbf{f}(k+2)-3\mathbf{f}(k+1)+\mathbf{f}(k)\end{aligned}$$

What is the dimension of the nullspace of an nth-order difference operator?

(*b*) Let $\mathbf{f}_1,\ldots,\mathbf{f}_n$ be infinite sequences of the form $\mathbf{f}_i(k)$, $k=0,1,2,\ldots$. The *Casorati matrix* of $\mathbf{f}_1,\ldots,\mathbf{f}_n$ is defined by

$$\mathbf{C}(k)\triangleq\begin{pmatrix}\mathbf{f}_1(k) & \cdots & \mathbf{f}_n(k)\\ \mathbf{E}\mathbf{f}_1(k) & \cdots & \mathbf{E}\mathbf{f}_n(k)\\ \vdots & & \vdots\\ \mathbf{E}^{n-1}\mathbf{f}_1(k) & \cdots & \mathbf{E}^{n-1}\mathbf{f}_n(k)\end{pmatrix}$$

The infinite sequences $\mathbf{f}_1,\ldots,\mathbf{f}_n$ are linearly independent if and only if $c(k) \neq 0$ for $k=0,1,2,\ldots$, where $c(k)$ is the *Casorati determinant*, $\det(\mathbf{C}(k))$. Use the Casorati determinant to show the independence of the basis vectors found in (*a*).

3.9 Use the power series method to find the complementary function for the differential operator $(\mathbf{D}-1)^2$.

3.10 Define $\mathbf{L}$: $\mathcal{C}^1(0,1) \to \mathcal{C}(0,1)$ by $\mathbf{L} \triangleq -\mathbf{D}-a\mathbf{I}$. Find the Green's function k and the inverse equation for the differential system $\mathbf{Lf}=\mathbf{u}$, $\mathbf{f}(0)=\mathbf{f}(1)$.

3.11 Define $\mathbf{L}$: $\mathcal{C}^2(0,b) \to \mathcal{C}(0,b)$ by $\mathbf{L} \triangleq \mathbf{D}^2-3\mathbf{D}+2\mathbf{I}$. Find the Green's function k, the boundary kernel ρ, and the inverse equation for the differential system $\mathbf{Lf}=\mathbf{u}$, $\mathbf{f}(0)=\alpha_1$, $\mathbf{f}(b)=\alpha_2$.

3.12 Find the inverse equation for each of the following differential systems:
(*a*) $\mathbf{f}''+6\mathbf{f}'+5\mathbf{f}=\mathbf{u}$, $\mathbf{f}(0)=\alpha_1$, $\mathbf{f}'(0)=\alpha_2$
(*b*) $\mathbf{f}''+2\mathbf{f}'+2\mathbf{f}=\mathbf{u}$, $\mathbf{f}(0)=\alpha_1$, $\mathbf{f}'(0)=\alpha_2$
(*c*) $\mathbf{f}'''+6\mathbf{f}''+5\mathbf{f}'=\mathbf{u}$, $\mathbf{f}(0)=\alpha_1$, $\mathbf{f}'(0)=\alpha_2$, $\mathbf{f}''(0)=\alpha_3$

3.13 The following differential system describes the steady-state temperature distribution along an insulated bar of length b: $-\mathbf{f}''=\mathbf{u}$, $\mathbf{f}(0)=\alpha_1$, $\mathbf{f}'(b)+\mathbf{f}(b)=\alpha_2$. (The second boundary condition implies that heat is removed by convection at point b.) Show that the inverse equation for this system is

$$\mathbf{f}(t)=\left(1-\frac{t}{1+b}\right)\int_0^t s\mathbf{u}(s)\,ds+t\int_t^b\left(1-\frac{s}{1+b}\right)\mathbf{u}(s)\,ds$$

$$+\alpha_1\left(1-\frac{t}{1+b}\right)+\alpha_2\left(\frac{t}{1+b}\right)$$

3.14 For the differential system $t\mathbf{f}'(t)-\mathbf{f}(t)=\mathbf{u}(t)$, $\mathbf{f}'(t_1)=\alpha$, $t_1>0$,
(*a*) Find the complementary function by the power series method;
(*b*) Find the Green's function $k(t,s)$;
(*c*) Find the boundary kernel $\rho_j(t)$;
(*d*) State explicitly the inverse equation.

*3.15 Let μ_1 and μ_2 be the roots of the characteristic equation for the differential system $\mathbf{f}''+a_1\mathbf{f}'+a_2\mathbf{f}=\mathbf{u}$, $\mathbf{f}(0)=\mathbf{f}'(0)=0$.
(*a*) Use (3.56) and (3.57) to find the Green's function k for this system. If $\mu_2 \approx \mu_1$, computed values of $\mu_2-\mu_1$ and $\exp(\mu_2 t)-\exp(\mu_1 t)$ will be badly in error. What is the effect of near equality of the roots on the numerical computation of $k(t,s)$ and $\int k(t,s)\mathbf{u}(s)\,ds$?

(*b*) If $\mu_2 \approx \mu_1$, the fundamental set $\{\exp(\mu_1 t), \exp(\mu_2 t)\}$ is nearly dependent. A better fundamental set (not nearly dependent) in this circumstance is

$$\mathbf{v}_1(t) = \frac{\exp(\mu_1 t) + \exp(\mu_2 t)}{2} \qquad \mathbf{v}_2(t) = \frac{\exp(\mu_1 t) - \exp(\mu_2 t)}{\mu_1 - \mu_2}$$

Derive a power series expansion of $\mathbf{v}_2$ which can be used to compute values of $\mathbf{v}_2$ without numerical division by the inaccurate quantity $\mu_2 - \mu_1$. Show that as $\mu_2 \to \mu_1$, $\{\mathbf{v}_1(t), \mathbf{v}_2(t)\} \to \{\exp(\mu_1 t), t\exp(\mu_1 t)\}$.

(*c*) Equation (3.52) expresses the Green's function k in terms of the functions $\{\mathbf{v}_i\}$ of (*b*). Evaluate the Wronskian determinant w in this expression in terms of exponentials. Values of $k(t,s)$ and $\int k(t,s)\mathbf{u}(s)\,ds$ can be computed accurately by using this expression for $k(t,s)$ together with computed values of $\mathbf{v}_1, \mathbf{v}_2$, and w. Show that this expression for $k(t,s)$ is a rearrangement of the expression for $k(t,s)$ found in (*a*).

3.16 One method for obtaining the Green's function for a constant-coefficient differential system is to solve (3.32) by means of one-sided Laplace transforms. Use this technique to show that the inverse of the differential equation $\ddot{\mathbf{f}} + \omega^2\mathbf{f} = \mathbf{u}$, with constant ω and given values of $\mathbf{f}(0)$ and $\dot{\mathbf{f}}(0)$, is

$$\mathbf{f}(t) = \mathbf{f}(0)\cos\omega t + \frac{\dot{\mathbf{f}}(0)}{\omega}\sin\omega t + \frac{1}{\omega}\int_0^t \sin\omega(t-s)\mathbf{u}(s)\,ds$$

3.17 The approximation of derivatives by finite-differences leads to the approximate representation of differential equations by difference equations. For instance, the use of a second-central difference plus a forward difference converts the second-order differential system $\phi'' + \phi' = \mathbf{u}$, $\phi(0) = \alpha_1$, $\phi'(0) = \alpha_2$ to the approximately equivalent second-order difference system $2\phi(i+2) - 3\phi(i+1) + \phi(i) = \mathbf{u}(i+1)$, $\phi(0) = \alpha_1$, $\phi(1) = \alpha_2 + \alpha_1$. A general form for the nth-order constant-coefficient difference system with initial conditions is

$$\mathbf{f}(i+n) + a_1\mathbf{f}(i+n-1) + \cdots + a_n\mathbf{f}(i) = \mathbf{v}(i)$$

$$\mathbf{f}(0) = \gamma_1, \quad \mathbf{f}(1) = \gamma_2 \quad \ldots, \quad \mathbf{f}(n-1) = \gamma_n$$

for $i = 0, 1, 2, \ldots$.

By analogy to the inverse equation for the nth-order differential system, we assume the inverse of the nth-order difference system is of the form

$$\mathbf{f}(i)=\sum_{j=0}^{\infty} k(i,j)\mathbf{v}(j)+\sum_{m=1}^{n} \rho_m(i)\mathbf{f}(m-1)$$

for $i=0,1,2,\ldots$.

(*a*) Show that the discrete Green's function $k(i,j)$ is specified by the difference system

$$k(i+n,j)+a_1k(i+n-1,j)+\cdots+a_nk(i,j)=\delta_{ij}$$
$$k(0,j)=k(1,j)=\cdots=k(n-1,j)=0$$

for $i=0,1,2,\ldots$ and $j=0,1,2,\ldots$.

(*b*) Show that the discrete boundary kernel $\rho_m(i)$ is specified by the difference system

$$\rho_m(i+n)+a_1\rho_m(i+n-1)+\cdots+a_n\rho_m(i)=0$$
$$\rho_m(p)=\delta_{m,p+1}$$

for $i=0,1,2,\ldots,\ m=1,\ldots,n$, and $p=0,\ldots,n-1$.

(*c*) Find the inverse of the second-order difference system mentioned above by solving the difference systems corresponding to those in (*a*) and (*b*). Hint: solutions to homogeneous constant-coefficient difference equations consist in sums of geometric sequences of the form $\mathbf{f}(i)=r^i,\ i=0,\pm 1,\pm 2,\ldots$.

3.18 The following pair of coupled differential equations relates a pair of system outputs $\{\mathbf{f}_i(t)\}$ to a pair of inputs $\{\mathbf{u}_i(t)\}$:

$$\mathbf{f}_1''+3\mathbf{f}_1'+2\mathbf{f}_2=\mathbf{u}_1,$$
$$\mathbf{f}_2''+\mathbf{f}_1'+\mathbf{f}_2=\mathbf{u}_2,$$
$$\mathbf{f}_1(0),\mathbf{f}_1'(0),\mathbf{f}_2(0),\mathbf{f}_2'(0)\text{ specified.}$$

(*a*) Find a first-order state equation of the form (3.65) which is equivalent to the set of coupled equations. (Hint: use as state variables the output functions and their first derivatives.) Is the state equation unique?

(*b*) The solution to the state equation is determined by the state transition matrix (3.73). How could this matrix function be computed for the system in (*a*)?

3.19 *Properties of state transition matrices:* the concept of a state transition matrix extends to time-varying dynamic systems [3.14]. Suppose a dynamic system satisfies $\dot{\mathbf{x}}(t)=\mathbf{A}(t)\mathbf{x}(t)$, where $\mathbf{x}(t_0)$ is given and $\mathbf{A}(t)$ is an $n \times n$ matrix. We can express the solution in the form $\mathbf{x}(t)=\mathbf{\Phi}(t,t_0)\mathbf{x}(t_0)$. We refer to the $n \times n$ matrix $\mathbf{\Phi}(t_0,t)$ as the **state transition matrix**. The state transition matrix has the following properties:

(*a*) $\frac{d}{dt}\mathbf{\Phi}(t,t_0)=\mathbf{A}(t)\mathbf{\Phi}(t,t_0)$, $\mathbf{\Phi}(t_0,t_0)=I$;

(*b*) $\mathbf{\Phi}(t_0,t_1)\mathbf{\Phi}(t_1,t_2)=\mathbf{\Phi}(t_0,t_2)$ for all t_0, t_1, and t_2;

(*c*) $\mathbf{\Phi}(t_1,t_0)^{-1}=\mathbf{\Phi}(t_0,t_1)$;

(*d*) If $\mathbf{A}(t)\int_{t_0}^{t}\mathbf{A}(s)\,ds=\int_{t_0}^{t}\mathbf{A}(s)\,ds\,\mathbf{A}(t)$, then $\mathbf{\Phi}(t,t_0)=\exp\int_{t_0}^{t}\mathbf{A}(s)\,ds$ (see P&C 4.29);

(*e*) $\det\mathbf{\Phi}(t,t_0)=\exp\int_{t_0}^{t}\text{trace}[\mathbf{A}(s)]\,ds$, where $\text{trace}[\mathbf{A}(s)]$ is the sum of the diagonal elements of $\mathbf{A}(s)$.

3.20 A certain system can be represented by a differential equation of the form $\ddot{\mathbf{f}}+a_1\dot{\mathbf{f}}+a_2\mathbf{f}=\mathbf{u}$. The values of the coefficients a_1 and a_2 are unknown. However, we have observed the response of the undriven system ($\mathbf{u}(t)=0$ for $t>0$) with various initial conditions. In particular, for $\mathbf{f}(0)=1$ and $\dot{\mathbf{f}}(0)=0$, we find that $\mathbf{f}(t)=2e^{-t}-e^{-2t}$ and $\dot{\mathbf{f}}(t)=e^{-2t}-e^{-t}$ for $t \geqslant 0$. Also, for $\mathbf{f}(0)=0$ and $\dot{\mathbf{f}}(0)=1$, we find that $\mathbf{f}(t)=e^{-t}+e^{-2t}$ and $\dot{\mathbf{f}}(t)=2e^{-2t}-e^{-t}$ for $t \geqslant 0$.

(*a*) Determine the state equation in terms of a_1 and a_2.

(*b*) Use the transient measurements to determine the state transition matrix and the precise inverse of the state equation.

3.21 *Discrete-time state equations*: by using finite-difference approximations for derivatives, an arbitrary nth-order linear constant-coefficient differential equation with initial conditions can be approximated by an nth-order linear constant-coefficient difference equation of the form

$$\mathbf{f}((k+n)\tau)+a_1\mathbf{f}((k+n-1)\tau)+\cdots+a_n\mathbf{f}(k\tau)=\mathbf{u}(k\tau)$$

for $k=0,1,2,\ldots$, with $\mathbf{f}(0),\mathbf{f}(\tau),\ldots,\mathbf{f}((n-1)\tau)$ given. The quantity τ is the time increment used in the finite-difference approximation.

(*a*) Put this nth-order difference equation in state-space form; that is, develop an equivalent first-order vector difference equation.

(*b*) Determine the form of the inverse of the discrete-time state equation.

3.6 References

[3.1] Bergman, Stefan and M. Schiffer, *Kernel Functions and Elliptic Differential Equations in Mathematical Physics*, Academic Press, New York, 1953.

[3.2] DeRusso, Paul M., Rob J. Roy, and Charles M. Close, *State Variables for Engineers*, Wiley, New York, 1966.

[3.3] Forsythe, George E. and Wolfgang R. Wasow, *Finite-Difference Methods for Partial Differential Equations*, Wiley, New York, 1960.

*[3.4] Friedman, Bernard, *Principles and Techniques of Applied Mathematics*, Wiley, New York, 1966.

[3.5] Greenberg, Michael D., *Applications of Green's Functions in Science and Engineering*, Prentice-Hall, Englewood Cliffs, N.J., 1971.

[3.6] Ince, E. L., *Ordinary Differential Equations*, Dover, New York, 1956.

[3.7] Kaplan, Wilfred, *Advanced Calculus*, Addison-Wesley, Reading, Mass., 1959.

[3.8] Morse, Philip M. and Herman Feshbach, *Methods of Theoretical Physics*, Parts I and II, McGraw-Hill, New York, 1953.

[3.9] Ralston, Anthony, *A First Course in Numerical Analysis*, McGraw-Hill, New York, 1965.

[3.10] Schwartz, L., *Théorie des Distributions*, Vols. 1 and 2, Hermann & Cie, Paris, 1951, 1957.

*[3.11] Stakgold, Ivar, *Boundary Value Problems of Mathematical Physics*, Volume I, Macmillan, New York, 1968.

[3.12] Varga, Richard S., *Matrix Iterative Analysis*, Prentice-Hall, Englewood Cliffs, N.J., 1962.

[3.13] Wylie, C. R., Jr., *Advanced Engineering Mathematics*, 3rd ed., McGraw-Hill, New York, 1966.

[3.14] Zadeh, Lofti A. and Charles A. Desoer, *Linear System Theory*, McGraw-Hill, New York, 1963.

4

Spectral Analysis of Linear Systems

In this chapter the central theme is the decomposition of the abstract linear equation $\mathbf{Tx}=\mathbf{y}$ into sets of simple linear equations which can be solved independently. Our initial purpose for exploring this decomposition is to obtain conceptual simplification of the system model. It is easier to think about the behavior of one scalar variable at a time than to think about the behavior of a vector variable. Furthermore, the solutions to the decomposed pieces of the original equation usually have physical meanings which provide insight into the behavior of the system. (See for example, P & C 4.7 or the discussion of the analysis of three-phase power systems by the method of symmetrical components.)

There are also computational reasons for examining the decomposition process. Generally speaking, decomposition provides an alternative to inversion as a technique for solving or analyzing the equations which describe a system. In particular, decomposition provides a practical technique for computing solutions to linear differential equations with arbitrary inputs (Section 5.5). In some instances decomposition provides both solutions and insight at no additional computational expense as compared to inversion. (Again, see the discussion of symmetrical components mentioned above.)

The ability to combine the solutions to small subproblems into a solution for the full system equation depends on the principle of linearity. Consequently, we restrict ourselves to linear models in this chapter in order to be able to fully develop the decomposition principle. We find that we can decompose most linear systems into sets of simple scalar multiplications. We refer to such "completely decomposable" systems as "diagonalizable" systems. A few systems are not diagonalizable or are so nearly nondiagonalizable that we cannot accurately compute fully decomposed solutions. We still split them into as small pieces as possible. Nondiagonalizable finite-dimensional systems are discussed in Sections 4.4 and 4.5. In Section 4.6 we explore the concept of functions of matrices for both the diagonalizable and nondiagonalizable cases. We encountered several such matrix functions in Chapter 3; we find the need for others in later chapters. The discussion of diagonalization of infinite-dimensional systems and of functions of linear operators on infinite dimensional spaces is begun in Section 4.6, but is not completed until Section 5.5.

4.1 System Decomposition

In this section we explore the subdivision of the system equation $\mathbf{Tx}=\mathbf{y}$ into a set of "smaller" equations which can be solved independently. Our ability to subdivide a linear equation in this manner is based partly on the fact that the effect of a linear transformation $\mathbf{T}$ on a basis determines the effect of $\mathbf{T}$ on all vectors in the space. In finding the matrix of a transformation, for instance, we simplified the process of determining the matrix elements by examining the effect of the transformation on the basis vectors. Consequently, we begin our investigation of decomposition by subdividing the vector space on which the transformation $\mathbf{T}$ acts. We can think of the space as a sum of smaller subspaces.

Definition. Let $\mathcal{W}_1$ and $\mathcal{W}_2$ be subspaces of the vector space $\mathcal{V}$. We call $\mathcal{V}$ the **direct sum** $\mathcal{W}_1 \oplus \mathcal{W}_2$ of $\mathcal{W}_1$ and $\mathcal{W}_2$ if*

(*a*) $\mathcal{V} = \mathcal{W}_1 + \mathcal{W}_2$ ($\mathcal{W}_1$ and $\mathcal{W}_2$ **span** $\mathcal{V}$) and
(*b*) $\mathcal{W}_1 \cap \mathcal{W}_2 = \boldsymbol{\theta}$ ($\mathcal{W}_1$ and $\mathcal{W}_2$ **are linearly independent**)

Example 1. Direct Sum in Arrow Space. The two-dimensional arrow space is the direct sum of two different lines which intersect at the origin (Figure 4.1). If the two lines are identical, they are not independent and do not span the arrow space.

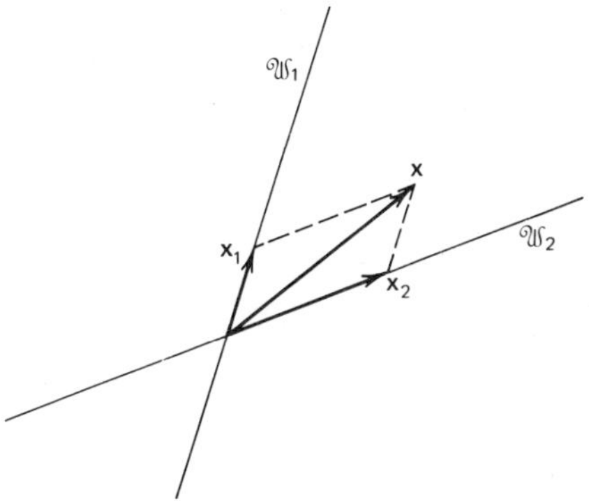

Figure 4.1. Direct sum in arrow space.

*See P&C 2.1 for definitions of the sum and intersection of subspaces.

This arrow space is also the sum of *three* lines which intersect at the origin. However, that sum is not direct; only two of the lines can be independent.

It is apparent from Figure 4.1 that for any finite-dimensional space every splitting of a basis into two parts determines a direct sum; that is, if $\{\mathbf{x}_1,\ldots,\mathbf{x}_n\}$ is a basis for $\mathcal{V}$, $\text{span}\{\mathbf{x}_1,\ldots,\mathbf{x}_n\}=\text{span}\{\mathbf{x}_1,\ldots,\mathbf{x}_k\}\oplus\text{span}\{\mathbf{x}_{k+1},\ldots,\mathbf{x}_n\}$. It is apparent that the two subspaces can also be subdivided. Although we have not yet defined a basis for an infinite-dimensional space, the concept of splitting a basis applies as well to direct sums in infinite-dimensional spaces (Sections 5.3–5.5).

Example 2. Direct Sum in a Function Space. Let $\mathcal{C}(-1,1)$ be the space of continuous functions defined on $[-1,1]$. Let $\mathcal{W}_1$ be the even functions in $\mathcal{C}(-1,1)$; $\mathbf{f}_e(-t)=\mathbf{f}_e(t)$. Let $\mathcal{W}_2$ be the odd functions in $\mathcal{C}(-1,1)$; $\mathbf{f}_o(-t)=-\mathbf{f}_o(t)$. Any function $\mathbf{f}$ in $\mathcal{C}(-1,1)$ decomposes into even and odd components:

$$\mathbf{f}(t)=\frac{\mathbf{f}(t)+\mathbf{f}(-t)}{2}+\frac{\mathbf{f}(t)-\mathbf{f}(-t)}{2}$$

Thus $\mathcal{W}_1$ and $\mathcal{W}_2$ span $\mathcal{C}(-1,1)$. The even and odd components of $\mathbf{f}$ are unique; for if $\mathbf{f}_e$ and $\mathbf{f}_o$ are even and odd functions, respectively, such that $\mathbf{f}=\mathbf{f}_e+\mathbf{f}_o$, then

$$\frac{\mathbf{f}(t)+\mathbf{f}(-t)}{2}=\frac{[\mathbf{f}_e(t)+\mathbf{f}_o(t)]+[\mathbf{f}_e(-t)+\mathbf{f}_o(-t)]}{2}=\mathbf{f}_e(t)$$

$$\frac{\mathbf{f}(t)-\mathbf{f}(-t)}{2}=\frac{[\mathbf{f}_e(t)+\mathbf{f}_o(t)]-[\mathbf{f}_e(-t)+\mathbf{f}_o(-t)]}{2}=\mathbf{f}_o(t)$$

Only the zero function is both even and odd; therefore, $\mathcal{W}_1\cap\mathcal{W}_2=\boldsymbol{\theta}$, and $\mathcal{C}(-1,1)=\mathcal{W}_1\oplus\mathcal{W}_2$.

Example 2 demonstrates an important property of the direct sum. Using bases for $\mathcal{W}_1$ and $\mathcal{W}_2$, it is easily shown that $\mathcal{V}=\mathcal{W}_1\oplus\mathcal{W}_2$ if and only if each $\mathbf{x}$ in $\mathcal{V}$ decomposes *uniquely* into a sum, $\mathbf{x}=\mathbf{x}_1+\mathbf{x}_2$, with $\mathbf{x}_1$ in $\mathcal{W}_1$ and $\mathbf{x}_2$ in $\mathcal{W}_2$.

It is a small step to extend the direct sum concept to several subspaces. We merely redefine independence of subspaces: $\mathcal{W}_1,\ldots,\mathcal{W}_p$ are linearly independent if each subspace is disjoint from the sum of the rest,

$$\mathcal{W}_i\cap\Big(\sum_{j\neq i}\mathcal{W}_j\Big)=\boldsymbol{\theta} \tag{4.1}$$

With the modification (4.1) we say $\mathcal{V}$ is the direct sum of $\{\mathcal{W}_i\}$ if the subspaces $\{\mathcal{W}_i\}$ are linearly independent and span $\mathcal{V}$. We denote the direct sum by

$$\mathcal{V}=\mathcal{W}_1\oplus\mathcal{W}_2\oplus\cdots\oplus\mathcal{W}_p \tag{4.2}$$

The previous comments concerning splitting of bases and unique decomposition of vectors also extend to the direct sum of several subspaces.

Exercise 1. Demonstrate in the two-dimensional arrow space that pairwise disjointness is not sufficient to guarantee independence of $\mathcal{W}_1, \ldots, \mathcal{W}_p$.

Example 3. Direct Sum of Three Subspaces. Let $\mathbf{f}_1(t) = 1+t$, $\mathbf{f}_2(t) = t + t^2$, and $\mathbf{f}_3(t) = 1 + t^2$ be a basis for $\mathcal{P}^3$. Define $\mathcal{W}_i = \text{span}\{\mathbf{f}_i\}$, $i = 1,2,3$. Then $\mathcal{P}^3 = \mathcal{W}_1 \oplus \mathcal{W}_2 \oplus \mathcal{W}_3$. Let $\mathbf{f}(t) \stackrel{\Delta}{=} \eta_1 + \eta_2 t + \eta_3 t^2$ be a specific vector in $\mathcal{P}^3$. By the process of determining coordinates of $\mathbf{f}$ relative to the basis $\{\mathbf{f}_1, \mathbf{f}_2, \mathbf{f}_3\}$ for $\mathcal{P}^3$, we decompose $\mathbf{f}$ uniquely into

$$\mathbf{f} = \left(\frac{\eta_1 + \eta_2 - \eta_3}{2}\right)\mathbf{f}_1 + \left(\frac{-\eta_1 + \eta_2 + \eta_3}{2}\right)\mathbf{f}_2 + \left(\frac{\eta_1 - \eta_2 + \eta_3}{2}\right)\mathbf{f}_3,$$

a sum of vectors from $\mathcal{W}_1$, $\mathcal{W}_2$, and $\mathcal{W}_3$, respectively.

Projection Operators

We can express the direct-sum decomposition of a space in terms of linear operators on the space. Suppose $\mathcal{V} = \mathcal{W}_1 \oplus \mathcal{W}_2$; any vector $\mathbf{x}$ in $\mathcal{V}$ can be written uniquely as $\mathbf{x} = \mathbf{x}_1 + \mathbf{x}_2$ with $\mathbf{x}_i$ in $\mathcal{W}_i$. We define the **projector** (or **projection operator**) $\mathbf{P}_1$ **on** $\mathcal{W}_1$ **along** $\mathcal{W}_2$ by $\mathbf{P}_1\mathbf{x} \stackrel{\Delta}{=} \mathbf{x}_1$ (see Figure 4.1). We call the vector $\mathbf{x}_1$ the **projection of x on** $\mathcal{W}_1$ **along** $\mathcal{W}_2$. Similarly $\mathbf{P}_2\mathbf{x} \stackrel{\Delta}{=} \mathbf{x}_2$ defines the projector on $\mathcal{W}_2$ along $\mathcal{W}_1$.

Example 4. Projectors on $\mathcal{P}^3$. Let $\mathbf{f}_1$, $\mathbf{f}_2$, and $\mathbf{f}_3$ be the functions defined in Example 3. Redefine $\mathcal{W}_1 \stackrel{\Delta}{=} \text{span}\{\mathbf{f}_1\}$ and $\mathcal{W}_2 \stackrel{\Delta}{=} \text{span}\{\mathbf{f}_2, \mathbf{f}_3\}$. Then $\mathcal{P}^3 = \mathcal{W}_1 \oplus \mathcal{W}_2$. In Example 3, the general vector $\mathbf{f}(t) = \eta_1 + \eta_2 t + \eta_3 t^2$ in $\mathcal{P}^3$ is decomposed into a linear combination of $\mathbf{f}_1$, $\mathbf{f}_2$, and $\mathbf{f}_3$. From that decomposition we see that the projections of $\mathbf{f}$ on $\mathcal{W}_1$ and $\mathcal{W}_2$, respectively, are

$$\mathbf{P}_1\mathbf{f} = \left(\frac{\eta_1 + \eta_2 - \eta_3}{2}\right)\mathbf{f}_1$$

$$\mathbf{P}_2\mathbf{f} = \left(\frac{-\eta_1 + \eta_2 + \eta_3}{2}\right)\mathbf{f}_2 + \left(\frac{\eta_1 - \eta_2 + \eta_3}{2}\right)\mathbf{f}_3$$

The bases for $\mathcal{W}_1$ and $\mathcal{W}_2$ combine to provide a basis which is particularly appropriate for matrix representation of the projectors. Using (2.48), the matrix of the projector $\mathbf{P}_1$ relative to the basis $\mathcal{F} \stackrel{\Delta}{=} \{\mathbf{f}_1, \mathbf{f}_2, \mathbf{f}_3\}$ is

$$\begin{aligned}
[\mathbf{P}_1]_{\mathcal{F}\mathcal{F}} &= ([\mathbf{P}_1\mathbf{f}_1]_{\mathcal{F}} \;\vdots\; [\mathbf{P}_1\mathbf{f}_2]_{\mathcal{F}} \;\vdots\; [\mathbf{P}_1\mathbf{f}_3]_{\mathcal{F}}) \\
&= ([\mathbf{f}_1]_{\mathcal{F}} \;\vdots\; [\boldsymbol{\theta}]_{\mathcal{F}} \;\vdots\; [\boldsymbol{\theta}]_{\mathcal{F}}) \\
&= \left(\begin{array}{c:cc} 1 & 0 & 0 \\ \hdashline 0 & 0 & 0 \\ 0 & 0 & 0 \end{array}\right)
\end{aligned}$$

Similarly, the matrix of $\mathbf{P}_2$ with respect to $\mathcal{F}$ is

$$[\mathbf{P}_2]_{\mathcal{F}\mathcal{F}} = \left(\begin{array}{c:cc} 0 & 0 & 0 \\ \hdashline 0 & 1 & 0 \\ 0 & 0 & 1 \end{array}\right)$$

Example 4 emphasizes the fact that a projector acts like the identity operator on its "own" subspace, the one *onto* which it projects, but like the zero operator on the subspace *along* which it projects. The following properties of projectors can be derived from the definition and verified by the matrices of Example 4. Assume $\mathcal{V} = \mathcal{W}_1 \oplus \mathcal{W}_2$. Let $\mathbf{P}_i$ be the projector on $\mathcal{W}_i$ along $\mathcal{W}_j\, (j \neq i)$, and $\mathbf{x}_i = \mathbf{P}_i \mathbf{x}$. Then

$$\begin{aligned} &(a) \quad \mathbf{P}_i \text{ is linear} \\ &(b) \quad \mathbf{P}_i^2 = \mathbf{P}_i \quad (\text{i.e., } \mathbf{P}_i \mathbf{x}_i = \mathbf{x}_i) \\ &(c) \quad \mathbf{P}_i \mathbf{P}_j = \boldsymbol{\Theta} \ (\text{i.e., } \mathbf{P}_i \mathbf{x}_j = \boldsymbol{\theta} \quad \text{for } j \neq i) \\ &(d) \quad \text{range}\,(\mathbf{P}_i) = \mathcal{W}_i \\ &(e) \quad \sum_i \mathbf{P}_i = \mathbf{I} \left(\text{i.e., } \sum_i \mathbf{P}_i \mathbf{x} = \mathbf{x} \right) \end{aligned} \tag{4.3}$$

If $\mathcal{V} = \mathcal{W}_1 \oplus \cdots \oplus \mathcal{W}_k$, we can define the projector $\mathbf{P}_i$ on $\mathcal{W}_i$ along $\Sigma_{j \neq i} \mathcal{W}_j$, for $i = 1, \ldots, k$. The properties (4.3) apply to this set of projectors as well.

Reduced Operators

The projectors in Example 4 act like scalar multiplication on certain vectors in $\mathcal{V}$; $\mathbf{P}_i$ acts like multiplication by 1 on all vectors in the subspace $\mathcal{W}_i$, and like multiplication by zero on $\mathcal{W}_j$, $j \neq i$. Other operators also act in a simple manner on certain subspaces. Define the nonlinear operator $\mathbf{G}$: $\mathcal{R}^2 \to \mathcal{R}^2$ by

$$\mathbf{G}(\xi_1, \xi_2) \triangleq \left((\xi_1 - \xi_2)^2 + 2\xi_2, 2\xi_2 \right)$$

On the subspace $\mathcal{W}_1 \triangleq \text{span}\{(1,0)\}$, $\mathbf{G}$ acts like the simple "squaring" operation, $\mathbf{G}(a,0) = (a^2, 0)$. On the subspace $\mathcal{W}_2 \triangleq \text{span}\{(1,1)\}$, $\mathbf{G}$ acts like the "doubling" operation $\mathbf{G}(b,b) = (2b, 2b)$. In point of fact, as far as vectors in $\mathcal{W}_1$ and $\mathcal{W}_2$ are concerned we can replace $\mathbf{G}$ by the "simpler" operators $\mathbf{G}_1$: $\mathcal{W}_1 \to \mathcal{W}_1$ defined by $\mathbf{G}_1(\xi, 0) \triangleq (\xi^2, 0)$ and $\mathbf{G}_2$: $\mathcal{W}_2 \to \mathcal{W}_2$ defined by $\mathbf{G}_2(\xi, \xi) \triangleq 2(\xi, \xi)$. We are able to reduce $\mathbf{G}$ to these simpler operators because the action of $\mathbf{G}$ on $\mathcal{W}_1$ produces only vectors in $\mathcal{W}_1$ and the action of $\mathbf{G}$ on $\mathcal{W}_2$ produces only vectors in $\mathcal{W}_2$.

Definition. Let $\mathbf{G}$ be an operator (perhaps nonlinear) on $\mathcal{V}$. The subspace $\mathcal{W}$ (of $\mathcal{V}$) is **invariant under G** if for each $\mathbf{x}$ in $\mathcal{W}$, $\mathbf{Gx}$ is also in $\mathcal{W}$; that is, if $\mathbf{G}(\mathcal{W})$ is contained in $\mathcal{W}$.

Example 5. Invariance of the Nullspace and Range. Let $\mathbf{G}$: $\mathcal{V}\to\mathcal{V}$. Then range $(\mathbf{G})$ is invariant under $\mathbf{G}$, for $\mathbf{G}$ takes all vectors in $\mathcal{V}$, including those in range($\mathbf{G}$), into range($\mathbf{G}$). By definition, $\mathbf{G}$ takes nullspace($\mathbf{G}$) into $\boldsymbol{\theta}$. If $\mathbf{G}(\boldsymbol{\theta})=\boldsymbol{\theta}$, then $\boldsymbol{\theta}$ is in nullspace($\mathbf{G}$). In this case, nullspace($\mathbf{G}$) is also invariant under $\mathbf{G}$. These subspaces are pictured abstractly in Figure 2.6.

If $\mathbf{G}$: $\mathcal{V}\to\mathcal{V}$, and $\mathcal{W}$ is a subspace of $\mathcal{V}$ which is invariant under $\mathbf{G}$, then we can define a **reduced operator** $\mathbf{G}_{\mathcal{W}}: \mathcal{W}\to\mathcal{W}$ by $\mathbf{G}_{\mathcal{W}}\mathbf{x} \triangleq \mathbf{Gx}$ for all $\mathbf{x}$ in $\mathcal{W}$. The operators $\mathbf{G}_1$ and $\mathbf{G}_2$ discussed earlier are examples of reduced operators. The following illustration shows that the reduced operator $\mathbf{G}_{\mathcal{W}}$ is truly different from $\mathbf{G}$.

Example 6. Reduced Linear Operators. We define $\mathbf{T}$: $\mathcal{R}^2\to\mathcal{R}^2$ by

$$\mathbf{T}(\xi_1,\xi_2) \triangleq (2\xi_1+3\xi_2, 4\xi_2) \tag{4.4}$$

The matrix of $\mathbf{T}$ relative to the standard basis $\mathcal{E}$ is

$$\begin{aligned}[\mathbf{T}]_{\mathcal{E}\mathcal{E}} &= ([\mathbf{T}\boldsymbol{\varepsilon}_1]_{\mathcal{E}} \vdots [\mathbf{T}\boldsymbol{\varepsilon}_2]_{\mathcal{E}})\\ &= \begin{pmatrix} 2 & 3\\ 0 & 4\end{pmatrix}\end{aligned}$$

The subspaces $\mathcal{W}_1 \triangleq \text{span}\{(1,0)\}$ and $\mathcal{W}_2 \triangleq \text{span}\{(3,2)\}$ are invariant under $\mathbf{T}$. Therefore, we can define the reduced operators $\mathbf{T}_1$: $\mathcal{W}_1\to\mathcal{W}_1$ by $\mathbf{T}_1(\xi,0) \triangleq \mathbf{T}(\xi,0) = 2(\xi,0)$ and $\mathbf{T}_2$: $\mathcal{W}_2\to\mathcal{W}_2$ by $\mathbf{T}_2(3\xi,2\xi) \triangleq \mathbf{T}(3\xi,2\xi)=4(3\xi,2\xi)$. Using $\mathcal{X} \triangleq \{(1,0)\}$ as a basis for $\mathcal{W}_1$ and $\mathcal{Y} \triangleq \{(3,2)\}$ as a basis for $\mathcal{W}_2$ we find

$$[\mathbf{T}_1]_{\mathcal{X}\mathcal{X}} = ([\mathbf{T}_1(1,0)]_{\mathcal{X}}) = (2)$$

$$[\mathbf{T}_2]_{\mathcal{Y}\mathcal{Y}} = ([\mathbf{T}_2(3,2)]_{\mathcal{Y}}) = (4)$$

The reduced operators $\mathbf{T}_1$ and $\mathbf{T}_2$ are scalar operators, represented by 1×1 matrices. They are very different from $\mathbf{T}$, which is represented by a 2×2 matrix. Clearly the domain and range of definition of a transformation are necessary parts of its definition.

Solution of Equations by Decomposition

The combination of three basic concepts—direct sum, invariance, and linearity—leads to the spectral decomposition, a decomposition of an operator or an equation into a set of scalar multipliers or scalar single-variable equations. The decomposition provides considerable insight into the nature of linear models. It also provides a technique for solving equations which is an alternative to inverting the equations.

If **T** is a linear operator on $\mathcal{V}$, if $\mathcal{V} = \mathcal{W}_1 \oplus \cdots \oplus \mathcal{W}_p$, and if each $\mathcal{W}_i$ is invariant under **T**, then the set $\{\mathcal{W}_i\}$ of subspaces **decomposes T** into a set of reduced linear operators $\mathbf{T}_i$: $\mathcal{W}_i \to \mathcal{W}_i$ defined by $\mathbf{T}_i\mathbf{x} \triangleq \mathbf{Tx}$ for all **x** in $\mathcal{W}_i$. Analysis of a system represented by **T** reduces to analysis of a set of *independent subsystems* represented by $\{\mathbf{T}_i\}$; that is, we can solve the equation $\mathbf{Tx}=\mathbf{y}$ by the following process.

The Spectral Decomposition Process (4.5)

1. Using the direct sum, decompose **y** into the unique combination

$$\mathbf{y}=\mathbf{y}_1+\cdots+\mathbf{y}_p \quad \text{with } \mathbf{y}_i \text{ in } \mathcal{W}_i$$

2. Using the invariance of $\mathcal{W}_i$ under **T**, solve the subsystems

$$\mathbf{Tx}_i=\mathbf{y}_i \qquad i=1,2,\ldots,p$$

(in effect solving the reduced equations $\mathbf{T}_i\mathbf{x}_i=\mathbf{y}_i$).

3. Using the linearity of **T**, get the solution **x** by adding

$$\mathbf{x}=\mathbf{x}_1+\cdots+\mathbf{x}_p$$

If the reduced operators $\mathbf{T}_i$ are simple scalar multipliers like those of Example 6, then solution of the subsystem equations is trivial; that is, if $\mathbf{Tx}_i=\lambda_i\mathbf{x}_i$ for each $\mathbf{x}_i$ in $\mathcal{W}_i$, then $\lambda_i\mathbf{x}_i=\mathbf{y}_i$ and parts (2) and (3) of (4.5) can be expressed as

$$\mathbf{x}=\left(\frac{1}{\lambda_1}\right)\mathbf{y}_1+\cdots+\left(\frac{1}{\lambda_p}\right)\mathbf{y}_p \tag{4.6}$$

If we know the invariant subspaces $\mathcal{W}_i$ and the scalars λ_i, the primary effort required to carry out this procedure is that in decomposing **y**.

Example 7. Solution of an Equation by Decomposition. Let $\mathbf{T}$: $\mathcal{R}^2 \to \mathcal{R}^2$ be as in (4.4):

$$\mathbf{T}(\xi_1,\xi_2) \triangleq (2\xi_1+3\xi_2, 4\xi_2)$$

From Example 6, we know the subspaces $\mathcal{W}_1 \triangleq \text{span}\{(1,0)\}$ and $\mathcal{W}_2 \triangleq \text{span}\{(3,2)\}$ are invariant under **T**; furthermore, **T** acts like $\mathbf{T}_1\mathbf{x} \triangleq 2\mathbf{x}$ for **x** in $\mathcal{W}_1$, and like $\mathbf{T}_2\mathbf{x} \triangleq 4\mathbf{x}$ for **x** in $\mathcal{W}_2$. Also $\mathcal{V} = \mathcal{W}_1 \oplus \mathcal{W}_2$. Therefore, we can solve the equation

$$\mathbf{Tx} = \mathbf{y} \triangleq (\eta_1, \eta_2)$$

by the process (4.5). We decompose $\mathbf{y}$ by solving $(\eta_1, \eta_2) = c_1(1,0) + c_2(3,2)$ to find

$$(\eta_1, \eta_2) = \left(\eta_1 - \frac{3\eta_2}{2}\right)(1,0) + \left(\frac{\eta_2}{2}\right)(3,2)$$

$$\triangleq \mathbf{y}_1 + \mathbf{y}_2$$

By (4.6)

$$(\xi_1, \xi_2) = \left(\frac{1}{2}\right)\left(\eta_1 - \frac{3\eta_2}{2}\right)(1,0) + \left(\frac{1}{4}\right)\left(\frac{\eta_2}{2}\right)(3,2)$$

$$= \left(\frac{\eta_1}{2} - \frac{3\eta_2}{8}, \frac{\eta_2}{4}\right)$$

The procedure (4.5) is essentially the one we use to determine the steady-state solution of a constant-coefficient differential equation by Fourier series. It is well known that a continuous function $\mathbf{f}$ can be expanded uniquely as a Fourier series of complex exponentials of the form $e^{i2\pi kt/b}$, where $i = \sqrt{-1}$ and b is the length of the interval over which $\mathbf{f}$ is defined. Each such exponential spans a subspace $\mathcal{W}_k$. The Fourier series expansion is possible because the space of continuous functions is in some sense the direct sum of $\{\mathcal{W}_k\}$. But each subspace $\mathcal{W}_k$ is invariant under any linear constant-coefficient differential operator; for instance, $(\mathbf{D}^2 + \mathbf{D})e^{\mu t} = (\mu^2 + \mu)e^{\mu t}$, a scalar multiple of $e^{\mu t}$. Thus the solution to certain differential equations can be found by an extension of (4.6). See P&C 5.35.

The Spectrum

The real goal of most systems analyses is insight into the system structure. Most linear models have a structure which permits decomposition into a set of scalar operations. It is not yet clear what effect the subdivision of a linear operator $\mathbf{T}$ has on the overall computation. In fact, since one result of the decomposition is valuable insight into the structure of the system represented by $\mathbf{T}$, perhaps we should expect an increase in total computation. Although this expectation is justified, we find that under certain circumstances the decomposition information is known a priori. Then decomposition can also lead to reduced computation (Section 5.2).

Definition. An **eigenvalue** (or **characteristic value**) of a linear operator $\mathbf{T}$ on a vector space $\mathcal{V}$ is a scalar λ such that $\mathbf{Tx} = \lambda\mathbf{x}$ for some nonzero vector $\mathbf{x}$ in $\mathcal{V}$. Any nonzero $\mathbf{x}$ for which $\mathbf{Tx}=\lambda\mathbf{x}$ is called an **eigenvector** of $\mathbf{T}$ associated with the eigenvalue λ.

The eigenvector $\mathbf{x}$ spans a subspace of $\mathcal{V}$. Each member of this subspace (or **eigenspace**) is also an eigenvector for the same eigenvalue. In fact, because $\mathbf{T}$ is linear, any one-dimensional subspace which is invariant under $\mathbf{T}$ must be an eigenspace of $\mathbf{T}$. The identity operator $\mathbf{I}$ clearly has only one eigenvalue; the whole space $\mathcal{V}$ is the eigenspace for $\lambda=1$. Similarly, for the zero operator $\mathbf{\Theta}$, $\mathcal{V}$ is the eigenspace for $\lambda=0$. If $\mathcal{V}=\mathcal{W}_1\oplus\mathcal{W}_2$, then for the projector $\mathbf{P}_i$ of (4.3), $\mathcal{W}_i$ is the eigenspace for $\lambda=1$ and $\mathcal{W}_j$ is the eigenspace for $\lambda=0$.

The eigenvectors of an operator which acts on a function space are often called **eigenfunctions**. We will refer to the eigenvalues and eigenvectors (or eigenfunctions) of $\mathbf{T}$ as the **eigendata for T**. The eigendata usually have some significant physical interpretation in terms of the system represented by $\mathbf{T}$.

Example 8. Eigendata for a Transformation in $\mathcal{R}^2$. The operator $\mathbf{T}$: $\mathcal{R}^2\rightarrow\mathcal{R}^2$ of (4.4) is

$$\mathbf{T}(\xi_1,\xi_2)\triangleq(2\xi_1+3\xi_2,4\xi_2)$$

It has two eigenvalues: $\lambda_1=2$ and $\lambda_2=4$. The corresponding eigenspaces are span$\{(1,0)\}$ for λ_1 and span$\{(3,2)\}$ for λ_2.

Example 9. Eigendata for Differential Operators. The exponential function $e^{\mu t}$ and its multiples form an eigenspace for any linear constant-coefficient differential operator *without boundary conditions*. For instance, since

$$\frac{d^n}{dt^n}e^{\mu t}+a_1\frac{d^{n-1}}{dt^{n-1}}e^{\mu t}+\cdots+a_n e^{\mu t}=(\mu^n+a_1\mu^{n-1}+\cdots+a_n)e^{\mu t}$$

for any complex scalar μ, the differential operator $\mathbf{D}^n+a_1\mathbf{D}^{n-1}+\cdots+a_n\mathbf{I}$ has the eigenfunction $e^{\mu t}$ corresponding to the eigenvalue $\lambda=\mu^n+a_1\mu^{n-1}+\cdots+a_n$. A differential operator without boundary conditions possesses a continuum of eigenvalues.

Example 10. An Operator Without Eigenvalues. A linear differential operator with homogeneous boundary conditions need not have any eigenvalues. For example, the only vector that satisfies

$$\frac{d\mathbf{f}(t)}{dt}=\lambda\mathbf{f}(t), \qquad \mathbf{f}(0)=0$$

is the zero function, regardless of the value we try for the eigenvalue λ. Thus the operator $\mathbf{D}$ acting on the space of differentiable functions $\mathbf{f}$ which satisfy $\mathbf{f}(0)=0$ has no eigenvalues. Furthermore, any nth order linear differential operator with n independent one-point homogeneous boundary conditions is without eigenvalues. [See the discussion following (3.28).]

The problem of finding eigenvalues for a linear operator $\mathbf{T}: \mathcal{V} \rightarrow \mathcal{V}$ is basically the problem of determining values of λ for which the equation

$$(\mathbf{T}-\lambda\mathbf{I})\mathbf{x}=\boldsymbol{\theta} \tag{4.7}$$

has nonzero solutions $\mathbf{x}$; that is, we seek the values of λ for which the operator $\mathbf{T}-\lambda\mathbf{I}$ is singular. Once we have a specific eigenvalue, say, λ_1, obtaining the corresponding eigenvectors involves the determination of nullspace($\mathbf{T}-\lambda_1\mathbf{I}$) —the solution of (4.7) with $\lambda=\lambda_1$. The determination of eigendata and the use of eigendata in practical analysis are explored for finite-dimensional systems in Section 4.2 and for infinite-dimensional systems in Section 4.3.

4.2 Spectral Analysis in Finite-Dimensional Spaces

In this section we convert (4.7) to a matrix equation for the case where $\mathcal{V}$ is finite-dimensional. We also examine the spectral (eigendata) properties of matrix equations. Practical computation of eigendata for finite-dimensional problems, a more difficult task than appears on the surface, is discussed at the end of the section.

In Section 2.5 we found we could convert any equation involving a linear operator on a finite-dimensional space into an equivalent matrix equation. If $\mathbf{T}: \mathcal{V} \rightarrow \mathcal{V}$, we simply pick a basis $\mathfrak{X}$ for $\mathcal{V}$. The basis converts the equation $\mathbf{Tx}=\mathbf{y}$ into the equation $[\mathbf{T}]_{\mathfrak{X}\mathfrak{X}}[\mathbf{x}]_{\mathfrak{X}}=[\mathbf{y}]_{\mathfrak{X}}$. We generally define $\mathbf{A} \triangleq [\mathbf{T}]_{\mathfrak{X}\mathfrak{X}}$, and use the simpler matrix notation $\mathbf{A}[\mathbf{x}]_{\mathfrak{X}}=[\mathbf{y}]_{\mathfrak{X}}$. The eigenvalues and eigenvectors for $\mathbf{T}$ are then specified by the matrix equivalent of (4.7):

$$(\mathbf{A}-\lambda\mathbf{I})[\mathbf{x}]_{\mathfrak{X}}=[\boldsymbol{\theta}]_{\mathfrak{X}} \tag{4.8}$$

The values of λ for which (4.8) has nonzero solutions constitute the eigenvalues of $\mathbf{T}$. We also refer to them as the **eigenvalues of the matrix A**.

From Section 1.5 we know that the square-matrix equation (4.8) has nonzero solutions if and only if

$$\det(\mathbf{A}-\lambda\mathbf{I})=0 \tag{4.9}$$

Equation (4.9) is known as the **characteristic equation of the matrix A** (or of the operator $\mathbf{T}$ which $\mathbf{A}$ represents). If $\mathbf{A}$ is an $n \times n$ matrix, then

$$c(\lambda) \triangleq \det(\lambda\mathbf{I}-\mathbf{A})=(-1)^n\det(\mathbf{A}-\lambda\mathbf{I}) \tag{4.10}$$

is an nth order polynomial in λ called the **characteristic polynomial of A** (or of **T**). An nth order polynomial has precisely n (possibly complex) roots. (This fact follows from the fundamental theorem of algebra.) The set $\{\lambda_1,\ldots,\lambda_n\}$ of roots of $c(\lambda)$ constitutes the complete set of eigenvalues of **A** (or **T**); the set is called the **spectrum of A** (or **T**). We often refer to an analysis which involves eigenvalues as a *spectral analysis*. Since $\lambda=\lambda_i$ makes $\mathbf{A}-\lambda\mathbf{I}$ singular, there must be at least one nonzero eigenvector for each different eigenvalue. A solution $[\mathbf{x}]_{\mathfrak{X}}$ of (4.8) for $\lambda=\lambda_i$ is an eigenvector of **A** for λ_i. The corresponding vector **x** is an eigenvector of **T** for λ_i.

***Example 1. Finding Eigendata from* [T].** Let $\mathbf{T}: \mathfrak{R}^2\rightarrow\mathfrak{R}^2$ be defined as in (4.4) by

$$\mathbf{T}(\xi_1,\xi_2)\triangleq(2\xi_1+3\xi_2,4\xi_2)$$

Using the standard basis $\mathcal{E}$ for $\mathfrak{R}^2$ as in Example 6, (4.8) becomes

$$\left(\begin{pmatrix}2 & 3\\ 0 & 4\end{pmatrix}-\lambda\begin{pmatrix}1 & 0\\ 0 & 1\end{pmatrix}\right)[\mathbf{x}]_{\mathcal{E}}=[\boldsymbol{\theta}]_{\mathcal{E}}$$

or

$$\begin{pmatrix}2-\lambda & 3\\ 0 & 4-\lambda\end{pmatrix}[\mathbf{x}]_{\mathcal{E}}=\begin{pmatrix}0\\ 0\end{pmatrix}$$

The characteristic equation is

$$\begin{vmatrix}2-\lambda & 3\\ 0 & 4-\lambda\end{vmatrix}=(2-\lambda)(4-\lambda)=0$$

The eigenvalues of **A** (and **T**) are $\lambda_1=2$ and $\lambda_2=4$. We find the eigenvectors of **A** for λ_i by solving (4.8) with $\lambda=\lambda_i$:

$$(\mathbf{A}-2\mathbf{I})\,[\mathbf{x}_1]_{\mathcal{E}}=\begin{pmatrix}0 & 3\\ 0 & 2\end{pmatrix}\begin{pmatrix}c_1\\ c_2\end{pmatrix}=\begin{pmatrix}0\\ 0\end{pmatrix}\quad\Rightarrow\quad[\mathbf{x}_1]_{\mathcal{E}}=\begin{pmatrix}c_1\\ 0\end{pmatrix}$$

$$(\mathbf{A}-4\mathbf{I})\,[\mathbf{x}_2]_{\mathcal{E}}=\begin{pmatrix}-2 & 3\\ 0 & 0\end{pmatrix}\begin{pmatrix}d_1\\ d_2\end{pmatrix}=\begin{pmatrix}0\\ 0\end{pmatrix}\quad\Rightarrow\quad[\mathbf{x}_2]_{\mathcal{E}}=\begin{pmatrix}3d_1\\ 2d_1\end{pmatrix}$$

The scalars c_1 and d_1 are arbitrary; there is a one-dimensional eigenspace for each eigenvalue. The eigenvectors of **T** for λ_i are found from the relationship between a vector and its coordinates relative to the basis $\mathcal{E}$:

$$[\mathbf{x}]_{\mathcal{E}}=\begin{pmatrix}c_1\\ c_2\end{pmatrix}\quad\Leftrightarrow\quad\mathbf{x}=c_1(1,0)+c_2(0,1)$$

Therefore, the eigenvectors of **T** corresponding to λ_1 and λ_2 are

$$\mathbf{x}_1=c_1(1,0)+0(0,1)=c_1(1,0)$$

$$\mathbf{x}_2=3d_1(1,0)+2d_1(0,1)=d_1(3,2)$$

In our previous discussions of vector spaces we have been able to allow freedom in the type of scalars which we use. We have thought primarily in terms of real numbers. However, in the discussion of eigenvalues this freedom in choice of scalars can cause difficulty. A real polynomial need not have real roots. Thus an operator on a space with real scalars may not have real eigenvalues; on the other hand, a complex eigenvalue has no meaning for such a space. The usual engineering practice is to accept the complex scalars whenever they appear, and assign them an appropriate meaning if necessary. We follow this approach, and assume, whenever we speak of eigenvalues, that the characteristic equation has a full set of roots.

Exercise 1. Define the operator **T** on $\mathcal{R}^2$ by

$$\mathbf{T}(\xi_1,\xi_2)=(\xi_1\cos\phi-\xi_2\sin\phi,\ \xi_2\cos\phi+\xi_1\sin\phi) \tag{4.11}$$

This operator describes "rotation through the angle ϕ" in $\mathcal{R}^2$. Show that the eigendata for **T** are

$$\lambda_1=\cos\phi+i\sin\phi=e^{i\phi},\qquad \mathbf{x}_1=(1,-i)$$

$$\lambda_2=\cos\phi-i\sin\phi=e^{-i\phi},\qquad \mathbf{x}_2=(1,i)$$

where $i=\sqrt{-1}$. The vector $(1,\pm i)$ is not a real 2-tuple; it is not in $\mathcal{R}^2$.

We could have used any basis in Example 1. The eigenvalues and eigenvectors of **T** are properties of **T**; they do not depend upon the basis. Suppose we use the invertible change of coordinate matrix $\mathbf{S}^{-1}$ to convert (4.8) from the $\mathcal{Z}$ coordinate system to a new coordinate system $\mathcal{X}$ as in (2.54):

$$[\mathbf{x}]_{\mathcal{X}}=\mathbf{S}^{-1}[\mathbf{x}]_{\mathcal{Z}}$$

The effect of the change of coordinates on the matrix of **T** is represented by the similarity transformation (2.62): $[\mathbf{T}]_{\mathcal{Z}\mathcal{Z}}=\mathbf{S}[\mathbf{T}]_{\mathcal{X}\mathcal{X}}\mathbf{S}^{-1}$. Recalling that $\mathbf{A}=[\mathbf{T}]_{\mathcal{Z}\mathcal{Z}}$, we find that (4.8) can be expressed as $([\mathbf{T}]_{\mathcal{Z}\mathcal{Z}}-\lambda\mathbf{I})[\mathbf{x}]_{\mathcal{Z}}=(\mathbf{S}[\mathbf{T}]_{\mathcal{X}\mathcal{X}}\mathbf{S}^{-1}-\lambda\mathbf{I})[\mathbf{x}]_{\mathcal{Z}}=\mathbf{S}([\mathbf{T}]_{\mathcal{X}\mathcal{X}}-\lambda\mathbf{I})\mathbf{S}^{-1}[\mathbf{x}]_{\mathcal{Z}}=[\boldsymbol{\theta}]_{\mathcal{Z}}$. Multiplying by the invertible matrix $\mathbf{S}^{-1}$, we find

$$([\mathbf{T}]_{\mathcal{X}\mathcal{X}}-\lambda\mathbf{I})[\mathbf{x}]_{\mathcal{X}}=[\boldsymbol{\theta}]_{\mathcal{X}} \tag{4.12}$$

Clearly, any λ which is an eigenvalue of **A** is also an eigenvalue of any other matrix $[\mathbf{T}]_{\mathcal{X}\mathcal{X}}$ which represents **T**. The similarity transformation, $[\mathbf{T}]_{\mathcal{X}\mathcal{X}}=\mathbf{S}^{-1}\mathbf{A}\mathbf{S}$, results in a change in the *coordinates* of the eigenvectors of **T** corresponding to λ, but it does not change either the eigenvectors of **T** or the characteristic polynomial of **T**.

Example 2. Invariance of Eigenvalues under a Change of Coordinates. The transformation $\mathbf{T}\colon \mathcal{R}^2 \to \mathcal{R}^2$ of Example 1 is

$$\mathbf{T}(\xi_1, \xi_2) \triangleq (2\xi_1 + 3\xi_2, 4\xi_2)$$

The eigenvectors (1,0) and (3,2) found for $\mathbf{T}$ in Example 1 form a basis for $\mathcal{R}^2$; denote this basis by $\mathcal{X}$. With respect to this basis,

$$[\mathbf{T}]_{\mathcal{X}\mathcal{X}} = \left([\mathbf{T}(1,0)]_{\mathcal{X}} \;\vdots\; [\mathbf{T}(3,2)]_{\mathcal{X}} \right)$$
$$= \begin{pmatrix} 2 & 0 \\ 0 & 4 \end{pmatrix}$$

Then

$$\det([\mathbf{T}]_{\mathcal{X}\mathcal{X}} - \lambda \mathbf{I}) = \begin{pmatrix} 2-\lambda & 0 \\ 0 & 4-\lambda \end{pmatrix}$$
$$= (2-\lambda)(4-\lambda)$$

The characteristic polynomial and the eigenvalues are those found in Example 1.

Diagonalization

It is apparent that the matrix of any linear operator $\mathbf{T}$ with respect to a basis of eigenvectors for $\mathbf{T}$ is of the form demonstrated in Example 2. If $\mathcal{X}$ is a basis of eigenvectors, $[\mathbf{T}]_{\mathcal{X}\mathcal{X}}$ has the eigenvalues of $\mathbf{T}$ on its diagonal; the rest of the matrix is zero. We call a linear operator $\mathbf{T}\colon \mathcal{V} \to \mathcal{V}$ **diagonalizable** if there is a basis $\mathcal{X}$ for $\mathcal{V}$ which is composed of eigenvectors of $\mathbf{T}$. We refer to the diagonal matrix $[\mathbf{T}]_{\mathcal{X}\mathcal{X}}$ as the **spectral matrix** of $\mathbf{T}$, and denote it by the symbol $\mathbf{\Lambda}$. If $\mathbf{A}$ is the matrix of $\mathbf{T}$ relative to some other basis, say $\mathcal{Z}$, for $\mathcal{V}$, we will also refer to $\mathbf{\Lambda}$ as the **diagonal form of A.**

A basis of eigenvectors converts the operator equation $\mathbf{Tx} = \mathbf{y}$ to the matrix equation

$$\mathbf{\Lambda}[\mathbf{x}]_{\mathcal{X}} = [\mathbf{y}]_{\mathcal{X}} \tag{4.13}$$

Equation (4.13) is actually a matrix version of the process (4.5) for solving an equation by decomposition. Finding an eigenvector basis $\mathcal{X}$ corresponds to finding a direct-sum decomposition of the space into subspaces $\mathcal{W}_i$ which are invariant under $\mathbf{T}$. Finding a coordinate matrix $[\mathbf{y}]_{\mathcal{X}}$ is equivalent to the decomposition of $\mathbf{y}$ in (4.5). Inverting the diagonal (or "uncoupled") matrix $\mathbf{\Lambda}$ amounts to solving the reduced equations, $\mathbf{T}_i\mathbf{x}_i = \lambda_i\mathbf{x}_i = \mathbf{y}_i$. When we find $\mathbf{x}$ from the coordinates $[\mathbf{x}]_{\mathcal{X}}$, we are merely combining the subsystem solutions as in (4.6). The process of computing eigenvalues and eigenvectors of matrices has been automated using a digital computer. Furthermore, the process of diagonalizing a matrix equation is more mnemonic than the decomposition process (4.5); the visual manner in which the eigenvalues and eigenvectors interact is easy to remember. Equation (4.13) is a clear and simple model for the system it represents.

What types of linear operators are diagonalizable? That is, for what finite-dimensional systems is there a basis of eigenvectors for the space? Since the existence of an eigenvalue λ_i implies the existence of a corresponding eigenvector $\mathbf{x}_i$, we expect the eigenvectors of an operator $\mathbf{T}$ on an n-dimensional space $\mathcal{V}$ to form a basis if its n eigenvalues are distinct. We verify that the n eigenvectors are independent if the eigenvalues are distinct by the test (2.11). Let

$$c_1\mathbf{x}_1 + c_2\mathbf{x}_2 + \cdots + c_n\mathbf{x}_n = \boldsymbol{\theta}$$

where $\mathbf{x}_i$ is an eigenvector of $\mathbf{T}$ for the eigenvalue λ_i. Operating with $(\mathbf{T}-\lambda_1\mathbf{I})$ we obtain

$$c_1\underbrace{(\lambda_1-\lambda_1)}_{0}\mathbf{x}_1 + c_2(\lambda_2-\lambda_1)\mathbf{x}_2 + \cdots + c_n(\lambda_n-\lambda_1)\mathbf{x}_n = \boldsymbol{\theta}$$

Successively operating with $(\mathbf{T}-\lambda_2\mathbf{I}),\ldots,(\mathbf{T}-\lambda_{n-1}\mathbf{I})$ eliminates all terms but

$$c_n(\lambda_n-\lambda_1)(\lambda_n-\lambda_2)\cdots(\lambda_n-\lambda_{n-1})\mathbf{x}_n = \boldsymbol{\theta}$$

since $\lambda_i \neq \lambda_j$, $c_n = 0$. By backtracking, we can successively show that $c_{n-1} = \cdots = c_1 = 0$; the eigenvectors are independent and form a basis for the n-dimensional space.

In the above proof we applied the operator $(\mathbf{T}-\lambda_1\mathbf{I})(\mathbf{T}-\lambda_2\mathbf{I})\cdots(\mathbf{T}-\lambda_{n-1}\mathbf{I})$ to a general vector in the space $\mathfrak{M}^{n\times 1}$ (i.e., to a linear combination, $\mathbf{x} = \sum c_i\mathbf{x}_i$, of the eigenvectors in the basis). Suppose we operate once more, using the factor $(\mathbf{T}-\lambda_n\mathbf{I})$. Then, for any $\mathbf{x}$, we obtain

$$c_n(\lambda_n-\lambda_1)(\lambda_n-\lambda_2)\cdots\underbrace{(\lambda_n-\lambda_n)}_{0}\mathbf{x}_n = \boldsymbol{\theta}$$

That is,

$$(\mathbf{T}-\lambda_1\mathbf{I})(\mathbf{T}-\lambda_2\mathbf{I})\cdots(\mathbf{T}-\lambda_n\mathbf{I}) = \boldsymbol{\Theta} \tag{4.14}$$

Recall from (4.10) that if $\mathbf{A}$ is a matrix of $\mathbf{T}$, the characteristic polynomial for $\mathbf{T}$ is $c(\lambda)=\det(\lambda\mathbf{I}-\mathbf{A})=(\lambda-\lambda_1)\cdots(\lambda-\lambda_n)$. Thus (4.14) is an operator analogue of $c(\lambda)$ which we denote by $c(\mathbf{T})$. The characteristic polynomial in $\mathbf{T}$ annihilates all vectors in the space. This fact is commonly known as the **Cayley–Hamilton theorem**. It applies as well to matrices—a square matrix satisfies its own characteristic equation:

$$c(\mathbf{A})=\boldsymbol{\Theta} \tag{4.15}$$

Although we have proved the theorem only for an operator with distinct roots, it holds for all square matrices [see (4.85)].

Example 3. A Nondiagonalizable Matrix. Suppose

$$[\mathbf{T}]_{\mathfrak{X}\mathfrak{X}}=\mathbf{A}=\begin{pmatrix}\lambda_1 & 2\\ 0 & \lambda_1\end{pmatrix}$$

Then

$$\begin{aligned}c(\lambda)&=\det(\lambda\mathbf{I}-\mathbf{A})\\&=(\lambda-\lambda_1)^2\end{aligned}$$

The only eigenvalue for $\mathbf{A}$ is $\lambda=\lambda_1$. Using (4.8) we solve for the associated eigenvectors of $\mathbf{A}$:

$$(\mathbf{A}-\lambda_1\mathbf{I})[\mathbf{x}_1]_{\mathfrak{X}}\triangleq\begin{pmatrix}0 & 2\\ 0 & 0\end{pmatrix}\begin{pmatrix}\xi_1\\ \xi_2\end{pmatrix}=\begin{pmatrix}0\\ 0\end{pmatrix}$$

or

$$[\mathbf{x}_1]_{\mathfrak{X}}=\text{span}\left\{\begin{pmatrix}1\\ 0\end{pmatrix}\right\}$$

There are not enough independent eigenvectors of $\mathbf{A}$ to form a basis for $\mathfrak{M}^{2\times 1}$. The characteristic polynomial in $\mathbf{A}$ is

$$\begin{aligned}c(\mathbf{A})&=(\mathbf{A}-\lambda_1\mathbf{I})^2\\&=\begin{pmatrix}0 & 2\\ 0 & 0\end{pmatrix}^2=\boldsymbol{\Theta}\end{aligned}$$

It is apparent that the Cayley–Hamilton theorem also applies to matrices which do not possess distinct eigenvalues.

Although repeated eigenvalues can signal difficulty, it is possible for the eigenvectors to form a basis even though the eigenvalues are not distinct. A notable example is the identity operator; any vector in the space is an eigenvector for the eigenvalue $\lambda = 1$. In Section 4.4 we discuss further those operators that are not diagonalizable.

Most matrices have distinct eigenvalues, and are thus diagonalizable. For a diagonalizable matrix **A**, the eigenvalues by themselves (or the equivalent spectral matrix $\mathbf{\Lambda}$) give a rough idea of the manner in which the system operates. However, in order to be specific about the operation of the system, we need to know what **A** does to specific vectors $[\mathbf{x}]_{\mathcal{E}}$ on which it operates. Thus we need the eigenvectors of **A**. In the process of finding the eigenvectors, we relate **A** and $\mathbf{\Lambda}$. A change of basis is the key. Let **T** act on a finite-dimensional space $\mathcal{V}$. Assume $\mathbf{A} = [\mathbf{T}]_{\mathcal{E}\mathcal{E}}$. Let $\mathcal{X} \stackrel{\Delta}{=} \{\mathbf{x}_1, \ldots, \mathbf{x}_n\}$ be a basis for $\mathcal{V}$ composed of eigenvectors of **T**. Let $\{[\mathbf{x}_1]_{\mathcal{E}}, \ldots, [\mathbf{x}_n]_{\mathcal{E}}\}$ be the corresponding basis for $\mathfrak{M}^{n\times 1}$ composed of eigenvectors of **A**. Define the change of basis matrix **S** by

$$\mathbf{S}[\mathbf{x}]_{\mathcal{X}} = [\mathbf{x}]_{\mathcal{E}} \tag{4.16}$$

Then, by (2.55),

$$\mathbf{S} = \left([\mathbf{x}_1]_{\mathcal{E}} \vdots \cdots \vdots [\mathbf{x}_n]_{\mathcal{E}} \right) \tag{4.17}$$

Furthermore, by (2.62),

$$\begin{aligned} [\mathbf{T}]_{\mathcal{X}\mathcal{X}} &= \mathbf{S}^{-1}[\mathbf{T}]_{\mathcal{E}\mathcal{E}}\mathbf{S} \\ &= \mathbf{S}^{-1}\mathbf{A}\mathbf{S} \\ &= \mathbf{\Lambda} \end{aligned} \tag{4.18}$$

We call the matrix **S**, the columns of which are eigenvectors of **A**, a **modal matrix** for **A**.* Of course, the definition of **S** in (4.16) is arbitrary; the roles of **S** and $\mathbf{S}^{-1}$ can be reversed. In order to help keep in mind which of the matrices **S** and $\mathbf{S}^{-1}$ is the modal matrix, we note that **A** in (4.18) multiplies the eigenvectors of **A** in the modal matrix.

An engineer often generates a system model directly in matrix form. The matrix form follows naturally from the use of standard models and standard physical units. When the underlying transformation is not explicitly stated, it becomes cumbersome to carry the coordinate notation $[\mathbf{x}]_{\mathcal{E}}$ for the vectors on which the $n \times n$ matrix **A** operates. Under these circumstances, we will change the notation in (4.8) to

$$(\mathbf{A} - \lambda\mathbf{I})\mathbf{x} = \boldsymbol{\theta} \tag{4.19}$$

*In some contexts the eigenvectors are referred to as modes of the system.

where $\mathbf{x}$ is a vector in $\mathfrak{M}^{n\times 1}$. This new notation can cause confusion—we are using the same notation $\mathbf{x}$ for both a vector (on which $\mathbf{T}$ operates) and its coordinate matrix (which $\mathbf{A}$ multiplies.) We must keep in mind that $\mathbf{A}$ and $\mathbf{x}$ may be representatives of an underlying transformation $\mathbf{T}$ and a vector $\mathbf{x}$ on which it operates.

Example 4. Diagonalization of a Matrix. Let

$$\mathbf{A}=\begin{pmatrix} 4 & -2 & 1\\ -2 & 1 & 2\\ 1 & 2 & 4\end{pmatrix}$$

Then $c(\lambda)=\det(\lambda\mathbf{I}-\mathbf{A})=(\lambda-5)^2(\lambda+1)=0$. The eigenvalues of $\mathbf{A}$ are $\lambda_1=5$, $\lambda_2=5$, $\lambda_3=-1$. The eigenvectors for $\lambda=5$ satisfy

$$(\mathbf{A}-5\mathbf{I})\mathbf{x}=\begin{pmatrix} -1 & -2 & 1\\ -2 & -4 & 2\\ 1 & 2 & -1\end{pmatrix}\begin{pmatrix}\xi_1\\ \xi_2\\ \xi_3\end{pmatrix}=\begin{pmatrix}0\\0\\0\end{pmatrix}$$

or $\xi_3=\xi_1+2\xi_2$. The eigenspace of $\mathbf{A}$ for $\lambda=5$ is two-dimensional; one basis for this space is

$$\mathbf{x}_1=\begin{pmatrix}1\\0\\1\end{pmatrix},\qquad \mathbf{x}_2=\begin{pmatrix}0\\1\\2\end{pmatrix}$$

The eigenvectors for $\lambda=-1$ satisfy

$$(\mathbf{A}+\mathbf{I})\mathbf{x}=\begin{pmatrix} 5 & -2 & 1\\ -2 & 2 & 2\\ 1 & 2 & 5\end{pmatrix}\begin{pmatrix}\xi_1\\ \xi_2\\ \xi_3\end{pmatrix}=\begin{pmatrix}0\\0\\0\end{pmatrix}$$

or, by row reduction, $\xi_1=-\xi_3$ and $\xi_2=-2\xi_3$. The eigenspace of $\mathbf{A}$ for $\lambda=-1$ is one-dimensional. We choose

$$\mathbf{x}_3=\begin{pmatrix}1\\2\\-1\end{pmatrix}$$

as a basis for this eigenspace. We use the eigenvectors $\mathbf{x}_1$, $\mathbf{x}_2$, and $\mathbf{x}_3$ of the matrix $\mathbf{A}$ as the columns of a modal matrix $\mathbf{S}$ for $\mathbf{A}$. We find $\mathbf{S}^{-1}$ from $\mathbf{S}$ by row reduction:

$$\mathbf{S}=\begin{pmatrix}1 & 0 & 1\\ 0 & 1 & 2\\ 1 & 2 & -1\end{pmatrix}\qquad \mathbf{S}^{-1}=\frac{1}{6}\begin{pmatrix}5 & -2 & 1\\ -2 & 2 & 2\\ 1 & 2 & -1\end{pmatrix}$$

The diagonal form of **A** is:

$$\boldsymbol{\Lambda}=\mathbf{S}^{-1}\mathbf{A}\mathbf{S}=\begin{pmatrix}5 & 0 & 0\\ 0 & 5 & 0\\ 0 & 0 & -1\end{pmatrix}$$

The eigenvalues appear on the diagonal of $\boldsymbol{\Lambda}$ in the same order as their corresponding eigenvectors appear in the modal matrix.

Eigendata and Inverse Operators

If **T** is an invertible operator and **x** is an eigenvector of **T** for the eigenvalue λ, it follows from the definition ($\mathbf{T}\mathbf{x}=\lambda\mathbf{x}$) that

$$\mathbf{T}^{-1}\mathbf{x}=\left(\frac{1}{\lambda}\right)\mathbf{x} \tag{4.20}$$

That is, **x** is also an eigenvector for $\mathbf{T}^{-1}$ corresponding to the eigenvalue $1/\lambda$. Furthermore, **T** is invertible if and only if $\lambda=0$ is not an eigenvalue of **T**. This fact is easily seen if **T** acts on a finite-dimensional space: suppose **A** is a matrix of **T** (relative to some basis). Then $\lambda=0$ is an eigenvalue of **T** if and only if

$$\det(\mathbf{A}-0\mathbf{I})=0 \tag{4.21}$$

But (4.21) is just the condition for noninvertibility of **A** (and **T**). If $\boldsymbol{\Lambda}$ is a diagonal form of **A**, the relationship between the eigenvalues and invertibility is even more transparent. If $\lambda=0$ is an eigenvalue of **A**, then $\boldsymbol{\Lambda}$ has a zero row, and **A** and **T** are not invertible.

Example 5. Eigendata for an Inverse Matrix. The inverse of the matrix **A** of Example 4 is

$$\mathbf{A}^{-1}=\frac{1}{5}\begin{pmatrix}0 & -2 & 1\\ -2 & -3 & 2\\ 1 & 2 & 0\end{pmatrix}$$

Using the spectral matrix $\boldsymbol{\Lambda}$ and the modal matrix **S** for **A** (from Example 4), we find the spectral matrix for $\mathbf{A}^{-1}$ by

$$\boldsymbol{\Lambda}_{\mathbf{A}^{-1}}=\mathbf{S}^{-1}\mathbf{A}^{-1}\mathbf{S}=(\mathbf{S}^{-1}\mathbf{A}\mathbf{S})^{-1}=\boldsymbol{\Lambda}_{\mathbf{A}}{}^{-1}$$

or

$$\boldsymbol{\Lambda}_{\mathbf{A}^{-1}}=\begin{pmatrix}\frac{1}{5} & 0 & 0\\ 0 & \frac{1}{5} & 0\\ 0 & 0 & -1\end{pmatrix}$$

Thus **A** and $\mathbf{A}^{-1}$ have inverse eigenvalues, but the same eigenvectors (modal matrices).

Computation of Eigendata for Matrices

Computation of the eigenvalues and eigenvectors of a square matrix appears straightforward. We need only solve for the roots λ_i of the characteristic polynomial, $c(\lambda)=\det(\lambda\mathbf{I}-\mathbf{A})$, then solve the equation $(\mathbf{A}-\lambda_i\mathbf{I})\mathbf{x}=\boldsymbol{\theta}$ for the eigenvectors associated with λ_i. For the selected low-order matrices used in the examples and in the Problems and Comments, the eigendata can be computed exactly using this approach. As a practical matter, however, the process is difficult for an arbitrary diagonalizable matrix. For a matrix larger than, say, 3×3, we resort to the digital computer.

Determination of the characteristic polynomial of the matrix by computing the determinant of $\lambda\mathbf{I}-\mathbf{A}$ is an expensive process. Computation of a simple $n\times n$ determinant requires $n^3/3$ multiplications, without the complication of the unspecified variable λ.* A more efficient approach for finding $c(\lambda)$ is **Krylov's method**, which is based on the Cayley–Hamilton theorem (4.15).† The characteristic equation for the $n\times n$ matrix **A** can be written

$$c(\lambda)=\lambda^n+b_1\lambda^{n-1}+\cdots+b_n=0 \tag{4.22}$$

where the coefficients $\{b_i\}$ are, as yet, unknown. By (4.15),

$$c(\mathbf{A})=\mathbf{A}^n+b_1\mathbf{A}^{n-1}+\cdots+b_n\mathbf{A}=\boldsymbol{\Theta}$$

Then for an arbitrary vector $\mathbf{x}$ in $\mathfrak{M}^{n\times 1}$,

$$\mathbf{A}^n\mathbf{x}+b_1\mathbf{A}^{n-1}\mathbf{x}+\cdots+b_n\mathbf{x}=\boldsymbol{\theta} \tag{4.23}$$

For a specific $\mathbf{x}$, the vector equation (4.23) can be solved by row reduction to obtain the coefficients $\{b_i\}$. Note that the powers of **A** need not be formed. Rather, $\mathbf{x}$ is multiplied by **A** n times. The method requires approximately n^3 multiplications to compute (4.23), then $n^3/3$ multiplications to solve for the coefficients $\{b_i\}$ by Gaussian elimination.

Example 6. Computing $c(\lambda)$ by Krylov's Method. Let **A** be the system matrix of Example 1, Section 3.4:

$$\mathbf{A}=\begin{pmatrix}0 & 1\\ 0 & -1\end{pmatrix}$$

*See Appendix 1 for a discussion of determinants and their evaluation.

†Ralston [4.13]. Refer also to P&C 1.3*c*.

The characteristic equation is second order:

$$c(\lambda)=\lambda^2+b_1\lambda+b_2=0$$

$$c(\mathbf{A})=\mathbf{A}^2+b_1\mathbf{A}+b_2\mathbf{I}=\boldsymbol{\Theta}$$

Let $\mathbf{x}=(1\ 1)^{\mathrm{T}}$. Then

$$\mathbf{A}^2\mathbf{x}+b_1\mathbf{A}\mathbf{x}+b_2\mathbf{x}=\boldsymbol{\theta}$$

or

$$\begin{pmatrix}-1\\1\end{pmatrix}+b_1\begin{pmatrix}1\\-1\end{pmatrix}+b_2\begin{pmatrix}1\\1\end{pmatrix}=\begin{pmatrix}0\\0\end{pmatrix}$$

The solution to these equations is $b_1=1$, $b_2=0$. Therefore,

$$c(\lambda)=\lambda^2+\lambda$$

Suppose that in Example 6 we had let $\mathbf{x}=(1\ -1)^{\mathrm{T}}$, the eigenvector of $\mathbf{A}$ for $\lambda=-1$. Then (4.23) would have been

$$\begin{pmatrix}1\\-1\end{pmatrix}+b_1\begin{pmatrix}-1\\1\end{pmatrix}+b_2\begin{pmatrix}1\\-1\end{pmatrix}=\begin{pmatrix}0\\0\end{pmatrix}$$

an underdetermined set of equations. The difficulty arises because $\mathbf{A}+\mathbf{I}$, one of the two factors of $c(\mathbf{A})$, is sufficient to annihilate $\mathbf{x}$. If we use an eigenvector of $\mathbf{A}$ in (4.23), we can determine only those factors of $c(\mathbf{A})$ that annihilate the eigenvector. Thus is it possible to make a poor choice for $\mathbf{x}$ in (4.23); try another! If the eigenvalues are not distinct, similar difficulties arise. (Try Krylov's method for $\mathbf{A}=\mathbf{I}$.)

Once we have $c(\lambda)$, we still need a scheme for finding its roots. A suitable method for finding the real roots is the iterative technique known as Newton's method. This method is discussed in detail in Section 8.1. If we need only the eigenvalues of $\mathbf{A}$ [as in evaluating functions of matrices by (4.108)], and if these eigenvalues are real, Krylov's method together with Newton's method is a reasonable approach to obtaining them.

Denote the eigenvalue of $\mathbf{A}$ which is of largest magnitude by λ_{L}. If λ_{L} is real, the **power method** obtains directly from $\mathbf{A}$ both its largest eigenvalue λ_{L} and a corresponding eigenvector $\mathbf{x}_{\mathrm{L}}$. The method relies on the "dominance" of the eigenvalue λ_{L}. Suppose eigenvectors of an $n\times n$ matrix $\mathbf{A}$ form a basis $\{\mathbf{x}_1,\ldots,\mathbf{x}_n\}$ for $\mathfrak{M}^{n\times 1}$. Then any vector $\mathbf{x}$ in $\mathfrak{M}^{n\times 1}$ can be expressed as $\mathbf{x}=\sum_{i=1}^{n}c_i\mathbf{x}_i$. Repeated multiplication of $\mathbf{x}$ by $\mathbf{A}$ yields $\mathbf{A}^k\mathbf{x}=\sum_{i=1}^{n}c_i\mathbf{A}^k\mathbf{x}_i=\sum_{i=1}^{n}c_i\lambda_i^k\mathbf{x}_i$. If one of the eigenvalues λ_{L} is larger in magnitude that the rest, then for large enough k, $\mathbf{A}^k\mathbf{x} \approx c_L\lambda_L^k\mathbf{x}_L$, an eigenvector for λ_L. Furthermore, λ_L is approximately equal to the ratio of the elements of $\mathbf{A}^{k+1}\mathbf{x}$ to those of $\mathbf{A}^k\mathbf{x}$. We explore the use of the power method in P&C 4.17. The method can be extended, by a process known as deflation, to obtain all the eigendata for **A**. However, computational errors accumulate; the method is practical only for a few dominant eigenvalues. See Wilkinson [4.19].

Practical computation of the full set of eigenvectors of an arbitrary matrix is more difficult than is computation of the eigenvalues. The eigenvalues $\{\lambda_i\}$, by whatever method they are obtained, will be inexact, if only because of computer roundoff. Therefore, $(\mathbf{A}-\lambda_i\mathbf{I})$ is not quite singular; we need to compute the "near nullspace" of $(\mathbf{A}-\lambda_i\mathbf{I})$ (i.e., the "near solution" to $(\mathbf{A}-\lambda_i\mathbf{I})\mathbf{x}=\boldsymbol{\theta}$). In Section 2.4 we describe the **inverse iteration method** for determining a vector in the "near nullspace" of a nearly singular matrix. We now justify that method. If a matrix **B** is nearly singular, its near nullspace is precisely the eigenspace for its smallest (least dominant) eigenvalue, λ_s. Then the near nullspace of **B** is also the eigenspace for the largest (dominant) eigenvalue $1/\lambda_s$ of $\mathbf{B}^{-1}$. If λ_s is real, we can determine an eigenvector $\mathbf{x}_s$ corresponding to λ_s by applying the power method to $\mathbf{B}^{-1}$. We pick an arbitrary vector $\mathbf{z}_0$, and repetitively determine $\mathbf{z}_{k+1}=\mathbf{B}^{-1}\mathbf{z}_k$; for large enough k, the vector $\mathbf{z}_k$ is a good approximation to $\mathbf{x}_s$; the ratio of the components of $\mathbf{z}_k$ to those of $\mathbf{z}_{k+1}$ is essentially λ_s. Thus the inverse iteration method is just the power method applied to the inverse matrix. In practice, rather than explicitly computing $\mathbf{B}^{-1}$, we would repetitively solve $\mathbf{B}\mathbf{z}_{k+1}=\mathbf{z}_k$, a less expensive operation.

The inverse iteration method can be used to obtain the eigenvectors of a matrix **A** which correspond to a previously computed real eigenvalue λ_i. Just repetitively solve $(\mathbf{A}-\lambda_i\mathbf{I})\mathbf{z}_{k+1}=\mathbf{z}_k$ for some initial vector $\mathbf{z}_0$; after several iterations, $\mathbf{z}_k$ will approximate an eigenvector $\mathbf{x}_i$ corresponding to λ_i. The ratio of the elements of $\mathbf{z}_{k+1}$ to those of $\mathbf{z}_k$ will approximate $1/\lambda_s$ where λ_s is the smallest eigenvalue of the matrix $\mathbf{B}=\mathbf{A}-\lambda_i\mathbf{I}$. The eigenvalue λ_s is a measure of the nonsingularity of **B** and, therefore, the inaccuracy in λ_i; a better approximation to the eigenvalue of **A** is $\lambda_i+\lambda_s$. A highly accurate value of λ_i implies a low value of λ_s and, consequently, rapid convergence. Of course, small λ_s also implies an ill-conditioned matrix $(\mathbf{A}-\lambda_i\mathbf{I})$; yet, as discussed in Section 2.4, the resulting uncertainty in the solution will be a vector in nullspace $(\mathbf{A}-\lambda_i\mathbf{I})$. The inverse iteration method works well as long as the eigenvalue λ_i is "isolated." *Any* method will have trouble distinguishing between eigenvectors corresponding to nearly equal eigenvalues.*

*Wilkinson [4.19].

Example 7. Computing Eigenvectors by Inverse Iteration. Let $\mathbf{A}$ be the following matrix

$$\mathbf{A}=\begin{pmatrix}1 & 0\\ 1 & -1\end{pmatrix}$$

The exact eigendata of $\mathbf{A}$ are

$$\lambda_1=1,\qquad \mathbf{x}_1=\begin{pmatrix}2\\ 1\end{pmatrix},\qquad \lambda_2=-1,\qquad \mathbf{x}_2=\begin{pmatrix}0\\ 1\end{pmatrix}$$

Suppose we have computed the eigenvalue $\hat{\lambda}_1=1+\epsilon$, perhaps by means of Krylöv's method and Newton's method. The equation $(\mathbf{A}-\hat{\lambda}_1\mathbf{I})\mathbf{x}=\boldsymbol{\theta}$ has no nonzero solution. We use inverse iteration with the matrix $(\mathbf{A}-\hat{\lambda}_1\mathbf{I})$ to approximate the true eigenvector $\mathbf{x}_1$. Denote $\mathbf{z}_k=(\eta_1\,\eta_2)^{\mathrm{T}}$ and $\mathbf{z}_{k+1}=(\xi_1\,\xi_2)^{\mathrm{T}}$. Then

$$(\mathbf{A}-\hat{\lambda}_1\mathbf{I})\mathbf{z}_{k+1}=\begin{pmatrix}-\epsilon & 0\\ 1 & -2-\epsilon\end{pmatrix}\begin{pmatrix}\xi_1\\ \xi_2\end{pmatrix}=\begin{pmatrix}\eta_1\\ \eta_2\end{pmatrix}=\mathbf{z}_k$$

has the exact solution

$$\mathbf{z}_{k+1}=\begin{pmatrix}\xi_1\\ \xi_2\end{pmatrix}=-\frac{1}{\epsilon}\begin{pmatrix}1 & 0\\ \dfrac{1}{2+\epsilon} & \dfrac{\epsilon}{2+\epsilon}\end{pmatrix}\begin{pmatrix}\eta_1\\ \eta_2\end{pmatrix}=(\mathbf{A}-\hat{\lambda}_1\mathbf{I})^{-1}\mathbf{z}_k$$

Let $\mathbf{z}_0=(1\ 1)^{\mathrm{T}}$. Then

$$\mathbf{z}_1=-\frac{1}{\epsilon}\begin{pmatrix}1\\ \dfrac{\epsilon+1}{\epsilon+2}\end{pmatrix},\qquad \mathbf{z}_2=\left(-\frac{1}{\epsilon}\right)^2\begin{pmatrix}1\\ \dfrac{\epsilon^2+2\epsilon+2}{(\epsilon+2)^2}\end{pmatrix}$$

This sequence rapidly approaches a true eigenvector for λ_1 even if the approximate eigenvalue $\hat{\lambda}_1$ contains significant error. If $\epsilon=0.1$, for instance, $\mathbf{z}_1=-10\ (1\ .52)^{\mathrm{T}}$ and $\mathbf{z}_2=100\ (1\ .501)^{\mathrm{T}}$. The smallest eigenvalue of $(\mathbf{A}-\hat{\lambda}_1\mathbf{I})$ is clearly $\hat{\lambda}_s=-\epsilon$, which approaches zero as the error in $\hat{\lambda}_1$ approaches zero. It is apparent that for small ϵ, the elements of $\mathbf{z}_k$ would soon become very large. Practical computer implementations of the inverse iteration method avoid large numbers by normalizing $\mathbf{z}_k$ at each iteration.

If $\mathbf{A}$ is symmetric, the eigenvalues of $\mathbf{A}$ are real (P&C 5.28) and there is a basis of eigenvectors for the space.* The most efficient and accurate algorithms for determination of the full set of eigendata for a symmetric matrix avoid computation of the characteristic polynomial altogether. Rather, they perform a series of similarity transformations on $\mathbf{A}$, reducing the matrix to its diagonal form $\boldsymbol{\Lambda}$; the eigenvalues appear on the diagonal. Since $\boldsymbol{\Lambda}=\mathbf{S}^{-1}\mathbf{A}\mathbf{S}$, where $\mathbf{S}$ is a matrix of eigenvectors, the sequence of

*See Section 5.4.

similarity transformations determines the eigenvectors of **A**. See P&C 4.11 for an example of such a method.

Because methods that produce the full set of eigendata for a matrix must, in effect, determine both **S** and $\mathbf{S}^{-1}$, we should expect the accuracy of the results to be related to the invertibility of the modal matrix **S**. In point of fact, it can be shown that if **S** is ill-conditioned, the eigenvalues are difficult to compute accurately; some of the eigenvalues are sensitive functions of the elements of **A**. As a general rule, symmetric matrices have easily determined eigenvalues, whereas unsymmetric matrices do not. For a full discussion of computer techniques for computing eigendata, see Wilkinson [4.19] and Forsythe [4.6].

Application of Spectral Decomposition—Symmetrical Components

Since a sinusoid of specified frequency is completely determined by two real numbers, its amplitude and phase, we can represent it by a single complex number; for example, the function $2\sin(\omega t+\phi)$ is equivalent to the complex number $2e^{i\phi}$, where $i=\sqrt{-1}$. Therefore, complex numbers adequately represent the steady-state 60-Hz sinusoidal voltages and currents in an electric power system (assuming physical units of volts and amperes, respectively).

Figure 4.2 is a simplified description of a three-phase electric power system. The complex amplitudes of the generated voltages, load voltages, and load currents are denoted by E_i, V_i, and I_i, respectively. These voltages and currents are related by the following matrix equations:

$$\mathbf{E}-\mathbf{V}=\mathbf{Z}\mathbf{I} \tag{4.24}$$

$$\mathbf{V}=\mathbf{W}\mathbf{I} \tag{4.25}$$

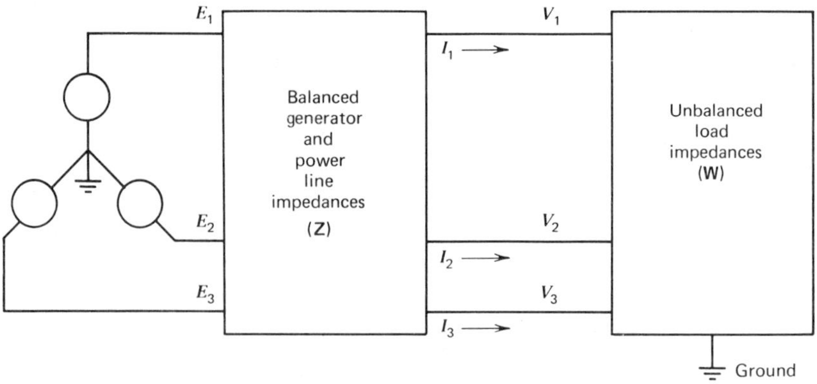

Figure 4.2. A three-phase electric power system.

where $\mathbf{E}=(E_1\ E_2\ E_3)^{\mathrm{T}}$, $\mathbf{V}=(V_1\ V_2\ V_3)^{\mathrm{T}}$, $\mathbf{I}=(I_1\ I_2\ I_3)^{\mathrm{T}}$, and $\mathbf{Z}$ and $\mathbf{W}$ are 3×3 impedance matrices. In a typical power system, the generating system is balanced; that is, $\mathbf{Z}$ has the form

$$\mathbf{Z}=\begin{pmatrix} z_1 & z_2 & z_2 \\ z_2 & z_1 & z_2 \\ z_2 & z_2 & z_1 \end{pmatrix} \tag{4.26}$$

A useful approach to analyzing a three-phase power system is to change coordinates in (4.24)–(4.25) in order to diagonalize (4.24). The method is known to power system engineers as the **method of symmetrical components.**

Exercise 2. Show (or verify) that the eigenvalues λ_i and corresponding eigenvectors $\mathbf{x}_i$ of $\mathbf{Z}$ are

$$\lambda_0=z_1+2z_2 \qquad \lambda_+=z_1-z_2 \qquad \lambda_-=z_1-z_2 \tag{4.27}$$

$$\mathbf{x}_0=\begin{pmatrix}1\\1\\1\end{pmatrix} \qquad \mathbf{x}_+=\begin{pmatrix}1\\a\\a^2\end{pmatrix} \qquad \mathbf{x}_-=\begin{pmatrix}1\\a^2\\a\end{pmatrix} \tag{4.28}$$

where $a=e^{i2\pi/3}$, a 120° counterclockwise rotation in the complex plane. (Note that $a^2+a+1=0$.) Let $\mathbf{S}=(\mathbf{x}_0 \ \vdots\ \mathbf{x}_+ \ \vdots\ \mathbf{x}_-)$. Show (or verify) that

$$\mathbf{S}^{-1}=\frac{1}{3}\begin{pmatrix}1 & 1 & 1\\ 1 & a^2 & a\\ 1 & a & a^2\end{pmatrix} \tag{4.29}$$

Each of the eigenvectors (4.28) represents the complex amplitudes of a symmetrical three-phase sinusoidal quantity (voltage or current). The subscripts indicate the relative placement of the elements of each vector in the complex plane. The generated voltage vector $\mathbf{E}$ typically has the form of $\mathbf{x}_+$. The eigenvalues (4.27) can be interpreted as impedances associated with the symmetrical (eigenvector) components of the voltage and current vectors.

The engineer usually needs to analyze the generation and distribution system under various loads. If the load impedance matrix $\mathbf{W}$ is an arbitrary matrix, it need not simplify during diagonalization. However, system loads are usually of a more specialized nature. For example, if the load is balanced (a goal of system planners), $\mathbf{W}$ is of the same form as $\mathbf{Z}$, both (4.24) and (4.25) diagonalize simultaneously, only positive sequence quantities appear in the equations, and the matrix equations reduce to two scalar equations. Certain unbalanced loads (such as a line-to-line fault) also lead to specialized forms of **W** for which symmetrical component analysis is useful. A more complete discussion of symmetrical component analysis can be found in Rothe [4.15].

4.3 Spectral Analysis in Function Spaces

Spectral analysis is at least as helpful for understanding differential systems as it is for matrix equations. Furthermore, for many distributed systems (those described by partial differential equations) it provides the only reasonable approach to the determination of solutions. This section is devoted primarily to a discussion of spectral analysis of differential systems. We found in Example 9 of Section 4.1 that for a differential operator without boundary conditions, every scalar is an eigenvalue. The differential operators of real interest, however, are the ones we use in modeling systems. These ordinarily possess an appropriate number of boundary conditions. Suppose

$$\begin{aligned} \mathbf{Lf} &\triangleq g_0(t)\frac{d^n\mathbf{f}(t)}{dt^n}+\cdots+g_n(t)\mathbf{f}(t)=\mathbf{u}(t) \\ \boldsymbol{\beta}_i(\mathbf{f}) &= \alpha_i \qquad i=1,\ldots,n \end{aligned} \tag{4.30}$$

It is convenient to decompose this differential system into two pieces:

$$\mathbf{Lf}=\mathbf{u} \quad \text{with } \boldsymbol{\beta}_i(\mathbf{f})=0, \qquad i=1,\ldots,n \tag{4.31}$$

and

$$\mathbf{Lf}=\boldsymbol{\theta} \quad \text{with } \boldsymbol{\beta}_i(\mathbf{f})=\alpha_i, \qquad i=1,\ldots,n \tag{4.32}$$

Equation (4.32) is essentially finite dimensional in nature—by substituting for **f** the complementary function $\mathbf{f}_c=c_1\mathbf{v}_1+\cdots+c_n\mathbf{v}_n$ of (3.19), we convert (4.32) to the matrix equation

$$\begin{pmatrix} \boldsymbol{\beta}_1(\mathbf{v}_1) & \cdots & \boldsymbol{\beta}_1(\mathbf{v}_n) \\ \vdots & & \vdots \\ \boldsymbol{\beta}_n(\mathbf{v}_1) & \cdots & \boldsymbol{\beta}_n(\mathbf{v}_n) \end{pmatrix}\begin{pmatrix} c_1 \\ \vdots \\ c_n \end{pmatrix}=\begin{pmatrix} \alpha_1 \\ \vdots \\ \alpha_n \end{pmatrix} \tag{4.33}$$

We examined the eigendata for matrix operators in Section 4.2. We focus now on the infinite-dimensional problem (4.31).

We seek the eigenvalues and eigenfunctions for the system **T** defined by **L** together with the homogeneous boundary conditions of (4.31). That is, we only allow **L** to operate on functions which satisfy these boundary conditions. The equation which defines the eigendata is (4.7); thus

$$(\mathbf{L}-\lambda\mathbf{I})\mathbf{f}=\boldsymbol{\theta}$$
$$\beta_i(\mathbf{f})=0 \qquad i=1,\ldots,n \tag{4.34}$$

We introduce, by means of an example, a procedure for obtaining from (4.34) the eigenvalues and eigenfunctions associated with (4.31). The armature-controlled motor of (3.40)–(3.41) is modeled by $\mathbf{L}\phi \triangleq \mathbf{D}^2\phi+\mathbf{D}\phi$, with $\beta_1(\phi) \triangleq \phi(0)$ and $\beta_2(\phi) \triangleq \phi(b)$. For this specific **L** and $\{\beta_i\}$, (4.34) becomes

$$\frac{d^2\phi(t)}{dt^2}+\frac{d\phi(t)}{dt}-\lambda\phi(t)=0$$
$$\phi(0)=\phi(b)=0 \tag{4.35}$$

We first obtain a fundamental set of solutions for $(\mathbf{L}-\lambda\mathbf{I})$. The characteristic equation for $(\mathbf{L}-\lambda\mathbf{I})$, found by inserting $\phi(t)=e^{\mu t}$, is

$$\mu^2+\mu-\lambda=0$$

with roots

$$\mu=\frac{-1\pm\sqrt{1+4\lambda}}{2}$$

If $\lambda=-\frac{1}{4}$, then the fundamental solutions are

$$\mathbf{v}_1(t)=e^{-t/2} \qquad \mathbf{v}_2(t)=te^{-t/2}$$

Any nonzero solutions to (4.35) for $\lambda=-\frac{1}{4}$ must be of the form $\mathbf{f}=c_1\mathbf{v}_1+c_2\mathbf{v}_2$ and must satisfy the boundary conditions:

$$\begin{pmatrix}\beta_1(\mathbf{f})\\ \beta_2(\mathbf{f})\end{pmatrix}=\begin{pmatrix}\beta_1(\mathbf{v}_1) & \beta_1(\mathbf{v}_2)\\ \beta_2(\mathbf{v}_1) & \beta_2(\mathbf{v}_2)\end{pmatrix}\begin{pmatrix}c_1\\ c_2\end{pmatrix}=\begin{pmatrix}1 & 0\\ e^{-b/2} & be^{-b/2}\end{pmatrix}\begin{pmatrix}c_1\\ c_2\end{pmatrix}=\begin{pmatrix}0\\ 0\end{pmatrix}$$

The boundary condition matrix is invertible; $c_1=c_2=0$. There are no nonzero solutions for $\lambda=-\frac{1}{4}$, and $\lambda=-\frac{1}{4}$ is not an eigenvalue.

If $\lambda\neq-\frac{1}{4}$, a pair of fundamental solutions is

$$\mathbf{g}_1(t)=e^{-t/2}\exp\left(\frac{(1+4\lambda)^{1/2}t}{2}\right), \qquad \mathbf{g}_2(t)=e^{-t/2}\exp\left(\frac{-(1+4\lambda)^{1/2}t}{2}\right)$$

A different but equivalent pair is

$$\mathbf{h}_1(t)=e^{-t/2}\cos\left(\frac{-i(1+4\lambda)^{1/2}t}{2}\right),\qquad \mathbf{h}_2(t)=e^{-t/2}\sin\left(\frac{-i(1+4\lambda)^{1/2}t}{2}\right)$$

We let $\mathbf{g}=c_1\mathbf{g}_1+c_2\mathbf{g}_2$, and again invoke the boundary conditions:

$$\begin{pmatrix}\beta_1(\mathbf{g})\\ \beta_2(\mathbf{g})\end{pmatrix}=\begin{pmatrix}\beta_1(\mathbf{g}_1) & \beta_1(\mathbf{g}_2)\\ \beta_2(\mathbf{g}_1) & \beta_2(\mathbf{g}_2)\end{pmatrix}\begin{pmatrix}c_1\\ c_2\end{pmatrix}$$

$$=\begin{pmatrix}1 & 1\\ e^{-b/2}\exp\left(\dfrac{(1+4\lambda)^{1/2}b}{2}\right) & e^{-b/2}\exp\left(\dfrac{-(1+4\lambda)^{1/2}b}{2}\right)\end{pmatrix}\begin{pmatrix}c_1\\ c_2\end{pmatrix}=\begin{pmatrix}0\\ 0\end{pmatrix}$$

There is a nonzero solution $\mathbf{g}$ (or nonzero coefficients $\{c_i\}$) if and only if the boundary condition matrix is singular; thus, denoting the boundary condition matrix by $\mathbf{B}(\lambda)$,

$$\det(\mathbf{B}(\lambda))=e^{-b/2}\exp\left(\frac{-(1+4\lambda)^{1/2}b}{2}\right)-e^{-b/2}\exp\left(\frac{(1+4\lambda)^{1/2}b}{2}\right)=0$$

or

$$\exp\left[(1+4\lambda)^{1/2}b\right]=1 \tag{4.36}$$

By analogy with the finite-dimensional case, we are inclined to refer to $\det(\mathbf{B}(\lambda))=0$ as the characteristic equation for the operator $\mathbf{T}$ ($\mathbf{L}$ with the homogeneous boundary conditions). However, the term characteristic equation is commonly used in reference to the equation (in the variable μ) used earlier to determine the fundamental solutions for $\mathbf{L}$. Therefore, we call $\det(\mathbf{B}(\lambda))=0$ the **eigenvalue equation** for $\mathbf{T}$. We may also refer to it as the eigenvalue equation for $\mathbf{L}$ if it is clear which homogeneous boundary conditions are intended. The eigenvalue equation (4.36) is a transcendental equation in λ. To find the roots, recall from the theory of complex variables that*

$$\ln(e^{\alpha+i\gamma})=\alpha+i\gamma+i2\pi k,\qquad k=0,\pm1,\pm2,\ldots$$

for real scalars α and γ. Thus (4.36) becomes

$$(1+4\lambda)^{1/2}b+i2\pi k=0\qquad k=0,\pm1,\pm2,\ldots$$

*See Chapter 14 of Wylie [4.18].

and the eigenvalues (for which nonzero solutions exist) are

$$\lambda_k = -\tfrac{1}{4} - \left(\frac{k\pi}{b}\right)^2 \qquad k = 1, 2, 3, \ldots \tag{4.37}$$

Note that $k=0$ has been deleted; it corresponds to $\lambda = -\frac{1}{4}$, for which case $\mathbf{g}_1$ and $\mathbf{g}_2$ are not a fundamental set of solutions. Since k is squared, the positive and negative values of k yield identical values of λ; thus, the positive values are sufficient.

We obtain the eigenfunctions ϕ_k corresponding to the eigenvalue λ_k by solving (4.35) with $\lambda = \lambda_k$. The solutions involve the roots μ_k of the characteristic equation:

$$\mu_k = \frac{-1 \pm (1+4\lambda_k)^{1/2}}{2} = -\tfrac{1}{2} \pm i\frac{k\pi}{b}$$

Since these roots are complex, we use the sinusoidal form $\{\mathbf{h}_i\}$ for the fundamental solutions:

$$\phi_k(t) = c_1 e^{-t/2}\cos\left(\frac{k\pi t}{b}\right) + c_2 e^{-t/2}\sin\left(\frac{k\pi t}{b}\right)$$

The boundary conditions yield

$$\begin{pmatrix}\beta_1(\phi_k)\\ \beta_2(\phi_k)\end{pmatrix} = \mathbf{B}(\lambda_k)\begin{pmatrix}c_1\\ c_2\end{pmatrix} = \begin{pmatrix} e^0\cos(0) & e^0\sin(0)\\ e^{-b/2}\cos(k\pi) & e^{-b/2}\sin(k\pi)\end{pmatrix}\begin{pmatrix}c_1\\ c_2\end{pmatrix} = \begin{pmatrix}0\\ 0\end{pmatrix}$$

It follows that $c_1 = 0$ and c_2 is arbitrary. Letting $c_2 = 1$, we obtain the eigenfunction

$$\phi_k(t) = e^{-t/2}\sin\left(\frac{k\pi t}{b}\right) \tag{4.38}$$

corresponding to the eigenvalue λ_k.

The eigenfunctions for the two-point boundary value operator of (4.35) are analogous to the modes of oscillation of a string which is tied at both ends. The modes are harmonics of the fundamental or lowest-order mode, $e^{-t/2}\sin(\pi t/b)$; that is, the frequencies of oscillation are integral multiples of the lowest-order frequency. The number μ_k is the complex "natural frequency" of the kth mode. The eigenvalue λ_k can be thought of as a "characteristic number" for the kth mode. It is not clear whether or not $\mathbf{T}$ is a diagonalizable operator. The eigenvalues are distinct; the set of eigenfunctions are suggestive of the terms of a Fourier series; however, we wait until Chapter 5 to determine that there are sufficient eigenfunctions $\{\boldsymbol{\phi}_k, k=1,2,\ldots\}$ to form a basis for the space of functions $\mathbf{f}$ on which $\mathbf{T}$ (or $\mathbf{L}$) operates. (See Example 3, Section 5.3.)

Finding Eigendata for Differential Operators

For general differential equations of the form (4.30) we find eigendata by following the procedure used for the specific operator of (4.35). We first seek values of λ (or **eigenvalues**) for which (4.34) has nonzero solutions (**eigenfunctions**). Then we determine the corresponding eigenfunctions. We occasionally refer to the eigendata for the differential equation when we really mean the eigendata for the differential operator which determines the equation. Let the functions $\mathbf{v}_1(\lambda),\ldots,\mathbf{v}_n(\lambda)$ be a fundamental set of solutions for $(\mathbf{L}-\lambda\mathbf{I})$; note that the functions depend on λ. The solutions to (4.34) consist in linear combinations

$$\mathbf{f}_c = c_1\mathbf{v}_1 + \cdots + c_n\mathbf{v}_n$$

which satisfy the boundary conditions. The coefficients are determined by the boundary condition matrix, whose λ dependency we denote explicitly by $\mathbf{B}(\lambda)$:

$$\mathbf{B}(\lambda)\begin{pmatrix} c_1 \\ \vdots \\ c_n \end{pmatrix} = \begin{pmatrix} \beta_1(\mathbf{v}_1) & \cdots & \beta_1(\mathbf{v}_n) \\ \vdots & & \vdots \\ \beta_n(\mathbf{v}_1) & \cdots & \beta_n(\mathbf{v}_n) \end{pmatrix}\begin{pmatrix} c_1 \\ \vdots \\ c_n \end{pmatrix} = \begin{pmatrix} 0 \\ \vdots \\ 0 \end{pmatrix} \tag{4.39}$$

There are nonzero solutions to (4.34) [or nonzero coefficients $\{c_i\}$ in (4.39)] only for λ such that

$$\det(\mathbf{B}(\lambda)) = 0 \tag{4.40}$$

As discussed beneath (4.36), we call (4.40) the **eigenvalue equation for T** (or for $\mathbf{L}$ with its boundary conditions). Its roots constitute the **spectrum of T** (or of $\mathbf{L}$ with its boundary conditions).

Determining the complementary function for $\mathbf{T}-\lambda\mathbf{I}$ is not necessarily a simple task. But it is the fundamental problem of differential equation analysis—standard techniques apply. The eigenvalue equation (4.40) is generally transcendental. Its solution, perhaps difficult, is a matter of algebra. Once we have determined a specific eigenvalue λ_k we return to (4.39) to determine those combinations of the fundamental solutions which are eigenfunctions for λ_k. The eigenfunctions are

$$\mathbf{f}_k = c_1\mathbf{v}_1(\lambda_k) + \cdots + c_n\mathbf{v}_n(\lambda_k) \tag{4.41}$$

where the scalars $c_1,\ldots,c_n$ satisfy

$$\mathbf{B}(\lambda_k)\begin{pmatrix} c_1 \\ \vdots \\ c_n \end{pmatrix} = \boldsymbol{\theta}$$

As noted in the discussion following (3.28), the boundary condition matrix for a one-point boundary value problem is always invertible. Thus if the boundary conditions for **L** are all initial conditions, (4.40) has no roots, and the system **T** has no eigenvalues.

Exercise 1. Seek the eigenvalues for the operator **L** of (4.35) with the initial conditions $\phi(0)=\phi'(0)=0$.

Example 1. Eigendata for a Heat-Flow Problem. Equation (3.1) is a steady-state description of a system wherein the heat generated within an insulated bar of length b diffuses toward heat sinks at the surfaces $t=0$ and $t=b$. We now modify the second boundary condition. At $t=b$ we withdraw heat from the system by convection. The equation and modified boundary conditions for the temperature distribution **f** are as follows:

$$(\mathbf{Lf})(t) \stackrel{\Delta}{=} -\frac{d^2\mathbf{f}(t)}{dt^2} = \mathbf{u}(t)$$
$$\beta_1(\mathbf{f}) \stackrel{\Delta}{=} \mathbf{f}(0)=\alpha_1, \qquad \beta_2(\mathbf{f}) \stackrel{\Delta}{=} \mathbf{f}'(b)+\mathbf{f}(b)=\alpha_2 \tag{4.42}$$

The characteristic equation for $(\mathbf{L}-\lambda\mathbf{I})$ is

$$-\mu^2-\lambda=0$$

with roots $\mu=\pm i\sqrt{\lambda}$. We pick as a fundamental set of solutions (for $\lambda\neq 0$):

$$\mathbf{v}_1(t)=\cos\sqrt{\lambda}\,t, \qquad \mathbf{v}_2(t)=\sin\sqrt{\lambda}\,t$$

The eigenvalue equation is

$$\det(\mathbf{B}(\lambda))=\begin{vmatrix} 1 & 0 \\ -\sqrt{\lambda}\,\sin(\sqrt{\lambda}\,b)+\cos(\sqrt{\lambda}\,b) & \sqrt{\lambda}\,\cos(\sqrt{\lambda}\,b)+\sin(\sqrt{\lambda}\,b) \end{vmatrix}$$
$$=\sqrt{\lambda}\,\cos(\sqrt{\lambda}\,b)+\sin(\sqrt{\lambda}\,b)=0$$

or

$$\tan\sqrt{\lambda}\,b=-\sqrt{\lambda} \tag{4.43}$$

Making the substitution $r \triangleq \sqrt{\lambda}\, b$, (4.43) becomes

$$\tan r = -\frac{r}{b} \tag{4.44}$$

Figure 4.3 shows the two halves of the eigenvalue equation plotted versus r for $b=2$. If $\{r_k, k=0,\pm 1,\pm 2,\ldots\}$ are the roots of (4.44), then the eigenvalues for (4.42) are

$$\lambda_k = \frac{r_k^2}{b^2} \qquad k=1,2,3,\ldots \tag{4.45}$$

The root r_0 has been eliminated. It corresponds to $\lambda=0$, for which the sinusoids are not a fundamental set of solutions. That $\lambda=0$ is not an eigenvalue is easily seen by repeating the above, using a fundamental set of solutions for $(\mathbf{L}-0\mathbf{I})$. Since

$$(r_{-k})^2 = (-r_k)^2 = r_k^2$$

the negative values of k are unnecessary. We find the eigenfunctions $\mathbf{f}_k$ for λ_k by (4.41):

$$\mathbf{B}(\lambda_k)\begin{pmatrix} c_1 \\ c_2 \end{pmatrix} = \begin{pmatrix} 1 & 0 \\ -\dfrac{r_k}{b}\sin r_k + \cos r_k & \dfrac{r_k}{b}\underbrace{\cos r_k + \sin r_k}_{0} \end{pmatrix}\begin{pmatrix} c_1 \\ c_2 \end{pmatrix} = \begin{pmatrix} 0 \\ 0 \end{pmatrix}$$

or $c_1=0$ and c_2 is arbitrary. Therefore, letting $c_2=1$, we obtain only one independent eigenvector,

$$\mathbf{f}_k(t) = \sin\left(\frac{r_k}{b}t\right) \tag{4.46}$$

for each eigenvalue $\lambda_k = r_k^2/b^2, k=1,2,3,\ldots$.
In this example, the modes are not harmonic; the frequencies r_k^2/b^2 are not integral multiples of the lowest frequency. Although the operator of (4.42) is diagonalizable (the eigenvectors (4.46) form a basis for the domain of $\mathbf{L}$), we are not presently prepared to show it.

Eigendata for Integral Operators

We found in (4.20) that if an operator $\mathbf{T}$ is invertible and $\mathbf{T}\mathbf{x}=\lambda\mathbf{x}$, then $\mathbf{T}^{-1}\mathbf{x}=(1/\lambda)\mathbf{x}$. That is, the eigenvectors of $\mathbf{T}$ and $\mathbf{T}^{-1}$ are identical and correspond to reciprocal eigenvalues. From (4.40) we know that a differential system $\mathbf{T}$ has the eigenvalue $\lambda=0$ if and only if $\det(\mathbf{B}(\lambda))=\det(\mathbf{B}(0))=0$. But this is just the opposite of the condition (3.28) for invertibility of $\mathbf{T}$. Thus a differential system $\mathbf{T}$ is invertible if and only if $\lambda=0$ is not an eigenvalue for $\mathbf{T}$. If we think in terms of a diagonalized $(\infty \times \infty)$ matrix representation of $\mathbf{T}$, it is clear that a zero eigenvalue is equivalent to singularity of the operator. Thus if $\lambda=0$ is an eigenvalue of $\mathbf{T}$, then the

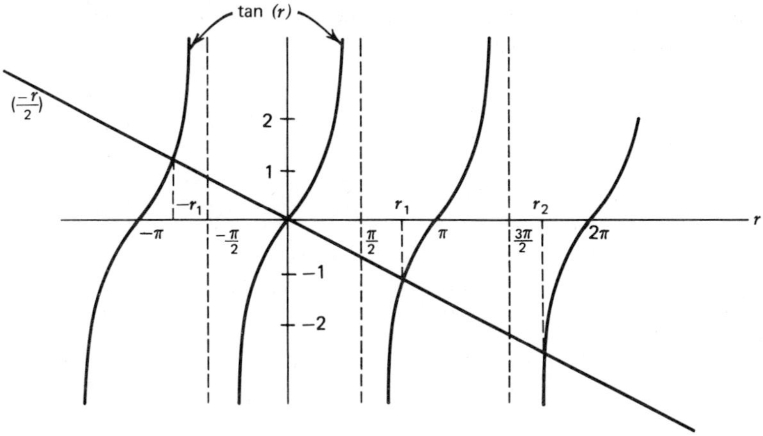

Figure 4.3. Roots of the eigenvalue equation (4.44) for $b=2$.

Green's function for **T** does not exist. Invertible differential and integral equations come in pairs, one the inverse of the other. Because the properties of integration are theoretically and computationally less troublesome than those of differentation, we use the integral form to derive useful information about the eigenfunctions of operators and the solutions of equations (Sections 5.4 and 5.5). We also use the integral form for approximate numerical solution of equations. Yet because integral equations are difficult to solve, we often return to the differential form and standard differential equation techniques to determine the eigenfunctions of specific operators or the solutions of specific equations. In the following example, we obtain the eigendata for an integral operator from its differential inverse.

Example 2. Eigendata for an Integral Operator. The eigendata for the system **T** represented by the differential operator $\mathbf{L}=\mathbf{D}^2+\mathbf{D}$ with $\phi(0)=0$ and $\phi(b)=0$ are given in (4.37) and (4.38). They are

$$\lambda_k=-\frac{1}{4}-\left(\frac{k\pi}{b}\right)^2,\quad \phi_k(t)=e^{-t/2}\sin\left(\frac{k\pi t}{b}\right),\qquad k=1,2,\ldots$$

Note that $\lambda=0$ is not an eigenvalue. The Green's function for this operator is (3.42). Using this Green's function, we write the inverse of the differential system as

$$\phi(t)=\frac{1-e^{b}e^{-t}}{e^{b}-1}\int_0^t(e^s-1)\mathbf{u}(s)\,ds+\frac{1-e^{-t}}{e^{b}-1}\int_t^b(e^s-e^b)\mathbf{u}(s)\,ds$$

$$=(\mathbf{T}^{-1}\mathbf{u})(t) \qquad (4.47)$$

We expect the eigenfunctions of $\mathbf{T}^{-1}$ to be the same as those of $\mathbf{T}$. Operating on ϕ_k with $\mathbf{T}^{-1}$, a complicated integration, we find

$$
\begin{aligned}
(\mathbf{T}^{-1}\phi_k)(t) &= \frac{1-e^{b}e^{-t}}{e^{b}-1}\int_0^t (e^{s}-1)e^{-s/2}\sin\left(\frac{k\pi s}{b}\right)ds \\
&\quad + \frac{1-e^{-t}}{e^{b}-1}\int_t^b (e^{s}-e^{b})e^{-s/2}\sin\left(\frac{k\pi s}{b}\right)ds \\
&= \frac{1}{-1/4-(k\pi/b)^2}e^{-t/2}\sin\left(\frac{k\pi t}{b}\right) \\
&= \left(\frac{1}{\lambda_k}\right)\phi_k(t) \quad k=1,2,3,\ldots
\end{aligned}
\tag{4.48}
$$

The eigenvalues of the integral operator $\mathbf{T}^{-1}$ are clearly $\{1/\lambda_k\}$.

Eigenvalue Problems in State Space

We introduced the state space model for dynamic systems in Section 3.4. We reproduce it here:

$$\dot{\mathbf{x}}(t)=\mathbf{A}\mathbf{x}(t)+\mathbf{B}\mathbf{u}(t) \quad \mathbf{x}(0)=\mathbf{x}_0 \tag{4.49}$$

where $\mathbf{A}$ is an $n\times n$ matrix multiplying the $n\times 1$ state vector $\mathbf{x}(t)$, and $\mathbf{B}$ is an $n\times m$ matrix multiplying the $m\times 1$ input vector $\mathbf{u}(t)$. We know the differential system of (4.49) has no eigenvalues—it is an initial-value problem.* However, there is a meaningful and interesting eigenvalue problem associated with (4.49). It has to do with the system matrix $\mathbf{A}$. We introduce the relationship between the eigendata for the system matrix and the solutions of (4.49) by examining the system matrix for the nth-order constant-coefficient differential equation, the companion matrix of (3.36). The eigenvalues of $\mathbf{A}$ are the roots of the equation $\det(\lambda\mathbf{I}-\mathbf{A})=0$.

Exercise 2. Show that if $\mathbf{A}$ is the companion matrix for the nth-order constant-coefficient differential equation

$$\mathbf{D}^n\mathbf{f}+a_1\mathbf{D}^{n-1}\mathbf{f}+\cdots+a_n\mathbf{f}=\mathbf{u} \tag{4.50}$$

then the characteristic equation for $\mathbf{A}$ is

$$\det(\lambda\mathbf{I}-\mathbf{A})=(\lambda^n+a_1\lambda^{n-1}+\cdots+a_n)=0 \tag{4.51}$$

* If the initial condition vector is $\mathbf{x}(0)=\boldsymbol{\theta}$, then $\dot{\mathbf{x}}(t)-\mathbf{A}\mathbf{x}(t)-\lambda\mathbf{x}(t)=0$ has only the zero solution, $\mathbf{x}(t)=\boldsymbol{\theta}$.

From (4.51), we see that if $\mathbf{A}$ is the system matrix corresponding to an nth-order constant-coefficient differential equation, the characteristic equation for $\mathbf{A}$ is the same as the characteristic equation (3.37) for the underlying nth-order differential equation. The eigenvalues of the system matrix are the exponents for a fundamental set of solutions to the differential equation. They are sometimes referred to as **poles** of the system. This relationship between the eigenvalues of the system matrix and the fundamental set of solutions to the underlying set of differential equations holds for any system matrix $\mathbf{A}$, not just for those in companion matrix form. [See the discussion below (4.94); refer also to P&C 4.16.] Thus in the state-space equation (4.49) the concepts of matrix transformations and differential operators merge in an interesting way. The origin of the term "characteristic equation for the differential equation" is apparent. Fortunately, the state-space formulation is not convenient for boundary value problems. Thus eigenvalues of a system matrix and eigenvalues of a differential equation usually do not appear in the same problem.

Suppose we use the eigenvectors of the system matrix $\mathbf{A}$ as a new basis for the state space, assuming, of course, that $\mathbf{A}$ is diagonalizable. We change coordinates as in (4.16)–(4.18). (We can think of the state vector $\mathbf{x}(t)$ in $\mathfrak{M}^{n\times 1}$ as representing itself relative to the standard basis for $\mathfrak{M}^{n\times 1}$.) If $\{\mathbf{x}_1,\ldots,\mathbf{x}_n\}$ is a basis of eigenvectors for $\mathbf{A}$ corresponding to the eigenvalues $\{\lambda_1,\ldots,\lambda_n\}$, we transform $\mathbf{x}(t)$ into the new coordinates $\mathbf{y}(t)$ by the transformation

$$\mathbf{y}(t)=\mathbf{S}^{-1}\mathbf{x}(t) \tag{4.52}$$

where $\mathbf{S}$ is the modal matrix for $\mathbf{A}$:

$$\mathbf{S}=(\mathbf{x}_1 \vdots \cdots \vdots \mathbf{x}_n) \tag{4.53}$$

Then, by (4.18), (4.49) becomes

$$\mathbf{S}\dot{\mathbf{y}}(t)=\mathbf{ASy}(t)+\mathbf{Bu}(t), \qquad \mathbf{Sy}(0)=\mathbf{x}_0$$

$$\dot{\mathbf{y}}(t)=\mathbf{S}^{-1}\mathbf{ASy}(t)+\mathbf{S}^{-1}\mathbf{Bu}(t)$$

$$=\mathbf{\Lambda y}(t)+\mathbf{S}^{-1}\mathbf{Bu}(t), \qquad \mathbf{y}(0)=\mathbf{S}^{-1}\mathbf{x}_0 \tag{4.54}$$

Equation (4.54) is a set of n uncoupled first-order differential equations which can be solved independently. The eigenvectors (or modes) of $\mathbf{A}$ in a sense express natural relationships among the state variables [the elements of $\mathbf{x}(t)$] at each instant t. By using these eigenvectors as a basis, we eliminate the interactions—the new state variables [the elements of $\mathbf{y}(t)$] do not affect each other.

Example 3. Diagonalizing a State Equation. The state equation for the armature controlled dc motor of (3.40) was obtained in Example 1 of Section 3.4. It is

$$\dot{\mathbf{x}}(t)=\begin{pmatrix}0 & 1\\ 0 & -1\end{pmatrix}\mathbf{x}(t)+\begin{pmatrix}0\\ 1\end{pmatrix}\mathbf{u}(t), \qquad \mathbf{x}(0)=\begin{pmatrix}\alpha_1\\ \alpha_2\end{pmatrix} \tag{4.55}$$

The eigendata for the system matrix are

$$\lambda_1=0, \qquad \lambda_2=-1 \qquad \mathbf{x}_1=\begin{pmatrix}1\\ 0\end{pmatrix} \qquad \mathbf{x}_2=\begin{pmatrix}1\\ -1\end{pmatrix} \tag{4.56}$$

The modal matrix is its own inverse

$$\mathbf{S}^{-1}=\begin{pmatrix}1 & 1\\ 0 & -1\end{pmatrix}=\mathbf{S} \tag{4.57}$$

The decoupled state equation is

$$\dot{\mathbf{y}}(t)=\begin{pmatrix}0 & 0\\ 0 & -1\end{pmatrix}\mathbf{y}(t)+\begin{pmatrix}1\\ -1\end{pmatrix}\mathbf{u}(t), \qquad \mathbf{y}(0)=\begin{pmatrix}\alpha_1+\alpha_2\\ -\alpha_2\end{pmatrix} \tag{4.58}$$

Denote the new state variables [elements of $\mathbf{y}(t)$] by $\mathbf{g}_1(t)$ and $\mathbf{g}_2(t)$. We can solve independently for $\mathbf{g}_1$ and $\mathbf{g}_2$. On the other hand, we can use (3.79) with $\mathbf{x}$, $\mathbf{A}$, and $\mathbf{B}$ replaced by $\mathbf{y}$, $\boldsymbol{\Lambda}$, and $\mathbf{S}^{-1}\mathbf{B}$, respectively. By either approach the result is

$$\mathbf{y}(t)\triangleq\begin{pmatrix}\mathbf{g}_1(t)\\ \mathbf{g}_2(t)\end{pmatrix}=\int_0^t\begin{pmatrix}1\\ -e^{-(t-s)}\end{pmatrix}\mathbf{u}(s)\,ds+\begin{pmatrix}1 & 0\\ 0 & e^{-t}\end{pmatrix}\begin{pmatrix}\alpha_1+\alpha_2\\ -\alpha_2\end{pmatrix} \tag{4.59}$$

Then

$$\begin{aligned}\mathbf{x}(t)&=\mathbf{S}^{-1}\mathbf{y}(t)\\ &=\int_0^t\begin{pmatrix}1 & 1\\ 0 & -1\end{pmatrix}\begin{pmatrix}1\\ -e^{-(t-s)}\end{pmatrix}\mathbf{u}(s)\,ds+\begin{pmatrix}1 & 1\\ 0 & -1\end{pmatrix}\begin{pmatrix}1 & 0\\ 0 & e^{-t}\end{pmatrix}\begin{pmatrix}\alpha_1+\alpha_2\\ -\alpha_2\end{pmatrix}\\ &=\int_0^t\begin{pmatrix}1-e^{-(t-s)}\\ e^{-(t-s)}\end{pmatrix}\mathbf{u}(s)\,ds+\begin{pmatrix}\alpha_1+\alpha_2-\alpha_2e^{-t}\\ \alpha_2e^{-t}\end{pmatrix}\end{aligned} \tag{4.60}$$

Compare this result with (3.80).

Note that the modal matrix in Example 3 is the Vandermond matrix for the system. Whenever the system matrix is in companion matrix form and the poles of the system are distinct, the Vandermond matrix is a modal matrix; then the eigenvectors of $\mathbf{A}$ need not be calculated, but follow directly from the eigenvalues. See P&C 4.16.

Eigenvalue Problems and Partial Differential Equations

As we found in Example 10 of Section 4.1, not all differential operators have eigenvalues. This statement applies to both ordinary and partial differential operators. However, the most common analytical method for solving partial differential equations, separation of variables, generally introduces an eigenvalue problem even if the partial differential operator itself does not have eigenvalues. In point of fact, an analytical solution to a partial differential equation and its associated boundary conditions is usually obtainable only by summing eigenfunctions of a related differential operator. See Wylie [4.18]. On the other hand, some partial differential operators do have eigenvalues. One example is the Laplacian operator ∇^2, defined by

$$\nabla^2\mathbf{f}(s,t) \stackrel{\Delta}{=} \frac{\partial^2\mathbf{f}(s,t)}{\partial s^2} + \frac{\partial^2\mathbf{f}(s,t)}{\partial t^2} \tag{4.61}$$

together with the "many-point" boundary conditions

$$\mathbf{f}(s,t)=0 \quad \text{on} \quad \Gamma \tag{4.62}$$

where Γ is a closed curve in the (s,t) plane,

Exercise 3. Let Γ be the boundary of the rectangle with sides at $s=0$, $s=a$, $t=0$, and $t=b$. Show (by separation of variables) or verify that the eigenvalues and eigenfunctions for ∇^2 together with the boundary conditions (4.62) are:

$$\lambda_{mk} = -\left(\frac{m\pi}{a}\right)^2 - \left(\frac{k\pi}{b}\right)^2$$

$$\mathbf{f}_{mk}(s,t) = \sin\left(\frac{m\pi s}{a}\right)\sin\left(\frac{k\pi t}{b}\right) \tag{4.63}$$

$$m=1,2,\ldots \qquad k=1,2,\ldots$$

Notice that $\lambda=0$ is not an eigenvalue of (4.61)–(4.62). Therefore the operator must be invertible, and we can expect to find a unique solution to Poisson's equation, $\nabla^2\mathbf{f}=\mathbf{u}$, together with the boundary conditions of Example 3.

4.4 Nondiagonalizable Operators and Jordan Form

Most useful linear transformations are diagonalizable. However, there occasionally arises in practical analysis a system which is best modeled by a nondiagonalizable transformation. Probably the most familiar example is a dynamic system with a pair of nearly equal poles. We use such an example to introduce the concept of nondiagonalizability.

Suppose we wish to solve the undriven differential equation $(\mathbf{D}+1)(\mathbf{D}+1+\epsilon)\mathbf{f}=\boldsymbol{\theta}$ with the boundary conditions $\mathbf{f}(0)=\alpha_1$ and $\mathbf{f}'(0)=\alpha_2$, where ϵ is a small constant. The solution is of the form

$$\mathbf{f}(t)=c_1e^{-t}+c_2e^{-(1+\epsilon)t} \tag{4.64}$$

Applying the boundary conditions, we find

$$\begin{pmatrix} 1 & 1 \\ -1 & -(1+\epsilon) \end{pmatrix}\begin{pmatrix} c_1 \\ c_2 \end{pmatrix}=\begin{pmatrix} \alpha_1 \\ \alpha_2 \end{pmatrix}$$

Since ϵ is small, this equation is ill-conditioned; it is difficult to compute accurately the multipliers c_1 and c_2 (see Section 1.5). The difficulty occurs because the poles of the system (or roots of the characteristic equation) are nearly equal; the functions e^{-t} and $e^{-(1+\epsilon)t}$ are nearly indistinguishable (see Figure 4.4). We resolve this computational difficulty by replacing e^{-t} and $e^{-(1+\epsilon)t}$ by a more easily distinguishable pair of functions; (4.64) becomes

$$\begin{aligned}\mathbf{f}(t)&=e^{-t}\left(c_1+c_2e^{-\epsilon t}\right)\\ &=e^{-t}\left[c_1+c_2\left(1-\epsilon t+\frac{(\epsilon t)^2}{2!}-\cdots\right)\right]\\ &\approx e^{-t}[(c_1+c_2)-c_2\epsilon t]\\ &=d_1e^{-t}+d_2te^{-t}\end{aligned} \tag{4.65}$$

where $d_1=c_1+c_2$ and $d_2=-\epsilon c_2$. Since ϵ is small, the functions e^{-t} and te^{-t} span essentially the same space as e^{-t} and $e^{-(1+\epsilon)t}$; yet this new pair of functions is clearly distinguishable (Figure 4.4b). The "new" function

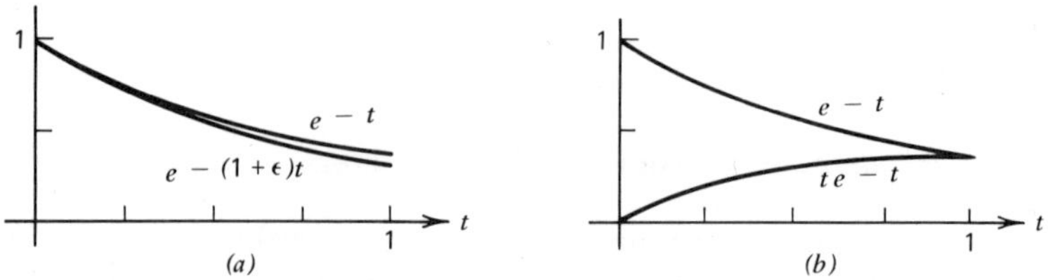

Figure 4.4. Alternative pairs of solutions to $(\mathbf{D}+1)(\mathbf{D}+1+\epsilon)\mathbf{f}=\boldsymbol{\theta}$.

te^{-t} is essentially the difference between the two nearly equal exponentials. The boundary conditions now require

$$\begin{pmatrix} 1 & 0 \\ -1 & 1 \end{pmatrix}\begin{pmatrix} d_1 \\ d_2 \end{pmatrix}=\begin{pmatrix} \alpha_1 \\ \alpha_2 \end{pmatrix}$$

or $\mathbf{f}(t)=\alpha_1 e^{-t}+(\alpha_1+\alpha_2)te^{-t}$. We have eliminated the computational difficulty by equating the nearly equal poles of the system. When the roots of the characteristic equation are equal, (4.65) is the exact complementary function for the differential operator.

It is enlightening to view the differential system in state-space form. By writing the differential equation in the form $(\mathbf{D}^2+(2+\epsilon)\mathbf{D}+(1+\epsilon)\mathbf{I})\mathbf{f}=\boldsymbol{\theta}$, we recognize from (3.63) that the state equation is

$$\dot{\mathbf{x}}(t)=\begin{pmatrix} 0 & 1 \\ -(1+\epsilon) & -(2+\epsilon) \end{pmatrix}\mathbf{x}(t), \qquad \mathbf{x}(0)=\begin{pmatrix} \alpha_1 \\ \alpha_2 \end{pmatrix}$$

The nearly equal poles of the system appear now as nearly equal eigenvalues of the system matrix, $\lambda_1=-1, \lambda_2=-(1+\epsilon)$. We know from P&C 4.16 that the modal matrix is the Vandermond matrix;

$$\mathbf{S}=\begin{pmatrix} 1 & 1 \\ \lambda_1 & \lambda_2 \end{pmatrix}=\begin{pmatrix} 1 & 1 \\ -1 & -(1+\epsilon) \end{pmatrix}$$

Since this matrix is ill-conditioned, we would have computational difficulty in finding $\mathbf{S}^{-1}$ in order to carry out a diagonalization of the system matrix $\mathbf{A}$. However, if we equate the eigenvalues (as we did above), the system matrix becomes

$$\mathbf{A}=\begin{pmatrix} 0 & 1 \\ -1 & -2 \end{pmatrix}$$

which is not diagonalizable. Moreover, the earlier computational difficulty arose because we tried to diagonalize a "nearly nondiagonalizable" matrix.

The above example has demonstrated the need for dealing with nondiagonalizable transformations. In this section we explore nondiagonalizable finite-dimensional operators in detail. We discover that they can be represented by simple, nearly diagonal matrices which have the eigenvalues on the diagonal. Thus the conceptual clarity associated with the decoupling of system equations extends, to a great extent, to general linear operators.

To avoid heavy use of the cumbersome coordinate matrix notation, we focus throughout this section on matrices. However, we should keep in mind that an $n\times n$ matrix $\mathbf{A}$ which arises in a system model usually represents an underlying linear operator **T**. The eigenvectors of **A** are the coordinates of the eigenvectors of **T**. Thus when we use a similarity transformation, $\mathbf{S}^{-1}\mathbf{A}\mathbf{S}$, to convert **A** to a new form, we are merely changing the coordinate system for the space on which **T** operates.

Generalized Nullspace and Range

Unlike a scalar, a linear operator **U** is generally neither invertible nor zero. It lies in a "gray region" in between; **U** takes some vectors to zero (acting like the zero operator); others it does not take to zero (thereby acting invertible). Perhaps even more significant is the fact that the nullspace and range of **U** may overlap. The second and higher operations by **U** may annihilate additional vectors. In some ways, the subspace annihilated by higher powers of **U** is more characteristic of the operator than is nullspace (**U**).

Example 1. Overlapping Nullspace and Range. Define the operator **U** on $\mathcal{M}^{3\times 1}$ by $\mathbf{U}\mathbf{x} \triangleq \mathbf{B}\mathbf{x}$, where

$$\mathbf{B}=\begin{pmatrix} 0 & 1 & 0 \\ 0 & 0 & 0 \\ 0 & 0 & 1 \end{pmatrix}$$

Then **U** has the following effect on a general vector in $\mathcal{M}^{3\times 1}$:

$$\underset{\mathcal{M}^{3\times 1}}{\begin{pmatrix} \xi_1 \\ \xi_2 \\ \xi_3 \end{pmatrix}} \xrightarrow{\ \mathbf{U}\ } \underset{\text{range}(\mathbf{U})}{\begin{pmatrix} \xi_2 \\ 0 \\ \xi_3 \end{pmatrix}} \xrightarrow{\ \mathbf{U}\ } \underset{\text{range}(\mathbf{U}^2)}{\begin{pmatrix} 0 \\ 0 \\ \xi_3 \end{pmatrix}} \circlearrowleft \mathbf{U}$$

The vectors annihilated by various powers of **U** are described by

$$\text{nullspace}(\mathbf{U})=\text{span}\left\{\begin{pmatrix} 1 \\ 0 \\ 0 \end{pmatrix}\right\}, \qquad \text{nullspace}(\mathbf{U}^2)=\text{span}\left\{\begin{pmatrix} 1 \\ 0 \\ 0 \end{pmatrix}, \begin{pmatrix} 0 \\ 1 \\ 0 \end{pmatrix}\right\}$$

The nullspace and range of $\mathbf{U}^k$ for $k>2$ are the same as the nullspace and range of $\mathbf{U}^2$.

Definition. The **generalized nullspace** $\mathcal{N}_g(\mathbf{U})$ of a linear operator **U** acting on an n-dimensional space $\mathcal{V}$ is the largest subspace of $\mathcal{V}$ annihilated by powers of **U**. Since $\mathcal{V}$ is finite dimensional, the annihilation must terminate. Let q be that power of **U** required for maximum annihilation.

We call q the **index of annihilation** for $\mathbf{U}$. Then $\mathfrak{N}_g(\mathbf{U}) = \text{nullspace}(\mathbf{U}^q)$. The generalized range $\mathfrak{R}_g(\mathbf{U})$ of the operator $\mathbf{U}$ is defined by $\mathfrak{R}_g(\mathbf{U}) = \text{range}(\mathbf{U}^q)$. Since multiplication by a square matrix is a linear operator, we speak also of the generalized nullspace and generalized range of square matrices.

In Example 1, the index of annihilation is $q=2$. The generalized range and generalized nullspace are

$$\mathfrak{R}_g(\mathbf{U}) = \text{span}\left\{\begin{pmatrix}0\\0\\1\end{pmatrix}\right\}, \qquad \mathfrak{N}_g(\mathbf{U}) = \text{span}\left\{\begin{pmatrix}1\\0\\0\end{pmatrix}, \begin{pmatrix}0\\1\\0\end{pmatrix}\right\}$$

Notice that $\mathcal{V}$ is the direct sum of the generalized range and the generalized nullspace of $\mathbf{U}$. It is proved in Theorem 1 of Appendix 3 that any linear operator on an n-dimensional space splits the space in this manner. It is further shown in that theorem that both $\mathfrak{N}_g(\mathbf{U})$ and $\mathfrak{R}_g(\mathbf{U})$ are invariant under $\mathbf{U}$, and that $\mathbf{U}$ acts like a reduced invertible operator on the generalized range of $\mathbf{U}$. These facts are verified by Example 1. An operator (or a square matrix) some power of which is zero is said to be **nilpotent**; $\mathbf{U}$ acts like a reduced nilpotent operator on the generalized nullspace of $\mathbf{U}$.

Exercise 1. Let $\mathbf{U}$ be the operator of Example 1. Define $\mathbf{U}_1: \mathfrak{R}_g(\mathbf{U}) \to \mathfrak{R}_g(\mathbf{U})$ by $\mathbf{U}_1\mathbf{x} \stackrel{\Delta}{=} \mathbf{U}\mathbf{x}$ for all $\mathbf{x}$ in $\mathfrak{R}_g(\mathbf{U})$; define $\mathbf{U}_2: \mathfrak{N}_g(\mathbf{U}) \to \mathfrak{N}_g(\mathbf{U})$ by $\mathbf{U}_2\mathbf{x} \stackrel{\Delta}{=} \mathbf{U}\mathbf{x}$ for all $\mathbf{x}$ in $\mathfrak{N}_g(\mathbf{U})$. Pick as bases for $\mathfrak{R}_g(\mathbf{U})$, $\mathfrak{N}_g(\mathbf{U})$, and $\mathfrak{M}^{3\times 1}$ the standard bases

$$\mathfrak{X}_1 = \left\{\begin{pmatrix}0\\0\\1\end{pmatrix}\right\}, \qquad \mathfrak{X}_2 = \left\{\begin{pmatrix}1\\0\\0\end{pmatrix}, \begin{pmatrix}0\\1\\0\end{pmatrix}\right\}, \qquad \text{and} \qquad \mathfrak{X} = \{\mathfrak{X}_1, \mathfrak{X}_2\}$$

respectively. Show that

$$[\mathbf{U}]_{\mathfrak{X}\mathfrak{X}} = \begin{pmatrix} [\mathbf{U}_1]_{\mathfrak{X}_1\mathfrak{X}_1} & \mathbf{O} \\ \mathbf{O} & [\mathbf{U}_2]_{\mathfrak{X}_2\mathfrak{X}_2} \end{pmatrix} = \left(\begin{array}{c:cc} 1 & 0 & 0 \\ \hdashline 0 & 0 & 1 \\ 0 & 0 & 0 \end{array}\right)$$

What are the characteristics of $\mathbf{U}_1$ and $\mathbf{U}_2$? Why is the matrix of $\mathbf{U}$ in "block-diagonal" form? (See P&C 4.3.)

Generalized Eigendata

The characteristic polynomial of an $n\times n$ matrix $\mathbf{A}$ can be expressed in the form

$$c(\lambda)=\det(\lambda\mathbf{I}-\mathbf{A})=(\lambda-\lambda_1)^{m_1}(\lambda-\lambda_2)^{m_2}\cdots(\lambda-\lambda_p)^{m_p} \tag{4.66}$$

where p is the number of distinct eigenvalues, and $m_1+\cdots+m_p=n$. We call m_i the **algebraic multiplicity** of λ_i. The eigenspace for λ_i is nullspace $(\mathbf{A}-\lambda_i\mathbf{I})$. The dimension of this eigenspace, the nullity of $(\mathbf{A}-\lambda_i\mathbf{I})$, we denote by k_i. We call k_i the **geometric multiplicity** of λ_i; it is the number of independent eigenvectors of $\mathbf{A}$ for λ_i. If the geometric multiplicity equals the algebraic multiplicity for each eigenvalue, it is reasonable to believe that there is a basis for $\mathfrak{M}^{n\times 1}$ composed of eigenvectors for $\mathbf{A}$, and that $\mathbf{A}$ is diagonalizable.

If λ_i is deficient in eigenvectors ($k_i<m_i$), we say $\mathbf{A}$ is **defective** at λ_i. If $\mathbf{A}$ has any defective eigenvalues, we must pick noneigenvectors to complete the basis. We seek (m_i-k_i) additional independent vectors from the subspace associated with λ_i—from the generalized nullspace of $(\mathbf{A}-\lambda_i\mathbf{I})$. Define

$$\begin{aligned}\mathcal{W}_i &\stackrel{\Delta}{=} \text{generalized nullspace of } (\mathbf{A}-\lambda_i\mathbf{I})\\ &=\text{nullspace}(\mathbf{A}-\lambda_i\mathbf{I})^{q_i}\end{aligned} \tag{4.67}$$

where q_i is the **index of annihilation** for $(\mathbf{A}-\lambda_i\mathbf{I})$. It is shown in Theorem 2 of Appendix 3 that

$$\dim(\mathcal{W}_i)=m_i \tag{4.68}$$

We will think of all vectors in the generalized nullspace of $(\mathbf{A}-\lambda_i\mathbf{I})$ as generalized eigenvectors of $\mathbf{A}$ for λ_i. Specifically, we call $\mathbf{x}_r$ a **generalized eigenvector of rank r** for λ_i if

$$\begin{aligned}(\mathbf{A}-\lambda_i\mathbf{I})^r\mathbf{x}_r&=\boldsymbol{\theta}\\ (\mathbf{A}-\lambda_i\mathbf{I})^{r-1}\mathbf{x}_r&\neq\boldsymbol{\theta}\end{aligned} \tag{4.69}$$

If $\mathbf{x}_r$ is a generalized eigenvector of rank r for λ_i, then $(\mathbf{A}-\lambda_i\mathbf{I})\mathbf{x}_r$ is a generalized eigenvector of rank $r-1$; for (4.69) can be rewritten

$$(\mathbf{A}-\lambda_i\mathbf{I})^{r-1}(\mathbf{A}-\lambda_i\mathbf{I})\mathbf{x}_r=\boldsymbol{\theta}$$

$$(\mathbf{A}-\lambda_i\mathbf{I})^{r-2}(\mathbf{A}-\lambda_i\mathbf{I})\mathbf{x}_r\neq\boldsymbol{\theta}$$

Thus each vector in $\mathcal{W}_i$ is a member of some chain of generalized eigenvectors generated by repeated multiplication with $(\mathbf{A}-\lambda_i\mathbf{I})$; the last member of each chain is a true eigenvector (of rank 1). We think of $\mathcal{W}_i$ as the **generalized eigenspace** for λ_i; $\mathcal{W}_i$ contains precisely the m_i independent vectors associated with λ_i that we intuitively expect in a basis for $\mathcal{M}^{n\times 1}$.

In Theorem 3 of Appendix 3 we show that

$$\mathcal{M}^{n\times 1}=\mathcal{W}_1\oplus\cdots\oplus\mathcal{W}_p \tag{4.70}$$

Therefore, any bases which we pick for $\{\mathcal{W}_i\}$ combine to form a basis for $\mathcal{M}^{n\times 1}$. Any basis for $\mathcal{W}_i$ consists in m_i generalized eigenvectors. Furthermore, k_i of these m_i generalized eigenvectors can be true eigenvectors for λ_i.

Jordan Canonical Form

If $\mathbf{A}$ is diagonalizable, we can diagonalize it by the similarity transformation $\mathbf{S}^{-1}\mathbf{A}\mathbf{S}$, where the columns of $\mathbf{S}$ are a basis for $\mathcal{M}^{n\times 1}$ composed of eigenvectors of $\mathbf{A}$. Suppose $\mathbf{A}$ is not diagonalizable. What form can we expect for the matrix $\mathbf{S}^{-1}\mathbf{A}\mathbf{S}$ if the columns of $\mathbf{S}$ are a basis of generalized eigenvectors of $\mathbf{A}$? It depends on the way we pick the bases for the subspaces $\{\mathcal{W}_i\}$. We demonstrate, by example, a way to pick the bases which results in as simple a form for the matrix $\mathbf{S}^{-1}\mathbf{A}\mathbf{S}$ as we can possibly get in the presence of multiple eigenvalues. In order that the form be as nearly diagonal as possible, we include, of course, the true eigenvectors for λ_i in the basis for $\mathcal{W}_i$.

Let

$$\mathbf{A}=\left(\begin{array}{ccc:cc:c} 2 & 3 & 0 & & & \\ 0 & 2 & 4 & & \bigcirc & \\ 0 & 0 & 2 & & & \\ \hdashline & & & 2 & -1 & \\ & \bigcirc & & 0 & 2 & \\ \hdashline & & & & & 3 \end{array}\right) \tag{4.71}$$

Then $c(\lambda)=(\lambda-2)^5(\lambda-3)$, or $p=2$, $\lambda_1=2$, $m_1=5$, $\lambda_2=3$, and $m_2=1$. Also,

$$(\mathbf{A}-2\mathbf{I})=\left(\begin{array}{ccc:cc:c} 0 & 3 & 0 & & & \\ 0 & 0 & 4 & & O & \\ 0 & 0 & 0 & & & \\ \hdashline & & & 0 & -1 & \\ & O & & 0 & 0 & \\ \hdashline & & & & & 1 \end{array}\right)$$

$$(\mathbf{A}-2\mathbf{I})^2=\left(\begin{array}{ccc:cc:c} 0 & 0 & 12 & & & \\ 0 & 0 & 0 & & O & \\ 0 & 0 & 0 & & & \\ \hdashline & & & 0 & 0 & \\ & O & & 0 & 0 & \\ \hdashline & & & & & 1 \end{array}\right)$$

$$(\mathbf{A}-2\mathbf{I})^3=\left(\begin{array}{ccc:cc:c} 0 & 0 & 0 & & & \\ 0 & 0 & 0 & & O & \\ 0 & 0 & 0 & & & \\ \hdashline & & & 0 & 0 & \\ & O & & 0 & 0 & \\ \hdashline & & & & & 1 \end{array}\right),$$

$$(\mathbf{A}-3\mathbf{I})=\left(\begin{array}{ccc:cc:c} -1 & 3 & 0 & & & \\ 0 & -1 & 4 & & O & \\ 0 & 0 & -1 & & & \\ \hdashline & & & -1 & -1 & \\ & O & & 0 & -1 & \\ \hdashline & & & & & 0 \end{array}\right)$$

It is apparent that

$$\begin{aligned} &\text{nullity}(\mathbf{A}-2\mathbf{I})=2=k_1\\ &\text{nullity}(\mathbf{A}-2\mathbf{I})^2=4\\ &\text{nullity}(\mathbf{A}-2\mathbf{I})^3=5\\ &\text{nullity}(\mathbf{A}-3\mathbf{I})=1=k_2 \end{aligned} \tag{4.72}$$

The indices of annihilation for $(\mathbf{A}-\lambda_1\mathbf{I})$ and $(\mathbf{A}-\lambda_2\mathbf{I})$, respectively, are $q_1=3$ and $q_2=1$. The five-dimensional subspace $\mathcal{W}_1$, the generalized eigenspace for λ_1, consists in vectors of the form $(\xi_1\ \xi_2\ \xi_3\ \xi_4\ \xi_5\ 0)^T$; vectors in $\mathcal{W}_2$, the generalized eigenspace for λ_2, are of the form $(0\ 0\ 0\ 0\ 0\ \xi_6)^T$. [Note that (4.68) and (4.70) are verified in this example.]

Any eigenvector for $\lambda = 3$ will form a basis $\mathcal{X}_2$ for $\mathcal{W}_2$. Clearly, a basis $\mathcal{X}_1$ for $\mathcal{W}_1$ must contain five vectors. Since there are only two independent true eigenvectors (of rank 1), three of the vectors in the basis must be generalized eigenvectors of rank greater than 1.

Assume we pick a basis which reflects the nullity structure of (4.72); that is, we pick two generalized eigenvectors of rank 1 for $\lambda = 2$, two of rank 2 for $\lambda = 2$, one of rank 3 for $\lambda = 2$, and one of rank 1 for $\lambda = 3$. Also assume we pick the basis vectors in chains; that is, if $\mathbf{x}$ is a vector of rank 3 for $\lambda = 2$, and $\mathbf{x}$ is in the basis, $(\mathbf{A} - 2\mathbf{I})\mathbf{x}$ and $(\mathbf{A} - 2\mathbf{I})^2\mathbf{x}$ will also be in the basis. We express both the nullity structure and chain structure by the following subscript notation:

$$\begin{array}{lllll} & \text{rank 1} & \text{rank 2} & \text{rank 3} & \\ & \downarrow & \downarrow & \downarrow & \\ \mathcal{X}_1 = & \left\{\begin{array}{l}\mathbf{x}_1 \\ \mathbf{x}_2\end{array}\right. & \begin{array}{l}\mathbf{x}_{12} \\ \mathbf{x}_{22}\end{array} & \mathbf{x}_{13} & \begin{array}{l}\leftarrow \text{chain 1} \\ \leftarrow \text{chain 2}\end{array} \\ \mathcal{X}_2 = & \{\mathbf{x}_3 & & & \leftarrow \text{chain 3} \end{array} \tag{4.73}$$

This nullity and chain structure is expressed mathematically by the following equations:

$$\begin{aligned} (\mathbf{A} - 2\mathbf{I})\mathbf{x}_{13} &= \mathbf{x}_{12} \\ (\mathbf{A} - 2\mathbf{I})\mathbf{x}_{12} &= \mathbf{x}_1 \\ (\mathbf{A} - 2\mathbf{I})\mathbf{x}_1 &= \boldsymbol{\theta} \\ (\mathbf{A} - 2\mathbf{I})\mathbf{x}_{22} &= \mathbf{x}_2 \\ (\mathbf{A} - 2\mathbf{I})\mathbf{x}_2 &= \boldsymbol{\theta} \\ (\mathbf{A} - 3\mathbf{I})\mathbf{x}_3 &= \boldsymbol{\theta} \end{aligned} \tag{4.74}$$

We propose the union of the sets $\mathcal{X}_i$ as a basis, denoted $\mathcal{X}$, for $\mathcal{M}^{6\times 1}$. It can be shown that a set of vectors of this form can be constructed and is a basis for $\mathcal{M}^{6\times 1}$ (see Friedman [4.7]). Using the basis $\mathcal{X}$, we form the change of coordinates matrix as in (4.17):

$$\mathbf{S} = (\mathbf{x}_1 \,\vdots\, \mathbf{x}_{12} \,\vdots\, \mathbf{x}_{13} \,\vdots\, \mathbf{x}_2 \,\vdots\, \mathbf{x}_{22} \,\vdots\, \mathbf{x}_3) \tag{4.75}$$

As in (4.18), this change of coordinates transforms $\mathbf{A}$ into the matrix $\mathbf{\Lambda} = \mathbf{S}^{-1}\mathbf{A}\mathbf{S}$. Recasting this similarity relation into the form $\mathbf{A}\mathbf{S} = \mathbf{S}\mathbf{\Lambda}$, we recognize that

$$\begin{aligned}
\mathbf{AS} &= \mathbf{A}(\mathbf{x}_1 \vdots \mathbf{x}_{12} \vdots \mathbf{x}_{13} \vdots \mathbf{x}_2 \vdots \mathbf{x}_{22} \vdots \mathbf{x}_3) \\
&= (2\mathbf{x}_1 \vdots 2\mathbf{x}_{12}+\mathbf{x}_1 \vdots 2\mathbf{x}_{13}+\mathbf{x}_{12} \vdots 2\mathbf{x}_2 \vdots 2\mathbf{x}_{22}+\mathbf{x}_2 \vdots 3\mathbf{x}_3) \\
&= (\mathbf{x}_1 \vdots \mathbf{x}_{12} \vdots \mathbf{x}_{13} \vdots \mathbf{x}_2 \vdots \mathbf{x}_{22} \vdots \mathbf{x}_3) \begin{pmatrix} 2 & 1 & 0 & & & \\ 0 & 2 & 1 & & \bigcirc & \\ 0 & 0 & 2 & & & \\ & & & 2 & 1 & \\ & \bigcirc & & 0 & 2 & \\ & & & & & 3 \end{pmatrix} \\
&= \mathbf{S\Lambda}
\end{aligned} \tag{4.76}$$

The form of $\mathbf{\Lambda}$ is as simple and as nearly diagonal a representation of $\mathbf{A}$ as we can expect to obtain. The eigenvalues are on the diagonal. The off-diagonal 1's specify in a simple manner the "rank structure" or "chain structure" inherent in $\mathbf{A}$.

It is apparent that whenever the columns of $\mathbf{S}$ form a basis for $\mathfrak{M}^{n\times 1}$ composed of generalized eigenvectors of $\mathbf{A}$, and these basis vectors consist in chains of vectors which express the nullity structure of $\mathbf{A}$ as in (4.73)–(4.74), then $\mathbf{S}^{-1}\mathbf{AS}$ will be of the simple form demonstrated in (4.76). It will consist in a series of blocks on the diagonal; each block will be of the form

$$\begin{pmatrix} \lambda_i & 1 & 0 & \cdots & 0 \\ 0 & \lambda_i & 1 & \cdots & 0 \\ \vdots & & & & \vdots \\ 0 & \cdots & & \lambda_i & 1 \\ 0 & \cdots & & 0 & \lambda_i \end{pmatrix}$$

By analogy with (4.16)–(4.18) in our discussion of diagonalization, we call $\mathbf{S}$ the **modal matrix for A**. We also call the near-diagonal matrix $\mathbf{\Lambda}$ the **spectral matrix for A** (or for the underlying transformation $\mathbf{T}$). The spectral matrix is also referred to as the **Jordan canonical form of A**. Each square block consisting in a repeated eigenvalue on the diagonal and an unbroken string of 1's above the diagonal is called a **Jordan block**. There is one Jordan block in $\mathbf{\Lambda}$ for each chain of generalized eigenvectors in the basis. The dimension of each block equals the length of the corresponding chain. Thus we can tell from the nullity structure (4.71) alone, the form of the basis (4.73) and the precise form of $\mathbf{\Lambda}$ (4.76). Observe that the Jordan form is not unique. We can choose arbitrarily the order of the Jordan blocks by choosing the order in which we place the generalized eigenvectors in the basis.

Example 2. Nullities Determine the Jordan Form. Suppose **A** is a 9×9 matrix for which

$$c(\lambda)=\det(\lambda\mathbf{I}-\mathbf{A})=(\lambda-\lambda_1)^6(\lambda-\lambda_2)^2(\lambda-\lambda_3)$$

$$\text{nullity}(\mathbf{A}-\lambda_1\mathbf{I})=3$$

$$\text{nullity}(\mathbf{A}-\lambda_1\mathbf{I})^2=5$$

$$\text{nullity}(\mathbf{A}-\lambda_1\mathbf{I})^3=6$$

$$\text{nullity}(\mathbf{A}-\lambda_2\mathbf{I})=1$$

$$\text{nullity}(\mathbf{A}-\lambda_2\mathbf{I})^2=2$$

$$\text{nullity}(\mathbf{A}-\lambda_3\mathbf{I})=1$$

From (4.68), the factored characteristic polynomial, and the nullities stated above, we know that

$$m_1=\dim(\mathcal{W}_1)=6, \qquad k_1=3$$

$$m_2=\dim(\mathcal{W}_2)=2, \qquad k_2=1$$

$$m_3=\dim(\mathcal{W}_3)=1, \qquad k_3=1$$

It follows that $q_1=3$, $q_2=2$, and $q_3=1$; higher powers than $(\mathbf{A}-\lambda_i\mathbf{I})^{q_i}$ do not have higher nullities. The form of the basis of generalized eigenvectors of **A** which will convert **A** to its Jordan form is

$$\mathscr{X} = \begin{cases} \mathscr{X}_1 = \begin{cases} \mathbf{x}_1 & \mathbf{x}_{12} & \mathbf{x}_{13} \\ \mathbf{x}_2 & \mathbf{x}_{22} & \\ \mathbf{x}_3 & & \end{cases} \\ \mathscr{X}_2 = \{\mathbf{x}_4 \quad \mathbf{x}_{42} \\ \mathscr{X}_3 = \{\mathbf{x}_5 \end{cases}$$

The Jordan form of **A** is

$$\mathbf{\Lambda}=\begin{pmatrix} \lambda_1 & 1 & 0 & & & & & & \\ 0 & \lambda_1 & 1 & & & & & & \\ 0 & 0 & \lambda_1 & & & & & & \\ & & & \lambda_1 & 1 & & & & \\ & & & 0 & \lambda_1 & & & & \\ & & & & & \lambda_1 & & & \\ & & & & & & \lambda_2 & 1 & \\ & & & & & & 0 & \lambda_2 & \\ & & & & & & & & \lambda_3 \end{pmatrix}$$

Bases of Generalized Eigenvectors

We now generate a specific basis for $\mathfrak{M}^{6\times 1}$ which is composed of generalized eigenvectors of the matrix $\mathbf{A}$ of (4.71). That is, we find a basis of the form (4.73) by satisfying (4.74). We use (4.69) to find the highest rank vector in each chain. We first seek the vector $\mathbf{x}_{13}$ of (4.73). All five of the basis vectors in $\mathcal{A}_1$ satisfy $(\mathbf{A}-2\mathbf{I})^3\mathbf{x}=\boldsymbol{\theta}$. But only $\mathbf{x}_{13}$ satisfies, in addition, $(\mathbf{A}-2\mathbf{I})^2\mathbf{x}\neq\boldsymbol{\theta}$. Therefoe, we let $\mathbf{x}_{13}=(c_1\ c_2\ c_3\ c_4\ c_5\ 0)^{\mathrm{T}}$, thegeneral solution to $(\mathbf{A}-2\mathbf{I})^3\mathbf{x}=\boldsymbol{\theta}$. Then

$$(\mathbf{A}-2\mathbf{I})^2\mathbf{x}_{13}=\begin{pmatrix}12c_3\\0\\0\\0\\0\\0\end{pmatrix}\neq\boldsymbol{\theta} \tag{4.77}$$

or $c_3\neq 0$. Thus any vector in $\mathfrak{M}^{6\times 1}$ which has a zero sixth element and a nonzero third element is a generalized eigenvector of rank 3 for $\lambda=2$. We have a lot of freedom in picking $\mathbf{x}_{13}$. Arbitrarily, we let $c_3=1$, and $c_1=c_2=c_4=c_5=0$. Then

$$\mathbf{x}_{13}=\begin{pmatrix}0\\0\\1\\0\\0\\0\end{pmatrix},\quad \mathbf{x}_{12}=(\mathbf{A}-2\mathbf{I})\mathbf{x}_{13}=\begin{pmatrix}0\\4\\0\\0\\0\\0\end{pmatrix},\quad \mathbf{x}_{1}=(\mathbf{A}-2\mathbf{I})\mathbf{x}_{12}=\begin{pmatrix}12\\0\\0\\0\\0\\0\end{pmatrix} \tag{4.78}$$

Notice that in (4.77) we looked at the eigenvector, $\mathbf{x}_1=(\mathbf{A}-2\mathbf{I})^2\mathbf{x}_{13}$, at the end of the chain in order to determine the vector $\mathbf{x}_{13}$ at the head of the chain.

To find the remaining vectors of $\mathcal{A}_1$, we look for the vector $\mathbf{x}_{22}$ at the head of the second chain. By (4.69), all vectors $(d_1\ d_2\ d_3\ d_4\ d_5\ d_6)^{\mathrm{T}}$ of rank 2 or less satisfy

$$(\mathbf{A}-2\mathbf{I})^2\begin{pmatrix}d_1\\d_2\\d_3\\d_4\\d_5\\d_6\end{pmatrix}=\begin{pmatrix}12d_3\\0\\0\\0\\0\\d_6\end{pmatrix}=\boldsymbol{\theta}$$

or $d_6 = d_3 = 0$. The vectors which are precisely of rank 2 also satisfy

$$(\mathbf{A}-2\mathbf{I})\begin{pmatrix} d_1 \\ d_2 \\ 0 \\ d_4 \\ d_5 \\ 0 \end{pmatrix} = \begin{pmatrix} 3d_2 \\ 0 \\ 0 \\ -d_5 \\ 0 \\ 0 \end{pmatrix} \neq \boldsymbol{\theta} \tag{4.79}$$

Again we are looking at the eigenvector at the end of the chain as we pick the constants. We must pick d_2 and d_5, not both zero, such that $\mathbf{x}_2$ is independent of the eigenvector $\mathbf{x}_1$ selected above (i.e., $d_2 = 1$, $d_5 = 0$ will not do). Arbitrarily, we let $d_5 = 1$, $d_1 = d_2 = d_4 = 0$; d_3 is already zero. Thus

$$\mathbf{x}_{22} = \begin{pmatrix} 0 \\ 0 \\ 0 \\ 0 \\ 1 \\ 0 \end{pmatrix}, \qquad \mathbf{x}_2 = (\mathbf{A}-2\mathbf{I})\mathbf{x}_{22} = \begin{pmatrix} 0 \\ 0 \\ 0 \\ -1 \\ 0 \\ 0 \end{pmatrix} \tag{4.80}$$

The five vectors of (4.78) and (4.80) satisfy (4.73), and they are a basis for $\mathcal{W}_1$. The equation $(\mathbf{A}-3\mathbf{I})\mathbf{x} = \boldsymbol{\theta}$ determines the form of eigenvectors for $\lambda = 3$: $\mathbf{x} = (0\ 0\ 0\ 0\ 0\ b_6)^{\mathrm{T}}$. We arbitrarily let $b_6 = 1$ to get

$$\mathbf{x}_3 = \begin{pmatrix} 0 \\ 0 \\ 0 \\ 0 \\ 0 \\ 1 \end{pmatrix}$$

a basis for $\mathcal{W}_2$. By (4.76), this basis of generalized eigenvectors generates the modal matrix $\mathbf{S}$:

$$\mathbf{S} = \begin{pmatrix} 12 & 0 & 0 & 0 & 0 & 0 \\ 0 & 4 & 0 & 0 & 0 & 0 \\ 0 & 0 & 1 & 0 & 0 & 0 \\ 0 & 0 & 0 & -1 & 0 & 0 \\ 0 & 0 & 0 & 0 & 1 & 0 \\ 0 & 0 & 0 & 0 & 0 & 1 \end{pmatrix} \qquad \mathbf{S}^{-1} = \begin{pmatrix} \frac{1}{12} & 0 & 0 & 0 & 0 & 0 \\ 0 & \frac{1}{4} & 0 & 0 & 0 & 0 \\ 0 & 0 & 1 & 0 & 0 & 0 \\ 0 & 0 & 0 & -1 & 0 & 0 \\ 0 & 0 & 0 & 0 & 1 & 0 \\ 0 & 0 & 0 & 0 & 0 & 1 \end{pmatrix}$$

The spectral matrix is

$$\boldsymbol{\Lambda}=\mathbf{S}^{-1}\mathbf{A}\mathbf{S}=\left(\begin{array}{ccc:cc:c} 2 & 1 & 0 & & & \\ 0 & 2 & 1 & & \bigcirc & \\ 0 & 0 & 2 & & & \\ \hdashline & & & 2 & 1 & \\ & \bigcirc & & 0 & 2 & \\ \hdashline & & & & & 3 \end{array}\right) \tag{4.81}$$

as we concluded earlier in (4.76).

Clearly, the chains of generalized eigenvectors which make up a basis are not unique. In fact, many different chains end in the same true eigenvector. It can be shown that any set of chains which possesses the structure of (4.73)–(4.74) will constitute a basis for $\mathfrak{M}^{6\times 1}$ if the eigenvectors at the ends of the chains are independent. Because of this fact, we might be led to find the true eigenvectors $\mathbf{x}_1$ and $\mathbf{x}_2$ first, and then find the rest of the basis by "backing up" each chain. This approach need not work. The vectors $\mathbf{x}_1=(1\ 0\ 0\ 1\ 0\ 0)^{\mathrm{T}}$ and $\mathbf{z}_2=(1\ 0\ 0\ -1\ 0\ 0)^{\mathrm{T}}$ are independent eigenvectors of $\mathbf{A}$. However, they are both of the form (4.79) of eigenvectors at the end of chains of length 2. Neither is of the form (4.77) of an eigenvector at the end of a chain of length 3. Although these two eigenvectors can be used as part of a basis for $\mathfrak{M}^{6\times 1}$, the basis cannot be of the form (4.73).

Exercise 2. Attempt to determine a basis for $\mathfrak{M}^{6\times 1}$ which is of the form (4.73) and yet includes the eigenvectors $\mathbf{x}_1=(1\ 0\ 0\ 1\ 0\ 0)^{\mathrm{T}}$ and $\mathbf{x}_2=(1\ 0\ 0\ -1\ 0\ 0)^{\mathrm{T}}$.

Procedure for Construction of the Basis

We summarize the procedure for generating a basis of generalized eigenvectors. Suppose the $n\times n$ matrix $\mathbf{A}$ has the characteristic polynomial (4.66). Associated with the eigenvalue λ_i is an m_i-dimensional subspace $\mathcal{W}_i$ (Theorem 1, Appendix 3). This subspace contains k_i independent eigenvectors for λ_i. Assume the basis vectors are ordered by decreasing chain length, with each chain ordered by increasing rank. We denote this basis for $\mathcal{W}_i$ by

$$\mathcal{Q}_i=\left\{\begin{array}{ccccc} \mathbf{x}_1 & \mathbf{x}_{12} & \cdots & & \mathbf{x}_{1q_i} \\ \mathbf{x}_2 & \mathbf{x}_{22} & \cdots & \mathbf{x}_{2l_2} & \\ \vdots & & & & \\ \mathbf{x}_{k_i} & \mathbf{x}_{k_i 2} & \cdots & \mathbf{x}_{k_i l_{k_i}} & \\ \uparrow & & & & \uparrow \\ \text{rank } 1 & & & & \text{rank } q_i \end{array}\right\} \begin{array}{l} \leftarrow\text{longest chain} \\ \\ \\ \leftarrow\text{shortest chain} \\ \\ \\ \end{array} \tag{4.82}$$

where l_j is the length of the jth chain for λ_i and q_i is the index of annihilation for $(\mathbf{A}-\lambda_i\mathbf{I})$; thus q_i is the length of the longest chain. The nullities of various powers of $(\mathbf{A}-\lambda_i\mathbf{I})$ determine the structure of (4.82) just as (4.73) is determined by (4.72). The procedure for construction of the basis $\mathcal{C}_i$ is as follows:

1. Determine the form of vectors of rank q_i or less by solving $(\mathbf{A}-\lambda_i\mathbf{I})^{q_i}\mathbf{x}=\boldsymbol{\theta}$.
2. Observe the true eigenvectors $(\mathbf{A}-\lambda_i\mathbf{I})^{q_i-1}\mathbf{x}$; choose from the vectors found in (1) a total of $(\text{nullity}(\mathbf{A}-\lambda_i\mathbf{I})^{q_i}-\text{nullity}(\mathbf{A}-\lambda_i\mathbf{I})^{q_i-1})$ vectors which lead to independent eigenvectors. These vectors are of rank q_i, and are the highest rank generalized eigenvectors in their respective chains.
3. Multiply each vector chosen in (2) by $(\mathbf{A}-\lambda_i\mathbf{I})$, thereby obtaining a set of generalized eigenvectors of rank(q_i-1), which is part of the set of basis vectors of rank (q_i-1).
4. Complete the set of basis vectors of rank(q_i-1) by adding enough vectors of rank (q_i-1) to obtain a total of $(\text{nullity}(\mathbf{A}-\lambda_i\mathbf{I})^{q_i-1}-\text{nullity}(\mathbf{A}-\lambda_i\mathbf{I})^{q_i-2})$ vectors which lead to independent eigenvectors. This step requires work equivalent to steps 1 and 2 with q_i replaced by (q_i-1). The vectors which are added are highest rank vectors in new chains.
5. Repeat steps 3 and 4 for lower ranks until a set of k_i eigenvectors is obtained.

Because $\mathfrak{M}^{n\times 1}=\mathcal{W}_1\oplus\cdots\oplus\mathcal{W}_p$, we can obtain a basis $\mathcal{C}$ for $\mathfrak{M}^{n\times 1}$ consisting of generalized eigenvectors of $\mathbf{A}$ by merely combining the bases for the subspaces $\mathcal{W}_i$:

$$\mathcal{C}=\{\mathcal{C}_1,\ldots,\mathcal{C}_p\}$$

Proceeding as in the example of (4.71), we can use the basis $\mathcal{C}$ to convert $\mathbf{A}$ to its nearly diagonal Jordan canonical form $\boldsymbol{\Lambda}$.

Example 3. A Basis of Generalized Eigenvectors. Let

$$\mathbf{A}=\begin{pmatrix} 2 & 1 & 0 & 0 & 0 & 0\\ -1 & 4 & 0 & 0 & 0 & 0\\ -1 & 1 & 2 & 1 & 0 & 0\\ -1 & 1 & -1 & 4 & 0 & 0\\ -1 & 1 & -1 & 1 & 3 & 0\\ -1 & 1 & -1 & 1 & 1 & 2 \end{pmatrix}$$

The process of finding and factoring the characteristic polynomial is complicated. We merely state it in factored form:

$$c(\lambda)=\det(\lambda\mathbf{I}-\mathbf{A})=(\lambda-3)^5(\lambda-2)$$

Therefore, $\lambda_1 = 3$, $m_1 = 5$, $\lambda_2 = 2$, and $m_2 = 1$. Furthermore,

$$(\mathbf{A}-3\mathbf{I}) = \begin{pmatrix} -1 & 1 & 0 & 0 & 0 & 0 \\ -1 & 1 & 0 & 0 & 0 & 0 \\ -1 & 1 & -1 & 1 & 0 & 0 \\ -1 & 1 & -1 & 1 & 0 & 0 \\ -1 & 1 & -1 & 1 & 0 & 0 \\ -1 & 1 & -1 & 1 & 1 & -1 \end{pmatrix}, \quad (\mathbf{A}-3\mathbf{I})^2 = \begin{pmatrix} 0 & 0 & 0 & 0 & 0 & 0 \\ 0 & 0 & 0 & 0 & 0 & 0 \\ 0 & 0 & 0 & 0 & 0 & 0 \\ 0 & 0 & 0 & 0 & 0 & 0 \\ 0 & 0 & 0 & 0 & 0 & 0 \\ 0 & 0 & 0 & 0 & -1 & 1 \end{pmatrix}$$

Clearly, $\text{nullity}(\mathbf{A}-3\mathbf{I}) = 3$ and $\text{nullity}(\mathbf{A}-3\mathbf{I})^2 = 5 = m_1$. It is also apparent that $\text{nullity}(\mathbf{A}-3\mathbf{I})^3 = 5$. Thus $k_1 = 3$, $q_1 = 2$, and $\dim(\mathcal{W}_1) = 5$. Moreover,

$$(\mathbf{A}-2\mathbf{I}) = \begin{pmatrix} 0 & 1 & 0 & 0 & 0 & 0 \\ -1 & 2 & 0 & 0 & 0 & 0 \\ -1 & 1 & 0 & 1 & 0 & 0 \\ -1 & 1 & -1 & 2 & 0 & 0 \\ -1 & 1 & -1 & 1 & 1 & 0 \\ -1 & 1 & -1 & 1 & 1 & 0 \end{pmatrix}$$

and $\text{nullity}(\mathbf{A}-2\mathbf{I}) = 1$. As a result, $k_2 = 1$, $q_2 = 1$, and $\dim(\mathcal{W}_2) = 1$. [Note that $\dim(\mathcal{W}_1) + \dim(\mathcal{W}_2) = \dim(\mathfrak{M}^{6\times 1})$.] From the nullity information above, we know that the Jordan form of $\mathbf{A}$ is

$$\boldsymbol{\Lambda} = \begin{pmatrix} 3 & 1 & & & & \\ 0 & 3 & & & \bigcirc & \\ & & 3 & 1 & & \\ & & 0 & 3 & & \\ & \bigcirc & & & 3 & \\ & & & & & 2 \end{pmatrix}$$

We find a basis $\mathscr{X}$ for $\mathfrak{M}^{6\times 1}$ consisting in chains of generalized eigenvectors with the following structure:

$$\mathscr{X} = \begin{cases} \mathscr{X}_1 = \begin{cases} \mathbf{x}_1 & \mathbf{x}_{12} \\ \mathbf{x}_2 & \mathbf{x}_{22} \\ \mathbf{x}_3 & \end{cases} \\ \mathscr{X}_2 = \{\mathbf{x}_4 \end{cases}$$

We first seek $\mathbf{x}_{12}$ and $\mathbf{x}_{22}$, the vectors at the heads of the two longest chains. All generalized eigenvectors for $\lambda = 3$ satisfy $(\mathbf{A}-3\mathbf{I})^2\mathbf{x} = \boldsymbol{\theta}$. The solutions to this equation are of the form $\mathbf{x} = (c_1\ c_2\ c_3\ c_4\ c_5\ c_5)^{\mathrm{T}}$. The vectors of rank 2 also satisfy

$$(\mathbf{A}-3\mathbf{I})\begin{pmatrix} c_1 \\ c_2 \\ c_3 \\ c_4 \\ c_5 \\ c_5 \end{pmatrix} = (c_2 - c_1)\begin{pmatrix} 1 \\ 1 \\ 1 \\ 1 \\ 1 \\ 1 \end{pmatrix} + (c_4 - c_3)\begin{pmatrix} 0 \\ 0 \\ 1 \\ 1 \\ 1 \\ 1 \end{pmatrix} \neq \begin{pmatrix} 0 \\ 0 \\ 0 \\ 0 \\ 0 \\ 0 \end{pmatrix}$$

We are looking at the true eigenvector at the end of the most general chain of length 2. We must select two different sets of constants in order to specify both $\mathbf{x}_{12}$ and $\mathbf{x}_{22}$. Furthermore, we must specify these constants in such a way that the eigenvectors $\mathbf{x}_1$ and $\mathbf{x}_2$ (which are derived from $\mathbf{x}_{12}$ and $\mathbf{x}_{22}$, respectively) are independent. It is clear by inspection of the above equation that precisely two independent eigenvectors are available. By choosing $c_2 = 1$ and $c_1 = c_3 = c_4 = c_5 = 0$, we make

$$\mathbf{x}_{12} = \begin{pmatrix} 0 \\ 1 \\ 0 \\ 0 \\ 0 \\ 0 \end{pmatrix} \quad \text{and} \quad \mathbf{x}_1 = \begin{pmatrix} 1 \\ 1 \\ 1 \\ 1 \\ 1 \\ 1 \end{pmatrix}$$

By selecting $c_4 = 1$ and $c_1 = c_2 = c_3 = c_5 = 0$ we get

$$\mathbf{x}_{22} = \begin{pmatrix} 0 \\ 0 \\ 0 \\ 1 \\ 0 \\ 0 \end{pmatrix} \quad \text{and} \quad \mathbf{x}_2 = \begin{pmatrix} 0 \\ 0 \\ 1 \\ 1 \\ 1 \\ 1 \end{pmatrix}$$

Of course, many other choices of $\mathbf{x}_{12}$ and $\mathbf{x}_{22}$ would yield the same $\mathbf{x}_1$ and $\mathbf{x}_2$. Furthermore, other choices of $\mathbf{x}_1$ and $\mathbf{x}_2$ would also have been appropriate. We now seek $\mathbf{x}_3$, a third true eigenvector for $\lambda = 3$ which is independent of $\mathbf{x}_1$ and $\mathbf{x}_2$. The eigenvectors for $\lambda = 3$ satisfy $(\mathbf{A} - 3\mathbf{I}) = \boldsymbol{\theta}$. From the matrix $\mathbf{A} - 3\mathbf{I}$ we recognize that $c_1 = c_2$ and $c_3 = c_4$, as well as $c_5 = c_6$ for all eigenvectors for $\lambda = 3$. Letting $c_1 = c_2 = c_3 = c_4 = 0$ and $c_5 = c_6 = 1$, we obtain

$$\mathbf{x}_3 = \begin{pmatrix} 0 \\ 0 \\ 0 \\ 0 \\ 1 \\ 1 \end{pmatrix}$$

an eigenvector independent of the other two. It is a simple matter to determine $\mathbf{x}_4$, an eigenvector for $\lambda = 2$; we choose

$$\mathbf{x}_4 = \begin{pmatrix} 0 \\ 0 \\ 0 \\ 0 \\ 0 \\ 1 \end{pmatrix}$$

Exercise 3. Continuing Example 3, let

$$\mathbf{S} = \left(\mathbf{x}_1 \ \vdots \ \mathbf{x}_{12} \ \vdots \ \mathbf{x}_2 \ \vdots \ \mathbf{x}_{22} \ \vdots \ \mathbf{x}_3 \ \vdots \ \mathbf{x}_4 \right)$$

Show that $\boldsymbol{\Lambda} = \mathbf{S}^{-1}\mathbf{A}\mathbf{S}$.

Generalized Eigenvectors in Function Spaces

Our discussion of generalized eigenvectors has been directed primarily toward matrices and, through matrices of transformations, toward any linear operator on an n-dimensional vector space. However, the concepts apply also to transformations on infinite-dimensional spaces. We have already noted that for the operator $\mathbf{D}$ acting on the space $\mathcal{C}^1(0,1)$, any scalar λ is an eigenvalue, and that $e^{\lambda t}$ is a corresponding eigenfunction. Furthermore, there is no other eigenfunction for λ which is independent from $e^{\lambda t}$—the geometric multiplicity of λ is one.

We have not to this point explored the generalized nullspace for λ. In point of fact, powers of $(\mathbf{D}-\lambda\mathbf{I})$ do annihilate additional functions. Specifically, $(\mathbf{D}-\lambda\mathbf{I})^r$ annihilates the r-dimensional subspace of functions of the form $c_1e^{\lambda t}+c_2te^{\lambda t}+c_3t^2e^{\lambda t}+\cdots+c_rt^{r-1}e^{\lambda t}$. The annihilation does not terminate as r increases; the index of annihilation is infinite. It is apparent that the following functions constitute an infinite chain of generalized eigenfunctions of $\mathbf{D}$ for the eigenvalue λ:

$$e^{\lambda t}, te^{\lambda t}, \frac{1}{2!}t^2e^{\lambda t}, \frac{1}{3!}t^3e^{\lambda t}, \ldots \tag{4.83}$$

Generally, differential operators are accompanied by boundary conditions. The eigenvalues of a differential operator $\mathbf{L}$ (with its boundary conditions) are the roots of the eigenvalue equation (4.40), $\det(\mathbf{B}(\lambda))=0$. As in (4.41), the eigenfunctions corresponding to the eigenvalue λ_i are linear combinations of a set of fundamental solutions for $\mathbf{L}$, where the multipliers in the linear combination satisfy

$$\mathbf{B}(\lambda_i)\begin{pmatrix} c_1 \\ \vdots \\ c_n \end{pmatrix}=\begin{pmatrix} 0 \\ \vdots \\ 0 \end{pmatrix}$$

The **algebraic multiplicity** of the eigenvalue λ_i is the multiplicity of λ_i as a root of the eigenvalue equation. The nullity of $\mathbf{B}(\lambda_i)$ equals the number of independent eigenfunctions of $\mathbf{L}$ for the single eigenvalue λ_i; we call this number the **geometric multiplicity** of λ_i. It can be shown that $k_i \leqslant m_i$, just as we found for matrices (see Ince [4.10]). In the above example, where no boundary conditions were applied to the operator $\mathbf{D}$, these definitions do not apply. However, it seems appropriate in that case to assume that $m_i=\infty$ and $k_i=1$ for each scalar λ_i, since there is an infinite string of generalized eigenfunctions associated with each λ_i. See P&C 4.12*d* for a differential operator (with boundary conditions) which possesses multiple eigenvalues.

The Minimal Polynomial

We showed in (4.15) that if an $n\times n$ matrix $\mathbf{A}$ has distinct roots, its characteristic polynomial in $\mathbf{A}$ is $\mathbf{\Theta}$; that is, $c(\mathbf{A})=(\mathbf{A}-\lambda_i\mathbf{I})\cdots(\mathbf{A}-\lambda_n\mathbf{I})=\mathbf{\Theta}$. We are now in a position to extend this result to all square matrices. The fact that $\mathfrak{M}^{n\times 1}=\mathfrak{W}_1\oplus\cdots\oplus\mathfrak{W}_p$ is proved in Theorem 3 of Appendix 3. By definition (4.67), $(\mathbf{A}-\lambda_i\mathbf{I})^{q_i}$ annihilates $\mathfrak{W}_i$. Furthermore, $\mathfrak{W}_j$ is invariant under $(\mathbf{A}-\lambda_i\mathbf{I})^{q_i}$ if $j\neq i$. Therefore, the matrix

$$(\mathbf{A}-\lambda_1\mathbf{I})^{q_1}\cdots(\mathbf{A}-\lambda_p\mathbf{I})^{q_p}$$

annihilates the whole space $\mathfrak{M}^{n\times 1}$. We call

$$m(\lambda)\triangleq(\lambda-\lambda_1)^{q_1}\cdots(\lambda-\lambda_p)^{q_p}$$

the **minimal polynomial** for $\mathbf{A}$. The minimal polynomial in $\mathbf{A}$ satisfies

$$m(\mathbf{A})\triangleq(\mathbf{A}-\lambda_1\mathbf{I})^{q_1}\cdots(\mathbf{A}-\lambda_p\mathbf{I})^{q_p}=\mathbf{\Theta} \tag{4.84}$$

If $r\triangleq q_1+\cdots+q_p$, then $m(\mathbf{A})=\mathbf{A}^r+a_1\mathbf{A}^{r-1}+\cdots+a_r\mathbf{I}$, an rth-order polynomial in $\mathbf{A}$. In fact, $m(\mathbf{A})$ is the lowest-order polynomial in $\mathbf{A}$ which annihilates the whole space. It is apparent that polynomials in $\mathbf{A}$ which include higher powers of $(\mathbf{A}-\lambda_i\mathbf{I})$ also annihilate the space. For instance, recalling that $m_i\geqslant q_i$, the characteristic polynomial in $\mathbf{A}$ satisfies

$$c(\mathbf{A})=(\mathbf{A}-\lambda_1\mathbf{I})^{m_1}\cdots(\mathbf{A}-\lambda_p\mathbf{I})^{m_p}=\mathbf{\Theta} \tag{4.85}$$

for any square matrix $\mathbf{A}$. Equation (4.85) is the **Cayley–Hamilton theorem**. Equations (4.84) and (4.85) find considerable use in computing. See, for example, Krylov's method (4.23) for finding the characteristic equation; see also the computation of functions of matrices via (4.108).

Example 4. A Minimal Polynomial. Let

$$\mathbf{A}=\begin{pmatrix}1&2&0\\0&1&0\\0&0&1\end{pmatrix}$$

Then $p=1$, $\lambda_1=1$, and $c(\lambda)=(\lambda-1)^3$. Since

$$(\mathbf{A}-\mathbf{I})=\begin{pmatrix}0&2&0\\0&0&0\\0&0&0\end{pmatrix}$$

and $(\mathbf{A}-\mathbf{I})^2=\mathbf{\Theta}$, $q_i=2$, and $m(\lambda)=(\lambda-1)^2$. It is apparent that $c(\mathbf{A})=m(\mathbf{A})=\mathbf{\Theta}$.

4.5 Applications of Generalized Eigendata

The concept of the Jordan form of a matrix is useful partly because it is mnemonic—it helps us remember and categorize the fundamental properties of the matrix (or the linear transformation which the matrix represents). The diagonal form of a diagonalizable matrix is merely a special case of the Jordan form. Whether an operator is diagonalizable or not, a complete eigenvalue analysis—obtaining eigenvalues and eigenvectors—is a computationally expensive process. Thus computational efficiency alone does not ordinarily justify the use of spectral decomposition (decomposition by means of eigenvectors) as a technique for solving an operator equation. However, our reason for analyzing an operator is usually to gain insight into the input–output relation which it describes. Spectral analysis of a model does develop intuitive insight concerning this input–output relation. In some instances a basis of eigenvectors is known a priori, and it need not be computed (e.g., the symmetrical components of (4.28), the Vandermond matrix of P&C 4.16, and the complex exponential functions of Fourier series expansions). In these instances, we gain the insight of spectral decomposition with little more effort than that involved in solution of the operator equation.

Nearly Equal Eigenvalues

True multiple eigenvalues rarely appear in physical systems. But nearly equal eigenvalues are often accompanied by near singularity of the linear operator and, therefore, by computational difficulty. This difficulty can sometimes be avoided by equating the nearly equal eigenvalues and computing generalized eigenvectors in the manner described earlier.

Example 1. Nearly Equal Eigenvalues. In the introduction to Section 4.4 we described a dynamic system with nearly equal poles: $(\mathbf{D}+1)(\mathbf{D}+1+\epsilon)\mathbf{f}=\boldsymbol{\theta}$ with $\mathbf{f}(0)=\alpha_1$ and $\mathbf{f}'(0)=\alpha_2$. As we found in our earlier discussion, the near equality of the poles causes computational difficulty which we remove by equating the poles. But equating the nearly equal poles is equivalent to replacing the nearly dependent set of solutions $\{e^{-t},e^{-(1+\epsilon)t}\}$ by the easily distinguishable pair of functions $\{e^{-t},te^{-t}\}$. Since the poles are made identical ($\epsilon=0$), the state-space representation of the system becomes $\dot{\mathbf{x}}=\mathbf{A}\mathbf{x}$, where

$$\mathbf{A}=\begin{pmatrix} 0 & 1 \\ -1 & -2 \end{pmatrix} \tag{4.86}$$

This system matrix is not diagonalizable. The pair of vectors $\mathbf{x}_1=(1\ \ -1)^{\mathrm{T}}$ and $\mathbf{x}_{12}=(\frac{1}{2}\ \ \frac{1}{2})^{\mathrm{T}}$ is a two-vector chain of generalized eigenvectors of $\mathbf{A}$ for the single eigenvalue $\lambda=-1$. This pair of vectors is a basis for the state space. Therefore, the matrix

$$\mathbf{S}=\begin{pmatrix} 1 & \frac{1}{2} \\ -1 & \frac{1}{2} \end{pmatrix} \tag{4.87}$$

is a modal matrix for the system. Note that **S** is well conditioned. There will be no computational difficulty in inverting **S**. The nondiagonal spectral matrix for the system is

$$\mathbf{\Lambda}=\mathbf{S}^{-1}\mathbf{A}\mathbf{S}=\begin{pmatrix} -1 & 1 \\ 0 & -1 \end{pmatrix} \tag{4.88}$$

Example 1 demonstrates the practical value of the concepts of generalized eigenvectors and Jordan form. Even though these concepts are important, the full generality of the Jordan form is seldom, if ever, needed. We are unlikely to encounter, in practice, a generalized eigenspace more complex than that characterized by the single two-vector chain of generalized eigenvectors of Example 1. In Example 1, the system matrix **A** is nondiagonalizable only for $\epsilon=0$. We focused on this nondiagonalizable case because it characterizes the situation for small ϵ better than does the true barely diagonalizable case.* It seems that diagonalizability is the rule in models which represent nature, except at the boundary between certain regions or at the limit of certain approximations. In Example 1, diagonalizability broke down completely only at the boundary between the two regions defined by $\epsilon>0$ and $\epsilon<0$. Yet from a practical point of view the boundary is a fuzzy, "small ϵ" transition region.

Pease [4.12, p. 81] presents a spectral analysis of the transmission of electrical signals through a 2-port system. His analysis illustrates the way that nondiagonalizability characterizes the boundary between different regions. The 2×2 system matrix which describes the transmission of signals through the 2-port network is diagonalizable for all sinusoidal signals except signals at the upper or lower cutoff frequencies. At these two frequencies the spectral analysis breaks down because of nondiagonalizability of the matrix of 2-port parameters. However, the analysis can be salvaged by using generalized eigenvectors. Even for frequencies *near* the cutoff frequencies, the spectral analysis is aided by the use of generalized eigenvectors because of the *near* nondiagonalizability of the system matrix.

* Forsythe [4.6] explores other problems in which accuracy is improved by treating near singularity as true singularity.

Application of Jordan Form—Feedback Control

The most common model for a linear time-invariant dynamic system is the state equation (3.67):

$$\dot{\mathbf{x}}(t)=\mathbf{A}\mathbf{x}(t)+\mathbf{B}\mathbf{u}(t), \qquad \mathbf{x}(0) \text{ given} \tag{4.89}$$

where $\mathbf{x}(t)$ is the state (or condition) of the system at time t, and $\mathbf{u}(t)$ is the control (or input) at time t; $\mathbf{A}$ and $\mathbf{B}$ are arbitrary $n\times n$ and $n\times m$ matrices, respectively. In (3.79) we inverted the state equation, obtaining

$$\mathbf{x}(t)=e^{\mathbf{A}t}\mathbf{x}(0)+\int_0^t e^{\mathbf{A}(t-s)}\mathbf{B}\mathbf{u}(s)\,ds \tag{4.90}$$

where the state transition matrix (or matrix exponential) $e^{\mathbf{A}t}$ is defined as the sum of an infinite series of matrices (3.72).

Equations (4.89) and (4.90) are generalizations of the simple first-order linear constant-coefficient differential equation

$$\dot{\mathbf{f}}(t)=a\mathbf{f}(t)+b\mathbf{u}(t), \qquad \mathbf{f}(0) \text{ given} \tag{4.91}$$

which has the solution

$$\mathbf{f}(t)=e^{at}\mathbf{f}(0)+\int_0^t e^{a(t-s)}b\mathbf{u}(s)\,ds$$

Another approach to the solution of (4.91) is through frequency domain analysis.* Taking the Laplace transform of (4.91), we obtain

$$s\mathbf{F}(s)-\mathbf{f}(0)=a\mathbf{F}(s)+b\mathbf{U}(s)$$

or

$$\mathbf{F}(s)=\left(\frac{1}{s-a}\right)\mathbf{f}(0)+\left(\frac{b}{s-a}\right)\mathbf{U}(s) \tag{4.92}$$

where the symbols $\mathbf{F}$ and $\mathbf{U}$ are the Laplace transforms of $\mathbf{f}$ and $\mathbf{u}$, respectively. The function $(s-a)^{-1}$ is known as the transfer function of the system (4.91). The pole of the tranfer function $(s=a)$ characterizes the time response of the system. In fact, the transfer function is the Laplace transform of the impulse response of the system, e^{at}.

The relationships among the variables in a linear equation can be represented pictorially by means of a signal flow graph. A signal flow

*For an introduction to frequency domain analysis, see Appendix 2. Refer also to Schwartz and Friedland [4.16] or DeRusso, Roy, and Close [4.3].

graph for (4.91) is shown in Figure 4.5. The variables in the system are associated with nodes in the graph. The arrows indicate the flow of information (or the relationships among the variables). The encircled symbols contained in each arrow are multipliers. Thus the variable $\mathbf{f}(t)$ is multiplied by a as it flows to the node labeled $\dot{\mathbf{f}}(t)$. The symbol $1/s$ represents an integration operation on the variable $\dot{\mathbf{f}}$ (multiplication of the Laplace transform of $\dot{\mathbf{f}}$ by $1/s$ yields the Laplace transform of $\mathbf{f}$). Nodes are treated as summing points for all incoming signals. Thus the node labeled $\dot{\mathbf{f}}(t)$ is a graphic representation of the differential equation (4.91). The primary information about the system, the position of the pole, is contained in the feedback path. The signal flow graph focuses attention on the feedback nature of the system represented by the differential equation.

We can also obtain a transformed equation and a signal flow graph corresponding to the vector state equation (4.89). Suppose the state variables [or elements of $\mathbf{x}(t)$] are denoted by $\mathbf{f}_i(t)$, $i=1,\ldots,n$. Then we define the Laplace transform of the vector $\mathbf{x}$ of (4.89) by

$$\mathbf{X} \triangleq \mathcal{L}(\mathbf{x}) \triangleq \begin{pmatrix} \mathcal{L}(\mathbf{f}_1) \\ \vdots \\ \mathcal{L}(\mathbf{f}_n) \end{pmatrix} \tag{4.93}$$

Exercise 1. Show that $\mathcal{L}(\mathbf{Ax})=\mathbf{A}\mathcal{L}(\mathbf{x})$ for any $n\times n$ matrix $\mathbf{A}$.

Using definition (4.93) and Exercise 1, we take the Laplace transform of (4.89):

$$s\mathbf{X}(s)-\mathbf{x}(0)=\mathbf{AX}(s)+\mathbf{BU}(s)$$

Solving for $\mathbf{X}(s)$, we obtain the following generalization of (4.92):

$$\mathbf{X}(s)=(s\mathbf{I}-\mathbf{A})^{-1}\mathbf{x}(0)+(s\mathbf{I}-\mathbf{A})^{-1}\mathbf{BU}(s) \tag{4.94}$$

The matrix $(s\mathbf{I}-\mathbf{A})^{-1}$ is called the **matrix transfer function** for the system represented by (4.89). The poles of the transfer function are those values of

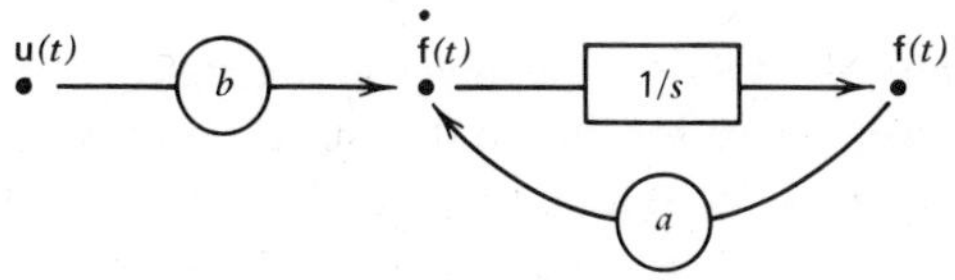

Figure 4.5. Signal flow graph for (4.91).

s for which $(s\mathbf{I}-\mathbf{A})$ is singular. Therefore, the poles of the system are the eigenvalues of the system matrix $\mathbf{A}$, a fact which we discovered for a restricted class of system matrices in (4.51). Because of the formal similarity between the results for the first-order system equation and for the n-dimensional state equation, we suspect that

$$\mathcal{L}(e^{\mathbf{A}t})=(s\mathbf{I}-\mathbf{A})^{-1} \tag{4.95}$$

Equation (4.95) is easily verified by comparing (4.90) and (4.94). We can think of the state transition matrix $e^{\mathbf{A}t}$ as a **matrix impulse response** [see (3.77)–(3.78)]. The vector signal flow graph is formally the same as that for the scalar equation (Figure 4.6). However, individual nodes now represent vector variables. Again, the feedback nature of the system is emphasized by the flow graph model. The feedback path in Figure 4.6 contains all the information peculiar to the particular system, although the poles of the system are stated only implicitly as the eigenvalues of $\mathbf{A}$. The graph would be more specific if we were to use a separate node for each element of each vector variable; however, the diagram would be much more complicated. We draw such a detailed flow graph for a special case in Figure 4.8.

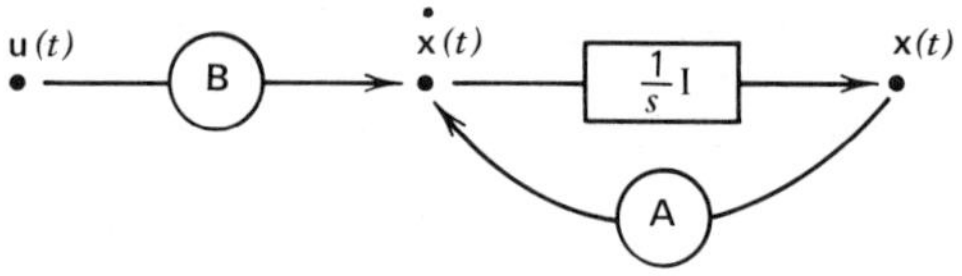

Figure 4.6. Vector signal flow graph for (4.89).

In order to obtain as much insight concerning the feedback nature of the state equation as we did for the scalar case, we change to a coordinate system which emphasizes the poles of the system. Let $\mathbf{x}=\mathbf{S}\mathbf{z}$, where $\mathbf{S}$ is an invertible $n\times n$ matrix. Then $\mathbf{z}(t)$ describes the state of the system relative to a new set of coordinates, and (4.89) becomes

$$\dot{\mathbf{z}}(t)=\mathbf{S}^{-1}\mathbf{A}\mathbf{S}\mathbf{z}(t)+\mathbf{S}^{-1}\mathbf{B}\mathbf{u}(t), \qquad \mathbf{z}(0)=\mathbf{S}^{-1}\mathbf{x}(0) \text{ given} \tag{4.96}$$

We choose $\mathbf{S}$ so that $\mathbf{S}^{-1}\mathbf{A}\mathbf{S}=\mathbf{\Lambda}$, the spectral matrix (or Jordan form) of $\mathbf{A}$. Thus $\mathbf{S}$ consists in a basis for the state space composed of generalized eigenvectors of $\mathbf{A}$ as in (4.76). The new signal flow graph is Figure 4.7.

In order to see that this new signal flow graph is particularly informative, we must examine the interconnections between the individual elements of $\mathbf{z}(t)$. We do so for a particular example.

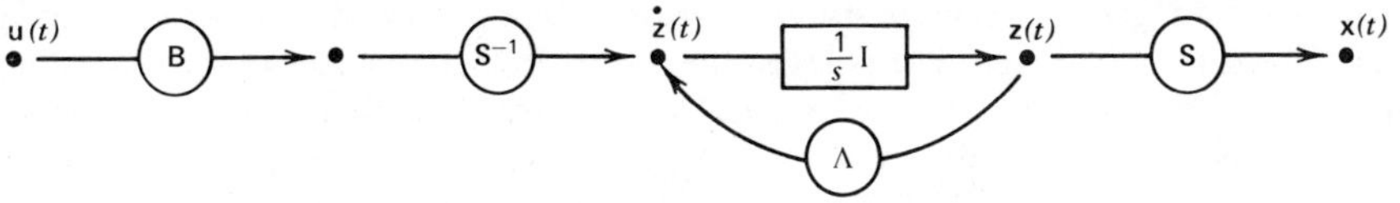

Figure 4.7. Signal flow graph for (4.96).

Example 2. A Specific Feedback System. Let the system and input matrices be

$$\mathbf{A}=\begin{pmatrix} 1 & 1 & 0 \\ -1 & 3 & 0 \\ -1 & 1 & 2 \end{pmatrix} \qquad \mathbf{B}=\begin{pmatrix} b_1 \\ b_2 \\ b_3 \end{pmatrix}$$

It is easily verified that the Jordan form of **A** is

$$\Lambda=\begin{pmatrix} 2 & 1 & 0 \\ 0 & 2 & 0 \\ 0 & 0 & 2 \end{pmatrix}$$

and that this nearly decoupled spectral matrix can be obtained using

$$\mathbf{S}=\begin{pmatrix} 1 & 0 & 1 \\ 1 & 1 & 1 \\ 1 & 0 & 0 \end{pmatrix} \qquad \mathbf{S}^{-1}=\begin{pmatrix} 0 & 0 & 1 \\ -1 & 1 & 0 \\ 1 & 0 & -1 \end{pmatrix}$$

There is only one element $\mathbf{u}(t)$ in the input vector ($\mathbf{B}$ is 3×1). Letting $\mathbf{f}_i$ and $\mathbf{v}_i$ represent the elements of $\mathbf{x}$ and $\mathbf{z}$, respectively, the flow graph corresponding to Figure 4.7 can be given in detail (Figure 4.8). We will refer to the new variables $\mathbf{v}_i(t)$ as the **canonical state variables** [as contrasted with the state variables $\mathbf{f}_i(t)$].

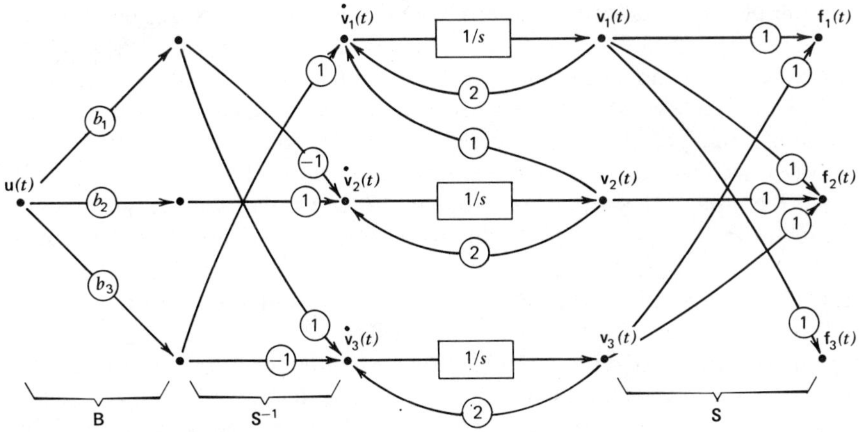

Figure 4.8. A detailed signal flow graph.

In the flow graph of Figure 4.8 the vector system is viewed as a set of nearly uncoupled scalar systems. The poles of the system (the eigenvalues of **A**) appear in the main feedback paths in the graph. The only other feedback paths are those corresponding to the off-diagonal 1's in $\mathbf{\Lambda}$. It is these off-diagonal 1's that give rise to nonexponential terms (te^{2t}) in the response of the system. Specifically, if the input function **u** is zero,

$$\mathbf{v}_3(t)=\mathbf{v}_3(0)e^{2t}$$
$$\mathbf{v}_2(t)=\mathbf{v}_2(0)e^{2t}$$
$$\mathbf{v}_1(t)=\mathbf{v}_1(0)e^{2t}+\mathbf{v}_2(0)te^{2t}$$

The extra term in $\mathbf{v}_1$ arises because the scalar system which determines $\mathbf{v}_1$ is driven by $\mathbf{v}_2$.

It is evident that the Jordan form of a system matrix is a convenient catalog of the information available concerning the system. The modal matrix **S** describes the interconnections between the canonical variables and the state variables. Suppose the above system is undriven [$\mathbf{u}(t)=0$] and the initial values of the canonical variables are $\mathbf{v}_1(0)=\mathbf{v}_2(0)=0$ and $\mathbf{v}_3(0)=1$. Then $\mathbf{v}_1(t)=\mathbf{v}_2(t)=0$ and $\mathbf{v}_3(t)=e^{2t}$. The corresponding output vector $\mathbf{x}(t)$ is

$$\mathbf{x}(t)=\begin{pmatrix}\mathbf{f}_1(t)\\ \mathbf{f}_2(t)\\ \mathbf{f}_3(t)\end{pmatrix}=\begin{pmatrix}e^{2t}\\ e^{2t}\\ 0\end{pmatrix}=e^{2t}\begin{pmatrix}1\\1\\0\end{pmatrix}$$

At each instant, the output vector is proportional to the third column of **S**, one of the eigenvectors of **A**. Under these circumstances, we say only one "mode of response" of the system has been excited. There is one mode of response corresponding to each canonical variable; corresponding to the variable $\mathbf{v}_i(t)$ is the mode where $\mathbf{x}(t)$ is proportional to the ith column of **S**.

We call the system represented by (4.89) **controllable** if there is some input $\mathbf{u}(t)$ that will drive the system [$\mathbf{z}(t)$ or $\mathbf{x}(t)$] from one arbitrary state to another arbitrary state in a finite amount of time. It should be apparent from Example 2 that in order to be able to control all the canonical state variables in the system, the input variables must be coupled to the inputs of each chain in the flow graph, namely, $\dot{\mathbf{v}}_2(t)$ and $\dot{\mathbf{v}}_3(t)$ in Figure 4.8. If in the above example $\mathbf{B}=(0\ 1\ 0)^{\mathrm{T}}$, $\mathbf{u}(t)$ is not coupled to (and has no influence on) $\mathbf{v}_3(t)$. On the other hand, if $\mathbf{B}=(1\ 0\ 0)^{\mathrm{T}}$, the input is coupled to all the canonical state variables; the system appears to be controllable. However, the variables $\mathbf{v}_2(t)$ and $\mathbf{v}_3(t)$ respond identically to **u**—they are associated with identical poles. As a result, $\mathbf{v}_2(t)$ and $\mathbf{v}_3(t)$ cannot be controlled independently. In point of fact, we cannot consider the single input system of Example 2 fully controllable regardless of which input matrix **B** we use. A system can be fully controlled only if we can influence identical subsystems independently. In Example 2, the use of a *pair* of inputs with the input matrix

$$\mathbf{B}=\begin{pmatrix}1&1\\0&1\\1&0\end{pmatrix}$$

yields a controllable system.

In physical systems we may not be able to measure the state variables directly. Perhaps we can only measure variables $\{\mathbf{g}_i(t)\}$ which are related to the state variables by

$$\mathbf{y}(t)=\mathbf{C}\mathbf{x}(t)$$

where $\mathbf{y}(t)=(\mathbf{g}_1(t)\cdots\mathbf{g}_p(t))^{\mathrm{T}}$ and $\mathbf{C}$ is $p\times n$. The matrix **C** would appear in the flow graph of Figure 4.8 as a set of connections between the state variables $\{\mathbf{f}_i(t)\}$ and the output (or measurable) variables $\{\mathbf{g}_i(t)\}$. Clearly, we cannot fully determine the state of the system from the measurements unless the output variables are coupled to the output of each chain; namely, $\mathbf{v}_1(t)$ and $\mathbf{v}_3(t)$. Furthermore, in this specific example, measurement of a single output variable $\mathbf{g}_i(t)$ is not sufficient to distinguish between the variables $\mathbf{v}_2(t)$ and $\mathbf{v}_3(t)$, because their behavior is identical. In general, we call a system **observable** if by observing the output $\mathbf{y}(t)$ of the undriven system for a finite interval of time, we get enough information to determine the initial state $\mathbf{x}(0)$. See Brown [4.2] or Zadeh and Desoer [4.20] for convenient tests for controllability and observability.

4.6 Functions of Matrices and Linear Operators

In previous examples we have encountered several functions of square matrices; namely, $\mathbf{A}^k$, $e^{\mathbf{A}t}$, and $(s\mathbf{I}-\mathbf{A})^{-1}$. In later sections we encounter additional matrix functions. The actual computation of such functions of matrices is a problem of practical importance, especially in the analysis of dynamic systems. In this section we develop a definition for functions of matrices which applies in essentially all situations where we might expect such functions to be meaningful. The definition applies to diagonalizable and nondiagonalizable matrices, and also to the linear operators that these matrices represent. (Functions of diagonalizable linear operators on infinite-dimensional spaces are considered in Section 5.5.) Much of this section is devoted to the development of techniques for analyzing and evaluating functions of matrices.

Two of the matrix functions mentioned above, $\mathbf{A}^k$ and $(s\mathbf{I}-\mathbf{A})^{-1}$, are defined in terms of ordinary matrix operations—addition, scalar multiplication, and inversion. The third matrix function, $e^{\mathbf{A}t}$, represents the sum of an infinite polynomial series in $\mathbf{A}$, as defined in (3.72). This latter function suggests an approach to the definition of general functions of the square matrix $\mathbf{A}$. Polynomial functions of matrices are clearly defined; they can be evaluated by matrix multiplications and additions. Suppose the nonpolynomial function f can be expanded in the power series*

$$f(\lambda)=\sum_{k=0}^{\infty} a_k\lambda^k$$

One reasonable way to define $f(\mathbf{A})$ is by using the same power series in $\mathbf{A}$,

$$f(\mathbf{A}) \triangleq \sum_{k=0}^{\infty} a_k\mathbf{A}^k \tag{4.97}$$

Each term of the series can be evaluated using ordinary matrix operations. Of course, the definition (4.97) is useful only if the series converges and we can evaluate the sum of the series. We explore the question of convergence of (4.97) shortly. The essential properties of $\mathbf{A}$ are displayed in its spectral matrix $\mathbf{\Lambda}$ and its modal matrix $\mathbf{S}$. Substituting $\mathbf{A}=\mathbf{S\Lambda S}^{-1}$ into (4.97) we find

$$\begin{aligned}
f(\mathbf{A})&=f(\mathbf{S\Lambda S}^{-1})\\
&=\sum_k a_k(\mathbf{S\Lambda S}^{-1})^k\\
&=\sum_k a_k\mathbf{S\Lambda}^k\mathbf{S}^{-1}\\
&=\mathbf{S}\left(\sum_k a_k\mathbf{\Lambda}^k\right)\mathbf{S}^{-1}\\
&=\mathbf{S}f(\mathbf{\Lambda})\mathbf{S}^{-1}
\end{aligned} \tag{4.98}$$

*The power series used in (4.97) is a Taylor series expansion about the origin. The matrix function could have been defined in terms of a Taylor series or Laurent series expansion about some other point in the complex plane. See Wylie [4.18] for a discussion of such power series expansions.

(We are able to take the similarity transformation outside the infinite sum because matrix multiplication is a continuous operator; see Section 5.4.) Thus if $f(\mathbf{A})$ as given in (4.97) is well-defined, then evaluation of $f(\mathbf{A})$ reduces to evaluation of $f(\mathbf{\Lambda})$. We again apply the power series definition to determine $f(\mathbf{\Lambda})$. If $\mathbf{A}$ is diagonalizable, then $\mathbf{\Lambda}$ is diagonal, and

$$\begin{aligned} f(\mathbf{\Lambda}) &= \sum_k a_k \begin{pmatrix} \lambda_1 & & \\ & \ddots & \\ & & \lambda_n \end{pmatrix}^k \\ &= \sum_k a_k \begin{pmatrix} \lambda_1^k & & \\ & \ddots & \\ & & \lambda_n^k \end{pmatrix} \\ &= \begin{pmatrix} \sum_k a_k \lambda_1^k & & \\ & \ddots & \\ & & \sum_k a_k \lambda_n^k \end{pmatrix} \\ &= \begin{pmatrix} f(\lambda_1) & & \\ & \ddots & \\ & & f(\lambda_n) \end{pmatrix} \end{aligned} \tag{4.99}$$

On the other hand, if $\mathbf{A}$ is not diagonalizable, $f(\mathbf{\Lambda})$ differs from (4.99) only as a result of the off-diagonal 1's in $\mathbf{\Lambda}$. By the same logic, we can express $f(\mathbf{\Lambda})$ as

$$f(\mathbf{\Lambda}) = \sum_k a_k \begin{pmatrix} \mathbf{J}_1 & & \\ & \ddots & \\ & & \mathbf{J}_r \end{pmatrix}^k = \begin{pmatrix} f(\mathbf{J}_1) & & \\ & \ddots & \\ & & f(\mathbf{J}_r) \end{pmatrix} \tag{4.100}$$

where $\mathbf{J}_i$ is the ith Jordan block in $\mathbf{\Lambda}$. Thus calculation of $f(\mathbf{A})$ reduces to the determination of $f(\mathbf{J}_i)$.

We explore $f(\mathbf{J}_i)$ by means of an example. For a 4×4 Jordan block we have

$$\mathbf{J}=\begin{pmatrix}\lambda & 1 & 0 & 0\\ 0 & \lambda & 1 & 0\\ 0 & 0 & \lambda & 1\\ 0 & 0 & 0 & \lambda\end{pmatrix}\qquad \mathbf{J}^2=\begin{pmatrix}\lambda^2 & 2\lambda & 1 & 0\\ 0 & \lambda^2 & 2\lambda & 1\\ 0 & 0 & \lambda^2 & 2\lambda\\ 0 & 0 & 0 & \lambda^2\end{pmatrix}$$

$$\mathbf{J}^3=\begin{pmatrix}\lambda^3 & 3\lambda^2 & 3\lambda & 1\\ 0 & \lambda^3 & 3\lambda^2 & 3\lambda\\ 0 & 0 & \lambda^3 & 3\lambda^2\\ 0 & 0 & 0 & \lambda^3\end{pmatrix},\qquad \mathbf{J}^4=\begin{pmatrix}\lambda^4 & 4\lambda^3 & 6\lambda^2 & 4\lambda\\ 0 & \lambda^4 & 4\lambda^3 & 6\lambda^2\\ 0 & 0 & \lambda^4 & 4\lambda^3\\ 0 & 0 & 0 & \lambda^4\end{pmatrix}$$

Observe that in each matrix the element which appears on the jth "superdiagonal" is $(1/j!)$ times the jth derivative (with respect to λ) of the element on the main diagonal. Thus, continuing the example,

$$f(\mathbf{J})=\sum_k a_k\mathbf{J}^k$$

$$=\begin{pmatrix}\sum_k a_k\lambda^k & \sum_k \frac{a_k}{1!}\frac{d\lambda^k}{d\lambda} & \sum_k \frac{a_k}{2!}\frac{d^2\lambda^k}{d\lambda^2} & \sum_k \frac{a_k}{3!}\frac{d^3\lambda^k}{d\lambda^3}\\ 0 & \sum_k a_k\lambda^k & \sum_k \frac{a_k}{1!}\frac{d\lambda^k}{d\lambda} & \sum_k \frac{a_k}{2!}\frac{d^2\lambda^k}{d\lambda^2}\\ 0 & 0 & \sum_k a_k\lambda^k & \sum_k \frac{a_k}{1!}\frac{d\lambda^k}{d\lambda}\\ 0 & 0 & 0 & \sum_k a_k\lambda^k\end{pmatrix}$$

Relying on the term-by-term differentiability of power series (Kaplan [4.11, p. 353]), we take all derivatives outside the summations to obtain

$$f(\mathbf{J})=\begin{pmatrix}f(\lambda) & \frac{f'(\lambda)}{1!} & \frac{f''(\lambda)}{2!} & \frac{f^{(3)}(\lambda)}{3!}\\ 0 & f(\lambda) & \frac{f'(\lambda)}{1!} & \frac{f''(\lambda)}{2!}\\ 0 & 0 & f(\lambda) & \frac{f'(\lambda)}{1!}\\ 0 & 0 & 0 & f(\lambda)\end{pmatrix}\qquad(4.101)$$

Corresponding to each Jordan block $\mathbf{J}_i$ of $\boldsymbol{\Lambda}$ (with eigenvalue λ_i), $f(\boldsymbol{\Lambda})$ contains a block which has $f(\lambda_i)$ on the main diagonal. The upper elements in the block are filled with appropriately scaled derivatives of f (evaluated at λ_i). The elements on the jth super-diagonal are

$$\frac{1}{j!}\frac{d^j f(\lambda_i)}{d\lambda^j}$$

Surprisingly, $f(\boldsymbol{\Lambda})$ is not in Jordan form.

Example 1. Matrix Inversion as a Matrix Function. Suppose $f(\lambda)=1/\lambda$. If $\mathbf{A}$ is an invertible $n\times n$ matrix, we use (4.98) and (4.99) to find

$$\begin{aligned}\mathbf{A}^{-1}&=\mathbf{S}\boldsymbol{\Lambda}^{-1}\mathbf{S}^{-1}\\ &=\mathbf{S}\begin{pmatrix}1/\lambda_1 & & \\ & \ddots & \\ & & 1/\lambda_n\end{pmatrix}\mathbf{S}^{-1}\end{aligned}$$

Suppose

$$\mathbf{A}=\begin{pmatrix}2&0&0\\0&1&0\\0&0&0\end{pmatrix}$$

Then $\mathbf{S}=\mathbf{S}^{-1}=\mathbf{I}$, and

$$\mathbf{A}^{-1}=\begin{pmatrix}1&0&0\\0&\frac{1}{2}&0\\0&0&\infty\end{pmatrix}$$

It is clear that $\mathbf{A}^{-1}$ does not exist if zero is an eigenvalue of $\mathbf{A}$. The function $1/\lambda$ is not defined at $\lambda=0$, and (4.99) cannot be evaluated.

Example 2. A Function of a Nondiagonalizable Matrix. As in Example 1, if $\mathbf{A}=\mathbf{S}\boldsymbol{\Lambda}\mathbf{S}^{-1}, \mathbf{A}^{-1}=\mathbf{S}\boldsymbol{\Lambda}^{-1}S^{-1}$. Suppose

$$\boldsymbol{\Lambda}=\left(\begin{array}{ccc:c:c}\lambda_1&1&0&&\\0&\lambda_1&1&&\\0&0&\lambda_1&&\\ \hdashline &&&\lambda_1&\\ \hdashline &&&&\lambda_2\end{array}\right)$$

Letting $f(\lambda)=1/\lambda$ we find that $f'(\lambda)=-1/\lambda^2$ and $f''(\lambda)/2!=1/\lambda^3$. Thus, using

(4.101) for each Jordan block,

$$\Lambda^{-1}=\left(\begin{array}{ccc:c:c} 1/\lambda_1 & -1/\lambda_1^2 & 1/\lambda_1^3 & & \\ 0 & 1/\lambda_1 & -1/\lambda_1^2 & & \\ 0 & 0 & 1/\lambda_1 & & \\ \hdashline & & & 1/\lambda_1 & \\ \hdashline & & & & 1/\lambda_2 \end{array}\right)$$

An Alternative Definition

Although we have used (4.97) to define $f(\mathbf{A})$, we have used (4.98) and (4.100) to perform the actual evaluation of $f(\mathbf{A})$. [Note that (4.99) is a special case of (4.100).] It can be shown that our original definition of $f(\mathbf{A})$, (4.97), converges if and only if f is analytic in a circle of the complex plane which contains all the eigenvalues of $\mathbf{A}$.* Yet (4.98) and (4.100), which we derived from (4.97), provide a correct evaluation of $f(\mathbf{A})$ in cases which do not satisfy this criterion. For example,

$$\text{if } \mathbf{A}=\begin{pmatrix} 2 & 0 \\ 0 & -2 \end{pmatrix} \qquad \text{then} \quad \mathbf{A}^{-1}=\begin{pmatrix} \frac{1}{2} & 0 \\ 0 & -\frac{1}{2} \end{pmatrix}$$

The function $f(\lambda)=\lambda^{-1}$ is not analytic at $\lambda=0$. No circle encloses the points 2 and -2 while excluding the point 0, yet (4.98) and (4.99) provide the correct inverse. It is apparent that (4.98)–(4.101) provide a *more general definition* of $f(\mathbf{A})$ than does (4.97).

The definition (4.98)–(4.101) applies to all functions f and matrices $\mathbf{A}$ for which $f(\mathbf{J}_i)$ can be evaluated for each Jordan block $\mathbf{J}_i$. If $\mathbf{A}$ is diagonalizable, this evaluation requires only that f be *defined on the spectrum*; that is, that f be defined at all the eigenvalues of $\mathbf{A}$. If $\mathbf{A}$ is not diagonalizable, the evaluation of $f(\mathbf{A})$ requires the existence of derivatives of f at some of the eigenvalues of $\mathbf{A}$. Thus the definition of $f(\mathbf{A})$ given in (4.98)–(4.101) certainly applies to all f and $\mathbf{A}$ for which f is not only defined on the spectrum of $\mathbf{A}$ but also analytic at those eigenvalues of $\mathbf{A}$ for which $\mathbf{A}$ is defective (i.e., for which the corresponding Jordan blocks $\mathbf{J}_i$ are larger than 1×1). In every case where the definition (4.97) applies, the evaluation of $f(\mathbf{A})$ which results is identical to the evaluation provided by (4.98)–(4.101).

As illustrated in (4.101), the actual evaluation of $f(\mathbf{A})$ leads to evaluation of

$$f(\lambda_i), f'(\lambda_i), \ldots, f^{(q_i-1)}(\lambda_i), \qquad i=1,\ldots,p \tag{4.102}$$

*Rinehart [4.14]. A function $f(\lambda)$ is said to be analytic at λ_1 if it is differentiable (as a function of a complex variable λ) in a neighborhood of λ_1 (see Wylie [4.18]).

We refer to this set of evaluations as **evaluation on the spectrum of A**. It is apparent that any two functions that have the same evaluation on the spectrum lead to the same function of **A**.

Exercise 1. Compare $f(\mathbf{A})$ and $g(\mathbf{A})$ for $f(\lambda) \triangleq 4\lambda - 8$, $g(\lambda) \triangleq \lambda^2 - 4$, and

$$\mathbf{A} = \begin{pmatrix} 2 & 1 \\ 0 & 2 \end{pmatrix}$$

Equations (4.98)–(4.101) provide a suitable definition of $f(\mathbf{A})$ for most choices of f and **A**. Rinehart [4.14] shows that with this definition of $f(\mathbf{A})$ and with single-valued functions g and h for which $g(\mathbf{A})$ and $h(\mathbf{A})$ exist,

1. If $f(\lambda) = c$ then $f(\mathbf{A}) = c\mathbf{I}$
2. If $f(\lambda) = \lambda$ then $f(\mathbf{A}) = \mathbf{A}$
3. If $f(\lambda) = g(\lambda) + h(\lambda)$ then $f(\mathbf{A}) = g(\mathbf{A}) + h(\mathbf{A})$
4. If $f(\lambda) = g(\lambda) \cdot h(\lambda)$ then $f(\mathbf{A}) = g(\mathbf{A}) \cdot h(\mathbf{A})$
5. If $f(\lambda) = g(h(\lambda))$ then $f(\mathbf{A}) = g(h(\mathbf{A}))$

If g or h is not single valued, then the matrix $f(\mathbf{A})$ depends upon which branches of g and h are used in the evaluation on the spectrum of **A**. From these properties it follows that scalar functional identities extend to matrices. For example, $\sin^2(\mathbf{A}) + \cos^2(\mathbf{A}) = \mathbf{I}$ and $e^{\ln \mathbf{A}} = \mathbf{A}$.

The Fundamental Formula for Matrices

Let **A** be a 3×3 diagonalizable matrix with only two distinct eigenvalues; that is, $c(\lambda) = (\lambda - \lambda_1)^2(\lambda - \lambda_2)$, and the eigenspace for λ_1 is two-dimensional. Suppose also that the function f is defined at λ_1 and λ_2. Then we can express $f(\mathbf{A})$ in the manner of Example 1:

$$f(\mathbf{A}) = \begin{pmatrix} f(\lambda_1) & 0 & 0 \\ 0 & f(\lambda_1) & 0 \\ 0 & 0 & f(\lambda_2) \end{pmatrix}$$

In order to express $f(\mathbf{A})$ in a manner that clearly separates the essential properties of **A** from those of f, we introduce the following notation. Let $\mathbf{E}_{i0}^{\mathbf{A}}$ be a matrix which has a one wherever $f(\mathbf{A})$ has $f(\lambda_i)$, and zeros elsewhere. (The second subscript, "0," is used only to provide consistency with the nondiagonalizable case introduced later.) Specifically,

$$\mathbf{E}_{10}^{\mathbf{A}} = \begin{pmatrix} 1 & 0 & 0 \\ 0 & 1 & 0 \\ 0 & 0 & 0 \end{pmatrix} \quad \text{and} \quad \mathbf{E}_{20}^{\mathbf{A}} = \begin{pmatrix} 0 & 0 & 0 \\ 0 & 0 & 0 \\ 0 & 0 & 1 \end{pmatrix}$$

Then we can express $f(\Lambda)$ by

$$f(\Lambda)=f(\lambda_1)\mathbf{E}_{10}^{\Lambda}+f(\lambda_2)\mathbf{E}_{20}^{\Lambda}$$

Since $f(\mathbf{A})=\mathbf{S}f(\mathbf{A})\mathbf{S}^{-1}$, to obtain $f(\mathbf{A})$ we simply perform the similarity transformations $\mathbf{E}_{i0}^{\mathrm{A}}=\mathbf{S}\mathbf{E}_{i0}^{\Lambda}\mathbf{S}^{-1}$ to obtain

$$f(\mathbf{A})=f(\lambda_1)\mathbf{E}_{10}^{\mathrm{A}}+f(\lambda_2)\mathbf{E}_{20}^{\mathrm{A}}$$

It is evident that we can express any well-defined function of the specific matrix $\mathbf{A}$ by means of this formula. Once we have the matrices $\mathbf{E}_{i0}^{\mathrm{A}}$, evaluation of $f(\mathbf{A})$ requires only evaluation of f on the spectrum of $\mathbf{A}$. By a derivation similar to that above, we can show that for any $n\times n$ diagonalizable matrix $\mathbf{A}$ and any f defined on the spectrum of $\mathbf{A}$, $f(\mathbf{A})$ can be expressed as

$$f(\mathbf{A})=\sum_{i=1}^{p} f(\lambda_i)\mathbf{E}_{i0}^{\mathrm{A}} \tag{4.103}$$

where p is the number of distinct eigenvalues of $\mathbf{A}$. We call (4.103) the **fundamental formula for** $f(\mathbf{A})$. The matrices $\mathbf{E}_{i0}^{\mathrm{A}}$ are called the **constituent matrices** (or **components**) of $\mathbf{A}$. (We drop the superscript A when confusion seems unlikely.) Notice that (4.103) separates the contributions of f and $\mathbf{A}$. In fact, (4.103) is a satisfactory definition of $f(\mathbf{A})$, equivalent to (4.98)–(4.99).

The definition of the fundamental formula (4.103) can be extended to nondiagonalizable matrices as well. Suppose f is analytic at λ_1 and defined at λ_2. Then we can write $f(\Lambda)$ for the matrix Λ of Example 2 as

$$f(\Lambda)=\left(\begin{array}{ccc:c:c} f(\lambda_1) & f'(\lambda_1) & \dfrac{f''(\lambda_1)}{2} & & \\ 0 & f(\lambda_1) & f'(\lambda_1) & & \\ 0 & 0 & f(\lambda_1) & & \\ \hdashline & & & f(\lambda_1) & \\ \hdashline & & & & f(\lambda_2) \end{array}\right)$$

In order to separate the essential properties of Λ from those of f, we define $\mathbf{E}_{ik}^{\Lambda}$ to be a matrix which has a one wherever $f(\Lambda)$ has $(1/k!)\,f^{(k)}(\lambda_i)$, $k=0,$

1, and 2, and zeros elsewhere. Thus

$$\mathbf{E}_{10}^{\Lambda}=\begin{pmatrix} 1 & 0 & 0 & & \\ 0 & 1 & 0 & & \\ 0 & 0 & 1 & & \\ & & & 1 & \\ & & & & 0 \end{pmatrix} \qquad \mathbf{E}_{11}^{\Lambda}=\begin{pmatrix} 0 & 1 & 0 & & \\ 0 & 0 & 1 & & \\ 0 & 0 & 0 & & \\ & & & 0 & \\ & & & & 0 \end{pmatrix}$$

$$\mathbf{E}_{12}^{\Lambda}=\begin{pmatrix} 0 & 0 & 1 & & \\ 0 & 0 & 0 & & \\ 0 & 0 & 0 & & \\ & & & 0 & \\ & & & & 0 \end{pmatrix} \qquad \mathbf{E}_{20}^{\Lambda}=\begin{pmatrix} 0 & 0 & 0 & & \\ 0 & 0 & 0 & & \\ 0 & 0 & 0 & & \\ & & & 0 & \\ & & & & 1 \end{pmatrix}$$

Then we can express $f(\mathbf{\Lambda})$ by

$$f(\mathbf{\Lambda})=f(\lambda_1)\mathbf{E}_{10}^{\Lambda}+f'(\lambda_1)\mathbf{E}_{11}^{\Lambda}+\frac{f''(\lambda_1)}{2!}\mathbf{E}_{12}^{\Lambda}+f(\lambda_2)\mathbf{E}_{20}^{\Lambda}$$

As in the diagonalizable case, we perform the similarity transformations $\mathbf{E}_{ij}^{A}=\mathbf{S}\mathbf{E}_{ij}^{\Lambda}\mathbf{S}^{-1}$ to obtain

$$f(\mathbf{A})=f(\lambda_1)\mathbf{E}_{10}^{A}+f'(\lambda_1)\mathbf{E}_{11}^{A}+\frac{f''(\lambda_1)}{2!}\mathbf{E}_{12}^{A}+f(\lambda_2)\mathbf{E}_{20}^{A}$$

We can compute any well-defined function of the matrix $\mathbf{A}$ of Example 2 by means of this formula. By a derivation similar to that above, we can show that for any $n\times n$ matrix $\mathbf{A}$ and any f which is defined on the spectrum of $\mathbf{A}$ and analytic at eigenvalues where $\mathbf{A}$ is defective, $f(\mathbf{A})$ can be expressed as

$$f(\mathbf{A})=\sum_{i=1}^{p}\left[f(\lambda_i)\mathbf{E}_{i0}^{A}+\frac{f'(\lambda_i)}{1!}\mathbf{E}_{i1}^{A}+\cdots+\frac{f^{(q_i-1)}(\lambda_i)}{(q_i-1)!}\mathbf{E}_{i(q_i-1)}^{A}\right] \tag{4.104}$$

where p is the number of distinct eigenvalues of $\mathbf{A}$, and q_i is the index of annihilation for λ_i [see (4.66) and (4.67)]. Equation (4.104) is the general form of the **fundamental formula** for $f(\mathbf{A})$. Again, we refer to the matrices $\mathbf{E}_{ij}^{A}$ as constituent matrices (or components) of A.*

*The constituent matrices are sometimes defined as $\mathbf{E}_{ij}^{A}/j!$.

The fundamental formula can be used to generate a **spectral decomposition of A**. If we let $f(\lambda)=\lambda$ in (4.104), we obtain

$$\mathbf{A}=\sum_{i=1}^{p}(\lambda_i\mathbf{E}_{i0}^{\mathbf{A}}+\mathbf{E}_{i1}^{\mathbf{A}}) \tag{4.105}$$

If $\mathbf{A}$ is diagonalizable, $q_i=1$ for each i, and (4.105) becomes

$$\mathbf{A}=\sum_{i=1}^{p}\lambda_i\mathbf{E}_{10}^{\mathbf{A}}$$

It is apparent that in the diagonalizable case $\mathbf{E}_{i0}^{\mathbf{A}}$ describes the projection onto the eigenspace associated with λ_i. That is, if $\mathbf{x}=\mathbf{x}_1+\cdots+\mathbf{x}_p$, where $\mathbf{x}_i$ is the component of $\mathbf{x}$ in the eigenspace for λ_i, then $\mathbf{x}_i=\mathbf{E}_{i0}^{\mathbf{A}}\mathbf{x}$ and $\mathbf{A}$ acts like λ_i on $\mathbf{x}_i$. In the nondiagonalizable case, $\mathbf{E}_{i0}^{\mathbf{A}}$ describes the projection onto the *generalized* eigenspace for λ_i. Furthermore, $\mathbf{E}_{ik}^{\mathbf{A}}$ acts like the nilpotent operator $(\mathbf{A}-\lambda_i\mathbf{I})^k$ on the generalized eigenspace for λ_i; that is, $\mathbf{E}_{ik}^{\mathbf{A}}=(\mathbf{A}-\lambda_i\mathbf{I})^k\mathbf{E}_{i0}^{\mathbf{A}}$.

Exercise 2. Verify that the matrices $\mathbf{E}_{10}^{\mathbf{A}}$ and $\mathbf{E}_{20}^{\mathbf{A}}$ of Example 2 satisfy the properties (4.3) for projectors. Show also that $\mathbf{E}_{ik}^{\mathbf{A}}=(\mathbf{A}-\lambda_i\mathbf{I})^k\mathbf{E}_{10}^{\mathbf{A}}$.

Functions of Linear Operators

The fundamental formula also serves to define functions of the underlying operator represented by $\mathbf{A}$. If $\mathbf{T}$ operates on an n-dimensional vector space $\mathcal{V}$, if $\mathbf{P}_{i0}$ is the operator which projects onto $\mathcal{W}_i$ (the generalized eigenspace for λ_i) along $\Sigma_{j\neq i}\mathcal{W}_j$, and if $\mathbf{P}_{ik}\triangleq(\mathbf{T}-\lambda_i\mathbf{I})^k\mathbf{P}_{i0}$, then the **fundamental formula for $f(\mathbf{T})$** is

$$f(\mathbf{T})\triangleq\sum_{i=1}^{p}\left[f(\lambda_i)\mathbf{P}_{i0}+\frac{1}{1!}f'(\lambda_i)\mathbf{P}_{i1}+\cdots+\frac{1}{(q_i-1)!}f^{(q_i-1)}(\lambda_i)\mathbf{P}_{i(q_i-1)}\right] \tag{4.106}$$

If $\mathcal{X}$ is a basis for $\mathcal{V}$ and we define $\mathbf{A}\triangleq[\mathbf{T}]_{\mathcal{X}\mathcal{X}}$, then $\mathbf{E}_{ij}^{\mathbf{A}}=[\mathbf{P}_{ij}]_{\mathcal{X}\mathcal{X}}$. As a result, (4.104) and (4.106) require that $[f(\mathbf{T})]_{\mathcal{X}\mathcal{X}}=f([\mathbf{T}]_{\mathcal{X}\mathcal{X}})$. For diagonalizable $\mathbf{T}$ ($\mathbf{T}$ for which there exists a basis for $\mathcal{V}$ composed of eigenvectors for $\mathbf{T}$), (4.106) simplifies to $f(\mathbf{T})=\Sigma_{i=1}^{p}f(\lambda_i)\mathbf{P}_{i0}$. This simple result is extended to certain infinite-dimensional operators in (5.90).

Example 3. A Function of a Linear Operator. Consider $\mathbf{D}: \mathcal{P}^3 \to \mathcal{P}^3$. We first find the eigendata for $\mathbf{D}$ (as an operator on $\mathcal{P}^3$). The set $\mathcal{N} \triangleq \{\mathbf{f}_i(t) = t^{i-1}, i = 1, 2, 3\}$ is a basis for $\mathcal{P}^3$. In Example 2 of Section 2.5 we found that

$$\mathbf{A} = [\mathbf{D}]_{\mathcal{N}\mathcal{N}} = \begin{pmatrix} 0 & 1 & 0 \\ 0 & 0 & 2 \\ 0 & 0 & 0 \end{pmatrix}$$

This matrix has only one eigenvalue, $\lambda_1 = 0$; a basis of generalized eigenvectors for $[\mathbf{D}]_{\mathcal{N}\mathcal{N}}$ is

$$\mathbf{x}_1 = \begin{pmatrix} 1 \\ 0 \\ 0 \end{pmatrix}, \quad \mathbf{x}_{12} = \begin{pmatrix} 0 \\ 1 \\ 0 \end{pmatrix}, \quad \mathbf{x}_{13} = \begin{pmatrix} 0 \\ 0 \\ \frac{1}{2} \end{pmatrix}$$

Thus

$$\Lambda = \begin{pmatrix} 0 & 1 & 0 \\ 0 & 0 & 1 \\ 0 & 0 & 0 \end{pmatrix}, \quad \mathbf{S} = \begin{pmatrix} 1 & 0 & 0 \\ 0 & 1 & 0 \\ 0 & 0 & \frac{1}{2} \end{pmatrix}, \quad \text{and} \quad \mathbf{S}^{-1} = \begin{pmatrix} 1 & 0 & 0 \\ 0 & 1 & 0 \\ 0 & 0 & 2 \end{pmatrix}$$

The generalized eigenfunctions of $\mathbf{D}$ corresponding to $\mathbf{x}_1$, $\mathbf{x}_{12}$, and $\mathbf{x}_{13}$ are

$$\mathbf{g}_1(t) = 1, \qquad \mathbf{g}_{12}(t) = t, \qquad \mathbf{g}_{13}(t) = \frac{t^2}{2}$$

Because the chain of generalized eigenvectors is of length 3, $q_1 = 3$. Therefore, in order to evaluate $\mathbf{f}(\mathbf{D})$, we must determine three operators: $\mathbf{P}_{10}$, $\mathbf{P}_{11}$, and $\mathbf{P}_{12}$. Since the generalized eigenspace of $\mathbf{D}$ for $\lambda_1 = 0$ is the whole space $\mathcal{P}^3$, the projector $\mathbf{P}_{10}$ onto the generalized eigenspace for λ_1 is $\mathbf{P}_{10} = \mathbf{I}$. We find the other two operators by

$$\mathbf{P}_{11} = (\mathbf{D} - \lambda_1 \mathbf{I})\mathbf{P}_{10} = \mathbf{D}\mathbf{I} = \mathbf{D}$$

$$\mathbf{P}_{12} = (\mathbf{D} - \lambda_1 \mathbf{I})^2 \mathbf{P}_{10} = \mathbf{D}^2 \mathbf{I} = \mathbf{D}^2$$

By (4.106), if $\mathbf{f}$ is analytic at $\lambda = 0$,

$$f(\mathbf{D}) = f(0)\mathbf{I} + f'(0)\mathbf{D} + \frac{f''(0)}{2}\mathbf{D}^2$$

Let $f(\lambda) = \lambda$. Then $f(\mathbf{D})$ reduces to

$$\mathbf{D} = (0)\mathbf{I} + (1)\mathbf{D} + (0)\mathbf{D}^2$$

which verifies the formula for $f(\mathbf{D})$. Let $f(\lambda) = e^{\lambda}$. Then

$$\begin{aligned} e^{\mathbf{D}} &= e^0 \mathbf{I} + e^0 \mathbf{D} + \tfrac{1}{2} e^0 \mathbf{D}^2 \\ &= \mathbf{I} + \mathbf{D} + \tfrac{1}{2}\mathbf{D}^2 \end{aligned}$$

Returning to $\mathbf{A}=[\mathbf{D}]_{\mathfrak{N}\mathfrak{N}}$, we generate those functions of $\mathbf{A}$ which correspond to the functions $f(\mathbf{D})$, $\mathbf{D}$, and $e^{\mathbf{D}}$ above. By inspection of $\mathbf{\Lambda}$ we find that

$$\mathbf{E}_{10}^{\Lambda}=\begin{pmatrix}1&0&0\\0&1&0\\0&0&1\end{pmatrix},\qquad \mathbf{E}_{11}^{\Lambda}=\begin{pmatrix}0&1&0\\0&0&1\\0&0&0\end{pmatrix},\qquad \mathbf{E}_{12}^{\Lambda}=\begin{pmatrix}0&0&1\\0&0&0\\0&0&0\end{pmatrix}$$

Using the similarity transformation $\mathbf{E}_{ij}^{\mathbf{A}}=\mathbf{S}\mathbf{E}_{ij}^{\Lambda}\mathbf{S}^{-1}$, we obtain

$$\mathbf{E}_{10}^{\mathbf{A}}=\begin{pmatrix}1&0&0\\0&1&0\\0&0&1\end{pmatrix}=\mathbf{I},\qquad \mathbf{E}_{11}^{\mathbf{A}}=\begin{pmatrix}0&1&0\\0&0&2\\0&0&0\end{pmatrix}=\mathbf{A},\qquad \mathbf{E}_{12}^{\mathbf{A}}=\begin{pmatrix}0&0&2\\0&0&0\\0&0&0\end{pmatrix}=\mathbf{A}^2$$

These constituents of $\mathbf{A}$ are $[\mathbf{P}_{10}]_{\mathfrak{N}\mathfrak{N}}$, $[\mathbf{P}_{11}]_{\mathfrak{N}\mathfrak{N}}$, and $[\mathbf{P}_{12}]_{\mathfrak{N}\mathfrak{N}}$, respectively. By (4.104),

$$f(\mathbf{A})=f(0)\mathbf{I}+f'(0)\mathbf{A}+\frac{f''(0)}{2}\mathbf{A}^2$$

If $f(\lambda)=\lambda$, we find

$$\mathbf{A}=(0)\mathbf{I}+(1)\mathbf{A}+(0)\mathbf{A}^2$$

Let $f(\lambda)=e^{\lambda}$. Then

$$\begin{aligned} e^{\mathbf{A}}&=e^0\mathbf{I}+e^0\mathbf{A}+\tfrac{1}{2}e^0\mathbf{A}^2\\ &=\mathbf{I}+\mathbf{A}+\tfrac{1}{2}\mathbf{A}^2\\ &=\begin{pmatrix}1&1&2\\0&1&2\\0&0&1\end{pmatrix}\end{aligned}$$

We easily verify that $e^{\mathbf{A}}=[\mathbf{e}^{\mathbf{A}}]_{\mathfrak{N}\mathfrak{N}}$. These results are consistent with the definition (3.72) of $e^{\mathbf{A}t}$, because $\mathbf{A}^k=\mathbf{\Theta}$ for $k>2$.

Computation of Functions of Matrices

We have already derived a method for computing $f(\mathbf{A})$ which relies on a complete eigenvalue analysis of $\mathbf{A}$. We summarize the method.

Computation of $f(\mathbf{A})$ by eigenvalue analysis of $\mathbf{A}$ (4.107)

1. Determine the Jordan form $\mathbf{\Lambda}$, the modal matrix $\mathbf{S}$, and $\mathbf{S}^{-1}$ such that $\mathbf{A}=\mathbf{S}\mathbf{\Lambda}\mathbf{S}^{-1}$.
2. Determine $\mathbf{E}_{ij}^{\Lambda}$ by inspection of $\mathbf{\Lambda}$.
3. Determine $\mathbf{E}_{ij}^{\mathbf{A}}$ by the similarity transformation $\mathbf{E}_{ij}^{\mathbf{A}}=\mathbf{S}\mathbf{E}_{ij}^{\Lambda}\mathbf{S}^{-1}$.
4. Evaluate f on the spectrum of $\mathbf{A}$.
5. Determine $f(\mathbf{A})$ from the fundamental formula, (4.103) or (4.104).

Example 4. Computing e^{At} Using Complete Eigenvalue Analysis. Let $f(\lambda)=e^{\lambda t}$. Let $\mathbf{A}$ be the matrix of Example 2, Section 4.5:

$$\mathbf{A}=\begin{pmatrix} 1 & 1 & 0 \\ -1 & 3 & 0 \\ -1 & 1 & 2 \end{pmatrix}$$

Then $f(\mathbf{A})=e^{\mathbf{A}t}$ is the state transition matrix for that example. We found in that example that

$$(1)\qquad \Lambda=\left(\begin{array}{cc:c} 2 & 1 & 0 \\ 0 & 2 & 0 \\ \hdashline 0 & 0 & 2 \end{array}\right),\quad \mathbf{S}=\begin{pmatrix} 1 & 0 & 1 \\ 1 & 1 & 1 \\ 1 & 0 & 0 \end{pmatrix},\quad \mathbf{S}^{-1}=\begin{pmatrix} 0 & 0 & 1 \\ -1 & 1 & 0 \\ 1 & 0 & -1 \end{pmatrix}$$

Following the other steps outlined above,

$$(2)\qquad \mathbf{E}_{10}^{\Lambda}=\left(\begin{array}{cc:c} 1 & 0 & 0 \\ 0 & 1 & 0 \\ \hdashline 0 & 0 & 1 \end{array}\right),\qquad \mathbf{E}_{11}^{\Lambda}=\left(\begin{array}{cc:c} 0 & 1 & 0 \\ 0 & 0 & 0 \\ \hdashline 0 & 0 & 0 \end{array}\right)$$

$$(3)\qquad \mathbf{E}_{10}^{\mathbf{A}}=\begin{pmatrix} 1 & 0 & 0 \\ 0 & 1 & 0 \\ 0 & 0 & 1 \end{pmatrix},\qquad \mathbf{E}_{11}^{\mathbf{A}}=\begin{pmatrix} -1 & 1 & 0 \\ -1 & 1 & 0 \\ -1 & 1 & 0 \end{pmatrix}$$

$$(4)\qquad f(2)=e^{2t},\qquad f'(2)=te^{2t}$$

$$(5)\qquad e^{\mathbf{A}t}=e^{2t}\mathbf{E}_{10}^{\mathbf{A}}+te^{2t}\mathbf{E}_{11}^{\mathbf{A}}$$

$$=\begin{pmatrix} e^{2t}-te^{2t} & te^{2t} & 0 \\ -te^{2t} & e^{2t}+te^{2t} & 0 \\ -te^{2t} & te^{2t} & e^{2t} \end{pmatrix}$$

Determination of $f(\mathbf{A})$ using complete eigenvalue analysis is lengthy and computationally expensive. The eigenvalue analysis serves only to determine constituents of $\mathbf{A}$. [Of course, it provides considerable insight into the structure of the matrix $\mathbf{A}$ in addition to producing $f(\mathbf{A})$]. We can eliminate most of this computation by employing the fundamental formula in evaluating the constituents. If we substitute several different functions into (4.103)–(4.104), we obtain several equations involving the constituents as unknowns. By a judicious choice of functions, we can obtain equations that allow us to determine each constituent independently. If the minimal polynomial $m(\lambda)$ is evaluated on the spectrum, the evaluations are all zero. If one factor is cancelled from $m(\lambda)$ and the resulting polynomial evaluated on the spectrum, precisely one evaluation is nonzero; if we evaluate this same polynomial in $\mathbf{A}$, precisely one constituent will remain in the fundamental formula. By successively cancelling factors from $m(\lambda)$, and evaluating the resulting polynomials in **A**, we obtain the constituents in an efficient manner.*

Computation of $f(\mathbf{A})$ by evaluating factors of $m(\lambda)$ (4.108)

1. Find and factor $m(\lambda)$, the minimal polynomial for **A**.
2. Cancel one factor from $m(\lambda)$. Denote the resulting polynomial $g_1(\lambda)$. Evaluating $g_1(\mathbf{A})$ will determine precisely one constituent matrix.
3. Cancel an additional factor from $m(\lambda)$. Let $g_i(\lambda)$ denote the polynomial which results from cancelling i factors from $m(\lambda)$. Evaluation of $g_i(\mathbf{A})$ determines precisely one constituent matrix in terms of previously determined constituents. This step is repeated until all the constituents E_{ij}^{A} are known.
4. Evaluate f on the spectrum of **A**.
5. Compute $f(\mathbf{A})$ from the fundamental formula, (4.103) or (4.104).

Example 5. Computing e^{At} by Evaluating Factors of the Minimal Polynomial. Let $f(\lambda)=e^{\lambda t}$. Assume **A** is the matrix given in Example 4. We compute the state transition matrix $e^{\mathbf{A}t}$ by the steps outlined above:

1. The characteristic polynomial for **A** is $c(\lambda)=\det(\lambda\mathbf{I}-\mathbf{A})=(\lambda-2)^3$. The only eigenvalue is $\lambda_1=2$. By investigating the nullities of $(\mathbf{A}-2\mathbf{I})$ and $(\mathbf{A}-2\mathbf{I})^2$, we find that $q_1=2$ and $m(\lambda)=(\lambda-2)^2$. Thus

$$f(\mathbf{A})=f(2)\mathbf{E}_{10}^{A}+f'(2)\mathbf{E}_{11}^{A}$$

2. $g_1(\lambda)=(\lambda-2)$ and $g_1'(\lambda)=1$. Therefore,

$$\begin{aligned} g_1(A) &\stackrel{\Delta}{=} (\mathbf{A}-2\mathbf{I})=g_1(2)\mathbf{E}_{10}^{A}+g_1'(2)\mathbf{E}_{11}^{A} \\ &=(0)\mathbf{E}_{10}^{A}+(1)\mathbf{E}_{11}^{A} \end{aligned}$$

and $\mathbf{E}_{11}^{A}=\mathbf{A}-2\mathbf{I}$.

3. $g_2(\lambda)=1$ and $g_2'(\lambda)=0$. Then,

$$\begin{aligned} g_2(\mathbf{A}) &\stackrel{\Delta}{=} \mathbf{I}=g_2(2)\mathbf{E}_{10}^{A}+g_2'(2)\mathbf{E}_{11}^{A} \\ &=(1)\mathbf{E}_{10}^{A}+(0)\mathbf{E}_{11}^{A} \end{aligned}$$

and $\mathbf{E}_{10}^{A}=\mathbf{I}$.

4. $f(2)=e^{2t}$, and $f'(2)=te^{2t}$.

5. $e^{\mathbf{A}t}=e^{2t}\mathbf{I}+te^{2t}(\mathbf{A}-2\mathbf{I})$

$$=\begin{pmatrix} e^{2t}-te^{2t} & te^{2t} & 0 \\ -te^{2t} & e^{2t}+te^{2t} & 0 \\ -te^{2t} & te^{2t} & e^{2t} \end{pmatrix}$$

*From Zadeh and Desoer [4.20].

Evaluating factors of $m(\lambda)$ is probably the most efficient known method for computing $f(\mathbf{A})$. A suitable sequence of functions can also be obtained by successively cancelling factors from the characteristic polynomial $c(\lambda)$, thereby avoiding determination of the nullities of powers of $(\mathbf{A}-\lambda_i\mathbf{I})$. If $c(\lambda)$ had been used in Example 5, we would have found that $\mathbf{E}_{12}^{\mathbf{A}}=\boldsymbol{\Theta}$.

From our computation of $f(\mathbf{A})$ by evaluating factors of the minimal polynomial, we recognize that each of the constituents $\mathbf{E}_{ij}^{\mathbf{A}}$ equals a polynomial in $\mathbf{A}$; the order of the polynomial is, in each case, less than that of the minimal polynomial. Therefore, by the fundamental formula, $f(\mathbf{A})$ is also equal to a polynomial in $\mathbf{A}$. Since powers of $\mathbf{A}$, and thus polynomials in $\mathbf{A}$, commute with each other, functions of $\mathbf{A}$ commute with each other also. See P&C 4.29 for properties of commuting matrices. Additional techniques for computing $f(\mathbf{A})$ are given in P&C 4.25–4.27.

Application of Functions of Matrices—Modes of Oscillation

Figure 4.9 is an idealized one-dimensional representation of a piece of spring-mounted equipment. The variables $\mathbf{v}_1$, $\mathbf{v}_2$, and $\mathbf{u}$ represent the positions, relative to their respective references, of the two identical masses (labeled m) and the frame which holds the equipment. The three springs have identical spring constants k. We treat the position (or vibration) of the frame as an independent variable; we seek the motions, $\mathbf{v}_1(t)$ and $\mathbf{v}_2(t)$, of the spring-mounted objects. The dynamic equations which describe these motions are

$$\begin{aligned} m\ddot{\mathbf{v}}_1(t) &= -2k\mathbf{v}_1(t)+k\mathbf{v}_2(t)+k\mathbf{u}(t) \\ m\ddot{\mathbf{v}}_2(t) &= k\mathbf{v}_1(t)-2k\mathbf{v}_2(t)+k\mathbf{u}(t) \end{aligned} \tag{4.109}$$

We could convert (4.109) to a four-dimensional first-order state equation. However, emboldened by the formal analogy which we found between the solution to the state equation and its scalar counterpart, we develop a second-order vector equation which is equivalent to (4.109) and which keeps explicit the second-order nature of the individual equations.

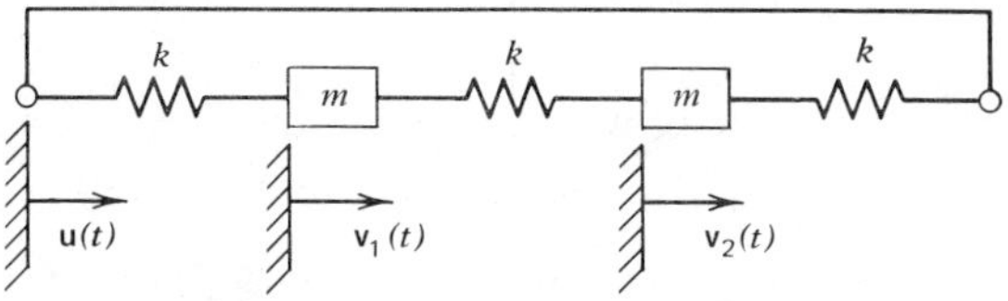

Figure 4.9. A model for spring-mounted equipment.

Let $\mathbf{x}=(\mathbf{v}_1\ \mathbf{v}_2)^{\mathrm{T}}$. Then (4.109) becomes

$$\ddot{\mathbf{x}}(t)+\begin{pmatrix} 2k/m & -k/m \\ -k/m & 2k/m \end{pmatrix}\mathbf{x}(t)=\begin{pmatrix} k/m \\ k/m \end{pmatrix}\mathbf{u}(t) \tag{4.110}$$

The 2×2 matrix in (4.110) is known as the **stiffness matrix** for the system. Equation (4.110) is a special case of the general vector equation

$$\ddot{\mathbf{x}}(t)+\mathbf{A}\mathbf{x}(t)=\mathbf{B}\mathbf{u}(t) \tag{4.111}$$

where $\mathbf{x}(t)$ is $n\times 1$, $\mathbf{u}(t)$ is $m\times 1$, $\mathbf{B}$ is $n\times m$, and $\mathbf{A}$ is an $n\times n$ diagonalizable matrix with positive eigenvalues.* Equation (4.111) is a convenient way to express many conservative systems; for example, a frictionless mechanical system which contains n masses coupled by springs; or a lossless electrical network containing interconnected inductors and capacitors. We solve (4.110) and (4.111) by analogy with the scalar case.

The scalar counterpart of (4.111) is

$$\ddot{\mathbf{f}}(t)+\omega^2\mathbf{f}(t)=\mathbf{u}(t) \tag{4.112}$$

We found in P&C 3.6 that the inverse of (4.112), in terms of the initial conditions $\mathbf{f}(0)$ and $\dot{\mathbf{f}}(0)$, is

$$\mathbf{f}(t)=\mathbf{f}(0)\cos\omega t+\frac{\dot{\mathbf{f}}(0)}{\omega}\sin\omega t+\int_0^t\frac{\sin\omega(t-s)}{\omega}\mathbf{u}(s)\,ds \tag{4.113}$$

The solution consists in an undamped oscillation of frequency ω plus a term affected by the input vibration $\mathbf{u}$.

Comparing (4.111) and (4.112), we recognize that $\mathbf{x}$ is the vector analog of $\mathbf{f}$, and $\mathbf{A}$ plays the same role as ω^2. Therefore, we expect the solution to (4.111) to be

$$\mathbf{x}(t)=\cos(\sqrt{\mathbf{A}}\,t)\mathbf{x}(0)+(\sqrt{\mathbf{A}}\,)^{-1}\sin(\sqrt{\mathbf{A}}\,t)\dot{\mathbf{x}}(0)$$
$$+\int_0^t(\sqrt{\mathbf{A}}\,)^{-1}\sin\left[\sqrt{\mathbf{A}}\,(t-s)\right]\mathbf{B}\mathbf{u}(s)\,ds \tag{4.114}$$

By $\sqrt{\mathbf{A}}$ we mean any matrix whose square equals $\mathbf{A}$. As with the scalar square root, $\sqrt{\mathbf{A}}$ is not unique. The fundamental formula (4.103) indicates that $\sqrt{\mathbf{A}}$ depends on the square roots of the eigenvalues of $\mathbf{A}$. We use in

*The matrix **A** is symmetric and positive definite. Such a matrix necessarily has positive real eigenvalues. See P&C 5.9 and 5.28.

(4.114) the principal square root of **A**—the one involving positive square roots of the eigenvalues (P&C 4.28). Recall from the discussion following Example 5 that functions of **A** commute with each other; the order of multiplication of $(\sqrt{\mathbf{A}})^{-1}$ and $\sin(\sqrt{\mathbf{A}}\,t)$ is arbitrary.

Equation (4.114) can be derived by finding a matrix Green's function and matrix boundary kernel for (4.111) (P&C 4.32). Or it can be verified by showing that it is a solution to the differential equation (4.111).

Exercise 3. Verify (4.114) by substituting $\mathbf{x}(t)$ into (4.111). Hint:

$$\frac{d}{dt} f(\mathbf{A}t) = \mathbf{A}\dot{f}(\mathbf{A}t) \qquad \text{(P\&C 4.30)}$$

$$\frac{d}{dt}\int_a^t g(t,s)\,ds = \int_a^t \frac{\partial}{\partial t} g(t,s)\,ds + g(t,t)$$

We now evaluate the solution (4.114) for the specific case (4.110) using the techniques derived for determining functions of matrices.

Exercise 4. Show that the eigendata for the 2×2 stiffness matrix **A** of (4.110) are

$$\lambda_1 = \frac{k}{m}, \qquad \lambda_2 = \frac{3k}{m}, \qquad \mathbf{x}_1 = \begin{pmatrix} 1 \\ 1 \end{pmatrix}, \qquad \mathbf{x}_2 = \begin{pmatrix} 1 \\ -1 \end{pmatrix}$$

Exercise 5. Show that for **A** of (4.110),

$$f(\mathbf{A}) = f\left(\frac{k}{m}\right)\begin{pmatrix} \frac{1}{2} & \frac{1}{2} \\ \frac{1}{2} & \frac{1}{2} \end{pmatrix} + f\left(\frac{3k}{m}\right)\begin{pmatrix} \frac{1}{2} & -\frac{1}{2} \\ -\frac{1}{2} & \frac{1}{2} \end{pmatrix}$$

It follows from Exercise 5 that

$$\cos\sqrt{\mathbf{A}}\,t = \begin{pmatrix} \dfrac{\cos\sqrt{k/m}\,t + \cos\sqrt{3k/m}\,t}{2} & \dfrac{\cos\sqrt{k/m}\,t - \cos\sqrt{3k/m}\,t}{2} \\[2ex] \dfrac{\cos\sqrt{k/m}\,t - \cos\sqrt{3k/m}\,t}{2} & \dfrac{\cos\sqrt{k/m}\,t + \cos\sqrt{3k/m}\,t}{2} \end{pmatrix}$$

$$(\sqrt{\mathbf{A}})^{-1}\sin\sqrt{\mathbf{A}}\,t = \begin{pmatrix} \dfrac{\sin\sqrt{k/m}\,t}{2\sqrt{k/m}} + \dfrac{\sin\sqrt{3k/m}\,t}{2\sqrt{3k/m}} & \dfrac{\sin\sqrt{k/m}\,t}{2\sqrt{k/m}} - \dfrac{\sin\sqrt{3k/m}\,t}{2\sqrt{3k/m}} \\[2ex] \dfrac{\sin\sqrt{k/m}\,t}{2\sqrt{k/m}} - \dfrac{\sin\sqrt{3k/m}\,t}{2\sqrt{3k/m}} & \dfrac{\sin\sqrt{k/m}\,t}{2\sqrt{k/m}} + \dfrac{\sin\sqrt{3k/m}\,t}{2\sqrt{3k/m}} \end{pmatrix}$$

$$(\sqrt{\mathbf{A}})^{-1}\sin\left[\sqrt{\mathbf{A}}\,(t-s)\right]\mathbf{B} = \sqrt{k/m}\,\sin\sqrt{k/m}\,(t-s)\begin{pmatrix} 1 \\ 1 \end{pmatrix}$$

These three matrices can be substituted into (4.114) to obtain $\mathbf{x}(t)$ explicitly as a complicated function of the input data $\mathbf{u}(t)$, $\mathbf{x}(0)$, and $\dot{\mathbf{x}}(0)$.

Even though the general form of $\mathbf{x}(t)$ is complicated, we can provide a simple physical interpretation of the eigendata of the stiffness matrix of (4.110). Let $\mathbf{x}(0)=\mathbf{x}_1$, $\dot{\mathbf{x}}(0)=\boldsymbol{\theta}$, and $\mathbf{u}(t)=0$. Then recalling that $\mathbf{A}$ and $f(\mathbf{A})$ have the same eigenvectors,

$$\mathbf{x}(t) \triangleq \begin{pmatrix} \mathbf{v}_1(t) \\ \mathbf{v}_2(t) \end{pmatrix} = \cos\sqrt{\mathbf{A}}\, t \begin{pmatrix} 1 \\ 1 \end{pmatrix} = \cos\sqrt{k/m}\, t \begin{pmatrix} 1 \\ 1 \end{pmatrix}$$

The first eigenvector initial condition excites a sinusoidal oscillation of frequency $\sqrt{k/m} = \sqrt{\lambda_1}$. In this first mode of oscillation, both masses move together—the center spring is not stressed. The system acts like a single mass with a spring–mass ratio of $2k/2m = k/m = \lambda_1$. A second mode of oscillation can be excited by the conditions $\mathbf{x}(0)=\mathbf{x}_2$, $\dot{\mathbf{x}}(0)=\boldsymbol{\theta}$, $\mathbf{u}(t)=0$;

$$\mathbf{x}(t) \triangleq \begin{pmatrix} \mathbf{v}_1(t) \\ \mathbf{v}_2(t) \end{pmatrix} = \cos\sqrt{\mathbf{A}}\, t \begin{pmatrix} 1 \\ -1 \end{pmatrix} = \cos\sqrt{3k/m}\, t \begin{pmatrix} 1 \\ -1 \end{pmatrix}$$

The second eigenvector initial condition excites a sinusoidal oscillation of frequency $\sqrt{3k/m} = \sqrt{\lambda_2}$. In this mode of oscillation, the masses move in opposite directions—the midpoint of the center spring does not move. The system acts like a pair of mirror images, each with a spring–mass ratio of $(k+2k)m = 3k/m = \lambda_2$. Thus the eigenvectors and eigenvalues of $\mathbf{A}$ are natural modes of oscillation and squares of natural frequencies of oscillation, respectively.

The initial conditions $\dot{\mathbf{x}}(0)=\mathbf{x}_1$ or $\dot{\mathbf{x}}(0)=\mathbf{x}_2$ also excite the above two natural modes of oscillation. We note that for this particular example $\mathbf{Bu}(t)$ is of the form of $\mathbf{x}_1$. The motion excited by the input vibration $\mathbf{u}(t)$ can only be proportional to $\mathbf{x}_1$. Whether or not the motion is a sinusoidal oscillation is determined by the form of $\mathbf{u}(t)$.

4.7 Problems and Comments

4.1 Let $\mathcal{W}_1 = \text{span}\{(1,0,1)\}$ and $\mathcal{W}_2 = \text{span}\{(1,0,0), (0,1,0)\}$ in $\mathcal{R}^3$.

(a) Show that an arbitrary vector $\mathbf{x}$ in $\mathcal{R}^3$ can be decomposed into a unique pair of components $\mathbf{x}_1$ and $\mathbf{x}_2$ from $\mathcal{W}_1$ and $\mathcal{W}_2$, respectively.

(b) Let $\mathbf{P}_1$ be the projector onto $\mathcal{W}_1$ along $\mathcal{W}_2$, and $\mathbf{P}_2$ the projector onto $\mathcal{W}_2$ along $\mathcal{W}_1$. Let $\mathcal{E}$ be the standard basis for $\mathcal{R}^3$. Find $[\mathbf{P}_1]_{\mathcal{X}\mathcal{X}}$ and $[\mathbf{P}_2]_{\mathcal{X}\mathcal{X}}$.

4.2 Let the linear operator $\mathbf{T}$ defined by $\mathbf{Tx} \triangleq \mathbf{Ax}$ operate on the space $\mathcal{M}^{n\times 1}$. Let the subspaces $\mathcal{W}_1$ and $\mathcal{W}_2$ of $\mathcal{M}^{n\times 1}$ be composed of vectors of the form

$$\begin{pmatrix} \xi_1 \\ \vdots \\ \xi_m \\ 0 \\ \vdots \\ 0 \end{pmatrix} \quad \text{and} \quad \begin{pmatrix} 0 \\ \vdots \\ 0 \\ \eta_{m+1} \\ \vdots \\ \eta_n \end{pmatrix}$$

respectively. Determine the form of $\mathbf{A}$ if
(*a*) $\mathcal{W}_1$ is invariant under $\mathbf{T}$.
(*b*) $\mathcal{W}_2$ is invariant under $\mathbf{T}$.
(*c*) Both $\mathcal{W}_1$ and $\mathcal{W}_2$ are invariant under $\mathbf{T}$.
Hint: investigate an example where $m=1$, $n=3$.

4.3 The *cartesian product* is useful for *building up* complicated vector spaces from simple ones. The *direct sum*, on the other hand, is useful for *subdividing* complicated vector spaces into smaller subspaces.

(*a*) Define $\mathbf{T}_a$: $\mathcal{R}^2 \to \mathcal{R}^2$ by $\mathbf{T}_a(\xi_1, \xi_2) \triangleq (\xi_1 - \xi_2, \xi_1)$.
Let $\mathcal{X}_a = \{(1,0), (0,1)\}$. Find $[\mathbf{T}_a]_{\mathcal{X}_a\mathcal{X}_a}$.
Define $\mathbf{T}_b$: $\mathcal{R}^1 \to \mathcal{R}^1$ by $\mathbf{T}_b(\xi_3) \triangleq (-\xi_3)$.
Let $\mathcal{X}_b = \{(1)\}$. Find $[\mathbf{T}_b]_{\mathcal{X}_b\mathcal{X}_b}$.

(*b*) If we do not distinguish between $((\xi_1, \xi_2), (\xi_3))$ and (ξ_1, ξ_2, ξ_3), then $\mathcal{R}^3 = \mathcal{R}^2 \times \mathcal{R}^1$. Define $\mathbf{T}$: $\mathcal{R}^3 \to \mathcal{R}^3$ by $\mathbf{T}((\xi_1, \xi_2), (\xi_3)) \triangleq (\mathbf{T}_a(\xi_1, \xi_2), \mathbf{T}_b(\xi_3))$. Let $\mathcal{X} = \{((1,0), (0)), ((0,1), (0)), ((0,0), (1))\}$. Find $[\mathbf{T}]_{\mathcal{X}\mathcal{X}}$. What is the relationship between $[\mathbf{T}]_{\mathcal{X}\mathcal{X}}$, $[\mathbf{T}_a]_{\mathcal{X}_a\mathcal{X}_a}$, $[\mathbf{T}_b]_{\mathcal{X}_b\mathcal{X}_b}$?

(*c*) Let $\mathcal{W}_1 = \mathcal{R}^2 \times \{(0)\}$ and $\mathcal{W}_2 = \{(0,0)\} \times \mathcal{R}^1$. Then $\mathcal{R}^3 = \mathcal{W}_1 \oplus \mathcal{W}_2$. Appropriate bases for $\mathcal{W}_1$ and $\mathcal{W}_2$ are $\mathcal{X}_1 = \{((1,0), (0)), ((0,1), (0))\}$ and $\mathcal{X}_2 = \{((0,0), (1))\}$. Define $\mathbf{T}_1$: $\mathcal{W}_1 \to \mathcal{W}_1$ by $\mathbf{T}_1(\xi_1, \xi_2, 0) \triangleq (\xi_1 - \xi_2, \xi_1, 0)$. Define $\mathbf{T}_2$: $\mathcal{W}_2 \to \mathcal{W}_2$ by $\mathbf{T}_2(0,0,\xi_3) \triangleq (0,0,-\xi_3)$. Find $[\mathbf{T}_1]_{\mathcal{X}_1\mathcal{X}_1}$ and $[\mathbf{T}_2]_{\mathcal{X}_2\mathcal{X}_2}$. What is the relationship between $[\mathbf{T}]_{\mathcal{X}\mathcal{X}}$, $[\mathbf{T}_1]_{\mathcal{X}_1\mathcal{X}_1}$, and $[\mathbf{T}_2]_{\mathcal{X}_2\mathcal{X}_2}$?

(*d*) In general, if $\mathcal{V}=\mathcal{W}_1\oplus\cdots\oplus\mathcal{W}_p$, with each subspace $\mathcal{W}_i$ invariant under $\mathbf{T}$, then $\{\mathcal{W}_i\}$ decomposes $\mathbf{T}$ into $\{\mathbf{T}_i\colon \mathcal{W}_i \to \mathcal{W}_i\}$. Let $\mathcal{X}_i$ be a basis for $\mathcal{W}_i$. Then $\mathcal{X}=\{\mathcal{X}_1,\ldots,\mathcal{X}_p\}$ is a basis for $\mathcal{V}$. If $\mathcal{V}$ is finite-dimensional, then

$$[\mathbf{T}]_{\mathcal{X}\mathcal{X}}=\begin{pmatrix}[\mathbf{T}_1]_{\mathcal{X}_1\mathcal{X}_1} & & \\ & \ddots & \\ & & [\mathbf{T}_p]_{\mathcal{X}_p\mathcal{X}_p}\end{pmatrix}$$

with zeros everywhere except in the blocks on the diagonal. Show that the transformation $\mathbf{T}\colon \mathcal{R}^3\to\mathcal{R}^3$ defined by $\mathbf{T}(\xi_1,\xi_2,\xi_3)\triangleq(\xi_1+\xi_2, 2\xi_1+\xi_2-\xi_3, \xi_1+\xi_3)$ is decomposed by $\mathcal{W}_1$ and $\mathcal{W}_2$, where $\mathcal{W}_1$ consists in vectors of the form $(\xi_1,\xi_2,\xi_1+\xi_2)$ and $\mathcal{W}_2$ consists in vectors of the form (ξ_1,ξ_1,ξ_1). Note that there is no cartesian product which corresponds to this invariant direct-sum decomposition in the same manner as (*b*) corresponds to (*c*).

4.4 Find the eigenvalues and eigenvectors of the following matrices:

$$(a)\quad \begin{pmatrix}-3 & 0 & 0\\ -5 & 2 & 0\\ -5 & 1 & 1\end{pmatrix} \qquad (b)\quad \begin{pmatrix}-2 & 0 & 0\\ -3 & 1 & 3\\ 0 & 0 & -2\end{pmatrix}$$

4.5 Let $\mathbf{A}$ be an $n\times n$ matrix. Denote the characteristic polynomial for $\mathbf{A}$ by $c(\lambda)=\lambda^n+b_1\lambda^{n-1}+\cdots+b_n$. The **trace** of a matrix is defined as the sum of its diagonal elements, an easily computed quantity. An iterative method based on the trace function has been proposed for computing the coefficients $\{b_i\}$ in the characteristic polynomial [4.3, p. 296]. The iteration is:

$$b_1=-\text{Trace}(\mathbf{A})$$

$$b_2=-\tfrac{1}{2}\left[b_1\,\text{Trace}(\mathbf{A})+\text{Trace}(\mathbf{A}^2)\right]$$

$$b_3=-\frac{1}{3}\left[b_2\,\text{Trace}(\mathbf{A})+b_1\,\text{Trace}(\mathbf{A}^2)+\text{Trace}(\mathbf{A}^3)\right]$$

$$\vdots$$

$$b_n=-\frac{1}{n}\left[b_{n-1}\,\text{Trace}(\mathbf{A})+\cdots+b_1\,\text{Trace}(\mathbf{A}^{n-1})+\text{Trace}(\mathbf{A}^n)\right]$$

(*a*) How many multiplications are required to compute the characteristic polynomial by means of this trace iteration? Compare the iteration with Krylov's method.

(*b*) Compute the characteristic polynomial by Krylov's method and by the trace iteration for the matrix

$$\mathbf{A}=\begin{pmatrix}1 & 2 & 3\\ 2 & 1 & 0\\ 1 & 1 & 1\end{pmatrix}$$

*4.6 Let $\mathbf{A}$ be an $n\times n$ matrix with eigenvalues $\lambda_1,\ldots,\lambda_n$. Then

(*a*) $\text{Det}(\mathbf{A})=\lambda_1\cdot\lambda_2\cdot\cdots\cdot\lambda_n$

(*b*) $\text{Trace}(\mathbf{A})\stackrel{\Delta}{=}a_{11}+a_{22}+\cdots+a_{nn}=\lambda_1+\lambda_2+\cdots+\lambda_n$

(*c*) If $\mathbf{A}$ is triangular (i.e, if all elements to one side of the main diagonal are zero), then the diagonal elements of $\mathbf{A}$ are $\mathbf{A}_{ii}=\lambda_i$.

4.7 Three men are playing ball. Every two seconds the one who has the ball tosses it to one of the others, with the probabilities shown in the diagram. Let $p_n(i)$ be the probability that the ball is held by the ith player (or is in the ith state) after the nth toss. Let p_{ij} be the probability with which player j throws the ball to player i. The theory of conditional probability requires that

$$p_n(i)=\sum_{j=1}^{3}p_{ij}p_{n-1}(j)\quad\text{for}\quad i=1,2,3$$

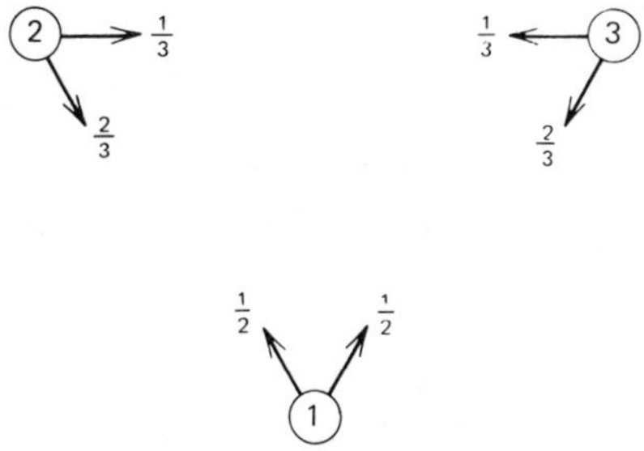

Let $\mathbf{x}_n\stackrel{\Delta}{=}(p_n(1)\ p_n(2)\ p_n(3))^{\mathrm{T}}$. We call $\mathbf{x}_n$ a state probability vector. Let Ω denote the set of all possible 3×1 state probability vectors. The elements of each vector in Ω are non-negative and sum to one. Note that Ω is a *subset* of $\mathfrak{M}^{n\times 1}$, rather than a subspace. The game is an example of a Markov process. The future state probability vectors depend only on the present state, and not on the past history.

(*a*) A matrix whose columns are members of Ω is called a *transition probability matrix*. Find the transition probability matrix $\mathbf{A}$ such that $\mathbf{x}_n = \mathbf{A}\mathbf{x}_{n-1}$. Note that $\mathbf{x}_n = \mathbf{A}^n\mathbf{x}_0$; we refer to $\mathbf{A}^n$ as the n-step transition probability matrix.

(*b*) Determine the eigenvalues and eigenvectors of $\mathbf{A}$. What do they tell us about the game? (Hint: $\lambda = 1$ is an eigenvalue.)

(*c*) Find the spectral matrix $\mathbf{\Lambda}$ and the modal matrix $\mathbf{S}$ such that $\mathbf{A} = \mathbf{S}\mathbf{\Lambda}\mathbf{S}^{-1}$. Show that every transition probability matrix has $\lambda = 1$ as an eigenvalue. (Hint: transpose the equation $\mathbf{A} = \mathbf{S}\mathbf{\Lambda}\mathbf{S}^{-1}$ for the matrix $\mathbf{A}$ obtained in (*a*). A matrix and its transpose have identical eigenvalues. What is the form of the eigenvectors of $\mathbf{A}^{\mathrm{T}}$ for $\lambda = 1$?)

(*d*) In the game described previously, the state probability vector $\mathbf{x}_n$ becomes independent of the initial state as n becomes large. Find the form of the limiting state probability vector. (Hint: find $\lim_{n\to\infty}\mathbf{A}^n$ using the substitution $\mathbf{A} = \mathbf{S}\mathbf{\Lambda}\mathbf{S}^{-1}$.) We note that the eigenvalues of every transition probability matrix satisfy $\lambda_i \leqslant 1$ [4.4, p. 429].

(*e*) A transition probability matrix wherein the elements of each row also sum to one is called a *stochastic matrix*. What is the limiting state probability vector, $\lim_{n\to\infty}\mathbf{x}_n$, if the transition probabilities in the above game are modified to yield a stochastic matrix?

4.8 Let

$$\mathbf{A} = \begin{pmatrix} 3 & -2 & 0 \\ 0 & 1 & 0 \\ 0 & 0 & 1 \end{pmatrix}$$

Find a matrix $\mathbf{S}$ for which $\mathbf{S}^{-1}\mathbf{A}\mathbf{S}$ is a diagonal matrix.

4.9 Find a nondiagonal matrix $\mathbf{A}$ which has as its diagonal form the matrix

$$\mathbf{\Lambda} = \begin{pmatrix} 1 & 0 & 0 \\ 0 & -1 & 0 \\ 0 & 0 & 3 \end{pmatrix}$$

What are eigenvectors of $\mathbf{A}$?

4.10 We wish to compute the eigendata of the matrix

$$\mathbf{A} = \begin{pmatrix} 1 & 0 \\ 1 & -1 \end{pmatrix}$$

Assume that numerical computations have produced the following approximations to the eigenvalues: $\lambda_1 \approx 0.99$ and $\lambda_2 \approx -1.01$. Use the inverse iteration method to compute more accurate eigenvalues and corresponding eigenvectors. Start the iterations with the initial vector $\mathbf{z}_0 = (1 \;\; -1)^{\mathrm{T}}$.

4.11 The **Jacobi method** for determining the eigenvalues and eigenvectors of a symmetric matrix $\mathbf{A}$ consists in performing a sequence of similarity transformations which reduce the off-diagonal elements of $\mathbf{A}$ to zero. In order to avoid a sequence of matrix inversions, we perform the similarity transformations with orthogonal matrices (matrices for which $\mathbf{S}^{-1} = \mathbf{S}^{\mathrm{T}}$). Thus we let $\mathbf{A}_1 = \mathbf{S}_1^{\mathrm{T}}\mathbf{A}\mathbf{S}_1$ and $\mathbf{A}_k = \mathbf{S}_k^{\mathrm{T}}\mathbf{A}_{k-1}\mathbf{S}_k$ for $k = 2, 3, \ldots$. The eigenvalues of a matrix are not changed by similarity transformations. Consequently, the resulting diagonal matrix must be the spectral matrix (with the eigenvalues of $\mathbf{A}$ on its diagonal); that is,

$$\lim_{k\to\infty} \mathbf{A}_k = \lim_{k\to\infty} (\mathbf{S}_1\mathbf{S}_2 \cdots \mathbf{S}_k)^{\mathrm{T}}\mathbf{A}(\mathbf{S}_1\mathbf{S}_2 \cdots \mathbf{S}_k) = \mathbf{\Lambda}$$

Furthermore, the matrix $\mathbf{S} = \lim_{k\to\infty}(\mathbf{S}_1\mathbf{S}_2 \cdots \mathbf{S}_k)$ must be a modal matrix for $\mathbf{A}$ (with the eigenvectors of $\mathbf{A}$ as its columns). Let $a_{ij} = (\mathbf{A}_{k-1})_{ij}$. It is shown in [4.13] that a_{ij} and a_{ji} can be driven to zero simultaneously by a similarity transformation which uses the orthogonal matrix $\mathbf{S}_k$ which differs from the identity matrix only in the following elements:

$$(\mathbf{S}_k)_{ii} = (\mathbf{S}_k)_{jj} = \sqrt{(\gamma + |\beta|)/2\gamma} = \cos\phi$$

$$(\mathbf{S}_k)_{ij} = -(\mathbf{S}_k)_{ji} = \alpha\,\mathrm{sign}(\beta)/(2\gamma\cos\phi) = \sin\phi$$

where $\alpha = -a_{ij}$, $\beta = (a_{ii} - a_{jj})/2$, and $\gamma = (\alpha^2 + \beta^2)^{1/2}$. (Multiplication by the matrix $\mathbf{S}_k$ can be interpreted as a rotation of the axes of the i and j coordinates through an angle ϕ.) In the Jacobi method we pick an $\mathbf{S}_k$ of the above form which drives the largest pair of off-diagonal elements of $\mathbf{A}_{k-1}$ to zero. Although later transformations will usually make these elements nonzero again, the sum of the squares of the off-diagonal elements is reduced at each iteration.

(*a*) Use the Jacobi method to compute (to slide rule accuracy) the eigenvalues and eigenvectors of the matrix

$$\mathbf{A} = \begin{pmatrix} 3 & 1 \\ 1 & 2 \end{pmatrix}$$

(*b*) Calculate the eigenvalues of **A** by solving the characteristic polynomial. Determine the corresponding eigenvectors. Compare the results with (*a*).

4.12 Let **L** be the differential operator defined by $\mathbf{Lf} \triangleq \mathbf{f}''$. Assume **L** acts on the subspace of functions in $\mathcal{C}^2(0,\pi)$ which satisfy the boundary conditions $\boldsymbol{\beta}_1(\mathbf{f})=\boldsymbol{\beta}_2(\mathbf{f})=0$. Find all the eigenvalues and corresponding eigenfunctions of **L** for each of the following definitions of the boundary conditions:
(*a*) $\boldsymbol{\beta}_1(\mathbf{f})=\mathbf{f}(0), \qquad \boldsymbol{\beta}_2(\mathbf{f})=\mathbf{f}(\pi)$
(*b*) $\boldsymbol{\beta}_1(\mathbf{f})=\mathbf{f}(0)+\mathbf{f}(\pi), \qquad \boldsymbol{\beta}_2(\mathbf{f})=\mathbf{f}'(0)-\mathbf{f}'(\pi)$
(*c*) $\boldsymbol{\beta}_1(\mathbf{f})=\mathbf{f}(0)+2\mathbf{f}(\pi), \qquad \boldsymbol{\beta}_2(\mathbf{f})=\mathbf{f}'(0)-2\mathbf{f}'(\pi)$
(*d*) $\boldsymbol{\beta}_1(\mathbf{f})=\mathbf{f}(0)-\mathbf{f}(\pi), \qquad \boldsymbol{\beta}_2(\mathbf{f})=\mathbf{f}'(0)-\mathbf{f}'(\pi)$

4.13 Find the eigenvalues and eigenfunctions associated with the differential system $\mathbf{f}''-c\mathbf{f}=\mathbf{u}$, $\mathbf{f}(0)=\mathbf{f}'(1)=0$. Hint: $\ln(-1)=i(\pi+2k\pi)$, $k=0, \pm 1, \pm 2, \ldots$. For what values of the constant c is the system invertible?

4.14 Let $\mathcal{V}$ be a space of functions **f** whose values $\mathbf{f}(n)$ are defined only for integer values of n. Define the forward difference operator $\boldsymbol{\Delta}$ on $\mathcal{V}$ by

$$(\boldsymbol{\Delta}\mathbf{f})(n) \triangleq \mathbf{f}(n+1)-\mathbf{f}(n)$$

(This operator can be used to approximate the differential operator **D**.) Find the eigenvalues and eigenfunctions of $\boldsymbol{\Delta}$.

4.15 Define $\nabla^2\mathbf{f}(s,t) \triangleq (\partial^2\mathbf{f}/\partial s^2)+(\partial^2\mathbf{f}/\partial t^2)$ in the rectangular region $0 \leqslant s \leqslant a$ and $0 \leqslant t \leqslant b$. Let **f** satisfy the boundary conditions

$$\frac{\partial \mathbf{f}}{\partial s}(0,t)=\frac{\partial \mathbf{f}}{\partial s}(a,t)=\frac{\partial \mathbf{f}}{\partial t}(s,0)=\frac{\partial \mathbf{f}}{\partial t}(s,b)=0$$

Show that the partial differential operator ∇^2 and the given boundary conditions have the eigendata

$$\lambda_{km}=-\left(\frac{m\pi}{a}\right)^2-\left(\frac{k\pi}{b}\right)^2$$

$$\mathbf{f}_{km}(s,t)=\cos\left(\frac{m\pi s}{a}\right)\cos\left(\frac{k\pi t}{b}\right)$$

for $k,m=0,1,2,\ldots$.

*4.16 Let **A** be the companion matrix for an nth order constant-coefficient differential operator. Denote the eigenvalues of **A** by $\lambda_1,\ldots,\lambda_n$.

(*a*) Show that the vector $\mathbf{z}_i = (1 \quad \lambda_i \quad \lambda_i^2 \quad \cdots \quad \lambda_i^{n-1})^{\mathrm{T}}$ is an eigenvector of $\mathbf{A}$ for the eigenvalue λ_i. Show further that there is only one independent eigenvector for each distinct eigenvalue.

(*b*) Show that the Vandermond matrix

$$\begin{pmatrix} 1 & \cdots & 1 \\ \lambda_1 & \cdots & \lambda_n \\ \vdots & \vdots & \vdots \\ \lambda_1^{n-1} & \cdots & \lambda_n^{n-1} \end{pmatrix}$$

is a modal matrix for $\mathbf{A}$ if and only if the eigenvalues of $\mathbf{A}$ are all distinct.

*4.17 *The Power method*: the inverse of the differential operator $\mathbf{L} = \mathbf{D}^2$ with the boundary conditions $\mathbf{f}(0) = \mathbf{f}(1) = 0$ is the integral operator $\mathbf{T}$ defined by

$$(\mathbf{Tu})(t) = \int_0^t (t-1)s\mathbf{u}(s)\,ds + \int_t^1 t(s-1)\mathbf{u}(s)\,ds$$

The functions $\mathbf{f}_n(t) = \sin n\pi t$, $n = 1, 2, \ldots$, are eigenfunctions for both the differential and integral operators. We can find the dominant eigenvalue and the corresponding eigenfunction of $\mathbf{T}$ by the power method. We just compute the sequence of functions $\mathbf{u}_k = \mathbf{T}^k\mathbf{u}_0$, for some initial function $\mathbf{u}_0$, until $\mathbf{u}_k$ is a sufficiently good approximation to the dominant eigenfunction.

(*a*) Let $\mathbf{u}_0(t) = 1$, and compute $\mathbf{u}_1$ and $\mathbf{u}_2$.

(*b*) Compare $\mathbf{u}_1$ and $\mathbf{u}_2$ with the true dominant eigenfunction. Use the iterates $\{\mathbf{u}_k\}$ to determine an approximation to the dominant eigenvalue.

4.18 (*a*) Determine an ordered basis of generalized eigenvectors for the matrix

$$\mathbf{A} = \begin{pmatrix} 5 & -1 & 1 & 1 & 0 & 0 \\ 1 & 3 & -1 & -1 & 0 & 0 \\ 0 & 0 & 4 & 0 & 1 & 1 \\ 0 & 0 & 0 & 4 & -1 & -1 \\ 0 & 0 & 0 & 0 & 3 & 1 \\ 0 & 0 & 0 & 0 & 1 & 3 \end{pmatrix}$$

Hint: $\det(\mathbf{A} - \lambda\mathbf{I}) = (4-\lambda)^5\,(2-\lambda)$.

(*b*) Determine the Jordan canonical form of $\mathbf{A}$ [relative to the basis found in (*a*)].

(*c*) Determine the "change of coordinates" matrix **S** which would be used in a similarity transformation on **A** in order to obtain the Jordan form found in (*b*). (Obtain only the obvious matrix, not its inverse.)

4.19 Find a matrix **S** such that $\mathbf{S}^{-1}\mathbf{B}\mathbf{S}$ is in Jordan form, for

$$\mathbf{B}=\begin{pmatrix} 3 & 0 & 0 & 1 \\ 0 & 2 & 0 & 0 \\ 1 & 1 & 3 & 1 \\ -1 & 0 & 0 & 1 \end{pmatrix}$$

Hint: $c(\lambda)=(2-\lambda)^3(3-\lambda)$.

4.20 The minimal polynomial $m(\lambda)$ and the characteristic polynomial $c(\lambda)$ are useful for reducing effort in matrix computations. Assume $f(\mathbf{A})$ is a polynomial in the $n\times n$ matrix **A**, and $f(\mathbf{A})$ includes powers of **A** higher than n. We divide $f(\lambda)$ by $m(\lambda)$ to determine a quotient $g(\lambda)$ and a remainder $r(\lambda)$; that is, $f(\lambda)=g(\lambda)\, m(\lambda)+r(\lambda)$. If we replace λ by **A**, and usė the fact that $m(\mathbf{A})=\boldsymbol{\Theta}$, we observe that $f(\mathbf{A})=r(\mathbf{A})$. The remainder $r(\mathbf{A})$ is of lower degree (in **A**) than $m(\mathbf{A})$, regardless of the degree of $f(\mathbf{A})$. Consequently, $r(\mathbf{A})$ is easier to compute than is $f(\mathbf{A})$. The same procedure can be carried out using the more easily determined characteristic polynomial rather than the minimal polynomial. Use this "remainder" method to compute the matrix $\mathbf{A}^5$ for

$$\mathbf{A}=\begin{pmatrix} 1 & 2 \\ 3 & 4 \end{pmatrix}$$

4.21 Assume f is analytic at the eigenvalues of the matrix **A**. Find the component matrices of **A** and express $f(\mathbf{A})$ as a linear combination of these components for:

(*a*) $\mathbf{A}=\begin{pmatrix} 1 & 1 & 1 \\ 0 & 2 & 1 \\ 0 & 0 & 3 \end{pmatrix}$ (*b*) $\mathbf{A}=\begin{pmatrix} 1 & 1 & 0 \\ 0 & 1 & 1 \\ 0 & 0 & 2 \end{pmatrix}$

4.22 The gamma function $\Gamma(p)$ is defined for all positive values of the scalar p. If p is a positive integer, $\Gamma(p)=(p-1)!$ Find $\Gamma(\mathbf{A})$, where

$$\mathbf{A}=\begin{pmatrix} 3 & 0 & -2 \\ 0 & 3 & -1 \\ 0 & 0 & 2 \end{pmatrix}$$

4.23 Let

$$f(\lambda) \overset{\Delta}{=} 0, \qquad \lambda \leqslant c$$
$$\overset{\Delta}{=} (\lambda - c)^2, \qquad \lambda \geqslant c$$

and

$$\mathbf{\Lambda} = \begin{pmatrix} 0 & 1 & 0 \\ 0 & 0 & 0 \\ 0 & 0 & 1 \end{pmatrix}$$

(*a*) Find $f(\mathbf{\Lambda})$.
(*b*) Consider various values of c. Is the resulting matrix what you would expect?

4.24 If $\mathbf{A}$ is invertible, the inverse can be computed by evaluating $f(\mathbf{A})$ for $f(\lambda) \overset{\Delta}{=} 1/\lambda$. By modifying f, we can compute a "pseudoinverse" for a matrix which has zero eigenvalues. We merely change the definition of f to

$$\hat{f}(\lambda) \overset{\Delta}{=} \frac{1}{\lambda}, \qquad \lambda \neq 0$$
$$\overset{\Delta}{=} 0, \qquad \lambda = 0$$

(See P&C 6.23 for an interpretation of this "pseudoinverse.")
(*a*) Find the inverse of the matrix $\mathbf{A}$ of P&C 4.21 *a* by evaluating $f(\mathbf{A})$.
(*b*) Find the "pseudoinverse" of the following matrix by evaluating $\hat{f}(\mathbf{B})$:

$$\mathbf{B} = \begin{pmatrix} 1 & 1 & 1 \\ 0 & 0 & 2 \\ 0 & 0 & 2 \end{pmatrix}$$

4.25 The constituent matrices of a square matrix $\mathbf{A}$ can be determined by partial fraction expansion of the *resolvant matrix*, $(s\mathbf{I} - \mathbf{A})^{-1}$ (the resolvant matrix is the Laplace transform of $e^{\mathbf{A}t}$). Let

$$\mathbf{A} = \begin{pmatrix} 1 & 1 & 0 \\ -1 & 3 & 0 \\ -1 & 1 & 2 \end{pmatrix}$$

(*a*) Determine the resolvant matrix $(s\mathbf{I}-\mathbf{A})^{-1}$ by inverting $(s\mathbf{I}-\mathbf{A})$.

(*b*) Perform a partial fraction expansion of $(s\mathbf{I}-\mathbf{A})^{-1}$; that is, perform a partial fraction expansion of each term of $(s\mathbf{I}-\mathbf{A})^{-1}$, and arrange the expansion into a sum of terms with multipliers which are constant 3×3 matrices.

(*c*) Let $f(\lambda)\triangleq 1/(s-\lambda)$; then $f(\mathbf{A})=(s\mathbf{I}-\mathbf{A})^{-1}$. Express the fundamental formula for $f(\mathbf{A})$ in terms of $\{\mathbf{E}_{ij}^{\mathbf{A}}\}$, the constituent matrices for $\mathbf{A}$. (The form of the fundamental formula is determined by the minimal polynomial for $\mathbf{A}$.) Determine the constituent matrices by comparing the fundamental formula for $f(\mathbf{A})$ with the partial fraction expansion obtained in (*b*).

(*d*) Use the fundamental formula and the constituent matrices to evaluate $\mathbf{A}^5$.

4.26 Let f be a scalar-valued function of a scalar variable. Assume f is defined on the spectrum of the $n\times n$ matrix $\mathbf{A}$. Then $f(\mathbf{A})$ can be expressed as a polynomial in $\mathbf{A}$ of lower degree than the minimal polynomial for $\mathbf{A}$. That is, if r is the degree of the minimal polynomial, then $f(\mathbf{A})=a_0\mathbf{I}+a_1\mathbf{A}+\cdots+a_{r-1}\mathbf{A}^{r-1}$. The coefficients $\{a_i\}$ can be determined by evaluating the corresponding scalar equation, $f(\lambda)=a_0+a_1\lambda+\cdots+a_{r-1}\lambda^{r-1}$, on the spectrum of $\mathbf{A}$; the resulting equations are always solvable.

(*a*) Find the minimal polynomial for the matrix

$$\mathbf{A}=\begin{pmatrix}1 & 1 & 0\\ -1 & 3 & 0\\ -1 & 1 & 2\end{pmatrix}$$

(*b*) For the matrix $\mathbf{A}$ introduced in (*a*), evaluate the matrix function $f(\mathbf{A})\triangleq\mathbf{A}^5$ by the technique described above.

4.27 Let the $n\times n$ matrix $\mathbf{A}$ be diagonalizable. Then, the fundamental formula is $f(\mathbf{A})=\sum_{i=1}^{p}f(\lambda_i)\mathbf{E}_{i0}^{\mathbf{A}}$, where p is the number of distinct eigenvalues. The constituent matrix $\mathbf{E}_{i0}^{\mathbf{A}}$ is the projector on the eigenspace for λ_i along the sum of the other eigenspaces. It can be expressed as

$$\mathbf{E}_{i0}^{\mathbf{A}}=\prod_{j\neq i}\left(\frac{\mathbf{A}-\lambda_j\mathbf{I}}{\lambda_i-\lambda_j}\right)$$

($\mathbf{E}_{i0}^{\mathbf{A}}$ acts like $\mathbf{I}$ on the eigenspace for λ_i and like $\boldsymbol{\Theta}$ on the eigenspace for λ_j.) The scalar equivalent of the fundamental formula,

$$f(\lambda)=\sum_{1=1}^{p} f(\lambda_i)\prod_{j\neq i}\left(\frac{\lambda-\lambda_j}{\lambda_i-\lambda_j}\right)$$

is known as the *Lagrange interpolation formula* for the data points $\lambda_1,\ldots,\lambda_p$.

(*a*) Let

$$\mathbf{A}=\begin{pmatrix}1 & 1 & 2\\ 0 & 2 & -1\\ 0 & 0 & 3\end{pmatrix}$$

Find the constituent matrices $\mathbf{E}_{i0}^{\mathbf{A}}$ by evaluating the polynomial expressions given above.

(*b*) Use the fundamental formula to evaluate the matrix exponential, $e^{\mathbf{A}t}$, for the matrix $\mathbf{A}$ given in (*a*).

4.28 Use the fundamental formula to find four square roots of the matrix

$$\mathbf{A}=\begin{pmatrix}20 & -8\\ 48 & -20\end{pmatrix}$$

*4.29 (*a*) *Commuting matrices*: if $\mathbf{A}$ and $\mathbf{B}$ commute (i.e., $\mathbf{AB}=\mathbf{BA}$), then

$$(\mathbf{A}+\mathbf{B})^n=\sum_{k=0}^{n}\binom{n}{k}\mathbf{A}^{n-k}\mathbf{B}^k, \qquad n=0,1,2,\ldots$$

where

$$\binom{n}{k}=\frac{n!}{k!(n-k)!}$$

That is, the binomial theorem is satisfied.

(*b*) The algebra of matrices is essentially the same as the algebra of scalars if the matrices commute with each other. Therefore, a functional relation which holds for scalars also holds for commuting matrices if the required matrix functions are defined. For example, $e^{\mathbf{A}+\mathbf{B}}=e^{\mathbf{A}}e^{\mathbf{B}}$, $\cos(\mathbf{A}+\mathbf{B})=\cos\mathbf{A}\cos\mathbf{B}-\sin\mathbf{A}\sin\mathbf{B}$; the binominal theorem is satisfied; etc.

(*c*) If $\mathbf{A}$ and $\mathbf{B}$ are diagonalizable, then they are commutable if and only if they are diagonalizable by the same similarity transformation (i.e., if and only if they have the same eigenvectors).

4.30 Use the fundamental formula to show that $(d/dt)f(\mathbf{A}t)=\mathbf{A}\dot{f}(\mathbf{A}t)$ for any square matrix $\mathbf{A}$ and any function f which is analytic on the spectrum of $\mathbf{A}$.

4.31 Let $\mathbf{f}''+6\mathbf{f}'+5\mathbf{f}=\mathbf{u}$, $\mathbf{f}(0)=\mathbf{f}'(0)=0$.

(*a*) Express the differential system in state-space form.

(*b*) Diagonalize the state equation found in (*a*).

(*c*) Draw a signal flow diagram which relates the original state variables, the canonical state variables, and the input.

(*d*) Find the state transition matrix and invert the state equation.

4.32 Let $\ddot{\mathbf{x}}+\mathbf{A}\mathbf{x}=\mathbf{B}\mathbf{u}$, where $\mathbf{x}(t)$ is $n\times 1$, $\mathbf{u}(t)$ is $m\times 1$, $\mathbf{B}$ is $n\times m$, and $\mathbf{A}$ is $n\times n$ with positive eigenvalues. Assume $\mathbf{x}(0)$ and $\dot{\mathbf{x}}(0)$ are known.

(*a*) Use the power series method of Frobenius to show that the complementary function for this vector differential equation is

$$\mathbf{F}_c(t)=\cos(\sqrt{\mathbf{A}}\,t)\mathbf{C}_0+(\sqrt{\mathbf{A}}\,)^{-1}\sin(\sqrt{\mathbf{A}}\,t)\mathbf{C}_1$$

where $\mathbf{C}_0$ and $\mathbf{C}_1$ are arbitrary $n\times n$ matrices.

(*b*) The inverse of the differential equation is of the form

$$\mathbf{x}(t)=\int_0^\infty \mathbf{K}(t,s)\mathbf{B}\mathbf{u}(s)\,ds+\mathbf{R}_1(t)\mathbf{x}(0)+\mathbf{R}_2(t)\dot{\mathbf{x}}(0)$$

Show that the Green's function $\mathbf{K}(t,s)$ and boundary kernel $\mathbf{R}_j(t)$ satisfy:

$$\frac{d^2}{dt^2}\mathbf{K}(t,s)+\mathbf{A}\mathbf{K}(t,s)=\delta(t-s)\mathbf{I}$$

$$\mathbf{K}(0,s)=\frac{d}{dt}\mathbf{K}(0,s)=\mathbf{\Theta}$$

$$\frac{d^2}{dt^2}\mathbf{R}_j(t)+\mathbf{A}\mathbf{R}_j(t)=\mathbf{\Theta},\qquad j=1,2$$

$$\mathbf{R}_1(0)=\mathbf{I},\qquad \dot{\mathbf{R}}_1(0)=\mathbf{\Theta}$$

$$\mathbf{R}_2(0)=\mathbf{\Theta},\qquad \dot{\mathbf{R}}_2(0)=\mathbf{I}$$

(c) Show that

$$\begin{aligned}\mathbf{K}(t,s) &= \mathbf{\Theta}, & t \leqslant s\\ &= (\sqrt{\mathbf{A}})^{-1}\sin\big(\sqrt{\mathbf{A}}\,(t-s)\big), & t \geqslant s\end{aligned}$$

$$\mathbf{R}_1(t) = \cos\sqrt{\mathbf{A}}\, t$$

$$\mathbf{R}_2(t) = (\sqrt{\mathbf{A}})^{-1}\sin(\sqrt{\mathbf{A}}\, t)$$

4.33 In optimal control problems we often need to solve a pair of simultaneous state equations. Suppose the equations are $\dot{\mathbf{x}} = \mathbf{A}\mathbf{x} - \mathbf{B}\mathbf{B}^{\mathrm{T}}\boldsymbol{\lambda}$ and $\dot{\boldsymbol{\lambda}} = -\mathbf{A}^{\mathrm{T}}\boldsymbol{\lambda}$, where

$$\mathbf{A} = \begin{pmatrix} 0 & 1 \\ 0 & -1 \end{pmatrix} \quad \text{and} \quad \mathbf{B} = \begin{pmatrix} 0 \\ 1 \end{pmatrix}$$

(a) Write the pair of equations as a single state equation $\dot{\mathbf{y}} = \mathbf{Q}\mathbf{y}$, where $\mathbf{y} \triangleq \begin{pmatrix} \mathbf{x} \\ \boldsymbol{\lambda} \end{pmatrix}$.

(b) Find the eigenvalues and constituent matrices of $\mathbf{Q}$.

(c) Find the solution $\mathbf{y}$ to the state equation as a function of $\mathbf{y}(0)$.

4.8 References

[4.1] Bocher, M., *Introduction to Higher Algebra*, Macmillan, New York, 1921.

[4.2] Brown, R. G., "Not Just Observable, But How Observable," *Proc. Natl. Elec. Conf.*, **22** (1966), 709–714.

[4.3] DeRusso, Paul M., Rob J. Roy, and Charles M. Close, *State Variables for Engineers*, Wiley, New York, 1966.

[4.4] Feller, W., *An Introduction to Probability Theory and its Applications*, 3rd Ed., Vol. I, Wiley, New York, 1968.

[4.5] Forsythe, George E., "Singularity and Near Singularity in Numerical Analysis," *Am. Math. Mon.*, **65** (1958), 229–240.

[4.6] Forsythe, George E., "Today's Computational Methods of Linear Algebra," *SIAM Rev.* **9** (July 1967), 489–515.

[4.7] Friedman, Bernard, *Principles and Techniques of Applied Mathematics*, Wiley, New York, 1966.

[4.8] Halmos, P. R., *Finite-Dimensional Vector Spaces*, Van Nostrand, Englewood Cliffs, N. J., 1958.

[4.9] Hoffman, Kenneth and Roy Kunze, *Linear Algebra*, Prentice-Hall, Englewood Cliffs, N. J., 1961.

[4.10] Ince, E. L., *Ordinary Differential Equations*, Dover, New York, 1956.

[4.11] Kaplan, Wilfred, *Advanced Calculus*, Addison-Wesley, Reading, Mass. 1959.

*[4.12] Pease, Marshall C., III, *Methods of Matrix Algebra*, Academic Press, New York, 1965.

[4.13] Ralston, Anthony, *A First Course in Numerical Analysis*, McGraw-Hill, New York, 1965.

[4.14] Rinehart, R. F., "The Equivalence of Definitions of A Matric Function," *Am. Math. Mon.*, **62** (1955), 395–414.

[4.15] Rothe, Frederich S., *An Introduction to Power System Analysis*, Wiley, New York, 1953.

[4.16] Schwartz, Ralph J. and Bernard Friedland, *Linear Systems*, McGraw-Hill, New York, 1965.

[4.17] Skilling, Hugh Hildreth, *Electrical Engineering Circuits*, Wiley, New York, 1958.

[4.18] Wylie, C. R., Jr., *Advanced Engineering Mathematics*, 3rd ed., McGraw-Hill, New York, 1966.

*[4.19] Wilkinson, J. H., *The Algebraic Eigenvalue Problem*, Clarenden Press, Oxford, 1965.

[4.20] Zadeh, Lofti A. and Charles A. Desoer, *Linear System Theory*, McGraw-Hill, New York, 1963.

5

Hilbert Spaces

Our previous discussions have been concerned with algebra. The representation of systems (quantities and their interrelations) by abstract symbols has forced us to distill out the most significant and fundamental properties of these systems. We have been able to carry our exploration much deeper for linear systems, in most cases decomposing the system models into sets of uncoupled scalar equations.

Our attention now turns to the geometric notions of length and angle. These concepts, which are fundamental to measurement and comparison of vectors, complete the analogy between general vector spaces and the physical three-dimensional space with which we are familiar. Then our intuition concerning the size and shape of objects provides us with valuable insight. The definition of length gives rigorous meaning to our previous heuristic discussions of an infinite sequence of vectors as a basis for an infinite-dimensional space. Length is also one of the most widely used optimization criteria. We explore this application of the concept of length in Chapter 6. The definition of orthogonality (or angle) allows us to carry even further our discussion of system decomposition. To this point, determination of the coordinates of a vector relative to a particular basis has required solution of a set of simultaneous equations. With orthogonal bases, each coordinate can be obtained independently, a much simpler process conceptually and, in some instances, computationally.

5.1 Inner Products

The dot product concept is familiar from analytic geometry. If $\mathbf{x}=(\xi_1,\xi_2)$ and $\mathbf{y}=(\eta_1,\eta_2)$ are two vectors from $\mathcal{R}^2$, the dot product $\mathbf{x}\cdot\mathbf{y}$ between $\mathbf{x}$ and $\mathbf{y}$ is defined by

$$\mathbf{x}\cdot\mathbf{y} \triangleq \xi_1\eta_1+\xi_2\eta_2 \tag{5.1}$$

The length $\|\mathbf{x}\|$ of the vector $\mathbf{x}$ is defined by

$$\|\mathbf{x}\| \triangleq \sqrt{\mathbf{x}\cdot\mathbf{x}} =\sqrt{\xi_1^2+\xi_2^2} \tag{5.2}$$

The angle between the vectors **x** and **y** is defined in terms of the dot product between the normalized vectors:

$$\cos\phi = \frac{\mathbf{x}}{\|\mathbf{x}\|} \cdot \frac{\mathbf{y}}{\|\mathbf{y}\|} = \frac{\mathbf{x}\cdot\mathbf{y}}{\|\mathbf{x}\|\,\|\mathbf{y}\|} \tag{5.3}$$

Example 1. The Dot Product in $\mathfrak{R}^2$. Let $\mathbf{x}=(1,1)$ and $\mathbf{y}=(2,0)$. Then $\mathbf{x}\cdot\mathbf{y}=2$, $\|\mathbf{x}\|=\sqrt{2}$, $\|\mathbf{y}|=2$, and $\cos\phi=1/\sqrt{2}$ (or $\phi=45°$)). Figure 5.1 is an arrow space equivalent of this example.

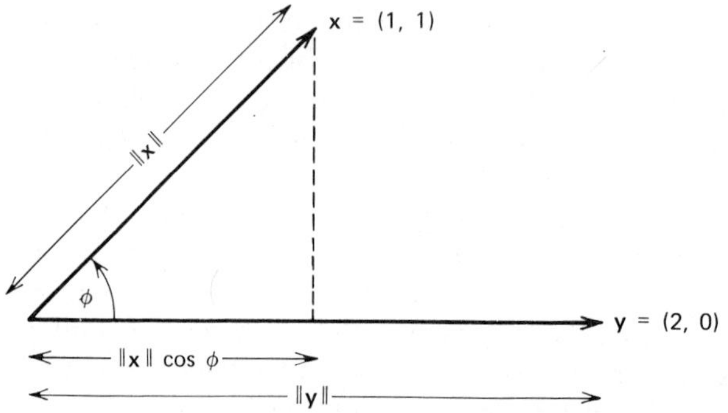

Figure 5.1. Arrow vectors corresponding to Example 1.

It is apparent from Example 1 that (5.3) can be interpreted, in terms of the natural correspondence to arrow space, as a definition of the dot product (as a function of the angle between the vectors):

$$\mathbf{x}\cdot\mathbf{y} \triangleq \|\mathbf{y}\|(\|\mathbf{x}\|\cos\phi) \tag{5.4}$$

where $\|\mathbf{x}\|\cos\phi$ is the length of the projection of **x** on **y** along the perpendicular to **y**. The following properties of the dot product seem fundamental:

1. Length is non-negative; that is,

$$\mathbf{x}\cdot\mathbf{x} \geqslant 0, \quad \text{with equality if and only if } \mathbf{x}=\boldsymbol{\theta}$$

2. The magnitude of ϕ (or $\cos\phi$) is independent of the order of **x** and **y**; that is,

$$\mathbf{x}\cdot\mathbf{y}=\mathbf{y}\cdot\mathbf{x}$$

3. The length of $c\mathbf{x}$ equals $|c|$ times the length of $\mathbf{x}$, for any scalar c; that is,

$$c\mathbf{x}\cdot c\mathbf{x}=c^2(\mathbf{x}\cdot\mathbf{x})$$

4. In order that (5.4) be consistent with the rules for addition of vectors, the dot product must be distributive over addition (see Figure 5.2); that is,

$$(\mathbf{x}_1+\mathbf{x}_2)\cdot\mathbf{y}=\mathbf{x}_1\cdot\mathbf{y}+\mathbf{x}_2\cdot\mathbf{y}$$

We now extend the dot product to arbitrary vector spaces with real or complex scalars in a manner which preserves these four properties.

Definition. An **inner product** (or **scalar product**) on a real or complex vector space $\mathcal{V}$ is a scalar-valued function $\langle\mathbf{x},\mathbf{y}\rangle$ of the ordered pair of vectors $\mathbf{x}$ and $\mathbf{y}$ such that:

1. $\langle\mathbf{x},\mathbf{x}\rangle\geqslant 0$, with equality if and only if $\mathbf{x}=\boldsymbol{\theta}$
2. $\langle\mathbf{x},\mathbf{y}\rangle=\overline{\langle\mathbf{y},\mathbf{x}\rangle}$ (the bar denotes complex conjugation).
3. $\langle c_1\mathbf{x}_1+c_2\mathbf{x}_2,\mathbf{y}\rangle=c_1\langle\mathbf{x}_1,\mathbf{y}\rangle+c_2\langle\mathbf{x}_2,\mathbf{y}\rangle$

It follows that $\langle\mathbf{y},c_1\mathbf{x}_1+c_2\mathbf{x}_2\rangle=\bar{c}_1\langle\mathbf{y},\mathbf{x}_1\rangle+\bar{c}_2\langle\mathbf{y},\mathbf{x}_2\rangle$. We describe these properties by saying that an inner product must be (1) **positive definite**, (2) **hermitian symmetric**, and (3) **conjugate bilinear**. Note that because of (2), $\langle\mathbf{x},\mathbf{x}\rangle$ is necessarily real, and the inequality (1) makes sense. If the scalars are real, the complex conjugation bar is superfluous.

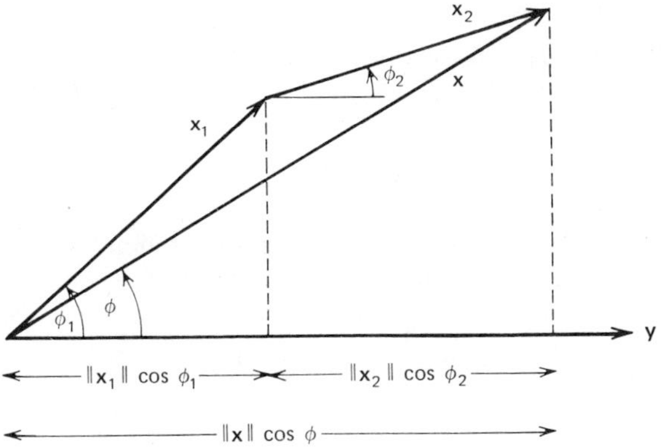

Figure 5.2. Dot products are distributive over addition.

We define the **norm** (or length) of **x** by

$$\|\mathbf{x}\| \triangleq \sqrt{\langle \mathbf{x}, \mathbf{x} \rangle} \tag{5.5}$$

When $\langle \mathbf{x}, \mathbf{y} \rangle$ is real, we can define the angle ϕ between **x** and **y** by

$$\cos\phi \triangleq \frac{\langle \mathbf{x}, \mathbf{y} \rangle}{\|\mathbf{x}\| \, \|\mathbf{y}\|}$$

Practically speaking, we are interested in the angle ϕ only in the following two cases:

$$\langle \mathbf{x}, \mathbf{y} \rangle = 0 \text{ (}\mathbf{x}\text{ and }\mathbf{y}\text{ are said to be } \textbf{orthogonal}\text{)} \tag{5.6}$$

$$\langle \mathbf{x}, \mathbf{y} \rangle = \pm \|\mathbf{x}\| \, \|\mathbf{y}\| \text{ (}\mathbf{x}\text{ and }\mathbf{y}\text{ are said to be } \textbf{collinear}\text{)} \tag{5.7}$$

Example 2. The Standard Inner Product for $\mathcal{C}^n$ and $\mathcal{R}^n$. The standard inner product for $\mathcal{C}^n$ (and $\mathcal{R}^n$) is defined by

$$\langle \mathbf{x}, \mathbf{y} \rangle \triangleq \sum_{i=1}^{n} \xi_i \bar{\eta}_i \tag{5.8}$$

where ξ_i and η_i are the elements of **x** and **y**, respectively. Of course, the complex conjugate bar is superfluous for $\mathcal{R}^n$. This inner product is simply the extension of the dot product to complex spaces and n dimensions. Consider the vector (i) in $\mathcal{C}^1$;

$$\|(i)\| = \sqrt{(i)(\bar{i})} = 1$$

The complex conjugation in (5.8) is needed in order to keep lengths non-negative for complex scalars.

Example 3. The Standard Inner Product for $\mathfrak{M}_c^{n\times 1}$ and $\mathfrak{M}^{n\times 1}$. The standard inner product for $\mathfrak{M}_c^{n\times 1}$ is defined by

$$\langle \mathbf{x}, \mathbf{y} \rangle \triangleq \bar{\mathbf{y}}^{\mathrm{T}} \mathbf{x} \tag{5.9}$$

Again, if only real scalars are involved, the conjugate is unnecessary. For instance, if $\mathbf{x} = (1 \;\; 2 \;\; 4)^{\mathrm{T}}$ and $\mathbf{y} = (-1 \;\; 3 \;\; 2)^{\mathrm{T}}$ in $\mathfrak{M}^{3\times 1}$, then, by (5.9),

$$\langle \mathbf{x}, \mathbf{y} \rangle = (-1 \;\; 3 \;\; 2)\begin{pmatrix} 1 \\ 2 \\ 4 \end{pmatrix} = 13$$

Example 4. The Standard Inner Product for Function Spaces. The standard inner product for a function space such as $\mathcal{P}(a,b)$ or $\mathcal{C}(a,b)$ is defined by

$$\langle \mathbf{f},\mathbf{g}\rangle \triangleq \int_a^b \mathbf{f}(t)\,\overline{\mathbf{g}(t)}\,dt \tag{5.10}$$

for each **f** and **g** in the space. We usually deal only with real functions and ignore the complex conjugation. Consider the function $\mathbf{f}(t)=1$ in $\mathcal{C}(0,1)$:

$$\|\mathbf{f}\| = \sqrt{\int_0^1 (1)^2\,dt} = 1$$

Any vector whose average value over the interval $[0,1]$ is zero is orthogonal to **f**; for then $\langle \mathbf{f},\mathbf{g}\rangle = \int_0^1 (1)\mathbf{g}(t)\,dt = 0$. We easily verify, for the case of continuous functions and real scalars, that (5.10) possesses the properties of an inner product; by the properties of integrals:

(*a*) $\langle \mathbf{f},\mathbf{f}\rangle = \int_a^b \mathbf{f}^2(t)\,dt \geqslant 0$, with equality if and only if $\mathbf{f}(t)=0$ for all t in $[a,b]$;
(*b*) $\int_a^b \mathbf{f}(t)\mathbf{g}(t)\,dt = \int_a^b \mathbf{g}(t)\mathbf{f}(t)\,dt$
(*c*) $\int_a^b [c_1\mathbf{f}_1(t) + c_2\mathbf{f}_2(t)]\mathbf{g}(t)\,dt = c_1\int_a^b \mathbf{f}_1(t)\mathbf{g}(t)\,dt + c_2\int_a^b \mathbf{f}_2(t)\mathbf{g}(t)\,dt$

Example 5. The Standard Inner Product for a Space of Two-Dimensional Functions. Let $\mathcal{C}^2(\Omega)$ denote the space of functions which are twice continuously differentiable over a two-dimensional region Ω. We define an inner product for $\mathcal{C}^2(\Omega)$ by

$$\langle \mathbf{f},\mathbf{g}\rangle \triangleq \int_\Omega \mathbf{f}(\mathbf{p})\mathbf{g}(\mathbf{p})\,d\mathbf{p} \tag{5.11}$$

where $\mathbf{p}=(s,t)$, an arbitrary point in Ω.

An inner product assigns a real number (or norm) to each vector in the space. The norm provides a simple means for comparing vectors in applications. Example 1 of Section 3.4 is concerned with the state (or position and velocity) of a motor shaft in the state space $\mathfrak{M}^{2\times 1}$. In a particular application we might require both the position and velocity to approach given values, say, zero. As a simple measure of the nearness of the state to the desired position ($\boldsymbol{\theta}$), we use the norm corresponding to (5.9):

$$\|\mathbf{x}(t)\| = \sqrt{\xi_1{}^2 + \xi_2{}^2}$$

where ξ_1 and ξ_2 are the angular position and velocity of the motor shaft at instant t. However, there is no inherent reason why position and velocity should be equally important. We might be satisfied if the velocity stayed large as long as the position of the shaft approached the target position

$\xi_1=0$. In this case, some other measure of the performance of the system would be more appropriate. The following measure weights ξ_1 more heavily than ξ_2.

$$\|\mathbf{x}(t)\|=\sqrt{100\xi_1{}^2+\xi_2{}^2}$$

This new measure is just the norm associated with the following weighted inner product for $\mathfrak{M}^{2\times 1}$:

$$\begin{aligned}\langle\mathbf{x},\mathbf{y}\rangle &\triangleq 100\xi_1\eta_1+\xi_2\eta_2\\ &=\mathbf{y}^{\mathrm{T}}\begin{pmatrix}100 & 0\\ 0 & 1\end{pmatrix}\mathbf{x}\end{aligned}$$

where $\mathbf{x}=(\xi_1\ \xi_2)^{\mathrm{T}}$ and $\mathbf{y}=(\eta_1\ \eta_2)^{\mathrm{T}}$. We generally select that inner product which is most appropriate to the purpose for which it is to be used.

Example 6. A Weighted Inner Product for Function Spaces. An inner product of the following form is often appropriate for such spaces as $\mathcal{P}(a,b)$ and $\mathcal{C}(a,b)$:

$$\langle\mathbf{f},\mathbf{g}\rangle=\int_a^b \omega(t)\mathbf{f}(t)\,\overline{\mathbf{g}(t)}\,dt,\qquad 0<\omega(t)<\infty \tag{5.12}$$

If the weight function is $\omega(t)=1$, (5.12) reduces to the standard inner product (5.10). The weight $\omega(t)=e^t$ might be used to emphasize the values of functions for large t and deemphasize the values for t small or negative.

Example 7. A Weighted Inner Product for $\mathfrak{R}^2$. Let $\mathbf{x}=(\xi_1,\xi_2)$ and $\mathbf{y}=(\eta_1,\eta_2)$ be arbitrary vectors in $\mathfrak{R}^2$. Define the inner product on $\mathfrak{R}^2$ by

$$\langle\mathbf{x},\mathbf{y}\rangle \triangleq \tfrac{1}{4}\xi_1\eta_1-\tfrac{1}{4}\xi_1\eta_2-\tfrac{1}{4}\xi_2\eta_1+\tfrac{5}{4}\xi_2\eta_2 \tag{5.13}$$

We apply this inner product to the vectors $\mathbf{x}=(1,1)$ and $\mathbf{y}=(2,0)$, the same vectors to which we previously applied the standard (or dot) inner product: $\langle\mathbf{x},\mathbf{y}\rangle=0$, $\|\mathbf{x}\|=1$, and $\|\mathbf{y}\|=1$. The same vectors which previously were displaced by 45° (Figure 5.1) are, by definition (5.13), orthogonal and of unit length. We see that (5.13) satisfies the properties required of an inner product:

1. By completing the square, we find

$$\langle\mathbf{x},\mathbf{x}\rangle=\tfrac{1}{4}(\xi_1-\xi_2)^2+\xi_2^2\geqslant 0$$

with equality if an only if $\xi_1=\xi_2=0$;

2. Since the coefficients for the cross-product terms are equal,

$$\langle\mathbf{x},\mathbf{y}\rangle=\langle\mathbf{y},\mathbf{x}\rangle$$

3. We rewrite (5.13) as

$$\langle \mathbf{x},\mathbf{y}\rangle=(\eta_1\ \eta_2)\begin{pmatrix}\frac{1}{4} & -\frac{1}{4}\\ -\frac{1}{4} & \frac{5}{4}\end{pmatrix}\begin{pmatrix}\xi_1\\ \xi_2\end{pmatrix}\triangleq \mathbf{y}^{\mathrm{T}}\mathbf{Q}\mathbf{x}$$

Then, by the linearity of matrix multiplication,

$$\begin{aligned}\langle c_1\mathbf{x}_1+c_2\mathbf{x}_2,\mathbf{y}\rangle&=\mathbf{y}^{\mathrm{T}}\mathbf{Q}(c_1\mathbf{x}_1+c_2\mathbf{x}_2)\\&=c_1\mathbf{y}^{\mathrm{T}}\mathbf{Q}\mathbf{x}_1+c_2\mathbf{y}^{\mathrm{T}}\mathbf{Q}\mathbf{x}_2\\&=c_1\langle\mathbf{x}_1,\mathbf{y}\rangle+c_2\langle\mathbf{x}_2,\mathbf{y}\rangle\end{aligned}$$

The last two examples suggest that we have considerable freedom in picking inner products. Length and orthogonality are, to a great extent, what we define them to be. Only if we use standard inner products in $\mathcal{R}^3$ do length and orthogonality correspond to physical length and 90° angles. Surprisingly, the concept suggested by (5.4) still holds in Example 7: $|\langle\mathbf{x},\mathbf{y}\rangle|$ is the product of $\|\mathbf{y}\|$ and the norm of the projection of $\mathbf{x}$ on $\mathbf{y}$ along the direction orthogonal [in the sense of (5.13)] to $\mathbf{y}$. The sign of $\langle\mathbf{x},\mathbf{y}\rangle$ is positive if the projection of $\mathbf{x}$ on $\mathbf{y}$ is in the same direction as $\mathbf{y}$; if the projection is in the opposite direction, the sign is negative.

Exercise 1. Let $\mathbf{x}=(0,1)$ and $\mathbf{y}=(1,0)$ in $\mathcal{R}^2$. Define the inner product in $\mathcal{R}^2$ by (5.13). Show that the projection of $\mathbf{x}$ on $\mathbf{y}$ along the direction orthogonal to $\mathbf{y}$ is the vector $(-1,0)$. Verify that $\langle\mathbf{x},\mathbf{y}\rangle$ is correctly determined by the above rule which uses the projection of $\mathbf{x}$ on $\mathbf{y}$.

An **inner product space** (or **pre-Hilbert space**) is a vector space on which a particular inner product is defined. A *real* inner product space is called a **Euclidean space**. A **unitary space** is an inner product space for which the scalars are the complex numbers. We will often employ the symbols $\mathcal{R}^n$ and $\mathfrak{M}^{n\times 1}$ to represent the Euclidean spaces consisting of the real vector spaces $\mathcal{R}^n$ and $\mathfrak{M}^{n\times 1}$ together with the standard inner products (5.8) and (5.9), respectively. Similarly, we use $\mathcal{P}(a,b)$, $\mathcal{C}(a,b)$, etc. to represent real Euclidean function spaces which make use of the standard inner product (5.10). Whereever we use a different (nonstandard) inner product, we mention it explicitly.

Matrices of Inner Products

To this point, we have not used the concept of a basis in our discussion of inner products. There is no particular basis inherent in any inner product space, although we will find some bases more convenient than others. We found in Chapter 2 that by picking a basis $\mathcal{X}$ for an n-dimensional space

$\mathcal{V}$ we can represent vectors $\mathbf{x}$ in $\mathcal{V}$ by their coordinates $[\mathbf{x}]_{\mathcal{X}}$ in the "standard" space $\mathfrak{M}^{n\times 1}$; moreover, we can represent a linear operator $\mathbf{T}$ on $\mathcal{V}$ by a matrix manipulation of $[\mathbf{x}]_{\mathcal{X}}$, multiplication by $[\mathbf{T}]_{\mathcal{X}\mathcal{X}}$. It seems only natural that by means of the same basis we should be able to convert the inner product operation to a matrix manipulation. We proceed by means of an example.

Let $\mathbf{x}=(\xi_1,\xi_2)$ and $\mathbf{y}=(\eta_1,\eta_2)$ be general vectors in the vector space $\mathcal{R}^2$. Let $\langle\mathbf{x},\mathbf{y}\rangle$ represent the inner product (5.13). We select $\mathcal{X} \triangleq \mathcal{E}$, the standard basis for $\mathcal{R}^2$. Then using the bilinearity of the inner product,

$$\begin{aligned}\langle\mathbf{x},\mathbf{y}\rangle &= \langle\xi_1(1,0)+\xi_2(0,1),\eta_1(1,0)+\eta_2(0,1)\rangle\\ &=\xi_1\langle(1,0),\eta_1(1,0)+\eta_2(0,1)\rangle+\xi_2\langle(0,1),\eta_1(1,0)+\eta_2(0,1)\rangle\\ &=\xi_1\eta_1\|(1,0)\|^2+\xi_1\eta_2\langle(1,0),(0,1)\rangle+\xi_2\eta_1\langle(0,1),(1,0)\rangle+\xi_2\eta_2\|(0,1)\|^2\\ &=\tfrac{1}{4}\xi_1\eta_1-\tfrac{1}{4}\xi_1\eta_2-\tfrac{1}{4}\xi_2\eta_1+\tfrac{5}{4}\xi_2\eta_2\end{aligned}$$

On the surface, we appear to have returned to the defining equation (5.13), but the meaning of the equation is now different; ξ_i and η_i now represent coordinates [or multipliers of the vectors (1,0) and (0,1)] rather than elements of the vectors $\mathbf{x}$ and $\mathbf{y}$. We rewrite the last line of the equation as

$$\begin{aligned}\langle\mathbf{x},\mathbf{y}\rangle &= (\eta_1\ \ \eta_2)\begin{pmatrix}\frac{1}{4} & -\frac{1}{4}\\ -\frac{1}{4} & \frac{5}{4}\end{pmatrix}\begin{pmatrix}\xi_1\\ \xi_2\end{pmatrix}\\ &\triangleq [\mathbf{y}]_{\mathcal{E}}^{\mathrm{T}}\mathbf{Q}_{\mathcal{E}}[\mathbf{x}]_{\mathcal{E}}\end{aligned}$$

We have converted the inner product operation to a matrix multiplication. We call $\mathbf{Q}_{\mathcal{E}}$ the matrix of the inner product relative to the basis $\mathcal{E}$. In similar fashion, any inner product on a finite-dimensional space can be represented by a matrix.

Let $\mathcal{X} \triangleq \{\mathbf{x}_1,\ldots,\mathbf{x}_n\}$ be a basis for an inner product space $\mathcal{V}$. Then

$$\mathbf{x}=\sum_{k=1}^{n} a_k\mathbf{x}_k \quad\text{and}\quad \mathbf{y}=\sum_{j=1}^{n} b_j\mathbf{x}_j$$

By the argument used for the special case above,

$$\begin{aligned}\langle \mathbf{x},\mathbf{y}\rangle &= \left\langle \sum_k a_k\mathbf{x}_k, \sum_j b_j\mathbf{x}_j\right\rangle\\ &= \sum_j \bar{b}_j\left\langle \sum_k a_k\mathbf{x}_k,\mathbf{x}_j\right\rangle\\ &= \sum_j \bar{b}_j\sum_k a_k\langle \mathbf{x}_k,\mathbf{x}_j\rangle\\ &= (\bar{b}_1\cdots\bar{b}_n)\begin{pmatrix}\langle\mathbf{x}_1,\mathbf{x}_1\rangle & \cdots & \langle\mathbf{x}_n,\mathbf{x}_1\rangle\\ \vdots & & \vdots\\ \langle\mathbf{x}_1,\mathbf{x}_n\rangle & \cdots & \langle\mathbf{x}_n,\mathbf{x}_n\rangle\end{pmatrix}\begin{pmatrix}a_1\\ \vdots\\ a_n\end{pmatrix}\\ &\stackrel{\Delta}{=} \overline{[\mathbf{y}]}_{\mathfrak{X}}^{\mathrm{T}}\mathbf{Q}_{\mathfrak{X}}[\mathbf{x}]_{\mathfrak{X}}\end{aligned} \tag{5.14}$$

We refer to $\mathbf{Q}_{\mathfrak{X}}$ as the **matrix of the inner product** $\langle\cdot,\cdot\rangle$ **relative to the basis** $\mathfrak{X}$. It is evident that

$$(\mathbf{Q}_{\mathfrak{X}})_{jk} = \langle \mathbf{x}_k,\mathbf{x}_j\rangle \tag{5.15}$$

We can use (5.15) directly to generate the matrix of a given inner product relative to a particular basis. The matrix (5.15) is also known as the **Gram matrix** for the basis $\mathfrak{X}$; the matrix consists in the inner products of all pairs of vectors from the basis.

Exercise 2. Use (5.15) to generate the matrix of the inner product (5.13) relative to the standard basis for $\mathfrak{R}^2$.

From (5.14), (5.15), and the definition of an inner product we deduce that a Gram matrix, or a matrix of an inner product, has certain special properties which are related to the properties of inner products:

1. Since $\langle \mathbf{x}_k,\mathbf{x}_j\rangle = \overline{\langle \mathbf{x}_j,\mathbf{x}_k\rangle}$, $\mathbf{Q}_{\mathfrak{X}} = \overline{\mathbf{Q}}_{\mathfrak{X}}{}^{\mathrm{T}}$.

2. The inner product is positive definite; denoting $\mathbf{z} \stackrel{\Delta}{=} [\mathbf{x}]_{\mathfrak{X}}$, we find $\bar{\mathbf{z}}^{\mathrm{T}}\mathbf{Q}_{\mathfrak{X}}\mathbf{z} \geqslant 0$ for all $\mathbf{z}$ in $\mathfrak{M}^{n\times 1}$, with equality if and only if $\mathbf{z} = \boldsymbol{\theta}$.

We describe these matrix properties by saying $\mathbf{Q}_{\mathfrak{X}}$ is (1) **hermitian symmetric*** and (2) **positive definite**. For a given basis, the set of all possible

*If $\mathbf{Q}_{\mathfrak{X}}$ is real, the complex conjugate is superfluous. Then, if $\mathbf{Q}_{\mathfrak{X}} = \mathbf{Q}_{\mathfrak{X}}{}^{\mathrm{T}}$, we say $\mathbf{Q}_{\mathfrak{X}}$ is symmetric.

inner products on an n-dimensional space $\mathcal{V}$ is equivalent to the set of positive-definite, hermitian symmetric $n \times n$ matrices. This fact indicates precisely how much freedom we have in picking inner products. In point of fact, (5.14) can be used in defining an inner product for $\mathcal{V}$. We will exploit it in our discussion of orthogonal bases in the next section. A method for determining whether or not a matrix is positive definite is described in P&C 5.9.

Exercise 3. Any inner product on the *real* space $\mathfrak{M}^{n\times 1}$ is of the form $\langle \mathbf{x},\mathbf{y}\rangle \triangleq \mathbf{y}^{\mathrm{T}}\mathbf{Q}\mathbf{x}$ for some symmetric positive-definite matrix $\mathbf{Q}$. The analogous definition for a real function space on the interval $[a,b]$ is

$$\langle \mathbf{f},\mathbf{g}\rangle \triangleq \int_a^b \int_a^b k(t,s)\mathbf{f}(t)\mathbf{g}(s)\,ds\,dt$$

What properties must the kernel function k possess in order that this equation define a valid inner product (see P&C 5.30)? Show that if $k(t,s)=\omega(t)\delta(t-s)$, then the inner product reduces to (5.12).

5.2 Orthogonality

The thrust of this section is that orthogonal sets of vectors are not only linearly independent, but also lead to independent computation of coordinates. A set $\mathbb{S}$ of vectors is an **orthogonal set** if the vectors are pairwise orthogonal. If, in addition, each vector in $\mathbb{S}$ has unit norm, the set is called **orthonormal**. The two vectors of Example 1 (below) form an orthonormal set relative to the inner product (5.13). The standard basis for $\mathcal{R}^n$ is an orthonormal set relative to the standard inner product. Suppose the set $\mathfrak{X} \triangleq \{x_1,\ldots,x_n\}$ is orthogonal. It follows that each vector in $\mathfrak{X}$ is orthogonal to (and linearly independent of) the space spanned by the other vectors in the set; for example,

$$\langle \mathbf{x}_1, c_2\mathbf{x}_2+\cdots+c_n\mathbf{x}_n\rangle = c_2\underbrace{\langle \mathbf{x}_1,\mathbf{x}_2\rangle}_{0}+\cdots+c_n\underbrace{\langle \mathbf{x}_1,\mathbf{x}_n\rangle}_{0}=0$$

If $\mathfrak{X}$ is an orthogonal basis for an n-dimensional space $\mathcal{V}$, then for any vector $\mathbf{x}$ in $\mathcal{V}$, $\mathbf{x}=\sum_{k=1}^{n} c_k\mathbf{x}_k$ and

$$\begin{aligned}\langle \mathbf{x},\mathbf{x}_k\rangle &= \langle c_1\mathbf{x}_1+\cdots+c_n\mathbf{x}_n,\mathbf{x}_k\rangle\\ &= c_1\langle \mathbf{x}_1,\mathbf{x}_k\rangle+\cdots+c_k\langle \mathbf{x}_k,\mathbf{x}_k\rangle+\cdots+c_n\langle \mathbf{x}_n,\mathbf{x}_k\rangle\\ &= c_k\langle \mathbf{x}_k,\mathbf{x}_k\rangle\end{aligned}$$

Thus the kth coordinate of $\mathbf{x}$ relative to the orthogonal basis $\mathcal{X}$ is

$$c_k = \frac{\langle \mathbf{x}, \mathbf{x}_k \rangle}{\langle \mathbf{x}_k, \mathbf{x}_k \rangle} \tag{5.16}$$

Each coordinate can be determined independently using (5.16). The set of simultaneous equations which, in previous chapters, had to be solved in order to find coordinates is not necessary in this case. Inherent in the "orthogonalizing" inner product is the computational decoupling of the coordinates. If, in fact, the vectors in $\mathcal{X}$ are *orthonormal*, the denominator in (5.16) is 1, and

$$\mathbf{x} = \sum_{k=1}^{n} \langle \mathbf{x}, \mathbf{x}_k \rangle \mathbf{x}_k \tag{5.17}$$

Equation (5.17) is known as a **generalized Fourier series expansion** (or orthonormal expansion) of $\mathbf{x}$ relative to the orthonormal basis $\mathcal{X}$. The kth coordinate, $\langle \mathbf{x}, \mathbf{x}_k \rangle$, is called the kth **Fourier coefficient** of $\mathbf{x}$ relative to the orthonormal basis $\mathcal{X}$. We will have little need to distinguish between (5.17) and the orthogonal expansion which uses the coefficients (5.16). We will also refer to the latter expansion as a Fourier series expansion, and to (5.16) as a Fourier coefficient.

Example 1. Independent Computation of Fourier Coefficients. From Example 7 of the previous section we know that the vectors $\mathbf{x}_1 \triangleq (1,1)$ and $\mathbf{x}_2 \triangleq (2,0)$ form a basis for $\mathcal{R}^2$ which is orthonormal relative to the inner product (5.13). Let $\mathbf{x} = (2,1)$. Then by (5.17) we know that

$$\mathbf{x} = (2,1) = c_1(1,1) + c_2(2,0)$$

where $c_1 = \langle \mathbf{x}, \mathbf{x}_1 \rangle = \langle (2,1),(1,1) \rangle = 1$ and $c_2 = \langle \mathbf{x}, \mathbf{x}_2 \rangle = \langle (2,1),(2,0) \rangle = \frac{1}{2}$.

Gram–Schmidt Orthogonalization Procedure

The Gram–Schmidt procedure is a technique for generating an orthonormal basis. Suppose $\mathbf{x}_1$ and $\mathbf{x}_2$ are independent vectors in the space $\mathcal{R}^2$ with the standard inner product (dot product). (See the arrow space equivalent in Figure 5.3.) We will convert this pair of vectors to an orthogonal pair of vectors which spans the same space. The vector $\mathbf{x}_2$ decomposes uniquely into a pair of components, one collinear with $\mathbf{x}_1$ and the other orthogonal to $\mathbf{x}_1$. The collinear component is $\|\mathbf{x}_2\| \cos\phi$ times the unit vector in the direction of $\mathbf{x}_1$; using the expression (5.4) for the dot

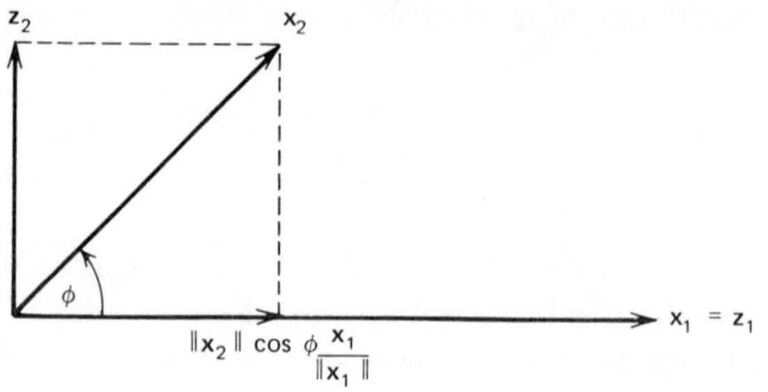

Figure 5.3. Gram–Schmidt orthogonalization in arrow space.

product, we convert this collinear vector to the form

$$\|\mathbf{x}_2\| \cos\phi \frac{\mathbf{x}_1}{\|\mathbf{x}_1\|} = \frac{\mathbf{x}_2 \cdot \mathbf{x}_1}{\|\mathbf{x}_1\|} \frac{\mathbf{x}_1}{\|\mathbf{x}_1\|}$$

Define $\mathbf{z}_1 \triangleq \mathbf{x}_1$ and $\mathbf{z}_2 \triangleq \mathbf{x}_2 - (\mathbf{x}_2 \cdot \mathbf{x}_1 / \|\mathbf{x}_1\|^2)\mathbf{x}_1$. Then $\mathbf{z}_2$ is orthogonal to $\mathbf{z}_1$, and $\{\mathbf{z}_1, \mathbf{z}_2\}$ is an orthogonal set which spans the same space as $\{\mathbf{x}_1, \mathbf{x}_2\}$. We can normalize these vectors to obtain an orthonormal set $\{\mathbf{y}_1, \mathbf{y}_2\}$ which also spans the same space: $\mathbf{y}_1 = \mathbf{z}_1 / \|\mathbf{z}_1\|$, and $\mathbf{y}_2 = \mathbf{z}_2 / \|\mathbf{z}_2\|$.

The procedure applied to the pair of vectors in $\mathcal{R}^2$ above can be used to orthogonalize a finite number of vectors in any inner product space. Suppose we wish to orthogonalize a set of vectors $\{\mathbf{x}_1, \ldots, \mathbf{x}_n\}$ from some inner product space $\mathcal{V}$. Assume we have already replaced $\mathbf{x}_1, \ldots, \mathbf{x}_k$ by an orthogonal set $\mathbf{z}_1, \ldots, \mathbf{z}_k$ which spans the same space as $\mathbf{x}_1, \ldots, \mathbf{x}_k$ (imagine $k = 1$). Then $\mathbf{x}_{k+1}$ decomposes uniquely into a pair of components, one in the space spanned by $\{\mathbf{z}_1, \ldots, \mathbf{z}_k\}$ and the other $(\mathbf{z}_{k+1})$ orthogonal to $\mathbf{z}_1, \ldots, \mathbf{z}_k$. Thus $\mathbf{z}_{k+1}$ must satisfy

$$\mathbf{x}_{k+1} = (c_1 \mathbf{z}_1 + \cdots + c_k \mathbf{z}_k) + \mathbf{z}_{k+1}$$

Since the set $\{\mathbf{z}_1, \ldots, \mathbf{z}_{k+1}\}$ must be orthogonal, and therefore a basis for the space it spans, the coefficients $\{c_j\}$ are determined by (5.16):

$$c_j = \frac{\langle \mathbf{x}_{k+1}, \mathbf{z}_j \rangle}{\langle \mathbf{z}_j, \mathbf{z}_j \rangle}$$

Therefore,

$$\mathbf{z}_{k+1} = \mathbf{x}_{k+1} - \sum_{j=1}^{k} \frac{\langle \mathbf{x}_{k+1}, \mathbf{z}_j \rangle}{\langle \mathbf{z}_j, \mathbf{z}_j \rangle} \mathbf{z}_j \tag{5.18}$$

Exercise 1. Verify that $\mathbf{z}_{k+1}$ as given in (5.18) is orthogonal to $\mathbf{z}_j$ for $j=1,\ldots,k$. How do we know the "orthogonal" decomposition of $\mathbf{x}_{k+1}$ is unique?

Starting with $\mathbf{z}_1=\mathbf{x}_1$ and using (5.18) for $k=1,\ldots,n-1$, we generate an orthogonal basis for the space spanned by $\{\mathbf{x}_1,\ldots,\mathbf{x}_n\}$. The procedure can be applied to any finite set of vectors, independent or not; any dependencies will be eliminated (P&C 5.14). Thus we can obtain an orthogonal basis for a vector space by applying (5.18) to any set of vectors which spans the space. The application of (5.18) is referred to as the **Gram–Schmidt orthogonalization procedure**. It requires no additional effort to normalize the vectors at each step, obtaining $\mathbf{y}_j=\mathbf{z}_j/\|\mathbf{z}_j\|$; then (5.18) becomes

$$\mathbf{z}_{k+1}=\mathbf{x}_{k+1}-\sum_{j=1}^{k}\langle\mathbf{x}_{k+1},\mathbf{y}_j\rangle\mathbf{y}_j \tag{5.19}$$

Numerical accuracy and techniques for retaining accuracy in Gram–Schmidt orthogonalization are discussed in Section 6.6.

Example 2. Gram–Schmidt Orthogonalization in a Function Space. Define $\mathbf{f}_k(t)\overset{\Delta}{=}t^k$ in the space $\mathcal{P}(-1,1)$ with the standard inner product

$$\langle\mathbf{f},\mathbf{g}\rangle=\int_{-1}^{1}\mathbf{f}(t)\,\mathbf{g}(t)\,dt$$

We will apply the Gram–Schmidt procedure to the first few functions in the set $\{\mathbf{f}_0,\mathbf{f}_1,\mathbf{f}_2,\ldots\}$. Using (5.18), with appropriate adjustments in notation, we let $\mathbf{g}_0(t)=\mathbf{f}_0(t)=1$ and

$$\mathbf{g}_1(t)=\mathbf{f}_1(t)-\frac{\langle\mathbf{f}_1,\mathbf{g}_0\rangle}{\langle\mathbf{g}_0,\mathbf{g}_0\rangle}\mathbf{g}_0(t)$$

But $\langle\mathbf{f}_1,\mathbf{g}_0\rangle=\int_{-1}^{1}(t)(1)\,dt=0$. Therefore, $\mathbf{g}_1(t)=\mathbf{f}_1(t)=t$, and $\mathbf{g}_1$ is orthogonal to $\mathbf{g}_0$. Again using (5.18),

$$\mathbf{g}_2(t)=\mathbf{f}_2(t)-\frac{\langle\mathbf{f}_2,\mathbf{g}_0\rangle}{\langle\mathbf{g}_0,\mathbf{g}_0\rangle}\mathbf{g}_0(t)-\frac{\langle\mathbf{f}_2,\mathbf{g}_1\rangle}{\langle\mathbf{g}_1,\mathbf{g}_1\rangle}\mathbf{g}_1(t)$$

The inner products are

$$\langle\mathbf{f}_2,\mathbf{g}_0\rangle=\int_{-1}^{1}(t^2)(1)\,dt=\tfrac{2}{3}$$

$$\langle\mathbf{g}_0,\mathbf{g}_0\rangle=\int_{-1}^{1}(1)(1)\,dt=2$$

$$\langle\mathbf{f}_2,\mathbf{g}_1\rangle=\int_{-1}^{1}(t^2)(t)\,dt=0$$

Therefore, $g_2(t)=t^2-\frac{1}{3}$, and $\mathbf{g}_2$ is orthogonal to $\mathbf{g}_1$ and $\mathbf{g}_0$. We could continue, if we wished, to generate additional vectors of the orthogonal set $\{\mathbf{g}_0,\mathbf{g}_1,\mathbf{g}_2,\ldots\}$. The functions $\{\mathbf{g}_k\}$ are known as *orthogonal polynomials*. Rather than normalize these orthogonal polynomials, we adjust their length as follows: define $\mathbf{p}_k \stackrel{\Delta}{=} \mathbf{g}_k/\mathbf{g}_k(1)$ so that $\mathbf{p}_k(1)=1$. The functions $\{\mathbf{p}_0,\mathbf{p}_1,\mathbf{p}_2,\ldots\}$ so defined are known as the **Legendre polynomials**. (These polynomials are useful for solving partial differential equations in spherical coordinates.) Thus $\mathbf{p}_0(t)=\mathbf{g}_0(t)=1$, $\mathbf{p}_1(t)=\mathbf{g}_1(t)=t$, and $\mathbf{p}_2(t)=\mathbf{g}_2(t)/\mathbf{g}_2(1)=(3t^2-1)/2$. A method of computing orthogonal polynomials which uses less computation than the Gram–Schmidt procedure is described in P&C 5.16.

Orthogonal Projection

The **orthogonal complement** of a set $\mathcal{S}$ of vectors in a vector space $\mathcal{V}$ is the set $\mathcal{S}^\perp$ of all vectors in $\mathcal{V}$ which are orthogonal to every vector in $\mathcal{S}$. For example, the orthogonal complement of the vector $\mathbf{x}_1$ of Figure 5.3 is the subspace spanned by $\mathbf{z}_2$. On the other hand, the orthogonal complement of $\text{span}\{\mathbf{z}_2\}$ is not the vector $\mathbf{x}_1$, but rather the space spanned by $\mathbf{x}_1$. An orthogonal complement is always a subspace.

Example 3. An Orthogonal Complement in $\mathcal{C}(0,1)$. Suppose the set $\mathcal{S}$ in the standard inner product space $\mathcal{C}(0,1)$ consists of the single function $\mathbf{f}_1(t)=1$. Then $\mathcal{S}^\perp$ is the set of all functions whose average is zero; that is, those functions $\mathbf{g}$ for which

$$\langle \mathbf{f}_1,\mathbf{g}\rangle=\int_0^1(1)\mathbf{g}(t)\,dt=0$$

As part of our discussion of the decomposition of a vector space $\mathcal{V}$ into a direct sum, $\mathcal{V}=\mathcal{W}_1\oplus\mathcal{W}_2$, we introduced the concept of a projection on one of the subspaces along the other (Section 4.1). In the derivation of (5.18) we again used this concept of projection. In particular, each time we apply (5.18), we project a vector $\mathbf{x}_{k+1}$ onto the space spanned by $\{\mathbf{z}_1,\ldots,\mathbf{z}_k\}$ along a direction orthogonal to $\mathbf{z}_1,\ldots,\mathbf{z}_k$ (Figure 5.3). Suppose we define $\mathcal{W} \stackrel{\Delta}{=} \text{span}\{\mathbf{z}_1,\ldots,\mathbf{z}_k\}$. Then any vector which is orthogonal to $\mathbf{z}_1,\ldots,\mathbf{z}_k$ is in $\mathcal{W}^\perp$, the orthogonal complement of $\mathcal{W}$. The only vector which is in both $\mathcal{W}$ and $\mathcal{W}^\perp$ is the vector $\boldsymbol{\theta}$. Since the vector $\mathbf{x}_{k+1}$ of (5.18) can be any vector in $\mathcal{V}$, the derivation of (5.18) constitutes a proof (for finite-dimensional $\mathcal{V}$)* that

$$\mathcal{V}=\mathcal{W}\oplus\mathcal{W}^\perp \tag{5.20}$$

*The projection theorem (5.20) also applies to certain infinite-dimensional spaces. Specifically, it is valid for any (complete) subspace $\mathcal{W}$ of a Hilbert space $\mathcal{V}$. See Bachman and Narici [5.2, p. 172]. These infinite-dimensional concepts (Hilbert space, subspace, and completeness) are discussed in Section 5.3.

That is, any vector in $\mathcal{V}$ can be decomposed uniquely into a pair of components, one in $\mathcal{W}$ and the other orthogonal to $\mathcal{W}$. The projection of a vector $\mathbf{x}$ on a subspace $\mathcal{W}$ along $\mathcal{W}^{\perp}$ is usually referred to as the **orthogonal projection of x on** $\mathcal{W}$. Equation (5.20), which guarantees the existence of orthogonal projections, is sometimes known as the **projection theorem.** This theorem is one of the keys to the solution of the least-square optimization problems explored in Chapter 6.

It is apparent from (5.18) that the orthogonal projection $\mathbf{x}_{\mathcal{W}}$ of an arbitrary vector $\mathbf{x}$ in $\mathcal{V}$ onto the subspace $\mathcal{W}$ spanned by the orthogonal set $\{\mathbf{z}_1,\ldots,\mathbf{z}_k\}$ is

$$\mathbf{x}_{\mathcal{W}} = \sum_{j=1}^{k} \frac{\langle \mathbf{x},\mathbf{z}_j\rangle}{\|\mathbf{z}_j\|^2}\mathbf{z}_j \tag{5.21}$$

We can also write (5.21) in terms of the normalized vectors $\{\mathbf{y}_1,\ldots,\mathbf{y}_k\}$ of (5.19):

$$\mathbf{x}_{\mathcal{W}} = \sum_{j=1}^{k} \langle \mathbf{x},\mathbf{y}_j\rangle \mathbf{y}_j \tag{5.22}$$

Equation (5.22) expresses $\mathbf{x}_{\mathcal{W}}$ as a partial Fourier series expansion, an "attempted" expansion of $\mathbf{x}$ in terms of an orthonormal basis for the subspace on which $\mathbf{x}$ is projected. If $\mathbf{x}-\sum_{j=1}^{k}\langle \mathbf{x},\mathbf{y}_j\rangle\mathbf{y}_j \neq \boldsymbol{\theta}$, we know that the orthonormal basis for $\mathcal{W}$ is not a basis for the whole space $\mathcal{V}$. It is evident that an orthonormal set $\{\mathbf{y}_i\}$ is a basis for a finite-dimensional space $\mathcal{V}$ if and only if there is no nonzero vector in $\mathcal{V}$ which is orthogonal to $\{\mathbf{y}_i\}$. We can compute the orthogonal projection of $\mathbf{x}$ on $\mathcal{W}$ without concerning ourselves with a basis for the orthogonal complement $\mathcal{W}^{\perp}$. We can do so because a description of $\mathcal{W}^{\perp}$ is inherent in the inner product. Clearly, Gram–Schmidt orthogonalization, orthogonal projection, and Fourier series are closely related. Equation (5.22), or its equivalent, (5.21), is a practical tool for computing orthogonal projections on finite-dimensional subspaces.

Example 4. Computation of an Orthogonal Projection. Let $\mathcal{W}$ be that subspace of the standard inner product space $\mathcal{R}^3$ which is spanned by $\{\mathbf{x}_1,\mathbf{x}_2\}$, where $\mathbf{x}_1=(1,0,1)$ and $\mathbf{x}_2=(0,1,1)$. We seek the orthogonal projection of $\mathbf{x}=(0,0,2)$ on $\mathcal{W}$. We first use the Gram–Schmidt procedure to orthogonalize the set $\{\mathbf{x}_1,\mathbf{x}_2\}$; then we apply (5.21). By (5.18), $\mathbf{z}_1=\mathbf{x}_1=(1,0,1)$ and

$$\mathbf{z}_2=\mathbf{x}_2-\frac{\langle \mathbf{x}_2,\mathbf{x}_1\rangle}{\|\mathbf{x}_1\|^2}\mathbf{x}_1=(0,1,1)-\tfrac{1}{2}(1,0,1)=(-\tfrac{1}{2},1,\tfrac{1}{2})$$

By (5.21),

$$\mathbf{x}_{\mathcal{W}} = \frac{\langle \mathbf{x}, \mathbf{z}_1 \rangle}{\|\mathbf{z}_1\|^2}\mathbf{z}_1 + \frac{\langle \mathbf{x}, \mathbf{z}_2 \rangle}{\|\mathbf{z}_2\|^2}\mathbf{z}_2 = \frac{2}{2}(1,0,1) + \frac{1}{(3/2)}\left(-\tfrac{1}{2},1,\tfrac{1}{2}\right) = \left(\tfrac{2}{3},\tfrac{2}{3},\tfrac{4}{3}\right)$$

Orthonormal Eigenvector Bases for Finite-Dimensional Spaces

In (4.13) we solved the operator equation $\mathbf{Tx}=\mathbf{y}$ by means of spectral decomposition (or diagonalization). By representing the input vector $\mathbf{y}$ in terms of its coordinates relative to a basis of eigenvectors, we converted the operator equation into a set of uncoupled scalar equations, and solution for the output $\mathbf{x}$ became simple. Of course, even when the eigendata were known, a set of simultaneous equations was required in order to decompose $\mathbf{y}$. We now explore the solution of equations by means of an *orthonormal* basis of eigenvectors. The orthonormality allows us to determine independently each eigenvector component of the input; the solution process is then completely decoupled.

Let $\mathbf{T}$ have eigendata $\{\lambda_i\}$ and $\{\mathbf{z}_i\}$, and let $\{\mathbf{z}_1,\ldots,\mathbf{z}_n\}$ be a basis for the space $\mathcal{V}$ on which $\mathbf{T}$ operates. (Then $\mathbf{T}$ must be diagonalizable.) Furthermore, suppose $\mathbf{T}$ is invertible; that is, $\lambda_i \neq 0$. We solve the operator equation $\mathbf{Tx}=\mathbf{y}$ as follows. The vectors $\mathbf{x}$ and $\mathbf{y}$ can be expanded as

$$\mathbf{y} = \sum_{i=1}^{n} c_i \mathbf{z}_i \quad \text{and} \quad \mathbf{x} = \sum_{i=1}^{n} d_i \mathbf{z}_i$$

The coordinates $\{c_i\}$ can be determined from $\mathbf{y}$; the numbers $\{d_i\}$ are coordinates of the unknown vector $\mathbf{x}$. Inserting these eigenvector expansions into the operator equation, we obtain

$$\mathbf{Tx} = \sum_i d_i \mathbf{T}\mathbf{z}_i = \sum_i d_i \lambda_i \mathbf{z}_i = \sum_i c_i \mathbf{z}_i = \mathbf{y}$$

or

$$\sum_i (d_i\lambda_i - c_i)\mathbf{z}_i = \boldsymbol{\theta}$$

Since the vectors $\mathbf{z}_i$ are independent, $d_i = c_i/\lambda_i$, and the solution to the operator equation is

$$\mathbf{x} = \sum_i \left(\frac{c_i}{\lambda_i}\right)\mathbf{z}_i \tag{5.23}$$

Suppose the eigenvector basis $\{\mathbf{z}_1,\ldots,\mathbf{z}_n\}$ is orthonormal relative to the inner product on $\mathcal{V}$. Then the eigenvector expansion of $\mathbf{y}$ can be expressed as the Fourier expansion

$$\mathbf{y}=\sum_{i=1}^{n}\langle\mathbf{y},\mathbf{z}_i\rangle\mathbf{z}_i$$

and the solution (5.23) becomes

$$\mathbf{x}=\sum_{i=1}^{n}\frac{\langle\mathbf{y},\mathbf{z}_i\rangle}{\lambda_i}\mathbf{z}_i \tag{5.24}$$

Each component of (5.24) can be evaluated independently.

If $\mathbf{T}$ is not invertible, (5.24) requires division by a zero eigenvalue. The eigenvectors for the zero eigenvalue form a basis for nullspace($\mathbf{T}$). The remaining eigenvectors are taken by $\mathbf{T}$ into range($\mathbf{T}$), and in fact form a basis for range($\mathbf{T}$). To avoid division by zero in (5.24), we split the space: $\mathcal{V}=\text{nullspace}(\mathbf{T})\oplus\text{range}(\mathbf{T})$. The equation $\mathbf{Tx}=\mathbf{y}$ has no solution unless $\mathbf{y}$ is in range($\mathbf{T}$) (or $c_i=0$ for i corresponding to a zero eigenvalue); since the eigenvectors are assumed to be orthonormal, an equivalent statement is that $\mathbf{y}$ must be orthogonal to nullspace($\mathbf{T}$). Treating the eigenvectors corresponding to zero eigenvalues separately, we replace the solution $\mathbf{x}$ in (5.23)–(5.24) by

$$\begin{aligned}\mathbf{x}&=\sum_{\text{nonzero }\lambda_i}d_i\mathbf{z}_i+\sum_{\text{zero }\lambda_i}d_i\mathbf{z}_i\\&=\sum_{\text{nonzero }\lambda_i}\frac{\langle\mathbf{y},\mathbf{z}_i\rangle}{\lambda_i}\mathbf{z}_i+\mathbf{x}_0\end{aligned} \tag{5.25}$$

where $\mathbf{x}_0$ is an arbitrary vector in nullspace($\mathbf{T}$). The first portion of (5.25) is a particular solution to the equation $\mathbf{Tx}=\mathbf{y}$. The second portion, $\mathbf{x}_0$, is the homogeneous solution. The undetermined coefficients d_i in the sum which constitutes $\mathbf{x}_0$ are indicative of the freedom in the solution owing to the noninvertibility of $\mathbf{T}$.

What fortunate circumstances will allow us to find an orthonormal basis of eigenvectors? The eigenvectors are properties of $\mathbf{T}$; they cannot be selected freely. Assume there are enough eigenvectors of $\mathbf{T}$ to form a basis for the space. Were we to orthogonalize an eigenvector basis using the Gram–Schmidt procedure, the resulting set of vectors would not be eigenvectors. However, we have considerable freedom in picking inner products.

In point of fact, since the space is finite dimensional, we can select the inner product to make any particular basis orthonormal.

The key to selection of inner products for finite-dimensional spaces is (5.14), the representation of inner products of vectors in terms of their coordinates. Let the basis $\mathfrak{X}$ be $\{\mathbf{z}_1,\ldots,\mathbf{z}_n\}$, the eigenvectors of $\mathbf{T}$. We select the matrix of the inner product, $\mathbf{Q}_{\mathfrak{X}}$, such that the basis vectors are orthonormal. By (5.15), if $\mathfrak{X}$ is to be orthonormal, $\mathbf{Q}_{\mathfrak{X}}$ satisfies

$$(\mathbf{Q}_{\mathfrak{X}})_{jk} = \langle \mathbf{z}_k, \mathbf{z}_j \rangle = 1, \quad j = k \\ = 0, \quad j \neq k \tag{5.26}$$

or $\mathbf{Q}_{\mathfrak{X}} = \mathbf{I}$. By (5.14), this matrix defines the following inner product on $\mathcal{V}$:

$$\langle \mathbf{x}, \mathbf{y} \rangle = \overline{[\mathbf{y}]}_{\mathfrak{X}}^{\mathrm{T}} [\mathbf{x}]_{\mathfrak{X}} \tag{5.27}$$

The expression (5.27) of an inner product in terms of coordinates relative to an orthonormal basis is called **Parseval's equation**. A basis $\mathfrak{X}$ is orthonormal if and only if (5.27) is satisfied; that is, if and only if the inner product between any two vectors equals the standard inner product (in $\mathfrak{M}^{n\times 1}$) between their coordinates relative to $\mathfrak{X}$.

Example 5. Solution of an Equation by Orthonormal Eigenfunction Expansion. Suppose we define $\mathbf{T}\colon \mathfrak{R}^2 \to \mathfrak{R}^2$ by

$$\mathbf{T}(\xi_1, \xi_2) \triangleq (2\xi_1 + 3\xi_2, 4\xi_2)$$

[The same operator is used in the decomposition of Example 7, Section 4.1.] The eigendata are $\lambda_1 = 2$, $\mathbf{z}_1 = (1,0)$, $\lambda_2 = 4$, and $\mathbf{z}_2 = (3,2)$. The pair of eigenvectors, $\mathfrak{X} \triangleq \{\mathbf{z}_1, \mathbf{z}_2\}$, is a basis for $\mathfrak{R}^2$. We define the inner product for $\mathfrak{R}^2$ by (5.27):

$$\langle \mathbf{x}, \mathbf{y} \rangle = [\mathbf{y}]_{\mathfrak{X}}^{\mathrm{T}} [\mathbf{x}]_{\mathfrak{X}}$$

To make this definition more explicit, we find the coordinates of $\mathbf{x}$ and $\mathbf{y}$; let $\mathbf{y} = a_1\mathbf{z}_1 + a_2\mathbf{z}_2$, or

$$\mathbf{y} = (\eta_1, \eta_2) = a_1(1,0) + a_2(3,2)$$

Solution (by row reduction) yields $a_1 = \eta_1 - 3\eta_2/2$ and $a_2 = \eta_2/2$. Similarly, the coordinates of $\mathbf{x} = (\xi_1, \xi_2)$ are $c_1 = \xi_1 - 3\xi_2/2$ and $c_2 = \xi_2/2$. Thus we can express the inner product as

$$\langle (\xi_1, \xi_2), (\eta_1, \eta_2) \rangle = \begin{pmatrix} \eta_1 - 3\eta_2/2 \\ \eta_2/2 \end{pmatrix}^{\mathrm{T}} \begin{pmatrix} \xi_1 - 3\xi_2/2 \\ \xi_2/2 \end{pmatrix}$$

$$= \xi_1\eta_1 - \tfrac{3}{2}\xi_1\eta_2 - \tfrac{3}{2}\xi_2\eta_1 + \tfrac{5}{2}\xi_2\eta_2$$

Relative to this inner product, the basis $\mathfrak{X}$ is orthonormal. We solve the equation $\mathbf{Tx}=\mathbf{T}(\xi_1,\xi_2)=(\eta_1,\eta_2)=\mathbf{y}$ using (5.24):

$$\begin{aligned}\mathbf{x}&=\frac{\langle \mathbf{y},\mathbf{z}_1\rangle}{\lambda_1}\mathbf{z}_1+\frac{\langle \mathbf{y},\mathbf{z}_2\rangle}{\lambda_2}\mathbf{z}_2\\ &=\frac{(\eta_1-3\eta_2/2)}{2}(1,0)+\frac{(\eta_2/2)}{4}(3,2)\\ &=(\eta_1/2-3\eta_2/8,\eta_2/4)\end{aligned}$$

We have developed two basic approaches for analyzing a finite-dimensional, invertible, diagonalizable, linear equation: (*a*) operator inversion and (*b*) spectral decomposition (or eigenvector expansion). Both methods give explicit descriptions of the input–output relationship of the system for which the equation is a model. The spectral decomposition yields a more detailed description; therefore, it provides more insight than does inversion. If the eigenvector expansion is orthonormal, we also obtain conceptual and computational independence of the individual terms in the expansion.

What price do we pay for the insight obtained by each of these approaches? We take as a measure of computational expense the approximate number of multiplications required to analyze an $n\times n$ matrix equation:

1. Inversion of an $n\times n$ matrix $\mathbf{A}$ (or solution of $\mathbf{Ax}=\mathbf{y}$ for an unspecified $\mathbf{y}$) by use of Gaussian elimination requires $4n^3/3$ multiplications. Actual multiplication of $\mathbf{y}$ by $\mathbf{A}^{-1}$ uses n^2 multiplications for each specific $\mathbf{y}$.

2. Analysis by the nonorthogonal eigenvector expansion (5.23) starts with computation of the eigendata. Determination of the characteristic equation, computation of its roots, and solution for the eigenvectors is considerably more expensive than matrix inversion (see Section 4.2). For each specific $\mathbf{y}$, determination of $\mathbf{x}$ requires $n^3/3$ multiplications to calculate the coordinates of $\mathbf{y}$ relative to the eigenvector basis. The number of multiplications needed to sum up the eigenvector components of $\mathbf{x}$ is relatively unimportant.

3. In order to express the solution $\mathbf{x}$ as the orthonormal eigenvector expansion (5.24), we need to determine the inner product which makes the basis of eigenvectors orthonormal. Determination of that inner product requires the solution of a vector equation with an unspecified right-hand side (see Example 5). Thus to fully define the expression (5.24), we need $4n^3/3$ multiplications in addition to the computation necessary to obtain

the eigendata. It is evident from Example 5 that evaluation of a single inner product in an n-dimensional space can require as few as n multiplications (if no cross-products terms appear and all coefficients are unity) and as many as $2n^2$ multiplications (if all cross-product terms appear). Therefore, for each specific $\mathbf{y}$, computation of $\mathbf{x}$ requires between n^2 and $2n^3$ multiplications to evaluate the inner products, and n^2+n multiplications to perform the linear combination.

The value of orthonormal eigenvector expansion as a vehicle for analyzing equations lies primarily in the insight provided by the complete decomposition (5.24). We pay for this insight by determining the eigendata. For certain classes of problems we are fortunate in that the eigendata is known a priori (e.g., the symmetrical components of (4.27)–(4.28), the Vandermond matrix of P&C 4.16, and the sinusoids or complex exponentials of classical Fourier series). Then the technique is computationally competitive with inversion. We note in Section 5.5 that for (infinite-dimensional) partial differential equations, eigenvector expansion is a commonly used analysis technique.

Infinite Orthonormal Expansions

We will find that most of the concepts we have discussed in this chapter apply in infinite-dimensional spaces. A significant characteristic of an infinite expansion of a vector (or function) is that the "first few" terms usually dominate. If the infinite expansion is also orthonormal, then we can not only approximate the vector by the first few terms of the expansion, but we can also compute these first few terms, ignoring the remainder—the individual terms of an orthonormal expansion are computationally independent. Thus the value of *orthonormal* eigenvector expansion is higher for infinite-dimensional systems than for finite-dimensional systems. Furthermore, for certain classes of models, orthonormal eigendata is standard—it is known a priori. (For example, all constant-coefficient linear differential operators with periodic boundary conditions have an easily determined set of orthogonal sine and cosine functions as eigenfunctions.) For these models, orthonormal eigenvector expansion is a computationally efficient analysis technique (P&C 5.35). In this section we examine briefly a few familiar infinite orthonormal expansions which are useful in the analysis of dynamic systems. A detailed general discussion of infinite orthonormal eigenvector expansions forms the subject of Section 5.5.

We noted in Section 4.3 that models of linear dynamic systems (linear differential operators with initial conditions) have no eigenfunctions because the boundary conditions all occur at one point in time. This fact would seem to preclude the use of eigenfunction expansions in analyzing dynamic systems. However, many practical dynamic systems, electric power systems for instance, are operated with periodic inputs. The output of a linear *time-invariant* dynamic system with a periodic input quickly approaches a steady-state form which is periodic with the same period as the input. The steady-state form depends only on the periodic input and not on the initial conditions. (Implicit in the term steady-state, however, is a set of **periodic boundary conditions**—the values of the solution $\mathbf{f}$ and its derivatives must be the same at the beginning and end of the period.) The transition from the initial conditions to the steady-state solution is described by a transient component of the solution. Suppose the system model is a differential equation, denoted by $\mathbf{Lf}=\mathbf{u}$, with initial conditions $\beta_i(\mathbf{f})=\alpha_i$. The **steady-state solution** $\mathbf{f}_1$ satisfies $\mathbf{Lf}_1=\mathbf{u}$ (with periodic boundary conditions). Define the **transient solution** $\mathbf{f}_2$ to be the solution of $\mathbf{Lf}_2=\boldsymbol{\theta}$ with $\beta_i(\mathbf{f}_1+\mathbf{f}_2)=\alpha_i$ (or $\beta_i(\mathbf{f}_2)=\alpha_i-\beta_i(\mathbf{f}_1)$). Then $\mathbf{f}\triangleq\mathbf{f}_1+\mathbf{f}_2$ satisfies both the differential equation and the initial conditions.

Example 6. Steady-State and Transient Solutions. The linear time-invariant electrical circuit of Figure 5.4 is described by the differential equation

$$\mathbf{e}(t)=L\frac{d\mathbf{i}(t)}{dt}+R\mathbf{i}(t),\qquad \mathbf{i}(0)=0 \tag{5.28}$$

Suppose the applied voltage (or input) is the periodic function $\mathbf{e}(t)=E\sin(\omega t+\phi_E)$. We can easily verify that the steady-state solution to the differential equation is

$$\mathbf{i}_1(t)=\frac{E}{\sqrt{(\omega L)^2+R^2}}\sin(\omega t+\phi_E-\phi_I),\qquad \phi_I=\tan^{-1}\left(\frac{\omega L}{R}\right)$$

Note that $\mathbf{i}_1$ does not satisfy the initial condition $\mathbf{i}_1(0)=0$ unless ϕ_E happens to equal ϕ_I. However, it does satisfy the periodic boundary condition $\mathbf{i}_1(2\pi/\omega)=\mathbf{i}_1(0)$. The transient solution (the solution of the homogeneous differential equation) is of the form

$$\mathbf{i}_2(t)=ce^{-(R/L)t}$$

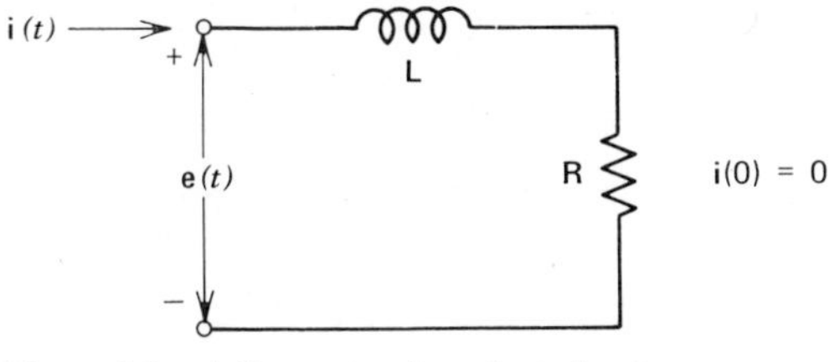

Figure 5.4. A linear time-invariant circuit.

We pick the constant c such that $\mathbf{i}_1(0)+\mathbf{i}_2(0)=0$:

$$c=-\frac{E\sin(\phi_E-\phi_I)}{\sqrt{(\omega L)^2+R^2}}$$

Then $\mathbf{i} \triangleq \mathbf{i}_1+\mathbf{i}_2$ satisfies (5.28).

Exercise 2. Verify that $\mathbf{i}_1$ of Example 6 satisfies the differential equation of (5.28), but not the initial condition. Hint:

$$a\cos\psi+b\sin\psi=\sqrt{a^2+b^2}\,\sin\left(\psi+\tan^{-1}\left(\frac{a}{b}\right)\right)$$

Steady-state analysis of a dynamic system is analysis of the system with periodic boundary conditions. A linear constant-coefficient differential operator with periodic boundary conditions *does* have eigenfunctions; namely, all sines, cosines, and complex exponentials which have the correct period. In point of fact, the steady-state solution to (5.28) was easy to determine only because the periodic input $\mathbf{e}(t)$ was an eigenfunction of the differential operator for periodic boundary conditions. The eigenvalue corresponding to that eigenfunction is the input impedance Z of the R-L circuit corresponding to the frequency ω of the applied voltage:

$$Z=R+i\omega L=\sqrt{(\omega L)^2+R^2}\,\exp\left(i\tan^{-1}\left(\frac{\omega L}{R}\right)\right)$$

where $i=\sqrt{-1}$.

It is well known that any "well-behaved" periodic function can be expanded in an orthonormal series of sines and cosines—eigenfunctions of linear constant-coefficient differential operators with periodic boundary conditions. Suppose $\mathbf{f}$ is a periodic function of period p; then*

$$\mathbf{f}(t)=a_0+a_1\cos\frac{2\pi t}{p}+a_2\cos\frac{4\pi t}{p}+\cdots$$
$$+b_1\sin\frac{2\pi t}{p}+b_2\sin\frac{4\pi t}{p}+\cdots \tag{5.29}$$

*This is the classical Fourier series expansion [5.5, p. 312].

where

$$a_0 = \frac{1}{p}\int_0^p \mathbf{f}(t)\,dt$$

$$a_j = \frac{2}{p}\int_0^p \mathbf{f}(t)\cos\frac{2\pi jt}{p}\,dt, \qquad j=1,2,\ldots$$

$$b_j = \frac{2}{p}\int_0^p \mathbf{f}(t)\sin\frac{2\pi jt}{p}\,dt, \qquad j=1,2,\ldots$$

We can replace the sinusoidal functions of (5.29) by the normalized functions

$$\begin{aligned} \mathbf{f}_0(t) &\triangleq \sqrt{1/p} \\ \mathbf{f}_k(t) &\triangleq \sqrt{2/p}\,\cos(2\pi kt/p), \quad k=-1,-2,\ldots \\ &\triangleq \sqrt{2/p}\,\sin(2\pi kt/p), \qquad k=1,2,\ldots \end{aligned} \tag{5.30}$$

Relative to the inner product

$$\langle \mathbf{f},\mathbf{g}\rangle \triangleq \int_0^p \mathbf{f}(t)\mathbf{g}(t)\,dt \tag{5.31}$$

the functions (5.30) form an orthonormal set. (Since the functions are periodic of period p, we concern ourselves only with values of the functions over a single period.) Therefore, we can write (5.29) in the standard form for a generalized Fourier series:

$$\mathbf{f} = \sum_{k=-\infty}^{\infty} \langle \mathbf{f},\mathbf{f}_k\rangle \mathbf{f}_k \tag{5.32}$$

Exercise 3. Show that the set of functions (5.30) is orthonormal relative to the standard inner product (5.31).

If $\mathbf{f}$ is any periodic function of period p, (5.32) is an orthonormal expansion of $\mathbf{f}$ in terms of the eigenfunctions of any linear constant-coefficient differential operator (assuming periodic boundary conditions of the same period p). Furthermore, since the eigenfunctions are known a priori, they need not be computed. Therefore, the Fourier series described by (5.29) or (5.32) is valuable in the steady-state analysis of linear time-invariant dynamic systems (P&C 5.35).

A sine or cosine can be expressed as the sum of a pair of complex exponentials with complex coefficients

$$\sin\psi = \frac{e^{i\psi} - e^{-i\psi}}{2i}, \qquad \cos\psi = \frac{e^{i\psi} + e^{-i\psi}}{2}, \qquad i = \sqrt{-1}$$

Therefore, the Fourier series (5.29) can be rewritten in terms of the functions

$$\mathbf{g}_k(t) \triangleq \frac{1}{\sqrt{p}} \exp\left(i\frac{2\pi kt}{p}\right), \qquad k = 0, \pm 1, \pm 2, \dots \tag{5.33}$$

Assume the inner product

$$\langle \mathbf{f}, \mathbf{g} \rangle \triangleq \int_0^p \mathbf{f}(t)\,\overline{\mathbf{g}(t)}\,dt \tag{5.34}$$

(We need the complex conjugation indicated in (5.34) because we are considering the complex-valued functions $\mathbf{g}_k$.) Then

$$\langle \mathbf{g}_k, \mathbf{g}_n \rangle = \frac{1}{p}\int_0^p \exp\left(i\frac{2\pi kt}{p}\right)\exp\left(-i\frac{2\pi nt}{p}\right)dt$$

$$= \left.\frac{\exp\left[i\dfrac{2\pi(k-n)t}{p}\right]}{i2\pi(k-n)}\right|_0^p = 0, \qquad k \neq n$$

$$\langle \mathbf{g}_k, \mathbf{g}_k \rangle = \frac{1}{p}\int_0^p \exp\left(i\frac{2\pi kt}{p}\right)\exp\left(-i\frac{2\pi kt}{p}\right)dt = 1,$$

The set (5.33) is orthonormal, and we can express (5.29) as the exponential Fourier series:

$$\mathbf{f} = \sum_{k=-\infty}^{\infty} \langle \mathbf{f}, \mathbf{g}_k \rangle \mathbf{g}_k$$

or

$$\mathbf{f}(t)=\sum_{k=-\infty}^{\infty}\frac{c_k}{\sqrt{p}}\exp\left(i\frac{2\pi kt}{p}\right),\qquad c_k=\frac{1}{\sqrt{p}}\int_0^p\mathbf{f}(s)\exp\left(-i\frac{2\pi ks}{p}\right)ds \tag{5.35}$$

The exponential series (5.35) is often used in place of (5.29). In some respects it is a more convenient series for use in analyzing constant-coefficient differential equations, because derivatives of exponentials are still exponentials.

We have discussed the applicability of an infinite eigenfunction expansion [the classical Fourier series in either of its forms, (5.29) or (5.35)] for steady-state analysis of dynamic systems. Surprisingly, the approach we have used for steady-state analysis can be applied to a dynamic system even if the system is not operated in a periodic fashion; we merely treat the system as if it were periodic with a single infinite period. We still seek a function $\mathbf{f}_1$ which satisfies the differential equation with periodic boundary conditions, then determine a "transient" solution $\mathbf{f}_2$ to the homogeneous differential system such that $\mathbf{f}_1+\mathbf{f}_2$ satisfies the initial conditions. Thus it still makes sense to work with exponentials, the eigenfunctions of linear constant-coefficient differential operators (ignoring the initial conditions). We could derive the expansion (in exponentials) of a nonperiodic function by changing variables and letting the period become large. However, we merely state the well-known result, known as the **Fourier integral theorem***:

$$\mathbf{f}(t)=\int_{-\infty}^{\infty}\mathbf{F}(s)e^{i2\pi st}\,ds \tag{5.36}$$

where

$$\mathbf{F}(s)=\int_{-\infty}^{\infty}\mathbf{f}(t)e^{-i2\pi st}\,dt$$

The expansion (5.36) applies for any "well-behaved" function **f** for which $\int_{-\infty}^{\infty}|\mathbf{f}(t)|\,dt<\infty$. The coefficient function **F** is known as the *Fourier integral* of **f**. The role of the discrete frequency variable k is taken over by the continuous real frequency variable s. The sum in (5.35) becomes an

*Churchill [5.5, pp. 88–90].

integral in (5.36). Let $\mathbf{q}(s,t) \stackrel{\Delta}{=} \exp(i2\pi st)$. Then defining the inner product by

$$\langle \mathbf{f},\mathbf{g}\rangle \stackrel{\Delta}{=} \int_{-\infty}^{\infty} \mathbf{f}(t)\,\overline{\mathbf{g}(t)}\,dt \tag{5.37}$$

we can express (5.36) as

$$\mathbf{f}(t)=\int_{-\infty}^{\infty} \langle \mathbf{f},\mathbf{q}(s,\cdot)\rangle \mathbf{q}(s,t)\,ds \tag{5.38}$$

It can be shown, by a limiting argument, that the infinite set $\{\mathbf{q}(s,\cdot), -\infty < s < \infty\}$ is an orthogonal set; however, $\|\mathbf{q}(s,\cdot)\|$ is not finite. **Parseval's theorem**, a handy tool in connection with Fourier integrals, states that

$$\int_{-\infty}^{\infty} \mathbf{f}(t)\,\overline{\mathbf{g}(t)}\,dt = \int_{-\infty}^{\infty} \mathbf{F}(s)\,\overline{\mathbf{G}(s)}\,ds \tag{5.39}$$

where $\mathbf{F}$ and $\mathbf{G}$ are the Fourier integrals of $\mathbf{f}$ and $\mathbf{g}$, respectively. This equation is a direct extension of (5.27). In effect, the "frequency domain" functions $\mathbf{F}$ and $\mathbf{G}$ constitute the coordinates of the "time domain" functions $\mathbf{f}$ and $\mathbf{g}$, respectively. Equations analogous to (5.39) can be written for the expansions (5.29) and (5.35).

It is interesting that restricting our concern to periodic functions (or, in effect, to the values of functions on the finite time interval $[0,p]$) reduces (5.36) to (5.35) and allows us to expand these functions in terms of a countable basis (a basis whose members can be numbered using only integer subscripts). Because of the duality exhibited in (5.36) and (5.39) between the time variable t and the frequency variable s, it should come as no surprise that restricting our interest to functions with finite "bandwidth" (functions whose transforms are nonzero only over a finite frequency interval) again allows us to expand the functions in terms of a countable basis. Limited bandwidth functions are fundamental to the analysis of periodic sampling. If $\mathbf{F}(s)=0$ for $|s|>w$, we say that $\mathbf{f}$ is band limited to w; or $\mathbf{f}$ has no frequency components as high as w. For such a function it is well known that the set of samples (values) of $\mathbf{f}$ at the points $t=k/2w$, $k=0,\pm 1,\pm 2,\ldots$ contains all the information possessed by $\mathbf{f}$. To be more specific, the **sampling theorem** states

$$\mathbf{f}(t)=\sum_{k=-\infty}^{\infty} \mathbf{f}\left(\frac{k}{2w}\right)\frac{\sin 2\pi w(t-k/2w)}{2\pi w(t-k/2w)} \tag{5.40}$$

for any function $\mathbf{f}$ which is band limited to w [5.18].

We define the functions $\{\mathbf{h}_k\}$ by

$$\mathbf{h}_k(t) \triangleq \sqrt{2w}\,\frac{\sin 2\pi w(t-k/2w)}{2\pi w(t-k/2w)}, \qquad k=0,\pm 1,\pm 2,\ldots$$

The function $\mathbf{h}_0$ is plotted in Figure 5.5; $\mathbf{h}_k$ is just $\mathbf{h}_0$ shifted by $t=k/2w$.

Exercise 4. Use (5.36), (5.39), and the inner product (5.37) to show that (*a*) $\{\mathbf{h}_k\}$ is an orthonormal set, and (*b*) $\langle \mathbf{f},\mathbf{h}_k\rangle = \mathbf{f}(k/2w)$. Hint: the Fourier integral of $\mathbf{h}_k$ is

$$\mathbf{H}_k(s) = \begin{cases} \dfrac{1}{\sqrt{2w}}\exp\left(-\dfrac{i\pi k s}{w}\right) & |s|<w \\ 0 & |s|\geqslant w \end{cases}$$

As a result of Exercise 4, we can express the sampling theorem as a generalized Fourier series:

$$\mathbf{f} = \sum_{k=-\infty}^{\infty} \langle \mathbf{f},\mathbf{h}_k\rangle \mathbf{h}_k \tag{5.41}$$

The coefficients of any orthonormal expansion can be computed independently. Thus the fact that the functions $\mathbf{h}_k$ are orthonormal is significant. Each coefficient in (5.41) can be obtained by physically sampling a single point of the function $\mathbf{f}$. It is common practice to sample functions in order to process them digitally. The samples of a function are the coordinates of that function relative to the orthonormal basis $\{\mathbf{h}_k\}$. The processes commonly used for physical reconstruction of functions from their samples are all, in some sense, approximations to the sum (5.41).

Extending Parseval's equation (5.27) to the set $\{\mathbf{h}_k\}$, we find that inner products of two band-limited functions can be computed in terms of the

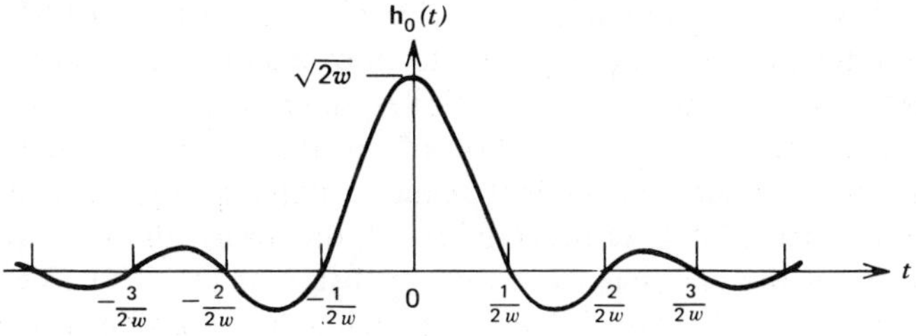

Figure 5.5. The function $\mathbf{h}_0$.

samples of the functions:

$$\int_{-\infty}^{\infty} \mathbf{f}(t)\,\overline{\mathbf{g}(t)}\,dt = \sum_{k=-\infty}^{\infty} \mathbf{f}(k/2w)\,\overline{\mathbf{g}(k/2w)} \tag{5.42}$$

If a function is both periodic and of finite bandwidth, then its Fourier series expansion, (5.29) or (5.35), contains only a finite number of terms; periodicity guarantees that discrete frequencies are sufficient to represent the function, whereas limiting the bandwidth to less than w guarantees that no (discrete) frequencies higher than w are required. Then, although (5.35) and (5.40) express the same function in different coordinates, the first set of coordinates is more efficient in the sense that it converges exactly in a finite number of terms. All the function samples are required in order to reconstruct the full function using (5.40). Yet (5.40) is dominated by its first few terms; only a "few" samples are required to accurately reconstruct the function over its first period. The remaining samples contain little additional information.

Exercise 5. Let $\mathbf{f}(t) = \sin 2\pi t$, a function which is periodic and band limited. ($\mathbf{F}(s) = 0$ for $|s| > 1$). Sample $\mathbf{f}$ at $t = 0, \pm\frac{1}{4}, \pm\frac{1}{2}, \pm\frac{3}{4}, \ldots$ (i.e., let $w = 2$). Then, by (5.40),

$$\mathbf{f}(t) = \sum_{k=-\infty}^{\infty} \mathbf{f}\left(\frac{k}{4}\right)\mathbf{h}_k(t) = \sum_{k=-\infty}^{\infty} \sin\left(\frac{k\pi}{2}\right)\frac{\sin 4\pi(t-k/4)}{4\pi(t-k/4)}$$

The samples are zero for $k = 0, \pm 2, \pm 4, \ldots$. Graphically combine the terms for $k = \pm 1, \pm 3, \pm 5$, and compare the sum with $\mathbf{f}$ over the interval $[0, 1]$.

5.3 Infinite-Dimensional Spaces

We developed the generalized Fourier series expansion (5.17) only for finite-dimensional spaces; yet we immediately recognized its extension to certain well-known infinite-dimensional examples, particularly (5.29). Our goal, ultimately, is to determine how to find orthonormal bases of eigenfunctions for linear operators on infinite-dimensional spaces. A basis of eigenfunctions permits decomposition of an infinite-dimensional operator equation into a set of independent scalar equations, just as in the finite-dimensional case (5.23). Orthogonality of the basis allows independent computation of the coefficients in the expansion as in (5.24). We will find this computational independence particularly valuable for infinite-dimensional problems because the "first few" terms in an infinite orthonormal expansion dominate that expansion; we can ignore the remaining terms.

To this point, wherever we have introduced infinite expansions of vectors, we have used well-known examples and avoided discussion of the meaning of an infinite sum. Thus we *interpret* the Taylor series expansion

$$\mathbf{f}(t)=\mathbf{f}(0)+\mathbf{f}'(0)t+\frac{\mathbf{f}''(0)t^2}{2!}+\cdots \tag{5.43}$$

of an infinitely differentiable function $\mathbf{f}$ as the expansion of $\mathbf{f}$ in terms of the "basis" $\{1,t,t^2,\ldots\}$. We consider the Fourier series expansion (5.29) as the expansion of a periodic function on the "orthonormal basis"

$$\left\{\sqrt{\frac{1}{p}}\,,\ \sqrt{\frac{2}{p}}\ \cos\frac{2\pi kt}{p},\ \sqrt{\frac{2}{p}}\ \sin\frac{2\pi kt}{p},\qquad k=1,2,\ldots\right\}$$

Yet the definition of linear combination does not pinpoint the meaning of $\Sigma_{k=1}^{\infty}c_k\mathbf{f}_k$ for an infinite set of functions $\{\mathbf{f}_k\}$. It seems natural and desirable to assume that such an infinite sum implies pointwise convergence of the partial sums. Certainly, the Taylor series (5.43) means that for each t,

$$\mathbf{f}_k(t) \triangleq \mathbf{f}(0)+\mathbf{f}'(0)t+\cdots+\frac{\mathbf{f}^{(k)}(0)t^k}{k!}\to\mathbf{f}(t)$$

as $k\to\infty$. However, it is well-known that the sequence of partial sums in the Fourier series expansion (5.29) of a discontinuous function is not pointwise convergent; the partial sums converge to the midpoints of any discontinuities (P&C 5.18). In an engineering sense, we do not care to which value the series converges at a discontinuity. The actual value of the function is usually defined arbitrarily at that point anyway. We define convergence of the partial sums in a way which ignores the value of the Fourier series at the discontinuities.

Convergence in Norm

Define $\mathbf{y}_n \triangleq \Sigma_{k=1}^{n}c_k\mathbf{x}_k$, the nth partial sum of the series $\Sigma_{k=1}^{\infty}c_k\mathbf{x}_k$. We can assign meaning to the infinite sum only if the partial sums $\mathbf{y}_n$ and $\mathbf{y}_m$ become more nearly alike in some sense as $n,m\to\infty$. The natural definition of "likeness" in an inner product space is likeness in norm. That is, $\mathbf{y}_n$ and $\mathbf{y}_m$ are alike if the norm $\|\mathbf{y}_n-\mathbf{y}_m\|$ of their difference is small. An infinite sequence $\{\mathbf{y}_n\}$ from an inner product space $\mathcal{V}$ is called a **Cauchy**

sequence if $\|\mathbf{y}_n - \mathbf{y}_m\| \to 0$ as $n,m \to \infty$; or, rigorously, if for each $\epsilon > 0$ there is an N such that $n,m > N$ implies $\|\mathbf{y}_n - \mathbf{y}_m\| < \epsilon$. Intuitively, a Cauchy sequence is a "convergent" sequence. By means of a Cauchy sequence we can discuss the fact of convergence without explicit reference to the limit vector. We say an infinite sequence $\{\mathbf{y}_n\}$ from an inner product space $\mathcal{V}$ **converges in norm** to the limit $\mathbf{x}$ if $\|\mathbf{x} - \mathbf{y}_n\| \to 0$ as $n \to \infty$.

Exercise 1. Use the triangle inequality (P&C 5.4) to show that a sequence from an inner product space $\mathcal{V}$ can converge in norm to a vector $\mathbf{x}$ in $\mathcal{V}$ only if it is a Cauchy sequence.

Assume the partial sums of a series, $\mathbf{y}_n \triangleq \sum_{k=1}^{n} c_k \mathbf{x}_k$, form a Cauchy sequence; by the infinite sum $\sum_{k=1}^{\infty} c_k \mathbf{x}_k$, we mean the vector $\mathbf{x}$ to which the partial sums converge in norm. We call $\mathbf{x}$ the **limit in norm** of the sequence $\{\mathbf{y}_n\}$. (Note that the limit of a Cauchy sequence need not be in $\mathcal{V}$. The mathematics literature usually does not consider a sequence *convergent* unless the limit *is* in $\mathcal{V}$.)

Let $\mathcal{V}$ be some space of functions defined on $[0,1]$ with the standard function space inner product. One of the properties of inner products guarantees that $\mathbf{f} = \boldsymbol{\theta}$ if $\|\mathbf{f}\| = 0$. We have assumed previously that $\mathbf{f} = \boldsymbol{\theta}$ meant $\mathbf{f}(t) = 0$ for all t in $[0,1]$. Suppose, however, that $\mathbf{f}$ is the discontinuous function shown in Figure 5.6. Observe that $\|\mathbf{f}\| = 0$, whereas $\mathbf{f}(t) \neq 0$ at $t = 0$, $\frac{1}{2}$, or 1. Changing the value of a function at a few points does not change its integral (or its norm). We are hard pressed to define any inner product for a space containing functions like the one in Figure 5.6 unless we ignore "slight" differences between functions.

We say $\mathbf{f} = \mathbf{g}$ **almost everywhere** if $\mathbf{f}(t) = \mathbf{g}(t)$ except at a finite number of points.* For most practical purposes we can consider convergence in norm to be pointwise convergence almost everywhere. (However, Bachman and

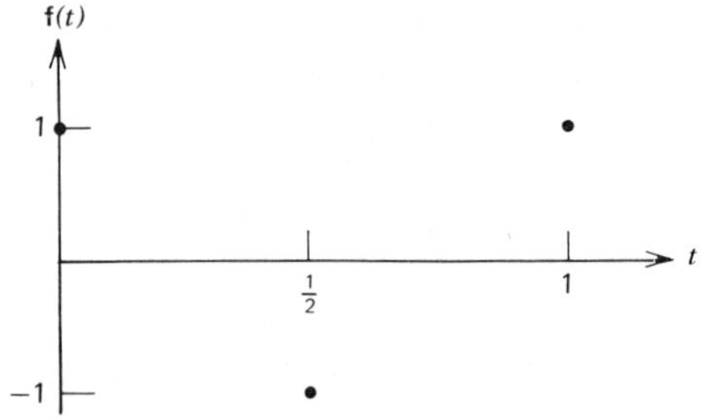

Figure 5.6. A nonzero function with zero norm.

*The definition of "almost everywhere" can be extended to except a countably infinite number of points.

Narici [5.2, p. 173] demonstrate that a sequence of functions can be convergent in norm, yet not converge at all in a pointwise sense.) Convergence in norm is sometimes called **convergence in the mean**. Convergence in norm is precisely the type of convergence which we need for discussion of Fourier series like (5.29). If $\mathbf{f}$ is a periodic function with period p, the Fourier series expansion (5.29) means

$$\int_0^p \left[\mathbf{f}(t) - a_0 - \sum_{k=1}^{n} \left(a_k \cos\frac{2\pi kt}{p} + b_k \sin\frac{2\pi kt}{p} \right) \right]^2 dt \to 0 \quad \text{as } n \to \infty \qquad (5.44)$$

That is, the sequence of partial sums converges in norm to the periodic function $\mathbf{f}$. The convergence is pointwise almost everywhere—pointwise except at discontinuities. It makes little practical difference how a function is defined at a finite number of points. Therefore we usually do not distinguish between functions which are equal almost everywhere. Of course, our focus on the convergence in norm of a series of functions does not preclude the possibility that the convergence is actually pointwise and, in fact, uniform.

Infinite-Dimensional Bases

We need to extend the n-dimensional concept of a basis to infinite-dimensional spaces. We naturally think in terms of extending a finite sum to an infinite sum. An infinite set is said to be **countable** if its elements can be numbered using only integer subscripts. We restrict ourselves to a discussion of inner product spaces which have countable bases.*

Definition. Let $\mathcal{V}$ be an infinite-dimensional inner product space. Let $\mathcal{X} \triangleq \{\mathbf{x}_1, \mathbf{x}_2, \ldots\}$ be a countable set in $\mathcal{V}$. Then $\mathcal{X}$ is said to be a **basis** for $\mathcal{V}$ if every vector $\mathbf{x}$ in $\mathcal{V}$ can be expressed uniquely as a convergent infinite series $\mathbf{x} = \sum_{k=1}^{\infty} c_k \mathbf{x}_k$; that is, if there is a unique set of coordinates $\{c_k\}$ such that $\|\mathbf{x} - \sum_{k=1}^{n} c_k \mathbf{x}_k\|$ can be made arbitrarily small by taking enough terms in the expansion.

Example 1. Bases for $\mathcal{P}(a,b)$. We denote by $\mathcal{P}(a,b)$ the infinite-dimensional space of all real polynomial functions defined on $[a,b]$. Since every polynomial is a (finite) linear combination of functions from the linearly independent set $\mathcal{F} \triangleq \{t^k, k=0,1,2,\ldots\}$, $\mathcal{F}$ is a basis for $\mathcal{P}(a,b)$. Observe that no norm is needed to define a basis for this particular infinite-dimensional space because no infinite sums are required. If we define an inner product on $\mathcal{P}(a,b)$, we can apply the

*A space which has a countable basis is said to be **separable**. Some spaces have only uncountable bases. See Bachman and Narici [5.2, p. 143] for an example.

Gram–Schmidt procedure to the set $\mathcal{F}$, and generate a basis for $\mathcal{P}(a,b)$ consisting of orthogonal polynomials. (See, for instance, the Legendre polynomials of Example 2, Section 5.2.) Each vector in $\mathcal{P}(a,b)$ is a finite linear combination of these orthogonal polynomials. Each different inner product leads to a different orthogonal basis. Of course, each such basis could also be normalized.

Any function that can be expanded in a Taylor series about the origin, as in (5.43), can be represented uniquely by the simple polynomial basis of Example 1. Many familiar functions (e^t, $\sin t$, rational functions, etc.) can be expanded in such a series. These functions are not in $\mathcal{P}(a,b)$, and true infinite sums are required. Thus $\mathcal{F}$ appears to serve as a basis for spaces larger than $\mathcal{P}(a,b)$. How do we tell whether or not $\mathcal{F}$ is a basis for any particular space $\mathcal{V}$ of functions? Of course, the coordinates of the function cannot be unique without independence of the basis vectors. Our previous concept of linear independence, which is based on addition and scalar multiplication, applies only to finite-dimensional spaces. We say an infinite set of vectors $\mathcal{X}$ is **linearly independent** if each finite subset of $\mathcal{X}$ is linearly independent. The vectors in a basis $\mathcal{X}$ must also span $\mathcal{V}$ in the sense that every $\mathbf{x}$ in $\mathcal{V}$ must be representable. But, merely making $\mathcal{X}$ a sufficiently large linearly independent set is not sufficient to guarantee that $\mathcal{X}$ is a basis. The set $\mathcal{F}$ of Example 1 is an infinite linearly independent set. Yet $\mathcal{F}$ is not a basis even for the "nice" space $\mathcal{C}^{\infty}(-1,1)$ of infinitely differentiable functions. For example, if we define the function $\mathbf{f}(t) \stackrel{\Delta}{=} \exp(-1/t^2)$ to have the value $\mathbf{f}(0)=0$ at the origin, it is infinitely differentiable; but it has the Taylor coefficients $\mathbf{f}(0)=\mathbf{f}'(0)=\mathbf{f}''(0)/2=\cdots=0$. Thus an attempted Taylor series expansion of $\mathbf{f}$ converges to the wrong (zero) function.

According to a famous theorem of Weierstrass [5.4], any function in $\mathcal{C}(a,b)$ can be represented arbitrarily closely in a pointwise sense (and in norm) by a polynomial. Yet this fact does not imply that $\mathcal{F}$ is a basis for $\mathcal{C}(a,b)$. We must still determine whether or not every $\mathbf{f}$ in $\mathcal{C}(a,b)$ is representable by a unique convergent expansion of the form $\sum_{k=0}^{\infty} c_k t^k$. In general, even though $\{\mathbf{x}_k\}$ is an infinite linearly independent set, there may be no approximation $\sum_{k=1}^{n} c_k \mathbf{x}_k$ that will approach a given vector $\mathbf{x}$ in norm unless the coefficients $\{c_k\}$ are modified as n increases. (See Naylor and Sell [5.17], pp. 315–316.) It is difficult to tell if a specific set is a basis without displaying and examining the coordinates of a general vector in the space. We will find that orthogonality of the vectors in a set eases considerably the task of determining whether or not the set is a basis.

Orthogonal Bases for Infinite-Dimensional Spaces

Actual determination of the coordinates of a specific vector relative to an arbitrary basis is not generally feasible in an infinite-dimensional space. It requires solving for the numbers c_k in the vector equation $\mathbf{x}=\sum_{k=1}^{\infty}c_k\mathbf{x}_k$; in effect, we must solve an infinite set of simultaneous equations. However, if the basis $\mathfrak{X}$ is orthogonal (or orthonormal), the coordinates c_k are the Fourier coefficients, which can be computed independently. This fact is one reason why we work almost exclusively with orthogonal (or orthonormal) bases in infinite-dimensional spaces. If $\mathfrak{X} \triangleq \{\mathbf{x}_k\}$ is a countable orthogonal basis for an inner product space $\mathcal{V}$, the Fourier series expansion of a vector $\mathbf{x}$ in $\mathcal{V}$ can be developed by an extension of the process used to obtain the finite-dimensional expansion (5.16)–(5.17). Let $\mathbf{x}_j$ be one of the first n vectors in the infinite dimensional basis $\mathfrak{X}$. Let c_i be the ith coordinate of $\mathbf{x}$ relative to $\mathfrak{X}$. The Cauchy–Schwartz inequality* shows that

$$\left|\left\langle \mathbf{x}-\sum_{k=1}^{n} c_k\mathbf{x}_k, \mathbf{x}_j\right\rangle\right| \leqslant \left\|\mathbf{x}-\sum_{k=1}^{n} c_k\mathbf{x}_k\right\| \|\mathbf{x}_j\|$$

The right side of this expression approaches zero as $n\to\infty$. Therefore, for each $j \leqslant n$,

$$\left|\left\langle \mathbf{x}-\sum_{k=1}^{n} c_k\mathbf{x}_k, \mathbf{x}_j\right\rangle\right| = \left|\langle \mathbf{x},\mathbf{x}_j\rangle - \sum_{k=1}^{n} c_k\langle \mathbf{x}_k,\mathbf{x}_j\rangle\right|$$
$$= \left|\langle \mathbf{x},\mathbf{x}_j\rangle - c_j\|\mathbf{x}_j\|^2\right| \to 0$$

as $n\to\infty$. Since the quantity approaching zero is independent of n, it must equal zero, and

$$c_j = \frac{\langle \mathbf{x},\mathbf{x}_j\rangle}{\|\mathbf{x}_j\|^2} \tag{5.45}$$

Thus the **Fourier series expansion of x** is

$$\mathbf{x} = \sum_{k=1}^{\infty} \frac{\langle \mathbf{x},\mathbf{x}_k\rangle}{\langle \mathbf{x}_k,\mathbf{x}_k\rangle}\mathbf{x}_k \tag{5.46}$$

Of course, if the basis is orthonormal, the kth coefficient in (5.46) is just $c_k = \langle \mathbf{x},\mathbf{x}_k\rangle$.

By an argument similar to the one above, we show that the coefficients $\{c_k\}$ in an orthogonal expansion are unique. Suppose $\mathbf{x}=\sum_{k=1}^{\infty}d_k\mathbf{x}_k$ is a

*P&C 5.4.

second expansion of $\mathbf{x}$. Then by the triangle inequality,*

$$\left\| \left(\mathbf{x} - \sum_{k=1}^{n} c_k \mathbf{x}_k\right) - \left(\mathbf{x} - \sum_{k=1}^{n} d_k \mathbf{x}_k\right) \right\| = \left\| \sum_{k=1}^{n} (d_k - c_k)\mathbf{x}_k \right\|$$

$$\leqslant \left\| \mathbf{x} - \sum_{k=1}^{n} c_k \mathbf{x}_k \right\| + \left\| \mathbf{x} - \sum_{k=1}^{n} d_k \mathbf{x}_k \right\| \to 0$$

as $n \to \infty$. Then if $\mathbf{x}_j$ is one of the vectors $\mathbf{x}_1, \ldots, \mathbf{x}_n$, we again employ the Cauchy–Schwartz inequality to find that as $n \to \infty$

$$\left| \left\langle \sum_{k=1}^{n} (d_k - c_k)\mathbf{x}_k, \mathbf{x}_j \right\rangle \right| = \left| \sum_{k=1}^{n} (d_k - c_k)\langle \mathbf{x}_k, \mathbf{x}_j \rangle \right| = |(d_j - c_j)| \|\mathbf{x}_j\|^2$$

$$\leqslant \left\| \sum_{k=1}^{n} (d_k - c_k)\mathbf{x}_k \right\| \|\mathbf{x}_j\| \to 0$$

It follows that $d_j = c_j$, and the coordinates of $\mathbf{x}$ with respect to an orthogonal basis are unique.

Thus the only question of concern, if $\mathcal{X}$ is an orthogonal set, is whether or not $\mathcal{X}$ is a large enough set to allow expansion of all vectors $\mathbf{x}$ in $\mathcal{V}$. If there is a vector $\mathbf{x}$ in $\mathcal{V}$ for which there is not a convergent expansion, then

$$\mathbf{z} \triangleq \mathbf{x} - \sum_{k=1}^{\infty} \frac{\langle \mathbf{x}, \mathbf{x}_k \rangle}{\langle \mathbf{x}_k, \mathbf{x}_k \rangle} \mathbf{x}_k$$

is nonzero. Furthermore, $\mathbf{z}$ is orthogonal to each vector $\mathbf{x}_j$ in $\mathcal{X}$, and could be added to $\mathcal{X}$ to make it more nearly complete (more nearly a basis).

Definition. We say **an orthogonal set is complete in the inner product space** $\mathcal{V}$ if there is no nonzero vector in $\mathcal{V}$ which is orthogonal to every vector in $\mathcal{X}$.

It follows from the discussion above that an *orthogonal set* $\mathcal{X}$ *is a basis for* $\mathcal{V}$ *if and only if it is complete in* $\mathcal{V}$. Any orthogonal set in a separable space $\mathcal{V}$ can be extended (by adding vectors) until it is complete in $\mathcal{V}$. A practical technique for testing an orthogonal set $\{\mathbf{x}_k\}$ to see if it is a basis consists in showing that the only vector orthogonal to each vector $\mathbf{x}_k$ is the zero vector $\boldsymbol{\theta}$. If $\mathcal{X}$ is an orthogonal basis for $\mathcal{V}$, then only for $\mathbf{x} = \boldsymbol{\theta}$ is it true that all the Fourier coefficients $\langle \mathbf{x}, \mathbf{x}_k \rangle$ are equal to zero. Thus this test

*P&C 5.4.

for completeness of the orthogonal set $\mathcal{X}$ is equivalent to a test for validity of the Fourier expansion (5.46) for each $\mathbf{x}$ in $\mathcal{V}$.

Example 2. Orthogonal bases for $\mathcal{C}(a,b)$. The Weierstrass approximation theorem [5.4] guarantees that any continuous function can be approximated arbitrarily closely in norm by a polynomial. We noted earlier that this fact is insufficient to guarantee that the set $\mathcal{F} \triangleq \{t^k,\ \ k=0,1,2,\ldots\}$ is a basis for $\mathcal{C}(a,b)$. On the other hand, suppose that $\mathcal{G} \triangleq \{\mathbf{p}_k\}$ is a basis for $\mathcal{P}(a,b)$ consisting in real polynomials $\mathbf{p}_k$ which are orthogonal relative to some inner product. (We could obtain $\mathcal{G}$ from $\mathcal{F}$ by the Gram–Schmidt procedure as in Example 2 of Section 5.2.) We now show that $\mathcal{G}$ is also a basis—an orthogonal basis—for $\mathcal{C}(a,b)$. Let $\mathbf{f}$ be a real continuous function on $[a,b]$. Assume $\langle \mathbf{f},\mathbf{p}_k\rangle=0$ for all polynomials $\mathbf{p}_k$ in $\mathcal{G}$. We show that $\mathbf{f}$ must be the zero vector. By the Weierstrass theorem, for each $\epsilon>0$ there is a polynomial $\mathbf{p}_\epsilon$ such that $\|\mathbf{f}-\mathbf{p}_\epsilon\|^2<\epsilon$. Furthermore, since $\mathcal{G}$ is a basis for $\mathcal{P}(a,b)$, $\mathbf{p}_\epsilon=\sum_{k=1}^{N} c_k\mathbf{p}_k$ for some finite number N. Then

$$\begin{aligned}\|\mathbf{f}-\mathbf{p}_\epsilon\|^2 &= \|\mathbf{f}\|^2-2|\langle \mathbf{f},\mathbf{p}_\epsilon\rangle|+\|\mathbf{p}_\epsilon\|^2\\ &= \|\mathbf{f}\|^2+\|\mathbf{p}_\epsilon\|^2-2\left|\sum_{k=1}^{N} c_k\langle \mathbf{f},\mathbf{p}_k\rangle\right|\\ &= \|\mathbf{f}\|^2+\|\mathbf{p}_\epsilon\|^2\end{aligned}$$

Since $\|\mathbf{f}\|^2+\|\mathbf{p}_\epsilon\|^2<\epsilon$ for an arbitrarily small number ϵ, $\|\mathbf{f}\|=0$, and the function $\mathbf{f}$ must be the zero vector. Thus the orthogonal set $\mathcal{G}$ is complete in $\mathcal{C}(a,b)$, and all orthogonal polynomial bases for $\mathcal{P}(a,b)$ are bases for $\mathcal{C}(a,b)$ as well.

Harmuth [5.13] describes an interesting orthogonal basis for $\mathcal{C}(a,b)$—the set of Walsh functions. These functions, which take on only the values 1 and -1, are extremely useful in digital signal processing; only additions and subtractions are needed to compute the Fourier coefficients.

The classical Fourier series expansion (5.29) for periodic functions applies to functions $\mathbf{f}$ in the standard inner product space $\mathcal{C}(a,b)$; we merely repeat the values of $\mathbf{f}$ on $[a,b]$ periodically outside of $[a,b]$ with period $p=b-a$. If we denote the set of sinusoidal functions (5.30) by $\mathcal{K}$, then the orthonormal set $\mathcal{K}$ is complete in $\mathcal{C}(a,b)$; it is an orthonormal basis for $\mathcal{C}(a,b)$.

Exercise 2. The Fourier series expansion (5.29) of a periodic function $\mathbf{f}$ contains only sine terms if $\mathbf{f}$ is an odd function and only cosine terms if $\mathbf{f}$ is an even function. Show that in addition to the sine–cosine expansion mentioned in Example 2, a function in $\mathcal{C}(0,b)$ can be expanded in two additional series of period $p=2b$, one involving only sines (the Fourier sine series), the other involving only cosines (the Fourier cosine series).

If $\{\mathbf{x}_k\}$ is an orthonormal basis for $\mathcal{V}$, the set of Fourier coefficients (or coordinates) $\{\langle \mathbf{x},\mathbf{x}_k\rangle\}$ is equivalent to the vector $\mathbf{x}$ itself, and operations on $\mathbf{x}$ can be carried out in terms of operations on the Fourier coefficients. For instance, we can compute inner products by means of **Parseval's equation**:

$$\langle \mathbf{x},\mathbf{y}\rangle = \left\langle \sum_{k=1}^{\infty} \langle \mathbf{x},\mathbf{x}_k\rangle \mathbf{x}_k, \sum_{j=1}^{\infty} \langle \mathbf{y},\mathbf{x}_j\rangle \mathbf{x}_j \right\rangle$$

$$= \sum_{k=1}^{\infty} \langle \mathbf{x},\mathbf{x}_k\rangle \langle \mathbf{y},\mathbf{x}_k\rangle \tag{5.47}$$

(Because we are concerned primarily with real spaces, we usually drop the complex conjugate.) If $\mathbf{y}=\mathbf{x}$, (5.47) becomes **Parseval's identity**:

$$\|\mathbf{x}\|^2 = \sum_{k=1}^{\infty} |\langle \mathbf{x},\mathbf{x}_k\rangle|^2 \tag{5.48}$$

Equation (5.48) is also a special case of the Pythagorean theorem. Furthermore, it is the limiting case (equality) of Bessel's inequality (P&C 5.4). In point of fact, Bessel's inequality becomes the identity (5.48) for each $\mathbf{x}$ in $\mathcal{V}$ if and only if the orthonormal set $\{\mathbf{x}_k\}$ is a basis for $\mathcal{V}$.

Of course, not all bases for infinite-dimensional spaces are orthogonal bases. Naylor and Sell [5.17, p. 317] describe one set of conditions which guarantees that a nonorthogonal countable set is a basis. However, rarely do we encounter in practical analysis the use of a nonorthogonal basis for an infinite-dimensional space.

In a finite-dimensional space we can pick an inner product to orthonormalize any basis; specifically, we pick the inner product defined by Parseval's equation (5.27). The infinite-dimensional equivalent (5.47) is less useful for this purpose because the unknown inner product is needed to find the coordinates in the equation. In an infinite-dimensional space, the choice of inner product still determines the orthonormality of a set of vectors; but the norm associated with the inner product also determines whether the vectors of an orthonormal set are complete in the space. Given a basis for an inner product space, what changes can we make in the inner product (in order to orthonormalize the basis) and still have a basis? For spaces of functions defined on a finite interval, a positive reweighting of the inner product does not destroy convergence. For example, if $\{\mathbf{f}_k\}$ is a basis for $\mathcal{C}(a,b)$ with the standard function space inner product, then for any $\mathbf{f}$ in $\mathcal{C}(a,b)$ (with unique coordinates $\{c_k\}$ relative to $\{\mathbf{f}_k\}$) and any $\epsilon>0$ there is a number N such that $\int_a^b|\mathbf{f}(t)-\sum_{k=1}^n c_k\mathbf{f}_k(t)|^2\,dt<\epsilon$ for $n>N$. Suppose we define a new inner product for the same space of continuous functions:

$$\langle \mathbf{f},\mathbf{g}\rangle_\omega \triangleq \int_a^b \omega(t)\mathbf{f}(t)\overline{\mathbf{g}(t)}\,dt \tag{5.49}$$

where $\omega(t)$ is bounded and positive for t in $[a,b]$. Then, using the same basis $\{\mathbf{f}_k\}$ and the same coefficients $\{c_k\}$,

$$\int_a^b \omega(t)\left|\mathbf{f}(t)-\sum_{k=1}^{n} c_k\mathbf{f}_k(t)\right|^2 dt \leqslant M\int_a^b \left|\mathbf{f}(t)-\sum_{k=1}^{n} c_k\mathbf{f}_k(t)\right|^2 dt < M\epsilon$$

where M is a positive bound on $\omega(t)$. Since ϵ is arbitrarily small, $M\epsilon$ is also arbitrarily small. Thus for large enough n the partial sum is still arbitrarily close to $\mathbf{f}$ in the new norm. We represent by $\mathcal{C}(\omega;\ a,b)$ the space of continuous functions with the inner product (5.49). It is evident that the choice of ω affects the definition of orthogonality, but does not affect the convergence or nonconvergence of sequences of vectors. Of course, the weighted inner product (5.49) does not represent all possible inner products on the function space $\mathcal{C}(a,b)$; it does not allow for "cross products" analogous to those in (5.13). Yet it is general enough to allow us to orthogonalize many useful bases.

Example 3. Orthogonalizing a Basis by Weighting the Inner Product. The shaft position ϕ of an armature-controlled motor as a function of armature voltage $\mathbf{u}$ is described by

$$(\mathbf{L}\phi)(t) \triangleq \frac{d^2\phi(t)}{dt^2}+\frac{d\phi(t)}{dt}=\mathbf{u}(t)$$

The eigenfunctions of $\mathbf{L}$ with the boundary conditions $\phi(0)=\phi(b)=0$ are given by (4.38):

$$\mathbf{f}_k(t)=e^{-t/2}\sin\left(\frac{\pi kt}{b}\right), \qquad k=1,2,\ldots$$

We pick the weight ω in the inner product (5.49) so that the set $\{\mathbf{f}_k\}$ is orthogonal:

$$\begin{aligned}\langle \mathbf{f}_k,\mathbf{f}_m\rangle_\omega &= \int_0^b \omega(t)e^{-t/2}\sin\left(\frac{\pi kt}{b}\right)e^{-t/2}\sin\left(\frac{\pi mt}{b}\right)dt\\ &= \int_0^b \omega(t)e^{-t}\sin\left(\frac{\pi kt}{b}\right)\sin\left(\frac{\pi mt}{b}\right)dt\\ &=0\end{aligned}$$

for $m\neq k$. The functions $\{\sin(\pi kt/b)\}$ form a well-known orthogonal basis for $\mathcal{C}(0,b)$ using the standard function space inner product, as we noted in Example 2. Therefore, the weight $\omega(t)=e^t$ makes the functions $\{\mathbf{f}_k\}$ orthogonal with respect to the weighted inner product. (The choice $\omega(t)=2e^t/b$ would make the set orthonormal. However, it is more convenient to normalize the eigenfunctions, multiplying each by $\sqrt{2/b}$).

We now demonstrate that the eigenfunctions $\{\mathbf{f}_k\}$ are a basis [complete in $\mathcal{C}(0,b)$] by showing that the only function orthogonal to all functions in the set is the zero function. Suppose $\langle \mathbf{f},\mathbf{f}_k\rangle_\omega = 0$ for all k. Then

$$\langle \mathbf{f},\mathbf{f}_k\rangle_\omega = \int_0^b e^t \mathbf{f}(t) e^{-t/2} \sin\left(\frac{\pi k t}{b}\right) dt = \int_0^b \mathbf{f}(t) e^{t/2} \sin\left(\frac{\pi k t}{b}\right) dt = 0$$

Since $\{\sin(\pi k t/b)\}$ is an orthogonal basis with respect to the standard function space inner product, and since the Fourier coefficients of $\mathbf{f}e^{t/2}$ relative to this basis are all zero, $\mathbf{f}e^{t/2} = \boldsymbol{\theta}$ and $\mathbf{f} = \boldsymbol{\theta}$. Therefore, $\{\mathbf{f}_k\}$ is an orthogonal basis for the space $\mathcal{C}(e^t;0,b)$. This orthogonal basis of eigenfunctions is used in Example 4, Section 5.5 to diagonalize and solve the differential equation described above.

Hilbert Spaces

From Example 2 it is evident that a single infinite set can be a basis for several different infinite-dimensional spaces. Suppose $\{\mathbf{x}_k\}$ is an orthonormal basis for an infinite-dimensional inner product space $\mathcal{V}$. Presumably there are vectors $\mathbf{x}$, not in $\mathcal{V}$, which can be expanded uniquely in terms of $\{\mathbf{x}_k\}$ (assuming we extend the inner product space operations to the additional vectors). What is the largest, most inclusive space for which $\{\mathbf{x}_k\}$ is a basis? We refer to the largest space $\mathcal{H}$ of vectors which can be represented in the form of $\mathbf{x} = \sum_{k=1}^{\infty} c_k \mathbf{x}_k$ as the space **spanned** (or **generated**) by the basis $\{\mathbf{x}_k\}$. (Because $\{\mathbf{x}_k\}$ is orthonormal, the coefficients in the expansion of $\mathbf{x}$ are necessarily unique.) We show that $\mathcal{H}$ is precisely the space of vectors $\mathbf{x}$ which are **square-summable** combinations of the basis vectors; that is, $\mathbf{x}$ such that $\mathbf{x} = \sum_{k=1}^{\infty} c_k \mathbf{x}_k$ with $\sum_{k=1}^{\infty} |c_k|^2 < \infty$.

Suppose a vector $\mathbf{x}$ in $\mathcal{H}$ can be expressed as $\mathbf{x} = \sum_{k=1}^{\infty} c_k \mathbf{x}_k$, where $\{\mathbf{x}_k\}$ is an orthonormal basis for the inner product space $\mathcal{V}$. Define $\mathbf{y}_n \triangleq \sum_{k=1}^{n} c_k \mathbf{x}_k$. Then $\{\mathbf{y}_n,\ n = 1,2,\ldots\}$ is a Cauchy sequence which approaches $\mathbf{x}$, and $\|\mathbf{y}_m - \mathbf{y}_n\| \to 0$ as $m,\ n \to \infty$. If we assume $n > m$ and use the orthonormality of $\{\mathbf{x}_k\}$, we find $\|\mathbf{y}_n - \mathbf{y}_m\|^2 = \|\sum_{k=1}^{n} c_k \mathbf{x}_k - \sum_{k=1}^{m} c_k \mathbf{x}_k\|^2 = \|\sum_{k=m+1}^{n} c_k \mathbf{x}_k\|^2 = \sum_{k=m+1}^{n} |c_k|^2$. Therefore, $\sum_{k=m+1}^{n} |c_k|^2 \to 0$ as $m,\ n \to \infty$. It follows that $\sum_{k=m+1}^{\infty} |c_k|^2 \to 0$ as $m \to \infty$; in other words, for each $\epsilon > 0$ there is a positive number M such that $m > M$ implies $\sum_{k=m+1}^{\infty} |c_k|^2 < \epsilon$. Pick a value of ϵ, and let m be a finite number greater than M. Then

$$\sum_{k=1}^{m} |c_k|^2 < \infty \quad \text{and} \quad \sum_{k=m+1}^{\infty} |c_k|^2 < \epsilon$$

Consequently, $\sum_{k=1}^{\infty} |c_k|^2 < \infty$, and $\mathbf{x}$ can be expanded on the basis $\{\mathbf{x}_k\}$ only if $\mathbf{x}$ is a square-summable combination of the basis vectors. Conversely, square summability of the coefficients $\{c_k\}$ implies that $\|\mathbf{y}_m - \mathbf{y}_n\|^2$

$\rightarrow 0$ as $m, n \rightarrow \infty$, and the sequence $\{\mathbf{y}_n\}$ is a Cauchy (convergent) sequence. Thus any square-summable combination of $\{\mathbf{x}_k\}$ must converge to some vector $\mathbf{x}$ in the space which we have denoted $\mathcal{H}$.

It is apparent that $\mathcal{H}$ may be more complete than $\mathcal{V}$. If we were to associate a single inner product space with the basis $\{\mathbf{x}_k\}$, the natural choice would be the largest space for which $\{\mathbf{x}_k\}$ is a basis, the space $\mathcal{H}$. If $\mathcal{V} \neq \mathcal{H}$, then $\mathcal{V}$ and $\mathcal{H}$ differ only in their "limit vectors." Suppose $\mathbf{x}$ satisfies $\mathbf{x} = \sum_{k=1}^{\infty} c_k \mathbf{x}_k$, and again denote the nth partial sum by $\mathbf{y}_n = \sum_{k=1}^{n} c_k \mathbf{x}_k$. The sequence of partial sums $\{\mathbf{y}_n\}$ is a Cauchy sequence with limit $\mathbf{x}$. Thus each $\mathbf{x}$ in $\mathcal{H}$ is the limit of a Cauchy sequence in $\mathcal{V}$. In point of fact, $\mathcal{H}$ differs from $\mathcal{V}$ only in that $\mathcal{H}$ contains the limits of more Cauchy sequences from $\mathcal{V}$ than does $\mathcal{V}$.

Example 4. A Cauchy Sequence in $\mathcal{C}(0,1)$ with no Limit in $\mathcal{C}(0,1)$. The functions $\{\mathbf{f}_k\}$ of Figure 5.7 form a Cauchy sequence in $\mathcal{C}(0,1)$ with the standard function space inner product (5.10); that is,

$$\int_0^1 (\mathbf{f}_m(t) - \mathbf{f}_n(t))^2 \, dt \rightarrow 0 \quad \text{as } n, m \rightarrow \infty$$

The limit in norm of the sequence $\{\mathbf{f}_k\}$ is the discontinuous function

$$\mathbf{f}(t) = \begin{cases} 1, & t \leq \frac{1}{2} \\ 0, & t > \frac{1}{2} \end{cases}$$

which is not in $\mathcal{C}(0,1)$. The limit vector $\mathbf{f}$ is a member of a space which is larger and more complete than $\mathcal{C}(0,1)$. Yet $\mathbf{f}$ can be expanded uniquely in the sine–cosine basis (5.30) for $\mathcal{C}(0,1)$.

Definition. Let $\mathcal{S}$ be set in an inner product space $\mathcal{V}$. A vector $\mathbf{x}$ in $\mathcal{V}$ is called a **point of closure** of $\mathcal{S}$ if for each $\epsilon > 0$ there is a vector $\mathbf{y}$ in $\mathcal{S}$ such

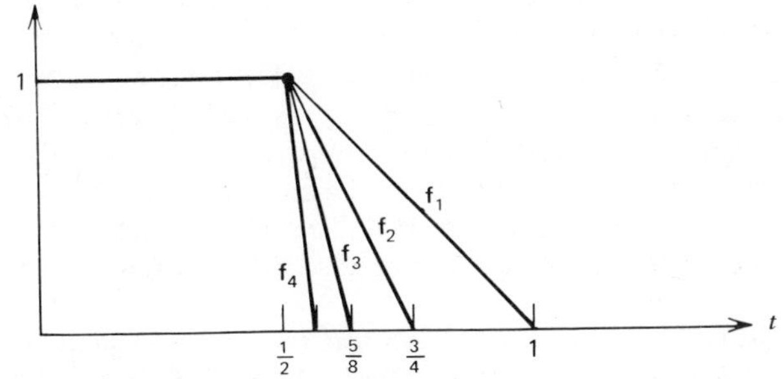

Figure 5.7. A Cauchy sequence in $\mathcal{C}(0,1)$.

that $\|\mathbf{x}-\mathbf{y}\|<\epsilon$; that is, $\mathbf{x}$ can be approximated arbitrarily closely in norm by vectors $\mathbf{y}$ in $\mathbb{S}$. The **closure** of $\mathbb{S}$, denoted $\overline{\mathbb{S}}$, consists in $\mathbb{S}$ together with all its points of closure. If $\mathbb{S}$ contains all its points of closure, it is said to be **closed**. A set $\mathbb{S}_1$ in $\mathbb{S}$ is said to be **dense** in $\mathbb{S}$ if $\mathbb{S}$ is the closure of $\mathbb{S}_1$; that is, if every vector in $\mathbb{S}$ can be approximated arbitrarily closely in norm by a vector in $\mathbb{S}_1$.

Definition. An inner produce space $\mathcal{H}$ is said to be **complete** if every Cauchy (convergent) sequence from $\mathcal{H}$ converges in norm to a limit in $\mathcal{H}$. A complete inner product space is called a **Hilbert space**.

The terms closed and complete, as applied to inner product spaces, are essentially equivalent concepts. The inner product space $\mathcal{V}$ discussed above is not complete, whereas the "enlarged" space $\mathcal{H}$ is complete; $\mathcal{H}$ is a Hilbert space. The space $\mathcal{V}$ is dense in $\mathcal{H}$; that is, $\mathcal{H}$ is only a slight enlargement of $\mathcal{V}$. We can complete any inner product space by extending its definition to include all of its limit vectors. Of course, the definitions of addition, scalar multiplication, and inner product must be extended to these additional limit vectors [5.11, p. 17].

Example 5. Finite-Dimensional Hilbert Spaces. Every finite-dimensional inner product space is complete [5.23, p. 143]. For instance, we cannot conceive of an infinite sequence of real n-tuples converging to anything but another real n-tuple; the ith components of a sequence of n-tuples constitute a sequence of real numbers, and the real numbers are complete.

Example 6. The Hilbert Space l_2. We denote by l_2^c the space of square-summable sequences of complex numbers with the inner product $\langle\mathbf{x},\mathbf{y}\rangle \triangleq \sum_{k=1}^{\infty}\xi_k\bar{\eta}_k$, where ξ_k and η_k are the kth elements of $\mathbf{x}$ and $\mathbf{y}$, respectively. (A square-summable sequence is a sequence for which $\|\mathbf{x}\|^2=\sum_{k=1}^{\infty}|\xi_k|^2<\infty$.) We use the symbol l_2 to represent the space of *real* square-summable sequences; then the complex conjugate in the inner product is superfluous. Both the real l_2 and the complex l_2^c are complete [5.23, p. 48]. The standard basis $\{\boldsymbol{\varepsilon}_i\}$, where $\boldsymbol{\varepsilon}_i=(0,\ldots,0,1_i,0,\ldots)$, is an orthonormal basis for both the real and complex cases.

Example 7. The Hilbert Space $\mathcal{L}_2(a,b)$. Let $\mathcal{L}_2^c(a,b)$ be the space of complex square-integrable* functions defined on the finite interval $[a,b]$ with the inner product $\langle\mathbf{f},\mathbf{g}\rangle \triangleq \int_a^b \mathbf{f}(t)\overline{\mathbf{g}(t)}\,dt$. (A square-integrable function is one for which $\|\mathbf{f}\|^2=\int_a^b|\mathbf{f}(t)|^2\,dt$ is finite.) The symbol $\mathcal{L}_2(a,b)$ is used to represent the space of *real* square-integrable functions; then the complex conjugate is unnecessary. We usually concern ourselves only with the real space. Both the real $\mathcal{L}_2(a,b)$ and the complex $\mathcal{L}_2^c(a,b)$ are complete [5.2, p. 115]. The space $\mathcal{L}_2(a,b)$ contains no delta functions. However, it does contain certain discontinuous functions, for example,

*The integral used in this definition is the Lebesgue integral. For all practical purposes, we can consider Lebesgue integration to be the same as the usual Riemann integration. Where the Riemann integral exists, the two integrals are equal. See Royden [5.21].

step functions. (Recall from the definition of equality in norm that we ignore isolated discontinuities. As a practical matter, we seldom encounter a function with more than a few discontinuities in a finite interval.) We can think of $\mathcal{L}_2(a,b)$ as essentially a space of functions which are piecewise continuous, but perhaps unbounded, in the finite interval $[a,b]$. Any set $\mathcal{G} \triangleq \{\mathbf{p}_k\}$ of orthogonal polynomials which forms a basis for $\mathcal{P}(a,b)$ is a basis for both the real and complex $\mathcal{L}_2(a,b)$. An orthonormal basis for both the real and complex $\mathcal{L}_2(a,b)$ is the set of sinusoids (5.30), with $p=b-a$. Another orthonormal basis for the complex $\mathcal{L}_2^{\mathcal{C}}(a,b)$ is the set of complex exponentials (5.33) with $p=b-a$.

Example 8. The Hilbert Space $\mathcal{L}_2(\omega;\ a,b)$. Let $\mathcal{L}_2(\omega;\ a,b)$ represent the set of all ω-square-integrable functions with the inner product (5.49). That is, $\mathcal{L}_2(\omega;\ a,b)$ contains those functions that have finite norm under the inner product (5.49). From the discussion associated with (5.49) it is apparent that $\mathcal{L}_2(\omega;\ a,b)$ differs from $\mathcal{L}_2(a,b)$ only in the inner product. Both spaces contain precisely the same functions, and completeness of $\mathcal{L}_2(\omega;\ a,b)$ follows from the completeness of $\mathcal{L}_2(a,b)$.

A Hilbert space possesses many subsets that are themselves inner product spaces (using the same inner product). These subsets may or may not be complete. If a subset is a complete inner product space, it is itself a Hilbert space, and we refer to it as a **subspace**. If a subset is a vector space, but is not necessarily complete, it is properly termed a **linear manifold**. Since all finite-dimensional vector spaces are complete, all finite-dimensional linear manifolds of $\mathcal{L}_2(a,b)$ are subspaces. However, $\mathcal{P}(a,b)$, $\mathcal{C}(a,b)$, $\mathcal{C}^n(a,b)$, $\mathcal{C}^\infty(a,b)$, and the space of piecewise-continuous functions on $[a,b]$ are (incomplete) linear manifolds of $\mathcal{L}_2(a,b)$. Each of these spaces is dense in $\mathcal{L}_2(a,b)$, and thus is nearly equal to $\mathcal{L}_2(a,b)$.

We note that if $\mathcal{H}$ is a Hilbert space and $\mathcal{S}$ is any set in $\mathcal{H}$, then the orthogonal complement $\mathcal{S}^\perp$ must be a subspace. For if $\{\mathbf{x}_n\}$ is a Cauchy sequence in $\mathcal{S}^\perp$ with limit $\mathbf{x}$ in $\mathcal{H}$, then $\langle \mathbf{x}_n, \mathbf{y}\rangle = 0$ for each $\mathbf{y}$ in $\mathcal{S}$; it follows that

$$\langle \mathbf{x},\mathbf{y}\rangle = \langle \lim_{n\to\infty} \mathbf{x}_n, \mathbf{y}\rangle = \lim_{n\to\infty} \langle \mathbf{x}_n, \mathbf{y}\rangle = 0$$

and the limit vector $\mathbf{x}$ is also orthogonal to $\mathcal{S}$. [In order to take the limit outside the inner product, we have relied on the continuity of inner products. See (5.56).]

Example 9. An Infinite-Dimensional (Complete) Subspace of $\mathcal{L}_2(a,b)$. Let $\mathcal{W}$ be the (one-dimensional) subspace of constant functions in $\mathcal{L}_2(a,b)$. By the previous paragraph, the orthogonal complement $\mathcal{W}^\perp$ is complete. But $\mathcal{W}^\perp$ consists in those functions $\mathbf{f}$ in $\mathcal{L}_2(a,b)$ which satisfy $\int_a^b c\mathbf{f}(t)\,dt = 0$ for all constants c. Thus the functions in $\mathcal{L}_2(a,b)$ whose average value is zero form a complete subspace of $\mathcal{L}_2(a,b)$. This subspace is itself a Hilbert space.

Why do we care whether or not a vector space is complete? One reason is that we wish to extend finite-dimensional concepts to infinite-dimensional cases. Some of these concepts extend only for a Hilbert space, the natural generalization of a finite-dimensional space. (Recall that finite-dimensional spaces are Hilbert spaces.) The only concept we have discussed thus far which applies only for Hilbert spaces is the **projection theorem** (5.20), $\mathcal{V} = \mathcal{W} \oplus \mathcal{W}^{\perp}$. The proof of (5.20) depends on the fact that repeated application of the Gram–Schmidt procedure can generate no more than n orthogonal vectors in an n-dimensional space. The theorem is valid in an infinite-dimensional space $\mathcal{V}$ if and only if $\mathcal{V}$ is a Hilbert space and $\mathcal{W}$ is a (complete) subspace of $\mathcal{V}$[5.2, p. 172]; of course, $\mathcal{W}^{\perp}$ is always complete.

Fortunately, the question of completeness of an inner product space is seldom of practical concern, since we can complete any inner product space by extending its definition. Suppose a linear transformation $\mathbf{T}$ has its domain and range in a separable Hilbert space $\mathcal{H}$ (a Hilbert space with a countable basis). Then if domain($\mathbf{T}$) is dense in $\mathcal{H}$, we refer to $\mathbf{T}$ as a *linear operator on the Hilbert space* $\mathcal{H}$. For instance, the completion of the space $\mathcal{C}^{\infty}(a,b)$ of real, infinitely differentiable functions on $[a,b]$ is just $\mathcal{L}_2(a,b)$ of Example 7. We apply a differential operator $\mathbf{L}$ to any space of "sufficiently differentiable" functions and still refer to $\mathbf{L}$ as a differential operator on $\mathcal{L}_2(a,b)$.

In our examination of finite-dimensional vector spaces we used the process of taking coordinates to equate every n-dimensional space to the matrix space $\mathfrak{M}^{n\times 1}$. We now equate inner product spaces, both finite and infinite-dimensional. Two inner product spaces, $\mathcal{V}$ with inner product $\langle\ ,\ \rangle_{\mathcal{V}}$ and $\mathcal{W}$ with inner product $\langle\ ,\ \rangle_{\mathcal{W}}$, are **isomorphic** (or **equivalent**) if there is an invertible linear transformation $\mathbf{T}$: $\mathcal{V}\rightarrow\mathcal{W}$ which preserves inner products; that is, for which $\langle \mathbf{x},\mathbf{y}\rangle_{\mathcal{V}} = \langle \mathbf{Tx},\mathbf{Ty}\rangle_{\mathcal{W}}$ for all $\mathbf{x}$ and $\mathbf{y}$ in $\mathcal{V}$. The process of taking coordinates relative to any orthonormal basis is just such a transformation.

Example 10. Coordinates for Real n-Dimensional Inner Product Spaces. For n-dimensional spaces, we take $\mathfrak{M}^{n\times 1}$ with its standard inner product as our space of coordinates. Let $\mathcal{V}$ be any real n-dimensional inner product space; let $\mathcal{X}$ be an orthonormal basis for $\mathcal{V}$. Define $\mathbf{T}$: $\mathcal{V}\rightarrow\mathfrak{M}^{n\times 1}$ as the invertible linear transformation which assigns to each vector $\mathbf{x}$ in $\mathcal{V}$ its set of Fourier coefficients (or coordinates) in $\mathfrak{M}^{n\times 1}$:

$$\mathbf{Tx} \triangleq (\langle \mathbf{x},\mathbf{x}_1\rangle_{\mathcal{V}} \cdots \langle \mathbf{x},\mathbf{x}_n\rangle_{\mathcal{V}})^{\mathrm{T}} = [\mathbf{x}]_{\mathcal{X}}$$

Since $\mathcal{X}$ is orthonormal, Parseval's equation (5.27) is satisfied:

$$\langle \mathbf{x},\mathbf{y}\rangle_{\mathcal{V}} = [\mathbf{y}]_{\mathcal{X}}^{\mathrm{T}}[\mathbf{x}]_{\mathcal{X}} = \langle \mathbf{Tx},\mathbf{Ty}\rangle_{\mathfrak{M}^{n\times 1}}$$

This is the standard inner product in the real space $\mathfrak{M}^{n\times 1}$. Clearly, each real n-dimensional inner product space (a Hilbert space) is equivalent to $\mathfrak{M}^{n\times 1}$ with its standard inner product. (By inserting a complex conjugate in the inner product, we can show that every complex n-dimensional inner product space is equivalent to $\mathfrak{M}_c^{n\times 1}$, the space of complex $n\times 1$ matrices with its standard inner product.)

Example 11. Coordinates for Real Separable Infinite-Dimensional Hilbert Spaces. The logic of Example 10 applies to all separable Hilbert spaces (Hilbert spaces which have countable bases). For separable infinite-dimensional spaces, we take l_2 with its standard inner product as our space of coordinates. A separable space has a countable basis. Any such basis can be orthonormalized by the Gram–Schmidt procedure. Suppose $\mathfrak{X}=\{\mathbf{x}_k\}$ is an orthonormal basis for a real separable Hilbert space $\mathcal{H}$. We define $\mathbf{T}$: $\mathcal{H}\to l_2$ as the process of assigning Fourier coefficients relative to this basis:

$$\mathbf{T}\mathbf{x} \triangleq (\langle\mathbf{x},\mathbf{x}_1\rangle_{\mathcal{H}},\langle\mathbf{x},\mathbf{x}_2\rangle_{\mathcal{H}},\ldots)=[\mathbf{x}]_{\mathfrak{X}} \tag{5.50}$$

From our discussion of the space spanned by an orthonormal basis, we know that the coordinates (Fourier coefficients) of vectors in $\mathcal{H}$ consist in the square-summable sequences which constitute l_2. Since Fourier expansions exist and are unique for each $\mathbf{x}$ in $\mathcal{H}$, $\mathbf{T}$ is invertible. Because the set $\{\mathbf{x}_k\}$ is orthonormal, Parseval's equation (5.47) applies:

$$\begin{aligned}\langle\mathbf{x},\mathbf{y}\rangle_{\mathcal{H}} &= \sum_{k=1}^{\infty}\langle\mathbf{x},\mathbf{x}_k\rangle_{\mathcal{H}}\,\langle\mathbf{y},\mathbf{x}_k\rangle_{\mathcal{H}}\\ &=\langle\mathbf{T}\mathbf{x},\mathbf{T}\mathbf{y}\rangle_{l_2}\end{aligned}$$

This is the standard inner product between the coordinates of $\mathbf{x}$ and $\mathbf{y}$ (in l_2). Therefore, every real separable infinite-dimensional Hilbert space is equivalent to the real space l_2 with its standard inner product. Thus the somewhat mysterious space $\mathcal{L}_2(a,b)$ is, in essence, no more complicated than l_2. In Example 11 of Section 5.4 we introduce $\mathcal{L}_2(\Omega)$, an inner product space of square-integrable functions defined on a finite two-dimensional domain Ω; $\mathcal{L}_2(\Omega)$ is also equivalent to l_2. (By inserting a complex conjugate in each inner product, we find that every complex separable infinite-dimensional Hilbert space is equivalent to the complex space l_2^c with its standard inner product.)

5.4 Adjoint Transformations

In the preceding section we developed separable Hilbert spaces as natural generalizations of n-dimensional inner product spaces. We know that we can generate a countable orthonormal basis for any such space. In (5.25) we diagonalized and computationally decoupled an operator equation in an n-dimensional space by means of an orthonormal basis of eigenvectors. We have also discussed the applicability of orthonormal eigenvectors to an infinite-dimensional example, the steady-state analysis of a dynamic system. Can we diagonalize a general linear operator on an infinite-dimensional space, a differential operator, for instance? We can if there is a countable orthonormal basis for the infinite-dimensional space which is composed of eigenvectors of the operator. In Example 3, Section 5.3 we orthogonalized a set of eigenfunctions by careful choice of the inner product. We would like to be able to make general statements which clearly characterize the existence of an orthonormal basis of eigenvectors for a given operator on a given infinite-dimensional space. Given a set of eigenvectors for an operator **T** on a given *vector space* $\mathcal{V}$, for what inner products are the eigenvectors an orthogonal (or orthonormal) basis? For what operators **T** on a given *inner product space* $\mathcal{V}$ do there exist orthonormal bases of eigenvectors? In sum, under what conditions can we diagonalize an operator equation by means of an orthonormal basis? The answers to these questions are to a great extent answered in the concept of "self-adjointness." We introduce the adjoint transformation in this section, and return to a discussion of orthonormal bases of eigenvectors for solving operator equations in Section 5.5.

We observed in Chapter 1 that we can interpret a matrix multiplication **Ax** either as a linear combination of the columns of **A** or as a set of standard inner products of **x** with the rows of **A**. Furthermore, we know that the number of independent rows of **A** equals the number of independent columns. Since the rows and columns of **A** possess much common information, we would be surprised if multiplication by the transposed matrix $\mathbf{A}^{\mathrm{T}}$ did not describe a transformation closely related to multiplication by **A**. Let $\mathbf{T}\colon \mathfrak{M}^{2\times 1}\to\mathfrak{M}^{3\times 1}$ be defined by

$$\mathbf{Tx} \stackrel{\Delta}{=} \mathbf{Ax} = \begin{pmatrix} 1 & 1 \\ 0 & 0 \\ 0 & 0 \end{pmatrix}\mathbf{x}$$

We define the "transpose" transformation $\mathbf{T}^{\mathrm{T}}\colon \mathfrak{M}^{3\times 1}\to\mathfrak{M}^{2\times 1}$ by

$$\mathbf{T}^{\mathrm{T}}\mathbf{y} \stackrel{\Delta}{=} \mathbf{A}^{\mathrm{T}}\mathbf{y} = \begin{pmatrix} 1 & 0 & 0 \\ 1 & 0 & 0 \end{pmatrix}\mathbf{y}$$

The range and nullspace of **T** are important indicators of its structure. They display the nonsolvability and nonuniqueness of solutions of the equation $\mathbf{Tx}=\mathbf{y}$. Observe that

$$\text{range}(\mathbf{T}) = \text{span}\left\{\begin{pmatrix} 1 \\ 0 \\ 0 \end{pmatrix}\right\}, \qquad \text{nullspace}(\mathbf{T}) = \text{span}\left\{\begin{pmatrix} 1 \\ -1 \end{pmatrix}\right\}$$

$$\text{range}(\mathbf{T}^{\mathrm{T}}) = \text{span}\left\{\begin{pmatrix} 1 \\ 1 \end{pmatrix}\right\}, \quad \text{nullspace}(\mathbf{T}^{\mathrm{T}}) = \text{span}\left\{\begin{pmatrix} 0 \\ 1 \\ 0 \end{pmatrix}, \begin{pmatrix} 0 \\ 0 \\ 1 \end{pmatrix}\right\}$$

Range(**T**) and nullspace(**T**) are, of course, in different spaces. However, range(**T**) and nullspace($\mathbf{T}^{\mathrm{T}}$) are in the same space; in fact,

$$\mathfrak{M}^{3\times 1} = \text{range}(\mathbf{T}) \oplus \text{nullspace}(\mathbf{T}^{\mathrm{T}})$$

Similarly,

$$\mathfrak{M}^{2\times 1} = \text{range}(\mathbf{T}^{\mathrm{T}}) \oplus \text{nullspace}(\mathbf{T})$$

Furthermore, these direct sums are orthogonal relative to the standard inner products for the two spaces. It is evident, at least for this example, that **T** and $\mathbf{T}^{\mathrm{T}}$ together characterize the transformation **T** more explicitly than does **T** alone.

We extend the transpose concept to general linear transformations. We find that the orthogonal decomposition illustrated above still applies. Recall that orthogonal decomposition is closely related to Fourier series expansion and, therefore, to orthonormal bases. The generalization of $\mathbf{T}^{\mathrm{T}}$, together with **T** itself, characterizes the existence or nonexistence of orthonormal bases of eigenvectors.

Bounded Linear Transformations

The generalization of the transpose matrix exists only for transformations which satisfy certain restrictions. We now define the concepts which we use to express these restrictions.

Definition. Let **T** be a (possibly nonlinear) transformation from an inner product space $\mathcal{V}$ into an inner product space $\mathcal{W}$. Then **T is bounded** if there is a positive number α such that

$$\|\mathbf{Tx}\|_{\mathcal{W}} \leqslant \alpha \|\mathbf{x}\|_{\mathcal{V}} \quad \text{for all } \mathbf{x} \text{ in } \mathcal{V} \tag{5.51}$$

We define the **norm of T** by*

$$\|\mathbf{T}\| \triangleq \inf\{\alpha : \|\mathbf{Tx}\|_{\mathcal{W}} \leqslant \alpha \|\mathbf{x}\|_{\mathcal{V}} \quad \text{for all } \mathbf{x} \text{ in } \mathcal{V}\} \tag{5.52}$$

We can think of $\|\mathbf{T}\|$ as the tightest bound for **T**. It follows that

$$\|\mathbf{Tx}\|_{\mathcal{W}} \leqslant \|\mathbf{T}\| \, \|\mathbf{x}\|_{\mathcal{V}} \tag{5.53}$$

*The term "inf" means **infimum** or greatest lower bound. If the bound is actually reached, the infimum is just the minimum. The term "sup" means **supremum** or least upper bound; if the bound is attained, the supremum is the maximum.

If $\mathbf{T}$ is linear, (5.52) can be expressed as

$$\|\mathbf{T}\| = \inf\left\{\alpha : \frac{\|\mathbf{Tx}\|_{\mathcal{W}}}{\|\mathbf{x}\|_{\mathcal{V}}} \leqslant \alpha\right\}$$

$$= \inf\{\alpha : \|\mathbf{Tx}\|_{\mathcal{W}} \leqslant \alpha, \quad \|\mathbf{x}\|_{\mathcal{V}} = 1\}$$

$$= \sup_{\|\mathbf{x}\|_{\mathcal{V}}=1}\{\|\mathbf{Tx}\|_{\mathcal{W}}\} \tag{5.54}$$

Example 1. A Norm is a Bounded Functional. Define $\mathbf{T}: \mathcal{V} \to \mathcal{R}$ by $\mathbf{Tx} \triangleq \|\mathbf{x}\|_{\mathcal{V}}$. Then $\|\mathbf{Tx}\|_{\mathcal{R}} = |\mathbf{Tx}| = \|\mathbf{x}\|_{\mathcal{V}}$. Clearly, the number 1 is a bound for $\mathbf{T}$ and $\|\mathbf{T}\| = 1$.

Example 2. Matrix Transformations are Bounded. Define $\mathbf{T}: \mathfrak{M}_c^{n\times 1} \to \mathfrak{M}_c^{m\times 1}$ by $\mathbf{Tx} \triangleq \mathbf{Ax}$, where $\mathbf{A}$ is a (possibly complex) $m \times n$ matrix. Assuming standard inner products, $\|\mathbf{Tx}\|^2 = \|\mathbf{Ax}\|^2 = \overline{\mathbf{x}}^{\mathrm{T}}\overline{\mathbf{A}}^{\mathrm{T}}\mathbf{Ax}$. Then, by (5.54), $\|\mathbf{T}\|^2 = \max_{\overline{\mathbf{x}}^{\mathrm{T}}\mathbf{x}=1} \overline{\mathbf{x}}^{\mathrm{T}}\overline{\mathbf{A}}^{\mathrm{T}}\mathbf{Ax}$. It can be shown that $\|\mathbf{T}\| = \sqrt{\lambda_m}$, where λ_m is the largest eigenvalue of the matrix $\overline{\mathbf{A}}^{\mathrm{T}}\mathbf{A}$ (see P&C 5.21). We call $\sqrt{\lambda_m}$ the **spectral radius of A** and denote it by $\sigma(\mathbf{A})$. We also refer to $\sigma(\mathbf{A})$ as the **norm of A**, denoted $\|\mathbf{A}\|$. Thus

$$\|\mathbf{T}\| = \|\mathbf{A}\| = \sigma(\mathbf{A}) = \sqrt{\lambda_m}$$

It is apparent that the bound $\sqrt{\lambda_m}$ is attained for $\mathbf{x}$ equal to a normalized eigenvector of $\overline{\mathbf{A}}^{\mathrm{T}}\mathbf{A}$ corresponding to the eigenvalue λ_m. The fact that matrix transformations are bounded implies that *all transformations on finite-dimensional spaces are bounded.*

If $\mathbf{A}$ is square, it makes sense to speak of the eigenvalues of $\mathbf{A}$ itself. If $\mathbf{A}$ is also real and symmetric, its spectral radius is just the largest eigenvalue of $\mathbf{A}$. That this statement is not true for every square matrix is demonstrated by the matrix

$$\mathbf{A} = \begin{pmatrix} 1 & 1 \\ 0 & 0 \end{pmatrix}$$

for which the largest eigenvalue is $\lambda = 1$, but for which $\|\mathbf{A}\| = \sigma(\mathbf{A}) = \sqrt{2}$. The bound $\|\mathbf{A}\|$ is attained for $\mathbf{x} = (1 \ 1)^{\mathrm{T}}$.

Example 3. Integral Operators are Bounded. Define the linear operator $\mathbf{T}$ on $\mathcal{L}^2(a,b)$ by $(\mathbf{Tf})(t) = \int_a^b k(t,s)\mathbf{f}(s)\,ds$, where the kernel k satisfies $\int_a^b \int_a^b k^2(t,s)\,ds\,dt < \infty$. Such a kernel is called a **Hilbert–Schmidt kernel** and $\mathbf{T}$ is known as a **Hilbert–Schmidt integral operator**. If k is bounded for t and s in $[a,b]$, for instance, then $\mathbf{T}$ is Hilbert–Schmidt. Many operators are of this type; for example, the inverses of most differential operators defined on a finite interval. We apply the

Cauchy–Schwartz inequality (P&C 5.4) to find

$$\|\mathbf{Tf}\|^2 = \int_a^b \left[\int_a^b k(t,s)\mathbf{f}(s)\,ds \right]^2 dt$$

$$\leqslant \int_a^b \left[\int_a^b k^2(t,s)\,ds \int_a^b \mathbf{f}^2(s)\,ds \right] dt$$

$$= \int_a^b \int_a^b k^2(t,s)\,ds\,dt\,\|\mathbf{f}\|^2$$

Therefore, by (5.54),

$$\|\mathbf{T}\| \leqslant \left[\int_a^b \int_a^b k^2(t,s)\,ds\,dt \right]^{1/2} \tag{5.55}$$

and $\mathbf{T}$ is bounded. Under what conditions is the bound (5.55) actually the norm of $\mathbf{T}$? The Cauchy–Schwartz inequality becomes an equality if and only if the two arguments $k(t,s)$ and $\mathbf{f}(s)$ are dependent functions of s. Therefore, if there is an $\mathbf{f}$ in $\mathcal{L}_2(a,b)$ such that $k(t,s)=\mathbf{g}(t)\mathbf{f}(s)$, then the bound which we have exhibited is actually $\|\mathbf{T}\|$. For many integral operators, $\|\mathbf{T}\|$ is equal to the magnitude of the largest eigenvalue of $\mathbf{T}$ (P&C 5.29).

Example 4. Differential Operators are not Bounded. Differential operators are among the most useful transformations, yet they are seldom bounded. For instance, let $\mathbf{D}$ operate on $\mathcal{L}_2(0,1)$. Let $\{\mathbf{f}_k\}$ be the sequence of functions shown in Figure 5.8*a*. In the $\mathcal{L}_2$ norm, $\|\mathbf{f}_1\|^2 = \int_0^1 t^2\,dt = \frac{1}{3}$ and $\|\mathbf{f}_\infty\|^2 = \int_0^1 dt = 1$. Therefore, the functions in the sequence satisfy

$$\|\mathbf{f}_1\|^2 = \tfrac{1}{3} \leqslant \|\mathbf{f}_k\|^2 < 1 = \|\mathbf{f}_\infty\|^2$$

Yet we recognize from Figure 5.8*b* that

$$\|\mathbf{Df}_k\|^2 = 2^{k-1} \to \infty$$

There is no number α which "bounds" $\mathbf{D}$ for all $\mathbf{f}$ in $\mathcal{L}_2(0,1)$. In the limit as $k \to \infty$, an equivalent statement is that the derivative of a discontinuous function contains a delta function, but delta functions are not square integrable; they are not in $\mathcal{L}_2(0,1)$.

Definition. A (possibly nonlinear) transformation $\mathbf{T}\colon \mathcal{V} \to \mathcal{W}$ is said to be **continuous at** $\mathbf{x}_0$ if for each $\epsilon > 0$ there is a $\delta > 0$ such that

$$\|\mathbf{x} - \mathbf{x}_0\|_{\mathcal{V}} < \delta \Rightarrow \|\mathbf{Tx} - \mathbf{Tx}_0\|_{\mathcal{W}} < \epsilon$$

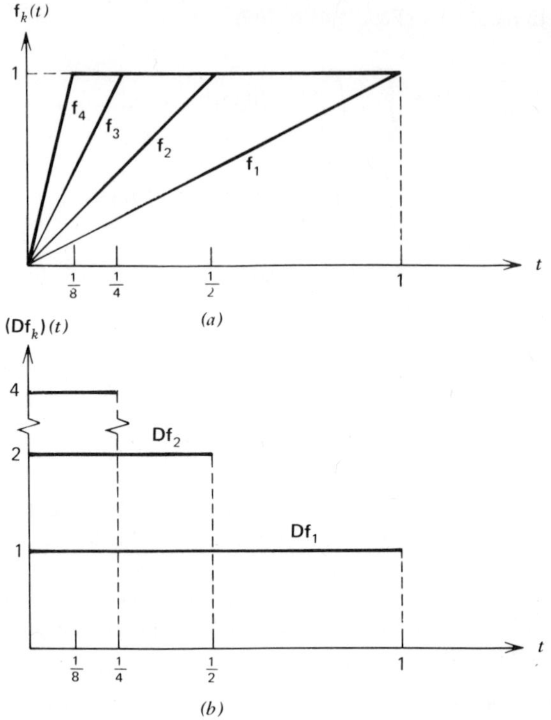

Figure 5.8. Differentiation of a sequence of functions.

That is, $\mathbf{T}$ is continuous at $\mathbf{x}_0$ if making $\|\mathbf{x}-\mathbf{x}_0\|_{\mathcal{V}}$ small will guarantee that $\|\mathbf{T}\mathbf{x}-\mathbf{T}\mathbf{x}_0\|_{\mathcal{W}}$ is small. If $\mathbf{T}$ is continuous for each $\mathbf{x}$ in $\mathcal{V}$, we just say **T is continuous**.

The nonlinear transformation $\mathbf{T}\mathbf{x} \triangleq \|\mathbf{x}\|_{\mathcal{V}}$, for example, is continuous. Suppose $\mathbf{T}$ is continuous and $\mathbf{x}_0 = \lim_{n\to\infty}\mathbf{x}_n$ in $\mathcal{V}$. If $\mathbf{x}_n$ approaches $\mathbf{x}_0$, then $\mathbf{T}\mathbf{x}_n$ approaches $\mathbf{T}\mathbf{x}_0$; in other words,

$$\lim_{n\to\infty}(\mathbf{T}\mathbf{x}_n) = \mathbf{T}\mathbf{x}_0 = \mathbf{T}\Big(\lim_{n\to\infty}\mathbf{x}_n\Big) \tag{5.56}$$

We will find this fact useful in the decoupling of equations on infinite-dimensional spaces. It is easy to show that *a linear transformation is continuous if and only if it is bounded*. Thus the linear transformations of Examples 2 and 3 are continuous transformations. It is apparent that bounded (or continuous) linear transformations are "well behaved." The linear differential operators of Example 4 are not continuous and are "poorly behaved." It is for bounded (continuous) linear transformations that we will show most of the useful results of this chapter. It is usually difficult, if not impossible, to extend these concepts to unbounded linear transformations in a rigorous manner. Yet the concepts will be shown, by example, to extend in certain instances.

Completely Continuous Transformations

We now introduce briefly the concept of "complete continuity" of a linear transformation in order to specify conditions which guarantee the existence of a *countable* basis of eigenvectors.

Definition. A set $\mathcal{S}$ in an inner product space $\mathcal{V}$ is said to be **bounded** if there is a constant M such that $\|\mathbf{x}\| \leqslant M$ for all $\mathbf{x}$ in $\mathcal{S}$. A set $\mathcal{S}$ is **compact** if each infinite sequence of vectors from $\mathcal{S}$ contains a subsequence that converges to a vector of $\mathcal{S}$. Every compact set is closed and bounded. In finite-dimensional spaces the converse is also true: a set is compact if and only if it is closed and bounded, and every bounded set is closed [5.22, p. 185].

It is easy to see that a linear transformation $\mathbf{T}: \mathcal{V} \rightarrow \mathcal{W}$ is bounded (continuous) if and only if it maps bounded sets in $\mathcal{V}$ into bounded sets in $\mathcal{W}$. A stronger restriction on $\mathbf{T}$, which guarantees the countability of the eigenvalues and eigenvectors of $\mathbf{T}$, is that of complete continuity.

Definition. A linear transformation $\mathbf{T}$ is **completely continuous** if it maps bounded sets into compact sets.

A completely continuous transformation is continuous, but the converse is not necessarily true. On an infinite-dimensional space, even the (continuous) identity operator is not completely continuous. Any bounded linear transformation whose range is finite-dimensional is completely continuous; thus any operator on a finite-dimensional space is completely continuous. The Hilbert–Schmidt integral operators of Example 3 are also completely continuous. Suppose $\mathbf{T}$ and $\mathbf{U}$ are completely continuous transformations mapping $\mathcal{V}$ into $\mathcal{W}$; then the linear combination $a\mathbf{T}+b\mathbf{U}$ is completely continuous. If $\mathbf{T}$ and $\mathbf{U}$ are linear operators on $\mathcal{V}$, one of which is bounded and the other completely continuous, then $\mathbf{TU}$ and $\mathbf{UT}$ are completely continuous. If $\mathbf{T}_k$ is the kth member of a sequence of completely continuous linear transformations mapping $\mathcal{V}$ into $\mathcal{W}$, then the limit operator $\mathbf{T}$ defined by

$$\lim_{k\rightarrow\infty} \|\mathbf{T}_k - \mathbf{T}\| = 0$$

is completely continuous. If a completely continuous transformation $\mathbf{T}$ is defined on a infinite-dimensional space, then $\mathbf{T}^{-1}$, if it exists, is unbounded. Detailed discussions of completely continuous transformations can be found in Bachman and Narici [5.2] and Stakgold [5.22]. We content ourselves with this brief introduction; however, we make occasional reference to the consequences of complete continuity.

Bounded Linear Functionals

The key theorem in the development of a generalization of the transpose matrix is the Riesz–Fréchet theorem. This theorem relates inner products and "bounded linear functionals." As indicated in Section 2.3, a functional is a scalar-valued transformation. Suppose $\mathcal{V}$ is an inner product space. The transformation $\mathbf{B}$: $\mathcal{V} \to \mathcal{C}$ defined by $\mathbf{Bx} \triangleq \langle \mathbf{x}, \mathbf{y} \rangle_{\mathcal{V}}$ for a fixed vector $\mathbf{y}$ in $\mathcal{V}$ is a linear functional. (Recall that an inner product is linear on the left.) By the Cauchy–Schwartz inequality (P&C 5.4), $\|\mathbf{Bx}\|_{\mathcal{C}} = |\langle \mathbf{x}, \mathbf{y} \rangle_{\mathcal{V}}| \leqslant \|\mathbf{y}\|_{\mathcal{V}} \|\mathbf{x}\|_{\mathcal{V}}$ and $\|\mathbf{y}\|_{\mathcal{V}}$ is a bound for the linear functional $\mathbf{B}$. Furthermore, the bound is attained with the normalized vector $\mathbf{x} = \mathbf{y}/\|\mathbf{y}\|_{\mathcal{V}}$, and thus $\|\mathbf{B}\| = \|\mathbf{y}\|_{\mathcal{V}}$. Each different $\mathbf{y}$ in $\mathcal{V}$ specifies a different bounded (or continuous) linear functional. If $\mathcal{V}$ is a Hilbert space, we can say more—any bounded linear functional on $\mathcal{V}$ can be represented by an inner product.

Riesz–Fréchet Theorem. Corresponding to any bounded linear functional $\mathbf{B}$ on a Hilbert space $\mathcal{H}$ there is a unique vector $\mathbf{y}$ in $\mathcal{H}$ such that

$$\mathbf{Bx} = \langle \mathbf{x}, \mathbf{y} \rangle_{\mathcal{H}} \quad \text{for all } \mathbf{x} \text{ in } \mathcal{H} \tag{5.57}$$

Furthermore, $\|\mathbf{B}\| = \|\mathbf{y}\|_{\mathcal{H}}$.

Proof. If $\mathbf{B}$ is the zero functional, it is obvious that $\mathbf{y} = \boldsymbol{\theta}$. Assume $\mathbf{B}$ is not the zero functional, and let $\mathcal{W}$ = nullspace($\mathbf{B}$). ($\mathcal{W}$ consists of all $\mathbf{x}$ in $\mathcal{H}$ for which $\mathbf{Bx} = \langle \mathbf{x}, \mathbf{y} \rangle = 0$. Thus the vector $\mathbf{y}$ which we seek spans $\mathcal{W}^{\perp}$.) Let $\mathbf{y}_0$ be a unit vector in $\mathcal{W}^{\perp}$, and $\mathbf{x}$ any vector in $\mathcal{H}$. The vector $(\mathbf{Bx})\mathbf{y}_0 - (\mathbf{By}_0)\mathbf{x}$ is in $\mathcal{W}$.* Therefore,

$$\langle (\mathbf{Bx})\mathbf{y}_0 - (\mathbf{By}_0)\mathbf{x}, \mathbf{y}_0 \rangle = (\mathbf{Bx}) - (\mathbf{By}_0)\langle \mathbf{x}, \mathbf{y}_0 \rangle = 0$$

or $\mathbf{Bx} = \langle \mathbf{x}, (\overline{\mathbf{By}_0})\mathbf{y}_0 \rangle$. The vector $\mathbf{y} = (\overline{\mathbf{By}_0})\mathbf{y}_0$ in $\mathcal{W}^{\perp}$ represents $\mathbf{B}$ as required by (5.57). To see that $\mathbf{y}$ is uniquely determined by $\mathbf{B}$, we assume both $\mathbf{y}$ and $\mathbf{z}$ will do. Then $\langle \mathbf{x}, \mathbf{y} \rangle = \langle \mathbf{x}, \mathbf{z} \rangle$ or $\langle \mathbf{x}, \mathbf{y} - \mathbf{z} \rangle = 0$ for all $\mathbf{x}$ in $\mathcal{H}$ including

*Since range($\mathbf{B}$) is one-dimensional, $\mathbf{Bx}$ and $\mathbf{By}_0$ are scalars; they can be used as scalar multipliers.

$\mathbf{x}=\mathbf{y}-\mathbf{z}$. If follows that $\mathbf{y}-\mathbf{z}=\boldsymbol{\theta}$ or $\mathbf{z}=\mathbf{y}$. We showed that $\|\mathbf{B}\|=\|\mathbf{y}\|_{\mathcal{V}}$ in the discussion prior to the theorem.

Example 5. Bounded Linear Functionals on $\mathfrak{M}_c^{n\times 1}$. Any linear functional $\mathbf{B}$ on the standard inner product space $\mathfrak{M}_c^{n\times 1}$ is bounded and representable as $\mathbf{Bx}=\bar{\mathbf{y}}^{\mathrm{T}}\mathbf{x}$ for some $\mathbf{y}$ in $\mathfrak{M}_c^{n\times 1}$. That is, a linear functional acting on $n\times 1$ matrices is necessarily the taking of a specific linear combination of the n elements of the matrices. Furthermore, $\|\mathbf{B}\|=\|\mathbf{y}\|=\sqrt{\bar{\mathbf{y}}^{\mathrm{T}}\mathbf{y}}=\sigma(\mathbf{y})$, the spectral radius of the $n\times 1$ matrix $\mathbf{y}$.

Example 6. Bounded Linear Functionals on $\mathcal{L}_2(a,b)$. The most general bounded linear functional $\mathbf{B}$ on the standard Hilbert space $\mathcal{L}_2(a,b)$ is $\mathbf{Bu}\triangleq\int_a^b\mathbf{u}(t)\mathbf{g}(t)\,dt$ for some specific $\mathbf{g}$ in $\mathcal{L}_2(a,b)$, and $\|\mathbf{B}\|=[\int_a^b|\mathbf{g}(t)|^2\,dt]^{1/2}$. For example, the response $\mathbf{f}$ of a single-input linear time-invariant dynamic system with zero initial conditions is the convolution of the system input $\mathbf{u}$ with the impulse response $\mathbf{g}$*:

$$\mathbf{f}(t)=\int_0^t\mathbf{u}(s)\mathbf{g}(t-s)\,ds$$

For the interval $0\leqslant t\leqslant b$, this function can be written

$$\mathbf{f}(t)=\int_0^b\mathbf{u}(s)\mathbf{k}_t(s)\,ds=\langle\mathbf{u},\mathbf{k}_t\rangle$$

where

$$\begin{aligned}\mathbf{k}_t(s)&=\mathbf{g}(t-s) && s\leqslant t\\ &=0 && s>t\end{aligned}$$

Thus for each instant t, $\mathbf{f}(t)$ is a bounded linear functional on $\mathcal{L}_2(0,b)$. The function in $\mathcal{L}_2(0,b)$ that represents the linear functional is $\mathbf{k}_t$. Treated as a function of t and s, $\mathbf{k}_t(s)$ is the Green's function for the dynamic system (see Chapter 3). A crude measure of the effect of the linear functional is the norm of the functional, $\|\mathbf{k}_t\|=[\int_0^t\mathbf{g}^2(t-s)\,ds]^{1/2}$.

Example 7. Function Evaluation, an Unbounded Linear Functional. Define $\mathbf{B}$: $\mathcal{L}_2(a,b)\to\mathfrak{R}$ by $\mathbf{Bf}\triangleq\mathbf{f}(t_0)$, where t_0 is in $[a,b]$. This linear functional, evaluation at t_0, is not bounded. For if $\mathbf{f}_k$ is a pulse of height k and width $1/k^2$, centered at t_0, then $\|\mathbf{f}_k\|=1$, but $\|\mathbf{Bf}_k\|=k\to\infty$. It is well known that $\mathbf{f}(t_0)=\int_a^b\mathbf{f}(t)\delta(t-t_0)\,dt$, where $\delta(t-t_0)$ is a Dirac delta function centered at t_0.† Thus in a sense the Riesz–Fréchet theorem extends to at least this unbounded linear functional. However, $\delta(t-t_0)$ does not have a finite norm and therefore is not in $\mathcal{L}_2(a,b)$.

*See Appendix 2 for a discussion of convolution and impulse response.
†See Appendix 2 for an introduction to the properties of delta functions.

The Adjoint

Let $\mathbf{T}: \mathcal{V} \rightarrow \mathcal{W}$ be a bounded linear transformation between *Hilbert spaces* $\mathcal{V}$ and $\mathcal{W}$. We now introduce the adjoint transformation $\mathbf{T}$, a generalization of the transposed matrix multiplication with which we introduced this section. The vector $\mathbf{Tx}$ is in $\mathcal{W}$. Since an inner product is a bounded linear functional of its left argument, $\mathbf{Ux} \stackrel{\Delta}{=} \langle \mathbf{Tx}, \mathbf{z} \rangle_{\mathcal{W}}$ is a bounded linear functional of the variable $\mathbf{x}$ in $\mathcal{V}$. In fact, by means of the Cauchy–Schwartz inequality (P&C 5.4) and the inequality (5.53), we can exhibit a bound:

$$\|\mathbf{Ux}\|_{\mathcal{W}} = |\langle \mathbf{Tx}, \mathbf{z} \rangle_{\mathcal{W}}| \leqslant \|\mathbf{Tx}\|_{\mathcal{W}} \|\mathbf{z}\|_{\mathcal{W}} \leqslant \|\mathbf{T}\| \, \|\mathbf{x}\|_{\mathcal{V}} \|\mathbf{z}\|_{\mathcal{W}}$$

or $\|\mathbf{U}\| \leqslant \|\mathbf{T}\| \, \|\mathbf{z}\|_{\mathcal{W}}$. The Riesz–Fréchet theorem (5.57) guarantees that there exists a unique vector $\mathbf{y}$ in $\mathcal{V}$ which represents this bounded linear functional $\mathbf{U}$ in the sense that

$$\mathbf{Ux} = \langle \mathbf{Tx}, \mathbf{z} \rangle_{\mathcal{W}} = \langle \mathbf{x}, \mathbf{y} \rangle_{\mathcal{V}}$$

It is evident from this equation that $\mathbf{y}$ in $\mathcal{V}$ and $\mathbf{z}$ in $\mathcal{W}$ are related. We define this relation to be the adjoint transformation, $\mathbf{y} = \mathbf{T}^*\mathbf{z}$.

Definition. The **adjoint transformation** $\mathbf{T}^*$ is defined by

$$\langle \mathbf{Tx}, \mathbf{z} \rangle_{\mathcal{W}} = \langle \mathbf{x}, \mathbf{T}^*\mathbf{z} \rangle_{\mathcal{V}} \tag{5.58}$$

for all $\mathbf{x}$ in $\mathcal{V}$ and $\mathbf{z}$ in $\mathcal{W}$.

The existence of $\mathbf{T}^*$ is guaranteed by the bounded linear nature of $\mathbf{T}$ and the completeness of $\mathcal{V}$. Uniqueness and linearity of $\mathbf{T}^*$ are easily verified. Furthermore, $\mathbf{T}^*$ is bounded; we recognize that

$$\|\mathbf{T}^*\mathbf{z}\|_{\mathcal{V}} = \|\mathbf{y}\|_{\mathcal{V}} = \|\mathbf{U}\| \leqslant \|\mathbf{T}\| \, \|\mathbf{z}\|_{\mathcal{W}}$$

By (5.54),

$$\|\mathbf{T}^*\| = \sup_{\|\mathbf{z}\|_{\mathcal{W}} = 1} \|\mathbf{T}^*\mathbf{z}\|_{\mathcal{V}} \leqslant \|\mathbf{T}\|$$

Since (5.58) is symmetric in $\mathbf{T}$ and $\mathbf{T}^*$, reversing the roles of $\mathbf{T}$ and $\mathbf{T}^*$ shows that $\|\mathbf{T}\| \leqslant \|\mathbf{T}^*\|$. Thus

$$\|\mathbf{T}\| = \|\mathbf{T}^*\| \tag{5.59}$$

An explicit description of $\mathbf{T}^*$ can be obtained from the defining equation (5.58) and a description of $\mathbf{T}$. The basic technique for obtaining the description of $\mathbf{T}^*$ is to write $\langle \mathbf{Tx},\mathbf{z}\rangle_{\mathcal{W}}$ and manipulate it into the form of $\langle \mathbf{x},\mathbf{T}^*\mathbf{z}\rangle_{\mathcal{V}}$; in effect, we work operations off of $\mathbf{x}$ and onto $\mathbf{z}$.

Example 8. The Adjoint of a Matrix Transformation. Let $\mathcal{V}=\mathfrak{M}_c^{n\times 1}$ and $\mathcal{W}=\mathfrak{M}_c^{m\times 1}$, each with its standard inner product. Define $\mathbf{T}$: $\mathcal{V}\to\mathcal{W}$ by $\mathbf{Tx}\triangleq\mathbf{Ax}$, where $\mathbf{A}$ is an $m\times n$ matrix. Then

$$\begin{aligned}\langle \mathbf{Tx},\mathbf{z}\rangle_{\mathcal{W}} &= \langle \mathbf{Ax},\mathbf{z}\rangle_{\mathcal{W}}\\ &= \overline{\mathbf{z}}^{\mathrm{T}}\mathbf{Ax}\\ &= \left(\overline{\overline{\mathbf{A}}^{\mathrm{T}}\mathbf{z}}\right)^{\mathrm{T}}\mathbf{x}\\ &= \langle \mathbf{x},\overline{\mathbf{A}}^{\mathrm{T}}\mathbf{z}\rangle_{\mathcal{V}} = \langle \mathbf{x},\mathbf{T}^*\mathbf{z}\rangle_{\mathcal{V}}\end{aligned}$$

Clearly,

$$\mathbf{T}^*\mathbf{z}=\overline{\mathbf{A}}^{\mathrm{T}}\mathbf{z} \tag{5.60}$$

Of course, if $\mathbf{A}$ is real (or if $\mathcal{V}$ and $\mathcal{W}$ are real) the conjugate is superfluous. It is apparent that the transposed matrix example with which we introduced Section 5.4 is just a special case of the general adjoint concept. If the inner products are not standard, multiplication by the conjugated transposed matrix is not the adjoint (P&C 5.19).

Example 9. The Adjoint of an Integral Operator. Let $\mathcal{V}=\mathcal{W}=\mathcal{L}_2^c(a,b)$. Define $\mathbf{T}$ by $(\mathbf{Tf})(t)\triangleq\int_a^b k(t,s)\mathbf{f}(s)\,ds$, where k is a Hilbert–Schmidt kernel; by Example 3, $\mathbf{T}$ is bounded. Then

$$\begin{aligned}\langle \mathbf{Tf},\mathbf{g}\rangle &= \int_a^b (\mathbf{Tf})(t)\,\overline{\mathbf{g}(t)}\,dt\\ &= \int_a^b\int_a^b k(t,s)\mathbf{f}(s)\,ds\,\overline{\mathbf{g}(t)}\,dt\\ &= \int_a^b \mathbf{f}(s)\int_a^b k(t,s)\,\overline{\mathbf{g}(t)}\,dt\,ds\\ &= \int_a^b \mathbf{f}(s)\,\overline{\int_a^b \overline{k(t,s)}\,\mathbf{g}(t)\,dt}\,ds\\ &= \int_a^b \mathbf{f}(s)\,\overline{(\mathbf{T}^*\mathbf{g})(s)}\,ds\end{aligned}$$

Therefore,

$$(\mathbf{T}^*\mathbf{g})(t)\triangleq\int_a^b \overline{k(s,t)}\,\mathbf{g}(s)\,ds \tag{5.61}$$

Whereas **T** requires integration with respect to the first variable in the kernel k, $\mathbf{T}^*$ requires integration with respect to the second variable. The kernel $\overline{k(s,t)}$ is the analogue of the conjugated transposed matrix of Example 8. Once again, if the spaces are real, the conjugations are superfluous.

Exercise 1. Let $\mathcal{V} = \mathcal{W} = \mathcal{L}_2(\omega;\ a,b)$, for which the inner product is (5.49). Assume the scalars are real. Define $\mathbf{T}\colon \mathcal{V} \to \mathcal{W}$ by

$$(\mathbf{Tf})(t) \triangleq \int_a^b k(t,s)\mathbf{f}(s)\,ds$$

where k is a Hilbert–Schmidt kernel. Show that

$$(\mathbf{T}^*\mathbf{g})(t) = \int_a^b \frac{\omega(s)k(s,t)}{\omega(t)}\mathbf{g}(s)\,ds \tag{5.62}$$

Adjoints of Differential Operators

Only for a bounded linear transformation on a Hilbert space does the preceding discussion guarantee the existence of an adjoint transformation which satisfies (5.58) and (5.59). In point of fact, if **T** is not bounded, it makes no sense to speak of $\|\mathbf{T}\|$. Yet among the most useful transformations are the linear differential operators, which are unbounded. Thus if the adjoint concept is useful, we have reason to attempt to apply the concept to differential operators. We shall see that differential operators do have adjoints.

Consider the simple differential operator **D** defined by $(\mathbf{Df})(t) \triangleq \mathbf{f}'(t)$ acting on functions defined over the interval $[a,b]$. Assume the standard function space inner product for both the domain and range of definition. Nothing prevents us from using the approach of Examples 8 and 9 to try to generate an adjoint for **D**. The natural technique for working differentiations off of one argument and onto another is integration by parts:

$$\begin{aligned}\langle \mathbf{Df},\mathbf{g}\rangle &= \int_a^b \mathbf{f}'(t)\mathbf{g}(t)\,dt\\ &= \mathbf{f}(t)\mathbf{g}(t)\big|_a^b - \int_a^b \mathbf{f}(t)\mathbf{g}'(t)\,dt\end{aligned} \tag{5.63}$$

It seems logical to define $\mathbf{D}^*$ by $(\mathbf{D}^*\mathbf{g})(t) \triangleq -\mathbf{g}'(t) = (-\mathbf{Dg})(t)$. However, this definition does not quite agree with the defining equation (5.58) unless the boundary term, $\mathbf{f}(b)\mathbf{g}(b) - \mathbf{f}(a)\mathbf{g}(a)$, is zero. We must not lose sight of the fact that differential operators usually have associated boundary conditions. Suppose the boundary condition associated with $\mathbf{D}$ is $\mathbf{f}(a)-\mathbf{f}(b)=0$. Then in order that the boundary term of (5.63) be zero, we must have $\mathbf{f}(b)[\mathbf{g}(b)-\mathbf{g}(a)]=0$, or $\mathbf{g}(b)-\mathbf{g}(a)=0$.

It should be apparent that for any ordinary linear differential operator $\mathbf{L}$ with a set of accompanying homogeneous boundary conditions we can use integration by parts to generate an adjoint differential operator $\mathbf{L}^*$ with accompanying adjoint homogeneous boundary conditions. The operators $\mathbf{L}$ and $\mathbf{L}^*$ satisfy the defining equation, (5.58), for all $\mathbf{f}$ and $\mathbf{g}$ in $\mathcal{L}_2(a,b)$ which satisfy the respective boundary conditions. [Of course, we have given up on (5.59).] If we were to change the homogeneous boundary conditions associated with $\mathbf{L}$, we would obtain a different set of adjoint boundary conditions, but the same adjoint differential operator $\mathbf{L}^*$. We refer to the adjoint differential operator $\mathbf{L}^*$ as the **formal adjoint of** $\mathbf{L}$. The formal adjoint is independent of boundary conditions. The definition of an operator always includes a definition of its domain. We use the homogeneous boundary conditions associated with $\mathbf{L}$ to restrict the domain of $\mathbf{L}$. The **adjoint boundary conditions**, which arise naturally out of the integration by parts, determine the restrictions on the domain of the formal adjoint $\mathbf{L}^*$ in order that it be a true adjoint of $\mathbf{L}$ [in the sense that it obeys (5.58)]. Thus the formal adjoint of $\mathbf{D}$ is $\mathbf{D}^*=-\mathbf{D}$. If the boundary condition associated with $\mathbf{D}$ is an initial condition, $\mathbf{f}(a)=0$, then by (5.63) the adjoint boundary condition is a final condition which requires that $\mathbf{g}(b)=0$ for each $\mathbf{g}$ in the domain of $\mathbf{D}^*$.

If we wished, we could further restrict the domains of $\mathbf{L}$ and $\mathbf{L}^*$ to include only differentiable functions, thereby eliminating delta functions and their derivatives from range($\mathbf{L}$) and range($\mathbf{L}^*$). However, formal use of integration by parts works for delta functions. Therefore, as a practical matter, we do not concern ourselves with this restriction.

Example 10. The Adjoint of $\mathbf{D}^n$. Let the nth derivative operator $\mathbf{D}^n$ act on a real space of functions defined over $[a,b]$. Assuming the standard inner product,

$$\begin{aligned}
\langle \mathbf{D}^n\mathbf{f},\mathbf{g}\rangle &= \int_a^b \mathbf{f}^{(n)}(t)\mathbf{g}(t)\,dt \\
&= \mathbf{f}^{(n-1)}(t)\mathbf{g}(t)|_a^b - \int_a^b \mathbf{f}^{(n-1)}(t)\mathbf{g}'(t)\,dt \\
&= \mathbf{f}^{(n-1)}(t)\mathbf{g}(t)|_a^b - \mathbf{f}^{(n-2)}(t)\mathbf{g}'(t)|_a^b + \int_a^b \mathbf{f}^{(n-2)}(t)\mathbf{g}^{(2)}(t)\,dt \\
&= \mathbf{f}^{(n-1)}(t)\mathbf{g}(t)|_a^b - \mathbf{f}^{(n-2)}(t)\mathbf{g}'(t)|_a^b + \cdots \\
&\quad + (-1)^{n-1}\mathbf{f}(t)\mathbf{g}^{(n-1)}(t)|_a^b + (-1)^n\int_a^b \mathbf{f}(t)\mathbf{g}^{(n)}(t)\,dt
\end{aligned}$$

(The intermediate terms are indicated only to show the pattern which the terms follow. Some of the terms shown are extraneous if $n=1$ or 2.) The formal adjoint of $\mathbf{D}^n$ is clearly $(-1)^n\mathbf{D}^n$. The adjoint boundary conditions depend upon the boundary conditions associated with $\mathbf{D}^n$. A specific set of boundary conditions does not determine a unique set of adjoint boundary conditions. However, the domain defined by the adjoint boundary conditions is unique.

Exercise 2. Show that the formal adjoint of the differential operator $\mathbf{L} \triangleq \mathbf{D}^n + a_1\mathbf{D}^{n-1} + \cdots + a_n\mathbf{I}$ acting on a space of functions with the *standard inner product* is $\mathbf{L}^* \triangleq (-1)^n\mathbf{D}^n + (-1)^{n-1}a_1\mathbf{D}^{n-1} + \cdots + a_n\mathbf{I}$.

Example 11. The Adjoint of a Partial Differential Operator. Let $\mathcal{L}_2(\Omega)$ be the space of real functions which are defined on a two-dimensional region Ω and which have finite norm under the inner product

$$\langle \mathbf{f},\mathbf{g}\rangle \triangleq \int_\Omega \mathbf{f}(\mathbf{p})\mathbf{g}(\mathbf{p})\,d\mathbf{p} \tag{5.64}$$

where $\mathbf{p}=(s,t)$, an arbitrary point in Ω. We define the Laplacian operator ∇^2 on $\mathcal{L}_2^2(\Omega)$ by

$$(\nabla^2\mathbf{f})(s,t) \triangleq \frac{\delta^2\mathbf{f}(s,t)}{\partial s^2} + \frac{\partial^2\mathbf{f}(s,t)}{\partial t^2} \qquad \text{for } (s,t) \text{ in } \Omega$$

For this problem, the **symmetric form of Green's theorem** is the equivalent of integration by parts; it states

$$\int_\Omega (\mathbf{f}\nabla^2\mathbf{g} - \mathbf{g}\nabla^2\mathbf{f})\,d\mathbf{p} = \oint_\Gamma (\mathbf{f}\mathbf{g}_n - \mathbf{g}\mathbf{f}_n)\,d\mathbf{p} \tag{5.65}$$

where Γ represents the boundary of the region Ω, and the subscript n indicates the derivative in a direction normal to Γ and directed out of Ω.* Using this theorem, we find

$$\begin{aligned}\langle \nabla^2\mathbf{f},\mathbf{g}\rangle &= \int_\Omega (\nabla^2\mathbf{f})(\mathbf{p})\mathbf{g}(\mathbf{p})\,d\mathbf{p} \\ &= \int_\Omega \mathbf{f}(\mathbf{p})(\nabla^2\mathbf{g})(\mathbf{p})\,d\mathbf{p} + \oint_\Gamma [\mathbf{g}(\mathbf{p})\mathbf{f}_n(\mathbf{p}) - \mathbf{f}(\mathbf{p})\mathbf{g}_n(\mathbf{p})]\,d\mathbf{p}\end{aligned}$$

Clearly, the formal adjoint of ∇^2 is just ∇^2 itself.

Not all boundary conditions are appropriate for a partial differential operator. For the Laplacian operator, one appropriate homogeneous boundary condition is $a\mathbf{f}(\mathbf{p}) + b\mathbf{f}_n(\mathbf{p}) = 0$ on the boundary. The adjoint boundary condition is selected such

*See Wylie [5.24, p. 575].

that the boundary integral in Green's theorem is zero. It is sufficient to make the integrand zero for each $\mathbf{p}$ on Γ:

$$\mathbf{g}(\mathbf{p})\mathbf{f}_n(\mathbf{p}) - \left(-\frac{b}{a}\mathbf{f}_n(\mathbf{p})\right)\mathbf{g}_n(\mathbf{p}) = \mathbf{f}_n(\mathbf{p})\left[\mathbf{g}(\mathbf{p}) + \frac{b}{a}\mathbf{g}_n(\mathbf{p})\right] = 0$$

Thus the adjoint boundary condition is $a\mathbf{g}(\mathbf{p}) + b\mathbf{g}_n(\mathbf{p}) = 0$ on Γ, the same as the original boundary condition associated with ∇^2.

Properties of Adjoints

Taking adjoints is similar to conjugation of complex numbers. Let $\mathcal{V}$, $\mathcal{W}$, and $\mathcal{U}$ be Hilbert spaces, and $\mathbf{I}$ the identity operator on $\mathcal{V}$. Suppose $\mathbf{T}$ and $\mathbf{U}$ are bounded linear transformations from $\mathcal{V}$ into $\mathcal{W}$, and $\mathbf{S}$ is a bounded linear transformation from $\mathcal{W}$ into $\mathcal{U}$. Then it is easy to show:

(*a*) $\mathbf{I}^* = \mathbf{I}$

(*b*) $(\mathbf{T}^*)^* = \mathbf{T}$

(*c*) $(a\mathbf{T} + b\mathbf{U})^* = \bar{a}\mathbf{T}^* + \bar{b}\mathbf{U}^*$ (5.66)

(*d*) $(\mathbf{ST})^* = \mathbf{T}^*\mathbf{S}^*$

(*e*) If $\mathbf{T}$ has a bounded inverse, $\mathbf{T}^*$ is invertible and $(\mathbf{T}^*)^{-1} = (\mathbf{T}^{-1})^*$

In fact, property (*e*) of (5.66) may be valid even if $\mathbf{T}^{-1}$ is not bounded. For example, let us define $\mathbf{T}$ on $\mathcal{L}_2(0,1)$ to be the bounded integral operator

$$(\mathbf{Tf})(t) \triangleq \int_0^t \mathbf{f}(s)\,ds$$

Then $\mathbf{T}^{-1}$ is the unbounded differential operator $\mathbf{Lf} \triangleq \mathbf{Df}$ with the homogeneous boundary condition $\mathbf{f}(0) = 0$. By (5.61) we know that $\mathbf{T}^*$ differs from $\mathbf{T}$ only in an interchange of the roles of t and s in the kernel function. In this instance the kernel function for $\mathbf{T}$ is

$$\begin{aligned} k(t,s) &= 1, \qquad s < t \\ &= 0, \qquad s > t \end{aligned}$$

Therefore,

$$(\mathbf{T}^*\mathbf{g})(t) \triangleq \int_0^1 k(s,t)\mathbf{g}(s)\,ds = \int_t^1 \mathbf{g}(s)\,ds$$

On the other hand, it follows from (5.63) that $(\mathbf{T}^{-1})^*$, the adjoint of $\mathbf{Lf} \stackrel{\Delta}{=} \mathbf{Df}$ with its homogeneous boundary condition $\mathbf{f}(0)=0$, is $\mathbf{L}^*\mathbf{g} \stackrel{\Delta}{=} -\mathbf{Dg}$ with the adjoint boundary condition $\mathbf{g}(1)=0$. But this differential system is also $(\mathbf{T}^*)^{-1}$, which we verify by acting on it with $\mathbf{T}^*$ to get the identity operator:

$$[\mathbf{T}^*(-\mathbf{Dg})](t)\Big|_{\mathbf{g}(1)=0} = \int_t^1 (-\mathbf{Dg})(s)\,ds\Big|_{\mathbf{g}(1)=0} = \mathbf{g}(t)$$

or $\mathbf{T}^*(\mathbf{T}^*)^{-1}\mathbf{g}=\mathbf{g}$.

One of the most valuable characteristics of the adjoint transformation is that it generates orthogonal decompositions of the domain and range of definition of $\mathbf{T}$. These decompositions are central to all forms of least-square optimization, to many iterative techniques for optimizing functionals, and to iterative techniques for solving nonlinear equations (Chapters 6–8).

Orthogonal Decomposition Theorem. If $\mathcal{V}$ and $\mathcal{W}$ are Hilbert spaces, and $\mathbf{T}: \mathcal{V} \to \mathcal{W}$ is a bounded linear transformation, then

$$\begin{aligned} \mathcal{V} &= \text{nullspace}(\mathbf{T}) \stackrel{\perp}{\oplus} \overline{\text{range}(\mathbf{T}^*)} \\ \mathcal{W} &= \text{nullspace}(\mathbf{T}^*) \stackrel{\perp}{\oplus} \overline{\text{range}(\mathbf{T})} \end{aligned} \tag{5.67}$$

The symbol $\stackrel{\perp}{\oplus}$ implies that these direct sums are orthogonal; the nullspaces and ranges are orthogonal complements. Theorem (5.67) is illustrated abstractly in Figure 5.9. The bars over range($\mathbf{T}$) and range($\mathbf{T}^*$) indicate the completion (or closure) of these linear manifolds. We have already seen this orthogonal decomposition demonstrated in the matrix example that introduced Section 5.4.

By the projection theorem, $\mathcal{V} = \mathcal{U} \stackrel{\perp}{\oplus} \mathcal{U}^\perp$ for any subspace $\mathcal{U}$ of a Hilbert space $\mathcal{V}$. Therefore, if we can show that nullspace($\mathbf{T}$)= [range($\mathbf{T}^*$)]$^\perp$, it follows that [nullspace($\mathbf{T}$)]$^\perp$ = [range($\mathbf{T}^*$)]$^{\perp\perp}$ = $\overline{\text{range}(\mathbf{T}^*)}$, and the first orthogonal decomposition of the theorem is proved. Let $\mathbf{y}$ be an arbitrary vector in $\mathcal{W}$; then $\mathbf{T}^*\mathbf{y}=\mathbf{z}$ is an arbitrary vector in range($\mathbf{T}^*$). The orthogonal complement of range($\mathbf{T}^*$) consists in all vectors $\mathbf{x}$ that are orthogonal to all $\mathbf{z}$ in range($\mathbf{T}^*$); that is, all $\mathbf{x}$ such that

$$0 = \langle \mathbf{x}, \mathbf{z} \rangle_{\mathcal{V}} = \langle \mathbf{x}, \mathbf{T}^*\mathbf{y} \rangle_{\mathcal{V}} = \langle \mathbf{Tx}, \mathbf{y} \rangle_{\mathcal{W}}$$

for all $\mathbf{y}$ in $\mathcal{W}$. Therefore, $\mathbf{Tx} = \boldsymbol{\theta}$, the vectors $\mathbf{x}$ constitute the nullspace of $\mathbf{T}$, and nullspace($\mathbf{T}$)=[range($\mathbf{T}^*$)]$^\perp$. The proof of the second orthogonal decomposition is parallel to that above.

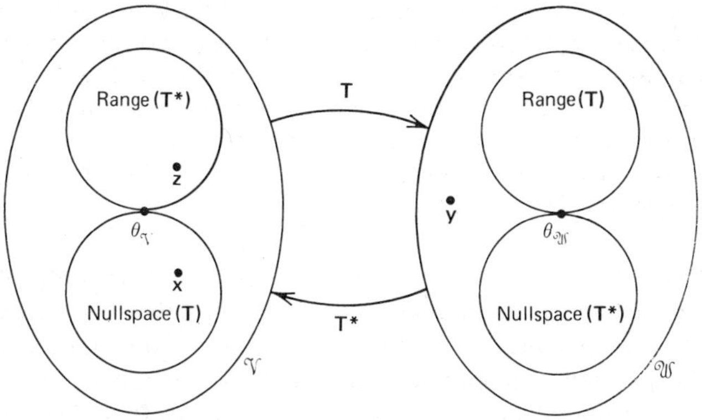

Figure 5.9. Orthogonal direct-sum decomposition described by **T***.

Since an orthogonal complement is always a (complete) subspace, we can see that the nullspace of any *bounded* linear transformation will be complete. On the other hand, the range need not be complete. For instance, let **T** on $\mathcal{L}_2(0,1)$ be defined by $(\mathbf{Tf})(t) \stackrel{\Delta}{=} \int_0^t \mathbf{f}(s)\,ds$. Then since $\mathcal{L}_2(0,1)$ contains no delta functions, range(**T**) contains only continuous functions; but the space of continuous functions [and thus, range(**T**)] is not complete under the $\mathcal{L}_2(0,1)$ norm. We usually assume range(**T**) and range(**T***) are complete, or ignore the difference between the ranges and their closures.

Although we have proved the orthogonal decomposition theorem (5.67) only for bounded linear transformations, it holds for many unbounded linear transformations as well. We use the theorem wherever the adjoint operator is defined. In particular, we apply the theorem to differential operators, even though they are not bounded.

Example 12. Orthogonal Decomposition for a Partial Differential Operator. Define ∇^2 on $\mathcal{L}_2(\Omega)$ as in Example 11. Let the boundary condition be $\mathbf{f}_n(\mathbf{p})=0$ on the boundary Γ. Then, by Example 11, the adjoint operator and adjoint boundary conditions are identical to the original operator and boundary conditions. The nullspace of ∇^2 with the boundary condition $\mathbf{f}_n=\boldsymbol{\theta}$ is the set of functions which are constant over Ω; that is, only if $\mathbf{f}(\mathbf{p})=c$ for $\mathbf{p}$ in Ω do we have $(\nabla^2\mathbf{f})(\mathbf{p})=0$ for $\mathbf{p}$ in Ω and $\mathbf{f}_n(\mathbf{p})=0$ for $\mathbf{p}$ on Γ. By the orthogonal decomposition theorem, we expect range(∇^2) to be the orthogonal complement of nullspace(∇^2). Therefore, if we wish to solve $\nabla^2\mathbf{f}=\mathbf{u}$ with $\mathbf{f}_n=\boldsymbol{\theta}$ on the boundary, we must be sure $\mathbf{u}$ is orthogonal to nullspace(∇^2), or

$$\begin{aligned}\langle \mathbf{u},\mathbf{f}\rangle &= \int_\Omega \mathbf{u}(\mathbf{p})\mathbf{f}(\mathbf{p})\,d\mathbf{p}\\ &= c\int_\Omega \mathbf{u}(\mathbf{p})\,d\mathbf{p}=0\end{aligned}$$

This result can be given a physical interpretation. If $\mathbf{u}(\mathbf{p})$ is the rate at which heat is introduced at the point $\mathbf{p}$ [with units of (heat)/(time)(area)], then the steady-state temperature distribution satisfies Poisson's equation $\nabla^2\mathbf{f}=\mathbf{u}$. The boundary condition $\mathbf{f}_n(\mathbf{p})=0$ says no heat is leaving Ω at the point $\mathbf{p}$ on the boundary. The orthogonal decomposition shows that we cannot find a steady-state temperature distribution such that no heat leaves the region unless the total heat generated per unit time, $\int_\Omega \mathbf{u}(\mathbf{p})\,d\mathbf{p}$, is zero.

It is apparent from Example 12 that the orthogonal decomposition does apply to at least some unbounded linear transformations. It often provides a useful way of checking whether or not a differential equation is solvable. If range($\mathbf{T}$) is closed, the operator equation $\mathbf{Tx}=\mathbf{y}$ is solvable if and only if $\mathbf{y}$ is orthogonal to nullspace($\mathbf{T}^*$). This nullspace is often easier to explore than is range($\mathbf{T}$); if $\mathbf{T}^*=\mathbf{T}$, as in Example 12, we find nullspace($\mathbf{T}^*$) by solving the homogeneous equation $\mathbf{Tx}=\boldsymbol{\theta}$. If the boundary condition in Example 12 were $\mathbf{f}(\mathbf{p})=0$ on the boundary Γ, the nullspace of ∇^2 would be empty. Then the orthogonal decomposition theorem would show that the operator was invertible.

If an operator $\mathbf{T}$ is not invertible, then the equation $\mathbf{Tx}=\mathbf{y}$ may have no solution or it may have many. The orthogonal decomposition theorem finds considerable use in solving such equations uniquely in a least square sense (Chapter 6). In point of fact, the decomposition pervades essentially all least-square optimization.

Let $\mathbf{T}$ be a bounded linear operator on a Hilbert space $\mathcal{H}$. Suppose $\mathbf{T}$ has eigenvalues and eigenvectors. We discovered in Section 4.3 that if $\mathbf{T}^{-1}$ exists, the eigendata for $\mathbf{T}$ and $\mathbf{T}^{-1}$ are related; $\mathbf{T}$ and $\mathbf{T}^{-1}$ have identical eigenvectors and inverse eigenvalues. Given the close relationship between $\mathbf{T}$ and $\mathbf{T}^*$, we also expect the eigendata for $\mathbf{T}$ to provide some information about the eigendata for $\mathbf{T}^*$. Let $\mathbf{x}_i$ be an eigenvector for $\mathbf{T}$ corresponding to the eigenvalue λ_i. Then, for any $\mathbf{y}$ in $\mathcal{V}$,

$$\langle \mathbf{Tx}_i,\mathbf{y}\rangle=\langle \lambda_i\mathbf{x}_i,\mathbf{y}\rangle=\langle \mathbf{x}_i,\bar{\lambda}_i\mathbf{y}\rangle=\langle \mathbf{x}_i,\mathbf{T}^*\mathbf{y}\rangle$$

Thus $\langle \mathbf{x}_i,(\mathbf{T}^*-\bar{\lambda}_i\mathbf{I})\mathbf{y}\rangle=0$ or range$(\mathbf{T}^*-\bar{\lambda}_i\mathbf{I})$ is orthogonal to $\mathbf{x}_i$. Consequently, range$(\mathbf{T}^*-\bar{\lambda}_i\mathbf{I})$ does not fill $\mathcal{H}$, and nullspace$(\mathbf{T}^*-\bar{\lambda}_i\mathbf{I})$ must be nonempty. In the finite-dimensional case we express this fact as

$$\text{nullity}\left(\mathbf{T}^*-\bar{\lambda}_i\mathbf{I}\right)=\dim\mathcal{H}-\text{rank}\left(\mathbf{T}^*-\bar{\lambda}_i\mathbf{I}\right)$$

We see that *if* λ_i *is an eigenvalue for* $\mathbf{T}$, $\bar{\lambda}_i$ *is an eigenvalue for* $\mathbf{T}^*$.

We will show that the eigenvectors of $\mathbf{T}$ and $\mathbf{T}^*$ are related as well. Suppose $\mathbf{x}_i$ and $\mathbf{x}_j$ are eigenvectors of $\mathbf{T}$ corresponding to the eigenvalues λ_i and λ_j, respectively; let $\mathbf{y}_i$ and $\mathbf{y}_j$ be the corresponding eigenvectors of $\mathbf{T}^*$.

Then

$$\begin{aligned} 0 &= \langle \mathbf{T}\mathbf{x}_i, \mathbf{y}_j \rangle - \langle \mathbf{x}_i, \mathbf{T}^*\mathbf{y}_j \rangle \\ &= (\lambda_i - \lambda_j)\langle \mathbf{x}_i, \mathbf{y}_j \rangle \end{aligned} \tag{5.68}$$

Clearly, *if* λ_i *and* λ_j *are different eigenvalues of* $\mathbf{T}$, *the eigenvectors* $\mathbf{x}_i$ *and* $\mathbf{y}_j$ (*of* $\mathbf{T}$ *and* $\mathbf{T}^*$, *respectively*) *are orthogonal.*

As a general rule, we expect $\mathbf{T}$ to be diagonalizable. That is, there is usually a countable (and perhaps orthonormal) basis for $\mathcal{H}$ composed of eigenvectors for $\mathbf{T}$. Nondiagonalizability is the exception. (Of course, in infinite-dimensional spaces we sometimes find there are no eigenvectors; any dynamic system with initial conditions is an example.) Suppose $\mathbf{T}$ is diagonalizable; let $\{\lambda_i\}$ be the eigenvalues of $\mathbf{T}$ and $\{\mathbf{x}_i\}$ a corresponding set of eigenvectors, a basis for $\mathcal{H}$. Then the numbers $\{\bar{\lambda}_i\}$ are the eigenvalues of $\mathbf{T}^*$; we denote the corresponding eigenvectors of $\mathbf{T}^*$ by $\{\mathbf{y}_i\}$. The eigenvectors $\mathbf{y}_i$ can be chosen such that

$$\langle \mathbf{x}_i, \mathbf{y}_j \rangle = \delta_{ij} \tag{5.69}$$

For those eigenspaces of $\mathbf{T}$ which are one-dimensional, (5.69) requires only normalization of the one available eigenvector $\mathbf{y}_i$ so that $\langle \mathbf{x}_i, \mathbf{y}_i \rangle = 1$. If $\mathbf{T}$ has several independent eigenvectors, say, $\mathbf{x}_1, \ldots, \mathbf{x}_m$, for a single eigenvalue λ_i, then the eigenvalue $\bar{\lambda}_i$ for $\mathbf{T}^*$ also has m independent eigenvectors, $\mathbf{y}_1, \ldots, \mathbf{y}_m$, which we choose by solving m^2 independent linear equations in m^2 unknowns, $\langle \mathbf{x}_k, \mathbf{y}_j \rangle = \delta_{kj}$, for $k, j = 1, \ldots, m$. The eigenvectors $\{\mathbf{y}_i\}$ of $\mathbf{T}^*$, chosen to satisfy (5.69), form a basis for $\mathcal{H}$ which we say is **biorthogonal** to the basis $\{\mathbf{x}_i\}$. We call $\{\mathbf{y}_i\}$ the **reciprocal basis** (P&C 5.31).

Since the eigenvectors of $\mathbf{T}$, $\{\mathbf{x}_i\}$, have been assumed to form a basis for $\mathcal{H}$, we can express any $\mathbf{x}$ in $\mathcal{H}$ in the form $\mathbf{x} = \sum_k c_k \mathbf{x}_k$. Then for any vector $\mathbf{y}_i$ in the reciprocal basis,

$$\langle \mathbf{x}, \mathbf{y}_i \rangle = \sum_k c_k \langle \mathbf{x}_k, \mathbf{y}_i \rangle = \sum_k c_k \delta_{ki} = c_i$$

where we have used the continuity of the inner product to take the infinite sum outside the inner product. Therefore any $\mathbf{x}$ in $\mathcal{H}$ has the representation

$$\mathbf{x} = \sum_k \langle \mathbf{x}, \mathbf{y}_k \rangle \mathbf{x}_k \tag{5.70}$$

The "biorthogonal" eigenvector expansion (5.70) is very much like an orthonormal eigenvector expansion. It can be used to diagonalize the operator equation, $\mathbf{Tx}=\mathbf{y}$. Furthermore, because of the biorthogonal nature of the reciprocal bases, the coefficients are computationally independent. Given a basis of eigenvectors $\{\mathbf{x}_i\}$, it is evident that finding the reciprocal eigenvector basis $\{\mathbf{y}_i\}$ is an alternative to finding the inner product which orthonormalizes $\{\mathbf{x}_i\}$.*

Exercise 3. Show that every $\mathbf{x}$ in $\mathcal{H}$ also has a biorthogonal expansion in the eigenvectors of $\mathbf{T}^*$:

$$\mathbf{x}=\sum_k \langle \mathbf{x},\mathbf{x}_k\rangle \mathbf{y}_k \tag{5.71}$$

5.5 Spectral Decomposition in Infinite-Dimensional Spaces

Because differential equations appear so frequently as models for real phenomena, we have a keen interest in the analysis of such equations. Motivated by the insight that comes from the decoupling of finite-dimensional equations, we seek to perform a similar decoupling of equations involving infinite-dimensional spaces. Suppose $\mathbf{T}$ is a linear transformation on an infinite-dimensional Hilbert space $\mathcal{H}$. In order to diagonalize (or decouple) the equation $\mathbf{Tx}=\mathbf{y}$, we search for a basis for $\mathcal{H}$ composed of eigenvectors of $\mathbf{T}$. Because the space is infinite dimensional, we naturally want to work only with an orthonormal basis; we will find that orthonormality of the eigenvectors requires that $\mathbf{T}$ be self-adjoint ($\mathbf{T}^*=\mathbf{T}$). Furthermore, we wish the orthonormal eigenvectors of $\mathbf{T}$ to be countable and complete in $\mathcal{H}$ in order that we can expand any vector in $\mathcal{V}$ as a unique, infinite sum of eigenvectors. If $\mathbf{T}$ has a countable, orthonormal set of eigenvectors $\{\mathbf{x}_k\}$ which is a basis for $\mathcal{H}$, then we can express any vector $\mathbf{x}$ in $\mathcal{H}$ uniquely as a Fourier series expansion in the eigenvectors:

$$\mathbf{x}=\sum_{k=1}^{\infty} \langle \mathbf{x},\mathbf{x}_k\rangle \mathbf{x}_k \tag{5.72}$$

Equivalent to the statement that any vector $\mathbf{x}$ in $\mathcal{H}$ can be expanded uniquely as in (5.72) is the following orthogonal direct-sum decomposition of $\mathcal{H}$ into one-dimensional eigenspaces:

$$\mathcal{H}=\text{span}(\mathbf{x}_1)\overset{\perp}{\oplus}\text{span}(\mathbf{x}_2)\overset{\perp}{\oplus}\cdots \tag{5.73}$$

If we sum those subspaces which are associated with identical eigenvalues,

*See Lamarsh [5.15, p. 549] for a practical function space example.

we can rewrite (5.73) as

$$\mathcal{H} = \sum_{j=1}^{\infty} \overset{\perp}{\oplus} \text{nullspace}(\mathbf{T} - \lambda_j \mathbf{I}) \tag{5.74}$$

where the set $\{\lambda_j\}$ consists of the distinct eigenvalues of **T** (which are not necessarily numbered in correspondence to the eigenvectors $\{\mathbf{x}_k\}$). Equation (5.74) is known as the **spectral theorem**. We will use (5.72) and its equivalents (5.73) and (5.74), to analyze (diagonalize or decouple) operator equations in infinite-dimensional spaces; in particular, differential equations.

Orthonormal Eigenvectors

Assume the bounded linear operator **T** on the Hilbert space $\mathcal{H}$ is diagonalizable; that is, $\mathcal{H}$ has a countable basis of eigenvectors of **T**. A logical place to begin exploration of *orthonormal* eigenvector bases for $\mathcal{H}$ is (5.69)–(5.70). It is clear that if $\mathbf{T}^* = \mathbf{T}$, the eigenvalues and eigenvectors of **T** and **T*** are identical. Then the eigenvalues λ_i are real, the eigenvectors corresponding to different eigenvalues are orthogonal, and the eigenvectors $\mathbf{x}_i$ can be selected so that they form a countable orthonormal basis. A linear operator for which $\mathbf{T}^* = \mathbf{T}$ is said to be **self-adjoint**. Self-adjointness is the key to orthonormality of eigenvectors. If the eigenvalues are real, self-adjointness of **T** is, in fact, necessary in order that there exist eigenvectors of **T** which form an orthonormal basis for $\mathcal{H}$. For if **T** is diagonalizable and $\mathfrak{X} \triangleq \{\mathbf{x}_i\}$ is an orthonormal eigenvector basis, then $[\mathbf{T}]_{\mathfrak{X}\mathfrak{X}}$ is a (possibly infinite) diagonal matrix;

$$[\mathbf{T}]_{\mathfrak{X}\mathfrak{X}} = \begin{pmatrix} \lambda_1 & & & \\ & \lambda_2 & & \\ & & \cdot & \\ & & & \cdot \\ & & & & \cdot \end{pmatrix}$$

But for any orthonormal basis, $[\mathbf{T}^*] = \overline{[\mathbf{T}]}^{\mathrm{T}}$ (P&C 5.27). Therefore,

$$[\mathbf{T}^*]_{\mathfrak{X}\mathfrak{X}} = \begin{pmatrix} \bar{\lambda}_1 & & & \\ & \bar{\lambda}_2 & & \\ & & \cdot & \\ & & & \cdot \\ & & & & \cdot \end{pmatrix}$$

and orthonormal eigenvectors for **T** are also orthonormal eigenvectors for $\mathbf{T}^*$. It follows that if the eigenvalues $\{\lambda_i\}$ are real, $\mathbf{T}^* = \mathbf{T}$. In sum, if the linear operator **T** is diagonalizable (a basis of eigenvectors exists) and the eigenvalues of **T** are real, then *there exists an orthonormal basis for the Hilbert space* $\mathcal{H}$ *consisting in eigenvectors of* **T** *if and only if* **T** *is self-adjoint*.

Exercise 1. Show that if **T** is diagonalizable (and the eigenvalues are not necessarily real) there is an orthonormal basis for $\mathcal{H}$ consisting in eigenvectors of **T** if and only if $\mathbf{TT}^* = \mathbf{T}^*\mathbf{T}$. A linear operator such that $\mathbf{TT}^* = \mathbf{T}^*\mathbf{T}$ is said to be a **normal operator**. Show that a normal operator which has real eigenvalues is self-adjoint.

In Section 5.2 we determined how to pick an inner product to orthonormalize a basis for a finite-dimensional space. The result, (5.27), was applied in Example 5 of that section to make the eigenvectors of a particular diagonalizable transformation orthonormal. Since the inner product was chosen to orthonormalize the eigenvector basis, it must also have made the operator self-adjoint. To see that this is the case, we find the adjoint of the operator **T** of Example 5, Section 5.2 relative to the orthonormalizing inner product. The operator **T** on $\mathcal{R}^2$ was defined by

$$\mathbf{T}(\xi_1, \xi_2) \stackrel{\Delta}{=} (2\xi_1 + 3\xi_2, 4\xi_2)$$

The orthonormalizing inner product was

$$\langle(\xi_1,\xi_2),(\eta_1,\eta_2)\rangle \stackrel{\Delta}{=} \xi_1\eta_1 - \tfrac{3}{2}\xi_1\eta_2 - \tfrac{3}{2}\xi_2\eta_1 + \tfrac{5}{2}\xi_2\eta_2$$

Let $\mathbf{x} \stackrel{\Delta}{=} (\xi_1, \xi_2)$ and $\mathbf{y} \stackrel{\Delta}{=} (\eta_1, \eta_2)$. Then

$$\begin{aligned}\langle \mathbf{Tx}, \mathbf{y}\rangle &= \langle(2\xi_1 + 3\xi_2, 4\xi_2), (\eta_1, \eta_2)\rangle \\ &= (2\xi_1 + 3\xi_2)\eta_1 - \tfrac{3}{2}(2\xi_1 + 3\xi_2)\eta_2 - \tfrac{3}{2}(4\xi_2)\eta_1 + \tfrac{5}{2}(4\xi_2)\eta_2 \\ &= 2\xi_1\eta_1 - 3\xi_1\eta_2 - 3\xi_2\eta_1 - \tfrac{11}{2}\xi_2\eta_2\end{aligned}$$

Since $\langle \mathbf{Tx}, \mathbf{y}\rangle$ is real and symmetric in **x** and **y**,

$$\langle \mathbf{Tx}, \mathbf{y}\rangle = \langle \mathbf{Ty}, \mathbf{x}\rangle = \langle \mathbf{x}, \mathbf{Ty}\rangle$$

Hence **T** is self-adjoint. In general, if we can pick an inner product to make a linear operator self-adjoint, we automatically guarantee that we can find an orthonormal basis consisting of eigenvectors of that operator (P&C 5.32).

Self-Adjoint Linear Operators on Real Function Spaces

Most models of physical systems have real eigenvalues. For these models, orthonormal bases of eigenvectors require self-adjointness of the model. It is because of the usefulness of orthonormal bases of eigenvectors for infinite-dimensional spaces that there is so much emphasis in the literature of physics and mathematics on self-adjoint differential and integral operators.

Let $(\mathbf{Tu})(t) \stackrel{\Delta}{=} \int_a^b k(t,s)\mathbf{u}(s)\,ds$ define an integral operator on a real function space. The adjoint of **T** relative to the standard inner product is given by (5.61). If the integral operator is to be self-adjoint, it is apparent that the kernel of the integral operator must be symmetric:

$$k(t,s) = k(s,t) \tag{5.75}$$

Suppose $\mathbf{Tu} = \mathbf{f}$. We can interpret $k(t,s)$ as a measure of the influence of the value, $\mathbf{u}(s)$, of the "input" function at point s on the value, $\mathbf{f}(t)$, of the "output" function at point t. Self-adjointness, (5.75), implies that the source point and observation point can be interchanged. That is, if $\mathbf{u}(s) = \delta(t_1 - s)$ then $\mathbf{f}(t_2) = \int_a^b k(t_2,s)\delta(t_1 - s)\,ds = k(t_2,t_1)$; interchanging the source and observation points, we find, $\mathbf{f}(t_1) = \int_a^b k(t_1,s)\delta(t_2 - s)\,ds = k(t_1,t_2) = k(t_2,t_1) = \mathbf{f}(t_2)$. This interchangeability of source and observation points is called **reciprocity**. Any system which can be described by an integral operator that is self-adjoint in the standard inner product exhibits reciprocity.

Example 1. A Self-Adjoint Integral Operator. The differential equation $-\mathbf{D}^2\mathbf{f} = \mathbf{u}$ with $\mathbf{f}(0) = \mathbf{f}(b) = 0$ describes the steady-state temperature distribution $\mathbf{f}$ along the length of an insulated bar of length b. The input function $\mathbf{u}$ represents the rate of heat generation, perhaps from induction heating, as a function of position within the bar. The temperature is fixed at the bar ends. We inverted this differential equation in Chapter 3. The Green's function (the kernel of the inverse operator), as given in (3.14), is

$$\begin{aligned} k(t,s) &= \frac{(b-s)t}{b}, & 0 \leqslant t \leqslant s \\ &= \frac{(b-t)s}{b}, & s \leqslant t \leqslant b \end{aligned}$$

for $0 \leqslant s \leqslant b$. By (5.61), the adjoint Green's function is

$$\begin{aligned} k(s,t) &= \frac{(b-t)s}{b}, & 0 \leqslant s \leqslant t \\ &= \frac{(b-s)t}{b}, & t \leqslant s \leqslant b \end{aligned}$$

for $0 \leqslant t \leqslant b$. Clearly, $k(t,s)=k(s,t)$, and the integral operator is self-adjoint. It is well known that steady-state heat flow problems exhibit reciprocity.

Suppose we use the weighted inner product $\langle \mathbf{f},\mathbf{g}\rangle = \int_a^b \omega(s)\mathbf{f}(s)\mathbf{g}(s)\,ds$ rather than the standard inner product. Then, by (5.62), in order that the integral operator $\mathbf{T}$ be self-adjoint we must have

$$k(t,s)=\frac{\omega(s)k(s,t)}{\omega(t)} \tag{5.76}$$

We will employ (5.76) when we discuss techniques for picking inner products for function spaces.

The adjoint of a differential system ($\mathbf{L}$ with its boundary conditions) consists in the formal adjoint $\mathbf{L}^*$ with the adjoint boundary conditions (or adjoint domain). Recall that the formal adjoint is independent of boundary conditions. If $\mathbf{L}^*=\mathbf{L}$, we say $\mathbf{L}$ is **formally self-adjoint.** We say *the differential system* ($\mathbf{L}$ *with its boundary conditions*) *is self-adjoint if* $\mathbf{L}$ *is formally self-adjoint and the adjoint boundary conditions are identical to the boundary conditions for* $\mathbf{L}$.† Exercise 2 of Section 5.4 shows that for the standard inner product the differential operators that are formally self-adjoint are those that contain only even derivatives. In the next example we explore various boundary conditions for the simplest differential operator which is formally self-adjoint with respect to the standard inner product.

Example 2. Self-Adjoint Boundary Conditions for $\mathbf{D}^2$**.** The operator $\mathbf{L}=-\mathbf{D}^2$ is formally self-adjoint. From Example 10 of Section 5.4, using the standard inner product, we find that

$$\begin{aligned}\langle -\mathbf{D}^2\mathbf{f},\mathbf{g}\rangle &= \langle \mathbf{f}, -\mathbf{D}^2\mathbf{g}\rangle - \mathbf{f}'(t)\mathbf{g}(t)|_a^b + \mathbf{f}(t)\mathbf{g}'(t)|_a^b \\ &= \langle \mathbf{f}, -\mathbf{D}^2\mathbf{g}\rangle + \mathbf{f}(b)\mathbf{g}'(b) - \mathbf{f}(a)\mathbf{g}'(a) - \mathbf{f}'(b)\mathbf{g}(b) + \mathbf{f}'(a)\mathbf{g}(a)\end{aligned}$$

Suppose the boundary conditions associated with $\mathbf{L}$ are $\mathbf{f}(a)=\mathbf{f}(b)=0$. Then the adjoint boundary conditions are $\mathbf{g}(a)=\mathbf{g}(b)=0$, and $\mathbf{L}=-\mathbf{D}^2$ is self-adjoint. This result is consistent with Example 1, wherein we showed the self-adjointness of the integral operator which is the inverse of this differential system (for the case where $a=0$). On the other hand, let the boundary conditions associated with $\mathbf{L}$ be the initial conditions $\mathbf{f}(a)=\mathbf{f}'(a)=0$. Then the adjoint boundary conditions are the final conditions $\mathbf{g}(b)=\mathbf{g}'(b)=0$, and $\mathbf{L}$ is not self-adjoint. We found in Chapter 3 that the Green's function $k(t,s)$ for an initial condition problem is always zero for $s \geqslant t$. Thus the integral operator which is the inverse of this initial condition problem is

†The adjoint boundary conditions are not necessarily unique, but the domain which they define is unique. To be precise, for self-adjointness we require the domain of $\mathbf{L}^*$ to be identical to the domain of $\mathbf{L}$.

not self-adjoint either. Furthermore, we found in Section 4.1 that initial condition problems have no eigenvalues, much less orthonormal eigenfunctions.

Exercise 2. Let $\mathbf{L} = -\mathbf{D}^2$. Verify the following adjoint boundary conditions. Assume $c_1 \neq c_2$.

Boundary Conditions on $\mathbf{L}$	Boundary Conditions on $\mathbf{L}^*$
$\mathbf{f}(a)+c_1\mathbf{f}'(a)=\mathbf{f}(b)+c_2\mathbf{f}'(b)=0$ (separated conditions)	$\mathbf{g}(a)+c_1\mathbf{g}'(a)=\mathbf{g}(b)+c_2\mathbf{g}'(b)=0$
$\mathbf{f}(a)+c_1\mathbf{f}(b)=\mathbf{f}'(a)+c_2\mathbf{f}'(b)=0$ (mixed conditions)	$\mathbf{g}(a)+\dfrac{1}{c_2}\mathbf{g}(b)=\mathbf{g}'(a)+\dfrac{1}{c_1}\mathbf{g}'(b)=0$
$\mathbf{f}(a)-\mathbf{f}(b)=\mathbf{f}'(a)-\mathbf{f}'(b)=0$ (periodic conditions)	$\mathbf{g}(a)-\mathbf{g}(b)=\mathbf{g}'(a)-\mathbf{g}'(b)=0$
No boundary conditions	$\mathbf{g}(a)=\mathbf{g}(b)=\mathbf{g}'(a)=\mathbf{g}'(b)=0$

Example 2 and Exercise 2 demonstrate a few general conclusions that can be drawn concerning self-adjointness of second-order differential operators, assuming the differential operator is formally self-adjoint:

1. Separated end-point conditions (wherein each boundary condition involves only one point of $[a,b]$) always yield a self-adjoint operator.
2. Mixed end-point conditions (wherein more than one point of $[a,b]$ can be involved in each boundary condition) seldom yield a self-adjoint operator.
3. Periodic boundary conditions (wherein the conditions at a equal the conditions at b) always yield a self-adjoint operator.
4. Initial conditions always lead to final adjoint boundary conditions. Thus dynamic initial condition problems are never self-adjoint.

We speak loosely of **self-adjoint boundary conditions** when we mean boundary conditions that lead to a self-adjoint operator.

Example 3. A Self-Adjoint Partial Differential Operator. In Example 11 of Section 5.4 we obtained the adjoint of the operator ∇^2 as it acted on the space of two-dimensional functions $\mathcal{L}_2(\Omega)$ with its standard inner product (5.64). We found that ∇^2 is formally self-adjoint. Furthermore, the boundary condition $a\mathbf{f}(\mathbf{p})+b\mathbf{f}_n(\mathbf{p})=0$ for $\mathbf{p}$ on the boundary Γ is also self-adjoint. (This boundary condition is an extension of the separated endpoint conditions illustrated above.) The inverse of the equation $\nabla^2\mathbf{f}=\mathbf{u}$ together with the above boundary condition is of the form

$$\mathbf{f}(\mathbf{p})=\int_\Omega k(\mathbf{p},\mathbf{q})\mathbf{u}(\mathbf{q})d\mathbf{q}$$

Since the differential system is self adjoint, we expect the Green's function k (which defines the inverse system) to be symmetric in $\mathbf{p}$ and $\mathbf{q}$. Consequently the inverse equation will exhibit reciprocity. Since the differential operator (and its inverse) is self adjoint, there exists an orthonormal basis for $\mathcal{L}_2(\Omega)$ consisting of eigenfunctions of ∇^2 which satisfy the above homogeneous boundary condition on Γ. We use these eigenfunctions later (Example 6) to derive the Green's function and solve the partial differential equation.

Exercise 3. Let Γ be the boundary of the rectangle $0\leqslant s\leqslant a$, $0\leqslant t\leqslant b$ in the (s,t) plane. The eigenfunctions corresponding to Example 3 are given in (4.63) for the boundary condition $\mathbf{f}(\mathbf{p})=0$ on Γ. Show that these eigenfunctions are orthogonal with respect to the standard inner product for $\mathcal{L}_2(\Omega)$.

Choosing Inner Products to Orthonormalize Eigenfunctions

Suppose $\mathbf{T}$ is a diagonalizable operator which acts on $\mathcal{L}_2(a,b)$. Then there is a basis for $\mathcal{L}_2(a,b)$ consisting in eigenfunctions for $\mathbf{T}$. The discussion associated with (5.49) showed that we can modify the inner product with a bounded positive weight function ω without changing the convergence of sequences of vectors; therefore, the eigenfunctions of $\mathbf{T}$ are also a basis for $\mathcal{L}_2(\omega;a,b)$. Although the weighted inner product (5.49) does not represent all possible inner products on the function space, in some circumstances we would expect to be able to make an eigenvector basis orthonormal (or at least orthogonal) by choice of the weight function. A given eigenvector basis can be orthogonal only if $\mathbf{T}$ is self-adjoint with respect to the weighted inner product. (Of course, we cannot make $\mathbf{T}$ self-adjoint unless the eigenvalues are real.)

In Example 3 of Section 5.3 we orthogonalized the eigenfunctions, $\mathbf{f}_k(t)=e^{-t/2}\sin(\pi kt/b)$, of the differential operator $\mathbf{D}^2+\mathbf{D}$ with the boundary conditions $\mathbf{f}(0)=\mathbf{f}(b)=0$ by choosing the weight function $\omega(t)=e^t$. Since the eigenfunction basis can be orthogonal only if the operator is self-adjoint, we could as well pick ω to assure self-adjointness. The adjoint of $\mathbf{D}^2+\mathbf{D}$ is determined by

$$\begin{aligned}\langle(\mathbf{D}^2+\mathbf{D})\mathbf{f},\mathbf{g}\rangle_\omega &\stackrel{\Delta}{=}\int_0^b\omega(\mathbf{f}''+\mathbf{f}')\mathbf{g}\,dt\\ &=\int_0^b\mathbf{f}[\omega\mathbf{g}''+(2\omega'-\omega)\mathbf{g}'+(\omega''-\omega')\mathbf{g}]\,dt\\ &\quad+(\omega\mathbf{g}\mathbf{f}'-\omega\mathbf{g}'\mathbf{f}-\omega'\mathbf{g}\mathbf{f}+\omega\mathbf{g}\mathbf{f})|_0^b\\ &\stackrel{\Delta}{=}\langle\mathbf{f},(\mathbf{D}^2+\mathbf{D})^*\mathbf{g}\rangle_\omega+\text{boundary terms}\end{aligned}$$

In order that the operator be formally self-adjoint, it must satisfy

$$\langle \mathbf{f}, (\mathbf{D}^2+\mathbf{D})^*\mathbf{g}\rangle_\omega = \int_0^b \omega \mathbf{f}(\mathbf{g}''+\mathbf{g}')\,dt$$

We choose ω so that the integrands in the above two expressions for $\langle \mathbf{f}, (\mathbf{D}^2+\mathbf{D})^*\mathbf{g}\rangle$ are identical; that is, so $2\omega'-\omega=\omega$ and $\omega''-\omega'=\boldsymbol{\theta}$. The common solutions to these two differential equations are the multiples of $\omega(t)=e^t$, the same weight function found earlier. Note that there is no additional freedom in the choice of ω with which to produce self-adjointness of the boundary conditions; the self-adjointness of the boundary conditions can be investigated after ω is determined.

Exercise 4. The Green's function for the differential operator $\mathbf{D}^2+\mathbf{D}$ with the boundary conditions $\mathbf{f}(0)=\mathbf{f}(b)=0$ is the function $k(t,s)$ of (3.42). Use the self-adjointness condition (5.76) to show again that $\omega(t)=e^t$.

We will demonstrate that every "nice" second-order differential operator is *formally* self-adjoint with respect to some weight function ω. As a consequence, since so many physical systems are representable by second-order differential equations, we can use orthonormal bases of eigenfunctions in analyzing an appreciable fraction of the differential equations which appear in practice. Suppose the differential operator $\mathbf{L}$ is defined for functions $\mathbf{f}$ which are twice continuously differentiable on $[a,b]$ by

$$(\mathbf{Lf})(t) \triangleq g_0(t)\mathbf{f}''(t)+g_1(t)\mathbf{f}'(t)+g_2(t)\mathbf{f}(t) \tag{5.77}$$

Assume $g_i(t)$ is continuous and $g_0(t)<0$ in the interval. We define the new variables p, q, and ω by

$$p(t) \triangleq \exp \int_a^t \frac{g_1(s)}{g_0(s)}\,ds, \qquad \omega(t) \triangleq -\frac{p(t)}{g_0(t)}, \qquad q(t) \triangleq g_2(t) \tag{5.78}$$

From (5.78) it follows that $g_1/g_0=p'/p$. (Furthermore, p', q, and ω are continuous; p and ω are bounded and positive.) Then the general second-order differential operator (5.77) can be expressed as

$$\begin{aligned}\mathbf{Lf} &= g_0\left(\mathbf{f}''+\frac{g_1}{g_0}\mathbf{f}'\right)+g_2\mathbf{f}\\ &= -\frac{p}{\omega}\left(\mathbf{f}''+\frac{p'}{p}\mathbf{f}'\right)+q\mathbf{f}\\ &= -\frac{1}{\omega}(p\mathbf{f}')'+q\mathbf{f}\end{aligned} \tag{5.79}$$

The operator (5.79) is commonly referred to as a **regular Sturm–Liouville operator.*** We now show via the form (5.79) that *the general second-order differential operator* **L** *is formally self-adjoint with respect to the positive weight function* ω *given in* (5.78):

$$\begin{aligned}
\langle \mathbf{Lf}, \mathbf{g} \rangle_\omega &= \int_a^b \omega \left[-\frac{1}{\omega}(p\mathbf{f}')' + q\mathbf{f} \right] \mathbf{g}\, dt \\
&= \int_a^b [-(p\mathbf{f}')'\mathbf{g} + \omega q \mathbf{f}\mathbf{g}]\, dt \\
&= -(p\mathbf{f}')\mathbf{g}|_a^b + \int_a^b p\mathbf{f}'\mathbf{g}'\, dt + \int_a^b \omega q \mathbf{f}\mathbf{g}\, dt \\
&= -p\mathbf{f}'\mathbf{g}|_a^b + \mathbf{f}p\mathbf{g}'|_a^b - \int_a^b \mathbf{f}(p\mathbf{g}')'\, dt + \int_a^b \omega q \mathbf{f}\mathbf{g}\, dt \\
&= p(\mathbf{f}\mathbf{g}' - \mathbf{f}'\mathbf{g})|_a^b + \int_a^b \omega \mathbf{f} \left[-\frac{1}{\omega}(p\mathbf{g}')' + q\mathbf{g} \right] dt \\
&= \text{boundary terms } + \langle \mathbf{f}, \mathbf{Lg} \rangle_\omega
\end{aligned} \tag{5.80}$$

If the boundary conditions are also self-adjoint, we expect to find an orthonormal set of eigenfunctions for **L**. In point of fact, this orthonormal set of eigenfunctions is complete in $\mathcal{L}_2(\omega; a, b)$, and we can diagonalize equations which involve the general second-order differential operator (5.77). See Birkhoff and Rota [5.3].

We experimented previously with the differential operator $\mathbf{L} = \mathbf{D}^2 + \mathbf{D}$ and boundary conditions $\mathbf{f}(0) = \mathbf{f}(b) = 0$, finding that self-adjointness of the operator and orthogonality of the eigenfunctions both require the weight function $\omega(t) = e^t$. We now treat this operator by means of our general result, (5.80). In order that $g_0(t)$ be negative as required for (5.77), we work with $-\mathbf{L} \triangleq -\mathbf{D}^2 - \mathbf{D}$ (which has the same eigenfunctions as does **L**). By the substitution (5.78) we find $p(t) = e^t$ and, once again, $\omega(t) = e^t$.

The eigenfunctions of a differential operator **L** satisfy the equation: $\mathbf{Lf} - \lambda\mathbf{f} = \boldsymbol{\theta}$. An equivalent equation for the second-order differential operator **L** of (5.79) is

$$-\omega(\mathbf{Lf} - \lambda\mathbf{f}) = (p\mathbf{f}')' + (\lambda\omega - \hat{q})\mathbf{f} = \boldsymbol{\theta} \tag{5.81}$$

* If the interval $[a, b]$ were infinite, if p or ω were equal to zero at some point, or if q were discontinuous, (5.79) would be a **singular Sturm–Liouville operator**. See Birkhoff and Rota [5.3] for examples.

where $\hat{q} \stackrel{\Delta}{=} \omega q$. Equation (5.81) is known as a **regular Sturm–Liouville equation.** For certain boundary conditions on **f**, **L** will have eigendata. If the boundary conditions are self-adjoint with weight ω, the eigenfunctions can be chosen so they are orthonormal with weight ω. We can obtain the eigendata from (5.81) and the boundary conditions. We call (5.81) together with self-adjoint boundary conditions a **regular Sturm–Liouville system.**

Decoupling of Equations By Means of Eigenvector Expansion

We wish to analyze the linear equation $\mathbf{Tx}=\mathbf{y}$, wherein $\mathbf{x}$ and $\mathbf{y}$ are members of a separable infinite-dimensional Hilbert space $\mathcal{H}$. We have found that we can have orthonormal eigenvectors of $\mathbf{T}$ only if $\mathbf{T}$ is self-adjoint. It can be shown that complete continuity of $\mathbf{T}$ is sufficient (but not necessary) to guarantee that the eigenvalues and eigenvectors of $\mathbf{T}$ are countable. Furthermore, complete continuity together with self-adjointness guarantees that eigendata exist and that the eigenvectors are complete in $\mathcal{H}$.* We assume $\mathbf{T}$ is self-adjoint and completely continuous; then the spectral theorem (5.74) applies, and $\mathbf{T}$ is diagonalizable by means of an orthonormal eigenvector basis $\{\mathbf{x}_k\}$. The vectors $\mathbf{x}$ and $\mathbf{y}$ can be expanded using (5.72):

$$\mathbf{y}=\sum_{k=1}^{\infty} c_k \mathbf{x}_k \quad \text{and} \quad \mathbf{x}=\sum_{k=1}^{\infty} d_k \mathbf{x}_k \tag{5.82}$$

where $c_k=\langle \mathbf{y},\mathbf{x}_k\rangle$, and can be computed using the known vector $\mathbf{y}$; d_k is the kth Fourier coefficient of the unknown solution vector $\mathbf{x}$. We substitute the expansions (5.82) into the equation $\mathbf{Tx}=\mathbf{y}$ to find

$$\begin{aligned}\boldsymbol{\theta}=\mathbf{y}-\mathbf{Tx} &= \sum_{k=1}^{\infty} c_k\mathbf{x}_k-\mathbf{T}\left(\sum_{k=1}^{\infty} d_k\mathbf{x}_k\right)\\ &= \sum_{k=1}^{\infty} c_k\mathbf{x}_k-\sum_{k=1}^{\infty} d_k\mathbf{T}\mathbf{x}_k\\ &= \sum_{k=1}^{\infty} (c_k-\lambda_k d_k)\mathbf{x}_k\end{aligned}$$

where we have relied on the continuity of $\mathbf{T}$ and (5.56) to take $\mathbf{T}$ inside the infinite sum. Then using the orthonormality of the basis $\{\mathbf{x}_k\}$, we find

$$0=\|\mathbf{y}-\mathbf{Tx}\|^2=\|\sum_{k=1}^{\infty}(c_k-\lambda_k d_k)\mathbf{x}_k\|^2=\sum_{k=1}^{\infty}|c_k-\lambda_k d_k|^2$$

* See Bachman and Narici [5.2, Chapter 24] and Stakgold [5.22, Chapter 3].

Since each term in the sum is non-negative, $c_k = \lambda_k d_k, k = 1,2,\ldots$. Then, if $\mathbf{T}$ is invertible (i.e., has no zero eigenvalues),

$$\mathbf{x} = \sum_{k=1}^{\infty} \frac{\langle \mathbf{y}, \mathbf{x}_k \rangle}{\lambda_k} \mathbf{x}_k \tag{5.83}$$

Equation (5.83) is an explicit expression of the solution $\mathbf{x}$ to the equation $\mathbf{Tx} = \mathbf{y}$ in terms of the eigendata for $\mathbf{T}$. The fact that $\mathbf{T}$ is assumed to act on a Hilbert space is not really a restriction. Were $\mathcal{H}$ an incomplete inner product space, it could be completed and the definition of $\mathbf{T}$ extended to the complete space. Furthermore, we are not significantly hampered by the boundedness (or complete continuity) used in the derivation of (5.83). Suppose, for example, that $\mathbf{T}$ represents a self-adjoint differential operator with its boundary conditions (an unbounded operator). If the boundary conditions are appropriate, $\mathbf{T}$ is invertible and $\mathbf{T}^{-1}$ is typically bounded (and perhaps completely continuous). We simply replace $\mathbf{Tx} = \mathbf{y}$ by $\mathbf{T}^{-1}\mathbf{y} = \mathbf{x}$, and repeat the above argument to find again that $d_k = c_k / \lambda_k$ and (5.83) is valid.

Example 4. Analysis of a Differential System by Eigenfunction Expansion. The shaft position $\phi(t)$ of a dc motor (with prescribed initial and final shaft positions) is related to the armature voltage $\mathbf{u}$ of the motor by the following differential equation and boundary conditions:

$$\mathbf{L}\phi \triangleq \ddot{\phi} + \dot{\phi} = \mathbf{u}, \qquad \phi(0) = \phi(b) = 0$$

The Green's function (3.42) for this system is bounded. Therefore, the inverse operator is Hilbert–Schmidt and, consequently, completely continuous. Furthermore, the differential system is self-adjoint relative to the weight function $\omega(t) = e^t$, as noted in Exercise 4. The eigenvalues and eigenfunctions for this differential system are given by (4.37) and (4.38):

$$\lambda_k = -\frac{1}{4} - \left(\frac{k\pi}{b}\right)^2, \qquad \mathbf{f}_k(t) = \sqrt{2/b}\, e^{-t/2} \sin\left(\frac{\pi k t}{b}\right), \qquad k = 1,2,\ldots$$

We determined in Example 3 of Section 5.3 that these eigenfunctions are orthogonal relative to the weight function which makes the differential system self-adjoint. (We have added the multiplier $\sqrt{2/b}$ in order to make the functions orthonormal.) We also showed in that example that these eigenfunctions form a basis for $\mathcal{C}(e^t; 0, b)$. [Of course, it is also a basis for the completion of that space, $\mathcal{L}_2(e^t; 0, b)$.] We now use (5.83) to express the solution to the differential system as an expansion in the eigenfunctions of the system; ϕ takes the role of $\mathbf{x}$, $\mathbf{y}$ becomes

$\mathbf{u}$, and $\mathbf{x}_k$ becomes $\mathbf{f}_k$:

$$\phi(t)=\frac{2}{b}\sum_{k=1}^{\infty}\frac{\int_0^b \mathbf{u}(s)e^{s/2}\sin(\pi ks/b)\,ds}{-\frac{1}{4}-\left(\frac{k\pi}{b}\right)^2}e^{-t/2}\sin\left(\frac{\pi kt}{b}\right) \tag{5.84}$$

The solution to this differential system was obtained by inversion in Section 3.3. Following (3.42) we used the inverse to determine the solution for the input $\mathbf{u}(t)=1$:

$$\phi(t)=t-\frac{be^b}{e^b-1}(1-e^{-t})$$

For this same input, $\mathbf{u}(t)=1$, (5.84) becomes

$$\phi(t)=\frac{2}{b}\sum_{k=1}^{\infty}\frac{(\pi k/b)[(-1)^k e^{b/2}-1]}{\left[\frac{1}{4}+(\pi k/b)^2\right]^2}e^{-t/2}\sin\left(\frac{\pi kt}{b}\right)$$

In Figure 5.10, we compare the exact solution with the first two terms of the eigenfunction expansion for $b=\pi$. It is apparent from the figure that for all practical purposes the first few terms of the series determine the solution.

Exercise 5. Using the eigenfunctions of Example 4, compute the first two terms of the eigenfunction expansion of the input $\mathbf{u}(t)=1$ with $b=\pi$. (Hint: multiply each curve of Figure 5.10 by the appropriate eigenvalue.) Note that the convergence of this series is considerably slower than the convergence of the output function $\phi(t)$. Furthermore, the series does not converge at the endpoints; convergence in the $\mathcal{L}_2$ norm does not imply convergence *everywhere*.

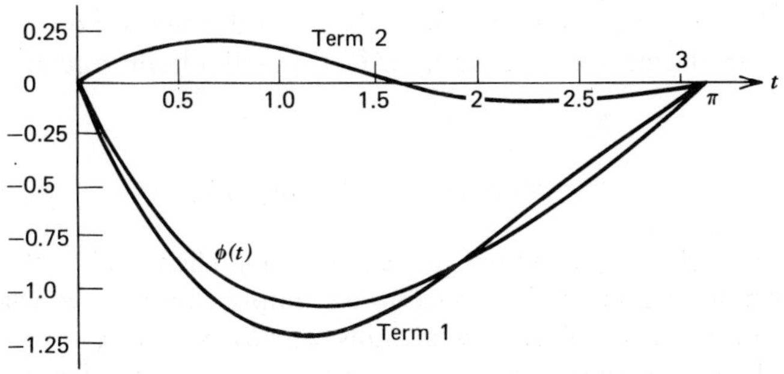

Figure 5.10. Convergence of (5.84) for $\mathbf{u}(t)=1$ and $b=\pi$.

Suppose that **T** is self-adjoint and completely continuous, but is not invertible. Then the equation $\mathbf{Tx}=\mathbf{y}$ has no solution unless **y** is in range(**T**). Furthermore, if the equation has a solution, we can add to it any vector in nullspace(**T**) to obtain another solution. Formal application of (5.83) would require division by a zero eigenvalue. To resolve this difficulty, we decompose the equation. By the orthogonal decomposition theorem (5.67) and the self-adjointness of **T**, $\mathcal{V}=\text{nullspace}(\mathbf{T}) \overset{\perp}{\oplus} \overline{\text{range}(\mathbf{T})}$. Those eigenvectors in the orthonormal basis which are associated with the zero eigenvalue form an orthonormal basis for nullspace(**T**). The remaining eigenvectors form an orthonormal basis for range(**T**). Moreover, the action of **T** on range(**T**) is one-to-one (the zero eigenvalues have been removed). We use (5.83) to obtain a particular solution to the equation $\mathbf{Tx}=\mathbf{y}$ by summing over only the nonzero eigenvalues; the resulting solution lies in range(**T**). Assuming **y** is in range(**T**), the general solution to the equation $\mathbf{Tx}=\mathbf{y}$, for a self-adjoint completely continuous transformation **T**, is expressed in terms of the eigendata for **T** by

$$\mathbf{x}=\sum_{\substack{\text{nonzero}\\ \lambda_k}} \frac{\langle \mathbf{y}, \mathbf{x}_k\rangle}{\lambda_k}\mathbf{x}_k+\mathbf{x}_0 \tag{5.85}$$

where $\mathbf{x}_0$ is an arbitrary vector in nullspace(**T**). As with (5.83), (5.85) may be valid even though **T** is not completely continuous. This fact is illustrated by the unbounded differential operator of Example 5. Rather than dwell further on conditions wherein (5.85) is valid, we adopt the (possibly risky) course of assuming its validity whenever the equation is useful.

In principle, in order to determine whether or not $\mathbf{Tx}=\mathbf{y}$ has solutions, we must solve explicitly for range(**T**) and see whether or not **y** is in range(**T**). Finding range(**T**) directly can be difficult for, say, a differential operator. It is simpler to apply the orthogonal decomposition which was introduced in the previous paragraph. If **T** is self-adjoint and range(**T**) is closed,

$$\mathcal{V}=\text{nullspace}\ (\mathbf{T}) \overset{\perp}{\oplus}\ \text{range}\ (\mathbf{T})$$

We solve $\mathbf{Tx}=\boldsymbol{\theta}$ for the vectors in nullspace(**T**). Then, rather than explicitly determine the vectors in range(**T**), we simply check to see whether or not **y** is orthogonal to all vectors in nullspace(**T**); **y** will be orthogonal to nullspace(**T**) if and only if it lies in range(**T**). Although the range of a differential operator is not closed, this orthogonality test for existence of a solution is used most often for differential equations. See Friedman [5.8, Chapter 3].

Example 5. Analysis by Eigenfunction Expansion—a Noninvertible Case. Let **T** represent the differential operator $\mathbf{L} \triangleq \mathbf{D}^2$ with the homogeneous boundary conditions $\mathbf{f}(1)-\mathbf{f}(0)=\mathbf{f}'(1)-\mathbf{f}'(0)=0$. The operator **T** is self-adjoint with respect to the standard function space inner product. However, **T** is degenerate; the differential system

$$\mathbf{f}''=\mathbf{u}, \qquad \mathbf{f}(1)=\mathbf{f}(0), \qquad \mathbf{f}'(1)=\mathbf{f}'(0)$$

has solutions only for a restricted set of functions **u**. Nullspace(**T**) consists in the solutions to the completely homogeneous system,

$$\mathbf{f}''(t)=0, \qquad \mathbf{f}(1)=\mathbf{f}(0), \qquad \mathbf{f}'(1)=\mathbf{f}'(0)$$

Thus nullspace(**T**) consists in the "constant functions," $\mathbf{f}_0(t)=c$. By the discussion above, **u** can be in range(**T**) only if **u** is orthogonal (with respect to the standard inner product) to nullspace(**T**). Therefore, in order that the differential system have a solution, **u** must satisfy

$$\langle \mathbf{u},\mathbf{f}_0\rangle = \int_0^1 \mathbf{u}(s)c\,ds=0$$

for all constants c; that is, $\int_0^1 \mathbf{u}(s)\,ds=0$. The constant functions are eigenfunctions of **T** for the eigenvalue $\lambda_0=0$. The nonzero eigenvalues and the corresponding eigenfunctions can be determined by the techniques of Section 4.3; they are

$$\lambda_k=-(2\pi k)^2, \qquad \mathbf{f}_k(t)=\sqrt{2}\,\cos 2\pi kt, \qquad \mathbf{g}_k(t)=\sqrt{2}\,\sin 2\pi kt$$

for $k=1,2,\ldots$. Note that there are two eigenfunctions, $\mathbf{f}_k$ and $\mathbf{g}_k$, for each eigenvalue. The orthogonality of the eigenfunctions for different eigenvalues follows from the self-adjointness of **T**. Each pair of eigenfunctions has been selected such that it forms an orthogonal pair. The whole set of eigenfunctions is essentially (5.30), the basis for the classical Fourier series; the constant function is missing since it is not in range(**T**). If $\int_0^1 \mathbf{u}(s)\,ds=0$, the solution to the differential system can be expressed by the eigenfunction expansion (5.85):

$$\mathbf{f}(t)=\mathbf{f}_0(t)+\sum_{k=1}^{\infty}\frac{\langle \mathbf{u},\mathbf{f}_k\rangle}{\lambda_k}\mathbf{f}_k(t)+\sum_{k=1}^{\infty}\frac{\langle \mathbf{u},\mathbf{g}_k\rangle}{\lambda_k}\mathbf{g}_k(t)$$

$$=c+2\sum_{k=1}^{\infty}\frac{\int_0^1 \mathbf{u}(s)\cos(2\pi ks)\,ds}{-(2\pi k)^2}\cos(2\pi kt)$$

$$+2\sum_{k=1}^{\infty}\frac{\int_0^1 \mathbf{u}(s)\sin(2\pi ks)\,ds}{-(2\pi k)^2}\sin(2\pi kt)$$

The arbitrary constant c expresses the freedom (or nonuniqueness) in the solution. As in Example 4. comparison of the exact solution with the sum of the first few terms of the series for a specific $\mathbf{u}$ demonstrates that convergence of the series is rapid.

We found earlier that the general second-order differential operator (5.77) is formally self-adjoint with respect to the weight function ω of (5.78). If the boundary conditions are also self-adjoint, the second-order differential operator (or regular Sturm–Liouville operator) has eigendata; these eigendata are determined by (5.81) together with the boundary conditions. It can be shown that the solutions (eigenfunctions) of any regular Sturm–Liouville system are complete; they form a countable basis, orthonormal with respect to weight ω, for the space $\mathcal{L}_2(a,b)$.* As exemplified by Example 4, it can also be shown that any regular Sturm–Liouville system has an infinite sequence of real eigenvalues $\lambda_1<\lambda_2<\lambda_3<\cdots$. (If the differential operator is invertible, its inverse is a Hilbert–Schmidt operator; that is, the inverse is completely continuous.) Furthermore, the eigenfunctions for a regular Sturm–Liouville system are similar to the sinusoidal functions in that the eigenfunctions for the nth eigenvalue have n zero crossings in the interval $[a,b]$. Sturm–Liouville problems are typically steady state (or standing-wave) problems. Examples of physical systems modeled by regular Sturm–Liouville operators and self-adjoint boundary conditions are vibrating strings, beams, and membranes. Steady-state heat flow in one dimension is another example. Less typical is the motor control problem introduced in (3.40) and solved by eigenfunction expansion in Example 4. In this problem, the standing-wave nature arises because conditions are placed on the future position of the motor shaft.

It follows from Parseval's identity (5.48) that the Fourier coefficients of a vector $\mathbf{x}$ relative to a countably infinite orthonormal basis $\{\mathbf{x}_k\}$ must approach zero: $\langle\mathbf{x},\mathbf{x}_k\rangle\rightarrow 0$ as $k\rightarrow\infty$. It is evident from Examples 4 and 5 that the convergence of eigenvector expansions of solutions is accounted for only in part by this property of the Fourier coefficients. In general, Fourier coefficients converge at least as fast as $1/k$. However, in these second-order examples a stronger influence on the convergence of the solutions is exerted by the eigenvalues, with $1/\lambda_k$ converging approximately at $1/k^2$. The "output expansion" (or solution) consists in a modification of the eigenfunction expansion of the input, wherein high-order eigenfunction components (or normal modes) of the input are attenuated more than are the low-order components. In analogy to a dynamic system with initial conditions, we can think of the systems of Examples 4 and 5 as "low-pass" systems; the systems emphasize (or pass) the low-order eigenfunctions.

* See Birkhoff and Rota [5.3].

Example 6. Solution of a Partial Differential Equation by Eigenfunction Expansion. The following partial differential equation is a model for problems in electrostatics or heat flow:

$$\nabla^2 \mathbf{f} = \mathbf{u}$$

Let the two-dimensional region Ω of the (s,t) plane on which $\mathbf{f}$ and $\mathbf{u}$ are defined be the rectangle $0 \leqslant s \leqslant a, 0 \leqslant t \leqslant b$. Let $\mathbf{f}(s,t)=0$ on the boundary Γ of this rectangle. In Example 11 of Section 5.4, we determined that ∇^2 is formally self-adjoint with respect to the standard inner product (5.64). Furthermore the boundary condition is also self-adjoint. Thus we expect to find a basis for $\mathcal{L}_2(\Omega)$ consisting in orthonormal eigenfunctions of the differential system. The eigendata for the system are given in (4.62) and (4.63); we express the eigenfunctions in normalized form:

$$\lambda_{mn} = -\left(\frac{m\pi}{a}\right)^2 - \left(\frac{n\pi}{b}\right)^2 = -\frac{\pi^2}{a^2b^2}(n^2a^2 + m^2b^2)$$

$$\mathbf{f}_{mn}(s,t) = \frac{2}{\sqrt{ab}} \sin\left(\frac{m\pi s}{a}\right) \sin\left(\frac{n\pi t}{b}\right)$$

It can be shown that these eigenfunctions are complete in $\mathcal{L}_2(\Omega)$ [5.22, p. 193]. By (5.83), the solution to the differential system, expressed as an eigenfunction expansion, is

$$\mathbf{f}(s,t)$$

$$= -\frac{4ab}{\pi^2} \sum_{m=1}^{\infty} \sum_{n=1}^{\infty} \frac{\int_0^b \int_0^a \mathbf{u}(\nu,\rho) \sin(m\pi\nu/a) \sin(n\pi\rho/b)\, d\nu\, d\rho}{n^2a^2 + m^2b^2} \sin\left(\frac{m\pi s}{a}\right) \sin\left(\frac{n\pi t}{b}\right)$$

We have no closed form solution with which to compare this result. Rather, techniques for finding Green's functions for partial differential operators are usually based upon eigenfunction expansions similar to the one used here. See (5.86).

Further Spectral Concepts

We have developed two different approaches to the solution of an invertible differential system, the inverse (3.35) and the eigenfunction expansion (5.83). We would be surprised if the two techniques were not closely related. Let the differential operator $\mathbf{L}$ act on a space of functions defined on $[a,b]$. Suppose $\mathbf{L}$ together with homogeneous boundary conditions has eigendata $\{\lambda_k, \mathbf{f}_k\}$. Further assume that the eigenfunctions form a basis for the function space which is orthonormal with respect to the standard inner product. We wish to explore the equation $\mathbf{Lf} = \mathbf{u}$ together with the boundary conditions. The solution $\mathbf{f}$ can be expressed in terms of the Green's function $k(t,s)$ as $\mathbf{f}(t) = \int_a^b k(t,s)\mathbf{u}(s)\, ds$. As discussed in Chapter 3, the Green's function is the solution corresponding to the input $\mathbf{u}(t)=\delta(t-s)$. Using (5.83) we express this solution in terms of the eigendata:

$$k(t,s)=\sum_{k=1}^{\infty}\frac{\int_a^b \delta(\tau-s)\mathbf{f}_k(\tau)\,d\tau}{\lambda_k}\mathbf{f}_k(t)$$

$$=\sum_{k=1}^{\infty}\frac{\mathbf{f}_k(s)\mathbf{f}_k(t)}{\lambda_k} \tag{5.86}$$

Equation (5.86) is known as the *bilinear expansion of the Green's function* for **L** in terms of the eigenfunctions of **L**. The Green's function for the Laplacian operator is in fact derivable from Example 6 using an extension of (5.86) to a space of two-dimensional functions.

Exercise 6. Find the bilinear expansion of the Green's function for Example 4 and compare the first term to the exact Green's function as expressed in (3.42).

The spectral theorem (5.74) can be used to define functions of linear operators analogous to the functions of matrix operators discussed in Section 4.6. Assume the linear operator **T** in the Hilbert space $\mathcal{V}$ is self-adjoint and completely continuous. Then, applying **T** to the expansion (5.72) of a general vector **x** in terms of an orthonormal set $\{\mathbf{x}_k\}$ of eigenvectors for **T**, we find

$$\mathbf{Tx}=\sum_{k=1}^{\infty}\lambda_k\langle\mathbf{x},\mathbf{x}_k\rangle\mathbf{x}_k \tag{5.87}$$

where we have used the continuity of **T** in order to take **T** inside the infinite sum. By combining all terms of (5.87) which are associated with identical eigenvalues, we reexpress (5.87) as

$$\mathbf{T}=\sum_{j=1}^{\infty}\lambda_j\mathbf{P}_j \tag{5.88}$$

where $\mathbf{P}_j$ is the orthogonal projector onto nullspace$(\mathbf{T}-\lambda_j\mathbf{I})$. Thus the effect of $\mathbf{P}_j$ on a general vector **x** in $\mathcal{V}$ can be expressed in terms of the orthonormal eigenvectors of **T**:

$$\mathbf{P}_j\mathbf{x}=\sum_{k}\langle\mathbf{x},\mathbf{x}_k\rangle\mathbf{x}_k \tag{5.89}$$

where the summation is over all values of k which correspond to the eigenvalue λ_j. Equation (5.88) is the **spectral decomposition of T**; it can be interpreted as a diagonalization of **T**. If f is a real continuous function which is defined at the eigenvalues of **T**, it can be shown that a suitable definition of a function of a transformation is provided by the **fundamental formula for** $f(\mathbf{T})$:

$$f(\mathbf{T}) = \sum_{j=1}^{\infty} f(\lambda_j)\mathbf{P}_j \tag{5.90}$$

Although we have defined (5.90) only for a self-adjoint, completely continuous **T**, the definition can be extended to any bounded normal linear transformation.* Furthermore, as we know from our examples, it can apply to unbounded differential operators. Equation (5.83), for instance, is essentially an expression of $\mathbf{x}=f(\mathbf{T})\mathbf{y}$ for the function $f(t) \triangleq t^{-1}$. We applied (5.83) to an unbounded differential operator in Example 4.

Throughout our examination of infinite-dimensional operator equations, we have restricted ourselves to operators for which there is a countable orthonormal set of eigenvectors which form a basis for the space. Self-adjoint, completely continuous transformations are of this type. We have restricted ourselves to these transformations in order to work with only the simplest infinite-dimensional extensions of matrix equations. More generality comes only with considerably increased abstraction. Let **T** be a linear operator on an inner product space $\mathcal{V}$. The eigenvalues and eigenvectors of **T** are determined by the equation $\mathbf{Tx}=\lambda\mathbf{x}$, or alternatively, by the **resolvant operator,** $(\mathbf{T}-\lambda\mathbf{I})^{-1}$; the eigenvalues of **T** are those values of λ for which the latter inverse does not exist. However the nonexistence of the inverse is only one of the ways in which the resolvant operator can be "irregular." Detailed discussions of the resolvant operator and general spectral concepts can be found in Bachman and Narici [5.2], Stakgold [5.22], Friedman [5.8], and Naylor and Sell [5.17].

Matched Filter Design–An Application of Spectral Decomposition

We wish to recognize the presence or absence of a signal $\mathbf{u}(t)$ of known shape (e.g., a radar return). Our measurement of the signal is corrupted by stationary noise $\mathbf{n}(t)$ whose autocorrelation function, $R(t,s) \triangleq \mathbf{E}[\mathbf{n}(t)\mathbf{n}(s)] = R(t-s)$, is known. Because $\mathbf{n}(t)$ is stationary, R is symmetric in t and s, and depends only on the time difference $t-s$; R is also finite and positive. We filter the noisy measurement in order to improve our estimate of the presence or absence of the signal (see Figure 5.11). We select the impulse

* Bachman and Narici [5.2].

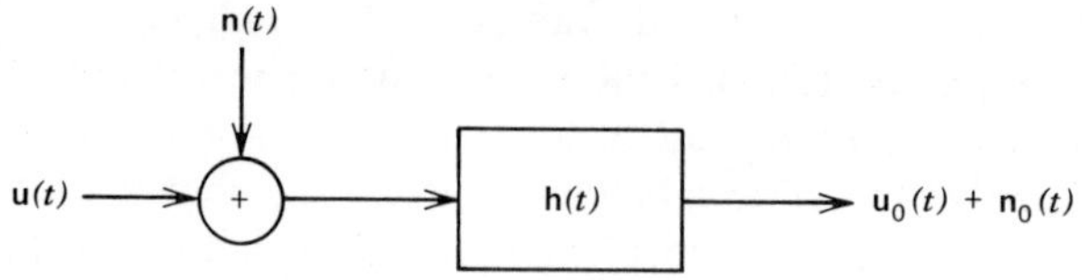

Figure 5.11. A linear filter.

response $\mathbf{h}(t)$ of the linear filter in such a way that the signal-to-noise ratio of the output, $\mathbf{u}_0^2(b)/\mathbf{E}[\mathbf{n}_0^2(b)]$, is maximized at some time $t=b$ units after measurement begins. [A circuit can then be synthesized which has the impulse response $\mathbf{h}(t)$.] The output signal and noise at time $t=b$ are, respectively,*

$$\mathbf{u}_0(b)=\int_0^b \mathbf{h}(s)\mathbf{u}(b-s)\,ds$$

$$\mathbf{n}_0(b)=\int_0^b \mathbf{h}(s)\mathbf{n}(b-s)\,ds$$

Then

$$\mathbf{E}\left[\mathbf{n}_0^2(b)\right]=\mathbf{E}\int_0^b \mathbf{h}(s)\mathbf{n}(b-s)\,ds\int_0^b \mathbf{h}(t)\mathbf{n}(b-t)\,dt$$

$$=\int_0^b\int_0^b \mathbf{h}(s)\mathbf{h}(t)R(s-t)\,dt\,ds$$

We use the concepts of P&C 5.30 and Exercise 3, Section 5.1 to interpret $\mathbf{E}[\mathbf{n}_0^2(b)]$ as the square of a norm. Define $\mathbf{T}$ by $(\mathbf{Th})(s)\triangleq\int_0^b\mathbf{h}(t)R(s-t)\,dt$. Then, since R is positive and symmetric in its variables, $\mathbf{T}$ is self-adjoint, completely continuous (Hilbert–Schmidt), and positive definite; $\mathbf{T}$ is diagonalizable by means of an orthonormal basis of eigenfunctions $\{\mathbf{f}_k\}$, and the eigenvalues $\{\lambda_k\}$ of $\mathbf{T}$ are positive (P&C 5.28). Therefore, the square roots $\{\sqrt{\lambda_k}\}$ exist, and a unique self-adjoint positive-definite operator $\sqrt{\mathbf{T}}$ is defined by (5.90). Thus

$$\mathbf{E}\left[\mathbf{n}_0^2(b)\right]=\int_0^b \mathbf{h}(s)(\mathbf{Th})(s)\,ds$$

$$=\langle\mathbf{h},\mathbf{Th}\rangle$$

$$=\|\sqrt{\mathbf{T}}\,\mathbf{h}\|^2$$

* Appendix 2 for a discussion of convolution and impulse response.

Let $\mathbf{u}_r$ denote the "reverse" of the signal shape $\mathbf{u}$; that is, $\mathbf{u}_r(s) \stackrel{\Delta}{=} \mathbf{u}(b-s)$. Then $\mathbf{u}_0(b) = \langle \mathbf{h}, \mathbf{u}_r \rangle$. Since the eigenvalues $\{\sqrt{\lambda_k}\}$ of $\sqrt{\mathbf{T}}$ are all positive, $\sqrt{\mathbf{T}}$ is invertible and range$(\sqrt{\mathbf{T}})$ is the whole function space, $\mathcal{L}_2(0,b)$. Therefore, we can assume $\mathbf{u}_r$ is in range $(\sqrt{\mathbf{T}})$; that is, $\mathbf{u}_r = \sqrt{\mathbf{T}}\,\mathbf{g}$ for some function $\mathbf{g}$. Then

$$\mathbf{u}_0^2(b) = |\langle \mathbf{h}, \mathbf{u}_r \rangle|^2 = |\langle \mathbf{h}, \sqrt{\mathbf{T}}\,\mathbf{g} \rangle|^2 = |\langle \sqrt{\mathbf{T}}\,\mathbf{h}, \mathbf{g} \rangle|^2$$

As a consequence, the signal-to-noise ratio satisfies

$$\frac{\mathbf{u}_0^2(b)}{\mathbf{E}\left[\mathbf{n}_0^2(b)\right]} = \frac{|\langle \sqrt{\mathbf{T}}\,\mathbf{h}, \mathbf{g} \rangle|^2}{\|\sqrt{\mathbf{T}}\,\mathbf{h}\|^2} \leqslant \|\mathbf{g}\|^2$$

The latter relationship is the Cauchy–Schwartz inequality (P&C 5.4); equality holds if $\sqrt{\mathbf{T}}\,\mathbf{h} = c\mathbf{g}$ for any constant c, or $\mathbf{Th} = c\sqrt{\mathbf{T}}\,\mathbf{g} = c\mathbf{u}_r$. We must solve this integral equation for $\mathbf{h}$. It is apparent that $\mathbf{h}$ depends only on the shape of the signal $\mathbf{u}$, but not its magnitude. We can express the solution to the equation in terms of eigendata for $\mathbf{T}$ by means of (5.83):

$$\mathbf{h} = c \sum_{k=1}^{\infty} \frac{\langle \mathbf{u}_r, \mathbf{f}_k \rangle}{\lambda_k} \mathbf{f}_k$$

Suppose the noise is "white"; that is, the autocorrelation function is the limiting case $R(s-t) = N\delta(s-t)$, where N is the noise power and $\delta(s-t)$ is the Dirac delta function. The integral equation becomes

$$(\mathbf{Th})(s) = N \int_0^b \mathbf{h}(t)\delta(s-t)\,dt = N\mathbf{h}(s) = c\mathbf{u}_r(s)$$

or $\mathbf{h}(s)$ is any multiple of $\mathbf{u}_r(s) = \mathbf{u}(b-s)$. The optimum impulse response for this case has the form of the signal running backward in time from the fixed time $t = b$. A filter with this characteristic is called a **matched filter**. We can also use the eigenfunction expansion to determine this solution: The eigendata are determined by

$$(\mathbf{Th})(s) = N\mathbf{h}(s) = \lambda\mathbf{h}(s)$$

The only eigenvalue is $\lambda = N$. Every function is an eigenfunction. Letting $\{\mathbf{f}_k\}$ be any orthonormal basis for the space, the eigenfunction expansion becomes

$$\mathbf{h} = \frac{c}{N} \sum_{k=1}^{\infty} \langle \mathbf{u}_r, \mathbf{f}_k \rangle \mathbf{f}_k = \frac{c}{N} \mathbf{u}_r$$

The eigenfunction expansion is just a Fourier series expansion of $\mathbf{u}_r$. Although we easily solved for this matched filter, solution of an integral equation and determination of the eigendata of an integral operator are usually difficult problems.

5.6 Problems and Comments

5.1 Let $\mathbf{A}$ be a real symmetric 2×2 matrix with positive eigenvalues.

(*a*) Show that the curve described by the quadratic equation $\langle \mathbf{x}, \mathbf{A}\mathbf{x}\rangle = \mathbf{x}^{\mathrm{T}}\mathbf{A}\mathbf{x} = 1$ is an ellipse in the $\mathbf{x}$ plane. Determine the relationship between the ellipse and the eigendata for $\mathbf{A}$. (Hint: a symmetric matrix has orthogonal eigenvectors. Therefore it can be diagonalized by means of the transformation $\boldsymbol{\Lambda} = \mathbf{S}^{\mathrm{T}}\mathbf{A}\mathbf{S}$.)

(*b*) Find the eigendata and sketch the ellipse for

$$\mathbf{A} = \begin{pmatrix} 5 & 3 \\ 3 & 5 \end{pmatrix}$$

5.2 Show that the following definition satisfies the rules for an inner product on $\mathcal{R}^2$:

$$\langle (\xi_1, \xi_2), (\eta_1, \eta_2) \rangle \triangleq 2\xi_1\eta_1 - \xi_1\eta_2 - \xi_2\eta_1 + \xi_2\eta_2$$

5.3 Let $\mathcal{V}$ and $\mathcal{W}$ be inner product spaces over the same scalar field with inner products denoted by $\langle\ ,\ \rangle_{\mathcal{V}}$ and $\langle\ ,\ \rangle_{\mathcal{W}}$, respectively. Let $\mathbf{u}$ and $\mathbf{v}$ be in $\mathcal{V}$; let $\mathbf{w}$ and $\mathbf{z}$ be in $\mathcal{W}$.

(*a*) Show that the following is an inner product on the cartesian product space $\mathcal{V}\times\mathcal{W}$: $\langle (\mathbf{u},\mathbf{w}), (\mathbf{v},\mathbf{z})\rangle \triangleq \langle \mathbf{u},\mathbf{v}\rangle_{\mathcal{V}} + \langle \mathbf{w},\mathbf{z}\rangle_{\mathcal{W}}$

(*b*) Let $\mathbf{x}$ and $\mathbf{y}$ denote vectors in $\mathcal{C}^2(0,1)\times\mathcal{C}^1(0,1)$. Express the elements of $\mathbf{x}$ and $\mathbf{y}$ as 2×1 matrices rather than as 2-tuples. (Then for each t in $[0,1]$, $\mathbf{x}(t)$ and $\mathbf{y}(t)$ are in the state space, $\mathfrak{M}^{2\times 1}$.) Show that the inner product $\langle \mathbf{x},\mathbf{y}\rangle \triangleq \int_0^1 \mathbf{y}^{\mathrm{T}}(t)\mathbf{x}(t)\,dt$ is essentially a special case of the inner product defined in (*a*).

*5.4 The following useful equalities and inequalities apply to the vectors in any inner product space $\mathcal{V}$:

(*a*) **Pythagorean theorem**: if $\langle \mathbf{x},\mathbf{y}\rangle = 0$, then

$$\|\mathbf{x}+\mathbf{y}\|^2 = \|\mathbf{x}\|^2 + \|\mathbf{y}\|^2$$

(*b*) **Bessel's inequality**: if $\{\mathbf{x}_i\}$ is an orthonormal set in $\mathcal{V}$, then

$$\|\mathbf{x}\|^2 \geqslant \sum_i |\langle \mathbf{x},\mathbf{x}_i\rangle|^2$$

(*c*) **Parseval's identity**: equality occurs in (*b*) if and only if $\{\mathbf{x}_i\}$ is a basis for $\mathcal{V}$;

(*d*) **Cauchy–Schwartz inequality**: $|\langle \mathbf{x},\mathbf{y}\rangle| \leqslant \|\mathbf{x}\|\|\mathbf{y}\|$, with equality if and only if $\mathbf{x}$ and $\mathbf{y}$ are collinear;

(*e*) **Triangle inequality**: $\|\mathbf{x}+\mathbf{y}\| \leqslant \|\mathbf{x}\| + \|\mathbf{y}\|$.

5.5 Equip the vector space $\mathcal{R}^2$ with the inner product

$$\langle \mathbf{x},\mathbf{y}\rangle \triangleq \xi_1\eta_1 - \xi_1\eta_2 - \xi_2\eta_1 + 4\xi_2\eta_2$$

where ξ_i and η_i are the components of $\mathbf{x}$ and $\mathbf{y}$, respectively.

(*a*) Find the matrix $\mathbf{Q}_{\mathcal{E}}$ of the inner product relative to the standard basis for $\mathcal{R}^2$;

(*b*) Find the matrix $\mathbf{Q}_{\mathcal{X}}$ of the inner product relative to the basis $\mathcal{X} \triangleq \{(1,0),(1,1)\}$. Explain the simple form of $\mathbf{Q}_{\mathcal{X}}$.

Let $\langle\ ,\ \rangle$ be an inner product defined on an n-dimensional space $\mathcal{V}$. Let $\mathbf{Q}_{\mathcal{X}}$ and $\mathbf{Q}_{\mathcal{Y}}$ be the matrices of this inner product relative to the bases $\mathcal{X}$ and $\mathcal{Y}$, respectively. Let $\mathbf{S}$ be the coordinate transformation matrix defined by $[\mathbf{x}]_{\mathcal{Y}} = \mathbf{S}[\mathbf{x}]_{\mathcal{X}}$.

(*c*) Determine the relationship between $\mathbf{Q}_{\mathcal{X}}$, $\mathbf{Q}_{\mathcal{Y}}$, and $\mathbf{S}$;

(*d*) What special property does $\mathbf{S}$ possess if $\mathcal{X}$ and $\mathcal{Y}$ are both orthonormal?

5.6 The set $\mathcal{X} \triangleq \{(1,1),(0,-1)\}$ is a basis for $\mathcal{R}^2$. Find an inner product which makes the basis $\mathcal{X}$ orthonormal. Determine the matrix $\mathbf{Q}_{\mathcal{E}}$ of this inner product relative to the standard basis for $\mathcal{R}^2$.

5.7 Let

$$\mathbf{x}_1 = \begin{pmatrix}1\\1\\0\end{pmatrix}, \qquad \mathbf{x}_2 = \begin{pmatrix}1\\0\\1\end{pmatrix}, \qquad \mathbf{x}_3 = \begin{pmatrix}0\\1\\1\end{pmatrix}$$

The set $\mathcal{X} = \{\mathbf{x}_1,\mathbf{x}_2,\mathbf{x}_3\}$ is a basis for $\mathfrak{M}^{3\times 1}$.

(*a*) Determine an inner product for $\mathfrak{M}^{3\times 1}$ with respect to which the basis is orthonormal.

(*b*) Find $\mathbf{Q}_{\mathcal{E}}$, the matrix of the inner product relative to the standard basis for $\mathfrak{M}^{3\times 1}$.

5.8 Let $\mathcal{W}_1$ be the subspace of $\mathcal{R}^3$ which is spanned by the pair of vectors $(1,0,1)$ and $(0,1,-1)$. Let $\mathcal{W}_2$ be the subspace of $\mathcal{R}^3$ which is spanned by the vector $(1,1,1)$. Pick an inner product for $\mathcal{R}^3$ which makes every vector in $\mathcal{W}_1$ orthogonal to every vector in $\mathcal{W}_2$.

*5.9 *Positive-definite matrices*: a symmetric $n\times n$ matrix $\mathbf{A}$ is called **positive definite** if $\mathbf{x}^{\mathrm{T}}\mathbf{A}\mathbf{x} \geqslant 0$ for all real $n\times 1$ vectors $\mathbf{x}$ and if equality occurs only for $\mathbf{x}=\boldsymbol{\theta}$. Suppose we pick the kth component of $\mathbf{x}$ equal to zero; it follows that the submatrix of $\mathbf{A}$ obtained by deleting the kth row and kth column of $\mathbf{A}$ must also be positive definite. In fact, any principal submatrix of $\mathbf{A}$ (obtained by deleting a set of rows and the corresponding columns of $\mathbf{A}$) must be positive definite. The determinant of a matrix equals the product of its eigenvalues (P&C 4.6). Furthermore, the eigenvalues of a positive definite matrix are all positive (P&C 5.28). Consequently, if $\mathbf{A}$ is positive definite, the determinant of $\mathbf{A}$ and of each principle submatrix of $\mathbf{A}$ must be positive.

Let $\mathbf{A}_r$ be obtained from $\mathbf{A}$ by deleting all but the first r rows and columns of $\mathbf{A}$; $\det(\mathbf{A}_r)$ is called the rth *leading principle minor* of $\mathbf{A}$. A symmetric $n\times n$ matrix $\mathbf{A}$ is positive definite if and only if the n leading principle minors of $\mathbf{A}$ are positive (see [5.14] and [5.25]). Checking the sign of the leading principle minors is a convenient test for positive definiteness of $\mathbf{A}$.

5.10 Show that the statement $\langle \mathbf{A},\mathbf{B}\rangle = \text{trace}\,(\mathbf{B}^{\mathrm{T}}\mathbf{A})$ defines a valid inner product on the real vector space $\mathfrak{M}^{n\times n}$; (the trace of a square matrix is defined to be the sum of the elements on its main diagonal).

5.11 Let $\mathfrak{X} \stackrel{\Delta}{=} \{\mathbf{x}_1,\ldots,\mathbf{x}_n\}$ be an orthonormal basis for a vector space $\mathcal{V}$. Let $\mathbf{T}$ be a linear operator on $\mathcal{V}$. Then the element in row i, column j of $[\mathbf{T}]_{\mathfrak{X}\mathfrak{X}}$ is $\langle \mathbf{T}\mathbf{x}_j,\mathbf{x}_i\rangle$ for $i,j=1,\ldots,n$.

5.12 Let $\mathfrak{X} = \{\mathbf{x}_1,\mathbf{x}_2,\ldots\}$ be an orthogonal basis for a real inner product space $\mathcal{V}$. Approximate a vector $\mathbf{x}$ of $\mathcal{V}$ by a linear combination, $\mathbf{x}_a=\Sigma_{k=1}^{n} c_k\mathbf{x}_k$, of the first n vectors of $\mathfrak{X}$ in such a way that $\|\mathbf{x}-\mathbf{x}_a\|^2$ is minimized. Show that the coefficients $\{c_k\}$ are the Fourier coefficients. How are the coefficients affected if we improve the approximation by adding more terms to $\mathbf{x}_a$ (increasing n)?

5.13 Let $\mathbf{A}$ be a 3×3 matrix with eigenvalues $\lambda_1=0$, $\lambda_2\neq 0$, $\lambda_3\neq 0$ and corresponding linearly independent eigenvectors $\mathbf{x}_1$, $\mathbf{x}_2$, and $\mathbf{x}_3$. Let $\langle\cdot,\cdot\rangle$ denote an inner product for which the above eigenvectors are orthonormal. We wish to solve the equation $\mathbf{A}\mathbf{x}=\mathbf{y}$.

(*a*) Assuming solutions exist, express the general solution $\mathbf{x}$ in terms of the eigendata and the inner product.

(*b*) Determine the conditions that $\mathbf{y}$ must satisfy in order that solutions exist. Express these conditions in terms of the eigendata and the inner product.

5.14 Equip $\mathfrak{M}^{3\times 1}$ with the standard inner product, $\langle \mathbf{x};\mathbf{y}\rangle \stackrel{\Delta}{=} \mathbf{y}^{\mathrm{T}}\mathbf{x}$. Let

$\mathbf{x}_1=\begin{pmatrix}1\\1\\0\end{pmatrix}$, $\mathbf{x}_2=\begin{pmatrix}2\\0\\1\end{pmatrix}$, $\mathbf{x}_3=\begin{pmatrix}4\\2\\1\end{pmatrix}$. Obtain an orthonormal basis for span $\{\mathbf{x}_1,\mathbf{x}_2,\mathbf{x}_3\}$ by applying the Gram–Schmidt procedure to $\{\mathbf{x}_1,\mathbf{x}_2,\mathbf{x}_3\}$.

5.15 Assign to $\mathfrak{M}^{3\times 1}$ the inner product $\langle \mathbf{x},\mathbf{y}\rangle \triangleq \mathbf{y}^{\mathrm{T}}\mathbf{Q}\mathbf{x}$, where

$$\mathbf{Q}=\begin{pmatrix}1&1&1\\1&5&0\\1&0&3\end{pmatrix}$$

Let $\mathbf{x}_1=(1\quad 0\quad 1)^{\mathrm{T}}$. Find an orthogonal basis for $\{\mathbf{x}_1\}^{\perp}$, the orthogonal complement of $\mathbf{x}_1$.

*5.16 *Recurrence formulas for orthogonal polynomials*: let $\{\mathbf{p}_0,\mathbf{p}_1,\mathbf{p}_2,\ldots\}$ be a set of polynomials orthogonal with respect to some inner product. Then $\mathbf{p}_n$ can be expressed in terms of $\mathbf{p}_{n-1}$ and $\mathbf{p}_{n-2}$ in the following fashion:

$$\mathbf{p}_n(t)=(c_nt+b_n)\mathbf{p}_{n-1}(t)-a_n\mathbf{p}_{n-2}(t)$$

Once the appropriate coefficients $\{a_n,b_n,c_n\}$ are known, the three-term recurrence formula allows successive determination of the orthogonal polynomials in a manner which is far less cumbersome than the Gram–Schmidt procedure [5.7]. Three-term recurrence formulas exist for other orthogonal sets as well: sine–cosine functions, Bessel functions, and various sets of functions defined on discrete domains. (See [5.24], p. 269] and [5.12]).

Verify for $n=2$ that the Legendre polynomials of Example 2, Section 5.2 obey the recurrence relation

$$\mathbf{p}_n(t)=\left(\frac{2n-1}{n}\right)t\mathbf{p}_{n-1}(t)-\left(\frac{n-1}{n}\right)\mathbf{p}_{n-2}(t)$$

Use this recurrence relation to compute $\mathbf{p}_3$ and verify that it is the next polynomial in the Legendre polynomial set; that is, show that $\mathbf{p}_3$ has the correct norm and is orthogonal to the lower-order polynomials in the set.

5.17 Let $\mathcal{P}^3(-1,1)$ be the space of real polynomial functions of degree less than three with the inner product

$$\langle \mathbf{f},\mathbf{g}\rangle \triangleq \int_{-1}^{1}(1+t^2)\mathbf{f}(t)\mathbf{g}(t)\,dt$$

Let $\mathcal{W}$ be the subspace of $\mathcal{P}^3(-1,1)$ spanned by $\mathbf{f}_0$, where $\mathbf{f}_0(t) \triangleq 1$. Find a basis for $\mathcal{W}^\perp$, the orthogonal complement of $\mathcal{W}$.

5.18 Let $\mathcal{V}$ be the space of complex-valued functions on $[0,1]$ which are bounded, piecewise continuous, and have no more than a finite number of maxima, minima, or discontinuities (these are called the Dirichlet conditions). A typical function in $\mathcal{V}$ is

$$\begin{aligned} \mathbf{h}(t) &= 1 \qquad 0 \leqslant t < \tfrac{1}{2} \\ &= -1 \qquad \tfrac{1}{2} \leqslant t \leqslant 1 \end{aligned}$$

Equip $\mathcal{V}$ with the inner product

$$\langle \mathbf{f}, \mathbf{g} \rangle \triangleq \int_0^1 \mathbf{f}(t)\, \overline{\mathbf{g}(t)}\, dt$$

Then the set of functions

$$\mathbf{g}_n(t) \triangleq e^{i2\pi n t} \qquad n = 0, \pm 1, \pm 2, \ldots$$

where $i = \sqrt{-1}$, is an orthonormal basis for the space.

(*a*) Determine the coordinates of the function $\mathbf{h}$ relative to this orthonormal basis; that is, expand $\mathbf{h}$ in its exponential Fourier series.

(*b*) To what value does the series converge at the discontinuities ($t = 0, \frac{1}{2}, 1$)? (Hint: combine the positive and negative nth order terms of the series.)

*5.19 Let $\mathbf{T}$ be the linear operator on $\mathfrak{M}_c^{n\times 1}$ defined by

$$\mathbf{T}\mathbf{x} \triangleq \mathbf{A}\mathbf{x}$$

where $\mathbf{A}$ is an $n \times n$ matrix. Let the inner product on $\mathfrak{M}_c^{n\times 1}$ be defined by

$$\langle \mathbf{x}, \mathbf{y} \rangle = \bar{\mathbf{y}}^{\mathrm{T}} \mathbf{Q} \mathbf{x},$$

where $\mathbf{Q}$ is a hermitian-symmetric, positive-definite matrix. Determine the form of $\mathbf{T}^*$.

5.20 Let $\mathbf{T}$ be the linear operator on the standard inner product space $\mathcal{L}_2(0,1)$ defined by

$$(\mathbf{T}\mathbf{f})(t) \triangleq \int_0^t b(s)\mathbf{f}(s)\, ds$$

Determine the form of $\mathbf{T}^*$. Hint: watch the limits of integration.

5.21 Let $\mathbf{T}\colon \mathcal{L}_2(0,1)\times\mathcal{L}_2(0,1)\rightarrow\mathfrak{M}^{2\times 1}$ be defined by

$$\mathbf{Tu}\triangleq\int_0^1\mathbf{Q}(s)\mathbf{u}(s)\,ds$$

where $\mathbf{Q}(s)$ is a 2×2 matrix and $\mathbf{u}(s)$ is a 2×1 matrix. Find the adjoint $\mathbf{T}^*$ for the inner products

$$\langle\mathbf{x},\mathbf{y}\rangle_{\mathfrak{M}^{2\times 1}}\triangleq\mathbf{y}^{\mathrm{T}}\mathbf{x}$$

$$\langle\mathbf{u},\mathbf{v}\rangle_{\mathcal{L}_2\times\mathcal{L}_2}\triangleq\int_0^1\mathbf{v}^{\mathrm{T}}(s)\mathbf{u}(s)\,ds$$

5.22 Let $\langle\mathbf{x},\mathbf{z}\rangle_n\triangleq\mathbf{z}^{\mathrm{T}}\mathbf{Q}\mathbf{x}$ and $\langle\mathbf{y},\mathbf{w}\rangle_m\triangleq\mathbf{w}^{\mathrm{T}}\mathbf{R}\mathbf{y}$ specify the inner products on the real spaces $\mathfrak{M}^{n\times 1}$ and $\mathfrak{M}^{m\times 1}$, respectively, where $\mathbf{Q}$ and $\mathbf{R}$ are symmetric, positive-definite matrices. Define $\mathbf{T}\colon\mathfrak{M}^{n\times 1}\rightarrow\mathfrak{M}^{m\times 1}$ by $\mathbf{Tx}\triangleq\mathbf{Ax}$.
(*a*) Find $\mathbf{T}^*$.
(*b*) Determine the properties which must be satisfied by $\mathbf{A}$, $\mathbf{Q}$, and $\mathbf{R}$ in order that $\mathbf{T}$ be self-adjoint.

5.23 Define $\mathbf{T}$ on $\mathfrak{M}^{2\times 1}$ by $\mathbf{Tx}\triangleq\mathbf{Ax}$, where $\mathbf{A}=\begin{pmatrix}1 & 2\\ 0 & 2\end{pmatrix}$. Define the inner product by $\langle\mathbf{x},\mathbf{y}\rangle=\mathbf{y}^{\mathrm{T}}\mathbf{Q}\mathbf{x}$. Pick $\mathbf{Q}$ such that $\mathbf{T}$ is self-adjoint.

5.24 Let $\mathbf{L}\triangleq\mathbf{D}^2+\mathbf{D}$ act on those functions $\mathbf{f}$ in $\mathcal{C}^2(a,b)$ which satisfy the boundary conditions $\mathbf{f}(a)=\mathbf{f}'(b)=0$. Assuming the standard inner product for $\mathcal{C}^2(a,b)$, find the formal adjoint $\mathbf{L}^*$ and the adjoint boundary conditions.

5.25 Define the differential operator $\mathbf{L}$ on $\mathcal{C}^2(a,b)$ by $\mathbf{Lf}\triangleq\mathbf{f}''-\mathbf{f}'$. Associate with $\mathbf{L}$ the boundary conditions $\mathbf{f}(a)+\mathbf{f}'(a)=\mathbf{f}(b)+\mathbf{f}'(b)=0$. Find the formal adjoint $\mathbf{L}^*$ and the adjoint boundary conditions relative to the standard inner product.

5.26 The wave equation is

$$\frac{\partial^2\mathbf{f}}{\partial s^2}+\frac{\partial^2\mathbf{f}}{\partial\sigma^2}-\frac{\partial^2\mathbf{f}}{\partial t^2}=0$$

where s and σ are space variables and t represents time. This equation can be represented in operator notation as

$$(\boldsymbol{\nabla}^2-\mathbf{D}^2)\mathbf{f}=\boldsymbol{\theta}$$

where the Laplacian operator $\boldsymbol{\nabla}^2$ acts only with respect to the space variables and the ordinary differential operator $\mathbf{D}^2$ acts only with respect to the time variable. Assume $\boldsymbol{\nabla}^2-\mathbf{D}^2$ acts on the space

$\mathcal{L}_2(\Omega)\times\mathcal{L}_2(0,b)$ with the inner product

$$\langle \mathbf{f},\mathbf{g}\rangle=\int_0^b\int_\Omega \mathbf{f}(\mathbf{p},t)\mathbf{g}(\mathbf{p},t)\,d\mathbf{p}\,dt$$

where $\mathbf{p}=(s,\sigma)$, Ω is the spatial domain (with boundary Γ), and t is in $[0,b]$. Show that the "wave operator" $\nabla^2-\mathbf{D}^2$ is formally self-adjoint. Hint: use Examples 10 and 11 of Section 5.4.

*5.24 Let $\mathcal{V}$ be an inner product space (perhaps infinite dimensional) with a basis $\mathcal{X}$. Let $\mathbf{T}$ be a linear operator on $\mathcal{V}$. Show that if $\mathcal{X}$ is orthonormal, $[\mathbf{T}^*]_{\mathcal{X}\mathcal{X}}=\overline{[\mathbf{T}]}^{\mathrm{T}}_{\mathcal{X}\mathcal{X}}$. Hint: express the inner product in terms of coordinates relative to the orthonormal basis.

*5.28 Let $\mathbf{T}$ be a linear operator on a complex inner product space $\mathcal{V}$.

(*a*) (1) If $\mathbf{T}^*\mathbf{T}=\mathbf{T}\mathbf{T}^*$, we call $\mathbf{T}$ a **normal** operator.

(2) If $\mathbf{T}^*\mathbf{T}=\mathbf{T}\mathbf{T}^*=\mathbf{I}$ (i.e., $\mathbf{T}^*=\mathbf{T}^{-1}$), we call $\mathbf{T}$ a **unitary** operator.

(3) We call $\mathbf{T}$ **non-negative** if $\langle \mathbf{Tx},\mathbf{x}\rangle\geqslant 0$ for all *complex* $\mathbf{x}$ in $\mathcal{V}$. If, in addition, $\langle \mathbf{Tx},\mathbf{x}\rangle=0$ only for $\mathbf{x}=\boldsymbol{\theta}$, we say $\mathbf{T}$ is **positive definite**.

(*b*) If $\mathbf{T}$ is (1) self-adjoint, (2) non-negative, (3) positive definite, (4) unitary, or (5) a projector, then the eigenvalues of $\mathbf{T}$ are, respectively, (1′) real, (2′) non-negative, (3′) positive, (4′) of absolute value 1, or (5′) equal to 1 or 0. If $\mathbf{T}$ is normal and $\mathcal{V}$ is finite dimensional, then (1′)–(5′) also imply (1)–(5). The inclusions among these classes of linear operators are illustrated by the following diagram.

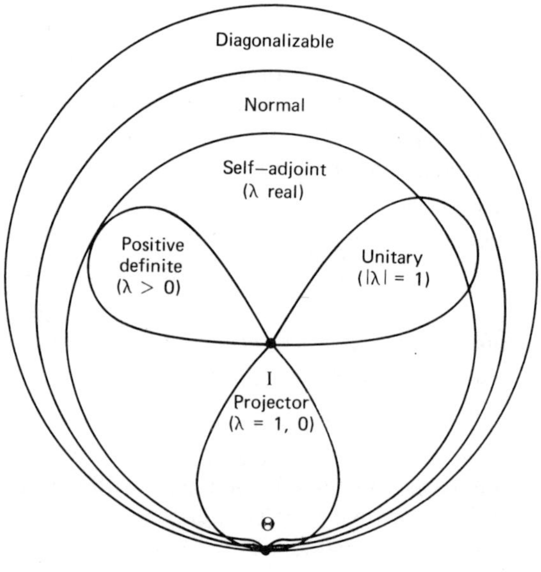

*5.29 *Norms of linear transformations*: define $\mathbf{T}$: $\mathfrak{M}_c^{n\times 1}\to\mathfrak{M}_c^{m\times 1}$ by $\mathbf{Tx} \triangleq \mathbf{Ax}$, where $\mathbf{A}$ is an $m\times n$ matrix. Assume the standard inner products. Then

$$\|\mathbf{T}\|^2=\|\mathbf{A}\|^2=\max_{\mathbf{x}^{\mathrm{T}}\mathbf{x}=1}\mathbf{x}^{\mathrm{T}}\mathbf{A}^{\mathrm{T}}\mathbf{Ax}=\lambda_{\mathrm{L}}$$

where λ_{L} is the eigenvalue of $\mathbf{A}^{\mathrm{T}}\mathbf{A}$ which is of largest magnitude.

(*a*) Find $\|\mathbf{A}\|$ for the following matrix by carrying out the maximization indicated above:

$$\mathbf{A}=\begin{pmatrix}1 & 2\\ 1 & 2\\ 1 & 2\end{pmatrix}$$

(*b*) Find $\|\mathbf{A}\|$ for the matrix $\mathbf{A}$ of (*a*) by determining the eigenvalue λ_{L}.

(*c*) A coarse, but easily computed, upper bound on $\|\mathbf{A}\|$ is the *Euclidean norm* of $\mathbf{A}$ defined by

$$\|\mathbf{A}\|_E^2 \triangleq \sum_{i,j}|a_{ij}|^2=\operatorname{trace}(\mathbf{A}^{\mathrm{T}}\mathbf{A})=\operatorname{trace}(\mathbf{AA}^{\mathrm{T}})=\sum_i|\lambda_i|^2$$

(The numbers $\{\lambda_i\}$ are the eigenvalues of $\mathbf{A}$.) Find $\|\mathbf{A}\|_E$ for the matrix $\mathbf{A}$ of (*a*).

(*d*) If $\mathbf{T}$ is a *bounded normal* operator on a *complex* Hilbert space $\mathcal{V}$, then

$$\|\mathbf{T}\|=\max_{\|\mathbf{x}\|=1}|\langle\mathbf{Tx},\mathbf{x}\rangle|=\max_i|\lambda_i|$$

where the numbers $\{\lambda_i\}$ are the eigenvalues of $\mathbf{T}$ [5.2, p. 382]. Use this relationship to find $\|\mathbf{T}^{-1}\|$ for $\mathbf{T}$ equal to the differential system of Example 2, Section 4.3. Note that this relationship between $\|\mathbf{T}\|$ and the largest eigenvalue of $\mathbf{T}$ can be used to determine $\|\mathbf{A}\|$ for any *symmetric* matrix $\mathbf{A}$; it cannot be used for the matrix $\mathbf{A}$ of (*a*).

*5.30 Let $\mathbf{T}$ be a bounded linear operator on a Hilbert space $\mathcal{V}$.

(*a*) Show that $\langle\mathbf{x},\mathbf{y}\rangle_{\mathrm{T}} \triangleq \langle\mathbf{x},\mathbf{Ty}\rangle$ is an inner product on $\mathcal{V}$ if and only if $\mathbf{T}$ is self-adjoint and positive definite.

(*b*) The operator $\mathbf{T}$ is self-adjoint and positive definite if and only if $\mathbf{T}$ can be decomposed as $\mathbf{T}=\mathbf{U}^2$ where $\mathbf{U}$ is a self-adjoint positive-definite linear operator on $\mathcal{V}$.

(*c*) Let $\mathcal{V}=\mathfrak{M}^{2\times 1}$ with the standard inner product. Let $\mathbf{Tx} \triangleq \mathbf{Qx}$

where

$$\mathbf{Q}=\begin{pmatrix} 13 & 5 \\ 5 & 13 \end{pmatrix}$$

Find a self-adjoint, positive-definite operator $\mathbf{U}$ on $\mathfrak{M}^{2\times 1}$ such that $\mathbf{T}=\mathbf{U}^2$.

*5.31 *Reciprocal bases*: let $\{\mathbf{x}_1,\ldots,\mathbf{x}_n\}$ be a basis for $\mathfrak{M}^{n\times 1}$ which is composed of eigenvectors for the invertible $n\times n$ matrix $\mathbf{A}$. Assume the standard inner product for $\mathfrak{M}^{n\times 1}$. The **reciprocal basis** $\{\mathbf{y}_1,\ldots,\mathbf{y}_n\}$ (reciprocal to $\{\mathbf{x}_i\}$) is defined by $\langle \mathbf{x}_i,\mathbf{y}_j\rangle=\mathbf{y}_j^{\mathrm{T}}\mathbf{x}_i=\delta_{ij}$.

(*a*) The vectors in the reciprocal basis are eigenvectors of $\mathbf{A}^{\mathrm{T}}$ (left-hand eigenvectors of $\mathbf{A}$).

(*b*) Every vector $\mathbf{x}$ in $\mathfrak{M}^{n\times 1}$ can be expressed as a biorthogonal expansion, $\mathbf{x}=\sum_{i=1}^{n}\langle \mathbf{x},\mathbf{y}_i\rangle \mathbf{x}_i$. Use this fact to show that the solution to $\mathbf{Ax}=\mathbf{y}$ can be expanded as

$$\mathbf{x}=\sum_{i=1}^{n}\frac{\langle \mathbf{y},\mathbf{y}_i\rangle}{\lambda_i}\mathbf{x}_i$$

Hint: follow the derivation of (5.24).

(*c*) The **outer product** of two vectors in $\mathfrak{M}^{n\times 1}$ is defined by $\mathbf{x}><\mathbf{y} \triangleq \mathbf{x}\mathbf{y}^{\mathrm{T}}$. Such a "backwards inner product" is sometimes referred to as a **dyad**. Use the dyad notation to convert the expansion in (*b*) to an explicit matrix multiplication of $\mathbf{y}$. Compare the resulting matrix to the fundamental formula for $\mathbf{A}^{-1}$,

$$\mathbf{A}^{-1}=\frac{1}{\lambda_1}\mathbf{E}_{10}^{\mathbf{A}}+\cdots+\frac{1}{\lambda_p}\mathbf{E}_{p0}^{\mathbf{A}}$$

where p is the number of *distinct* eigenvalues of $\mathbf{A}$. How are the constituent matrices $\{\mathbf{E}_{i0}^{\mathbf{A}}\}$ related to the pair of biorthogonal bases $\{\mathbf{x}_i\}$ and $\{\mathbf{y}_i\}$?

(*d*) Let $\mathbf{A}=\begin{pmatrix} 0 & -1 \\ -2 & -1 \end{pmatrix}$. Find a basis for $\mathfrak{M}^{2\times 1}$ consisting in eigenvectors for $\mathbf{A}$; find the reciprocal basis; use the pair of biorthogonal bases to find the constituents of $\mathbf{A}$; use the constituents to compute $\mathbf{A}^{-1}$.

5.32 Let $\mathbf{A}$ be an $n\times n$ matrix. Then $\mathbf{Tx} \triangleq \mathbf{Ax}$ defines a linear operator $\mathbf{T}$ on $\mathfrak{M}^{n\times 1}$. Let $\mathbf{Q}$ be an $n\times n$ symmetric positive-definite matrix.

(*a*) Show that if $\mathbf{T}$ is self-adjoint with respect to the inner product

$\langle \mathbf{x},\mathbf{y}\rangle_Q \triangleq \mathbf{y}^{\mathrm{T}}\mathbf{Q}\mathbf{x}$, then the operator $\mathbf{U}$ defined by $\mathbf{U}\mathbf{x} \triangleq \mathbf{Q}\mathbf{A}\mathbf{x}$ is self-adjoint with respect to the standard inner product. The matrix equation $\mathbf{A}\mathbf{x}=\mathbf{y}$ can be replaced by an equivalent equation, $\mathbf{Q}\mathbf{A}\mathbf{x}=\mathbf{Q}\mathbf{y}$; the latter equation can be analyzed in terms of a set of eigenvectors (of $\mathbf{Q}\mathbf{A}$) which is orthonormal with respect to the standard inner product.

(*b*) Let $\mathbf{A}=\begin{pmatrix} 2 & 3 \\ 0 & 4 \end{pmatrix}$. Find a matrix $\mathbf{Q}$ such that $\mathbf{T}$ is self-adjoint with respect to the inner product $\langle \mathbf{x},\mathbf{y}\rangle_Q=\mathbf{y}^{\mathrm{T}}\mathbf{Q}\mathbf{x}$. Hint: a test for positive definiteness is given in P&C 5.9.

5.33 By Example 2, Section 5.5, the differential operator $\mathbf{L} \triangleq -\mathbf{D}^2$ and the boundary conditions $\mathbf{f}(0)=\mathbf{f}(b)=0$ are self-adjoint with respect to the standard inner product on the interval $[0,b]$.

(*a*) The eigendata for $\mathbf{L}$ with the given boundary conditions are

$$\lambda_k=\left(\frac{k\pi}{b}\right)^2, \qquad \mathbf{f}_k(t)=\sin\left(\frac{k\pi t}{b}\right), \qquad k=1,2,3,\ldots$$

Show that the eigenfunctions form an orthogonal set.

(*b*) Express the solution to the differential system $-\mathbf{f}''=\mathbf{u}, \mathbf{f}(0)=\mathbf{f}(\mathrm{b})=0$ as an eigenfunction expansion.

(*c*) Compare the first term of the eigenfunction expansion in (*b*) to the exact solution for the specific input function $\mathbf{u}(t)=1$ and $b=1$.

5.34 The (nonharmonic) eigendata for the differential operator $-\mathbf{D}^2$ with the boundary conditions $\mathbf{f}(0)=\mathbf{f}(b)+\mathbf{f}'(b)=0$ are derived in Example 1, Section 4.3.

(*a*) Show that the eigenfunctions are orthogonal with respect to the standard inner product on the interval $[0,b]$ (and, consequently, that $\mathbf{L}$ and the boundary conditions are self-adjoint).

(*b*) Express the solution to the differential system $-\mathbf{f}''=\mathbf{u}, \mathbf{f}(0)=\mathbf{f}(b)+\mathbf{f}'(b)=0$ as an eigenfunction expansion.

(*c*) Compare the first term of the eigenfunction expansion in (*b*) to the exact solution for the specific input $\mathbf{u}(t)=1$ and $b=1$. Hint: $\tan(2.0288)\approx -2.0288$. (The exact solution for the differential system is given in P&C 3.13. Note that the symmetry of the Green's function again implies the self-adjointness of the system with respect to the standard inner product.)

5.35 A (nonsinusoidal) periodic voltage $\mathbf{e}$ of frequency ω (period $2\pi/\omega$)

is applied to the terminals of the R–L circuit of Figure 5.4. The steady-state current $\mathbf{i}_1$ satisfies the differential equation $L\mathbf{i}_1' + R\mathbf{i}_1 = \mathbf{e}$ with the periodic boundary condition $\mathbf{i}_1(2\pi/\omega) = \mathbf{i}_1(0)$.

(*a*) Find the eigendata for the differential operator $L\mathbf{D} + R\mathbf{I}$ with periodic boundary conditions.

(*b*) Show that the eigenfunctions are orthogonal with respect to the standard *complex* inner product on the interval $[0, 2\pi/\omega]$.

(*c*) Determine the eigenfunction expansion of the steady-state current for an arbitrary periodic voltage. Verify the result by applying the voltage $\mathbf{e}(t) = \sin\omega t$.

5.36 Define $\nabla^2\mathbf{f}(s,t) \stackrel{\Delta}{=} (\partial^2\mathbf{f}/\partial s^2) + (\partial^2\mathbf{f}/\partial t^2)$ on the rectangle $0 \leqslant s \leqslant a$, $0 \leqslant t \leqslant b$. Let $\mathbf{f}$ satisfy the boundary conditions

$$\frac{\partial \mathbf{f}}{\partial s}(0,t) = \frac{\partial \mathbf{f}}{\partial s}(a,t) = \frac{\partial \mathbf{f}}{\partial t}(s,0) = \frac{\partial \mathbf{f}}{\partial t}(s,b) = 0$$

(*a*) The eigendata for ∇^2 with the given boundary conditions are displayed in P&C 4.15. Show that the eigenfunctions are orthogonal with respect to the inner product

$$\langle \mathbf{f}, \mathbf{g} \rangle \stackrel{\Delta}{=} \int_0^b \int_0^a \mathbf{f}(s,t)\mathbf{g}(s,t)\,ds\,dt$$

(*b*) Note that one of the eigenvalues is zero. The range of the operator was derived in Example 12, Section 5.4. Express as an eigenfunction expansion the general solution to the partial differential system $\nabla^2\mathbf{f} = \mathbf{u}$ with the given boundary conditions.

*5.37 *A Hilbert space of random variables*: let $\mathcal{V}$ be a vector space of real-valued random variables defined on a particular experiment (Example 11, Section 2.1). An inner product can be defined on $\mathcal{V}$ in terms of the expected value operation (P&C 2.23):

$$\langle \mathbf{x}, \mathbf{y} \rangle \stackrel{\Delta}{=} \mathbf{E}(\mathbf{xy}) = \int xy\,\omega_{\mathbf{xy}}(x,y)\,dx\,dy$$

(*a*) Show that $\mathbf{E}(\mathbf{xy})$ is a valid inner product on $\mathcal{V}$.

(*b*) We refer to $\mathbf{E}(\mathbf{x})$ as the *mean* of the random variable $\mathbf{x}$. The *variance* of $\mathbf{x}$ is defined by

$$\operatorname{var}(\mathbf{x}) \stackrel{\Delta}{=} \|\mathbf{x} - \mathbf{E}(\mathbf{x})\|^2 = \mathbf{E}(\mathbf{x}^2) - \mathbf{E}^2(\mathbf{x})$$

The *covariance* between $\mathbf{x}$ and $\mathbf{y}$ is defined by

$$\operatorname{cov}(\mathbf{x}, \mathbf{y}) \stackrel{\Delta}{=} \langle \mathbf{x} - \mathbf{E}(\mathbf{x}), \mathbf{y} - \mathbf{E}(\mathbf{y}) \rangle = \mathbf{E}(\mathbf{xy}) - \mathbf{E}(\mathbf{x})\mathbf{E}(\mathbf{y})$$

The random variables $\mathbf{x}$ and $\mathbf{y}$ are said to be *uncorrelated* if $\text{cov}(\mathbf{x},\mathbf{y})=0$. Show that if $\mathbf{x}$ and $\mathbf{y}$ are uncorrelated, then $\text{var}(\mathbf{x}+\mathbf{y})=\text{var}(\mathbf{x})+\text{var}(\mathbf{y})$. Show that if $\mathbf{x}$ and $\mathbf{y}$ are orthogonal, then $\mathbf{E}((\mathbf{x}+\mathbf{y})^2)=\mathbf{E}(\mathbf{x}^2)+\mathbf{E}(\mathbf{y}^2)$ (Pythagorean theorem). If either $\mathbf{x}$ or $\mathbf{y}$ has zero mean, then $\mathbf{x}$ and $\mathbf{y}$ are orthogonal if and only if they are uncorrelated.

(*c*) The vector space $\mathcal{H}$ which consists in all random variables (defined on the experiment) of finite norm is a Hilbert space. Show that $\mathcal{H}$ consists in precisely those random variables which have finite mean and finite variance.

5.38 *Karhunen–Loève Expansion*: let $\mathbf{x} \triangleq (\mathbf{x}(1)\cdots\mathbf{x}(n))^{\mathrm{T}}$ be a discrete finite random process with zero mean; that is, $\mathbf{x}$ consists in a sequence of n random variables $\{\mathbf{x}(i)\}$, all defined on a single experiment,* and $\mathbf{E}(\mathbf{x}(i))=0, i=1,\ldots,n$. (For notational convenience we treat the n elements of $\mathbf{x}$ as an $n\times 1$ column vector.) A particular running of the underlying experiment yields a sample function $\bar{\mathbf{x}}$, a specific column of n numbers. The sample function $\bar{\mathbf{x}}$ has a unique Fourier series expansion $\bar{\mathbf{x}}=\sum_{j=1}^{n}\langle\bar{\mathbf{x}},\mathbf{y}_j\rangle\mathbf{y}_j$ corresponding to each orthonormal basis $\mathcal{Y}=\{\mathbf{y}_j\}$ for the standard inner product space $\mathfrak{M}^{n\times 1}$. We can also expand the random process itself in a Fourier series, $\mathbf{x}=\sum_{j=1}^{n}c_j\mathbf{y}_j$, where $\mathbf{c}_j=\langle\mathbf{x},\mathbf{y}_j\rangle=\sum_{p=1}^{n}\mathbf{x}(p)\mathbf{y}_j(p)$, and $\mathbf{y}_j(p)$ is the pth element of $\mathbf{y}_j$. However, since the elements $\{\mathbf{x}(p)\}$ of $\mathbf{x}$ are random variables, the Fourier coefficients c_j are also random variables. We wish to pick the basis $\mathcal{Y}$ for $\mathfrak{M}^{n\times 1}$ in such a way that the random variables $\{c_j\}$ are statistically orthogonal ($\mathbf{E}(c_jc_k)=0$). The resulting Fourier series expansion is known as the *Karhunen–Loêve expansion* of the random process.† The sequence of random variables $\{\mathbf{x}(i)\}$ can be represented by the sequence of random variables $\{c_j\}$; the latter are uncorrelated.

(*a*) If we substitute into $\mathbf{E}(\mathbf{x}(i)c_k)$ the Fourier series expansion $\mathbf{x}=\sum_{j=1}^{n}c_j\mathbf{y}_j$, we find $\mathbf{E}(\mathbf{x}(i)c_k)=\sum_{j=1}^{n}\mathbf{E}(c_jc_k)\mathbf{y}_j(i)$. On the other hand, if we substitute the Fourier coefficient expansion $c_k=\sum_{p=1}^{n}\mathbf{x}(p)\mathbf{y}_k(p)$, we obtain $\mathbf{E}(\mathbf{x}(i)c_k)=\sum_{p=1}^{n}\mathbf{E}(\mathbf{x}(i)\mathbf{x}(p))\mathbf{y}_k(p)$. By equating these two expansions, show that the random variables $\{c_k\}$ are orthogonal if and only if the basis functions $\{\mathbf{y}_k\}$ satisfy

$$\mathbf{R}\mathbf{y}_k=\mathbf{E}(c_k^2)\mathbf{y}_k, \qquad k=1,\ldots,n$$

where $\mathbf{R}$ is the autocorrelation matrix for the random process;

* See Example 11, Section 2.1.

† See Papoulis [5.19] for a discussion of the Karhunen–Loève expansion for continuous random processes.

R is defined by

$$\mathbf{R}(i,p) = \mathbf{E}(\mathbf{x}(i)\mathbf{x}(p)), \qquad i,p = 1,\ldots,n$$

(*b*) Let the autocorrelation matrix of a two-element random process **x** be

$$\mathbf{R} = \begin{pmatrix} 2 & 1 \\ 1 & 2 \end{pmatrix}$$

Find an orthonormal basis $\{\mathbf{y}_1, \mathbf{y}_2\}$ for the standard inner product space $\mathfrak{M}^{2\times 1}$ relative to which the coordinates of **x** are statistically orthogonal. Verify your results by computing the coordinates, c_1 and c_2.

5.7 References

[5.1] Apostol, Tom M., *Mathematical Analysis*, Addison-Wesley, Reading, Mass., 1957.

[5.2] Bachman, George and Lawrence Narici, *Functional Analysis*, Academic Press, New York, 1966.

[5.3] Birkhoff, Garrett and Gian-Carlo Rota, *Ordinary Differential Equations*, 2nd ed., Blaisdell, Waltham, Mass., 1969.

[5.4] Buck, R., *Advanced Calculus*, McGraw-Hill, New York, 1956.

[5.5] Churchill, R. V., *Fourier Series and Boundary Value Problems*, McGraw-Hill, New York, 1941.

[5.6] Davenport, Wilbur B., Jr. and William L. Root, *Random Signals and Noise*, McGraw-Hill, New York, 1958.

[5.7] Forsythe, George E., "Generation and Use of Orthogonal Polynomials for Data Fitting with a Digital Computer," *J. Soc. Ind. Appl. Math.*, **5**, 2 (June 1957).

[5.8] Friedman, Bernard, *Principles and Techniques of Applied Mathematics*, Wiley, New York, 1956.

[5.9] Goffman, Casper and George Pedrick, *First Course in Functional Analysis*, Prentice-Hall, Englewood Cliffs, N.J., 1965.

*[5.10] Halmos, P. R., *Finite-Dimensional Vector Spaces*, Van Nostrand, Princeton, N.J., 1958.

[5.11] Halmos, P. R., *Introduction to Hilbert Space*, 2nd ed., Chelsea, New York, 1957.

[5.12] Hamming, R. W., *Numerical Methods for Scientists and Engineers*, McGraw-Hill, New York, 1962.

[5.13] Harmuth, H. F., *Transmission of Information by Orthogonal Functions*, Springer-Verlag, Berlin/New York, 1970.

*[5.14] Hoffman, Kenneth and Ray Kunze, *Linear Algebra*, Prentice-Hall, Englewood Cliffs, N.J., 1961.

[5.15] Lamarsh, John R., *Introduction to Nuclear Reactor Theory*, Addison–Wesley, Reading, Mass., 1966.

[5.16] Lanczos, Cornelius, *Linear Differential Operators*, Van Nostrand, Princeton, N.J., 1961.

*[5.17] Naylor, Arch W. and George R. Sell, *Linear Operator Theory in Engineering and Science*, Holt, Rinehart, and Winston, New York, 1971.

[5.18] Oliver, B. M., J. R. Pierce, and C. E. Shannon, "The Philosophy of Pulse Code Modulation," *Proc. IRE*, **36**, 11 (November 1948), 1324–1331.

[5.19] Papoulis, Athanasios, *Probability, Random Variables, and Stochastic Processes*, McGraw-Hill, New York, 1965.

[5.20] Ralston, Anthony, *A First Course in Numerical Analysis*, McGraw-Hill, New York, 1965.

[5.21] Royden, H. L., *Real Analysis*, 2nd ed., Macmillan, New York, 1968.

[5.22] Stakgold, Ivar, *Boundary Value Problems of Mathematical Physics*, Vol. I, Macmillan, New York, 1967.

[5.23] Vulikh, B. Z., *Introduction to Functional Analysis*, Pergamon Press, Oxford, Addison-Wesley, Reading, Mass., 1963.

[5.24] Wylie, C. R., Jr., *Advanced Engineering Mathematics*, 3rd ed., McGraw-Hill, New York, 1966.

[5.25] Wilkinson, J. H., *The Algebraic Eigenvalue Problem*, Clarendon Press, Oxford, 1965.

6

Least-Square Minimization

The basic purpose for modeling is to determine how to design and/or operate a system without experimenting with the actual system. Chapters 1–5 are concerned primarily with describing, categorizing, and analyzing linear mathematical models of systems. In the process of constructing and using a mathematical model we must specify values for the model parameters. For instance, if we postulate that the capital cost c of an electric power plant (in dollars) and the size s of the plant (in megawatts) are related by an algebraic relation of the form $c = a_1 + a_2 s + a_3 s^2$, we must specify the values of the constants a_1, a_2, and a_3. Naturally, we select parameter values for which the model acts most nearly like the system it represents. We must choose an optimization criterion. Model parameters often are chosen to minimize the sum of squares of the errors between a set of measurements and a corresponding set of values predicted by the model. Occasionally some other criterion is used, such as minimization of the largest of the errors.

Once we have an adequate model, we use it to determine how to operate the system represented by the model. One approach is to use the model to determine the system input which makes the output behave in some desired fashion. For example, we might wish to determine the voltage that should be applied to the armature of a dc motor, as a function of time, so that the motor will position a piece of equipment satisfactorily. Once again, we need an optimization criterion. The voltage function might be chosen to minimize the time it takes to position the piece of equipment. The model for the motor acts as a constraint on the choice of the voltage function; the dynamic equations which describe the motor relate the voltage function to the prescribed final position of the piece of equipment.

Optimization problems like those described above arise frequently. The most widely used minimization criterion for such problems is a quadratic function of the variables. Any constraint equations associated with the minimization are usually treated as being linear. The primary reason for the extensive use of a quadratic minimization criterion with linear constraints is clear—derivatives of quadratic functions are linear functions;

the equations that must be solved in order to find the minimizing values of the variables are all linear, and can be solved with relative ease. Most minimizations carried out in detail in the literature are of this type. Few other types of minimization problems have a structure that is nice enough to be thoroughly understood. Fortunately, adequate motivation usually can be given for use of the quadratic minimization criterion.

In this chapter we begin our exploration of minimization problems in a vector space context. Now that we have introduced inner products, we are able to use the norm associated with an inner product to compare vectors. We consider two vectors $\mathbf{x}$ and $\mathbf{y}$ to be nearly alike if $\|\mathbf{x}-\mathbf{y}\|$ (or $\|\mathbf{x}-\mathbf{y}\|^2$) is small; the norm provides a "quadratic" measure of the size of $\mathbf{x}-\mathbf{y}$. This chapter explores in detail those minimization problems which are expressible as the minimization of an inner product norm. We restrict ourselves to linear constraint equations. More general minimization problems are the subject of Chapters 7 and 8.

We have already encountered norm (or least-square) minimization in connection with truncated Fourier series expansions; the Fourier coefficients provide a least-square approximation to a function (P&C 5.12). We noted in Section 5.2 the close relationship that exists between Fourier series, orthogonal projection, and Gram–Schmidt orthogonalization. This relationship pervades all of least-square minimization. The first two sections of this chapter introduce, in an orthogonal projection context, two basic least-square problems: (*a*) the least-square error solution of an overdetermined equation; and (*b*) the smallest (or least-effort) solution of an underdetermined equation. Sections 6.3 and 6.4 constitute a second view of these same two problems. This second view is expressed in terms of "pseudoinverse" operators; it is based on the orthogonal decomposition associated with a linear transformation and its adjoint [see (5.67)]. Section 6.5 combines the concepts of the four previous sections in order to handle degenerate inconsistent equations. Computational aspects of least-square problems are considered in Section 6.6.

6.1 Least-Error Problems by Orthogonal Projection

Consider the three-dimensional space $\mathcal{V}$ spanned by the mutually perpendicular vectors $\{\mathbf{y}_1, \mathbf{y}_2, \mathbf{y}_3\}$ of Figure 6.1. Let $\mathcal{W}$ be the plane spanned by $\{\mathbf{y}_1, \mathbf{y}_2\}$, and let $\mathbf{x}$ be a vector not in $\mathcal{W}$. Which vector $\mathbf{x}_a$ in $\mathcal{W}$ best approximates $\mathbf{x}$ in the sense that the length (norm) of the error, $\|\mathbf{x}-\mathbf{x}_a\|$, is minimized? The minimizing vector $\mathbf{x}_a$ is the projection of $\mathbf{x}$ on $\mathcal{W}$ along the direction perpendicular to $\mathcal{W}$. Stated another way, the minimum error vector $\mathbf{x}-\mathbf{x}_a$ is perpendicular to $\mathcal{W}$.

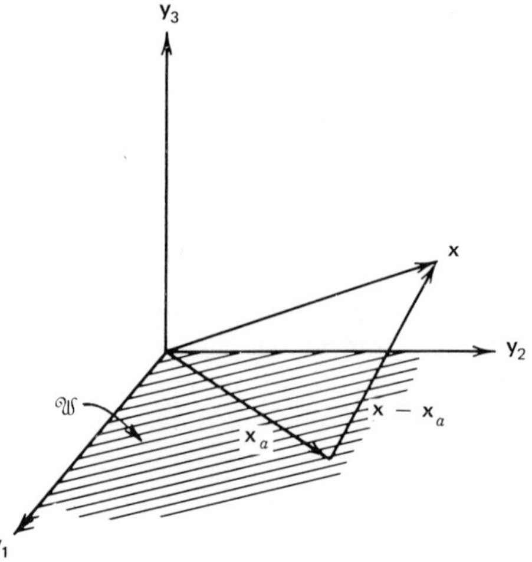

Figure 6.1. A least-error problem.

Many well-known minimization problems possess a mathematical structure analogous to this geometrical minimization problem. We generalize the geometrical problem by expressing it in the terminology of Hilbert space.

Definition. Let $\mathcal{W}$ be a (complete) subspace of a separable Hilbert space $\mathcal{V}$. Let $\mathbf{x}$ be in $\mathcal{V}$ and $\mathbf{x}_a$ in $\mathcal{W}$. The **least-square error problem** (or approximation problem) consists in determining that vector $\mathbf{x}_a$ for which $\|\mathbf{x}-\mathbf{x}_a\|$ is minimum.

We solve the least-square error problem as follows. The space $\mathcal{V}$ can be decomposed into the direct sum (5.20), $\mathcal{V}=\mathcal{W}\oplus\mathcal{W}^{\perp}$. Suppose we combine orthonormal bases for $\mathcal{W}$ and $\mathcal{W}^{\perp}$ to obtain an orthonormal basis $\{\mathbf{y}_k\}$ for $\mathcal{V}$. We can express the vector $\mathbf{x}$ and *any* approximating vector $\mathbf{x}_a$ in the pair of Fourier series:

$$\mathbf{x}=\sum_{\substack{\mathbf{y}_k \text{ in } \mathcal{W}\\ \text{and } \mathcal{W}^{\perp}}}\langle\mathbf{x},\mathbf{y}_k\rangle\mathbf{y}_k,\qquad \mathbf{x}_a=\sum_{\mathbf{y}_k \text{ in } \mathcal{W}}\langle\mathbf{x}_a,\mathbf{y}_k\rangle\mathbf{y}_k \tag{6.1}$$

Then

$$\mathbf{x}-\mathbf{x}_a=\sum_{\mathbf{y}_k \text{ in } \mathcal{W}}(\langle\mathbf{x},\mathbf{y}_k\rangle-\langle\mathbf{x}_a,\mathbf{y}_k\rangle)\mathbf{y}_k+\sum_{\mathbf{y}_k \text{ in } \mathcal{W}^{\perp}}\langle\mathbf{x},\mathbf{y}_k\rangle\mathbf{y}_k \tag{6.2}$$

Because the basis $\{\mathbf{y}_k\}$ is orthonormal, the Pythagorean theorem (P&C

5.4) yields

$$\|\mathbf{x}-\mathbf{x}_a\|^2 = \sum_{\mathbf{y}_k \text{ in } \mathcal{W}} \|(\langle \mathbf{x},\mathbf{y}_k\rangle - \langle \mathbf{x}_a,\mathbf{y}_k\rangle)\mathbf{y}_k\|^2 + \sum_{\mathbf{y}_k \text{ in } \mathcal{W}^\perp} \|\langle \mathbf{x},\mathbf{y}_k\rangle \mathbf{y}_k\|^2$$

$$= \sum_{\mathbf{y}_k \text{ in } \mathcal{W}} (\langle \mathbf{x},\mathbf{y}_k\rangle - \langle \mathbf{x}_a,\mathbf{y}_k\rangle)^2 + \sum_{\mathbf{y}_k \text{ in } \mathcal{W}^\perp} (\langle \mathbf{x},\mathbf{y}_k\rangle)^2 \tag{6.3}$$

The second sum of (6.3) is independent of $\mathbf{x}_a$. The first sum is minimized by picking $\mathbf{x}_a$ such that the Fourier coefficients of $\mathbf{x}$ and $\mathbf{x}_a$ are identical for $\mathbf{y}_k$ in $\mathcal{W}$,

$$\langle \mathbf{x},\mathbf{y}_k\rangle = \langle \mathbf{x}_a,\mathbf{y}_k\rangle \quad \text{for } \mathbf{y}_k \text{ in } \mathcal{W}$$

This equality of coefficients can also be expressed in the form

$$\langle \mathbf{x}-\mathbf{x}_a,\mathbf{y}_k\rangle = 0, \quad \mathbf{y}_k \text{ in } \mathcal{W} \tag{6.4}$$

showing that the minimum norm error vector $\mathbf{x}-\mathbf{x}_a$ is orthogonal to $\mathcal{W}$. Furthermore, by replacing the coefficients in the Fourier series expansion (6.1) of $\mathbf{x}_a$ with the identical coefficients for $\mathbf{x}$, we find that

$$\mathbf{x}_a = \sum_{\mathbf{y}_k \text{ in } \mathcal{W}} \langle \mathbf{x},\mathbf{y}_k\rangle \mathbf{y}_k \tag{6.5}$$

That is, $\mathbf{x}_a$ is the "partial" Fourier series expansion of $\mathbf{x}$ relative to any orthonormal basis for $\mathcal{W}$. From (5.22) we recognize (6.5) as an expression of the orthogonal projection of $\mathbf{x}$ onto $\mathcal{W}$. It follows that the minimum norm error vector $\mathbf{x}-\mathbf{x}_a$ is the orthogonal projection of $\mathbf{x}$ onto $\mathcal{W}^\perp$. The statements (6.4) and (6.5) constitute the **classical projection theorem**; they specify the solution to the least-square error problem. The obvious geometrical conclusion associated with the problem of Figure 6.1 is a special case.

Obtaining Least-Error Solutions

If $\mathcal{W}$ is n-dimensional, the two conditions (6.4) and (6.5) provide *two methods* for calculating the "least-error" approximation $\mathbf{x}_a$. One method amounts to *direct orthogonal projection*. Suppose $\mathcal{W}$ is described by a basis $\{\mathbf{x}_1,\ldots,\mathbf{x}_n\}$. If we apply the Gram–Schmidt procedure to the set $\{\mathbf{x}_1,\ldots,\mathbf{x}_n,\mathbf{x}\}$, the first n steps yield an orthonormal basis $\{\mathbf{y}_1,\ldots,\mathbf{y}_n\}$ for $\mathcal{W}$. The last step yields both the minimizing solution (6.5) and the corresponding error vector $\mathbf{x}-\mathbf{x}_a$.

Exercise 1. Use the Gram–Schmidt procedure graphically to solve the geometrical problem of Figure 6.1, wherein the orthonormal basis $\{\mathbf{y}_1, \mathbf{y}_2\}$ is already determined.

In the second method, we determine the orthogonal projection $\mathbf{x}_a$ indirectly using (6.4). Once again, let $\mathbf{x}_1, \ldots, \mathbf{x}_n$ be a basis for $\mathcal{W}$. It follows from (6.4) that $\mathbf{x}_a$ must satisfy the *orthogonality conditions*

$$\langle \mathbf{x} - \mathbf{x}_a, \mathbf{x}_k \rangle = 0, \qquad k = 1, \ldots, n \tag{6.6}$$

We put these conditions in a more standard form by substituting $\mathbf{x}_a = \sum_{j=1}^{n} \xi_j \mathbf{x}_j$ to obtain the **normal equations**:

$$\langle \mathbf{x}, \mathbf{x}_k \rangle = \sum_{j=1}^{n} \xi_j \langle \mathbf{x}_j, \mathbf{x}_k \rangle, \qquad k = 1, \ldots, n$$

or

$$\begin{pmatrix} \langle \mathbf{x}_1, \mathbf{x}_1 \rangle & \cdots & \langle \mathbf{x}_n, \mathbf{x}_1 \rangle \\ \vdots & & \vdots \\ \langle \mathbf{x}_1, \mathbf{x}_n \rangle & \cdots & \langle \mathbf{x}_n, \mathbf{x}_n \rangle \end{pmatrix} \begin{pmatrix} \xi_1 \\ \vdots \\ \xi_n \end{pmatrix} = \begin{pmatrix} \langle \mathbf{x}, \mathbf{x}_1 \rangle \\ \vdots \\ \langle \mathbf{x}, \mathbf{x}_n \rangle \end{pmatrix} \tag{6.7}$$

These equations can be solved by elimination to yield the minimizing coordinates $\{\xi_j\}$. In many practical problems the goal is this set of coordinates itself (Example 1); if not, we compute $\mathbf{x}_a = \sum_j \xi_j \mathbf{x}_j$. The matrix of (6.7) is familiar; it is known as the **Gram matrix** for the basis $\{\mathbf{x}_j\}$. We encountered the Gram matrix earlier, in (5.15), as the matrix of an inner product. From the independence of the basis vectors, it follows that the Gram matrix is invertible.

Exercise 2. Use the normal equations to determine the solution to the geometrical problem of Figure 6.1, using the orthonormal basis $\{\mathbf{y}_1, \mathbf{y}_2\}$.

Suppose we orthogonalize the basis $\{\mathbf{x}_1, \ldots, \mathbf{x}_n\}$, producing $\{\mathbf{y}_1, \ldots, \mathbf{y}_n\}$. Then we can express the orthogonality conditions (6.6) as $\langle \mathbf{x} - \mathbf{x}_a, \mathbf{y}_k \rangle = 0$, $k = 1, \ldots, n$. Substituting $\mathbf{x}_a = \sum_j c_j \mathbf{y}_j$ again leads to a set of normal equations. However, these equations are diagonal, and the computations are precisely equivalent to the orthogonal projection procedure described earlier.

The orthogonality conditions (6.6) can also be used if $\mathcal{V}$ is infinite-dimensional, although techniques for solution are not simple. The Weiner–Hopf equation of classical least-square filtering is essentially the set of normal equations for such an infinite-dimensional problem.* If $\mathcal{W}$ is infinite-dimensional, but $\mathcal{W}^{\perp}$ is finite-dimensional, we can use Gram–Schmidt orthogonalization to find the projection, $\mathbf{x}-\mathbf{x}_a$, of $\mathbf{x}$ on $\mathcal{W}^{\perp}$. Then $\mathbf{x}_a=\mathbf{x}-(\mathbf{x}-\mathbf{x}_a)$.

Exercise 3. Apply the Gram–Schmidt procedure graphically to the vectors $\{\mathbf{y}_3,\mathbf{x}\}$ of Figure 6.1 in order to determine the projection of $\mathbf{x}$ on $\mathcal{W}^{\perp}$.

It is not important that $\mathcal{V}$ be a complete space. But the completeness of the subspace $\mathcal{W}$ is crucial in that it guarantees the existence of a minimizing vector in $\mathcal{W}$. Yet if $\mathcal{W}$ were not complete, we could still find a least-error approximation $\mathbf{x}_a$ in $\mathcal{W}$ for *most* vectors $\mathbf{x}$ in $\mathcal{V}$.

Example 1. Inconsistent Linear Equations of Full Rank. Suppose that in the following set of equations there are more equations than unknowns ($m>n$) and the equations are inconsistent:

$$\begin{matrix} a_{11}\xi_1+\cdots+a_{1n}\xi_n=\eta_1 \\ \vdots \qquad\qquad \vdots \quad \vdots \\ a_{m1}\xi_1+\cdots+a_{mn}\xi_n=\eta_m \end{matrix} \tag{6.8}$$

These equations can be expressed in the form

$$\xi_1\mathbf{x}_1+\cdots+\xi_n\mathbf{x}_n=\mathbf{x} \tag{6.9}$$

where $\mathbf{x}_i$ is the ith column of the matrix of coefficients and $\mathbf{x}$ is the column on the right-hand side of (6.8). Assume the equations (6.8) are of full rank (i.e., the vectors $\{\mathbf{x}_1,\ldots,\mathbf{x}_n\}$ are independent). The fact that the equations are inconsistent means that $\mathbf{x}$ is not in span $\{\mathbf{x}_1,\ldots,\mathbf{x}_n\}$. Although the equations have no true solution, we can find a least-square error solution, one which minimizes the sum of the squares of the "residuals" defined by

$$r_i \triangleq \eta_i-(a_{i1}\xi_1+\cdots+a_{in}\xi_n), \qquad i=i,\ldots,m$$

Let $\mathcal{V}=\mathfrak{M}^{m\times 1}$ with its standard inner product. Let $\mathcal{W}\triangleq$ span $\{\mathbf{x}_1,\ldots,\mathbf{x}_n\}$ and denote $\mathbf{x}_a\triangleq\sum_{i=1}^{n}\xi_i\mathbf{x}_i$. Then $\mathcal{V}$, $\mathcal{W}$, $\mathbf{x}$, and $\mathbf{x}_a$ fit the structure of the least-square error problem. The least-square error solution to (6.8) is the solution to the normal equations (6.7). The orthogonal projection procedure could also be used here, but it determines only the vector $\mathbf{x}_a$; we must then solve the consistent equations

$$\xi_1\mathbf{x}_1+\cdots+\xi_n\mathbf{x}_n=\mathbf{x}_a$$

to obtain the unknowns $\xi_1,\ldots,\xi_n$.

*See Papoulis [6.15].

Example 2. Inconsistent Equations—Weighted Inner Product. There is no strong reason for using the standard inner product in Example 1. Suppose that the following three measurements were taken of a single quantity ξ_1:

$$\xi_1 = 1.2$$
$$\xi_1 = 1.1$$
$$\xi_1 = 1.3$$

Also suppose that we have less confidence in the last measurement than we do in the other two. We seek ξ_1 to minimize

$$2(1.2-\xi_1)^2 + 2(1.1-\xi_1)^2 + (1.3-\xi_1)^2$$

The smaller weight on the third residual expresses our lack of confidence in the third measurement; we are just minimizing $\|\mathbf{x}-\xi_1\mathbf{x}_1\|^2$, with $\mathbf{x} \triangleq (1.2 \quad 1.1 \quad 1.3)^{\mathrm{T}}$, $\mathbf{x}_1 \triangleq (1 \quad 1 \quad 1)^{\mathrm{T}}$, and the weighted inner product defined as

$$\left\langle \begin{pmatrix} c_1 \\ c_2 \\ c_3 \end{pmatrix}, \begin{pmatrix} d_1 \\ d_2 \\ d_3 \end{pmatrix} \right\rangle \triangleq 2c_1d_1 + 2c_2d_2 + c_3d_3$$

For this problem, (6.7) is the single normal equation

$$(\langle \mathbf{x}_1, \mathbf{x}_1 \rangle)\xi_1 = \langle \mathbf{x}, \mathbf{x}_1 \rangle$$

Thus $\xi_1 = \langle \mathbf{x}, \mathbf{x}_1 \rangle / \|\mathbf{x}_1\|^2 = 5.9/5 = 1.18$, a *weighted* average of the measurements. In this simple example, the orthogonal projection procedure requires computation of essentially the same quantities.

Application to Data Fitting

In modeling a physical system, we often represent the relationship between two variables by a polynomial and choose the coefficients of the polynomial such that the model fits actual data. In order to compensate for errors in the data, we use more data points than are needed to determine unique values for the coefficients; the coefficients are chosen such that the polynomial is the best approximation to the data in the sense that the sum of the squared errors is minimized. This curve-fitting process is called **linear regression.**

As a simple example of a curve-fitting problem, assume we have measured the following data,

s	$f(s)$
1	0
3	2
4	1

where s is the length of a spring when it supports a weight $f(s)$. We wish to describe the relationship between the variables s and $f(s)$ by a first-degree polynomial $f_a(s)=\xi_1+\xi_2 s$. The data and a possible f_a are shown in Figure 6.2. We pick ξ_1 and ξ_2 (the parameters of the line in the figure) to minimize the sum of the squared errors at the data points, $\epsilon_1^2+\epsilon_3^2+\epsilon_4^2$.

The problem reduces to finding the least-error solution to the set of overdetermined (inconsistent) equations

$$f_a(s)=f(s), \qquad s=1,3,4$$

These equations can be written as

$$\xi_1+\xi_2(1)=0$$
$$\xi_1+\xi_2(3)=2$$
$$\xi_1+\xi_2(4)=1$$

or, corresponding to (6.9),

$$\xi_1\mathbf{x}_1+\xi_2\mathbf{x}_2=\mathbf{x} \tag{6.10}$$

where $\mathbf{x}_1=(1 \quad 1 \quad 1)^{\mathrm{T}}$, $\mathbf{x}_2=(1 \quad 3 \quad 4)^{\mathrm{T}}$, and $\mathbf{x}=(0 \quad 2 \quad 1)^{\mathrm{T}}$. In the notation

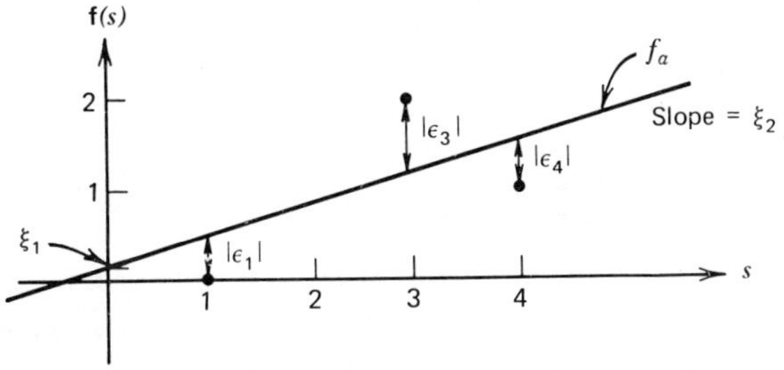

Figure 6.2. Straight-line approximation to measured data.

of (6.10), the quantity to be minimized is $\|\mathbf{x}-\mathbf{x}_a\|^2$, where $\mathbf{x}_a \stackrel{\Delta}{=} \xi_1\mathbf{x}_1+\xi_2\mathbf{x}_2$ and the norm is the one associated with the standard inner product for $\mathfrak{M}^{3\times 1}$. Applying the normal equations (6.7), we find that

$$\begin{pmatrix} \langle \mathbf{x}_1,\mathbf{x}_1\rangle & \langle \mathbf{x}_2,\mathbf{x}_1\rangle \\ \langle \mathbf{x}_1,\mathbf{x}_2\rangle & \langle \mathbf{x}_2,\mathbf{x}_2\rangle \end{pmatrix} \begin{pmatrix} \xi_1 \\ \xi_2 \end{pmatrix} = \begin{pmatrix} \langle \mathbf{x},\mathbf{x}_1\rangle \\ \langle \mathbf{x},\mathbf{x}_2\rangle \end{pmatrix} \tag{6.11}$$

Evaluation of the inner products yields

$$\begin{pmatrix} 3 & 8 \\ 8 & 26 \end{pmatrix} \begin{pmatrix} \xi_1 \\ \xi_2 \end{pmatrix} = \begin{pmatrix} 3 \\ 10 \end{pmatrix}$$

A simple row reduction shows that $\xi_1=-\frac{1}{7}$, $\xi_2=\frac{3}{7}$, and

$$\begin{aligned} f_a(s) &= -\tfrac{1}{7}+\tfrac{3}{7}s \\ &= \tfrac{3}{7}\left(s-\tfrac{1}{3}\right) \end{aligned} \tag{6.12}$$

According to the "straight-line" model (6.12), the unstretched length of the spring is $\frac{1}{3}$ and the spring constant is $\frac{3}{7}$. The difference between the measurements and the model predictions at the data points (exaggerated for this example) can be explained partly by errors in measurement and partly by inadequacy of the straight-line model in representing the nonlinearity of the spring.

We have computed the solution to (6.10) by means of the normal equations (6.11) rather than by the orthogonal projection procedure because the normal equations produce the unknowns $\{\xi_i\}$ directly. However, the normal equations can be very ill-conditioned, and therefore difficult to solve accurately. Forsythe [6.4] demonstrates this ill conditioning in the fitting of high-order polynomials $f_a(s)=\xi_1+\xi_2 s+\cdots+\xi_n s^{n-1}$ with equally spaced data. In Section 6.6 we demonstrate a practical computational scheme, based on the orthogonal projection procedure, which is not hampered by the ill conditioning which affects the normal equations. This computational procedure not only provides accurate solutions, but it also produces the coefficients in the model directly.

Linear Regression Models

It is evident from the above example that uniformly spaced data points are not necessary. Nor must we use the standard inner product; we can use an inner product that weights errors according to our confidence in the measurements. Each inner product constitutes a different definition of orthogonality and leads to a different orthogonal projection of the data vector $\mathbf{x}$ onto $\mathcal{W}$.

In some systems the measured quantities vary to such an extent that we treat them as random. Rather than try to fit the data accurately in such a case, we merely attempt to find a trend (or correlation) among the variables; a straight-line fit like that of Figure 6.2 describes such a trend.* In fact, the term "regression" came from just such a correlation analysis which had to do with biological regression.

On the other hand, in other systems the measured relationships are essentially reproducible. For these systems a more complicated model is justified. We wish the model to fit the data well except for slight errors resulting from inaccurate measurement. Thus we may want to use a relatively high-order polynomial for the model f_a; of course, the number of data points must be larger than the order of the polynomial. We still refer to f_a as a *linear* model, because it is linear in the unknown coefficients; we still perform *linear* regression. What order polynomial should we use? In the absence of some theoretical reason to expect a certain order polynomial, we simply try different orders, taking as our model the lowest order polynomial that seems to fit the data well.†

There is no reason not to try other functions for the model f_a besides polynomials. In some systems we have reason to expect other functional relationships. As long as the unknowns appear linearly, we are performing linear regression—row-reduction algorithms will suffice to compute the least-square error solutions. We can handle several independent variables by means of multivariable linear regression—the fitting of data with linear combinations of functions of several variables. The equations still reduce to the form of (6.9) (P&C 6.3). An infinite-dimensional linear regression is used in the identification (modeling) of a dynamic system in Pearson [6.16],

Application to Mean-Square Estimation Problems

In the linear regression example of Figure 6.2 we approximated a function f by a linear combination of functions $\xi_1 f_1 + \xi_2 f_2$, wherein $f_1(s) = 1$ and $f_2(s) = s$. Although we compared the function and its approximation only at the discrete data points, we could have taken data for an interval of values of s and picked ξ_1 and ξ_2 to minimize

$$\|f - f_a\|^2 = \int_a^b [f(s) - f_a(s)]^2\, ds \tag{6.13}$$

In this section we carry out a function space minimization of this sort; our function space is a space of random variables.

*See Sprent [6.21].
†See Forsythe [6.4].

The taking of a sequence of n measurements of a random quantity constitutes a statistical experiment. The measurements could be numbers resulting from throws of a single die; or they could be heights of electrical pulses emitted from a radar receiver. We define the Hilbert space $\mathcal{H}$ to be the set of all random variables $\mathbf{x}$ defined on this experiment which have a finite mean and variance; the inner product of two random variables $\mathbf{x}$ and $\mathbf{y}$ is given by*

$$\langle \mathbf{x},\mathbf{y}\rangle \stackrel{\Delta}{=} \mathbf{E}(\mathbf{xy}) \tag{6.14}$$

We focus on the random variables (functions) $\mathbf{x}_1,\dots,\mathbf{x}_n$ in $\mathcal{H}$ which are defined by

$$\mathbf{x}_k(\sigma_1,\dots,\sigma_n) \stackrel{\Delta}{=} \sigma_k \tag{6.15}$$

for all possible outcomes $(\sigma_1,\dots,\sigma_n)$ of the experiment. Each function $\mathbf{x}_k$ is defined on the whole experiment, but takes its values from only part of the experiment. The random variable $\mathbf{x}_k$ can be thought of as the outcome of the kth measurement of the experiment. However, we must not lose sight of the fact that the variables σ_i of (6.15) are *dummy* variables, *possible* outcomes of the experiment. A single running of the experiment yields *sample values* of the random variables $\mathbf{x}_1,\dots,\mathbf{x}_n$ (denoted $\bar{x}_1,\dots,\bar{x}_n$), but it does not *fix* the values of $\mathbf{x}_1,\dots,\mathbf{x}_n$.† The random variable $\mathbf{x}_k$ remains a *function* with a set of possible sample values. Each possible sample value has a certain likelihood of occurring at *each* running of the experiment.

Associated with the experiment is a joint probability mass function

$$\omega(\sigma_1,\dots,\sigma_n) \stackrel{\Delta}{=} \begin{array}{l}\text{probability that the outcome of}\\ \text{the experiment will be } (\sigma_1,\dots,\sigma_n)\end{array} \tag{6.16}$$

Therefore, there is a probability mass function associated with each of the random variables $\mathbf{x}_1,\dots,\mathbf{x}_n$, and a *joint* probability mass function associated with each *set* of these random variables. We note that the random variables $\mathbf{x}_1,\dots,\mathbf{x}_n$ are linearly independent functions. (If they

*The concept of a random variable as a function defined on the outcomes of an experiment is introduced in Example 11, Section 2.1. See P&C 5.37 for a description of Hilbert spaces of random variables. Papoulis [6.15] discusses random variables and mean-square estimation in detail. Luenberger [6.13] gives an extensive discussion of mean-square estimation from a vector space point of view.

† The notation used for sample values is merely an abbreviation for the correct notation, $\bar{x}_1 = \mathbf{x}_1(\sigma_1,\dots,\sigma_n),\dots,\bar{x}_n = \mathbf{x}_n(\sigma_1,\dots,\sigma_n)$, where $(\sigma_1,\dots,\sigma_n)$ is the outcome of the specific experiment.

were linearly dependent, their sample values would bear the same linear relationship to each other on every trial of the experiment.)

We wish to approximate the random variable $\mathbf{x}_n$ by another random variable, $\mathbf{x}_{na}$, which is a linear combination of the random variables $\{\mathbf{1},\mathbf{x}_1,\ldots,\mathbf{x}_{n-1}\}$:

$$\mathbf{x}_{na}=\xi_0\mathbf{1}+\xi_1\mathbf{x}_1+\cdots+\xi_{n-1}\mathbf{x}_{n-1} \tag{6.17}$$

The random variable $\mathbf{1}$ is defined by

$$\mathbf{1}(\sigma_1,\ldots,\sigma_n)=1 \quad \text{for all } (\sigma_1,\ldots,\sigma_n) \tag{6.18}$$

We pick the coefficients $\{\xi_i\}$ such that the approximation is best in a least-square sense; we minimize

$$\|\mathbf{x}_n-\mathbf{x}_{na}\|^2 \triangleq \mathbf{E}\left[(\mathbf{x}_n-\mathbf{x}_{na})^2\right] \tag{6.19}$$

The process of least-square approximation of a random variable by a linear combination of other random variables is known as **linear mean-square estimation**. The random variable $\mathbf{x}_{na}$ is called an **estimator**. In this particular setting, where we use a combination of a *sequence* of random variables to estimate the next random variable in the sequence, the estimation is sometimes referred to as **prediction**.

The vectors $\mathbf{1}$, $\mathbf{x}_1,\ldots,\mathbf{x}_{n-1}$ form a basis for a subspace $\mathcal{W}$ of $\mathcal{H}$. We must find a vector $\mathbf{x}_{na}$ in $\mathcal{W}$ which best approximates the vector $\mathbf{x}_n$ in $\mathcal{H}$ in a minimum norm sense. We can determine $\mathbf{x}_{na}$ (or the coefficients $\{\xi_i\}$ which specify $\mathbf{x}_{na}$) either by the normal equations (6.7) or by orthogonal projection using (6.5). The normal equations are

$$\begin{pmatrix} \langle \mathbf{1},\mathbf{1}\rangle & \cdots & \langle \mathbf{x}_{n-1},\mathbf{1}\rangle \\ \vdots & & \vdots \\ \langle \mathbf{1},\mathbf{x}_{n-1}\rangle & \cdots & \langle \mathbf{x}_{n-1},\mathbf{x}_{n-1}\rangle \end{pmatrix} \begin{pmatrix} \xi_0 \\ \vdots \\ \xi_{n-1} \end{pmatrix} = \begin{pmatrix} \langle \mathbf{x}_n,\mathbf{1}\rangle \\ \vdots \\ \langle \mathbf{x}_n,\mathbf{x}_{n-1}\rangle \end{pmatrix}$$

or, in terms of the specific inner product (6.14),

$$\begin{pmatrix} 1 & \mathbf{E}(\mathbf{x}_1) & \cdots & \mathbf{E}(\mathbf{x}_{n-1}) \\ \mathbf{E}(\mathbf{x}_1) & \mathbf{E}(\mathbf{x}_1^2) & \cdots & \mathbf{E}(\mathbf{x}_1\mathbf{x}_{n-1}) \\ \vdots & \vdots & & \vdots \\ \mathbf{E}(\mathbf{x}_{n-1}) & \mathbf{E}(\mathbf{x}_1\mathbf{x}_{n-1}) & \cdots & \mathbf{E}(\mathbf{x}_{n-1}^2) \end{pmatrix} \begin{pmatrix} \xi_0 \\ \xi_1 \\ \vdots \\ \xi_{n-1} \end{pmatrix} = \begin{pmatrix} \mathbf{E}(\mathbf{x}_n) \\ \mathbf{E}(\mathbf{x}_1\mathbf{x}_n) \\ \vdots \\ \mathbf{E}(\mathbf{x}_{n-1}\mathbf{x}_n) \end{pmatrix} \tag{6.20}$$

It is apparent from (6.20) that in order to compute the linear mean-square estimate $\mathbf{x}_{na}$, we need to know all the first and second order statistics of the random variables—the means $\mathbf{E}(\mathbf{x}_i)$ and the second moments $\mathbf{E}(\mathbf{x}_i\mathbf{x}_j)$. Suppose the random process from which the measurements are taken is *stationary*; that is, the statistics do not vary with time. Then the means and second moments can be found from a "long" sequence of measurements $a_1,\ldots,a_n,\ldots,a_{n+m}$ of the random quantity upon which the experiment is based. (Any n consecutive measurements from this list constitutes a set of sample values $\bar{x}_1,\ldots,\bar{x}_n$.) The means and second moments are approximately equal to the following averages:

$$\mathbf{E}(\mathbf{x}_1)=\cdots=\mathbf{E}(\mathbf{x}_n)\approx\frac{a_1+\cdots+a_m}{m}$$

$$\mathbf{E}(\mathbf{x}_i\mathbf{x}_j)\approx\frac{a_1a_{1+j-i}+a_2a_{2+j-i}+\cdots+a_ma_{m+j-i}}{m} \tag{6.21}$$

$$i=1,\ldots,n, \qquad j=i,i+1,\ldots,n$$

The numbers computed by (6.21) completely specify the normal equations (6.20) and determine the mean-square estimate. We note that the *best* mean-square estimate of $\mathbf{x}_n$, given $\mathbf{x}_1,\ldots,\mathbf{x}_{n-1}$, is the conditional expected value $\mathbf{E}(\mathbf{x}_n|\mathbf{x}_1,\ldots,\mathbf{x}_{n-1})$. However, this quantity is usually too difficult to determine. We are fortunate if we have available even the second moments. A major justification for the use of *linear* mean-square estimation is the fact that the estimate can be computed using *relatively* little statistical information.

Example 3. Estimating the Throw of a Die. Suppose the experiment described above consists in the outcomes of n throws of a single die. Each throw is statistically independent of the others. It follows that

$$\mathbf{E}(\mathbf{x}_i)=\mathbf{E}(\mathbf{x}_n) \quad \text{and} \quad \mathbf{E}(\mathbf{x}_i^2)=\mathbf{E}(\mathbf{x}_n^2) \quad \text{for each } i$$

$$\mathbf{E}(\mathbf{x}_i\mathbf{x}_j)=\mathbf{E}(\mathbf{x}_i)\mathbf{E}(\mathbf{x}_j)=\mathbf{E}^2(\mathbf{x}_n) \quad \text{for } i\neq j$$

(The latter equation is equivalent to the statement that $\mathbf{x}_i$ and $\mathbf{x}_j$ are **uncorrelated**: $\mathbf{E}[(\mathbf{x}_i-\mathbf{E}(\mathbf{x}_i))(\mathbf{x}_j-\mathbf{E}(\mathbf{x}_j))]=0$.) The normal equations (6.20) become

$$\begin{pmatrix} 1 & \mathbf{E}(\mathbf{x}_n) & \mathbf{E}(\mathbf{x}_n) & \cdots & \mathbf{E}(\mathbf{x}_n) \\ \mathbf{E}(\mathbf{x}_n) & \mathbf{E}(\mathbf{x}_n^2) & \mathbf{E}^2(\mathbf{x}_n) & \cdots & \mathbf{E}^2(\mathbf{x}_n) \\ \mathbf{E}(\mathbf{x}_n) & \mathbf{E}^2(\mathbf{x}_n) & \mathbf{E}(\mathbf{x}_n^2) & \cdots & \mathbf{E}^2(\mathbf{x}_n) \\ \vdots & \vdots & \vdots & & \vdots \\ \mathbf{E}(\mathbf{x}_n) & \mathbf{E}^2(\mathbf{x}_n) & \mathbf{E}^2(\mathbf{x}_n) & \cdots & \mathbf{E}(\mathbf{x}_n^2) \end{pmatrix} \begin{pmatrix} \xi_0 \\ \xi_1 \\ \xi_2 \\ \vdots \\ \xi_{n-1} \end{pmatrix} = \begin{pmatrix} \mathbf{E}(\mathbf{x}_n) \\ \mathbf{E}^2(\mathbf{x}_n) \\ \vdots \\ \mathbf{E}^2(\mathbf{x}_n) \end{pmatrix} \tag{6.22}$$

A surprisingly simple row reduction of (6.22) leads to $\xi_1 = \cdots = \xi_{n-1} = 0$ and $\xi_0 = \mathbf{E}(\mathbf{x}_n)$. The best estimate of $\mathbf{x}_n$ is the *random variable* $\mathbf{x}_{na} = \mathbf{E}(\mathbf{x}_n)$—the function whose value is $\mathbf{E}(\mathbf{x}_n)$ for any outcome of the experiment. This result should have been expected. The random variables $\mathbf{x}_1, \ldots, \mathbf{x}_{n-1}$, being statistically independent of $\mathbf{x}_n$, provide no information about $\mathbf{x}_n$.

Estimating the Distance to a Target

A radar processor emits a sequence of electrical pulses. Suppose the height of each pulse represents the distance to a target at a particular instant. We wish to design an estimator which predicts each pulse height in terms of the known heights of several previous pulses (Figure 6.3).

Consider the observation of n consecutive pulse heights as an experiment. The random variable $\mathbf{x}_k$ of (6.15) can be thought of as the height of the kth pulse in the sequence. We estimate the random variable $\mathbf{x}_n$ by a linear combination of the random variables $\mathbf{1}, \mathbf{x}_1, \ldots, \mathbf{x}_{n-1}$:

$$\mathbf{x}_{na} \triangleq \xi_0 \mathbf{1} + \xi_1 \mathbf{x}_1 + \cdots + \xi_{n-1} \mathbf{x}_{n-1} \tag{6.23}$$

The coefficients $\{\xi_i\}$ which lead to the best mean-square estimator $\mathbf{x}_{na}$ are given by the normal equations (6.20). The sample values of $\mathbf{x}_1, \ldots, \mathbf{x}_{n-1}$ determine a sample value of the random variable $\mathbf{x}_{na}$. The purpose for finding the estimator $\mathbf{x}_{na}$ is to allow us to compute $\bar{x}_{na}$ from $\bar{x}_1, \ldots, \bar{x}_{n-1}$ (Figure 6.3). The sample value $\bar{x}_{na}$ is the best mean-square estimate of the as yet unmeasured sample value $\bar{x}_n$.

We obtain the estimator $\mathbf{x}_{na}$ for the case $n = 2$. The random variables $\mathbf{x}_1$ and $\mathbf{x}_2$ are correlated. (The distance to the target at any instant cannot be very different from the distance at the immediately preceding instant.) We assume stationarity: $\mathbf{E}(\mathbf{x}_1) = \mathbf{E}(\mathbf{x}_2)$ and $\mathbf{E}(\mathbf{x}_1^2) = \mathbf{E}(\mathbf{x}_2^2)$. Rather than solve the normal equations, we find $\mathbf{x}_{2a}$ by orthogonal projection of $\mathbf{x}_2$ onto the subspace $\mathcal{W}$ spanned by $\{\mathbf{1}, \mathbf{x}_1\}$ as indicated by (6.5). We first apply the Gram–Schmidt procedure to $\{\mathbf{1}, \mathbf{x}_1\}$ to find an orthogonal basis for $\mathcal{W}$: let $\mathbf{z}_0 \triangleq \mathbf{1}$; then $\|\mathbf{z}_0\|^2 = \mathbf{E}(\mathbf{1}^2) = 1$. Let $\mathbf{z}_1 \triangleq \mathbf{x}_1 - (\langle \mathbf{x}_1, \mathbf{z}_0 \rangle / \|\mathbf{z}_0\|^2)\mathbf{z}_0 = \mathbf{x}_1 - \mathbf{E}(\mathbf{x}_1 \mathbf{1})\mathbf{1}$ $= \mathbf{x}_1 - \mathbf{E}(\mathbf{x}_1)\mathbf{1}$; then

$$\|\mathbf{z}_1\|^2 = \mathbf{E}\left[(\mathbf{x}_1 - \mathbf{E}(\mathbf{x}_1)\mathbf{1})^2\right] = \mathbf{E}(\mathbf{x}_1^2) - \mathbf{E}^2(\mathbf{x}_1) \triangleq \text{variance}(\mathbf{x}_1)$$

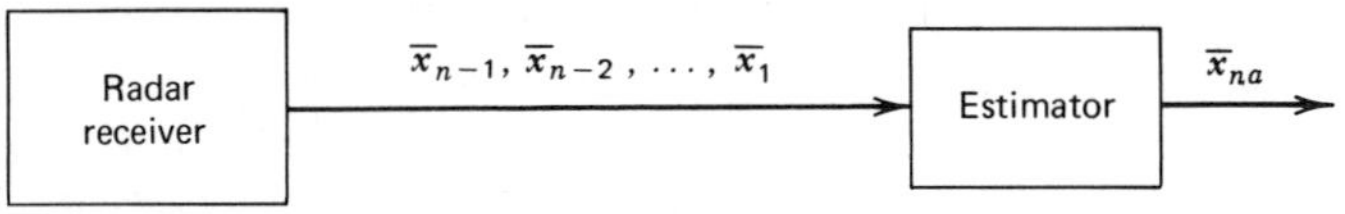

Figure 6.3. Predicting the sample value of $\mathbf{x}_n$.

Thus $\|\mathbf{z}_1\|$ is the standard deviation of $\mathbf{x}_1$ (the square root of the variance). The orthogonal projection of $\mathbf{x}_2$ onto $\mathcal{W}$ is given in terms of the unit vectors $\mathbf{y}_0=\mathbf{z}_0/\|\mathbf{z}_0\|$ and $\mathbf{y}_1=\mathbf{z}_1/\|\mathbf{z}_1\|$ by (6.5). Using the fact that $\mathbf{E}(\mathbf{x}_1)=\mathbf{E}(\mathbf{x}_2)$, and defining the covariance of $\mathbf{x}_2$ and $\mathbf{x}_1$ by

$$\operatorname{cov}(\mathbf{x}_2,\mathbf{x}_1) \stackrel{\Delta}{=} \mathbf{E}(\mathbf{x}_2\mathbf{x}_1)-\mathbf{E}(\mathbf{x}_2)\mathbf{E}(\mathbf{x}_1)$$

we find that

$$\begin{aligned}
\mathbf{x}_{2a} &= \frac{\langle \mathbf{x}_2,\mathbf{z}_0\rangle}{\|\mathbf{z}_0\|^2}\mathbf{z}_0+\frac{\langle \mathbf{x}_2,\mathbf{z}_1\rangle}{\|\mathbf{z}_1\|^2}\mathbf{z}_1\\
&= \mathbf{E}(\mathbf{x}_2\mathbf{1})\mathbf{1}+\frac{\mathbf{E}\left[\mathbf{x}_2(\mathbf{x}_1-\mathbf{E}(\mathbf{x}_1)\mathbf{1})\right]}{\operatorname{var}(\mathbf{x}_1)}(\mathbf{x}_1-\mathbf{E}(\mathbf{x}_1)\mathbf{1})\\
&= \mathbf{E}(\mathbf{x}_2)\mathbf{1}+\frac{\operatorname{cov}(\mathbf{x}_2,\mathbf{x}_1)}{\operatorname{var}(\mathbf{x}_1)}(\mathbf{x}_1-\mathbf{E}(\mathbf{x}_1)\mathbf{1})\\
&= \left(1-\frac{\operatorname{cov}(\mathbf{x}_2,\mathbf{x}_1)}{\operatorname{var}(\mathbf{x}_1)}\right)\mathbf{E}(\mathbf{x}_2)\mathbf{1}+\frac{\operatorname{cov}(\mathbf{x}_2,\mathbf{x}_1)}{\operatorname{var}(\mathbf{x}_1)}\mathbf{x}_1\\
&= (1-\rho)\mathbf{E}(\mathbf{x}_2)\mathbf{1}+\rho\mathbf{x}_1
\end{aligned} \tag{6.24}$$

where $\rho \stackrel{\Delta}{=} \operatorname{cov}(\mathbf{x}_2,\mathbf{x}_1)/\operatorname{var}(\mathbf{x}_1)$ is the correlation coefficient for $\mathbf{x}_1$ and $\mathbf{x}_2$.

Implementation of the estimator consists in multiplying the sample value $\bar{x}_1$ by the constant ρ and adding the constant $(1-\rho)\mathbf{E}(\mathbf{x}_2)$ to obtain $\bar{x}_{2a}$. Observe that if adjacent samples are uncorrelated, $\operatorname{cov}(\mathbf{x}_2,\mathbf{x}_1)=0$, and $\bar{x}_{2a}=\mathbf{E}(\mathbf{x}_2)$.

The estimator $\mathbf{x}_{2a}$ of (6.24) has the property that $\mathbf{E}(\mathbf{x}_{2a})=\mathbf{E}(\mathbf{x}_2)$. An estimator that has this property is said to be an **unbiased estimator.** Intuitively speaking, if the estimator were not unbiased, its sample values would not, even on the average, act like the sample values of the random variable it is estimating.

Exercise 4. Estimate the random variable $\mathbf{x}_2$ of the above example by a scalar multiple of $\mathbf{x}_1$ (i.e., eliminate the "unit" random variable, $\mathbf{1}$). Show that this estimator is biased.

The orthogonal projection technique for obtaining the least-square estimate has produced in the third line of (6.24) a description of $\mathbf{x}_{2a}$ as a sum of orthogonal (uncorrelated) random variables, $\mathbf{z}_0=\mathbf{1}$ and $\mathbf{z}_1=\mathbf{x}_1-\mathbf{E}(\mathbf{x}_1)\mathbf{1}$. It is because these random variables are orthogonal that we are able to determine their coefficients independently as in (6.5). Because of this independence of the coefficients, we can determine the best *constant* mean-square estimator for $\mathbf{x}_2$ by dropping the "$\mathbf{z}_1$" term in the third line of (6.24). The result is $\mathbf{x}_{2a} = \mathbf{E}(\mathbf{x}_2)\mathbf{1}$, an intuitively obvious result. Note that it is *not* correct to use the constant term from the last line of (6.24); $\mathbf{1}$ and $\mathbf{x}_1$ are correlated (not orthogonal).

We can also compute the mean-square error in the estimator. If $\mathbf{x}_{na}$ is the best mean-square estimator (6.23), then, by the orthogonality of $\mathbf{x}_{na}$ and $\mathbf{x}_n - \mathbf{x}_{na}$, $\|\mathbf{x}_n\|^2 = \|\mathbf{x}_{na}\|^2 + \|\mathbf{x}_n - \mathbf{x}_{na}\|^2$, or

$$\mathbf{E}\left[(\mathbf{x}_n - \mathbf{x}_{na})^2\right] = \mathbf{E}(\mathbf{x}_n^2) - \mathbf{E}(\mathbf{x}_{na}^2) \tag{6.25}$$

For the estimator (6.24), where $n=2$ and the sequence of pulse heights is stationary, we find

$$\begin{aligned}\mathbf{E}(\mathbf{x}_{na}^2) &= \mathbf{E}\left[((1-\rho)\mathbf{E}(\mathbf{x}_2)\mathbf{1} + \rho\mathbf{x}_1)^2\right]\\ &= (1-\rho)^2\mathbf{E}^2(\mathbf{x}_2) + 2\rho(1-\rho)\mathbf{E}(\mathbf{x}_1)\mathbf{E}(\mathbf{x}_2) + \rho^2\mathbf{E}(\mathbf{x}_1^2)\\ &= (1-\rho^2)\mathbf{E}^2(\mathbf{x}_2) + \rho^2\mathbf{E}(\mathbf{x}_2^2)\end{aligned} \tag{6.26}$$

Therefore, corresponding to $\mathbf{x}_{2a}$ of (6.24) is the squared error

$$\begin{aligned}\mathbf{E}\left[(\mathbf{x}_2 - \mathbf{x}_{2a})^2\right] &= \mathbf{E}(\mathbf{x}_2^2) - (1-\rho^2)\mathbf{E}^2(\mathbf{x}_2) - \rho^2\mathbf{E}(\mathbf{x}_2^2)\\ &= (1-\rho^2)\left(\mathbf{E}(\mathbf{x}_2^2) - \mathbf{E}^2(\mathbf{x}_2)\right)\\ &= (1-\rho^2)\,\mathrm{var}(\mathbf{x}_2)\end{aligned} \tag{6.27}$$

If we estimate the same pulse height $\mathbf{x}_2$ without sampling the previous pulse height, the estimator (6.24) must be replaced by $\mathbf{x}_{2a} = \mathbf{E}(\mathbf{x}_2)\mathbf{1}$. The corresponding squared error is

$$\mathbf{E}\left[(\mathbf{x}_2 - \mathbf{x}_{2a})^2\right] = \mathbf{E}(\mathbf{x}_2^2) - \mathbf{E}(\mathbf{x}_{2a}^2) = \mathbf{E}(\mathbf{x}_2^2) - \mathbf{E}^2(\mathbf{x}_2) = \mathrm{var}(\mathbf{x}_2)$$

The increase in squared error over the estimator (6.24) is $\mathrm{var}(\mathbf{x}_2) - (1 - \rho^2)\,\mathrm{var}(\mathbf{x}_2) = \rho^2\,\mathrm{var}(\mathbf{x}_2)$.

6.2 Least-Effort Problems by Orthogonal Projection

As demonstrated by the examples of the last section, the least-square error problem is essentially a problem of overdetermined equations. The antithesis of the least-square error problem is a set of consistent underdetermined equations. Consider the following simple example

$$\xi_1 + (0)\xi_2 + (0)\xi_3 = 1$$
$$(0)\xi_1 + \quad \xi_2 + (0)\xi_3 = 1$$

Let $\mathbf{x} = (\xi_1, \xi_2, \xi_3)$, $\mathbf{x}_1 = (1,0,0)$, and $\mathbf{x}_2 = (0,1,0)$. Using the standard inner product for $\mathcal{R}^3$, we express the pair of equations as

$$\langle \mathbf{x}, \mathbf{x}_1 \rangle = 1 \quad \text{and} \quad \langle \mathbf{x}, \mathbf{x}_2 \rangle = 1$$

By means of a natural correspondence between $\mathcal{R}^3$ and the arrow space of Figure 6.4, we can interpret the locus of the solutions to each of these equations as a plane. The line where the two planes intersect is the locus of solutions to the pair of simultaneous equations.

In the absence of other criteria dictated by the physical problem from which the equations arise, we focus on the one solution that appears to be unique—the "smallest" solution $\mathbf{x}_s$. Intuitively, we can think of $\mathbf{x}_s$ as the most efficient or "least-effort" solution. Let $\mathcal{W} = \text{span}\{\mathbf{x}_1, \mathbf{x}_2\}$, where $\mathbf{x}_1$ and $\mathbf{x}_2$ are the vectors which generate the equations. Then it is clear that $\mathbf{x}_s$ is the orthogonal projection onto $\mathcal{W}$ of any solution $\mathbf{x}$ to the set of

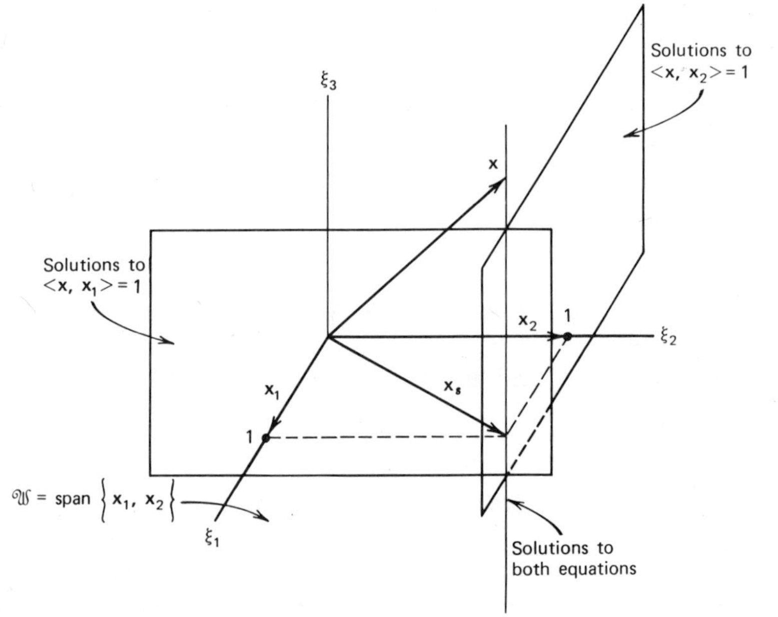

Figure 6.4. Solutions to an underdetermined set of equations.

equations. Thus $\mathbf{x}_s$ is the unique intersection of the plane $\mathcal{W}$ with the planes that represent the equations.

We will show that every set of indeterminate linear equations has the same geometrical structure as the example of Figure 6.4. Furthermore, the minimum norm solution appears to be a useful optimum solution in many practical problems. The solution of minimum norm always appears as an orthogonal projection, and can be obtained in a manner analogous to solving for the intersection of $\mathcal{W}$ with the line that satisfies the equations.

Linear Equations and Hyperplanes

The pair of equations used in Figure 6.4 defines planes and a line in the three-dimensional space. It is apparent that more equations in a higher-dimensional space will define, by their various intersections, higher-dimensional analogues of lines and planes. We include all of these lines and planes in the concept of a hyperplane.

Definition. A shifted subspace is called a **hyperplane**. That is, if $\mathcal{W}$ is a subspace of a vector space $\mathcal{V}$, and $\mathbf{x}_p$ is a particular vector of $\mathcal{V}$, then the set of vectors $\mathbf{x}_p + \mathcal{W}$ is a hyperplane. By the **dimension** of the hyperplane we mean the dimension of the subspace which defines it. We refer to the dimension of the orthogonal complement of this subspace as the **co-dimension** of the hyperplane.

In Figure 6.4, the plane defined by the equation $\langle \mathbf{x}, \mathbf{x}_1 \rangle = 1$ is a hyperplane of dimension 2 and co-dimension 1. It is a shifted version of the subspace $\{\mathbf{x}_1\}^{\perp}$. The vector $\mathbf{x}_1$ serves as a "shift vector" $\mathbf{x}_p$. The line determined by the intersection of the planes in Figure 6.4 is a hyperplane of dimension 1 and co-dimension 2. It is a shifted version of $\mathcal{W}^{\perp}$; any solution vector $\mathbf{x}$, and $\mathbf{x}_s$ in particular, will serve as a shift vector $\mathbf{x}_p$. It is evident that the shift vector $\mathbf{x}_p$ used to define a hyperplane is not unique; any vector in the hyperplane will do. However, the subspace of which the hyperplane is a shifted version is unique. Furthermore, the subspace itself is a hyperplane (with the shift vector $\mathbf{x}_p = \boldsymbol{\theta}$). Rather than denote a hyperplane by a specific symbol, we will denote it by the notation $\mathbf{x}_p + \mathcal{W}$, where $\mathbf{x}_p$ is a vector in the hyperplane and $\mathcal{W}$ is the subspace which defines the hyperplane.

Example 1. Consistent Underdetermined Linear Equations. Any consistent linear equation or set of linear equations which has more than one solution has as its set of solutions a hyperplane. The set of homogeneous solutions is a subspace. The addition of a particular solution and the set of homogeneous solutions defines the hyperplane. Thus we sometimes refer to an indeterminate set of linear equations as a hyperplane.

Exercise 1. Let $\mathbf{T}$ be a linear transformation. Assuming the equation

$\mathbf{Tx}=\mathbf{y}$ is consistent, it defines a hyperplane. Show for a finite-dimensional domain $\mathcal{V}$ that nullity(**T**) is the dimension of the hyperplane and rank(**T**) is its co-dimension.

Example 2. Linear Functionals. Every linear functional **T** on a vector space $\mathcal{V}$ defines hyperplanes of co-dimension 1. If **T** is a linear functional, the linear equation $\mathbf{Tx}=c$ is necessarily consistent. By Example 1, this equation defines a different hyperplane for each scalar c. By Exercise 1, the co-dimension of the hyperplane is rank(**T**), which equals one. As a specific example, consider the equation $\int_a^b \mathbf{f}(t)\,dt=1$, a hyperplane of co-dimension 1. The subspace $\mathcal{W}$ which defines this hyperplane is the infinite-dimensional space of functions with average zero. The orthogonal complement $\mathcal{W}^{\perp}$ is the space of constant functions. The function $\mathbf{f}_p(t)=1/(b-a)$ will serve as a shift vector. In this example, the co-dimension of the hyperplane is a useful quantity, whereas the dimension is not.

Exercise 2. Let $\mathbf{x}_1,\ldots,\mathbf{x}_m$ be independent vectors in a vector space $\mathcal{V}$, and let $c_1,\ldots,c_m$ be specified scalars. Show that the set of vectors which satisfy the equations

$$\langle \mathbf{x},\mathbf{x}_j\rangle=c_j, \qquad j=1,\ldots,m$$

is a hyperplane of co-dimension m. Hint: let $\mathbf{x}_s$ be a fixed vector which satisfies the equations, and investigate $\mathbf{x}-\mathbf{x}_s$.

The Least-Effort Problem

Definition. Let the independent vectors $\{\mathbf{x}_1,\ldots,\mathbf{x}_m\}$ span an m-dimensional subspace $\mathcal{W}$ of a Hilbert space $\mathcal{V}$. Let $\{c_1,\ldots,c_m\}$ be scalars. Then the underdetermined equations

$$\langle \mathbf{x},\mathbf{x}_j\rangle=c_j, \qquad j=1,\ldots,m \tag{6.28}$$

specify a hyperplane of co-dimension m in $\mathcal{V}$; this hyperplane is a shifted version of $\mathcal{W}^{\perp}$. The **least-effort problem** consists in determining that vector $\mathbf{x}$ in the hyperplane for which $\|\mathbf{x}\|$ is minimum.

The solution to the least-effort problem involves orthogonal projections. Suppose $\mathbf{x}$ is any vector in the hyperplane (6.28). By (5.20), we can decompose $\mathbf{x}$ into unique orthogonal projections on $\mathcal{W}$ and $\mathcal{W}^{\perp}$: $\mathbf{x}=\mathbf{x}_{\mathcal{W}}+\mathbf{x}_{\mathcal{W}^{\perp}}$. Substituting this decomposed form into (6.28), we find

$$\langle \mathbf{x}_{\mathcal{W}},\mathbf{x}_j\rangle+\underbrace{\langle \mathbf{x}_{\mathcal{W}^{\perp}},\mathbf{x}_j\rangle}_{0}=c_j, \qquad j=1,\ldots,m$$

Clearly, $\mathbf{x}_{\mathcal{W}}$ is in the hyperplane. Furthermore, $\mathbf{x}_{\mathcal{W}^{\perp}}$ can be adjusted freely without $\mathbf{x}$ leaving the hyperplane. (The uniqueness of the orthogonal projection $\mathbf{x}_{\mathcal{W}}$ is intuitively clear from the fact that it lies at the intersection of the hyperplane and the subspace $\mathcal{W}$; the m independent equations which define the hyperplane are just sufficient to eliminate the m degrees of freedom in $\mathcal{W}$.)

Exercise 3. Show that any two vectors in the hyperplane (6.28) have the same orthogonal projection on $\mathcal{W}$.

By adjusting only $\mathbf{x}_{\mathcal{W}^\perp}$, we are able to obtain any vector in the hyperplane. To solve the least-effort problem, we adjust $\mathbf{x}_{\mathcal{W}^\perp}$ (thereby adjusting $\mathbf{x}$) to obtain the minimum of $\|\mathbf{x}\|^2$. According to the Pythagorean theorem (P&C 5.4),

$$\|\mathbf{x}\|^2 = \|\mathbf{x}_{\mathcal{W}}\|^2 + \|\mathbf{x}_{\mathcal{W}^\perp}\|^2$$

Obviously, the minimum occurs for $\mathbf{x}_{\mathcal{W}^\perp} = \boldsymbol{\theta}$. Thus if $\mathbf{x}$ is a solution to the underdetermined equations (6.28), then the minimum norm (or smallest) solution $\mathbf{x}_s$ equals $\mathbf{x}_{\mathcal{W}}$, the orthogonal projection of $\mathbf{x}$ on $\mathcal{W}$.

We have used the completeness of $\mathcal{V}$ only to simplify the discussion of $\mathcal{W}^\perp$. The results apply to any vector space $\mathcal{V}$. One approach to obtaining the least-effort solution $\mathbf{x}_s$ to (6.28) is to obtain first a particular solution $\mathbf{x}$, then compute the orthogonal projection of $\mathbf{x}$ onto $\mathcal{W}$. The orthogonal projection consists essentially in applying the Gram–Schmidt orthogonalization procedure to the set $\{\mathbf{x}_1,\ldots,\mathbf{x}_m,\mathbf{x}\}$.

A second approach is to solve for the intersection between $\mathcal{W}$ and the hyperplane. Since $\mathbf{x}_s$ is in $\mathcal{W}$, it can be expressed as

$$\mathbf{x}_s = \sum_{i=1}^{m} \xi_i \mathbf{x}_i \tag{6.29}$$

Substituting this expansion into the hyperplane equations (6.28) we obtain

$$\left\langle \sum_{i=1}^{m} \xi_i \mathbf{x}_i, \mathbf{x}_j \right\rangle = \sum_{i=1}^{m} \xi_i \langle \mathbf{x}_i, \mathbf{x}_j \rangle = c_j, \qquad j = 1,\ldots,m$$

or

$$\begin{pmatrix} \langle \mathbf{x}_1,\mathbf{x}_1\rangle & \cdots & \langle \mathbf{x}_m,\mathbf{x}_1\rangle \\ \vdots & & \vdots \\ \langle \mathbf{x}_1,\mathbf{x}_m\rangle & \cdots & \langle \mathbf{x}_m,\mathbf{x}_m\rangle \end{pmatrix} \begin{pmatrix} \xi_1 \\ \vdots \\ \xi_m \end{pmatrix} = \begin{pmatrix} c_1 \\ \vdots \\ c_m \end{pmatrix} \tag{6.30}$$

We call (6.30) the **Gram equations** for the least-effort problem. The independence of the vectors $\{\mathbf{x}_1,\ldots,\mathbf{x}_m\}$ guarantees the invertibility of the

Gram matrix. We substitute the coefficients (ξ_i) obtained from (6.30) back into (6.29) to yield $\mathbf{x}_s$.

Example 3. Underdetermined Linear Equations of Full Rank. Suppose the following equations are consistent and there are more unknowns than equations ($n > m$):

$$\begin{aligned} a_{11}\xi_1 + \cdots + a_{1n}\xi_n &= \eta_1 \\ \vdots \qquad\qquad \vdots \quad &\ \ \vdots \\ a_{m1}\xi_1 + \cdots + a_{mn}\xi_n &= \eta_m \end{aligned} \tag{6.31}$$

These equations can be expressed in the inner product form

$$\langle \mathbf{x}, \mathbf{x}_j \rangle = \eta_j, \qquad j = 1, \ldots, m$$

where $\mathbf{x}_j$ is the jth row of the matrix of coefficients, and $\mathbf{x}$ is a $1 \times n$ row vector consisting in the unknowns $\{\xi_i\}$. Assume the equations are of full rank; that is, the vectors $\mathbf{x}_1, \ldots, \mathbf{x}_m$ are independent. Let $\mathcal{V} = \mathfrak{M}^{1 \times n}$ with its standard inner product. Let $\mathcal{W} = \text{span}\{\mathbf{x}_1, \ldots, \mathbf{x}_m\}$. If we let $c_j = \eta_j$, the inner product version of (6.31) corresponds precisely to (6.28); the equations represent a hyperplane of co-dimension m in $\mathcal{V}$. The equations have many solutions. We can get a solution vector $\mathbf{x}$ by row reducing (6.31) and picking $n - m$ variables arbitrarily. Projecting $\mathbf{x}$ on $\mathcal{W}$ would yield the least-effort solution $\mathbf{x}_s$, the one for which $\xi_1^2 + \cdots + \xi_n^2$ is minimum. Alternatively, we can obtain $\mathbf{x}_s$ directly by solving the Gram equations (6.30) with c_j replaced by η_j.

Example 4. Underdetermined Equations in a Function Space. The functions $\mathbf{f}$ which satisfy

$$\int_0^1 \mathbf{f}(s)\omega(s)\,ds = c_1 \quad \text{and} \quad \int_0^1 s\mathbf{f}(s)\omega(s)\,ds = c_2$$

form an infinite-dimensional hyperplane of co-dimension 2 in $\mathcal{C}(\omega; 0, 1)$, the space of continuous functions with the weighted inner product $\langle \mathbf{f}, \mathbf{g} \rangle \triangleq \int_0^1 \mathbf{f}(s)\mathbf{g}(s)\omega(s)\,ds$. We find that solution which minimizes $\int_0^1 \mathbf{f}^2(s)\omega(s)\,ds$. Define $\mathbf{f}_1(s) \triangleq 1$ and $\mathbf{f}_2(s) \triangleq s$. Then the integral equations become $\langle \mathbf{f}, \mathbf{f}_1 \rangle = c_1$ and $\langle \mathbf{f}, \mathbf{f}_2 \rangle = c_2$. The functions $\{\mathbf{f}_1, \mathbf{f}_2\}$ form a basis for a subspace $\mathcal{W}$. We express the minimum norm solution $\mathbf{f}_s$ as a vector in $\mathcal{W}$: $\mathbf{f}_s = \xi_1\mathbf{f}_1 + \xi_2\mathbf{f}_2$. The Gram equations (6.30) which determine ξ_1 and ξ_2 are

$$\begin{pmatrix} \langle \mathbf{f}_1, \mathbf{f}_1 \rangle & \langle \mathbf{f}_2, \mathbf{f}_1 \rangle \\ \langle \mathbf{f}_1, \mathbf{f}_2 \rangle & \langle \mathbf{f}_2, \mathbf{f}_2 \rangle \end{pmatrix} \begin{pmatrix} \xi_1 \\ \xi_2 \end{pmatrix} = \begin{pmatrix} c_1 \\ c_2 \end{pmatrix}$$

The coefficients in the solution are easily found:

$$\xi_1 = \frac{\langle \mathbf{f}_2, \mathbf{f}_2 \rangle c_1 - \langle \mathbf{f}_2, \mathbf{f}_1 \rangle c_2}{\langle \mathbf{f}_2, \mathbf{f}_2 \rangle \langle \mathbf{f}_1, \mathbf{f}_1 \rangle - \langle \mathbf{f}_2, \mathbf{f}_1 \rangle \langle \mathbf{f}_1, \mathbf{f}_2 \rangle}$$

$$\xi_2 = \frac{\langle \mathbf{f}_1, \mathbf{f}_1 \rangle c_2 - \langle \mathbf{f}_1, \mathbf{f}_2 \rangle c_1}{\langle \mathbf{f}_2, \mathbf{f}_2 \rangle \langle \mathbf{f}_1, \mathbf{f}_1 \rangle - \langle \mathbf{f}_2, \mathbf{f}_1 \rangle \langle \mathbf{f}_1, \mathbf{f}_2 \rangle}$$

We cannot compute the inner products without knowing ω.

This example has an interesting interpretation. Suppose $\omega(s)$ is the probability density function for a random variable s; that is, the probability that $a \leqslant s \leqslant b$ is $\int_a^b \omega(s)\,ds$ for a and b in $[0, 1]$, and $\int_0^1 \omega(s)\,ds = 1$. Then the expected value of the product of two functions of the random variable, $\mathbf{f}(s)$ and $\mathbf{g}(s)$, is*

$$\mathbf{E}[\mathbf{f}(s)\mathbf{g}(s)] = \int_0^1 \mathbf{f}(s)\mathbf{g}(s)\omega(s)\,ds = \langle \mathbf{f}, \mathbf{g} \rangle$$

By this random variable interpretation, we have derived a "straight-line" relationship between a random variable s with known density function and another random variable $\mathbf{f}_s(s)$ with stated statistical properties. We have minimized $\mathbf{E}(\mathbf{f}_s^2(s))$ while assuring that $\mathbf{E}(\mathbf{f}_s(s)) = c_1$ and $\mathbf{E}(s\mathbf{f}_s(s)) = c_2$; in essence, $\mathbf{f}_s(s)$ has minimum variance subject to specification of its mean and its correlation with s. With this interpretation, the minimum norm solution can be expressed as

$$\mathbf{f}_s(s) = \frac{\mathbf{E}(s^2)c_1 - \mathbf{E}(s)c_2}{\mathrm{var}(s)} + \frac{c_2 - \mathbf{E}(s)c_1}{\mathrm{var}(s)} s$$

The derivation from a simple random variable (say, uniformly distributed or Gaussian) of another random variable with specified statistical properties is useful for computer simulation of physical systems.

Application to Minimum Energy Control

A loaded dc motor with armature voltage $\mathbf{u}(t)$ is described by the differential system

$$\ddot{\phi}(t) + \dot{\phi}(t) = \mathbf{u}(t) \quad \text{with } \dot{\phi}(0) = \phi(0) = 0 \tag{6.32}$$

where ϕ is the angular position of the shaft. We wish to apply an input $\mathbf{u}$ which will cause the motor to move within 1 second to the new position $\phi(1) = 1$, $\dot{\phi}(1) = 0$, thereby positioning a piece of equipment. There are an infinite number of inputs which will accomplish this task. Should we

*The expected value operation is discussed in P&C 2.23.

carefully ease the motor to its new position? Should we drive it hard and then decelerate quickly? We will pick that motor control function $\mathbf{u}$ which consumes the least energy, $\int_0^1 \mathbf{u}^2(t)\,dt$.

In Chapter 3 we put the differential system (6.32) in state-space form. The inverse of the state equation is given in (3.80); we express it as the pair of equations

$$\phi(t) = \int_0^t (1 - e^{-t}e^{s})\mathbf{u}(s)\,ds$$

$$\dot{\phi}(t) = \int_0^t e^{-t}e^{s}\mathbf{u}(s)\,ds$$

In order that the motor meet the terminal conditions, the inverse equations must satisfy

$$\phi(1) = \int_0^1 (1 - e^{-1}e^{s})\mathbf{u}(s)\,ds$$

$$\dot{\phi}(1) = \int_0^1 e^{-1}e^{s}\mathbf{u}(s)\,ds$$

We simplify this pair of equations slightly, adding the second to the first, then multiplying the second by e:

$$\begin{aligned} \int_0^1 \mathbf{u}(s)\,ds &= \phi(1) + \dot{\phi}(1) \\ \int_0^1 e^{s}\mathbf{u}(s)\,ds &= e\dot{\phi}(1) \end{aligned} \tag{6.33}$$

The control functions $\mathbf{u}$ which satisfy (6.33) for specific final conditions form an infinite-dimensional hyperplane of co-dimension 2 in the space $\mathcal{C}(0,1)$ with its *standard inner product*; we write (6.33) in the inner product notation

$$\langle \mathbf{u}, \mathbf{f}_1 \rangle = \phi(1) + \dot{\phi}(1) \quad \text{and} \quad \langle \mathbf{u}, \mathbf{f}_2 \rangle = e\dot{\phi}(1)$$

where $\mathbf{f}_1(s) \triangleq 1$ and $\mathbf{f}_2(s) \triangleq e^{s}$. The quantity $\int_0^1 \mathbf{u}^2(t)\,dt$, which must be minimized, is just $\|\mathbf{u}\|^2$ in the space $\mathcal{C}(0,1)$. Thus the minimization problem is of the same geometrical structure as the least-effort problem; the term "least-effort" is certainly appropriate in this instance. Following the procedure of (6.29) and (6.30), we express the least-effort control $\mathbf{u}_s$ as a vector in the space $\mathcal{W}$ spanned by $\{\mathbf{f}_1, \mathbf{f}_2\}$: $\mathbf{u}_s = \xi_1\mathbf{f}_1 + \xi_2\mathbf{f}_2$. The Gram equations (6.30) become

$$\begin{pmatrix} \langle \mathbf{f}_1, \mathbf{f}_1 \rangle & \langle \mathbf{f}_2, \mathbf{f}_1 \rangle \\ \langle \mathbf{f}_1, \mathbf{f}_2 \rangle & \langle \mathbf{f}_2, \mathbf{f}_2 \rangle \end{pmatrix} \begin{pmatrix} \xi_1 \\ \xi_2 \end{pmatrix} = \begin{pmatrix} \phi(1) + \dot{\phi}(1) \\ e\dot{\phi}(1) \end{pmatrix}$$

The inner products in the Gram matrix are evaluated by simple integrations; the Gram matrix and its inverse are, respectively,

$$\begin{pmatrix} 1 & e-1 \\ e-1 & \dfrac{e^2-1}{2} \end{pmatrix} \quad \text{and} \quad \begin{pmatrix} \dfrac{e+1}{3-e} & \dfrac{-2}{3-e} \\ \dfrac{-2}{3-e} & \dfrac{2}{(3-e)(e-1)} \end{pmatrix}$$

For the specific final conditions $\phi(1)=1$, $\dot{\phi}(1)=0$, the solution is $\xi_1 = (e+1)/(3-e)$, $\xi_2 = -2/(3-e)$, and

$$\mathbf{u}_s(t) = \left(\frac{e+1}{3-e}\right) - \left(\frac{2}{3-e}\right)e^t \tag{6.34}$$

This minimum energy control function is displayed in Figure 6.5.

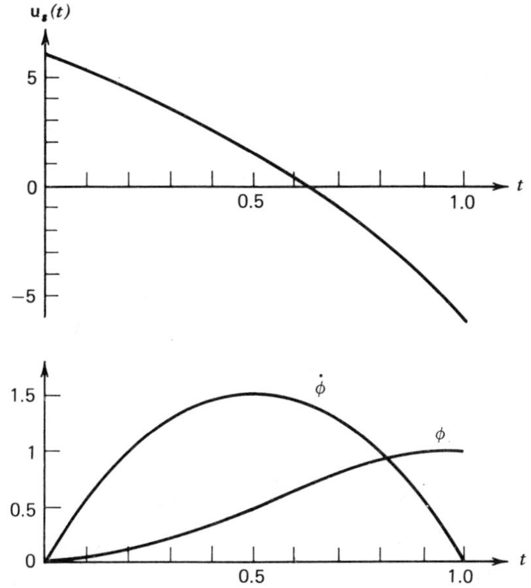

Figure 6.5. Minimum energy control function and state trajectories.

A least-effort control can be determined for any linear dynamic system by the approach used for (6.32). If the cost of control action for a given system varies throughout the control period (owing, perhaps, to variations in the price of fuel), this variation can be accounted for by use of a weighted inner product*: $\int_0^1 \mathbf{u}^2(t)\omega(t)\,dt$. The approach demonstrated here applies as well to time-varying linear systems with multiple inputs.

The minimum effort control (6.34) is an "open-loop" control. That is, it is calculated prior to operation of the system and applied regardless of the actual system behavior. In some situations it is desireable to express the control $\mathbf{u}$ as a function of the state variables (ϕ and $\dot{\phi}$) which represent the system behavior (P&C 7.14). If we use this "closed-loop" form, then random influences on the system and errors in the model (6.32) used to derive the control are not ignored; through their effect on the state of the system, they modify the control function.

6.3 Problem 1—Resolution of Incompatibility by Adjoints

Sections 6.1 and 6.2 have explored two quadratic minimization problems, the least-error problem and the least-effort problem. Most of the examples and applications of these two sections have consisted primarily in "solving" noninvertible linear equations. This fact should not be surprising, since most system models consist in sets of equations. Throughout Chapters 1–5 our approach to the analysis of a system has been to treat the system equations as special cases of the general vector equation $\mathbf{Tx}=\mathbf{y}$. This approach helps us to catalog concepts in a clear geometrical structure which aids our memory and our intuition. We now examine the least-square solution of linear equations in the context of linear transformations. We find that the adjoint transformation is the key to the analysis and understanding of these least-square problems.

Suppose $\mathbf{T}$: $\mathcal{V}\rightarrow\mathcal{W}$ is a linear transformation between two Hilbert spaces. According to the orthogonal decomposition theorem (5.67), if $\mathbf{T}$ is bounded,†

$$\begin{aligned}\mathcal{V}&=\text{nullspace}(\mathbf{T})\overset{\perp}{\oplus}\text{range}(\mathbf{T}^*)\\ \mathcal{W}&=\text{nullspace}(\mathbf{T}^*)\overset{\perp}{\oplus}\text{range}(\mathbf{T})\end{aligned}\tag{6.35}$$

*See P&C 6.12.

†We have specified the boundedness of $\mathbf{T}$ and the completeness of $\mathcal{V}$ and $\mathcal{W}$ only to *guarantee* the existence of $\mathbf{T}^*$ and the validity of (6.35). However, we have already seen in Sections 5.4 and 5.5 that these concepts can apply to unbounded linear transformations also. We assume range($\mathbf{T}$) and range($\mathbf{T}^*$) are complete; therefore we ignore the closure bars of (5.67).

This decomposition is illustrated abstractly in Figure 5.9. By definition of range(**T**), an equation **Tx**=**y** is compatible (has a solution) if and only if **y** is in range(**T**). If **T** is *onto* $\mathcal{W}$, then **y** *must* be in range(**T**), and a solution must exist. It is apparent from (6.35) that **T** is onto $\mathcal{W}$ if and only if nullspace(**T***)=$\boldsymbol{\theta}$. In turn, nullspace(**T***)=$\boldsymbol{\theta}$ if and only if **T*** is one-to-one. These three conditions on **T** and **T*** are equivalent ways of stating that the equation **Tx**=**y** is compatible for all **y**. If these conditions are satisfied, we say the equation **Tx**=**y** satisfies the **compatibility condition**. We emphasize that the compatibility condition implies the solvability of **Tx**=**y** for *any* **y**.

Complementary to the compatibility condition is the uniqueness condition. We say the equation **Tx**=**y** satisfies the **uniqueness condition** if **T** is one-to-one, for then any solution to the equation is unique. We learned in Section 2.4 that **T** is one-to-one if and only if nullspace(**T**)=$\boldsymbol{\theta}$. From (6.35), it follows that nullspace(**T**)=$\boldsymbol{\theta}$ if and only if **T*** is onto $\mathcal{V}$. Thus we have three equivalent ways of establishing the uniqueness condition for **Tx**=**y**.

Some equations satisfy the uniqueness condition but not the compatibility condition. In this section we treat these equations as least-square error problems. Equations which satisfy the compatibility condition but not the uniqueness condition are treated as least-effort problems in Section 6.4. Occasionally a transformation **T** satisfies neither of these conditions. For such transformations we develop the concept of a "pseudoinverse" $\mathbf{T}^{\dagger}$ which uses aspects of both the least-error and least-effort problems (Section 6.5). We have already learned that orthogonal projections are central to least squares. We now find that for the abstract equation **Tx**=**y**, the necessary orthogonal projections are intimately related to the orthogonal decomposition (6.35) generated by **T** and **T***. The concepts that we explore apply to all linear equations: algebraic, differential, integral, etc.

Solution to Problem 1

Definition. Let $\mathcal{V}$ and $\mathcal{W}$ be Hilbert spaces and **T**: $\mathcal{V} \to \mathcal{W}$ a bounded linear transformation. Assume nullspace(**T**)=$\boldsymbol{\theta}$, but nullspace(**T***)$\neq \boldsymbol{\theta}$. **Problem 1** consists in solving the equation **Tx**=**y** for that vector **x** which minimizes $\|\mathbf{y}-\mathbf{Tx}\|_{\mathcal{W}}$. This problem is illustrated in Figure 6.6.

The transformation **T** of Problem 1 is one-to-one but not onto $\mathcal{W}$. Problem 1 is a special case of the least-square error problem of Section 6.1. If **y** is in range(**T**), the equation has a unique solution **x**, and the minimum error norm is zero. If **y** is not in range(**T**), we can solve Problem 1 by applying the conclusion (6.5) of the least-square error problem. We wish, in effect, to approximate **y** in $\mathcal{W}$ by that vector **Tx** in range(**T**) which

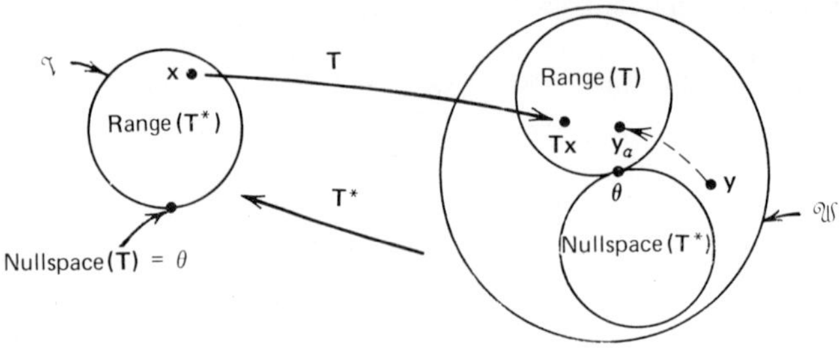

Figure 6.6. Abstract description of Problem 1.

minimizes $\|\mathbf{y}-\mathbf{Tx}\|_{\mathcal{W}}$. The best approximation $(\mathbf{Tx})_a$ is the orthogonal projection of $\mathbf{y}$ onto the subspace range($\mathbf{T}$); rather than carry the notation $(\mathbf{Tx})_a$, we denote this orthogonal projection by $\mathbf{y}_a$. We denote by $\mathbf{x}_a$ the solution to Problem 1; it is the unique solution to the equation $\mathbf{Tx}_a=\mathbf{y}_a$, where $\mathbf{y}_a$ is the orthogonal projection of $\mathbf{y}$ onto range($\mathbf{T}$). We can compute $\mathbf{x}_a$ by performing the projection and solving the equation. Since the orthogonal complement of range($\mathbf{T}$) is nullspace($\mathbf{T}^*$), the least-square error vector $\mathbf{y}-\mathbf{Tx}_a$ is the orthogonal projection of $\mathbf{y}$ onto nullspace($\mathbf{T}^*$).

By employing the properties of $\mathbf{T}^*$ we can express the least-error solution $\mathbf{x}_a$ explicitly. From the orthogonal decomposition (6.35) of $\mathcal{W}$, we recognize that $\mathbf{y}-\mathbf{y}_a$ is in nullspace($\mathbf{T}^*$). Therefore,

$$\mathbf{T}^*\mathbf{y}=\mathbf{T}^*\mathbf{y}_a+\underbrace{\mathbf{T}^*(\mathbf{y}-\mathbf{y}_a)}_{\boldsymbol{\theta}}=\mathbf{T}^*\mathbf{y}_a \tag{6.36}$$

Substituting $\mathbf{Tx}_a$ for $\mathbf{y}_a$ yields

$$\mathbf{T}^*\mathbf{Tx}_a=\mathbf{T}^*\mathbf{y} \tag{6.37}$$

Although (6.37) may not seem familiar because of the abstractness of the transformations, it is just the expression of the **normal equations** for Problem 1. We know from the above derivation that (6.37) has a unique solution for each $\mathbf{y}$ in $\mathcal{W}$. Therefore, $\mathbf{T}^*\mathbf{T}$ must be invertible, and

$$\mathbf{x}_a=(\mathbf{T}^*\mathbf{T})^{-1}\mathbf{T}^*\mathbf{y} \tag{6.38}$$

The transformation $(\mathbf{T}^*\mathbf{T})^{-1}\mathbf{T}^*$ is a left inverse for $\mathbf{T}$; that is, $((\mathbf{T}^*\mathbf{T})^{-1}\mathbf{T}^*)\mathbf{T}=\mathbf{I}$. We call it the **pseudoinverse $\mathbf{T}^\dagger$ for Problem 1**.

Exercise 1. Show that the uniqueness condition, nullspace($\mathbf{T}$)$=\boldsymbol{\theta}$, is necessary and sufficient to guarantee that $\mathbf{T}^*\mathbf{T}$ is invertible.

Example 1. An Inconsistent Matrix Equation of Full Rank. Assume $\mathbf{A}$ is a real $m\times n$ matrix, $m>n$, and rank($\mathbf{A}$)$=n$. Then, $\mathbf{Ax}=\mathbf{y}$ represents more equations than unknowns, but the columns of **A** are independent. The equation is identical to (6.8) of Example 1, Section 6.1. We find the least-square error solution by determining the pseudoinverse expressed in (6.38). We minimize $\|\mathbf{y}-\mathbf{Ax}\|$ where the norm is derived from the standard inner product for $\mathfrak{M}^{m\times 1}$. We can pick *any* inner product for $\mathfrak{M}^{n\times 1}$ (P&C 6.14); we choose, again, the standard inner product. Define $\mathbf{T}$: $\mathfrak{M}^{n\times 1}\to\mathfrak{M}^{m\times 1}$ by $\mathbf{Tx} \triangleq \mathbf{Ax}$. Because the columns of **A** are independent, both **A** and **T** are nonsingular [i.e., nullspace$(\mathbf{T})=\boldsymbol{\theta}$]. We know from (5.60) that *for standard inner products* in both spaces, $\mathbf{T}^*\mathbf{y} \triangleq \mathbf{A}^{\mathrm{T}}\mathbf{y}$. Then the least-square error solution (6.38) becomes

$$\mathbf{x}_a=(\mathbf{A}^{\mathrm{T}}\mathbf{A})^{-1}\mathbf{A}^{\mathrm{T}}\mathbf{y} \tag{6.39}$$

We refer to the matrix $(\mathbf{A}^{\mathrm{T}}\mathbf{A})^{-1}\mathbf{A}^{\mathrm{T}}$ as the *pseudoinverse of the matrix* **A** for this problem.

As a specific example, we take the linear regression equations of Section 6.1:

$$\mathbf{Ax}=\begin{pmatrix}1&1\\1&3\\1&4\end{pmatrix}\begin{pmatrix}\xi_1\\\xi_2\end{pmatrix}=\begin{pmatrix}0\\2\\1\end{pmatrix}=\mathbf{y} \tag{6.40}$$

The matrix $\mathbf{A}^{\mathrm{T}}\mathbf{A}$ is just the Gram matrix of (6.11):

$$\mathbf{A}^{\mathrm{T}}\mathbf{A}=\begin{pmatrix}1&1&1\\1&3&4\end{pmatrix}\begin{pmatrix}1&1\\1&3\\1&4\end{pmatrix}=\begin{pmatrix}3&8\\8&26\end{pmatrix} \tag{6.41}$$

The solution (6.39) becomes

$$\mathbf{x}_a=\frac{1}{14}\begin{pmatrix}18&2&-6\\-5&1&4\end{pmatrix}\begin{pmatrix}0\\2\\1\end{pmatrix}=\begin{pmatrix}-1/7\\3/7\end{pmatrix} \tag{6.42}$$

Example 2. Inconsistent Matrix Equation–Weighted Inner Product. The equations of Example 2, Section 6.1 are

$$\mathbf{Ax}=\begin{pmatrix}1\\1\\1\end{pmatrix}(\xi_1)=\begin{pmatrix}1.2\\1.1\\1.3\end{pmatrix}=\mathbf{y} \tag{6.43}$$

In that example, we used the inner product $\langle\mathbf{y},\mathbf{z}\rangle \triangleq \mathbf{z}^{\mathrm{T}}\mathbf{Qy}$, where

$$\mathbf{Q}=\begin{pmatrix}2&0&0\\0&2&0\\0&0&1\end{pmatrix} \tag{6.44}$$

for the space $\mathfrak{M}^{3\times 1}$. We found the "solution" which minimized $\|\mathbf{y}-\mathbf{Ax}\|$ where the norm is the one associated with the weighted inner product. We now find the same solution using the pseudoinverse given in (6.38). Define $\mathbf{T}\colon \mathfrak{M}^{1\times 1}\to\mathfrak{M}^{3\times 1}$ by $\mathbf{Tx}\triangleq\mathbf{Ax}$. We must determine $\mathbf{T}^*$. We pick the standard inner product $\langle\mathbf{x},\mathbf{w}\rangle\triangleq\mathbf{wx}$ for the "scalar" space $\mathfrak{M}^{1\times 1}$. Using the defining equation (5.58),

$$\langle\mathbf{Tx},\mathbf{y}\rangle=\langle\mathbf{Ax},\mathbf{y}\rangle=\mathbf{y}^{\mathrm{T}}\mathbf{QAx}=(\mathbf{A}^{\mathrm{T}}\mathbf{Qy})^{\mathrm{T}}\mathbf{x}=(\mathbf{T}^*\mathbf{y})^{\mathrm{T}}\mathbf{x}=\langle\mathbf{x},\mathbf{T}^*\mathbf{y}\rangle,$$

Therefore, $\mathbf{T}^*\mathbf{y}\triangleq\mathbf{A}^{\mathrm{T}}\mathbf{Qy}$. Then the weighted least-square error solution (6.38) becomes

$$\begin{aligned}(\xi_1)=\mathbf{x}_a&=(\mathbf{A}^{\mathrm{T}}\mathbf{QA})^{-1}\mathbf{A}^{\mathrm{T}}\mathbf{Qy}\\&=\tfrac{1}{5}(2\ 2\ 1)\begin{pmatrix}1.2\\1.1\\1.3\end{pmatrix}\\&=(1.18)\end{aligned}\qquad(6.45)$$

The weighted inner product amounts to a "nonstandard" definition of orthogonal projection.

6.4 Problem 2—Resolution of Nonuniqueness by Adjoints

In Section 6.3 we treated the least-square solution of incompatible linear system equations in the context of linear transformations. We now explore the least-square solution of underdetermined linear system equations from the same viewpoint, thereby fitting the concepts associated with this class of systems into the intuitive structure and terminology developed in Chapters 1–5. Once again, we find that the adjoint transformation provides the key to the analysis.

Definition. Let $\mathcal{V}$ and $\mathcal{W}$ be Hilbert spaces and $\mathbf{T}\colon\mathcal{V}\to\mathcal{W}$ a bounded linear transformation. Assume nullspace$(\mathbf{T}^*)=\boldsymbol{\theta}$, but nullspace$(\mathbf{T})\neq\boldsymbol{\theta}$. **Problem 2** consists in finding that vector $\mathbf{x}$ for which $\mathbf{Tx}=\mathbf{y}$ and for which $\|\mathbf{x}\|_{\mathcal{V}}$ is minimum. This problem is illustrated in Figure 6.7.

The transformation $\mathbf{T}$ of Problem 2 is onto $\mathcal{W}$ but not one-to-one. Problem 2 is an extension of the least-effort problem of Section 6.2. The equation always has solutions; we must find the smallest. Problem 2 and Problem 1 are *duals*; if we interchange the roles of $\mathbf{T}$ and $\mathbf{T}^*$, $\mathbf{x}$ and $\mathbf{y}$, $\mathbf{x}_s$ and $\mathbf{y}_a$, Problem 1 becomes Problem 2, and vice versa. This duality appeared previously in the dual (row and column) interpretations of matrix multiplication used in (6.8) and (6.31).

We solve Problem 2 by repeating the steps of the solution to the least-effort problem. The solutions to the equation $\mathbf{Tx}=\mathbf{y}$ form a hyperplane (Example 1 and Exercise 1 of Section 6.2). Let $\mathfrak{N}=$nullspace$(\mathbf{T})$.

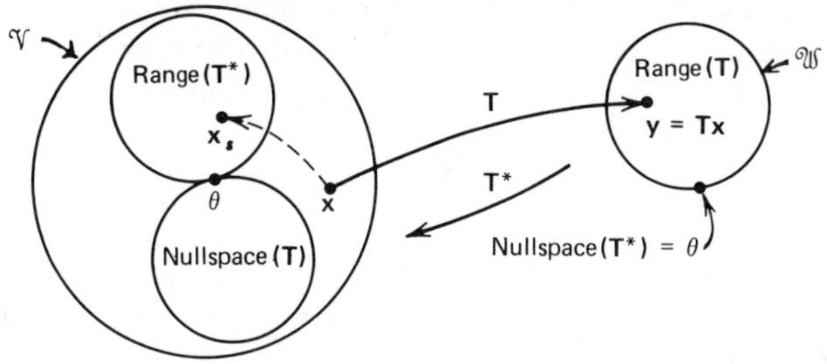

Figure 6.7. Abstract representation of Problem 2.

Then any solution $\mathbf{x}$ in the hyperplane can be decomposed into $\mathbf{x}=\mathbf{x}_{\mathfrak{N}}+\mathbf{x}_{\mathfrak{N}^\perp}$, where $\mathbf{x}_{\mathfrak{N}}$ is the orthogonal projection onto nullspace(**T**) and $\mathbf{x}_{\mathfrak{N}^\perp}$ is the orthogonal projection onto range(**T***). Substituting $\mathbf{x}$ into the equation, we find

$$\mathbf{Tx}=\underbrace{\mathbf{Tx}_{\mathfrak{N}}}_{\boldsymbol{\theta}}+\mathbf{Tx}_{\mathfrak{N}^\perp}=\mathbf{y}$$

Obviously $\mathbf{x}_{\mathfrak{N}^\perp}$ is a particular vector in the hyperplane and $\mathbf{x}_{\mathfrak{N}}$ can be adjusted freely without $\mathbf{x}$ leaving the hyperplane. (The hyperplane is $\mathbf{x}_{\mathfrak{N}^\perp}+\mathfrak{N}$.) By the Pythagorean theorem, $\|\mathbf{x}\|^2=\|\mathbf{x}_{\mathfrak{N}}\|^2+\|\mathbf{x}_{\mathfrak{N}^\perp}\|^2$. In order to minimize $\|\mathbf{x}\|$ without $\mathbf{x}$ leaving the hyperplane, we must pick $\mathbf{x}_{\mathfrak{N}}=\boldsymbol{\theta}$. Consequently, the minimum norm (or smallest) solution $\mathbf{x}_s$ to the equation $\mathbf{Tx}=\mathbf{y}$ is the orthogonal projection of an arbitrary solution $\mathbf{x}$ onto range (**T***). The uniqueness of $\mathbf{x}_s$ can be demonstrated by showing that any two solutions have the same orthogonal projection onto range(**T***). We can compute $\mathbf{x}_s$ by solving the equation and performing the projection.

We use the properties of **T*** to obtain an explicit expression for $\mathbf{x}_s$. In geometric terms, $\mathbf{x}_s$ lies at the intersection of the hyperplane $\mathbf{Tx}=\mathbf{y}$ and the subspace range(**T***). In order for $\mathbf{x}_s$ to lie in range(**T***), it must satisfy $\mathbf{x}_s=\mathbf{T}^*\boldsymbol{\lambda}$ for some $\boldsymbol{\lambda}$ in $\mathcal{W}$. Since **T*** is nonsingular, $\boldsymbol{\lambda}$ is unique. Then, since $\mathbf{x}_s$ must also satisfy the original equation $\mathbf{Tx}_s=\mathbf{y}$, the unique intersection $\mathbf{x}_s$ is determined by the equations

$$\begin{aligned}\mathbf{x}_s&=\mathbf{T}^*\boldsymbol{\lambda}\\ \mathbf{TT}^*\boldsymbol{\lambda}&=\mathbf{y}\end{aligned}\tag{6.46}$$

The first equation is an extension of the result (6.29) for the least-effort problem. The second equation extends the **Gram equation** (6.30). Finding the vector $\boldsymbol{\lambda}$ appears to be merely an unavoidable step toward the solution $\mathbf{x}_s$. However, in Chapter 7 we rederive (6.46) in a manner wherein $\boldsymbol{\lambda}$ is a "Lagrange multiplier"; $\boldsymbol{\lambda}$ specifies the sensitivity of the solution $\mathbf{x}_s$ to slight variations in the constraining equation $\mathbf{Tx}=\mathbf{y}$. We sometimes refer to the first of the equations (6.46) as the **adjoint equation** and call $\boldsymbol{\lambda}$ the **adjoint variable**. Because $\boldsymbol{\lambda}$ is uniquely determined for any $\mathbf{y}$, $\mathbf{TT}^*$ must be invertible. Therefore we can combine the equations (6.46) to produce

$$\mathbf{x}_s = \mathbf{T}^*(\mathbf{TT}^*)^{-1}\mathbf{y} \tag{6.47}$$

The transformation $\mathbf{T}^*(\mathbf{TT}^*)^{-1}$ is a right inverse for $\mathbf{T}$; that is, $\mathbf{T}(\mathbf{T}^*(\mathbf{TT}^*)^{-1}) = \mathbf{I}$. We call it the **pseudoinverse $\mathbf{T}^\dagger$ for Problem 2.**

Exercise 1. Show that the compatibility condition, nullspace$(\mathbf{T}^*) = \boldsymbol{\theta}$, is necessary and sufficient to guarantee the invertibility of $\mathbf{TT}^*$.

Example 1. An Underdetermined Matrix Equation of Full Rank. Suppose $\mathbf{A}$ is a real $m \times n$ matrix, $m < n$, and rank$(\mathbf{A}) = m$. Then $\mathbf{Ax} = \mathbf{y}$ represents fewer equations than unknowns, but the rows of $\mathbf{A}$ are independent. The equation is identical to (6.31) of Example 3, Section 6.2. We find the minimum norm solution by evaluating the pseudoinverse expressed in (6.47). Assume the norm which must be minimized is that associated with the standard inner product for $\mathfrak{M}^{n\times 1}$. Being free to pick any inner product for $\mathfrak{M}^{m\times 1}$, we again use the standard inner product. Define $\mathbf{T}: \mathfrak{M}^{n\times 1} \to \mathfrak{M}^{m\times 1}$ by $\mathbf{Tx} \triangleq \mathbf{Ax}$. For *standard inner products*, $\mathbf{T}^*$ is given by (5.60): $\mathbf{T}^*\mathbf{x} = \mathbf{A}^{\mathrm{T}}\mathbf{x}$. Because the rows of $\mathbf{A}$ are independent, the columns of $\mathbf{A}^{\mathrm{T}}$ are independent, and both $\mathbf{A}^{\mathrm{T}}$ and $\mathbf{T}^*$ are nonsingular; $\mathbf{T}$ satisfies the compatibility condition. The minimum norm solution (6.47) becomes

$$\mathbf{x}_s = \mathbf{A}^{\mathrm{T}}(\mathbf{AA}^{\mathrm{T}})^{-1}\mathbf{y} \tag{6.48}$$

We call the matrix $\mathbf{A}^{\mathrm{T}}(\mathbf{AA}^{\mathrm{T}})^{-1}$ the *pseudoinverse of the matrix* $\mathbf{A}$ for this problem.

As a specific example, let

$$\mathbf{A} = \begin{pmatrix} 2 & 1 & 1 \\ 1 & 2 & 2 \end{pmatrix} \tag{6.49}$$

The matrix $\mathbf{AA}^{\mathrm{T}}$ is the Gram matrix of the row vectors of $\mathbf{A}$:

$$\mathbf{AA}^{\mathrm{T}} = \begin{pmatrix} 2 & 1 & 1 \\ 1 & 2 & 2 \end{pmatrix}\begin{pmatrix} 2 & 1 \\ 1 & 2 \\ 1 & 2 \end{pmatrix} = \begin{pmatrix} 6 & 6 \\ 6 & 9 \end{pmatrix}$$

The minimum norm solution to the equation $\mathbf{Ax} = \mathbf{y}$ is (6.48):

$$\mathbf{x}_s = \frac{1}{6}\begin{pmatrix} 4 & -2 \\ -1 & 2 \\ -1 & 2 \end{pmatrix}\mathbf{y} \tag{6.50}$$

for any vector $\mathbf{y}$. As a check on the pseudoinverse of $\mathbf{A}$, we can verify that it is a right inverse of $\mathbf{A}$.

Application to Minimum Energy Control

In Section 6.2 we found the armature voltage function $\mathbf{u}$ which would consume minimum energy $\int_0^1 \mathbf{u}^2(t)\,dt$ in driving a motor shaft from the condition $\phi(0)=\dot{\phi}(0)=0$ to the condition $\phi(1)=1$, $\dot{\phi}(1)=0$. We now repeat that optimization using the concepts developed in the discussion of Problem 2. The state-space model for the differential system (6.32) which describes the motor is derived in Example 1 of Section 3.4; it is

$$\dot{\mathbf{x}}(t)=\mathbf{A}\mathbf{x}(t)+\mathbf{B}\mathbf{u}(t), \qquad \mathbf{x}(0)=\boldsymbol{\theta}$$
$$\mathbf{x}=\begin{pmatrix}\phi\\ \dot{\phi}\end{pmatrix}, \qquad \mathbf{A}=\begin{pmatrix}0 & 1\\ 0 & -1\end{pmatrix}, \qquad \mathbf{B}=\begin{pmatrix}0\\ 1\end{pmatrix} \tag{6.51}$$

The inverse of this state equation is given in (3.80); for the given initial conditions, the result is

$$\mathbf{x}(t)=\int_0^t \begin{pmatrix}1-e^{-(t-s)}\\ e^{-(t-s)}\end{pmatrix}\mathbf{u}(s)\,ds \tag{6.52}$$

In order that the final state of the system be as specified, we need

$$\mathbf{y} \triangleq \mathbf{x}(1)=\begin{pmatrix}1\\ 0\end{pmatrix}=\int_0^1 \begin{pmatrix}1-e^{-1}e^{s}\\ e^{-1}e^{s}\end{pmatrix}\mathbf{u}(s)\,ds \triangleq \mathbf{T}\mathbf{u} \tag{6.53}$$

Assume the operator $\mathbf{T}$ of (6.53) acts on $\mathcal{C}(0,1)$ with the standard function space inner product. The range of $\mathbf{T}$ is $\mathfrak{M}^{2\times 1}$. We *arbitrarily* pick the standard inner product for $\mathfrak{M}^{2\times 1}$. The minimum energy control problem consists in finding that control function $\mathbf{u}$ which satisfies (6.53) and minimizes $\|\mathbf{u}\|^2$. Although we have not yet verified that nullspace$(\mathbf{T}^*)=\boldsymbol{\theta}$, this problem clearly has the structure of Problem 2. We solve for the minimum energy control $\mathbf{u}_s$ by determining $\mathbf{T}^*$ and evaluating the pseudoinverse (6.47). By the defining equation (5.58) for $\mathbf{T}^*$,

$$\begin{aligned}\langle \mathbf{T}\mathbf{u},\mathbf{z}\rangle_{\mathfrak{M}^{2\times 1}} &= \mathbf{z}^{\mathrm{T}}\mathbf{T}\mathbf{u}\\
&= \mathbf{z}^{\mathrm{T}}\int_0^1 \begin{pmatrix}1-e^{-1}e^{s}\\ e^{-1}e^{s}\end{pmatrix}\mathbf{u}(s)\,ds\\
&= \int_0^1 (1-e^{-1}e^{s} \quad e^{-1}e^{s})\mathbf{z}\mathbf{u}(s)\,ds\\
&= \int_0^1 (\mathbf{T}^*\mathbf{z})(s)\mathbf{u}(s)\,ds\\
&= \langle \mathbf{u},\mathbf{T}^*\mathbf{z}\rangle_{\mathcal{C}(0,1)}\end{aligned} \tag{6.54}$$

for each $\mathbf{u}$ in $\mathcal{C}(0,1)$ and $\mathbf{z}$ in $\mathfrak{M}^{2\times 1}$. (We are able to transpose the quantity

$$\mathbf{z}^{\mathrm{T}}\begin{pmatrix} 1-e^{-1}e^{s} \\ e^{-1}e^{s} \end{pmatrix} \tag{6.55}$$

because it is a scalar.) Therefore,

$$(\mathbf{T}^*\mathbf{z})(s) \stackrel{\Delta}{=} (1-e^{-1}e^{s} \quad e^{-1}e^{s})\mathbf{z} \tag{6.56}$$

Suppose $\mathbf{z} \stackrel{\Delta}{=} (\eta_1\ \eta_2)^{\mathrm{T}}$ is an arbitrary vector in $\mathfrak{M}^{2\times 1}$. Then nullspace($\mathbf{T}^*$) consists in those $\mathbf{z}$ which satisfy $(\mathbf{T}^*\mathbf{z})(s) = \eta_1 + (\eta_2 - \eta_1)e^{-1}e^{s} = 0$ for $0 \leqslant s \leqslant 1$. It is evident that $\eta_1 = \eta_2 = 0$, $\mathbf{z} = \boldsymbol{\theta}$, and nullspace($\mathbf{T}^*$) $= \boldsymbol{\theta}$; this is indeed Problem 2.

The operator $\mathbf{TT}^*$ is given by

$$\begin{aligned}
\mathbf{TT}^*\mathbf{z} &= \int_0^1 \begin{pmatrix} 1-e^{-1}e^{s} \\ e^{-1}e^{s} \end{pmatrix} (\mathbf{T}^*\mathbf{z})(s)\,ds \\
&= \int_0^1 \begin{pmatrix} 1-e^{-1}e^{s} \\ e^{-1}e^{s} \end{pmatrix} (1-e^{-1}e^{s} \quad e^{-1}e^{s})\mathbf{z}\,ds \\
&= \frac{1}{2e^2}\begin{pmatrix} -e^2+4e-1 & (e-1)^2 \\ (e-1)^2 & e^2-1 \end{pmatrix}\mathbf{z}
\end{aligned} \tag{6.57}$$

for all $\mathbf{z}$ in $\mathfrak{M}^{2\times 1}$. Using (6.46) we find

$$\begin{aligned}
\boldsymbol{\lambda} &= (\mathbf{TT}^*)^{-1}\mathbf{y} \\
&= \frac{1}{(e-1)(3-e)}\begin{pmatrix} (e-1)(e+1) & -(e-1)^2 \\ -(e-1)^2 & -e^2+4e-1 \end{pmatrix}\mathbf{y}
\end{aligned} \tag{6.58}$$

Finally,

$$\mathbf{u}_s = \mathbf{T}^*\boldsymbol{\lambda} = \mathbf{T}^*(\mathbf{TT}^*)^{-1}\mathbf{y} = \mathbf{T}^{\dagger}\mathbf{y} \tag{6.59}$$

a *function* in $\mathcal{C}(0,1)$; we use (6.56) to express $\mathbf{u}_s$ in terms of its values

$$\begin{aligned}\mathbf{u}_s(t) &= (\mathbf{T}^*\boldsymbol{\lambda})(t)\\ &= (1-e^{-1}e^t \quad e^{-1}e^t)\boldsymbol{\lambda}\\ &= \left(\frac{e+1}{3-e} - \frac{2e^t}{3-e} \quad -\frac{e-1}{3-e} + \frac{2e^t}{(e-1)(3-e)}\right)\mathbf{y}\\ &= (\mathbf{T}^\dagger\mathbf{y})(t)\end{aligned} \tag{6.60}$$

For the specific vector $\mathbf{y}=\mathbf{x}(1)=(1 \quad 0)^{\mathrm{T}}$ of (6.53), the minimum energy control is

$$\mathbf{u}_s(t) = \frac{e+1}{3-e} - \frac{2}{3-e}e^t \tag{6.61}$$

Quadratic Objective Functions for Optimal Control

In section 7.2 we determine how to compute the minimum energy control (6.61) directly from the state equation (6.51), thereby avoiding the need to invert the state equation. The method involves the adjoint of the state equation. A more general quadratic objective function than the minimum energy criterion used to determine (6.60) is the function

$$\|\mathbf{u}-\hat{\mathbf{u}}\|_{\mathcal{V}}^2 + \|\mathbf{x}-\hat{\mathbf{x}}\|_{\mathcal{W}}^2 \tag{6.62}$$

where $\hat{\mathbf{x}}$ and $\hat{\mathbf{u}}$ are specified functions. (The subscripts $\mathcal{V}$ and $\mathcal{W}$ refer to the control space and the state space, respectively.) A control system designer might minimize this objective function with $\hat{\mathbf{u}}=\boldsymbol{\theta}$ in order to have the system follow closely some trajectory $\mathbf{x}(t)$ and yet to avoid large energy consumption. Kuo and Kazada [6.10] show that (6.62) can be rewritten as $\|\mathbf{u}\|_F$, where the subscript refers to a new inner product. Thus (6.62) is not conceptually different from the minimum energy problem solved above. Kuo and Kazda also describe a technique for computing the optimal control $\mathbf{u}$ and the corresponding trajectory $\mathbf{x}$ by solving the state equation together with an adjoint of the state equation.

6.5 The Pseudoinverse

For bounded linear transformations $\mathbf{T}$: $\mathcal{V}\rightarrow\mathcal{W}$ we have "resolved" the linear equation $\mathbf{Tx}=\mathbf{y}$ for two important special cases:

Problem 1. If $\mathbf{T}$ is one-to-one but not onto, $\|\mathbf{y}-\mathbf{Tx}\|_{\mathcal{W}}$ is minimized by $\mathbf{x}_a=(\mathbf{T}^*\mathbf{T})^{-1}\mathbf{T}^*\mathbf{y}$. Although $\mathbf{T}^*$ is determined by the inner products in both $\mathcal{V}$ and $\mathcal{W}$, the inner product in $\mathcal{V}$ is arbitrary. Each inner product for $\mathcal{V}$ specifies a different $\mathbf{T}^*$, but all such adjoints lead to the same vector $\mathbf{x}_a$ in $\mathcal{V}$.

Problem 2. If $\mathbf{T}$ is onto but not one-to-one, the solution which minimizes $\|\mathbf{x}\|_{\mathcal{V}}$ is $\mathbf{x}_s=\mathbf{T}^*(\mathbf{TT}^*)^{-1}\mathbf{y}$. Here only the inner product in $\mathcal{V}$ is fixed. Different inner products for $\mathcal{W}$ lead to different adjoint operators, but the same vector $\mathbf{x}_s$. In this section we combine the results of the two previous sections to yield least-square solutions for *arbitrary* linear system equations; thus we fit all linear systems into the framework of least-square analysis.

If $\mathbf{T}$ is both one-to-one and onto, the solutions of Problem 1 and Problem 2 both reduce to $\mathbf{x}=\mathbf{T}^{-1}\mathbf{y}$. Suppose, however, that $\mathbf{T}$ is neither one-to-one nor onto. The normal equations

$$\mathbf{T}^*\mathbf{Tx}=\mathbf{T}^*\mathbf{y}$$

of (6.37) still specify those vectors $\mathbf{x}$ which minimize $\|\mathbf{y}-\mathbf{Tx}\|_{\mathcal{W}}$. However, because $\mathbf{T}$ is not one-to-one, $\mathbf{T}^*\mathbf{T}$ is not invertible; there are many least-error solutions $\mathbf{x}$ which result in the same minimum error. This is the case even if $\mathbf{y}$ is in range($\mathbf{T}$) and $\|\mathbf{y}-\mathbf{Tx}\|_{\mathcal{W}}=0$. Of course, if $\mathbf{y}$ is in range($\mathbf{T}$), the unique solution of minimum norm $\|\mathbf{x}\|_{\mathcal{V}}$ is still determined by (6.46):

$$\mathbf{x}=\mathbf{T}^*\boldsymbol{\lambda} \quad \text{with} \quad \mathbf{TT}^*\boldsymbol{\lambda}=\mathbf{y}$$

However, because $\mathbf{T}$ is not onto, $\mathbf{TT}^*$ is not invertible; we cannot solve the Gram equation $\mathbf{TT}^*\boldsymbol{\lambda}=\mathbf{y}$ uniquely. There are many adjoint vectors $\boldsymbol{\lambda}$ which lead to the same least-effort solution $\mathbf{x}$.

Example 1. Incompatibility and Nonuniqueness. The equation

$$\mathbf{Ax} \stackrel{\Delta}{=} \begin{pmatrix} 1 & 2 \\ 1 & 2 \end{pmatrix}\begin{pmatrix} \xi_1 \\ \xi_2 \end{pmatrix} = \begin{pmatrix} \eta_1 \\ \eta_2 \end{pmatrix} \stackrel{\Delta}{=} \mathbf{y}$$

does not satisfy either the compatibility condition, nullspace($\mathbf{A}^{\mathrm{T}}$)$=\boldsymbol{\theta}$, or the uniqueness condition, nullspace($\mathbf{A}$)$=\boldsymbol{\theta}$. Any solution to the normal equations

$$\mathbf{A}^{\mathrm{T}}\mathbf{Ax}=\begin{pmatrix} 2 & 4 \\ 4 & 8 \end{pmatrix}\begin{pmatrix} \xi_1 \\ \xi_2 \end{pmatrix}=\begin{pmatrix} \eta_1+\eta_2 \\ 2\eta_1+2\eta_2 \end{pmatrix}=\mathbf{A}^{\mathrm{T}}\mathbf{y} \tag{6.63}$$

is a least-square error solution (with respect to the standard norm for $\mathfrak{M}^{2\times 1}$). Although the matrix $\mathbf{A}^{\mathrm{T}}\mathbf{A}$ is not invertible, the normal equations can be solved by elimination to obtain

$$\mathbf{x}=\begin{pmatrix} \xi_1 \\ \dfrac{\eta_1+\eta_2}{4}-\dfrac{\xi_1}{2} \end{pmatrix} \tag{6.64}$$

In keeping with the sense of this chapter, we seek the smallest of these least-error solutions. That is, if $\mathbf{B} \triangleq \mathbf{A}^T\mathbf{A}$ and $\mathbf{z} \triangleq \mathbf{A}^T\mathbf{y}$, the solution to $\mathbf{Bx}=\mathbf{z}$ which is of minimum norm (in the standard inner product for $\mathfrak{M}^{2\times 1}$) is $\mathbf{x}=\mathbf{B}^T\boldsymbol{\lambda}$, where $\mathbf{BB}^T\boldsymbol{\lambda}=\mathbf{z}$. Although $\mathbf{BB}^T$ is not invertible, the Gram equations

$$\mathbf{BB}^T\boldsymbol{\lambda}=\begin{pmatrix} 20 & 40 \\ 40 & 80 \end{pmatrix}\begin{pmatrix} \zeta_1 \\ \zeta_2 \end{pmatrix}=\begin{pmatrix} \eta_1+\eta_2 \\ 2\eta_1+2\eta_2 \end{pmatrix}=\mathbf{z} \tag{6.65}$$

can be solved by elimination to yield

$$\boldsymbol{\lambda}=\begin{pmatrix} \zeta_1 \\ \dfrac{\eta_1+\eta_2}{40}-\dfrac{\zeta_1}{2} \end{pmatrix} \tag{6.66}$$

The solutions $\boldsymbol{\lambda}$ are many, yet the minimum norm vector $\mathbf{x}$ is unique:

$$\mathbf{x}=\mathbf{B}^T\boldsymbol{\lambda}=\begin{pmatrix} 2 & 4 \\ 4 & 8 \end{pmatrix}\begin{pmatrix} \zeta_1 \\ \dfrac{\eta_1+\eta_2}{40}-\dfrac{\zeta_1}{2} \end{pmatrix}=\begin{pmatrix} \dfrac{\eta_1+\eta_2}{10} \\ \dfrac{\eta_1+\eta_2}{5} \end{pmatrix} \tag{6.67}$$

Practical situations do arise where $\mathbf{T}$ is neither one-to-one nor onto. Furthermore, nearly degenerate equations are *nearly* in this category. (As we have noted in earlier chapters, nearly degenerate problems are often handled more accurately by treating them as precisely degenerate.) We resolve the general linear equation $\mathbf{Tx}=\mathbf{y}$ in a least-square sense by combining our approaches to Problems 1 and 2.

Smallest Least-Error Solutions

Definition. Let $\mathbf{T}$: $\mathcal{V}\to\mathcal{W}$ be a bounded linear transformation between Hilbert spaces.* Assume $\mathbf{T}$ is neither one-to-one nor onto. We wish to find the **smallest least-error solution** to $\mathbf{Tx}=\mathbf{y}$. That is, of all $\mathbf{x}$ which minimize

*We should also assume range($\mathbf{T}$) is complete. It then follows that range($\mathbf{T}^*$) is complete and the orthogonal decompositions (6.35) are assured. See Dunford and Schwartz [6.3, p. 488]. In practice, of course, we may ignore some of these restrictions.

$\|\mathbf{y}-\mathbf{T}\mathbf{x}\|_{\mathcal{W}}$, we wish to determine that vector $\mathbf{x}_s$ for which $\|\mathbf{x}\|_{\mathcal{V}}$ is minimum. We define the **pseudoinverse** $\mathbf{T}^{\dagger}$ of $\mathbf{T}$ as that transformation which produces $\mathbf{x}_s$ from $\mathbf{y}$ for all $\mathbf{y}$ in $\mathcal{W}$.

The basic principle of Problem 1 can be used to minimize the error; $\|\mathbf{y}-\mathbf{T}\mathbf{x}\|_{\mathcal{W}}$ is minimum if $\mathbf{T}\mathbf{x}$ is the orthogonal projection $\mathbf{y}_a$ of $\mathbf{y}$ onto range($\mathbf{T}$). We can compute $\mathbf{y}_a$. To obtain the minimum norm solution to $\mathbf{T}\mathbf{x}=\mathbf{y}_a$, we apply the fundamental principle of Problem 2; the solution which minimizes $\|\mathbf{x}\|_{\mathcal{V}}$ is the orthogonal projection $\mathbf{x}_s$ of any particular solution onto range($\mathbf{T}^*$). The process of projecting $\mathbf{y}$ onto range($\mathbf{T}$), solving $\mathbf{T}\mathbf{x}=\mathbf{y}_a$, then projecting a solution $\mathbf{x}$ onto range($\mathbf{T}^*$) constitutes $\mathbf{T}^{\dagger}$. See Figure 6.8. The relationship $\mathbf{T}^{\dagger}$ between $\mathbf{y}$ and $\mathbf{x}_s$ exists for every $\mathbf{T}$ of the type described above. Moreover, because we have simply performed an orthogonal projection followed by a "Problem 2" solution, it is intuitively clear that $\mathbf{T}^{\dagger}$ is unique and linear; it is also bounded. Problem 1, Problem 2, and the problem where $\mathbf{T}$ is invertible are special cases of the smallest least-error problem. The pseudoinverse solves all linear equations in a least-square sense.

Example 2. Pseudoinverse for Problems 1 and 2. Suppose $\mathbf{T}_1$ is one-to-one [or nullspace($\mathbf{T}_1$)$=\boldsymbol{\theta}$]; then the problem of determining the smallest least-error solution reduces to Problem 1. By comparing the pseudoinverse operation (Figure 6.8) with the solution process for Problem 1 (Figure 6.6), we recognize that the pseudoinverse of $\mathbf{T}_1$ is given by (6.38):

$$\mathbf{T}_1^{\dagger}=(\mathbf{T}_1^*\mathbf{T}_1)^{-1}\mathbf{T}_1^* \tag{6.68}$$

The pseudoinverse of an operator $\mathbf{T}$ is of the form (6.68) if and only if $\mathbf{T}^*\mathbf{T}$ is invertible.

If $\mathbf{T}_2$ is onto [or nullspace($\mathbf{T}_2^*$)$=\boldsymbol{\theta}$], the problem of finding the smallest least-error solution reduces to Problem 2. Comparison of the pseudoinverse process with the solution process for Problem 2 (Figure 6.7) shows that $\mathbf{T}_2^{\dagger}$ is given by (6.47):

$$\mathbf{T}_2^{\dagger}=\mathbf{T}_2^*(\mathbf{T}_2\mathbf{T}_2^*)^{-1} \tag{6.69}$$

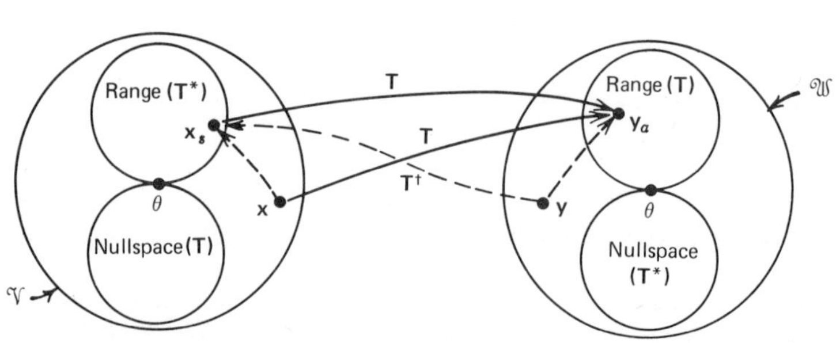

Figure 6.8. Construction of the pseudoinverse.

The pseudoinverse of an operator **T** is of the form (6.69) if and only if $\mathbf{TT}^*$ is invertible.

Example 3. The Pseudoinverse of a Matrix. We solve the smallest least-error problem of Example 1 by performing orthogonal projections. We carry out the operations for an arbitrary vector **y**, thereby finding the pseudoinverse.

$$\mathbf{Ax} \triangleq \begin{pmatrix} 1 & 2 \\ 1 & 2 \end{pmatrix}\begin{pmatrix} \xi_1 \\ \xi_2 \end{pmatrix} = \begin{pmatrix} \eta_1 \\ \eta_2 \end{pmatrix} \triangleq \mathbf{y}$$

Assume we use the standard inner product for $\mathfrak{M}^{2\times 1}$; if $\mathbf{Tx} \triangleq \mathbf{Ax}$, then $\mathbf{T}^*\mathbf{y} \triangleq \mathbf{A}^{\mathrm{T}}\mathbf{y}$. Note that **T** is neither one-to-one nor onto.

We need an orthogonal basis for range(**A**). We find one by row reducing $(\mathbf{A} \vdots \mathbf{y})$‡:

$$\left(\begin{array}{cc:c} 1 & 2 & \eta_1 \\ 1 & 2 & \eta_2 \end{array}\right) \rightarrow \left(\begin{array}{cc:c} 1 & 2 & \eta_1 \\ 0 & 0 & \eta_2-\eta_1 \end{array}\right)$$

The vectors **y** in range(**A**) satisfy $\eta_2-\eta_1=0$. The vector $\mathbf{y}_1=(1 \quad 1)^{\mathrm{T}}$ is obviously a basis for range(**A**). (Consisting in only one vector, the basis need not be orthogonalized.) We find $\mathbf{y}_a$, the orthogonal projection of **y** onto range(**A**), by the partial Fourier series expansion

$$\mathbf{y}_a = \frac{\langle \mathbf{y},\mathbf{y}_1\rangle}{\|\mathbf{y}_1\|^2}\mathbf{y}_1 = \frac{\eta_1+\eta_2}{2}\begin{pmatrix} 1 \\ 1 \end{pmatrix}$$

Although the remainder of the computation process is essentially in the form of Problem 2, we cannot use (6.69) to compute it; $\mathbf{AA}^{\mathrm{T}}$ is not invertible. We solve the equation $\mathbf{Ax}=\mathbf{y}_a$ by elimination to obtain

$$\mathbf{x} = \begin{pmatrix} \xi_1 \\ \dfrac{\eta_1+\eta_2}{4} - \dfrac{\xi_1}{2} \end{pmatrix}$$

We need a basis for range($\mathbf{A}^{\mathrm{T}}$). Since range($\mathbf{A}^{\mathrm{T}}$) consists in linear combinations of rows of **A** (columns of $\mathbf{A}^{\mathrm{T}}$), we can find a basis from the previous row reduction of $(\mathbf{A} \vdots \mathbf{y})$; a basis, again orthogonal, is $\mathbf{x}_1=(1 \quad 2)^{\mathrm{T}}$. The smallest least-error solution is the projection of **x** onto range($\mathbf{A}^{\mathrm{T}}$):

$$\mathbf{x}_s = \frac{\langle \mathbf{x},\mathbf{x}_1\rangle}{\|\mathbf{x}_1\|^2}\mathbf{x}_1 = \begin{pmatrix} \dfrac{\eta_1+\eta_2}{10} \\ \dfrac{\eta_1+\eta_2}{5} \end{pmatrix} = \begin{pmatrix} \dfrac{1}{10} & \dfrac{1}{10} \\ \dfrac{1}{5} & \dfrac{1}{5} \end{pmatrix}\mathbf{y} \triangleq \mathbf{A}^{\dagger}\mathbf{y}$$

‡See P&C 2.19.

The pseudoinverse operation is based upon two *orthogonal* projections. However, in solving $\mathbf{Tx}=\mathbf{y}$, we *could* project $\mathbf{y}$ onto range($\mathbf{T}$) along *any* particular subspace $\mathcal{U}$ which is a complement to range($\mathbf{T}$); that is, along a subspace $\mathcal{U}$ such that $\mathcal{W} = \mathcal{U} \oplus$ range($\mathbf{T}$). Similarly, the projection of vectors in $\mathcal{V}$ *could* be along nullspace($\mathbf{T}$) and onto *any* particular subspace $\mathcal{Z}$ which is a complement of nullspace($\mathbf{T}$). The process of projecting $\mathbf{y}$ to get $\mathbf{y}_a$, solving $\mathbf{Tx}=\mathbf{y}_a$, then projecting $\mathbf{x}$ to get $\mathbf{x}_s$ defines a **generalized inverse** of $\mathbf{T}$. The process is useful in that it specifies a unique $\mathbf{x}_s$ for each $\mathbf{y}$ and acts to some extent like an inverse. However, a generalized inverse does not possess the minimization properties of the pseudoinverse unless the projections are precisely the orthogonal projections which constitute the pseudoinverse. Of course, we can always *pick* the inner products so these orthogonalities occur. Then $\mathcal{U}$ = nullspace($\mathbf{T}^*$) and $\mathcal{Z}$ = range($\mathbf{T}^*$). It follows that any generalized inverse yields solutions that are smallest least-error solutions for *some* pair of inner products for $\mathcal{V}$ and $\mathcal{W}$. The pseudoinverse, for which the projections are orthogonal, is sometimes referred to as the Moore–Penrose inverse.*

The Pseudoinverse of a Differential Operator

The equation of motion of a simple mass–spring system is

$$\mathbf{Lf} \stackrel{\Delta}{=} \ddot{\mathbf{f}} + a^2\mathbf{f} = \mathbf{g} \tag{6.70}$$

where $\mathbf{f}(t)$ is the displacement of the mass relative to some fixed position, $\mathbf{g}(t)$ is the force on the mass at time t, and a^2 is the spring-constant-to-mass ratio. We attempt to analyze the spring-mass system as a vibration problem by seeking a solution to (6.70) which is periodic in time (with the natural period, $2\pi/a$):

$$\mathbf{f}\left(\frac{2\pi}{a}\right) = \mathbf{f}(0), \qquad \dot{\mathbf{f}}\left(\frac{2\pi}{a}\right) = \dot{\mathbf{f}}(0) \tag{6.71}$$

If we look only at the interval $0 \leqslant t \leqslant 2\pi/a$, any forcing function $\mathbf{g}$ can be considered periodic with period $2\pi/a$. However, not all such periodic functions $\mathbf{g}$ lead to periodic oscillations of the system; if the system is forced by a sinusoidal function of the natural frequency, *growing* oscillations occur. Even if a periodic solution $\mathbf{f}$ to (6.70)–(6.71) does exist for a particular $\mathbf{g}$, the oscillation $\mathbf{f}$ is not unique; the natural modes of oscillation of the system superimpose themselves on any particular solution. Many realistic vibration problems, although not this simple, can be analyzed in

*See Gillies [6.6] for a concise classification of generalized inverses. An alternate, but equivalent, definition of the pseudoinverse is contained in P&C 6.22.

this fashion; for instance, the vertical vibration of a railcar rolling on imperfect rails. We seek the smallest least-effort solution to the differential system. By finding the solution for an unspecified forcing function **g**, we generate the pseudoinverse of that differential operator **T** which consists in **L** with the periodic boundary conditions (6.71).

Our procedure imitates the simple matrix problem of Example 3; we project **g** onto range(**T**), solve the equation $\mathbf{Lf}=\mathbf{g}_a$ with (6.71), then project **f** onto range(**T***). We assume the standard function space inner product. Rather than find a basis for the infinite-dimensional range(**T**), we seek a basis for its orthogonal complement, nullspace(**T***). The differential operator **T** is self-adjoint; $\mathbf{L}^*=\mathbf{L}$, and the adjoint boundary conditions are identical to (6.71). Consequently, nullspace(**T**) equals nullspace(**T***), and each has as a basis a set of fundamental solutions for (6.70)–(6.71); that is, solutions for $\mathbf{g}=\boldsymbol{\theta}$. These solutions are the sinusoids of frequency a. An orthogonal basis for nullspace(**T***) is

$$\mathbf{f}_1(t)=\cos at \qquad \mathbf{f}_2(t)=\sin at \tag{6.72}$$

The projection $\mathbf{g}_a$ of **g** onto range(**T**) equals the function **g** minus the projection of **g** onto nullspace(**T***):

$$\begin{aligned}\mathbf{g}_a(t)&=\mathbf{g}(t)-\frac{\langle \mathbf{g},\mathbf{f}_1\rangle}{\|\mathbf{f}_1\|^2}\mathbf{f}_1(t)-\frac{\langle \mathbf{g},\mathbf{f}_2\rangle}{\|\mathbf{f}_2\|^2}\mathbf{f}_2(t)\\ &=\mathbf{g}(t)-\frac{a}{\pi}\cos at\int_0^{2\pi/a}\mathbf{g}(s)\cos as\,ds\\ &\quad-\frac{a}{\pi}\sin at\int_0^{2\pi/a}\mathbf{g}(s)\sin as\,ds\end{aligned} \tag{6.73}$$

In the approximation $\mathbf{g}_a$, the "natural frequencies" have been removed from the forcing function; the equation $\mathbf{Lf}=\mathbf{g}_a$ does have periodic solutions. Moreover, we have minimized $\|\mathbf{g}-\mathbf{g}_a\|$.

In P&C 3.16, the solutions to (6.70) are derived in terms of initial conditions:

$$\mathbf{f}(t)=\mathbf{f}(0)\cos at+\frac{\dot{\mathbf{f}}(0)}{a}\sin at+\frac{1}{a}\int_0^t \sin a(t-\sigma)\mathbf{g}(\sigma)\,d\sigma \tag{6.74}$$

Because the differential equation is invertible with initial conditions, this expression is a general solution to the differential equation. The boundary conditions (6.71) restrict the functions **g** for which it is the general solution. Since $\mathbf{g}_a$ is the projection of **g** onto range(**T**), the general solution **f** satisfies the periodic boundary conditions if we replace $\mathbf{g}$ by $\mathbf{g}_a$. We pick a *particular* solution which satisfies the boundary conditions by picking values for $\mathbf{f}(0)$ and $\dot{\mathbf{f}}(0)$; we let them both be zero. Then

$$\begin{aligned}
\mathbf{f}(t) &= \frac{1}{a}\int_0^t \sin a(t-\sigma)\mathbf{g}_a(\sigma)\,d\sigma \\
&= \frac{1}{a}\int_0^t \sin a(t-\sigma)\Big[\mathbf{g}(\sigma) - \frac{a}{\pi}\cos a\sigma \int_0^{2\pi/a} \mathbf{g}(s)\cos as\,ds \\
&\qquad - \frac{a}{\pi}\sin a\sigma \int_0^{2\pi/a} \mathbf{g}(s)\sin as\,ds\Big]\,d\sigma \\
&= \frac{1}{a}\int_0^t \sin a(t-\sigma)\mathbf{g}(\sigma)\,d\sigma - \frac{1}{2\pi}t\sin at \int_0^{2\pi/a} \mathbf{g}(s)\cos as\,ds \\
&\quad + \frac{1}{2\pi}\Big[t\cos at - \frac{\sin at}{a}\Big]\int_0^{2\pi/a} \mathbf{g}(s)\sin as\,ds
\end{aligned} \tag{6.75}$$

It should be apparent that the addition of any natural mode, $c_1 \cos at + c_2 \sin at$, to $\mathbf{f}(t)$ yields another periodic solution to $\mathbf{Lf} = \mathbf{g}_a$. We pick the solution $\mathbf{f}_s$ of smallest norm $\|\mathbf{f}_s\|$ by projecting (6.75) onto range($\mathbf{T}^*$). Since $\mathbf{T}$ is self-adjoint, the basis $\{\mathbf{f}_1, \mathbf{f}_2\}$ for nullspace($\mathbf{T}$) is also a basis for nullspace($\mathbf{T}^*$). Therefore,

$$\begin{aligned}
\mathbf{f}_s(t) &= \mathbf{f}(t) - \frac{\langle \mathbf{f}, \mathbf{f}_1\rangle}{\|\mathbf{f}_1\|^2}\mathbf{f}_1(t) - \frac{\langle \mathbf{f}, \mathbf{f}_2\rangle}{\|\mathbf{f}_2\|^2}\mathbf{f}_2(t) \\
&= \mathbf{f}(t) - \frac{a}{\pi}\cos at \int_0^{2\pi/a} \mathbf{f}(s)\cos as\,ds \\
&\quad - \frac{a}{\pi}\sin at \int_0^{2\pi/a} \mathbf{f}(s)\sin as\,ds
\end{aligned} \tag{6.76}$$

A straightforward but lengthy substitution of (6.75) into this equation gives

$$\begin{aligned}
\mathbf{f}_s(t) &= \int_0^t \mathbf{g}(\sigma)\Big[\frac{1}{a}\sin a(t-\sigma)\Big]\,d\sigma \\
&\quad - \int_0^{2\pi/a} \mathbf{g}(\sigma)\Big[\frac{1}{2a}\sin a(t-\sigma) + \frac{1}{4\pi a}\cos a(t-\sigma) \\
&\qquad + \frac{1}{2\pi}(t-\sigma)\sin a(t-\sigma)\Big]\,d\sigma
\end{aligned} \tag{6.77}$$

The smallest least-effort solution (6.77) can be expressed as

$$\mathbf{f}_s(t)=\int_0^{2\pi/a} k^\dagger(t,\sigma)\mathbf{g}(\sigma)\,d\sigma=(\mathbf{T}^\dagger\mathbf{g})(t) \tag{6.78}$$

where

$$k^\dagger(t,\sigma)=-\frac{1}{4\pi a}\cos a(t-\sigma)-\frac{1}{2\pi}(t-\sigma)\sin a(t-\sigma)$$
$$\pm\frac{1}{2a}\sin a(t-\sigma) \tag{6.79}$$

with the upper sign used if $\sigma<t$ and the lower used if $\sigma>t$. We call the kernel $k^\dagger$ of the pseudoinverse operator $\mathbf{T}^\dagger$ the **generalized Green's function** for the differential operator $\mathbf{T}$.

It should be clear from this example that any regular linear differential operator together with associated homogeneous boundary conditions has a generalized inverse and a generalized Green's function. Loud [6.12] explores in detail the pseudoinverse for differential operators. He also describes a technique for determining the generalized Green's function of an arbitrary differential system which is analogous to the method we use for invertible differential systems in Section 3.3.

Decomposition of the Pseudoinverse

In practical computing, determination of a solution to a matrix equation for a specific right-hand side is cheaper than inversion of the matrix. The same principle holds true for least-square problems. Direct computation of the orthogonal projections that are inherent in the pseudoinverse is a suitable means for getting a specific solution. By using an unspecified right-hand side, the full pseudoinverse can also be determined via projections.

The projection method seems somewhat informal in that we find the solution without expressing explicitly the operations that determine $\mathbf{T}^\dagger$. Our intuitive grasp of the pseudoinverse operator would be improved, perhaps, if we could express $\mathbf{T}^\dagger$ for a general transformation $\mathbf{T}$ in a simple explicit form as we have done for Problems 1 and 2 in (6.68) and (6.69). As we carry out the projections, we determine bases for range($\mathbf{T}$) and range($\mathbf{T}^*$) or their orthogonal complements, nullspace($\mathbf{T}^*$) and nullspace($\mathbf{T}$), respectively. Sufficient information is obtained in the process of finding these bases to express $\mathbf{T}^\dagger$ explicitly. Consider the matrix equations of Example 3, for instance. In the single row reduction

$$\left(\mathbf{A}\ \vdots\ \mathbf{y}\right)=\left(\begin{array}{cc:c}1&2&\eta_1\\1&2&\eta_2\end{array}\right)\rightarrow\left(\begin{array}{cc:c}1&2&\eta_1\\0&0&\eta_2-\eta_1\end{array}\right)$$

we determine bases for both range(**T**) and range(**T***). The vector $(1\ 2)^T$ is a basis for range(**T***). The vector $(1\ 1)^T$ [which satisfies $\eta_2-\eta_1=0$] is a basis for range(**T**). We explore further in the next section the use of row reduction as a technique for decomposing a general matrix least-squares problem into a combination of Problems 1 and 2.

We return to the abstract equation $\mathbf{Tx}=\mathbf{y}$, where **T** is neither one-to-one nor onto, and explore a decomposition of **T** which copes with the degeneracy in **T**. The key to the decomposition is displayed in the abstract diagram of the operations **T** and $\mathbf{T}^\dagger$ in Figure 6.8. By a simple redrawing of Figure 6.8, we convert **T** to a pair of operations which are, respectively, one-to-one and onto (Figure 6.9). Define $\mathbf{T}_2$: $\mathcal{V}\rightarrow$range(**T**) by

$$\mathbf{T}_2\mathbf{x} \stackrel{\Delta}{=} \mathbf{Tx} \quad \text{for } \mathbf{x} \text{ in } \mathcal{V} \tag{6.80}$$

That is, we remove from the definition of **T** the "unused" part of the range of definition, creating thereby a "Problem 2" structure; $\mathbf{T}_2$ is onto. To complete the operation **T** [because nullspace(**T***) is still a part of the initial problem], we define a second transformation $\mathbf{T}_1$: range(**T**)$\rightarrow\mathcal{W}$ by

$$\mathbf{T}_1\mathbf{z}=\mathbf{z} \quad \text{for } \mathbf{z} \text{ in range}(\mathbf{T}) \tag{6.81}$$

The operator $\mathbf{T}_1$ has a "Problem 1" structure; it is one-to-one. Then, $\mathbf{T}=\mathbf{T}_1\mathbf{T}_2$, and we seek a least-square solution to the equation $\mathbf{Tx}=\mathbf{T}_1\mathbf{T}_2\mathbf{x}=\mathbf{y}$.

Define $\mathbf{z}\stackrel{\Delta}{=}\mathbf{T}_2\mathbf{x}$. Then we seek a least-error solution to $\mathbf{T}_1\mathbf{z}=\mathbf{y}$, an equation in the form of Problem 1. The solution is $\mathbf{z}_a=\mathbf{T}_1^\dagger\mathbf{y}$, where $\mathbf{T}_1^\dagger$ has the explicit representation (6.68); $\mathbf{z}_a$ minimizes the error norm, $\|\mathbf{y}-\mathbf{T}_1\mathbf{z}\|_{\mathcal{W}}=\|\mathbf{y}-\mathbf{T}_1\mathbf{T}_2\mathbf{x}\|_{\mathcal{W}}$. The equation $\mathbf{T}_2\mathbf{x}=\mathbf{z}_a$ remains; it is in the form of Problem 2. The solution $\mathbf{x}_s=\mathbf{T}_2^\dagger\mathbf{z}_a$, where $\mathbf{T}_2$ is given in (6.69), minimizes

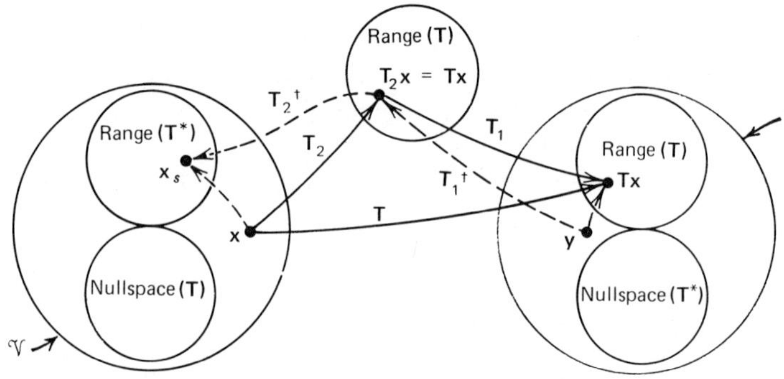

Figure 6.9. Decomposition of the pseudoinverse.

$\|\mathbf{x}\|_{\mathcal{V}}$. The smallest least-error solution to the equation $\mathbf{Tx}=\mathbf{y}$ is given by

$$\mathbf{x}_s=\mathbf{T}_2^\dagger\mathbf{z}_a=\mathbf{T}_2^\dagger\mathbf{T}_1^\dagger\mathbf{y}=\mathbf{T}^\dagger\mathbf{y}$$

Therefore $\mathbf{T}^\dagger$ can be expressed explicitly in the form

$$\begin{aligned}\mathbf{T}^\dagger &= (\mathbf{T}_1\mathbf{T}_2)^\dagger \\ &= \mathbf{T}_2^\dagger\mathbf{T}_1^\dagger \\ &= \mathbf{T}_2^*(\mathbf{T}_2\mathbf{T}_2^*)^{-1}(\mathbf{T}_1^*\mathbf{T}_1)^{-1}\mathbf{T}_1^* \end{aligned} \tag{6.82}$$

Generally speaking, $(\mathbf{TU})^\dagger \neq \mathbf{U}^\dagger\mathbf{T}^\dagger$. However, (6.82) holds for *any* factorization $\mathbf{T}=\mathbf{T}_1\mathbf{T}_2$ where $\mathbf{T}_1$ is one-to-one and $\mathbf{T}_2$ is onto; in particular, it holds for the factorization given in (6.80)–(6.81).* The factorization (6.82) is valuable primarily because it unifies all least-square problems and aids in their conceptualization. In addition, it can be used for computing the pseudoinverse. The ranges and nullspaces which must be determined in order to carry out the factorization would also have to be determined in order to compute solutions by orthogonal projection.

Matrix Factorization

As we suggested in the introduction to the decomposition of the pseudoinverse, we can determine a suitable factorization of a matrix $\mathbf{A}$ from the row-reduced form of $\mathbf{A}$. We demonstrate the technique by an example. However, the technique applies to any matrix. Suppose we perform the following row reduction:

$$(\mathbf{A} \,\vdots\, \mathbf{I}) \triangleq \begin{pmatrix} 1 & 2 & 3 & 2 & \vdots & 1 & 0 & 0 \\ 2 & 3 & 4 & 2 & \vdots & 0 & 1 & 0 \\ 3 & 5 & 7 & 4 & \vdots & 0 & 0 & 1 \end{pmatrix} \rightarrow$$

$$\begin{pmatrix} 1 & 0 & -1 & -2 & \vdots & -3 & 2 & 0 \\ 0 & 1 & 2 & 2 & \vdots & 2 & -1 & 0 \\ 0 & 0 & 0 & 0 & \vdots & -1 & -1 & 1 \end{pmatrix} \triangleq (\hat{\mathbf{A}} \,\vdots\, \mathbf{B})$$

The matrix $\mathbf{B}$ records the row operations which produce $\hat{\mathbf{A}}$; that is, $\mathbf{BA}=\hat{\mathbf{A}}$. Therefore $\mathbf{B}^{-1}$ tells how to reconstruct $\mathbf{A}$, $\mathbf{A}=\mathbf{B}^{-1}\hat{\mathbf{A}}$. In this instance,

$$\mathbf{B}^{-1}=\begin{pmatrix} 1 & 2 & 0 \\ 2 & 3 & 0 \\ 3 & 5 & 1 \end{pmatrix}$$

*Greville [6.8] describes necessary and sufficient conditions for the validity of $(\mathbf{TU})^\dagger=\mathbf{U}^\dagger\mathbf{T}^\dagger$.

Thus **A** can be factored as

$$\mathbf{A}=\begin{pmatrix}1&2&3&2\\2&3&4&2\\3&5&7&4\end{pmatrix}=\begin{pmatrix}1&2&0\\2&3&0\\3&5&1\end{pmatrix}\begin{pmatrix}1&0&-1&-2\\0&1&2&2\\0&0&0&0\end{pmatrix}$$

$$=\begin{pmatrix}1&2\\2&3\\3&5\end{pmatrix}\begin{pmatrix}1&0&-1&-2\\0&1&2&2\end{pmatrix}\triangleq\mathbf{A}_1\mathbf{A}_2 \tag{6.83}$$

The circled row and column can be dropped because the circled row is zero. The dropping of these unnecessary rows and columns produces matrices $\mathbf{A}_1$ and $\mathbf{A}_2$ which are of full rank. If we think of $\mathbf{A}_1$ and $\mathbf{A}_2$ as defining matrix transformations, $\mathbf{A}_2$ is onto and $\mathbf{A}_1$ is one-to-one.

It is perhaps surprising to note that we need not even record **B** or find $\mathbf{B}^{-1}$ in order to obtain the matrix factorization. All we need is the row reduction of **A**. The columns of $\mathbf{A}_1$ are also columns of **A**. The "standard basis" columns in the row-reduced matrix $\hat{\mathbf{A}}$ indicate which columns of **A** should appear in $\mathbf{A}_1$. We view the matrix product $\mathbf{A}=\mathbf{A}_1\mathbf{A}_2$ in the following way. Each column of $\mathbf{A}_2$ consists in a set of multipliers for the columns of $\mathbf{A}_1$. The multipliers in the jth column of $\mathbf{A}_2$ indicate which linear combination of the columns of $\mathbf{A}_1$ produces the jth column of **A**; the "basis" columns of $\mathbf{A}_2$ merely "place" columns of $\mathbf{A}_1$ in the appropriate column of **A**. Therefore, if the ith basis vector appears as the jth column of the row-reduced $\hat{\mathbf{A}}$, then the jth column of **A** appears as the ith column of $\mathbf{A}_1$.

Example 4. Matrix Factorization by Row Reduction. Consider the following row reduction:

$$\mathbf{A}=\begin{pmatrix}2&-2&-2&0&-4\\2&-2&1&6&5\\1&-1&1&4&4\end{pmatrix}\rightarrow\begin{pmatrix}1&-1&0&2&1\\0&0&1&2&3\\0&0&0&0&0\end{pmatrix}=\hat{\mathbf{A}}$$

The nonzero rows of $\hat{\mathbf{A}}$ constitute $\mathbf{A}_2$,

$$\mathbf{A}_2=\begin{pmatrix}1&-1&0&2&1\\0&0&1&2&3\end{pmatrix}$$
$$\uparrow\qquad\quad\uparrow$$

We pick $\mathbf{A}_1$ so that $\mathbf{A}_1\mathbf{A}_2=\mathbf{A}$. The basis columns of $\mathbf{A}_2$ indicate that the first and second columns of $\mathbf{A}_1$ are the first and third columns of **A**,

$$\mathbf{A}_1=\begin{pmatrix}2&-2\\2&1\\1&1\end{pmatrix}$$

The matrix factorization (6.83) is easily related to the transformation decomposition defined by (6.80) and (6.81). Suppose $\mathbf{T}$: $\mathfrak{M}^{4\times 1}\to\mathfrak{M}^{3\times 1}$ is given by $\mathbf{Tx}=\mathbf{Ax}$. Then if $\mathcal{E}_4$ and $\mathcal{E}_3$ are the standard bases for $\mathfrak{M}^{4\times 1}$ and $\mathfrak{M}^{3\times 1}$, respectively, $[\mathbf{T}]_{\mathcal{E}_4\mathcal{E}_3}=\mathbf{A}$. The number of nonzero rows of the row-reduced matrix $\hat{\mathbf{A}}$ equals the rank of $\mathbf{A}$. There is necessarily one basis column in $\mathbf{A}_2$ corresponding to each nonzero row of $\hat{\mathbf{A}}$. Therefore, the number of basis columns of $\mathbf{A}_2$ equals the number of independent columns of $\mathbf{A}$ [the dimension of range($\mathbf{T}$)]. Furthermore, the basis columns of $\mathbf{A}_2$ point to a basis for range($\mathbf{T}$), namely, the columns of $\mathbf{A}_1$. The other columns of $\mathbf{A}_2$ construct the remaining (dependent) columns of $\mathbf{A}$ from the "chosen" independent columns of $\mathbf{A}$.* Define $\mathbf{T}_2$: $\mathfrak{M}^{4\times 1}\to$range($\mathbf{T}$) and $\mathbf{T}_1$: range($\mathbf{T}$)$\to\mathfrak{M}^{3\times 1}$ by (6.80) and (6.81). Let $\mathcal{Y}\triangleq\{\mathbf{y}_1,\mathbf{y}_2\}$ denote the columns of $\mathbf{A}_1$, a basis for range($\mathbf{T}$). Then $[\mathbf{T}_2]_{\mathcal{E}_4\mathcal{Y}}=\mathbf{A}_2$ and $[\mathbf{T}_1]_{\mathcal{Y}\mathcal{E}_3}=\mathbf{A}_1$. A similar relationship holds for the factorization of an arbitrary matrix.

Exercise 1. Use (6.80), (6.81), and the technique (2.48) for determining the matrix of a transformation to verify that $[\mathbf{T}_2]_{\mathcal{E}_4\mathcal{Y}}=\mathbf{A}_2$ and $[\mathbf{T}_1]_{\mathcal{Y}\mathcal{E}_3}=\mathbf{A}_1$ for (6.83).

Since there are many different bases for the range of a matrix transformation, it should be clear from the above discussion that a matrix $\mathbf{A}$ has many different factorizations; there is one factorization for each basis. A second factorization for the matrix $\mathbf{A}$ of (6.83) is

$$\mathbf{A}=\begin{pmatrix}1&2&3&2\\2&3&4&2\\3&5&7&4\end{pmatrix}=\begin{pmatrix}1&0\\0&1\\1&1\end{pmatrix}\begin{pmatrix}1&2&3&2\\2&3&4&2\end{pmatrix}=\mathbf{B}_1\mathbf{B}_2 \qquad (6.84)$$

The factorization (6.84) can be obtained by row reducing $\mathbf{A}^{\mathrm{T}}$ (or column reducing $\mathbf{A}$). The columns of $\mathbf{B}_1$ [a basis for range($\mathbf{A}$)] are the nonzero rows of the row-reduced transposed matrix. The rows of $\mathbf{B}_2$ are independent rows of $\mathbf{A}$.

After having obtained a factorization of a matrix $\mathbf{A}$ into a product $\mathbf{A}_1\mathbf{A}_2$ with the columns of $\mathbf{A}_1$ and the rows of $\mathbf{A}_2$ independent, we can employ (6.82) to compute the pseudoinverse of $\mathbf{A}$:

$$\begin{aligned}\mathbf{A}^{\dagger}&=\mathbf{A}_2^{\mathrm{T}}(\mathbf{A}_2\mathbf{A}_2^{\mathrm{T}})^{-1}(\mathbf{A}_1^{\mathrm{T}}\mathbf{A}_1)^{-1}\mathbf{A}_1^{\mathrm{T}}\\&=\mathbf{A}_2^{\mathrm{T}}(\mathbf{A}_1^{\mathrm{T}}\mathbf{A}\mathbf{A}_2^{\mathrm{T}})^{-1}\mathbf{A}_1^{\mathrm{T}}\end{aligned} \qquad (6.85)$$

*Frame [6.5] refers to the independent columns of **A** which have been selected as a basis for range(**T**) as **distinguished columns of A**.

Note that we need perform only one matrix inversion in the simplification (6.85). If $\mathbf{Tx} \overset{\Delta}{=} \mathbf{Ax} = \mathbf{y}$, it follows that $\mathbf{x}_s = \mathbf{T}^\dagger\mathbf{y} = \mathbf{A}^\dagger\mathbf{y}$ is the smallest least-error solution.

Example 5. Computation of $\mathbf{A}^\dagger$. Let $\mathbf{A}$ be the matrix of (6.83). Then we compute (6.85) as follows:

$$(\mathbf{A}_1^T\mathbf{A}\mathbf{A}_2^T) = \begin{pmatrix} -54 & 123 \\ -90 & 204 \end{pmatrix}, \quad (\mathbf{A}_1^T\mathbf{A}\mathbf{A}_2^T)^{-1} = \frac{1}{18}\begin{pmatrix} 68 & -41 \\ 30 & -18 \end{pmatrix}$$

$$\mathbf{A}^\dagger = \mathbf{A}_2^T(\mathbf{A}_1^T\mathbf{A}\mathbf{A}_2^T)^{-1}\mathbf{A}_1^T = \frac{1}{18}\begin{pmatrix} -14 & 13 & -1 \\ -6 & 6 & 0 \\ 2 & -1 & 1 \\ 16 & -14 & 2 \end{pmatrix}$$

The algebraic definition of $\mathbf{A}^\dagger$ given in P&C 6.22 can be used as a numerical check on the accuracy of the computations.

6.6 Practical Computation of Least-Square Solutions

The data used in numerical computations often include empirical error. Even if the data are exact, computer manipulations quickly introduce small roundoff errors. If we subtract two nearly equal quantities, the relative error in the difference is far greater than the relative error in the quantities before subtraction. A sequence of computations can magnify and propagate errors to such an extent that the results are meaningless.

Suppose we apply the Gram–Schmidt orthogonalization procedure to the pair of vectors $\mathbf{x}_1$ and $\mathbf{x}_2$ of Figure 6.10. We obtain

$$\mathbf{z}_1 = \mathbf{x}_1, \qquad \mathbf{z}_2 = \mathbf{x}_2 - \frac{\langle \mathbf{x}_2, \mathbf{z}_1 \rangle}{\|\mathbf{z}_1\|^2}\mathbf{z}_1 \tag{6.86}$$

Because $\mathbf{z}_2$ results from the subtraction of two nearly equal vectors, a numerical computation of $\mathbf{z}_2$ is likely to be significantly in error (in both

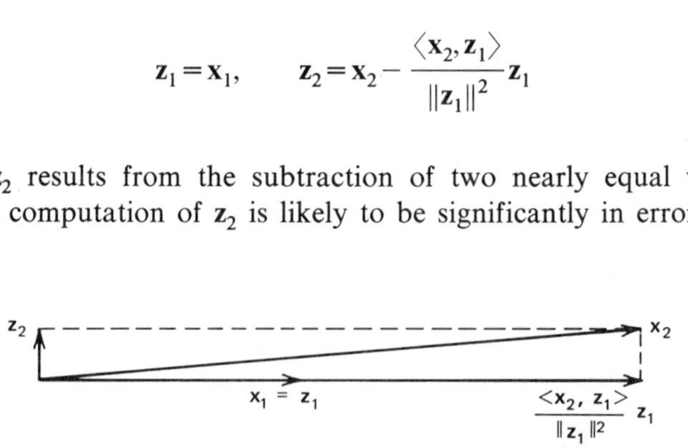

Figure 6.10. Orthogonalization of nearly dependent vectors.

magnitude and direction). The application of the Gram–Schmidt procedure to a nearly dependent set of vectors invariably results in such a subtraction of nearly equal quantities. Near dependency is common in practical least-square problems. Yet Gram–Schmidt orthogonalization is the primary computational tool for such problems. What can we do?

If the row vectors of a matrix $\mathbf{A}$ are nearly dependent, then elimination algorithms for solving the equation $\mathbf{Ax}=\mathbf{y}$ encounter the same magnification and propagation of errors that we described above. The standard way of handling ill-conditioned sets of equations, as discussed in Section 1.5, is to use scaling, pivoting for size, and double-precision arithmetic. By rearranging the equations, we are able to perform the necessary divisions using the largest available data elements at each stage; the small elements are likely to have arisen from the subtraction of nearly equal quantities at an earlier stage of the elimination.

The pivoting concept can also be used to minimize the propagation of errors in the Gram–Schmidt procedure. The orthogonalization consists in making a sequence of one-dimensional orthogonal projections and subtracting vectors (Figure 6.10). At each stage, a vector of small norm (e.g., $\mathbf{z}_2$ above) is likely to be the inaccurate result of a previous subtraction. We avoid using these vectors in the orthogonalization until near the end of the procedure. Osborne [6.14] recommends the following sequence of steps for orthogonalizing a set $\{\mathbf{x}_1,\ldots,\mathbf{x}_n\}$. Begin with the vector of largest norm (the pivot vector), say, $\mathbf{x}_p$; let $\mathbf{z}_1=\mathbf{x}_p$, and make all other vectors orthogonal to $\mathbf{z}$ by the subtractions

$$\mathbf{x}_i - \frac{\langle \mathbf{x}_i, \mathbf{z}_1\rangle}{\|\mathbf{z}_1\|^2}\mathbf{z}_1, \qquad i\neq p \tag{6.87}$$

From the now modified vectors $\{\hat{\mathbf{x}}_i, i\neq p\}$, pick the vector of largest norm (the second pivot vector), say, $\hat{\mathbf{x}}_q$; let $\mathbf{z}_2=\hat{\mathbf{x}}_q$, and make all other vectors orthogonal to $\mathbf{z}_2$ by the subtractions

$$\hat{\mathbf{x}}_i - \frac{\langle \hat{\mathbf{x}}_i, \mathbf{z}_2\rangle}{\|\mathbf{z}_2\|^2}\mathbf{z}_2, \qquad i\neq p,q \tag{6.88}$$

The procedure is continued until we have the orthogonal set $\{\mathbf{z}_1,\ldots,\mathbf{z}_n\}$. If at the $(m+1)$st stage the search for $\mathbf{z}_{m+1}$ finds no vectors whose norms are above a given threshold, assume the remaining vectors are zero, and take $\{\mathbf{z}_1,\ldots,\mathbf{z}_m\}$ as an orthogonal basis for span$\{\mathbf{x}_1,\ldots,\mathbf{x}_n\}$. Any computer algorithm for orthogonalizing vectors should use some form of pivoting in order to maintain as much accuracy as is possible; even with careful computing, significant error will result for nearly dependent sets of vectors.

The Condition Number

A computational problem is said to be **ill-conditioned** if the values to be computed are sensitive functions of the data; a **condition number** is an estimate of the sensitivity of the computed values to the data. Suppose $\mathbf{A}$ is an *invertible* square matrix, and we wish to solve $\mathbf{Ax}=\mathbf{y}$. We scale the equation so that the magnitude of the largest element of $\mathbf{A}$ is approximately one. Then the magnitude of the largest element of $\mathbf{A}^{-1}$ is a condition number c for the problem. If we are computing $\mathbf{A}^{-1}$, there is at least one element of $\mathbf{A}^{-1}$ which has sensitivity c to data elements of $\mathbf{A}$; if we seek a solution $\mathbf{x}$, there is at least one element of $\mathbf{x}$ which has sensitivity c to data elements of $\mathbf{y}$. As an example, let $\mathbf{A}=\mathbf{\Lambda}$, a diagonal matrix which has its eigenvalues on the diagonal. Let λ_L and λ_s be the eigenvalues of $\mathbf{\Lambda}$ which are of largest and smallest magnitude, respectively. The scaled matrix $\mathbf{\Lambda}_s \triangleq |1/\lambda_L|\mathbf{\Lambda}$ has as the magnitude of its largest element, $|\lambda_L/\lambda_L| = 1$. The eigenvalues of $\mathbf{\Lambda}_s^{-1}$ are the inverses of the eigenvalues of $\mathbf{\Lambda}_s$. Consequently, the matrix $\mathbf{\Lambda}_s^{-1}$ has as the magnitude of its largest element the number $|\lambda_L/\lambda_S|$. Therefore, a condition number of the matrix $\mathbf{\Lambda}$ is $c=|\lambda_L/\lambda_S|$. It can be shown that for any symmetric matrix the ratio of magnitudes of the largest and smallest eigenvalues is a condition number for the matrix. A slightly modified result applies for nonsymmetric matrices.*

In least-square problems the matrix $\mathbf{A}$ is not invertible and not necessarily square. Thus $\mathbf{A}^{-1}$ cannot be used to determine a condition number for the problem. However, it can be shown that $c=\sqrt{\lambda_L}\,/\sqrt{\lambda_s}$ serves as a condition number, where λ_L and λ_s are the eigenvalues of largest and smallest magnitude, respectively, for the smaller (in dimension) of the matrices $\mathbf{A}^T\mathbf{A}$ and $\mathbf{AA}^T$. (The numbers $\sqrt{\lambda_i}$ are called *singular values* of $\mathbf{A}$.) For any matrix equation, square or not, a high condition number is equivalent to near dependency of the row (and column) vectors of the matrix. Such dependencies are common in least-square problems.

Exercise 1. The vectors $\mathbf{x}_1=(1\ 1)^T$ and $\mathbf{x}_2=(1\ a)^T$ are nearly dependent if $a\cong 1$. Explore the ratio $|\lambda_L/\lambda_S|$ as a function of a for the matrix $(\mathbf{x}_1 \vdots \mathbf{x}_2)$.

Nearly Degenerate Equations

Suppose we wish to solve the equation

$$\mathbf{Ax}=\begin{pmatrix}1 & 0 & 0\\ 0 & \frac{1}{2} & 0\\ 0 & 0 & \epsilon\end{pmatrix}\begin{pmatrix}\xi_1\\ \xi_2\\ \xi_3\end{pmatrix}=\begin{pmatrix}1\\ 1\\ 1\end{pmatrix}=\mathbf{y} \tag{6.89}$$

*See Wilkinson [6.24] for a thorough discussion of condition numbers for matrices.

where the number ϵ is very small. Perhaps ϵ is nonzero only because of empirical error in the data. The matrix **A** is invertible, its condition number is $c=|1/\epsilon|$, and the solution to the equation is $\mathbf{x}=(1\quad 2\quad 1/\epsilon)^{\mathrm{T}}$. The solution is completely dominated by the unsure element $\xi_3=1/\epsilon$. The least we should do is determine the near- nullspace of **A**, the eigenspace for the eigenvalue $\lambda=\epsilon$.* This is the space of uncertainty or arbitrariness in the solution.

An alternative approach for (6.89) is to assume ϵ should be zero, in which event the equations exhibit both incompatibility and nonuniqueness. It is reasonable, then, to seek a smallest least-error solution. The pseudoinverse of **A**, assuming $\epsilon=0$, is

$$\mathbf{A}^{\dagger}=\begin{pmatrix}1 & 0 & 0\\ 0 & 2 & 0\\ 0 & 0 & 0\end{pmatrix}$$

Using the pseudoinverse, the solution to (6.89) is $\mathbf{x}=(1\ 2\ 0)^{\mathrm{T}}$. Note that this solution differs from the previous solution only in that the unsure element has been set to a convenient value, $\xi_3=0$.

A *complete* solution to (6.89) probably consists in a combination of the two approaches described above, a smallest least-error solution together with a description of the near nullspace and the measure $c=|1/\epsilon|$ of its sensitivity. The first point to be made is that blind acceptance of a straightforward elimination solution is foolish if the equations are ill-conditioned. The second point is that we can use the least-square concept to remove the arbitrariness in a nearly degenerate equation.

Treated as a least-square problem, the condition number of a nearly degenerate matrix **A** is $\sqrt{\lambda_L}\,/\sqrt{\lambda_S}$ where λ_L is the largest eigenvalue of $\mathbf{A}^{\mathrm{T}}\mathbf{A}$ (or $\mathbf{A}\mathbf{A}^{\mathrm{T}}$) and λ_S is the smallest of the eigenvalues *excluding those considered small enough to be zero*. Golub and Kahan [6.7] describe a technique for eliminating near degeneracy in a nonsquare matrix by "diagonalizing" the matrix and setting to zero its nearly zero singular values. They then find the pseudoinverse of the "improved" matrix.

Condition Number for the Normal Equations

It is apparent from the above discussion of condition numbers that the condition number for the matrix $\mathbf{A}^{\mathrm{T}}\mathbf{A}$ (or $\mathbf{A}\mathbf{A}^{\mathrm{T}}$) is c^2, the square of the condition number of **A**. If **A** is nearly degenerate, then $\mathbf{A}^{\mathrm{T}}\mathbf{A}$ (or $\mathbf{A}\mathbf{A}^{\mathrm{T}}$) is considerably more so. Consequently, the normal equations $\mathbf{A}^{\mathrm{T}}\mathbf{A}\mathbf{x}=\mathbf{A}^{\mathrm{T}}\mathbf{y}$ of Problem 1 and the Gram equations $\mathbf{A}\mathbf{A}^{\mathrm{T}}\boldsymbol{\lambda}=\mathbf{y}$ of Problem 2 provide poor

*See Section 2.4.

computational approaches to obtaining least-square solutions for ill-conditioned problems. The orthogonal projection approach, which attacks the original equations, is more accurate.

The difficulty that can occur with the normal equations is particularly apparent in polynomial linear regression (Section 6.1). On the surface, the matching (in a least-square sense) of a set of data at equally spaced points by means of a polynomial of the form

$$\mathbf{f}_a(s)=\xi_1+\xi_2 s+\cdots+\xi_n s^{n-1} \tag{6.90}$$

seems straightforward. However, for $n=10$, the condition number for the normal equations is approximately 3×10^{12} (see Forsythe [6.4]); an error of 10^{-10} in one of the data points can lead to errors of size 300 in the computed least-error coefficients ξ_i. With most computers, a sixth degree polynomial ($n=7$) is about the highest that can be handled accurately using the normal equations.

The solution to the polynomial regression problem can be computed more accurately by orthogonal projection; in the process we would orthogonalize the polynomials $\{1,t,\ldots,t^{n-1}\}$ in a discrete sense (over the data points). A cheaper approach is to *start* with the model

$$\mathbf{f}_a(s)=\xi_1\mathbf{p}_0(s)+\xi_2\mathbf{p}_1(s)+\cdots+\xi_n\mathbf{p}_{n-1}(s) \tag{6.91}$$

where $\mathbf{p}_k$ is a kth degree polynomial, and the set $\{\mathbf{p}_k\}$ is orthogonal over the data points. The normal equations associated with the regression model (6.91) are diagonal and easily solved. The orthogonal polynomials $\{\mathbf{p}_k\}$ can be computed by means of a recurrence relationship. Furthermore, after computing the least-error coefficients $\{\xi_k\}$, we can increase the order of the polynomial and compute the added coefficients without recomputing the old coefficients. The coefficients are computationally independent. See Ralston [6.17] for a complete discussion of the technique.

A Projection Method for Matrix Pseudoinverses

Numerous techniques have been proposed for computing the pseudoinverse of an arbitrary matrix **A**. Decell [6.2] has proposed an algorithm which is based on the Cayley–Hamilton theorem (P&C 6.21). Stallings and Boullion [6.22] describe a technique for minimizing the effect of roundoff with Decell's method. An infinite iteration method has been developed by Tanabe [6.23]. We mentioned earlier a method of Golub and Kahan [6.7] which is based on the singular values of **A**. We explore here a method which is based on Gram–Schmidt orthogonalization. We assume that a computer algorithm based on this method would use some form of pivoting in performing the Gram–Schmidt orthogonalization.

We introduce the method by means of an example. We wish to solve for the smallest least-error solution $\mathbf{x}_s$ to the equation $\mathbf{Ax}=\mathbf{y}$ (using the standard inner product), where

$$\mathbf{A}=\begin{pmatrix}2&0&1\\0&1&1\\2&1&2\end{pmatrix},\qquad \mathbf{y}=\begin{pmatrix}1\\0\\0\end{pmatrix} \tag{6.92}$$

The matrix $\mathbf{A}$ has rank two. The basic idea is to project $\mathbf{y}$ onto range($\mathbf{A}$) to obtain $\mathbf{y}_a$, solve the equation $\mathbf{Ax}=\mathbf{y}_a$, then project $\mathbf{x}$ onto range($\mathbf{A}^{\mathrm{T}}$) to obtain $\mathbf{x}_s$. We need an orthogonal basis for range($\mathbf{A}$). Since the columns of $\mathbf{A}$ span range($\mathbf{A}$), we obtain the orthogonal basis by applying the Gram–Schmidt procedure to these columns, $\mathbf{A}^{(1)}$, $\mathbf{A}^{(2)}$, and $\mathbf{A}^{(3)}$, obtaining the vectors $\mathbf{z}_1,\mathbf{z}_2,\mathbf{z}_3$:

$$\mathbf{z}_1 \triangleq \mathbf{A}^{(1)}=\begin{pmatrix}2\\0\\2\end{pmatrix}$$

$$\mathbf{z}_2 \triangleq \mathbf{A}^{(2)}-\frac{\langle \mathbf{A}^{(2)},\mathbf{z}_1\rangle}{\|\mathbf{z}_1\|^2}\mathbf{z}_1=\begin{pmatrix}0\\1\\1\end{pmatrix}-\tfrac{1}{4}\begin{pmatrix}2\\0\\2\end{pmatrix}=\begin{pmatrix}-\frac{1}{2}\\1\\\frac{1}{2}\end{pmatrix}$$

$$\mathbf{z}_3=\mathbf{A}^{(3)}-\frac{\langle \mathbf{A}^{(3)},\mathbf{z}_1\rangle}{\|\mathbf{z}_1\|^2}\mathbf{z}_1-\frac{\langle \mathbf{A}^{(3)},\mathbf{z}_2\rangle}{\|\mathbf{z}_2\|^2}\mathbf{z}_2$$

$$=\begin{pmatrix}1\\1\\2\end{pmatrix}-\tfrac{3}{4}\begin{pmatrix}2\\0\\2\end{pmatrix}-1\begin{pmatrix}-\frac{1}{2}\\1\\\frac{1}{2}\end{pmatrix}=\begin{pmatrix}0\\0\\0\end{pmatrix}$$

The vectors $\mathbf{z}_1$ and $\mathbf{z}_2$ form an orthogonal basis for range($\mathbf{A}$). In order to keep a record of the operations on the columns of $\mathbf{A}$, we perform the same operations on an identity matrix [this idea is somewhat like the row reduction of a matrix $(\mathbf{A} \,\vdots\, \mathbf{I})$ to obtain $(\mathbf{I} \,\vdots\, \mathbf{A}^{-1})$]:

$$\begin{pmatrix}\mathbf{A}\\\cdots\\\mathbf{I}\end{pmatrix}=\begin{pmatrix}2&0&1\\0&1&1\\2&1&2\\\cdots&\cdots&\cdots\\1&0&0\\0&1&0\\0&0&1\end{pmatrix}\xrightarrow{\text{Gram–Schmidt}}\begin{pmatrix}2&-\frac{1}{2}&0\\0&1&0\\2&\frac{1}{2}&0\\\cdots&\cdots&\cdots\\1&-\frac{1}{4}&-\frac{1}{2}\\0&1&-1\\0&0&1\end{pmatrix}\triangleq\begin{pmatrix}\mathbf{B}\\\cdots\\\mathbf{Q}\end{pmatrix} \tag{6.93}$$

Because $\mathbf{Q}$ records the operations performed on $\mathbf{A}$ to obtain $\mathbf{B}$, $\mathbf{AQ}=\mathbf{B}$. (Right multiplication by $\mathbf{Q}$ corresponds to operations on the columns of $\mathbf{A}$.) Each column of $\mathbf{Q}$ is a solution to $\mathbf{Ax}=\mathbf{y}$ for $\mathbf{y}$ equal to the corresponding column of $\mathbf{B}$. Consequently, the last column of $\mathbf{Q}$, the one corresponding to the zero column of $\mathbf{B}$, is a basis for nullspace($\mathbf{A}$). Denote that column of $\mathbf{Q}$ by $\mathbf{x}_1$. If there were more vectors $\{\mathbf{x}_1, \mathbf{x}_2, \ldots, \mathbf{x}_m\}$ in the basis for nullspace($\mathbf{A}$), we would orthogonalize them; that is, we would perform the following orthogonalization:

$$\begin{pmatrix} \mathbf{A} \\ \cdots \\ \mathbf{I} \end{pmatrix} \overset{\text{Gram–Schmidt}}{\longrightarrow} \begin{pmatrix} \mathbf{z}_1 \cdots \mathbf{z}_m & \vdots & \boldsymbol{\theta} \cdots \boldsymbol{\theta} \\ \cdots\cdots & & \cdots\cdots \\ \mathbf{y}_1 \cdots \mathbf{y}_m & \vdots & \mathbf{x}_1 \cdots \mathbf{x}_\rho \end{pmatrix} = \begin{pmatrix} \mathbf{B} \\ \cdots \\ \mathbf{Q} \end{pmatrix} \tag{6.94}$$

Continuing the example of (6.92)–(6.93), we denote by $\mathbf{w}$ the vector

$$\mathbf{w} = \begin{pmatrix} \dfrac{\langle \mathbf{y}, \mathbf{z}_1 \rangle}{\|\mathbf{z}_1\|^2} & \dfrac{\langle \mathbf{y}, \mathbf{z}_2 \rangle}{\|\mathbf{z}_2\|^2} & 0 \end{pmatrix}^{\mathrm{T}} \tag{6.95}$$

Then the orthogonal projection of $\mathbf{y}$ onto range($\mathbf{A}$) becomes

$$\mathbf{y}_a = \frac{\langle \mathbf{y}, \mathbf{z}_1 \rangle}{\|\mathbf{z}_1\|^2} \mathbf{z}_1 + \frac{\langle \mathbf{y}, \mathbf{z}_2 \rangle}{\|\mathbf{z}_2\|^2} \mathbf{z}_2 = \mathbf{Bw} = \mathbf{AQw} \tag{6.96}$$

The vector $\mathbf{w}$ is easy to compute, and a solution to $\mathbf{Ax}=\mathbf{y}_a$ is clearly $\mathbf{x}=\mathbf{Qw}$, a simple multiplication; thus

$$\mathbf{w} = \begin{pmatrix} \frac{1}{4} \\ -\frac{1}{3} \\ 0 \end{pmatrix}, \qquad \mathbf{x} = \begin{pmatrix} \frac{1}{3} \\ -\frac{1}{3} \\ 0 \end{pmatrix} \tag{6.97}$$

(If we had used a number other than zero for the last element of $\mathbf{w}$, $\mathbf{x}$ would change only by a vector in nullspace($\mathbf{A}$). The result would still be a solution to $\mathbf{Ax}=\mathbf{y}_a$.)

Only the orthogonal projection of $\mathbf{x}$ onto range($\mathbf{A}^{\mathrm{T}}$) remains to be carried out. We have in the right side of $\mathbf{Q}$ an orthogonal basis $\{\mathbf{x}_1\}$ for nullspace($\mathbf{A}$), the orthogonal complement of range($\mathbf{A}^{\mathrm{T}}$). As a consequence,

$$\mathbf{x}_s = \mathbf{x} - \frac{\langle \mathbf{x}, \mathbf{x}_1 \rangle}{\|\mathbf{x}_1\|^2} \mathbf{x}_1 = \begin{pmatrix} \frac{1}{3} \\ -\frac{1}{3} \\ 0 \end{pmatrix} - \frac{2}{27} \begin{pmatrix} -\frac{1}{2} \\ -1 \\ 1 \end{pmatrix} = \begin{pmatrix} \frac{10}{27} \\ -\frac{7}{27} \\ -\frac{2}{27} \end{pmatrix} \tag{6.98}$$

Equation (6.98) is the smallest least-error solution to the equation $\mathbf{Ax}=\mathbf{y}$ for the data of (6.92).

The computational procedure of (6.93)–(6.98) is easily automated. It is apparent that the technique can be used to find a smallest least-error solution for any $m\times n$ matrix equation. If we desire the pseudoinverse explicitly, we carry out the operations for a sequence of right-hand sides, $\boldsymbol{\varepsilon}_1,\boldsymbol{\varepsilon}_2,\ldots,\boldsymbol{\varepsilon}_m$, where $\boldsymbol{\varepsilon}_i$ is the ith standard basis vector for $\mathfrak{M}^{m\times 1}$. The corresponding solutions are columns of $\mathbf{A}^{\dagger}$. [Equation (6.98) is the first column of $\mathbf{A}^{\dagger}$ for the data of (6.92).] Note that the Gram–Schmidt orthogonalization (6.94) has to be performed only once. The remaining operations are simple.

Exercise 2. Show that for the general orthogonalization (6.94),

$$\mathbf{w}=\mathbf{Zy} \qquad \text{and} \qquad \mathbf{x}_s=\mathbf{Px} \tag{6.99}$$

where

$$\mathbf{P}=\mathbf{I}-\sum_{i=1}^{\rho}\frac{\mathbf{x}_i\mathbf{x}_i^{\mathrm{T}}}{\|\mathbf{x}_i\|^2}$$

and

$$\mathbf{Z}=\left(\frac{\mathbf{z}_1}{\|\mathbf{z}_1\|^2}\cdots\frac{\mathbf{z}_m}{\|\mathbf{z}_m\|^2}\quad\boldsymbol{\theta}\cdots\boldsymbol{\theta}\right)^{\mathrm{T}}$$

Consequently, the smallest least-error solution to $\mathbf{Ax}=\mathbf{y}$ is

$$\mathbf{x}_s=\mathbf{PQZy} \tag{6.100}$$

and $\mathbf{A}^{\dagger}$ can be written explicitly as

$$\mathbf{A}^{\dagger}=\mathbf{PQZ} \tag{6.101}$$

6.7 Problems and Comments

6.1 Find a solution to the following set of equations which is best in the least-square sense:

$$\begin{aligned}\xi_1+\xi_2&=1\\ \xi_1&=1\\ \xi_2&=1\end{aligned}$$

Sketch the equations and the least-square solution in the (ξ_1, ξ_2) plane.

6.2 Let $\mathcal{W}$ be the one-dimensional subspace of $\mathcal{R}^n$ spanned by $\mathbf{x}_1 = (1, 1, \ldots, 1)$, the vector whose entries are all 1. Let $\mathbf{x} = (\xi_1, \xi_2, \ldots, \xi_n)$ be an arbitrary vector in $\mathcal{R}^n$. Find the vector $\mathbf{x}_a$ in $\mathcal{W}$ which is closest to $\mathbf{x}$ in the least-square sense. Note that the vector $\mathbf{x}$ in $\mathcal{R}^n$ can be considered as a set of n measurements of a quantity ξ. We can think of $\mathcal{W}$ as the subspace of perfect results (where all measurements are the same). Thus $\mathbf{x}_a$ is that perfect result which is nearest to the actual set of measurements.

6.3 *Multivariable linear regression*: a planner has developed the following manpower model of a government agency:

$$\mathbf{f}(\xi_1, \xi_2) = b_0 + b_1\xi_1 + b_2\xi_2$$

The quantity $\mathbf{f}(\xi_1, \xi_2)$ is the total manpower in the agency, b_0 is the manpower required for overhead (payroll, building maintenance, planning, etc.), ξ_1 is the number of grant applications processed by the agency in a year, b_1 is the manpower (man-years) required to process one application, ξ_2 is the number of technical consultations provided by the agency in a year, and b_2 is the manpower required to provide one consultation. The quantities b_0, b_1, and b_2 are measures of the productivity of the agency. The following historical data is available from the agency:

Year	Employees	Applications processed	Consultations provided
1	93	5804	1750
2	100	6250	1878
3	102	6200	1852
4	105	6225	1860

(*a*) Indicate how to determine those values b_0, b_1, and b_2 for which the model best fits this data in a least-square sense.
(*b*) Compute the solution by means of a computer program.

6.4 Let $\mathbf{A}$ be a real $n \times n$ matrix in the space $\mathcal{M}^{n \times n}$. Assume the standard inner product, $\langle \mathbf{A}, \mathbf{B} \rangle \triangleq \text{trace}(\mathbf{B}^T\mathbf{A})$, where trace($\mathbf{C}$) is the sum of the diagonal elements of $\mathbf{C}$. Find the "constant matrix," $c\mathbf{I}$, which best approximates $\mathbf{A}$ in the least-square sense.

6.5 Find a least-square error solution (relative to the standard inner product) for the following set of equations

$$2\xi_1+3\xi_2=1$$

$$-\xi_1+\xi_2=2$$

$$4\xi_1+6\xi_2=2$$

(*a*) By means of orthogonal projection (the Gram–Schmidt procedure).

(*b*) By means of the normal equations.

6.6 Define $\mathbf{T}: \mathcal{R}^2 \to \mathcal{C}(0,1)$ by

$$[\mathbf{T}(\xi_1,\xi_2)](t) \triangleq (\xi_2-\xi_1)+\xi_2 t$$

Use orthogonal projection to find the least-error solution to the equation $\xi_2-\xi_1+\xi_2 t=e^t$, assuming the standard inner product for $\mathcal{C}(0,1)$. That is, pick ξ_1 and ξ_2 to minimize $\int_0^1[e^t-(\xi_2-\xi_1)-\xi_2 t]^2\,dt$.

6.7 A system puts out a sequence of electrical pulses of magnitude $\mathbf{x}_i$, where $0 \leqslant \mathbf{x}_i \leqslant 2$. The probability density for these pulses is unknown. However, observation of past pulses has shown that average values of $\mathbf{x}_i$, $\mathbf{x}_i^2$, $\mathbf{x}_i\mathbf{x}_{i-1}$, and $\mathbf{x}_i\mathbf{x}_{i-2}$ are 1, 9/4, 7/4, and 5/4, respectively. Estimate the future pulse height $\mathbf{x}_n$ by a "linear" function of the two previous pulse heights, $\mathbf{x}_{na} \triangleq \xi_0\mathbf{1}+\xi_1\mathbf{x}_{n-1}+\xi_2\mathbf{x}_{n-2}$. Pick the multipliers ξ_0, ξ_1, and ξ_2 to minimize the expected mean-square error, $\mathbf{E}[(\mathbf{x}_n-\mathbf{x}_{na})^2]$. (Note that ξ_0 multiplies a random variable whose sample values are always 1.)

*6.8 *Galerkin's method*: suppose we wish to solve the equation $\mathbf{Tx}=\mathbf{y}$, where $\mathbf{T}: \mathcal{V} \to \mathcal{W}$ is a bounded linear transformation between two Hilbert spaces. If the equation is difficult to solve exactly, we may prefer to approximate the solution by a finite linear combination of known vectors from $\mathcal{V}$; that is, we let $\mathbf{x}=c_1\mathbf{v}_1+\cdots+c_n\mathbf{v}_n$, and we choose the coefficients $\{c_i\}$ so that $\mathbf{T}(\Sigma_i c_i\mathbf{v}_i)=\Sigma_i c_i\mathbf{T}\mathbf{v}_i \approx \mathbf{y}$. There are various criteria for making the two sides of the "equation" nearly equal. Suppose $\{\mathbf{w}_1,\ldots,\mathbf{w}_n\}$ is a set of independent "weighting" functions. **Galerkin's method** is to choose the $\{c_i\}$ so that the orthogonal projections of the two sides of the equation onto span$\{\mathbf{w}_j\}$ are identical. The *Galerkin equations* are $\langle \mathbf{w}_j, \Sigma_i c_i\mathbf{T}\mathbf{v}_i\rangle=\Sigma_i c_i\langle \mathbf{w}_j,\mathbf{T}\mathbf{v}_i\rangle=\langle \mathbf{w}_j,\mathbf{y}\rangle$, $j=1,\ldots,n$, where the inner product is the one associated with $\mathcal{W}$.

(*a*) Determine the choice of weighting functions $\{\mathbf{w}_j\}$ for which the approximation is best in a least-square sense. That is, pick $\{\mathbf{w}_j\}$ so that $\|\mathbf{y}-\Sigma_i c_i\mathbf{T}\mathbf{v}_i\|^2$ is minimized.

(*b*) A suitable alternative to the least-square choice of coefficients considered in (*a*) is the set of coefficients determined by the weighting functions $\mathbf{w}_j = \mathbf{v}_j$, $j = 1, \ldots, n$. Use this set of weighting functions to approximate the solution to the nonlinear integral equation

$$(\mathbf{Tf})(t) \triangleq \int_0^1 (t-s)\mathbf{f}^2(s)\,ds = t^3 - \frac{t^2}{2} - \frac{t}{2} + \frac{1}{4} \triangleq \mathbf{g}(t)$$

Let $\mathbf{f}(t) \approx c_1 + c_2 t$. Use the standard inner product for $\mathcal{L}_2(0,1)$.

6.9 Find the minimum norm solution to the following set of equations:

$$\xi_1 + 2\xi_2 + 3\xi_3 = 1$$

$$2\xi_1 + \xi_3 = 1$$

6.10 Find the continuous function $\mathbf{f}$ for which $\int_0^1 \mathbf{f}(t)\,dt = 1$ and $\int_0^1 \mathbf{f}^2(t)\,dt$ is minimum.

6.11 Let $\ddot{\mathbf{f}} = \mathbf{u}$, $\mathbf{f}(0) = 0$ describe the one-dimensional control of a frictionless unit mass by means of a force $\mathbf{u}$.

(*a*) Express this equation in the state-space form

$$\dot{\mathbf{x}} = \mathbf{Ax} + \mathbf{Bu}, \qquad \mathbf{x}(0) = \boldsymbol{\theta}$$

(*b*) According to (3.79), the solution to the state equation can be expressed in the inverse form:

$$\mathbf{x}(t) = e^{\mathbf{A}t}\mathbf{x}(0) + \int_0^t e^{\mathbf{A}(t-s)}\mathbf{Bu}(s)\,ds$$

Evaluate the expression at $t = 1$, and rewrite it as a pair of hyperplanes (in the form $\langle \mathbf{u}, \mathbf{g}_i \rangle = c_i$) in the Hilbert space $\mathcal{L}_2(0,1)$.

(*c*) Use the Gram equations to determine the input that will drive the system to the state $\mathbf{f}(1) = 1$, $\dot{\mathbf{f}}(1) = 0$ with minimum energy $\int_0^1 \mathbf{u}^2(t)\,dt$.

6.12 Define a weighted inner product on $\mathcal{L}_2(0,1)$ by $\langle \mathbf{f}, \mathbf{g} \rangle = \int_0^1 e^t \mathbf{f}(t)\,\mathbf{g}(t)\,dt$. Using this inner product to define the norm, find the control voltage $\mathbf{u}_s$ of smallest norm which will drive the loaded dc motor of (6.32) from the state $\phi(0) = \dot{\phi}(0) = 0$ to the state $\phi(1) = 1$, $\dot{\phi}(1) = 0$. Compare this control to (6.34), the minimum energy control. What is the effect of adding the weighting function to the inner product?

6.13 Use a pseudoinverse matrix to find the least-square error solution (relative to the standard inner product) for the following set of equations:

$$\xi_1+2\xi_2=3$$

$$2\xi_1+\xi_2=3$$

$$4\xi_1+5\xi_2=10$$

6.14 Let $\mathbf{A}$ be a real $m\times n$ matrix with n independent columns ($m\geqslant n$). Let the inner products for $\mathfrak{M}^{n\times 1}$ and $\mathfrak{M}^{m\times 1}$, respectively, be $\langle \mathbf{x},\mathbf{z}\rangle_Q \overset{\Delta}{=} \mathbf{z}^{\mathrm{T}}\mathbf{Q}\mathbf{x}$ and $\langle \mathbf{y},\mathbf{w}\rangle_R \overset{\Delta}{=} \mathbf{w}^{\mathrm{T}}\mathbf{R}\mathbf{y}$, where $\mathbf{Q}$ and $\mathbf{R}$ are real, symmetric, positive-definite matrices.

(*a*) Find the pseudoinverse of $\mathbf{A}$ relative to the given inner products. Show that $\mathbf{A}^{\dagger}$ does not depend upon $\mathbf{Q}$.

(*b*) Let $\mathbf{A}=\binom{1}{2}$, $\mathbf{y}=\binom{1}{1}$, and $\mathbf{R}=\begin{pmatrix}2&1\\1&1\end{pmatrix}$. Find the vector $\mathbf{x}$ which minimizes $\|\mathbf{y}-\mathbf{A}\mathbf{x}\|_R$.

6.15 Use a pseudoinverse matrix to find the minimum norm solution (relative to the standard inner product) for the following set of equations:

$$2\xi_1+\xi_2+\xi_3=1$$

$$3\xi_1-\xi_2+\xi_3=2$$

6.16 Let $\mathbf{A}$ be a real $m\times n$ matrix with m independent rows ($m\leqslant n$). Let the inner products for $\mathfrak{M}^{n\times 1}$ and $\mathfrak{M}^{m\times 1}$, respectively, be $\langle \mathbf{x},\mathbf{z}\rangle_Q \overset{\Delta}{=} \mathbf{z}^{\mathrm{T}}\mathbf{Q}\mathbf{x}$ and $\langle \mathbf{y},\mathbf{w}\rangle_R \overset{\Delta}{=} \mathbf{w}^{\mathrm{T}}\mathbf{R}\mathbf{y}$, where $\mathbf{Q}$ and $\mathbf{R}$ are real, symmetric, positive-definite matrices.

(*a*) Find the pseudoinverse of $\mathbf{A}$ relative to the given inner products. Show that $\mathbf{A}^{\dagger}$ does not depend upon $\mathbf{R}$.

(*b*) Calculate the minimum "Q-norm" solution to the underdetermined equation $\mathbf{A}\mathbf{x}=\mathbf{y}$ for $\mathbf{A}=(1\ \ 2)$, $\mathbf{y}=(2)$, and $\mathbf{Q}=\begin{pmatrix}2&1\\1&1\end{pmatrix}$. Sketch the solution in the two-dimensional plane.

6.17 Let $\mathbf{x}(t)$ and $\mathbf{u}(t)$ denote an $n\times 1$ state vector and an $m\times 1$ control vector, respectively, at time t. Let $\mathbf{A}$ and $\mathbf{B}$ be real $n\times n$ and $n\times m$ matrices, respectively. The state equation $\dot{\mathbf{x}}=\mathbf{A}\mathbf{x}+\mathbf{B}\mathbf{u}$ with $\mathbf{x}(0)=\mathbf{x}_0$ describes an nth-order dynamic system with m control inputs. We seek the control vector $\mathbf{u}$ which drives the system to the final state $\mathbf{x}(t_f)=\mathbf{x}_f$ and minimizes $\int_0^{t_f}\mathbf{u}^{\mathrm{T}}(s)\mathbf{Q}\mathbf{u}(s)\,ds$, where $\mathbf{Q}$ is a real, symmetric, positive-definite $m\times m$ matrix. According to (3.79), the system equation can be expressed at the final time t_f in the inverse form

$$\mathbf{x}_f - e^{\mathbf{A}t_f}\mathbf{x}_0 = \int_0^{t_f} \exp\left[\mathbf{A}(t_f - s)\right]\mathbf{Bu}(s)\,ds$$

Let $\mathcal{L}_2^m(0,t_f)$ denote the space of $m\times 1$ vectors each of whose elements is a function from $\mathcal{L}_2(0,t_f)$; $\mathcal{L}_2^m(0,t_f)$ is a Hilbert space under the inner product $\langle \mathbf{u},\mathbf{v}\rangle_Q \triangleq \int_0^{t_f}\mathbf{v}^{\mathrm{T}}(s)\mathbf{Qu}(s)\,ds$. Assign the standard inner product to $\mathfrak{M}^{n\times 1}$. If we define $\mathbf{T}$: $\mathcal{L}_2^m(0,t_f)\to\mathfrak{M}^{n\times 1}$ by $\mathbf{Tu} \triangleq \int_0^{t_f}\exp[\mathbf{A}(t_f - s)]\mathbf{Bu}(s)\,ds$ and let $\mathbf{y} \triangleq \mathbf{x}_f - \exp(\mathbf{A}t_f)\mathbf{x}_0$, then the problem consists in determining the control $\mathbf{u}$ which satisfies $\mathbf{Tu}=\mathbf{y}$ and minimizes $\|\mathbf{u}\|_Q^2$. The solution is $\mathbf{u}=\mathbf{T}^*(\mathbf{TT}^*)^{-1}\mathbf{y}$.

(*a*) Describe $\mathbf{T}^*$ in terms of its action on an arbitrary vector $\mathbf{z}$ in $\mathfrak{M}^{n\times 1}$.

(*b*) Describe the operator $\mathbf{TT}^*$ by finding the $n\times n$ matrix $\mathbf{C}$ such that $\mathbf{TT}^*\mathbf{z}=\mathbf{Cz}$ for each $\mathbf{z}$ in $\mathfrak{M}^{n\times 1}$.

(*c*) Let $(\mathbf{f}_1(t),\mathbf{f}_2(t))$ denote the position of a unit mass at time t in a two-dimensional coordinate system. If $(\mathbf{u}_1(t),\mathbf{u}_2(t))$ is the force applied to the mass at time t, then the motion of the mass is described by $\ddot{\mathbf{f}}_1=\mathbf{u}_1, \mathbf{f}_1(0)=\dot{\mathbf{f}}_1(0)=0, \ddot{\mathbf{f}}_2=\mathbf{u}_2, \mathbf{f}_2(0)=\dot{\mathbf{f}}_2(0)=0$. Find the four-dimensional state equation $\dot{\mathbf{x}}=\mathbf{Ax}+\mathbf{Bu}$ using the state vector $\mathbf{x} \triangleq (\mathbf{f}_1\,\dot{\mathbf{f}}_1\,\mathbf{f}_2\,\dot{\mathbf{f}}_2)^{\mathrm{T}}$ and the control vector $\mathbf{u} \triangleq (\mathbf{u}_1\,\mathbf{u}_2)^{\mathrm{T}}$. Find the state transition matrix $e^{\mathbf{A}t}$. Let $\mathbf{Q}=\begin{pmatrix}1 & 0\\ 0 & .5\end{pmatrix}$ (denoting more concern with $\mathbf{u}_1$ than $\mathbf{u}_2$). Let $t_f=1$, $\mathbf{f}_1(1)=1$, $\dot{\mathbf{f}}_1(1)=0$, $\mathbf{f}_2(1)=1$, and $\dot{\mathbf{f}}_2(1)=0$. Determine $\mathbf{T}$, $\mathbf{y}$, $\mathbf{T}^*$, and $\mathbf{C}$ for this specific system.

(*d*) Find the pseudoinverse $\mathbf{T}^\dagger$ corresponding to the system given in (*c*). Use $\mathbf{T}^\dagger$ to determine the minimum effort control.

6.18 The steady-state temperature distribution $\mathbf{f}$ in an insulated bar is related to the rate of heat generation $\mathbf{u}$, as a function of position s along the bar, by the differential equation $-\mathbf{f}''(s)=\mathbf{u}(s)$. The function $\mathbf{u}$ is specified. We seek a temperature distribution $\mathbf{f}$ for which the rod ends, $s=0$ and $s=1$, are held at the temperature $\mathbf{f}(0)=\mathbf{f}(1)=0$, and the total heat flowing out of the ends of the bar is $\mathbf{f}'(1)-\mathbf{f}'(0)=0$. Physical understanding of steady-state heat flow shows that this problem is overspecified.

(*a*) Solve the differential equation and boundary conditions in a least-square sense. That is, pick $\mathbf{f}$ to minimize

$$\int_0^1 \left[\mathbf{u}(s)+\mathbf{f}''(s)\right]^2 ds$$

while satisfying the three boundary conditions. Hint: Let $-\mathbf{D}^2$ act only on functions which satisfy the boundary conditions. Project $\mathbf{u}$ onto either the range or the nullspace of this operator.

(*b*) Find the generalized Green's function $k^{\dagger}$, in terms of which the solution to (*a*) can be expressed as

$$\mathbf{f}(s)=\int_0^1 k^{\dagger}(s,\sigma)\mathbf{u}(\sigma)\,d\sigma$$

6.19 Use matrix factorization to find the pseudoinverses of the following matrices (relative to the standard inner products for $\mathfrak{M}^{4\times 1}$ and $\mathfrak{M}^{3\times 1}$):

(*a*) $$\mathbf{A}=\begin{pmatrix}2&3&2\\3&5&3\\1&1&1\\1&2&1\end{pmatrix}$$ (*b*) $$\mathbf{A}=\begin{pmatrix}1&1&1&1\\1&0&-1&1\\3&2&1&3\\0&1&2&0\end{pmatrix}$$

6.20 Let

$$\mathbf{A}=\begin{pmatrix}0&-1&2\\2&1&0\\1&0&1\end{pmatrix}$$

(*a*) Find $\mathbf{A}^{\dagger}$ (relative to the standard inner product) by matrix factorization.

(*b*) Find $\mathbf{A}^{\dagger}$ (relative to the standard inner product) by the projection method of (6.93)–(6.101).

*6.21 Let $\mathbf{A}$ be a real (nonzero) $m\times n$ matrix. Then $\mathbf{AA}^{\mathrm{T}}$ is $m\times m$. Denote the characteristic polynomial of $\mathbf{AA}^{\mathrm{T}}$ by $c(\lambda)=(-1)^m(\lambda^m+a_1\lambda^{m-1}+\cdots+a_m)$. Let k be the largest subscript for which $a_k\neq 0$. Decell [6.2] shows that $c(\lambda)$ and $\mathbf{A}^{\dagger}$ can be determined by the following algorithm:

$$\begin{aligned}
a_0&=1 & \mathbf{B}_0&=\mathbf{I}\\
a_1&=-\operatorname{trace}(\mathbf{AA}^{\mathrm{T}}), & \mathbf{B}_1&=\mathbf{AA}^{\mathrm{T}}+a_1\mathbf{I}\\
a_2&=-\tfrac{1}{2}\operatorname{trace}(\mathbf{AA}^{\mathrm{T}}\mathbf{B}_1), & \mathbf{B}_2&=\mathbf{AA}^{\mathrm{T}}\mathbf{B}_1+a_2\mathbf{I}\\
&\;\vdots & &\;\vdots\\
a_k&=-\frac{1}{k}\operatorname{trace}(\mathbf{AA}^{\mathrm{T}}\mathbf{B}_{k-1}), & \mathbf{B}_k&=\mathbf{AA}^{\mathrm{T}}\mathbf{B}_{k-1}+a_k\mathbf{I}
\end{aligned}$$

$$\mathbf{A}^{\dagger}=-\frac{1}{a_k}\mathbf{A}^{\mathrm{T}}\mathbf{B}_{k-1}$$

(The notation trace($\mathbf{C}$) denotes the sum of the diagonal elements of the square matrix $\mathbf{C}$.) It turns out that $\mathbf{A}\mathbf{A}^{\mathrm{T}}\mathbf{B}_k=\boldsymbol{\Theta}$. Therefore, the value of k becomes obvious during the course of the iteration. If $k=0$, then $\mathbf{A}^{\dagger}=\boldsymbol{\Theta}$. Carry out the algorithm for

$$(a)\quad \mathbf{A}=\begin{pmatrix}1 & 1\\ 1 & 1\end{pmatrix} \qquad (b)\quad \mathbf{A}=\begin{pmatrix}2 & 3 & 1 & 1\\ 3 & 5 & 1 & 2\\ 2 & 3 & 1 & 1\end{pmatrix}$$

*6.22 Let $\mathcal{V}$ and $\mathcal{W}$ be Hilbert spaces, and $\mathbf{T}: \mathcal{V}\to\mathcal{W}$ a bounded linear transformation. [We assume range($\mathbf{T}$) is closed.] The pseudoinverse can be defined algebraically as that bounded linear transformation $\mathbf{T}^{\dagger}: \mathcal{W}\to\mathcal{V}$ which satisfies:

(1) $\mathbf{T}\mathbf{T}^{\dagger}\mathbf{T}=\mathbf{T}$

(2) $\mathbf{T}^{\dagger}\mathbf{T}\mathbf{T}^{\dagger}=\mathbf{T}^{\dagger}$

(3) $(\mathbf{T}^{\dagger}\mathbf{T})^*=\mathbf{T}^{\dagger}\mathbf{T}$

(4) $(\mathbf{T}\mathbf{T}^{\dagger})^*=\mathbf{T}\mathbf{T}^{\dagger}$

This definition of $\mathbf{T}^{\dagger}$ is equivalent to the definition depicted in Figure 6.8 [6.1].

*6.23 Let $\mathbf{T}: \mathcal{V}\to\mathcal{W}$ be a bounded linear transformation from the Hilbert space $\mathcal{V}$ into the Hilbert space $\mathcal{W}$. (We assume $\mathbf{T}$ has closed range.) The pseudoinverse $\mathbf{T}^{\dagger}$ has the following properties:

(1) Range($\mathbf{T}^{\dagger}$) = range($\mathbf{T}^*$)

(2) Nullspace($\mathbf{T}^{\dagger}$) = nullspace($\mathbf{T}^*$)

(3) $\mathbf{T}^{\dagger}\mathbf{T}$ is the orthogonal projector onto range($\mathbf{T}^*$)

(4) $\mathbf{T}\mathbf{T}^{\dagger}$ is the orthogonal projector onto range($\mathbf{T}$)

(5) $(\mathbf{T}^{\dagger})^{\dagger}=\mathbf{T}$

(6) $(\mathbf{T}^*)^{\dagger}=(\mathbf{T}^{\dagger})^*$

(7) $\mathbf{T}^{\dagger}=(\mathbf{T}^*\mathbf{T})^{\dagger}\mathbf{T}^*=\mathbf{T}^*(\mathbf{T}\mathbf{T}^*)^{\dagger}$

*6.24 Let $\mathbf{A}$ be a real, symmetric, $n\times n$ matrix. The eigenvectors of a symmetric matrix are orthogonal relative to the standard inner product for $\mathcal{M}^{n\times 1}$. Consequently, $\mathbf{A}$ can be diagonalized by a similarity transformation $\boldsymbol{\Lambda}=\mathbf{S}^{-1}\mathbf{A}\mathbf{S}$, where $\mathbf{S}^{-1}=\mathbf{S}^{\mathrm{T}}$. Let the diagonal form of **A** be

$$\Lambda=\begin{pmatrix} \lambda_1 & & & & & & \\ & \cdot & & & & & \\ & & \cdot & & & & \\ & & & \lambda_m & & & \\ & & & & 0 & & \\ & & & & & \ddots & \\ & & & & & & 0 \end{pmatrix}$$

(a) Find $\Lambda^\dagger$ (relative to the standard inner product).
(b) Show that $\mathbf{A}^\dagger=\mathbf{S}\Lambda^\dagger\mathbf{S}^{-1}$.

6.25 Use the projection method of (6.93)–(6.101) to find $\mathbf{A}^\dagger$ for

$$\mathbf{A}=\begin{pmatrix} 2 & 0 & 1 \\ 0 & 1 & 1 \\ 2 & 1 & 2 \end{pmatrix}$$

Hint: the necessary Gram–Schmidt orthogonalization is carried out in (6.93).

6.8 References

[6.1] Ben-Israel, A. and A. Charnes, "Contributions to the Theory of Generalized Inverses," *J. SIAM Appl. Math.*, **11** (1963), 667–699.

[6.2] Decell, Henry P., "An Application of the Cayley–Hamilton Theorem to Generalized Matrix Inversion," *SIAM Rev.*, **7**, 4 (October 1965), 526–528.

[6.3] Dunford, N. and J. T. Schwartz, *Linear Operators*, Part I, Interscience, New York, 1958.

[6.4] Forsythe, George E., "Generation and Use of Orthogonal Polynomials for Data-Fitting with a Digital Computer," *J. Soc. Indust. Appl. Math.*, **5**, 2 (June 1957), 74–88.

[6.5] Frame, J. S., "Matrix Functions and Applications I: Matrix Operations and Generalized Inverses," *IEEE Spectrum*, **1**, 3 (March 1964), 209–220.

[6.6] Gillies, A. W., "On the Classification of Matrix Generalized Inverses," *SIAM Rev.*, **12**, 4 (October 1970), 573–576.

[6.7] Golub, G. and W. Kahan, "Calculating the Singular Values and Pseudoinverse of a Matrix," *J. SIAM Numer. Anal., Ser. B*, **2**, 2 (1965), 205–224.

[6.8] Greville, T.N.E., "Note on the Generalized Inverse of a Matrix Product," *SIAM Rev.*, **8**, 4 (October 1966), 518–521.

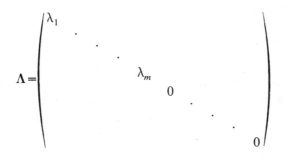

[6.9] Harmuth, H. F., *Transmission of Information by Orthogonal Functions*, Springer-Verlag, Berlin/New York, 1970.

[6.10] Kuo, Marshall C. and Louis F. Kazda, "Minimum Energy Problems in Hilbert Function Space," *J. Franklin Inst.*, **283**, 1 (January 1967), 38–54.

[6.11] Kwakernaak, Huibert and Raphael Sivan, *Linear Optimal Control Systems*, Wiley, New York, 1972.

[6.12] Loud, Warren S., "Some Examples of Generalized Green's Functions and Generalized Green's Matrices," *SIAM Rev.*, **12**, 2 (April 1970), 194–210.

*[6.13] Luenberger, David G., *Optimization by Vector Space Methods*, Wiley, New York, 1969.

[6.14] Osborne, E. E., "Smallest Least-Squares Solutions of Linear Equations," *J. SIAM Numer. Anal., Ser. B*, **2**, 2 (1965), 300–307.

[6.15] Papoulis, Athanasios, *Probability, Random Variables, and Stochastic Processes*, McGraw-Hill, New York, 1965.

[6.16] Pearson, A. E., "A Regression Analysis in Function Space for the Identification of Multi-Variable Linear Systems," *IFAC Conference*, Paper 46.B, June 1966.

[6.17] Ralston, Anthony, *A First Course in Numerical Analysis*, McGraw-Hill, New York, 1965.

[6.18] Sage, Andrew P., Jr., and George W. Masters, Jr., "Least-Squares Curve Fitting and Discrete Optimum Filtering," *IEEE Trans. Educ.*, **E-10**, 1 (March 1967).

[6.19] Sage, Andrew P. and James L. Melsa, *Estimation Theory with Applications to Communications and Control*, McGraw-Hill, New York, 1971.

[6.20] Scroggs, James E. and Patrick L. Odell, "An Alternate Definition of a Pseudoinverse of a Matrix," *J. SIAM Appl. Math.*, **14**, 4 (July 1966), 796–810.

[6.21] Sprent, Peter, *Models in Regression*, Methuen, London, 1969.

[6.22] Stallings, W. T. and T. L. Boullion, "Computation of Pseudoinverse Matrices using Residue Arithmetic," *SIAM REV.* **14**, 1 (January 1972), 152–163.

[6.23] Tanabe, Kunio, "Projection Method for Solving a Singular System of Linear Equations and its Applications," *Numer. Math.*, **17**, 1971, 203–214.

[6.24] Wilkinson, J. H., *The Algebraic Eigenvalue Problem*, Clarenden Press, Oxford, 1965.

7

Characterizing the Optimum: Linearization in a Hilbert Space

Mathematical optimization consists in determining values of variables to maximize or minimize an objective function. In many optimization problems the variables are also required to satisfy constraining equations or inequalities. Optimization problems often arise from the need to determine parameters of a model so that the model best fits measured data according to some criterion. An elementary example of optimum data fitting is the linear regression described in Section 6.1 (Figure 6.2). In that example, the polynomial structure of the model determines constraining equations which the model parameters must satisfy. Many optimization problems also arise from the desire to determine a system operating policy which is in some sense best. In this setting, the equations which constitute the model of the system act as constraints on the optimization. The minimum energy control problem treated in Section 6.2 (Figure 6.5) illustrates the selection of an optimum operating policy. We shall find that physical limitations sometimes introduce additional constraints into optimization problems, often in the form of inequalities. For example, temperatures in a system must be kept below the melting point of most materials in that system.

Unconstrained least-square minimization and least-square minimization with linear equality constraints are explored in detail in Chapter 6. The Hilbert space setting of that chapter provides considerable insight into the implications of least-square objective functions; it also provides a computational technique, orthogonal projection, for solving least-square problems. In this chapter we consider optimization problems that are considerably more general than least-square problems. We seek the minima and maxima of real-valued functionals, possibly subject to equality and/or inequality constraints. The objective functions and the constraint functions may be arbitrary nonlinear functions, subject to the condition that they be "smooth" near the extrema.* We use the term *extremum* to mean either a maximum or a minimum.

The main concept that we use in understanding and computing extrema is *linearization*. (This concept is central to almost all nonlinear analyses.) As a consequence, in Chapters 7 and 8 we lean heavily on the linear transformation notions of Chapters 1–5. In this chapter the use of linearization leads to conditions that are necessary for the existence of extrema—equations that can be solved to *determine* the extrema. Most classical optimization results are special cases of these necessary conditions. We explore computational techniques for solving nonlinear equations and for computing extrema in Chapter 8. These computational techniques, again, make heavy use of linearization. In Chapters 7 and 8 we work primarily in inner product space (or Hilbert space) settings in order to use the simple geometrical concepts that are associated with inner products.†

7.1 Local Linearization and Unconstrained Extrema

We introduce the concept of linearization by means of the simplest case, a function of a single variable. Let f be some smooth nonlinear real-valued function of the scalar variable s. It is common practice to approximate such a function in the neighborhood of interest by a straight line which is tangent to the function at some point s_0 in that neighborhood (Figure 7.1). Define the "excursion variable" ds by $ds = s - s_0$. The line tangent to f at s_0 is described, as a function of ds, by $f(s_0)+f'(s_0)\,ds$. Clearly, the equation of the tangent line consists in the first two terms of the Taylor series expansion of f about s_0:

$$f(s)=f(s_0+ds)=f(s_0)+f'(s_0)\,ds+\epsilon \tag{7.1}$$

If we are interested only in s near s_0, ϵ is small and the tangent is a good approximation to the function. We focus on the second term of the expansion, $f'(s_0)\,ds$; in effect, we shift the origin in Figure 7.1 to the point

*The minimax (or Chebyshev) criterion—minimization of the maximum of the errors—is a frequently used objective function. The techniques of Chapter 7 do not apply to minimax problems because minimax objective functions are not sufficiently differentiable. A minimax problem can be treated as a minimum norm problem, but the norm which is required is not associated with an inner product (P&C 7.1); thus the techniques of Chapter 6 do not apply. Nonquadratic minimum norm problems can be handled by a generalization of the orthogonal projection techniques of Chapter 6. See Luenberger [7.11] and Porter [7.15].

†A discussion of linearization and necessary conditions for extrema in normed spaces is contained in Luenberger [7.11].

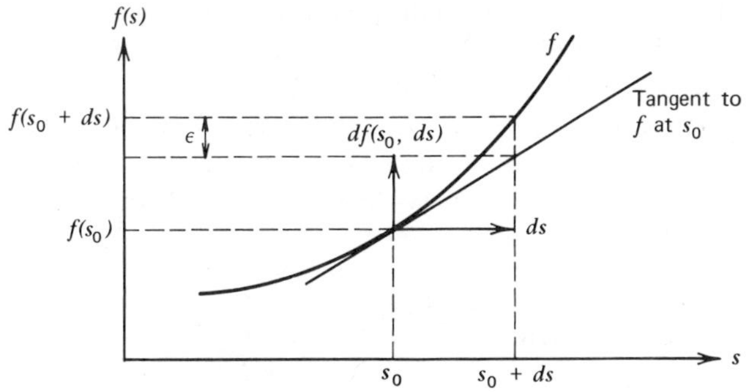

Figure 7.1. A tangent approximation.

$(s_0, f(s_0))$. We denote the new function by $df(s_0, ds) \triangleq f'(s_0)\,ds$, and refer to it as the *differential* of f at s_0; df is linear in ds.

One of the simplest results of differential calculus is that for a relative maximum or minimum of a differentiable function f to occur at s_0, we must have $f'(s_0)=0$. An equivalent condition is that the tangent to f at s_0 must be horizontal; that is, $f(s_0)+df(s_0, ds)$ must be constant, or $df(s_0, ds) = f'(s_0)\,ds = 0$, for all excursions ds.

The concept of a tangent and the associated necessary condition for the existence of extrema generalize easily to functions of more than one variable. Let $\mathbf{F}$: $\mathcal{R}^2 \to \mathcal{R}$ be a smooth (differentiable) nonlinear function of the vector $\mathbf{x} = (\xi_1, \xi_2)$. We approximate $\mathbf{F}$ in the neighborhood of $\mathbf{x}_0$ by a tangent function of the form $\mathbf{F}(\mathbf{x}_0) + \mathbf{dF}(\mathbf{x}_0, \mathbf{dx})$, where the function $\mathbf{dF}$, again referred to as the differential of $\mathbf{F}$, is linear in the excursion variable $\mathbf{dx}$. In the example of Figure 7.1, there are many straight lines that pass through the point $(s_0, f(s_0))$; each of these straight lines is linear in the variable ds. The tangent line is unique among these straight line approximations to f in that the error ϵ in the tangent approximation goes to zero faster than ds goes to zero. Consequently, for the two-variable case we pick as $\mathbf{dF}(\mathbf{x}_0, \mathbf{dx})$ that unique, linear function of $\mathbf{dx}$ for which the error, $\boldsymbol{\epsilon} \triangleq \mathbf{F}(\mathbf{x}_0 + \mathbf{dx}) - \mathbf{F}(\mathbf{x}_0) - \mathbf{dF}(\mathbf{x}_0, \mathbf{dx})$, approaches zero faster than $\|\mathbf{dx}\|$ approaches zero; that is,

$$\lim_{\|\mathbf{dx}\| \to 0} \left(\frac{\boldsymbol{\epsilon}}{\|\mathbf{dx}\|} \right) = 0 \tag{7.2}$$

It is a property of norms that $\|\mathbf{dx}\| \to 0$ implies $\mathbf{dx} \to \boldsymbol{\theta}$. However, $\mathbf{dx}$ can approach $\boldsymbol{\theta}$ by many different paths. According to the criterion (7.2), $\boldsymbol{\epsilon}/\|\mathbf{dx}\|$ must approach zero for all paths of the $\mathbf{dx}$ vector. The function $\mathbf{F}$

can be graphed in the cartesian product space $\mathcal{R}\times\mathcal{R}^2$. The linear approximation $\mathbf{F}(\mathbf{x}_0)+\mathbf{dF}(\mathbf{x}_0,\mathbf{dx})$ is a hyperplane in $\mathcal{R}\times\mathcal{R}^2$ which is tangent to the graph of $\mathbf{F}$ at $\mathbf{x}_0$ (Figure 7.2).

In the example of Figure 7.2, it is apparent that the equation of the tangent plane is determined by the slopes (or directional derivatives) of $\mathbf{F}$ at $\mathbf{x}_0$ in the directions of the two axes, ξ_1 and ξ_2. In point of fact, in Example 2 we show that

$$\mathbf{dF}(\mathbf{x}_0,\mathbf{dx})=\frac{\partial\,\mathbf{F}(\mathbf{x}_0)}{\partial\,\xi_1}d\xi_1+\frac{\partial\,\mathbf{F}(\mathbf{x}_0)}{\partial\,\xi_2}d\xi_2 \tag{7.3}$$

where the components of the excursion variable are denoted by $\mathbf{dx}=(d\xi_1,d\xi_2)$. The function $\mathbf{dF}$ of (7.3) is certainly linear in $\mathbf{dx}$. To be sure it qualifies as the differential of $\mathbf{F}$ at $\mathbf{x}_0$, we must check to see that it satisfies (7.2). We do so for a specific function $\mathbf{F}$.

Example 1. Testing a Differential. Define $\mathbf{F}\colon \mathcal{R}^2\to\mathcal{R}$ by $\mathbf{F}(\xi_1,\xi_2)=\xi_1^2+\xi_2^2$. According to (7.3), $\mathbf{dF}(\mathbf{x},\mathbf{dx})=2\xi_1 d\xi_1+2\xi_2 d\xi_2$. We apply the test (7.2) to the differential $\mathbf{dF}$, using the norm associated with the standard inner product on $\mathcal{R}^2$.

$$\lim_{\|\mathbf{dx}\|\to 0}\left(\frac{\mathbf{F}(\mathbf{x}+\mathbf{dx})-\mathbf{F}(\mathbf{x})-\mathbf{dF}(\mathbf{x},\mathbf{dx})}{\|\mathbf{dx}\|}\right)$$

$$=\lim_{\|\mathbf{dx}\|\to 0}\left(\frac{(\xi_1+d\xi_1)^2+(\xi_2+d\xi_2)^2-\xi_1^2-\xi_2^2-2\xi_1 d\xi_1-2\xi_2 d\xi_2}{\sqrt{(d\xi_1)^2+(d\xi_2)^2}}\right)$$

$$=\lim_{\|\mathbf{dx}\|\to 0}(\|\mathbf{dx}\|)=0$$

The determination of the differential, at least in this example, reduces to the determination of the partial derivatives of $\mathbf{F}$ at $\mathbf{x}$. The set of partial derivatives forms a vector $\nabla\mathbf{F}(\mathbf{x})=(\partial\,\mathbf{F}(\mathbf{x})/\partial\,\xi_1,\partial\,\mathbf{F}(\mathbf{x})/\partial\,\xi_2)$ which we call the *gradient* of $\mathbf{F}$ at $\mathbf{x}$. It is evident that (7.3) is in the form of a standard inner product on $\mathcal{R}^2$:

$$\mathbf{dF}(\mathbf{x}_0,\mathbf{dx})=\langle\mathbf{dx},\nabla\mathbf{F}(\mathbf{x}_0)\rangle \tag{7.4}$$

One way to view the relationship between $\mathbf{dF}(\mathbf{x}_0,\mathbf{dx})$ and $\nabla\mathbf{F}(\mathbf{x}_0)$ is to project level surfaces (horizontal slices) of the graphs of $\mathbf{F}(\mathbf{x}_0)$ and $\mathbf{dF}(\mathbf{x}_0,\mathbf{dx})$ onto the domain $\mathcal{R}^2$. Figure 7.3 shows the surface at the level $\mathbf{F}(\mathbf{x}_0)$ for the function $\mathbf{F}$ of Figure 7.2. The equation of the level surface of the tangent function is $\mathbf{F}(\mathbf{x}_0)+\mathbf{dF}(\mathbf{x}_0,\mathbf{dx})=\mathbf{F}(\mathbf{x}_0)$, or $\mathbf{dF}(\mathbf{x}_0,\mathbf{dx})=0$. The gradient vector is orthogonal (in the standard inner product) to excursions

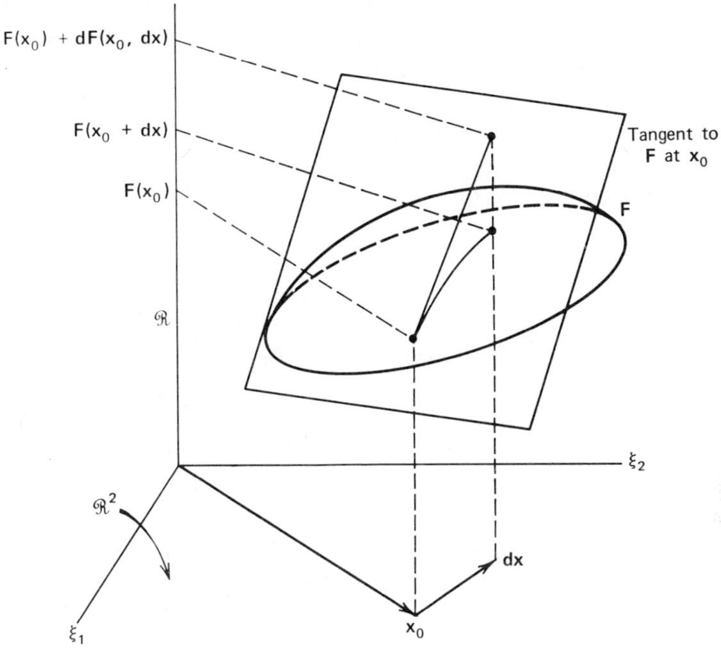

Figure 7.2. A tangent approximation in $\mathcal{R} \times \mathcal{R}^2$.

dx along the line $\mathbf{dF}(\mathbf{x}_0, \mathbf{dx}) = 0$. On the other hand, an excursion in the direction $\mathbf{dx} = \nabla\mathbf{F}(\mathbf{x}_0)$ yields a greater rate of increase in **F** than an excursion in any other direction from $\mathbf{x}_0$.

It is evident from Figure 7.2 that for a minimum or a maximum of a two-variable function **F** to occur at $\mathbf{x}_0$, it is necessary that the plane tangent to **F** at $\mathbf{x}_0$ be horizontal. Equivalent conditions are $\mathbf{dF}(\mathbf{x}_0, \mathbf{dx}) = 0$ or

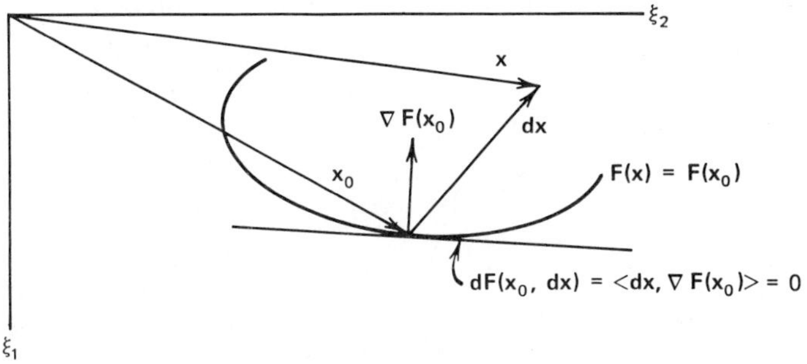

Figure 7.3. Level surfaces of **F**(**x**) and $\mathbf{dF}(\mathbf{x}_0, \mathbf{dx})$.

$\nabla \mathbf{F}(\mathbf{x}_0) = \boldsymbol{\theta}$. Applying any of these criteria to Example 1, we find the point $\mathbf{x} = (0,0)$ is a *critical point* of $\mathbf{F}$; that is, it is a candidate for a maximum or minimum of the function $\mathbf{F}$ of Example 1. We must use some other means to determine that the point is, in this instance, a minimum. Of course, the condition, $\mathbf{dF}(\mathbf{x}_0, \mathbf{dx}) = 0$, is not sufficient to *guarantee* the presence of a maximum or minimum. The tangent plane would also be horizontal at a horizontal inflection point of $\mathbf{F}$.

Fréchet Differentials

In order to be able to analyze infinite-dimensional optimization problems, we now extend the linearization concept to general nonlinear transformations. Let $\mathcal{V}$ and $\mathcal{W}$ be Hilbert spaces, $\mathbf{G}$: $\mathcal{V} \rightarrow \mathcal{W}$ a nonlinear real, vector-valued transformation, and $\mathbf{x}$ a point in $\mathcal{V}$.* There are many "linear approximations" to $\mathbf{G}$ at $\mathbf{x}$; that is, there are many hyperplanes which pass through the point $(\mathbf{x}, \mathbf{G}(\mathbf{x}))$ in $\mathcal{V} \times \mathcal{W}$. From among these linear approximations, we wish to pick the tangent to $\mathbf{G}$ at $\mathbf{x}$.

Definition. The **Fréchet differential** $\mathbf{dG}(\mathbf{x}, \cdot)$ of the function $\mathbf{G}$ at the point $\mathbf{x}$ is that bounded (continuous) linear transformation from $\mathcal{V}$ into $\mathcal{W}$ for which

$$\lim_{\|\mathbf{dx}\|_{\mathcal{V}} \rightarrow 0} \left(\frac{\|\boldsymbol{\epsilon}\|_{\mathcal{W}}}{\|\mathbf{dx}\|_{\mathcal{V}}} \right) = 0 \tag{7.5}$$

where the error vector $\boldsymbol{\epsilon}$ is given by

$$\boldsymbol{\epsilon} = \mathbf{G}(\mathbf{x} + \mathbf{dx}) - \mathbf{G}(\mathbf{x}) - \mathbf{dG}(\mathbf{x}, \mathbf{dx}) \tag{7.6}$$

Of course, there may be some points $\mathbf{x}$ at which $\mathbf{G}$ is not smooth enough to have a tangent. If the Fréchet differential exists at a point $\mathbf{x}$, we say $\mathbf{G}$ is *Fréchet differentiable* at $\mathbf{x}$. We assume all the functions we deal with are Fréchet differentiable for all $\mathbf{x}$ of interest. It can be shown that a Fréchet differential is unique.† We emphasize that $\mathbf{dG}(\mathbf{x}, \cdot)$ is linear in its second argument rather than its first. We often denote this linear relationship explicitly by the notation $\mathbf{G}'(\mathbf{x})$: $\mathcal{V} \rightarrow \mathcal{W}$, where

$$\mathbf{G}'(\mathbf{x})\mathbf{dx} \stackrel{\Delta}{=} \mathbf{dG}(\mathbf{x}, \mathbf{dx}) \tag{7.7}$$

*We concern ourselves primarily with problems in which $\mathcal{V}$ and $\mathcal{W}$ are Hilbert spaces. However, the definition of a Fréchet differential applies as well to problems in which $\mathcal{V}$ and $\mathcal{W}$ are inner product spaces or normed linear spaces. See P&C 7.1.

†Luenberger [7.11].

We refer to the linear transformation $\mathbf{G}'(\mathbf{x})$ as the **Fréchet derivative of G at x**.

Observe that the definition of $\mathbf{dG}(\mathbf{x},\cdot)$ depends upon the norms in $\mathcal{V}$ and $\mathcal{W}$, both for the meaning of boundedness and for the meaning of norm convergence in (7.5). We can change a norm arbitrarily without destroying convergence if the space on which the norm is defined is finite dimensional; all norms are equivalent in such spaces. Nor is the meaning of convergence in a function space changed by reweighting the norm with a bounded positive weighting function.

We found the Fréchet differential for a scalar-valued function of two variables in (7.3). The problem of finding an explicit description of the Fréchet differential for a more general vector-valued function $\mathbf{G}$ appears considerably more formidable. We need a general technique for finding such differentials. If the differential exists at $\mathbf{x}$, the limit (7.5) must be zero for all paths (e.g., spirals) by which $\mathbf{dx}$ approaches $\boldsymbol{\theta}$. Therefore, the limit must be zero, in particular, if we let $\mathbf{dx}$ approach zero along a fixed direction vector $\mathbf{h}$. We let $\mathbf{dx}=\beta\mathbf{h}$, and let the real scalar $\beta\to 0$. The limit (7.5) becomes

$$
\begin{aligned}
0 &= \lim_{\|\mathbf{dx}\|_{\mathcal{V}}\to 0}\left(\frac{\|\boldsymbol{\epsilon}\|_{\mathcal{W}}}{\|\mathbf{dx}\|_{\mathcal{V}}}\right)\\
&= \lim_{\|\beta\mathbf{h}\|_{\mathcal{V}}\to 0}\left(\frac{\|\mathbf{G}(\mathbf{x}+\beta\mathbf{h})-\mathbf{G}(\mathbf{x})-\mathbf{dG}(\mathbf{x},\beta\mathbf{h})\|_{\mathcal{W}}}{\|\beta\mathbf{h}\|_{\mathcal{V}}}\right)\\
&= \frac{1}{\|\mathbf{h}\|_{\mathcal{V}}}\lim_{\beta\to 0}\left\|\frac{\mathbf{G}(\mathbf{x}+\beta\mathbf{h})-\mathbf{G}(\mathbf{x})}{\beta}-\frac{\mathbf{dG}(\mathbf{x},\beta\mathbf{h})}{\beta}\right\|_{\mathcal{W}}
\end{aligned}
$$

Since $\mathbf{dG}(\mathbf{x},\beta\mathbf{h})$ is linear in $\beta\mathbf{h}$, we can cancel β from the last term. We can also drop the constant $1/\|\mathbf{h}\|_{\mathcal{V}}$. Therefore,

$$
\lim_{\beta\to 0}\left\|\frac{\mathbf{G}(\mathbf{x}+\beta\mathbf{h})-\mathbf{G}(\mathbf{x})}{\beta}-\mathbf{dG}(\mathbf{x},\mathbf{h})\right\|_{\mathcal{W}}=0 \qquad (7.8)
$$

or

$$
\mathbf{dG}(\mathbf{x},\mathbf{h})=\lim_{\beta\to 0}\frac{\mathbf{G}(\mathbf{x}+\beta\mathbf{h})-\mathbf{G}(\mathbf{x})}{\beta} \qquad (7.9)
$$

where the limit in (7.9) is taken in the sense of norm convergence as indicated in (7.8). Equation (7.9), although involving a limiting operation, can be used to calculate $\mathbf{dG}(\mathbf{x},\mathbf{h})$. Note that the excursion variable in (7.9) is denoted **h** rather than **dx**. We will use this new notation from this point forward.

Example 2. Finding a Differential by Means of (7.9). Let $\mathbf{F}(\xi_1,\xi_2)=\xi_1^2+\xi_2^2$. If we denote $\mathbf{x}=(\xi_1,\xi_2)$ and $\mathbf{h}=(h_1,h_2)$, then (7.9) is

$$\begin{aligned}\mathbf{dF}(\mathbf{x},\mathbf{h}) &= \lim_{\beta\to 0}\left(\frac{(\xi_1+\beta h_1)^2+(\xi_2+\beta h_2)^2-\xi_1^2-\xi_2^2}{\beta}\right)\\ &= \lim_{\beta\to 0}\left(2\xi_1 h_1+2\xi_2 h_2+\beta\left(h_1^2+h_2^2\right)\right)\\ &=2\xi_1 h_1+2\xi_2 h_2\end{aligned}$$

This function is clearly linear in **h**.

We should keep in mind that (7.9) is not a definition of $\mathbf{dG}(\mathbf{x},\mathbf{h})$. The limit in (7.9) can exist for all **h** even if $\mathbf{dG}(\mathbf{x},\mathbf{h})$ does not exist. Consider, for example, the function

$$\begin{aligned}\mathbf{F}(\xi_1,\xi_2) &\triangleq 0, && \xi_1\geqslant 0 \text{ and } \xi_2\geqslant 0\\ &\triangleq 0, && \xi_1\leqslant 0 \text{ and } \xi_2\leqslant 0\\ &\triangleq \xi_1+\xi_2, && \text{otherwise}\end{aligned}$$

By (7.9),

$$\begin{aligned}\mathbf{dF}(0,0;\,h_1,h_2) &=0, && h_1\geqslant 0 \text{ and } h_2\geqslant 0\\ &=0, && h_1\leqslant 0 \text{ and } h_2\leqslant 0\\ &=h_1+h_2, && \text{otherwise}\end{aligned}$$

The function $\mathbf{dF}(\boldsymbol{\theta},\mathbf{h})$ is not linear in **h**, and therefore it is not a Fréchet differential. The limit in (7.9), if it exists for a specific **h**, is known as the **weak differential** (or **Gateaux differential**) at **x** with increment **h**. If the Fréchet differential exists, it follows from the derivation of (7.9) that it is the same function of **h** as is the weak differential. Thus we are safe in using (7.9) to determine the explicit form of a Fréchet differential; verification that the result is, in fact, a Fréchet differential consists in showing that $\mathbf{dG}(\mathbf{x},\mathbf{h})$ is bounded (continuous) and linear in **h**.

Differentials of Functionals

Suppose we apply (7.9) to a real-valued functional $\mathbf{F}: \mathcal{V} \to \mathcal{R}$. If we define $f(\beta) \triangleq \mathbf{F}(\mathbf{x}+\beta\mathbf{h})$, then

$$\mathbf{dF}(\mathbf{x},\mathbf{h}) = \lim_{\beta\to 0}\left(\frac{f(\beta)-f(0)}{\beta}\right) = \left.\frac{df(\beta)}{d\beta}\right|_{\beta=0}$$

$$= \frac{d}{d\beta}\mathbf{F}(\mathbf{x}+\beta\mathbf{h})|_{\beta=0} \tag{7.10}$$

Equation (7.10) is extremely valuable for determining differentials of functionals because it replaces the evaluation of a limit by the use of standard differentiation rules.

Example 3. The Differential of a Functional on $\mathcal{R}^n$. Let $\mathbf{F}$ be a real-valued functional on $\mathcal{R}^n$. Denote the elements of $\mathbf{x}$ and $\mathbf{h}$ by $\mathbf{x}=(\xi_1,\ldots,\xi_n)$ and $\mathbf{h} =(h_1,\ldots,h_n)$. According to (7.10),

$$\mathbf{dF}(\mathbf{x},\mathbf{h}) = \frac{d}{d\beta}\mathbf{F}(\xi_1+\beta h_1,\ldots,\xi_n+\beta h_n)|_{\beta=0}$$

$$= \frac{\partial \mathbf{F}(\mathbf{x})}{\partial \xi_1}h_1 + \cdots + \frac{\partial \mathbf{F}(\mathbf{x})}{\partial \xi_n}h_n \tag{7.11}$$

Equation (7.11) extends the earlier result (7.3); the function is certainly linear in $\mathbf{h}$.

Example 4. The Differential of a Functional on $\mathcal{L}_2(a,b)$. Let $\mathbf{f}$ be in the standard Hilbert space $\mathcal{L}_2(a,b)$. Define $\mathbf{F}(\mathbf{f}) = \|\mathbf{f}\|^2 = \int_a^b \mathbf{f}^2(s)\,ds$. Then, by (7.10),

$$\mathbf{dF}(\mathbf{f},\mathbf{h}) = \frac{d}{d\beta}\int_a^b (\mathbf{f}(s)+\beta\mathbf{h}(s))^2\,ds|_{\beta=0}$$

$$= \int_a^b 2\mathbf{f}(s)\mathbf{h}(s)\,ds$$

According to the Riesz–Fréchet theorem (5.57), this integral is a bounded linear functional of $\mathbf{h}$.

The differential of a functional $\mathbf{F}: \mathcal{V}\to\mathcal{R}$ is a bounded linear functional of the excursion variable $\mathbf{h}$. If $\mathcal{V}$ is a Hilbert space, this bounded linear functional [denoted $\mathbf{F}'(\mathbf{x})$] can be expressed as an inner product*:

$$\mathbf{F}'(\mathbf{x})\mathbf{h} = \mathbf{dF}(\mathbf{x},\mathbf{h}) = \langle \mathbf{h}, \nabla \mathbf{F}(\mathbf{x})\rangle \tag{7.12}$$

The vector $\nabla\mathbf{F}(\mathbf{x})$ which represents the functional $\mathbf{F}'(\mathbf{x})$ is called the **gradient of F at x**. The inner product on $\mathcal{V}$ can be changed (reweighted) without changing the convergence of vectors in $\mathcal{V}$. A change in the inner product on $\mathcal{V}$ results in a change in the form of $\nabla\mathbf{F}(\mathbf{x})$, even though the differential $\mathbf{dF}(\mathbf{x},\mathbf{h})$ which the gradient represents is not changed. The gradient corresponding to the differential (7.11), relative to the standard inner product for $\mathcal{R}^n$, is

$$\nabla \mathbf{F}(\mathbf{x}) = \left(\frac{\partial \mathbf{F}(\mathbf{x})}{\partial \xi_1}, \ldots, \frac{\partial \mathbf{F}(\mathbf{x})}{\partial \xi_n} \right) \tag{7.13}$$

The gradient of the functional $\mathbf{F}$ of Example 4 is

$$\nabla \mathbf{F}(\mathbf{f}) = 2\mathbf{f}$$

Differentials with m-Dimensional Range

Suppose $\mathbf{G}$: $\mathcal{V} \to \mathcal{M}^{m\times 1}$. In other words, the values of $\mathbf{G}$ are of the form

$$\mathbf{G}(\mathbf{x}) = (\mathbf{G}_1(\mathbf{x}) \cdots \mathbf{G}_m(\mathbf{x}))^{\mathrm{T}} \tag{7.14}$$

where $\mathbf{G}_i$ is a functional on $\mathcal{V}$. To find the differential of $\mathbf{G}$ we apply (7.9):

$$\begin{aligned}
\mathbf{dG}(\mathbf{x},\mathbf{h}) &= \lim_{\beta\to 0} \left(\frac{(\mathbf{G}_1(\mathbf{x}+\beta\mathbf{h}) \cdots \mathbf{G}_m(\mathbf{x}+\beta\mathbf{h}))^{\mathrm{T}} - (\mathbf{G}_1(\mathbf{x}) \cdots \mathbf{G}_m(\mathbf{x}))^{\mathrm{T}}}{\beta} \right) \\
&= \left(\lim_{\beta\to 0} \frac{\mathbf{G}_1(\mathbf{x}+\beta\mathbf{h}) - \mathbf{G}_1(\mathbf{x})}{\beta} \cdots \lim_{\beta\to 0} \frac{\mathbf{G}_m(\mathbf{x}+\beta\mathbf{h}) - \mathbf{G}_m(\mathbf{x})}{\beta} \right)^{\mathrm{T}} \\
&= (\mathbf{dG}_1(\mathbf{x},\mathbf{h}) \cdots \mathbf{dG}_m(\mathbf{x},\mathbf{h}))^{\mathrm{T}}
\end{aligned} \tag{7.15}$$

The differential of the vector-valued function $\mathbf{G}$ is the vector of differentials of the scalar-valued functions which make up $\mathbf{G}$. The conclusion (7.15) is valid regardless of which norm is used on $\mathcal{M}^{m\times 1}$, since all

*The existence of a vector $\nabla\mathbf{F}(\mathbf{x})$ which represents $\mathbf{F}'(\mathbf{x})$ is guaranteed by the Riesz–Fréchet theorem. Of course, it is quite possible that $\nabla\mathbf{F}(\mathbf{x})$ would exist in $\mathcal{V}$ even if $\mathcal{V}$ were not complete.

norms provide equivalent definitions of convergence in a finite-dimensional space. Moreover, all m-dimensional spaces are equivalent to $\mathfrak{M}^{m\times 1}$. Consequently, the differential of any function $\mathbf{G}$ with an m-dimensional range can be expressed in a form equivalent to (7.15).

Exercise 1. Let $\mathbf{G}: \mathcal{V}\to\mathcal{W}$, where $\dim(\mathcal{W})=m$. Let $\mathcal{Y}=\{\mathbf{y}_1,\ldots,\mathbf{y}_m\}$ be a basis for $\mathcal{W}$. Denote the coordinates of $\mathbf{G}(\mathbf{x})$ by $[\mathbf{G}(\mathbf{x})]_{\mathcal{Y}} = (\mathbf{H}_1(\mathbf{x})\cdots\mathbf{H}_m(\mathbf{x}))^{\mathrm{T}}$, where the $\mathbf{H}_i$ are scalar-valued functions of $\mathbf{x}$. Show that

$$\mathbf{dG}(\mathbf{x},\mathbf{h})=\mathbf{dH}_1(\mathbf{x},\mathbf{h})\mathbf{y}_1+\cdots+\mathbf{dH}_m(\mathbf{x},\mathbf{h})\mathbf{y}_m \tag{7.16}$$

Example 5. A Differential from $\mathfrak{M}^{n\times 1}$ into $\mathfrak{M}^{m\times 1}$. Let $\mathbf{G}$: $\mathfrak{M}^{n\times 1}\to\mathfrak{M}^{m\times 1}$. Then $\mathbf{G}(\mathbf{x})$ is in the form of (7.14). The ith element of $\mathbf{G}(\mathbf{x})$ is a scalar-valued function of $\mathbf{x}$. Therefore its differential can be expressed as in (7.11) or (7.12):

$$\mathbf{dG}_i(\mathbf{x},\mathbf{h})=\frac{\partial\,\mathbf{G}_i(\mathbf{x})}{\partial\,\xi_1}h_1+\cdots+\frac{\partial\,\mathbf{G}_i(\mathbf{x})}{\partial\,\xi_n}h_n=\nabla\mathbf{G}_i(\mathbf{x})^{\mathrm{T}}\mathbf{h}$$

Then, according to (7.15), the differential of $\mathbf{G}$ is

$$\mathbf{dG}(\mathbf{x},\mathbf{h})=\begin{pmatrix}\nabla\mathbf{G}_1(\mathbf{x})^{\mathrm{T}}\mathbf{h}\\ \vdots\\ \nabla\mathbf{G}_m(\mathbf{x})^{\mathrm{T}}\mathbf{h}\end{pmatrix}=(\nabla\mathbf{G}_1(\mathbf{x})\cdots\nabla\mathbf{G}_m(\mathbf{x}))^{\mathrm{T}}\mathbf{h}$$

$$=\begin{pmatrix}\dfrac{\partial\,\mathbf{G}_1(\mathbf{x})}{\partial\,\xi_1} & \cdots & \dfrac{\partial\,\mathbf{G}_1(\mathbf{x})}{\partial\,\xi_n}\\ \vdots & & \vdots\\ \dfrac{\partial\,\mathbf{G}_m(\mathbf{x})}{\partial\,\xi_1} & \cdots & \dfrac{\partial\,\mathbf{G}_m(\mathbf{x})}{\partial\,\xi_n}\end{pmatrix}\mathbf{h}\triangleq\frac{\partial\,\mathbf{G}}{\partial\,\mathbf{x}}\mathbf{h} \tag{7.17}$$

The matrix of partial derivatives denoted $\partial\,\mathbf{G}/\partial\,\mathbf{x}$ in (7.17) is known as the **Jacobian matrix of G**. Multiplication by this matrix is clearly a bounded linear transformation from $\mathcal{V}$ into $\mathcal{W}$ as required by the definition of $\mathbf{dG}(\mathbf{x},\mathbf{h})$.

Observe that the notation $\mathbf{G}'(\mathbf{x})$ defined in (7.7) can conflict with the standard notation for the derivative of a function. In the scalar case (7.1), $f'(s_0)$ is a number which multiplies ds. The Fréchet derivative of f at s_0, which we also denote by $f'(s_0)$, consists in *multiplication by* the number $f'(s_0)$. The analogue of the *number* $f'(s_0)$ in the vector case (7.17) is $\partial\,\mathbf{G}/\partial\,\mathbf{x}$, the Jacobian matrix of $\mathbf{G}$ at $\mathbf{x}$. Thus $\mathbf{G}'(\mathbf{x})\mathbf{h}=(\partial\,\mathbf{G}/\partial\,\mathbf{x})\mathbf{h}$; in other words, the transformation $\mathbf{G}'(\mathbf{x})$ consists in *multiplication by* the matrix $\partial\,\mathbf{G}/\partial\,\mathbf{x}$.

Unconstrained Extrema

We now turn to the problem that is the focus of this chapter, the minimization or maximization of an objective function. We apply the term **objective function** to any real-valued transformation (or functional) for which we seek a maximum or minimum. We usually denote functionals by the symbol **F**. We use the term **extremum** in reference to either a maximum or minimum of $\mathbf{F}(\mathbf{x})$. We acknowledge the distinction between local (relative) extrema and global (absolute) extrema. We content ourselves with a search for local extrema. A discussion of conditions under which extrema are guaranteed to be global may be found in Hadley [7.7].

Suppose **F** is a nonlinear functional on some space $\mathcal{V}$. Assume $\mathbf{F}(\mathbf{x})$ is smooth (has a Fréchet differential) at $\mathbf{x}_0$, and has a local minimum at $\mathbf{x}_0$. Then as $\mathbf{x}$ approaches $\mathbf{x}_0$ along any direction vector $\mathbf{h}$, the rate of change of $\mathbf{F}(\mathbf{x})$ must approach zero. That is,

$$\lim_{\beta\to 0}\frac{d}{d\beta}\mathbf{F}(\mathbf{x}+\beta\mathbf{h})=0$$

for all $\mathbf{h}$ in $\mathcal{V}$. But this is just the requirement that the differential [as determined by (7.10)] be zero. The same conclusion holds for a relative maximum. We conclude, then, that the maxima and minima of an arbitrary functional **F** occur at points where the tangent plane is horizontal, a conclusion which is not surprising in the light of the one and two-dimensional examples of Figures 7.1 and 7.2.

Definition. Let $\mathcal{V}$ be an inner product space and let $\mathbf{F}: \mathcal{V}\to\mathcal{R}$ be Fréchet differentiable.* A **critical point** (or stationary point) of **F** is a point $\mathbf{x}$ in $\mathcal{V}$ such that

$$\mathbf{dF}(\mathbf{x},\mathbf{h})=0 \quad \text{for all } \mathbf{h} \text{ in } \mathcal{V} \tag{7.18}$$

A search for extrema of the objective function **F** begins with a search for the critical points of **F**. The necessary condition $\mathbf{dF}(\mathbf{x},\mathbf{h})=0$ merely provides an equation to be solved for the critical points. Because the equation is usually nonlinear, obtaining the solution can be difficult. (See Section 8.1.) Not all critical points are extrema; the tangent to **F** will also be horizontal at a horizontal inflection point of **F**. One simple approach for distinguishing between minima, maxima, and inflection points is to check the values of **F** for points $\mathbf{x}$ near each critical point. Or, since we usually seek a global maximum or minimum, we may be able to just select the largest or smallest of the critical points. We should also note that an extremum of **F** can lie at a point $\mathbf{x}$ where **F** is not smooth. Under this

*The definition applies as well if $\mathcal{V}$ is a normal linear space. See P&C 7.1.

circumstance, the necessary condition $\mathbf{dF}(\mathbf{x},\mathbf{h})=0$ does not apply. An example of such a situation is the function $f(s)=|s|$. The minimum of f occurs at $s=0$, a point where f has no tangent.

If $\mathcal{V}$ is a Hilbert space, then the differential $\mathbf{dF}(\mathbf{x},\mathbf{h})$ can be expressed in terms of the gradient $\nabla\mathbf{F}(\mathbf{x})$. The condition

$$\mathbf{dF}(\mathbf{x},\mathbf{h})=\langle\mathbf{h},\nabla\mathbf{F}(\mathbf{x})\rangle=0 \quad \text{for all } \mathbf{h} \text{ in } \mathcal{V}$$

is equivalent to

$$\nabla\mathbf{F}(\mathbf{x})=\boldsymbol{\theta} \tag{7.19}$$

We can think of (7.19) as an alternate definition of the critical points in a Hilbert space. Using this definition, the critical points of the function $\mathbf{F}(\mathbf{f})=\|\mathbf{f}\|^2$ of Example 4 are the solutions of $\nabla\mathbf{F}(\mathbf{f})=2\mathbf{f}=\boldsymbol{\theta}$. The only solution is $\mathbf{f}=\boldsymbol{\theta}$, the obvious minimum of $\mathbf{F}$.

Exercise 2. Define $\mathbf{F}$: $\mathcal{L}_2(\omega; a,b)\rightarrow\mathcal{R}$ by

$$\mathbf{F}(\mathbf{f})=\|\mathbf{f}\|_\omega^2=\int_a^b \omega(t)\mathbf{f}^2(t)\,dt$$

Find the gradient $\nabla_\omega\mathbf{F}(\mathbf{f})$; use it to show that the minimum of $\mathbf{F}$ occurs at $\mathbf{f}=\boldsymbol{\theta}$.

Example 6. A Least-Square Minimization. Let $\mathbf{A}$ be a real $m\times n$ matrix. Let $\mathbf{Tx}\triangleq\mathbf{Ax}$ define a linear transformation $\mathbf{T}$ between the Hilbert spaces $\mathfrak{M}^{n\times 1}$ and $\mathfrak{M}^{m\times 1}$. Assume the standard inner product on these spaces. We seek the minimum of the functional $\mathbf{F}(\mathbf{x})=\frac{1}{2}\|\mathbf{y}-\mathbf{Ax}\|^2$. If $\operatorname{rank}(\mathbf{A})=n$, this is an example of Problem 1 (Section 6.3). The minimum must satisfy the necessary condition

$$\mathbf{dF}(\mathbf{x},\,\mathbf{h})=\frac{d}{d\beta}\frac{1}{2}\langle\mathbf{y}-\mathbf{A}(\mathbf{x}+\beta\mathbf{h}),\mathbf{y}-\mathbf{A}(\mathbf{x}+\beta\mathbf{h})\rangle|_{\beta=0}=0$$

for all $\mathbf{h}$ in $\mathfrak{M}^{n\times 1}$. In order to evaluate the differential, we need to take the derivative inside both the inner product and the matrix multiplication. According to Exercises 3 and 4 (below), we find that

$$\begin{aligned}\mathbf{dF}(\mathbf{x},\mathbf{h})&=\tfrac{1}{2}\langle\mathbf{y}-\mathbf{A}(\mathbf{x}+\beta\mathbf{h}),\frac{d}{d\beta}(\mathbf{y}-\mathbf{A}(\mathbf{x}+\beta\mathbf{h}))\rangle|_{\beta=0}\\&\quad+\tfrac{1}{2}\langle\frac{d}{d\beta}(\mathbf{y}-\mathbf{A}(\mathbf{x}+\beta\mathbf{h})),\mathbf{y}-\mathbf{A}(\mathbf{x}+\beta\mathbf{h})\rangle|_{\beta=0}\\&=\tfrac{1}{2}\langle\mathbf{y}-\mathbf{Ax},-\mathbf{Ah}\rangle+\tfrac{1}{2}\langle-\mathbf{Ah},\mathbf{y}-\mathbf{Ax}\rangle\\&=\langle-\mathbf{Ah},\mathbf{y}-\mathbf{Ax}\rangle\\&=\langle\mathbf{h},-\mathbf{A}^{\mathrm{T}}(\mathbf{y}-\mathbf{Ax})\rangle=0\end{aligned}$$

Clearly, $\nabla \mathbf{F}(\mathbf{x}) = -\mathbf{A}^T(\mathbf{y}-\mathbf{A}\mathbf{x}) = \boldsymbol{\theta}$, and the minimum $\mathbf{x}$ satisfies the normal equations $\mathbf{A}^T\mathbf{A}\mathbf{x} = \mathbf{A}^T\mathbf{y}$. If $\mathbf{A}^T\mathbf{A}$ is invertible, the solution is unique, and is expressed explicitly by $\mathbf{x} = (\mathbf{A}^T\mathbf{A})^{-1}\mathbf{A}^T\mathbf{y}$. We obtained the same result in (6.39).

Exercise 3. Let $\mathbf{A}$ be an $m \times n$ matrix. Let $\mathbf{x}(\beta)$ be an $n \times 1$ column vector with elements that depend upon the scalar β. Assume that the first derivative of $\mathbf{x}$, element by element, exists and is denoted by $\mathbf{x}'(\beta)$. Show that

$$\frac{d}{d\beta}\mathbf{A}\mathbf{x}(\beta) = \mathbf{A}\mathbf{x}'(\beta) \tag{7.20}$$

Exercise 4. Suppose that $\mathbf{y}(\beta)$ and $\mathbf{z}(\beta)$ are $n \times 1$ column vectors with elements that depend upon the scalar β. Assume that first derivatives of $\mathbf{y}$ and $\mathbf{z}$, element by element, exist and are denoted by $\mathbf{y}'(\beta)$ and $\mathbf{z}'(\beta)$, respectively. Show that*

$$\frac{d}{d\beta}\langle \mathbf{y}(\beta), \mathbf{z}(\beta)\rangle = \langle \mathbf{y}(\beta), \mathbf{z}'(\beta)\rangle + \langle \mathbf{y}'(\beta), \mathbf{z}(\beta)\rangle \tag{7.21}$$

A Problem in the Calculus of Variations

The calculus of variations is concerned with the minimization of functionals defined on a function space. Our previous minimization of $\|\mathbf{f}\|^2$ is a trivial example. Suppose we wish to determine the shape $\mathbf{f}(s)$ of a playground slide that will provide the fastest ride for a "frictionless" child (Figure 7.4). Assume the child has mass m and zero initial velocity. During the child's ride, his potential energy (relative to the top of the slide) is converted to kinetic energy; that is, at each instant of time,

$$\tfrac{1}{2}mv^2(t) = mg(\mathbf{f}(0) - \mathbf{f}(t))$$

where v is the velocity of the child and g is the acceleration due to gravity. The notation $\mathbf{f}(t)$ is intended to mean $\mathbf{f}(s(t))$; at each instant t the child is at some position $(s(t), \mathbf{f}(s(t)))$ on the slide. We specify that the top and bottom of the slide must be at the points $(0,0)$ and $(1,-1)$, respectively. Then $v(t) = (-2g\mathbf{f}(s(t)))^{1/2}$.

We seek the function (or slide) $\mathbf{f}$ which minimizes the total time of descent, $t_f = \int_0^{t_f} dt$, where the final time t_f is unknown. But $v(t) = dl/dt$, where l denotes position along the slide. Therefore in the integral we can

*Equation (7.20) extends to arbitrary continuous linear transformations between normed spaces. Equation (7.21) extends to arbitrary inner product spaces. See Collatz [7.3].

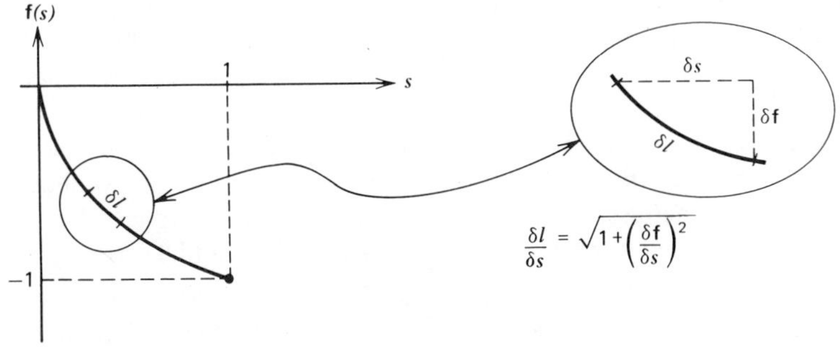

Figure 7.4. Determination of slide shape for a fast ride.

make the substitution

$$dt = \frac{dl}{v(t)} = \frac{dl}{\sqrt{-2g\mathbf{f}(s(t))}}$$

Furthermore, it is apparent from Figure 7.4 that we can make the additional substitution

$$dl = \sqrt{1+\left(\frac{d\mathbf{f}}{ds}\right)^2}\, ds$$

Consequently, we seek the function $\mathbf{f}$ which satisfies $\mathbf{f}(0)=0$ and $\mathbf{f}(1)=-1$, and which minimizes the integral

$$\mathbf{G}(\mathbf{f}) \triangleq \frac{1}{\sqrt{2g}} \int_0^1 \sqrt{\frac{1+(\mathbf{f}'(s))^2}{-\mathbf{f}(s)}}\, ds \tag{7.22}$$

where the superscript (′) denotes differentiation with respect to s. In the calculus of variations literature, this problem is referred to as a *brachistochrone* (shortest time) problem. We solve a generalization of this problem by setting a gradient equal to zero. We use the brachistochrone problem as an illustration.

Functionals defined on a function space are usually written as integrals. A fairly general form of functional which allows for the explicit appearance of $\mathbf{f}(t), \mathbf{f}'(t)$, and t in the integral is

$$\mathbf{G}(\mathbf{f}) \triangleq \int_a^b \mathbf{F}(\mathbf{f}(t), \mathbf{f}'(t), t)\, dt \tag{7.23}$$

The integral (7.22) is an example. Suppose we search some function space $\mathcal{V}$ defined on $[a,b]$ for a function $\mathbf{f}$ which has fixed end points $\mathbf{f}(a)$ and $\mathbf{f}(b)$, and which minimizes (7.23). [For the integral (7.22) it is apparent that the functions in $\mathcal{V}$ must be at least differentiable.] Because the fixed end points need not be zero, the functions to be searched are constrained to lie in a hyperplane in $\mathcal{V}$. We modify the problem slightly so that we can perform an unconstrained minimization in a vector space. Let $\mathbf{f}_1$ be an arbitrarily chosen function for which $\mathbf{f}_1(a)=\mathbf{f}(a)$ and $\mathbf{f}_1(b)=\mathbf{f}(b)$. Define $\mathbf{g} \stackrel{\Delta}{=} \mathbf{f}-\mathbf{f}_1$. Then $\mathbf{g}(a)=\mathbf{g}(b)=0$. Let $\mathcal{W}$ be the manifold of functions $\mathbf{g}$ in $\mathcal{V}$ for which $\mathbf{g}(a)=\mathbf{g}(b)=0$. Rather than seek $\mathbf{f}$ in the hyperplane to minimize $\mathbf{G}(\mathbf{f})$, we seek $\mathbf{g}$ in the vector space $\mathcal{W}$ to minimize $\hat{\mathbf{G}}(\mathbf{g}) \stackrel{\Delta}{=} \mathbf{G}(\mathbf{f}_1+\mathbf{g})=\mathbf{G}(\mathbf{f})$. According to the optimization principle expressed in (7.18), $\hat{\mathbf{G}}(\mathbf{g})$ can be minimum only if $\mathbf{d}\hat{\mathbf{G}}(\mathbf{g},\mathbf{h})=0$ for all $\mathbf{h}$ in $\mathcal{W}$. The differential is

$$
\begin{aligned}
\mathbf{d}\hat{\mathbf{G}}(\mathbf{g},\mathbf{h}) &= \frac{d}{d\beta}\hat{\mathbf{G}}(\mathbf{g}+\beta\mathbf{h})\big|_{\beta=0} = \frac{d}{d\beta}\mathbf{G}(\mathbf{f}+\beta\mathbf{h})\big|_{\beta=0} \\
&= \frac{d}{d\beta}\int_a^b \mathbf{F}(\mathbf{f}+\beta\mathbf{h},\mathbf{f}'+\beta\mathbf{h}',t)\,dt\big|_{\beta=0} \\
&= \int_a^b \left[\mathbf{F}_{\mathbf{f}}(\mathbf{f},\mathbf{f}',t)\mathbf{h}(t)+\mathbf{F}_{\mathbf{f}'}(\mathbf{f},\mathbf{f}',t)\mathbf{h}'(t)\right]dt \\
&= \int_a^b \left[\mathbf{F}_{\mathbf{f}}(\mathbf{f},\mathbf{f}',t)-\frac{d}{dt}\mathbf{F}_{\mathbf{f}'}(\mathbf{f},\mathbf{f}',t)\right]\mathbf{h}(t)\,dt+\underbrace{\mathbf{F}_{\mathbf{f}'}(\mathbf{f},\mathbf{f}',t)\mathbf{h}(t)\big|_a^b}_{0} \\
&= \langle \mathbf{h},\nabla\hat{\mathbf{G}}(\mathbf{g})\rangle
\end{aligned}
\tag{7.24}
$$

To make the notation more compact, we have abbreviated the arguments of $\mathbf{F}$ rather than use the full expression (7.23). The symbol $\mathbf{F}_{\mathbf{f}}(\mathbf{f},\mathbf{f}',t)$ is a standard abbreviation for $[\partial\mathbf{F}/\partial\mathbf{f}(t)](\mathbf{f}(t),\mathbf{f}'(t),t)$. Since $d\hat{\mathbf{G}}(\mathbf{g},\mathbf{h})=0$ for all $\mathbf{h}$ in $\mathcal{W}$, the gradient of $\hat{\mathbf{G}}(\mathbf{g})$ must be zero at the minimizing vector $\mathbf{g}$.* Because $\mathbf{d}\hat{\mathbf{G}}(\mathbf{g},\mathbf{h})=\mathbf{dG}(\mathbf{f},\mathbf{h})$, it follows that $\nabla\hat{\mathbf{G}}(\mathbf{g})=\nabla\mathbf{G}(\mathbf{f})$. Therefore

$$
\nabla\mathbf{G}(\mathbf{f})=\mathbf{F}_{\mathbf{f}}(\mathbf{f},\mathbf{f}',t)-\frac{d}{dt}\mathbf{F}_{\mathbf{f}'}(\mathbf{f},\mathbf{f}',t)=\boldsymbol{\theta}
\tag{7.25}
$$

Equation (7.25) is known as the **Euler–Lagrange equation** for the fixed end-point minimization problem (7.23). Notice that there is no need to

*This statement assumes that $\nabla\hat{\mathbf{G}}(\mathbf{g})$ (or $\nabla\mathbf{G}(\mathbf{f})$) is in $\mathcal{W}$. Because $\mathcal{W}$ is usually not complete, this assumption need not be valid. The calculus of variations literature usually restricts $\mathbf{F}$ in such a manner that $\nabla\mathbf{G}(\mathbf{f})$ lies in $\mathcal{W}$. See Gelfand and Fomin [7.5].

concern ourselves with $\mathbf{f}_1$ or $\mathbf{g}$ in (7.25); this equation is expressed in terms of the gradient of the original function $\mathbf{G(f)}$. As a practical matter, we can ignore the fixed end-point conditions on $\mathbf{f}$ during the determination of the necessary conditions (7.25). However, we must use these end-point conditions as boundary conditions for the Euler–Lagrange equation in order to solve for the minimizing vector $\mathbf{f}$. In the calculus of variations literature the differential $\mathbf{dG(f,h)}$ is called the **first variation** of $\mathbf{G}$. The space $\mathfrak{W}$ over which $\mathbf{h}$ is allowed to range is known as the space of **admissible variations.** See Gelfand and Fomin [7.5] for a discussion of similar problems with variable end-point conditions.

Example 7. The Brachistochrone Problem. The function $\mathbf{f}$ which minimizes the integral (7.22) specifies the shape of the playground slide which provides the fastest ride for a "frictionless" child. The integrand (ignoring the constant $1/\sqrt{2g}$) is

$$\mathbf{F}(\mathbf{f},\mathbf{f}',s)=\sqrt{\frac{1+(\mathbf{f}'(s))^2}{-\mathbf{f}(s)}}$$

The quantities required by the Euler–Lagrange equation (7.25) are

$$\frac{\partial \mathbf{F}}{\partial \mathbf{f}}=\tfrac{1}{2}\cdot\frac{\sqrt{1+(\mathbf{f}')^2}}{(-\mathbf{f})^{3/2}},\qquad \frac{\partial \mathbf{F}}{\partial \mathbf{f}'}=\frac{\mathbf{f}'}{\sqrt{-\mathbf{f}(1+(\mathbf{f}')^2)}}$$

$$\frac{d}{ds}\frac{\partial \mathbf{F}}{\partial \mathbf{f}'}=\frac{-2\mathbf{f}\mathbf{f}''+(\mathbf{f}')^2+(\mathbf{f}')^4}{2\left(-\mathbf{f}(1+(\mathbf{f}')^2)\right)^{3/2}}$$

The Euler–Lagrange equation, $\partial\mathbf{F}/\partial\mathbf{f}=(d/ds)(\partial\mathbf{F}/\partial\mathbf{f}')$, reduces to

$$2\mathbf{f}\mathbf{f}''+(\mathbf{f}')^2+1=0$$

a nonlinear differential equation which must be solved to determine the optimum shape of the slide. The solution of such equations is not straightforward. In this instance, the solution can be expressed in terms of a parameter ϕ:

$$\mathbf{f}(\phi)=a(\cos\phi-1),\qquad s(\phi)=s_0+a(\sin\phi-\phi)$$

This solution can be verified by substituting it into the differential equation and using the identities

$$\mathbf{f}'=\frac{d\mathbf{f}/d\phi}{ds/d\phi},\qquad \mathbf{f}''=\frac{d\mathbf{f}'/d\phi}{ds/d\phi}$$

The solution has the form of a cycloid generated by the motion of a fixed point on the circumference of a circle of radius a as it rolls along the line $\mathbf{f}(s)=0$. For the boundary conditions $\mathbf{f}(0)=0$ and $\mathbf{f}(1)=-1$, the constants are $s_0=0$ and $a\approx0.575$. The solution is displayed in Figure 7.4.

7.2 Optimization with Equality Constraints

The optimization problems of the preceding section are straightforward. In this section we consider the effect on the optimization process of requiring, as a side condition to the maximization or minimization, that certain equations be satisfied; that is, we explore optimization problems which contain equality constraints.

Let $\mathbf{x}=(\xi_1\ \xi_2\ \xi_3)^{\mathrm{T}}$ denote an arbitrary point in $\mathfrak{M}^{3\times 1}$. Suppose we drop a ball of weight w onto the terrain determined by the paraboloid $\xi_3=\xi_1^2+\xi_2^2$ and the plane $\xi_2=1$ as shown in Figure 7.5. If friction damps out the motion of the ball, it will eventually come to rest at some point which lies on both constraining surfaces. The point where it rests is that point where $w\xi_3$, its potential energy relative to the origin, is minimized. Obviously, the minimizing point is $\mathbf{x}=(0\ 1\ 1)^{\mathrm{T}}$.

From physical considerations we know the weight of the ball must be balanced by the forces exerted by the constraining surfaces. Mathemati-

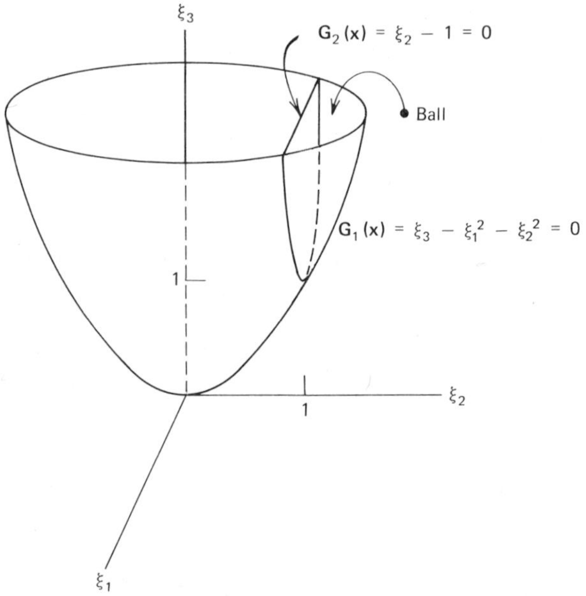

Figure 7.5. An energy minimization problem with constraints.

cally, we express this fact as follows. Define $\mathbf{F}(\mathbf{x}) \triangleq w\xi_3$, $\mathbf{G}_1(\mathbf{x}) \triangleq \xi_3-\xi_1^2-\xi_2^2$, and $\mathbf{G}_2(\mathbf{x}) \triangleq \xi_2-1$. The force exerted by constraint i must act in a direction collinear with $\nabla\mathbf{G}_i(\mathbf{x})$, the direction perpendicular to the surface $\mathbf{G}_i(\mathbf{x})=0$. The weight of the ball acts in the direction of $-\nabla\mathbf{F}(\mathbf{x})$, the direction of the greatest rate of decrease in $\mathbf{F}(\mathbf{x})$. Therefore the balance of forces requires that

$$\nabla\mathbf{F}(\mathbf{x})=\lambda_1\nabla\mathbf{G}_1(\mathbf{x})+\lambda_2\nabla\mathbf{G}_2(\mathbf{x}) \tag{7.26}$$

for some numbers λ_1 and λ_2 (Figure 7.6). Also, the variable $\mathbf{x}$ must lie on the constraint surfaces $\mathbf{G}_i(\mathbf{x})=0$. Therefore, we have five equations which determine the five unknowns:

$$\begin{aligned}\begin{pmatrix}0\\0\\w\end{pmatrix}&=\lambda_1\begin{pmatrix}-2\xi_1\\-2\xi_2\\1\end{pmatrix}+\lambda_2\begin{pmatrix}0\\1\\0\end{pmatrix}\\ \xi_3&=\xi_1^2+\xi_2^2\\ \xi_2&=1\end{aligned} \tag{7.27}$$

The solution to (7.27) is $\mathbf{x}=(0\ 1\ 1)^{\mathrm{T}}$, $\lambda_1=w$, and $\lambda_2=2w$.

The above analysis has produced a pair of numbers λ_1 and λ_2, known as *Lagrange multipliers*, which have as much practical significance as the elements of the minimizing position vector $\mathbf{x}$. In this particular instance, the Lagrange multipliers specify the magnitudes of the forces exerted by the constraints; that is, they indicate how hard each constraint is working. As we investigate further the properties of Lagrange multipliers, we find

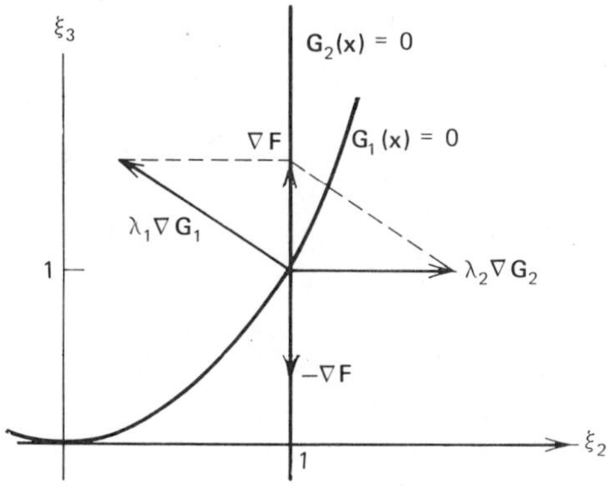

Figure 7.6. Balance of forces.

that the multipliers indicate the sensitivity of the minimum value of the objective function to changes in each of the constraints.

The method of analysis used above provides a general technique, known as the *method of Lagrange multipliers*, for finding local minima of real-valued functionals subject to equality constraints. If $\mathbf{F}$ is a real-valued functional acting on vectors $\mathbf{x}$ in a Hilbert space $\mathcal{V}$, if $\mathbf{G}_i(\mathbf{x})=0$, $i=1,\ldots,m$, and if at a constrained local minimum $\mathbf{x}$ the constraints act independently (i.e., the set $\{\nabla\mathbf{G}_i(\mathbf{x})\}$ is linearly independent), then the minimum $\mathbf{x}$ satisfies

$$\begin{aligned} \nabla\mathbf{F}(\mathbf{x}) &= \sum_{i=1}^{m} \lambda_i \nabla\mathbf{G}_i(\mathbf{x}) \\ \mathbf{G}_i(\mathbf{x}) &= 0, \qquad i=1,\ldots,m \end{aligned} \tag{7.28}$$

for some finite set of numbers $\lambda_1,\ldots,\lambda_m$. [We assume that the functional $\mathbf{F}$ and the constraint functions $\{\mathbf{G}_i\}$ are smooth (Fréchet differentiable) in the neighborhood of the extrema.] Of course, the method applies to maximization problems as well. As in the unconstrained examples of Section 7.1, the method converts the problem of finding extrema to the problem of solving a set of equations, (7.28). These equations are typically nonlinear and difficult to solve. Furthermore, a particular solution to (7.28) is not necessarily a maximum or minimum of $\mathbf{F}$; it may be a horizontal inflection point of the constrained optimization problem. As a practical matter, we often use some ad hoc scheme for determining whether a constrained critical point is a local or global extremum.

The Adjoint Equation

The Lagrange multiplier method can be derived in a manner that applies to problems which are more general than the gravity problem discussed above. Suppose the problem is to

$$\text{minimize } \mathbf{F}(\mathbf{x}) \text{ subject to } \mathbf{G}(\mathbf{x}) = \boldsymbol{\theta} \tag{7.29}$$

where $\mathbf{F}\colon \mathcal{V}\to\mathcal{R}$ and $\mathbf{G}\colon \mathcal{V}\to\mathcal{W}$ are transformations (usually nonlinear) that are Fréchet differentiable near the minima. Assume $\mathcal{V}$ and $\mathcal{W}$ are Hilbert spaces. If $\mathbf{F}$ has a constrained minimum at a particular point $\mathbf{x}$, then excursions $\mathbf{h}$ about $\mathbf{x}$ which do not violate the constraints should not lower the value of $\mathbf{F}$. If we restrict our discussion to very small excursions about the minimizing $\mathbf{x}$, it is reasonable to replace $\mathbf{F}$ and $\mathbf{G}$ by their linearizations (or tangent functions); we pick $\mathbf{h}$ to

$$\begin{aligned} &\text{minimize} && \mathbf{F}(\mathbf{x}+\mathbf{h}) \approx \mathbf{F}(\mathbf{x}) + \mathbf{dF}(\mathbf{x},\mathbf{h}) \\ &\text{subject to} && \mathbf{G}(\mathbf{x}+\mathbf{h}) \approx \mathbf{G}(\mathbf{x}) + \mathbf{dG}(\mathbf{x},\mathbf{h}) = \boldsymbol{\theta} \end{aligned} \tag{7.30}$$

Thus excursions $\mathbf{h}$ which lie along the tangent to the constraint, $\mathbf{dG}(\mathbf{x},\mathbf{h}) = \boldsymbol{\theta}$, must not cause $\mathbf{F}(\mathbf{x})+\mathbf{dF}(\mathbf{x},\mathbf{h})$ to vary from the value $\mathbf{F}(\mathbf{x})$. Consequently, in order for $\mathbf{x}$ to be a solution to (7.29), it is necessary that

$$\mathbf{dF}(\mathbf{x},\mathbf{h})=0 \quad \text{for all } \mathbf{h} \text{ such that } \mathbf{dG}(\mathbf{x},\mathbf{h})=\boldsymbol{\theta} \tag{7.31}$$

We refer to the vectors $\mathbf{x}$ which satisfy (7.31) as **constrained critical points** of $\mathbf{F}$. The necessary condition (7.31) can be put in a more easily recognizable form if we express the differentials in the form $\mathbf{dF}(\mathbf{x},\mathbf{h})=\langle \mathbf{h}, \nabla\mathbf{F}(\mathbf{x})\rangle_{\mathcal{V}}$ and $\mathbf{dG}(\mathbf{x},\mathbf{h})=\mathbf{G}'(\mathbf{x})\mathbf{h}$. The permissible excursions $\mathbf{h}$, which satisfy $\mathbf{G}'(\mathbf{x})\mathbf{h} = \boldsymbol{\theta}$, are the vectors in nullspace($\mathbf{G}'(\mathbf{x})$). The necessary condition (7.31) is $\langle \mathbf{h}, \nabla\mathbf{F}(\mathbf{x})\rangle_{\mathcal{V}}=0$ for all $\mathbf{h}$ in nullspace($\mathbf{G}'(\mathbf{x})$); that is, $\nabla\mathbf{F}(\mathbf{x})$ must be orthogonal to nullspace($\mathbf{G}'(\mathbf{x})$). But by the orthogonal decomposition theorem (5.67), the orthogonal complement of nullspace($\mathbf{G}'(\mathbf{x})$) is range($\mathbf{G}'(\mathbf{x})^*$).‡ As a result, $\nabla\mathbf{F}(\mathbf{x})=\mathbf{G}'(\mathbf{x})^*\boldsymbol{\lambda}$ for some vector $\boldsymbol{\lambda}$ in $\mathcal{W}$. Thus if (7.29) has a solution $\mathbf{x}$, if the linearized versions of $\mathbf{F}$ and $\mathbf{G}$ adequately represent $\mathbf{F}$ and $\mathbf{G}$ near $\mathbf{x}$, and if the constraint function $\mathbf{G}$ is not "degenerate" at $\mathbf{x}$, there necessarily exists a unique $\boldsymbol{\lambda}$ in $\mathcal{W}$, which together with $\mathbf{x}$ in $\mathcal{V}$, satisfies

$$\begin{aligned} &\nabla\mathbf{F}(\mathbf{x})=\mathbf{G}'(\mathbf{x})^*\boldsymbol{\lambda} \quad &&\textbf{(Adjoint equation)}\\ &\mathbf{G}(\mathbf{x})=\boldsymbol{\theta} &&\textbf{(Primal equation)} \end{aligned} \tag{7.32}$$

We call the vector $\boldsymbol{\lambda}$ the **Lagrange multiplier vector**; thus we occasionally refer to the adjoint equation as the *Lagrange multiplier equation*.† We note that the sign on $\boldsymbol{\lambda}$ is arbitrary. Some authors write the adjoint equation as $\nabla\mathbf{F}(\mathbf{x})= -\mathbf{G}'(\mathbf{x})^*\boldsymbol{\lambda}$. This statement concerning the sign convention on $\boldsymbol{\lambda}$ applies also to (7.28) and to later developments which involve Lagrange multipliers. A reversal in the sign convention results in a change in the sign of all Lagrange multipliers.

Equations (7.32) apply unchanged if we seek a maximum rather than a minimum in (7.29). The process of determining the solutions to (7.29) typically consists in solving the equations (7.32), then employing some scheme to distinguish the constrained local maxima and minima from other constrained critical points which may also satisfy (7.32). To find a description of the operator $\mathbf{G}'(\mathbf{x})^*$ which appears in the adjoint equation,

‡According to (5.67), $(\text{nullspace}(\mathbf{G}'(\mathbf{x})))^{\perp}=\text{range}(\mathbf{G}'(\mathbf{x})^*)$. Thus if range($\mathbf{G}'(\mathbf{x})^*$) were not closed, the gradient $\nabla\mathbf{F}(\mathbf{x})$ could be in the closure of range($\mathbf{G}'(\mathbf{x})^*$) but not in range($\mathbf{G}'(\mathbf{x})^*$) itself. We assume range($\mathbf{G}'(\mathbf{x})^*$) is closed.

†The fact that $\mathcal{V}$ and $\mathcal{W}$ are assumed to be Hilbert spaces is not a significant restriction in practice. In point of fact, inner products are not essential. Luenberger [7.11] derives a replacement for the adjoint equation of (7.32) which applies in normed spaces.

we use the inner product equation which defines the adjoint:

$$\langle \mathbf{dG}(\mathbf{x},\mathbf{h}),\boldsymbol{\lambda}\rangle_{\mathcal{W}} = \langle \mathbf{G}'(\mathbf{x})\mathbf{h},\boldsymbol{\lambda}\rangle_{\mathcal{W}} = \langle \mathbf{h},\mathbf{G}'(\mathbf{x})^*\boldsymbol{\lambda}\rangle_{\mathcal{V}} \tag{7.33}$$

Example 1. The Adjoint Equation for Finite Dimensional $\mathcal{V}$ ***and*** $\mathcal{W}$**.** Let $\mathbf{F},\mathbf{G}_1,\ldots,\mathbf{G}_m$ be real-valued functionals which act on the standard inner product space $\mathfrak{M}^{n\times 1}$. Define $\mathbf{G}$: $\mathfrak{M}^{n\times 1}\to\mathfrak{M}^{m\times 1}$ by $\mathbf{G}(\mathbf{x}) \overset{\Delta}{=} (\mathbf{G}_1(\mathbf{x})\cdots\mathbf{G}_m(\mathbf{x}))^{\mathrm{T}}$. Assume the standard inner product on $\mathfrak{M}^{m\times 1}$. Using (7.33) and (7.17), we find

$$\begin{aligned}\langle \mathbf{dG}(\mathbf{x},\mathbf{h}),\boldsymbol{\lambda}\rangle &= \boldsymbol{\lambda}^{\mathrm{T}}(\nabla\mathbf{G}_1(\mathbf{x})\cdots\nabla\mathbf{G}_m(\mathbf{x}))^{\mathrm{T}}\mathbf{h}\\ &= \langle \mathbf{h},(\nabla\mathbf{G}_1(\mathbf{x})\cdots\nabla\mathbf{G}_m(\mathbf{x}))\boldsymbol{\lambda}\rangle\end{aligned}$$

Consequently, the necessary equations (7.32) are

$$\begin{aligned}&\nabla\mathbf{F}(\mathbf{x}) = (\nabla\mathbf{G}_1(\mathbf{x})\cdots\nabla\mathbf{G}_m(\mathbf{x}))\boldsymbol{\lambda}\\ &\mathbf{G}_i(\mathbf{x}) = 0, \qquad i=1,\ldots,m\end{aligned}$$

These equations are essentially the same as those in (7.28). Equations (7.27) are a special case.

Example 2. Least-Square Minimization with Equality Constraints. Let $\mathbf{A}$ be a real $m\times n$ matrix. Define $\mathbf{T}$: $\mathfrak{M}^{n\times 1}\to\mathfrak{M}^{m\times 1}$ by $\mathbf{Tx} \overset{\Delta}{=} \mathbf{Ax}$. Assume standard inner products. We determine the minimum of $\mathbf{F}(\mathbf{x}) \overset{\Delta}{=} \frac{1}{2}\|\mathbf{x}\|^2$ subject to the equality constraints $\mathbf{G}(\mathbf{x}) \overset{\Delta}{=} \mathbf{Ax}-\mathbf{y}=\boldsymbol{\theta}$. If rank$(\mathbf{A})=m$, this is an example of Problem 2 (Section 6.4). We employ (7.20) and (7.21) to determine $\mathbf{dF}$ and $\mathbf{dG}$:

$$\mathbf{dF}(\mathbf{x},\mathbf{h}) = \frac{1}{2}\frac{d}{d\beta}\langle \mathbf{x}+\beta\mathbf{h},\mathbf{x}+\beta\mathbf{h}\rangle|_{\beta=0} = \langle \mathbf{h},\mathbf{x}\rangle = \langle \mathbf{h},\nabla\mathbf{F}(\mathbf{x})\rangle$$

$$\mathbf{dG}(\mathbf{x},\mathbf{h}) = \frac{d}{d\beta}(\mathbf{A}(\mathbf{x}+\beta\mathbf{h})-\mathbf{y})|_{\beta=0} = \mathbf{Ah} = \mathbf{G}'(\mathbf{x})\mathbf{h}$$

We find $\mathbf{G}'(\mathbf{x})^*$ by (7.33):

$$\begin{aligned}\langle \mathbf{G}'(\mathbf{x})\mathbf{h},\boldsymbol{\lambda}\rangle &= \langle \mathbf{Ah},\boldsymbol{\lambda}\rangle = \boldsymbol{\lambda}^{\mathrm{T}}\mathbf{Ah}\\ &= (\mathbf{A}^{\mathrm{T}}\boldsymbol{\lambda})^{\mathrm{T}}\mathbf{h} = \langle \mathbf{h},\mathbf{A}^{\mathrm{T}}\boldsymbol{\lambda}\rangle = \langle \mathbf{h},\mathbf{G}'(\mathbf{x})^*\boldsymbol{\lambda}\rangle\end{aligned}$$

Then the necessary conditions (7.32) are

$$\begin{aligned}&\nabla\mathbf{F}(\mathbf{x}) = \mathbf{x} = \mathbf{A}^{\mathrm{T}}\boldsymbol{\lambda} = \mathbf{G}'(\mathbf{x})^*\boldsymbol{\lambda}\\ &\mathbf{G}(\mathbf{x}) = \mathbf{Ax}-\mathbf{y} = \boldsymbol{\theta}\end{aligned}$$

Substituting $\mathbf{x}$ from the adjoint equation into the primal equation, we obtain

$$\mathbf{AA}^{\mathrm{T}}\boldsymbol{\lambda}=\mathbf{y} \quad \text{and} \quad \mathbf{x}=\mathbf{A}^{\mathrm{T}}\boldsymbol{\lambda}$$

If $\mathbf{AA}^{\mathrm{T}}$ is invertible, then the Lagrange multiplier vector $\boldsymbol{\lambda}$ is unique and $\mathbf{x} = \mathbf{A}^{\mathrm{T}}(\mathbf{AA}^{\mathrm{T}})^{-1}\mathbf{y}$. We obtained the same result in (6.48).

Degenerate Constraints

We indicated previously that the adjoint equation (7.32) specifies a unique $\boldsymbol{\lambda}$ in the Hilbert space $\mathcal{W}$ if "nothing is wrong" with the constraint function $\mathbf{G}$ at the extremum $\mathbf{x}$. If there is no $\boldsymbol{\lambda}$ that satisfies (7.32) or if the $\boldsymbol{\lambda}$ vector is not unique, we say the constraint $\mathbf{G}$ is **degenerate** at $\mathbf{x}$. In what ways might the constraint function $\mathbf{G}$ be degenerate at the minimizing vector $\mathbf{x}$ of problem (7.29)? We first explore this question for a finite number of constraints, using the notation of (7.28). If the constraint gradients $\{\nabla G_i(\mathbf{x}), i=1,\ldots,m\}$ are linearly independent at the constrained minimum $\mathbf{x}$, it can be shown that there exists a set of Lagrange multipliers $\{\lambda_i\}$ which satisfies the adjoint equation, $\nabla F(\mathbf{x})=\sum_{i=1}^{m}\lambda_i\nabla G_i(\mathbf{x})$, at the constrained minimum; that is, $\nabla F(\mathbf{x})$ is in span$\{\nabla G_i(\mathbf{x})\}$.‡ Furthermore, since the gradients $\{\nabla G_i(\mathbf{x})\}$ are independent, the Lagrange multipliers must be unique as well.

If the constraint gradients $\{\nabla G_i(\mathbf{x})\}$ are not independent, the Lagrange multipliers which satisfy the adjoint equation of (7.28) need not be unique. For example, if we replace the vertical constraint of the minimum potential energy problem of Figure 7.5 by the horizontal constraint $G_2(\mathbf{x}) \triangleq \xi_3=0$, then the constraint surfaces are tangent at the minimum [$\nabla G_1(\boldsymbol{\theta})$ and $\nabla G_2(\boldsymbol{\theta})$ are collinear], and one of the two Lagrange multipliers can be chosen arbitrarily.

As another example of degeneracy in the constraint equations, denote $\mathbf{x}=(\xi_1\ \xi_2\ \xi_3)^{\mathrm{T}}$ in $\mathfrak{M}^{3\times 1}$, and seek to minimize $F(\mathbf{x}) \triangleq \xi_2$ subject to $G_1(\mathbf{x}) \triangleq (\xi_2-\xi_3^2)^3+\xi_1=0$ and $G_2(\mathbf{x}) \triangleq (\xi_2-\xi_3^2)^3-\xi_1=0$. Since the pair of constraints is equivalent to the single condition $\xi_2-\xi_3^2=0$, the minimum of $F(\mathbf{x})$ is clearly at $\mathbf{x}=\boldsymbol{\theta}$. But at $\mathbf{x}=\boldsymbol{\theta}$, $\nabla G_1(\boldsymbol{\theta})=(1\ 0\ 0)^{\mathrm{T}}$, $\nabla G_2(\boldsymbol{\theta})=(-1\ 0\ 0)^{\mathrm{T}}$, and $\nabla F=(0\,1\,0)^{\mathrm{T}}$. Consequently, the Lagrange multiplier equation, $\nabla F(\mathbf{x}) =\lambda_1\nabla G_1(\mathbf{x})+\lambda_2\nabla G_2(\mathbf{x})$, has no finite solution, λ_1 and λ_2, at $\mathbf{x}=\boldsymbol{\theta}$; the constraint gradients $\{\nabla G_i(\boldsymbol{\theta})\}$ are linearly dependent, and $\nabla F(\boldsymbol{\theta})$ is not in span$\{\nabla G_i(\boldsymbol{\theta})\}$.

If we wish to delineate those minimization problems for which there exists a unique Lagrange multiplier vector $\boldsymbol{\lambda}$ which, together with a

‡See Hadley [7.7, p. 66] for a rigorous proof of this statement for the case where $\mathbf{x}$ is an n-element vector.

minimizing vector $\mathbf{x}$, satisfies (7.32), it is apparent that we must determine some restriction upon the constraint transformation $\mathbf{G}$ to guarantee that the effect of $\mathbf{G}$ is not degenerate at the constrained minimum. For a finite set of constraints, we have noted that the linear independence of the constraint gradients, $\{\nabla \mathbf{G}_i(\mathbf{x})\}$, is a sufficient constraint qualification. For the more general problem (7.29), wherein $\mathcal{W}$ can be infinite dimensional, we express this same constraint qualification by saying that $\mathbf{G}'(\mathbf{x})^*$ is one-to-one at the constrained minimum $\mathbf{x}$. An equivalent statement concerning $\mathbf{G}$ is that $\mathbf{G}'(\mathbf{x})$ is onto $\mathcal{W}$.* Thus if the $m \times n$ matrix $\mathbf{A}$ of Example 2 has rank m, then $\mathbf{G}'(\mathbf{x})$ is onto $\mathfrak{M}^{m\times 1}$, $\mathbf{AA}^{\mathrm{T}}$ is invertible, and a unique $\boldsymbol{\lambda}$ exists.

If the functions $\mathbf{F}$ and $\mathbf{G}$ are not sufficiently smooth at the solutions to (7.29), the adjoint equation of (7.32) does not apply. Consider the minimization of $\mathbf{F}(\xi_1,\xi_2) \triangleq \xi_2$ subject to the constraint $\mathbf{G}(\xi_1,\xi_2) \triangleq \xi_2 - |\xi_1| = 0$. At the constrained minimum, $\mathbf{x}=(0,0)$, $\mathbf{G}$ has no gradient (or tangent). Problems of this type often can be solved by the iterative techniques of Chapter 8.

A Minimum Energy Control Problem

In Section 6.4 we use the concept of an adjoint transformation to find the armature voltage function $\mathbf{u}$ which would consume minimum energy, $\int_0^1 \mathbf{u}^2(t)\,dt$, in driving a motor shaft from the condition $\phi(0)=\dot{\phi}(0)=0$ to the condition $\phi(1)=1$, $\dot{\phi}(1)=0$. The state-space form of the differential equation which describes the system is given in (6.51). By inverting the differential equation and using the required final state of the system, we previously obtained the following integral description of the system:

$$\mathbf{Tu} \triangleq \int_0^1 \begin{pmatrix} 1-e^{-1}e^s \\ e^{-1}e^s \end{pmatrix} \mathbf{u}(s)\,ds = \begin{pmatrix} \phi(1) \\ \dot{\phi}(1) \end{pmatrix} \triangleq \mathbf{y} \tag{7.34}$$

We now restate this constrained minimum energy problem in the terminology of this chapter. Define $\mathbf{F}$: $\mathcal{C}(0,1)\rightarrow \mathcal{R}$ and $\mathbf{G}$: $\mathcal{C}(0,1)\rightarrow \mathfrak{M}^{2\times 1}$ by

$$\begin{aligned} \mathbf{F}(\mathbf{u}) &\triangleq \tfrac{1}{2}\int_0^1 \mathbf{u}^2(t)\,dt = \tfrac{1}{2}\|\mathbf{u}\|^2 \\ \mathbf{G}(\mathbf{u}) &\triangleq \mathbf{Tu}-\mathbf{y} = \int_0^1 \begin{pmatrix} 1-e^{-1}e^s \\ e^{-1}e^s \end{pmatrix} \mathbf{u}(s)\,ds - \begin{pmatrix} \phi(1) \\ \dot{\phi}(1) \end{pmatrix} = \begin{pmatrix} 0 \\ 0 \end{pmatrix} \end{aligned} \tag{7.35}$$

*See Luenberger [7.11, p. 242] for a proof (for infinite-dimensional $\mathcal{W}$) that if $\mathbf{G}'(\mathbf{x})$ is onto $\mathcal{W}$ vector $\boldsymbol{\lambda}$ exists which satisfies (7.32) at the constrained minimum $\mathbf{x}$. It can also be shown that $\boldsymbol{\lambda}$ is unique.

We seek the control function $\mathbf{u}$ that minimizes $\mathbf{F}$ subject to $\mathbf{G}(\mathbf{u})=\boldsymbol{\theta}$. This is precisely the problem (7.29).

Because $\mathbf{T}$ is a linear transformation, this minimization problem has precisely the same mathematical structure as the matrix problem of Example 2; we should expect to go through precisely the same formal steps in solving both problems. The only significant difference between the two problems is that the space $\mathcal{C}(0,1)$ is infinite dimensional, whereas $\mathfrak{M}^{n\times 1}$ is finite dimensional. We use the standard inner products for $\mathcal{C}(0,1)$ and $\mathfrak{M}^{2\times 1}$. The differentials of $\mathbf{F}$ and $\mathbf{G}$ are

$$\mathbf{dF}(\mathbf{u},\mathbf{h})=\frac{1}{2}\frac{d}{d\beta}\int_0^1(\mathbf{u}(t)+\beta\mathbf{h}(t))^2\,dt\Big|_{\beta=0}$$

$$=\int_0^1\mathbf{u}(t)\mathbf{h}(t)\,dt=\langle\mathbf{h},\nabla\mathbf{F}(\mathbf{u})\rangle$$

$$\mathbf{dG}(\mathbf{u},\mathbf{h})=\frac{d}{d\beta}\left[\int_0^1\begin{pmatrix}1-e^{-1}e^{s}\\ e^{-1}e^{s}\end{pmatrix}(\mathbf{u}(s)+\beta\mathbf{h}(s))\,ds-\begin{pmatrix}\phi(1)\\ \dot{\phi}(1)\end{pmatrix}\right]_{\beta=0}$$

$$=\int_0^1\begin{pmatrix}1-e^{-1}e^{s}\\ e^{-1}e^{s}\end{pmatrix}\mathbf{h}(s)\,ds=\mathbf{G}'(\mathbf{u})\mathbf{h}$$

Thus $\nabla\mathbf{F}(\mathbf{u})=\mathbf{u}$ and $\mathbf{G}'(\mathbf{u})=\mathbf{T}$; the results are identical to the matrix example. Then $\mathbf{G}'(\mathbf{u})^*=\mathbf{T}^*$, and the necessary conditions (7.32) are

$$\nabla\mathbf{F}(\mathbf{u})=\mathbf{u}=\mathbf{T}^*\boldsymbol{\lambda}=\mathbf{G}'(\mathbf{u})^*\boldsymbol{\lambda}$$

$$\mathbf{G}(\mathbf{u})=\mathbf{T}\mathbf{u}-\mathbf{y}=\boldsymbol{\theta}$$

Substituting $\mathbf{u}$ from the adjoint equation into the primal equation, we obtain

$$\mathbf{T}\,\mathbf{T}^*\boldsymbol{\lambda}=\mathbf{y}\quad\text{and}\quad\mathbf{u}=\mathbf{T}^*\boldsymbol{\lambda}\tag{7.36}$$

The quantity $\mathbf{T}\,\mathbf{T}^*$ is invertible in this instance; we computed $\mathbf{T}^*$ and $(\mathbf{T}\,\mathbf{T}^*)^{-1}$ for this problem in (6.56) and the succeeding computations. For the final conditions $\phi(1)=1$, $\dot{\phi}(1)=0$, these computations show that

$$\begin{aligned}\boldsymbol{\lambda}&=\begin{pmatrix}\dfrac{e+1}{3-e} & -\dfrac{e-1}{3-e}\end{pmatrix}^{\mathrm{T}}\\ \mathbf{u}(t)&=\frac{e+1}{3-e}-\frac{2}{3-e}e^{t}\end{aligned}\tag{7.37}$$

During the derivation of the pseudoinverse for Problem 2, we found that we could not avoid determining the vector $\boldsymbol{\lambda}$ as an intermediate step in the computation of a least-effort solution. In the context of the present discussion of constrained minimization, we find that this intermediate vector is the Lagrange multiplier vector for the problem. In the next section we discover that the Lagrange multiplier vector contains useful information concerning the sensitivity of the minimum value of $\mathbf{F}(\mathbf{u})$ to changes in the constraints.

The Lagrangian

A useful contribution to the theory of constrained optimization is the discovery that the solutions $(\mathbf{x},\boldsymbol{\lambda})$ to the constrained minimization problem (7.29) are also *unconstrained* critical points of another functional known as the Lagrangian. Suppose $\mathbf{x}$ is a vector in $\mathcal{R}^n$, and we wish to minimize the functional $\mathbf{F}(\mathbf{x})$ subject to the constraints $\mathbf{G}_i(\mathbf{x})=0, i=1,\ldots,m$. We define the Lagrangian function for this problem by

$$\mathcal{L}(\mathbf{x},\boldsymbol{\lambda}) \triangleq \mathbf{F}(\mathbf{x}) - \sum_{i=1}^{m} \lambda_i \mathbf{G}_i(\mathbf{x}) \tag{7.38}$$

where λ_i denotes the ith element of $\boldsymbol{\lambda}$. The Lagrangian is a function of the $n+m$ element vector $(\mathbf{x},\boldsymbol{\lambda})$. By (7.19), the critical points of the unconstrained functional $\mathcal{L}(\mathbf{x},\boldsymbol{\lambda})$ are those vectors $(\mathbf{x},\boldsymbol{\lambda})$ for which the gradient of $\mathcal{L}$ is zero. The gradient is zero where the partial derivatives of $\mathcal{L}$ with respect to all the variables are zero. Denote the elements of $\mathbf{x}$ by $\mathbf{x}=(\xi_1,\ldots,\xi_n)$. Then the critical points of (7.38) satisfy

$$\begin{aligned} \frac{\partial \mathcal{L}(\mathbf{x},\boldsymbol{\lambda})}{\partial \xi_j} &= \frac{\partial \mathbf{F}(\mathbf{x})}{\partial \xi_j} - \sum_{i=1}^{m} \lambda_i \frac{\partial \mathbf{G}_i(\mathbf{x})}{\partial \xi_j} = 0, \qquad j=1,\ldots,n \\ \frac{\partial \mathcal{L}(\mathbf{x},\boldsymbol{\lambda})}{\partial \lambda_i} &= -\mathbf{G}_i(\mathbf{x}) = 0, \qquad i=1,\ldots,m \end{aligned} \tag{7.39}$$

The equations (7.39) are identical to the necessary conditions (7.28) for the constrained minimization of $\mathbf{F}$. As a result, a constrained minimum of $\mathbf{F}$ is necessarily a critical point of the unconstrained Lagrangian, (7.38). We extend this conclusion to all constrained minimization problems of the form (7.29).

Let $\mathbf{F}\colon \mathcal{V}\to\mathcal{R}$ and $\mathbf{G}\colon \mathcal{V}\to\mathcal{W}$ be as in (7.29). We define the **Lagrangian function** associated with the constrained minimization by

$$\mathcal{L}(\mathbf{x},\boldsymbol{\lambda}) \triangleq \mathbf{F}(\mathbf{x}) - \langle \mathbf{G}(\mathbf{x}),\boldsymbol{\lambda}\rangle_{\mathcal{W}} \tag{7.40}$$

Some authors use a plus sign in the definition of the Lagrangian. However, the choice of sign is arbitrary; it merely affects the sign of $\boldsymbol{\lambda}$. We sometimes refer to $\mathbf{x}$ and $\boldsymbol{\lambda}$, respectively, as the **primal variable** and the **dual variable** for the minimization problem (7.29).

The vector $(\mathbf{x},\boldsymbol{\lambda})$ is in $\mathcal{V}\times\mathcal{W}$. The differential of $\mathcal{L}(\mathbf{x},\boldsymbol{\lambda})$ is

$$\begin{aligned}
\mathbf{d}\mathcal{L}((\mathbf{x},\boldsymbol{\lambda}),(\mathbf{h}_x,\mathbf{h}_\lambda)) &= \frac{d}{d\beta}\mathcal{L}(\mathbf{x}+\beta\mathbf{h}_x,\lambda+\beta\mathbf{h}_\lambda)|_{\beta=0} \\
&= \frac{d}{d\beta}\mathbf{F}(\mathbf{x}+\beta\mathbf{h}_x)|_{\beta=0} - \frac{d}{d\beta}\langle\mathbf{G}(\mathbf{x}+\beta\mathbf{h}_x,\lambda+\beta\mathbf{h}_\lambda)\rangle_{\mathcal{W}}|_{\beta=0} \\
&= \mathbf{dF}(\mathbf{x},\mathbf{h}_x) - \langle\frac{d}{d\beta}\mathbf{G}(\mathbf{x}+\beta\mathbf{h}_x)|_{\beta=0},\lambda\rangle_{\mathcal{W}} - \langle\mathbf{G}(\mathbf{x}),\mathbf{h}_\lambda\rangle_{\mathcal{W}} \\
&= \mathbf{dF}(\mathbf{x},\mathbf{h}_x) - \langle\mathbf{dG}(\mathbf{x},\mathbf{h}_x),\lambda\rangle_{\mathcal{W}} - \langle\mathbf{G}(\mathbf{x}),\mathbf{h}_\lambda\rangle_{\mathcal{W}} \\
&= \langle\mathbf{h}_x,\nabla\mathbf{F}(\mathbf{x})-\mathbf{G}'(\mathbf{x})^*\boldsymbol{\lambda}\rangle_{\mathcal{V}} + \langle\mathbf{h}_\lambda,-\mathbf{G}(\mathbf{x})\rangle_{\mathcal{W}} \\
&= \langle(\mathbf{h}_x,\mathbf{h}_\lambda),(\nabla\mathbf{F}(\mathbf{x})-\mathbf{G}'(\mathbf{x})^*\lambda,-\mathbf{G}(\mathbf{x}))\rangle_{\mathcal{V}\times\mathcal{W}} \qquad (7.41)
\end{aligned}$$

The third line of the derivation of (7.41), wherein the derivative is taken inside the inner product, relies on (7.20). In the last line we use an inner product for the cartesian product space $\mathcal{V}\times\mathcal{W}$ as discussed in P&C 5.3. The gradient of $\mathcal{L}(\mathbf{x},\boldsymbol{\lambda})$ is clearly $\nabla\mathcal{L}(\mathbf{x},\boldsymbol{\lambda})=(\nabla\mathbf{F}(\mathbf{x})-\mathbf{G}'(\mathbf{x})^*\boldsymbol{\lambda},-\mathbf{G}(\mathbf{x}))$. The differential of $\mathcal{L}(\mathbf{x},\boldsymbol{\lambda})$ equals zero for all increments $(\mathbf{h}_x,\mathbf{h}_\lambda)$ in $\mathcal{V}\times\mathcal{W}$ if and only if the gradient of $\mathcal{L}(\mathbf{x},\boldsymbol{\lambda})$ equals zero. Consequently, the unconstrained critical points of $\mathcal{L}(\mathbf{x},\boldsymbol{\lambda})$ are precisely identical to the constrained critical points of $\mathbf{F}$, the solutions to (7.32).*

We use the Lagrangian to explore the effect of changes in the constraints on the minimum value of $\mathbf{F}(\mathbf{x})$. In the process of this exploration, we find that the Lagrange multipliers possess precisely the information we seek. Assume $\mathbf{F}$: $\mathcal{R}^n\to\mathcal{R}$ and $\mathbf{G}$: $\mathcal{R}^n\to\mathcal{R}^m$. Denote the elements of the primal and dual variables, respectively, by $\mathbf{x}=(\xi_1,\ldots,\xi_n)$ and $\boldsymbol{\lambda}=(\lambda_1,\ldots,\lambda_m)$. Let $\mathbf{x}_0$ denote a constrained minimum for (7.29), and let $\boldsymbol{\lambda}_0$ denote the corresponding Lagrange multiplier vector. Figure 7.7 illustrates a typical constraint, $\mathbf{G}_i(\mathbf{x})=0$, as viewed near the constrained minimum in the space $\mathcal{R}^n$. A slight tightening or loosening of the constraint (a motion of the surface perpendicular to the surface itself) can be expressed mathematically by the equation $\mathbf{G}_i(\mathbf{x})=v_i$ for some scalar v_i. The Lagrangian corresponding to slightly perturbed constraints is

$$\mathcal{L}(\mathbf{x},\boldsymbol{\lambda},\mathbf{v})=\mathbf{F}(\mathbf{x})-\sum_{i=1}^{m}\lambda_i(\mathbf{G}_i(\mathbf{x})-v_i)$$

*See Hadley [7.7] for a discussion of the nature of the critical points of the Lagrangian.

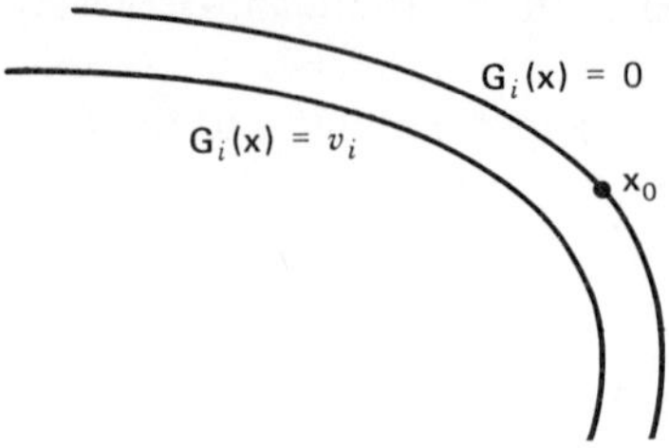

Figure 7.7. A perturbed constraint.

Define $\mathbf{F}_{\min}$ by

$$\mathbf{F}_{\min}(\mathbf{v}) \stackrel{\Delta}{=} \min_{\mathbf{G}(\mathbf{x})=\mathbf{v}} \mathbf{F}(\mathbf{x}) = \mathcal{L}(\mathbf{x},\boldsymbol{\lambda},\mathbf{v})\Big|_{\substack{\mathbf{x}=\mathbf{x}_v\\ \boldsymbol{\lambda}=\boldsymbol{\lambda}_v}} \tag{7.42}$$

where $\mathbf{v}=(v_1,\ldots,v_m)$, and the subscript v denotes the solution to the problem "minimize $\mathbf{F}(\mathbf{x})$ with $\mathbf{G}(\mathbf{x})=\mathbf{v}$." Then the value of $\mathbf{F}$ at the constrained minimum is

$$\mathbf{F}(\mathbf{x}_0)=\mathbf{F}_{\min}(\mathbf{v})\big|_{\mathbf{v}=\boldsymbol{\theta}}=\mathcal{L}(\mathbf{x},\boldsymbol{\lambda},\mathbf{v})\Big|_{\substack{\mathbf{x}=\mathbf{x}_v\\ \boldsymbol{\lambda}=\boldsymbol{\lambda}_v\\ \mathbf{v}=\boldsymbol{\theta}}}$$

Using (7.39), we determine the change in the minimum value of $\mathbf{F}$ which results from a slight shifting of the jth constraint from its original position:

$$\frac{\partial \mathbf{F}_{\min}(\mathbf{v})}{\partial v_j}\Big|_{\mathbf{v}=\boldsymbol{\theta}} = \frac{\partial \mathcal{L}(\mathbf{x},\boldsymbol{\lambda},\mathbf{v})}{\partial v_j}\Big|_{\substack{\mathbf{x}=\mathbf{x}_v\\ \boldsymbol{\lambda}=\boldsymbol{\lambda}_v\\ \mathbf{v}=\boldsymbol{\theta}}}$$

$$= \left[\sum_{k=1}^{n}\left(\underbrace{\frac{\partial \mathcal{L}}{\partial \xi_k}}_{0}\frac{\partial \xi_k}{\partial v_j}\right)+\sum_{k=1}^{m}\left(\underbrace{\frac{\partial \mathcal{L}}{\partial \lambda_k}}_{0}\frac{\partial \lambda_k}{\partial v_j}\right)+\underbrace{\frac{\partial \mathcal{L}}{\partial v_j}}_{\lambda_j}\right]\Bigg|_{\substack{\mathbf{x}=\mathbf{x}_v\\ \boldsymbol{\lambda}=\boldsymbol{\lambda}_v\\ \mathbf{v}=\boldsymbol{\theta}}}$$

$$=\lambda_j\big|_{\boldsymbol{\lambda}=\boldsymbol{\lambda}_0} \tag{7.43}$$

We see that jth Lagrange multiplier is the sensitivity of the constrained minimum value of $\mathbf{F}(\mathbf{x})$ to perturbations in the specified value (or position) of the jth constraint. These sensitivities provide important information for system design. If $\mathbf{F}(\mathbf{x})$ is a performance index, the magnitude of λ_j indicates how seriously the constraint hampers us in our attempt to optimize performance. Thus the dual variables are fundamental to constrained minimization. The solution to the problem with perturbed constraints, described by (7.42), can be expressed in terms of the solution to the unperturbed problem if the perturbations $\{v_i\}$ are small:

$$\mathbf{F}_{\min}(\mathbf{v}) \approx \mathbf{F}_{\min}(\boldsymbol{\theta}) + \sum_{i=1}^{m} \frac{\partial \mathbf{F}_{\min}(\boldsymbol{\theta})}{\partial v_i} v_i = \mathbf{F}(\mathbf{x}_0) + \sum_{i=1}^{m} \lambda_i v_i \tag{7.44}$$

The sign of the ith Lagrange multiplier indicates, via (7.44), which sign v_i must have in order to tighten the ith constraint. The ith constraint will be tightened (the minimum of $\mathbf{F}(\mathbf{x})$ increased) if the sign on v_i is the same as the sign on λ_i. Definition (7.42) and the sensitivity results (7.43) and (7.44) apply for maximization of $\mathbf{F}(\mathbf{x})$ if we replace $\mathbf{F}_{\min}(v)$ by $\mathbf{F}_{\max}(v)$. For maximization, however, the ith constraint will be tightened (the maximum of $\mathbf{F}(\mathbf{x})$ decreased) if the sign on v_i is opposite to the sign on λ_i. Although (7.43) and (7.44) are derived only for $\mathbf{G}$: $\mathfrak{R}^n \rightarrow \mathfrak{R}^m$, they apply more generally. We demonstrate their use in the following two examples.

Example 3. Sensitivity to Constraints in a Finite-Dimensional Problem. Consider the potential energy minimization illustrated in Figure 7.5, in which $\mathbf{F}(\mathbf{x}) = w\xi_3$, $\mathbf{G}_1(\mathbf{x}) = \xi_3 - \xi_1^2 - \xi_2^2$, and $\mathbf{G}_2(\mathbf{x}) = \xi_2 - 1$. From (7.27) we found the constrained minimum: $\mathbf{x} = (0\ 1\ 1)^{\mathrm{T}}$, $\lambda_1 = w$, and $\lambda_2 = 2w$. From (7.43) we know that

$$\frac{\partial \mathbf{F}_{\min}(\boldsymbol{\theta})}{\partial v_1} = \lambda_1 = w \quad \text{and} \quad \frac{\partial \mathbf{F}_{\min}(\boldsymbol{\theta})}{\partial v_2} = \lambda_2 = 2w$$

Thus the potential energy, $\mathbf{F}(\mathbf{x}_0)$, is twice as sensitive to slight perturbations in constraint 2 as it is to perturbations in constraint 1. A greater reduction in energy (altitude) can be obtained by relaxing constraint 2 than by relaxing constraint 1. These sensitivities are illustrated for a ball of weight $w=1$ in Figure 7.8. Note that the sensitivities predict precisely the effect of perturbations of the *linearized* constraints. Thus the sensitivities apply only for small perturbations.

Example 4. Sensitivity to Constraints in a Function Space. Earlier in this section we used the adjoint equations to determine the solution to a minimum energy control problem. The quantity minimized, $\mathbf{F}(\mathbf{u}) = \frac{1}{2}\int_0^1 \mathbf{u}^2(t)\,dt$, is the energy required to drive the shaft position ϕ of an electric motor from the condition $\phi(0) = \dot{\phi}(0) = 0$ to the condition $\phi(1) = 1$, $\dot{\phi}(1) = 0$. The system constraints, (7.34), can be expressed as

$$\mathbf{G}_1(\mathbf{u}) \triangleq \int_0^1 (1 - e^{-1}e^{s})\mathbf{u}(s)\,ds - \phi(1) = 0$$

$$\mathbf{G}_2(\mathbf{u}) \triangleq \int_0^1 e^{-1}e^{s}\mathbf{u}(s)\,ds - \dot{\phi}(1) = 0$$

For $\phi(1) = 1$ and $\dot{\phi}(1) = 0$, the minimum energy control and the corresponding Lagrange multiplier vector are given in (7.37). Thus the Lagrange multipliers

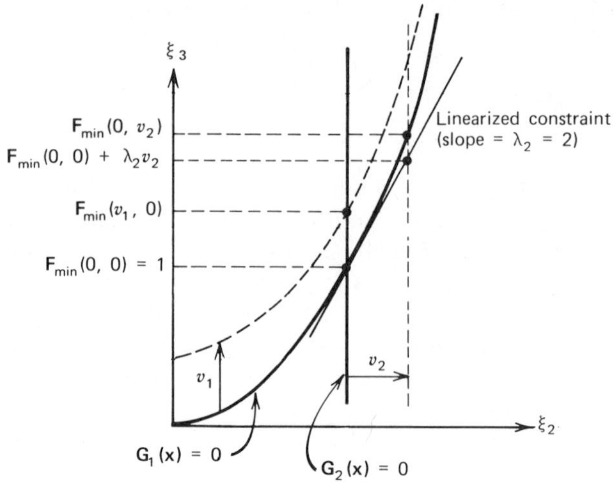

Figure 7.8. Sensitivity of $\mathbf{F}_{\min}$ to constraint perturbations (Example 3).

corresponding to $\mathbf{G}_1$ and $\mathbf{G}_2$ are $\lambda_1 = (e+1)/(3-e)$ and $\lambda_2 = -(e-1)/(3-e)$. The minimum energy, which results from use of the minimum energy control, is $\mathbf{F}(\mathbf{u}) = (\frac{1}{2})(e+1)/(3-e)$. Suppose we slightly perturb the final conditions, requiring instead that $\phi(1) = 1 + v_1$ and $\dot{\phi}(1) = v_2$. Then, by (7.44), the new minimum energy is

$$\mathbf{F}_{\min}(\mathbf{v}) = (\tfrac{1}{2})\left(\frac{e+1}{3-e}\right) + \left(\frac{e+1}{3-e}\right)v_1 - \left(\frac{e-1}{3-e}\right)v_2$$

An increase in $\phi(1)$ requires more energy—the motor must turn farther within the given time. An increase in $\dot{\phi}(1)$, on the other hand, reduces the energy requirement; less energy is consumed if we do not fully stop the motor by the final instant. Note that the energy consumed is more than twice as sensitive to changes in the requirement on $\phi(1)$ as it is to changes in the requirement to $\dot{\phi}(1)$.

An Isoperimetric Problem

A classical problem in the calculus of variations is the determination of the shape of a piece of string which will cause it to enclose maximum area. One version of this "fixed-perimeter" maximization problem is illustrated in Figure 7.9. Let $\mathbf{f}(t)$ describe the shape of a string of length l. Assume $\mathbf{f}(a) = \mathbf{f}(b) = 0$. We seek the continuous function $\mathbf{f}$ which encloses the largest area in the "top half-plane." The area under the curve is related to $\mathbf{f}$ by

$$\text{Area} = \mathbf{F}(\mathbf{f}) = \int_a^b \mathbf{f}(t)\,dt \qquad (7.45)$$

The relationship between the length of the string and the shape of the string follows from Figure 7.4:

$$\mathbf{G}(\mathbf{f})=\int_a^b \sqrt{1+(\mathbf{f}'(t))^2}\,dt-l=0 \tag{7.46}$$

Thus we must find the function **f** which maximizes **F(f)** subject to **G(f)**=0 and $\mathbf{f}(a)=\mathbf{f}(b)=0$.

The above problem can be phrased in more general terms: Pick **f** in $\mathcal{C}(a,b)$ to

$$\begin{aligned} &\text{maximize} \quad \mathbf{F}(\mathbf{f})=\int_a^b \hat{\mathbf{F}}(\mathbf{f},\mathbf{f}',t)\,dt \\ &\text{with} \quad \mathbf{G}(\mathbf{f})=\int_a^b \hat{\mathbf{G}}(\mathbf{f},\mathbf{f}',t)\,dt-l=0 \\ &\qquad \mathbf{f}(a) \text{ and } \mathbf{f}(b) \text{ fixed} \end{aligned} \tag{7.47}$$

Function space problems of this type are referred to as *isoperimetric* problems with fixed end points. We discovered during the derivation of the Euler–Lagrange equation (7.25) for an unconstrained calculus of variations problem that fixed end-point conditions can be ignored during the derivation of the necessary conditions. The end-point conditions need only be applied as boundary conditions at the time the necessary conditions are solved for the optimum **f**. Maximizing **F(f)** is equivalent to minimizing −**F(f)**. According to (7.32), the necessary conditions for a minimum of −**F(f)** are

$$-\nabla\mathbf{F}(\mathbf{f})=\mathbf{G}'(\mathbf{f})^*\lambda=\lambda\nabla\mathbf{G}(\mathbf{f})$$
$$\mathbf{G}(\mathbf{f})=\boldsymbol{\theta}$$

where the gradients are defined in terms of the standard function space

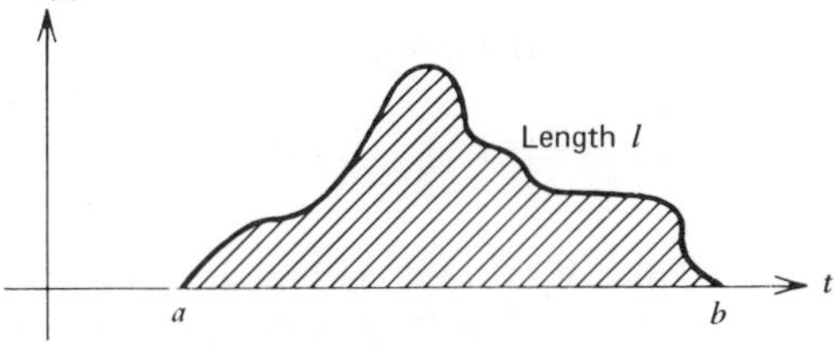

Figure 7.9. Maximization of the area under a curve.

inner product.* We obtain differentials and gradients in the same manner as in (7.24):

$$\mathbf{dF}(\mathbf{f},\mathbf{h}) = \int_a^b \left[\hat{\mathbf{F}}_{\mathbf{f}}(\mathbf{f},\mathbf{f}',t) - \frac{d}{dt}\hat{\mathbf{F}}_{\mathbf{f}'}(\mathbf{f},\mathbf{f}',t) \right] \mathbf{h}(t)\,dt$$

$$= \langle \mathbf{h}, \nabla \mathbf{F}(\mathbf{f}) \rangle$$

$$\mathbf{dG}(\mathbf{f},\mathbf{h}) = \int_a^b \left[\hat{\mathbf{G}}_{\mathbf{f}}(\mathbf{f},\mathbf{f}',t) - \frac{d}{dt}\hat{\mathbf{G}}_{\mathbf{f}'}(\mathbf{f},\mathbf{f}',t) \right] \mathbf{h}(t)\,dt$$

$$= \langle \mathbf{h}, \nabla \mathbf{G}(\mathbf{f}) \rangle$$

Consequently, the necessary conditions become

$$-\hat{\mathbf{F}}_{\mathbf{f}} + \frac{d}{dt}\hat{\mathbf{F}}_{\mathbf{f}'} = \lambda\left(\hat{\mathbf{G}}_{\mathbf{f}} - \frac{d}{dt}\hat{\mathbf{G}}_{\mathbf{f}'}\right)$$

$$\mathbf{G}(\mathbf{f}) = 0 \tag{7.48}$$

$$\mathbf{f}(a) \text{ and } \mathbf{f}(b) \text{ fixed}$$

Equation (7.48) can also be derived by means of the Lagrangian

$$\mathcal{L}(\mathbf{f},\lambda) = -\mathbf{F}(\mathbf{f}) - \lambda\mathbf{G}(\mathbf{f}) = \int_a^b (-\hat{\mathbf{F}} - \lambda\hat{\mathbf{G}})\,dt$$

By analogy to the unconstrained minimization treated in (7.23)–(7.25), we see that an unconstrained critical point of $\mathcal{L}(\mathbf{f},\lambda)$ must satisfy $\nabla_{\mathbf{f}}\mathcal{L}(\mathbf{f},\lambda) = (-\hat{\mathbf{F}} - \lambda\hat{\mathbf{G}})_{\mathbf{f}} - (d/dt)\ (-\hat{\mathbf{F}} - \lambda\hat{\mathbf{G}})_{\mathbf{f}'} = \boldsymbol{\theta}$. But this is the first equation of (7.48). We refer to it as the *Euler–Lagrange equation* for the constrained maximization problem (7.47).

Example 5. Maximization of the Area in Figure 7.9. For the isoperimetric problem of Figure 7.9, the functions $\hat{\mathbf{F}}$ and $\hat{\mathbf{G}}$ can be determined from (7.45) and (7.46):

$$\hat{\mathbf{F}}(\mathbf{f},\mathbf{f}',t) = \mathbf{f}$$

$$\hat{\mathbf{G}}(\mathbf{f},\mathbf{f}',t) = \sqrt{1 + (\mathbf{f}')^2}$$

*Actually the condition $-\nabla\mathbf{F}(\mathbf{f}) = \lambda\nabla\mathbf{G}(\mathbf{f})$ need not be a necessary condition unless certain restrictions are placed on the functions **F** and **G**. See the footnote associated with (7.24).

We obtain the Euler–Lagrange equation for this problem from the first equation of (7.48):

$$-\frac{\partial\,(\mathbf{f})}{\partial \mathbf{f}}+\frac{d}{dt}\underbrace{\frac{\partial (\mathbf{f})}{\partial \mathbf{f}'}}_{0}=\lambda\left(\frac{\partial}{\partial \mathbf{f}}\sqrt{1+(\mathbf{f}')^2}-\frac{d}{dt}\frac{\partial}{\partial \mathbf{f}'}\sqrt{1+(\mathbf{f}')^2}\right)$$

or

$$\frac{1}{\lambda}=\frac{d}{dt}\left(\frac{\mathbf{f}'}{\sqrt{1+(\mathbf{f}')^2}}\right)$$

The solutions to this differential equation are circles of the form $(\mathbf{f}(t)-c)^2+(t-d)^2=r^2$. The constants c, d, and r can be determined by applying the boundary conditions $\mathbf{f}(a)=\mathbf{f}(b)=0$ and the constraint equation, $\int_a^b[1+(\mathbf{f}')^2]^{1/2}dt=l$. The manipulations are messy; they show that

$$d=\frac{a+b}{2},\qquad c=r\cos\left(\frac{l}{2r}\right)$$

and r is determined by the equation

$$\sin\left(\frac{l}{2r}\right)=\frac{b-a}{2r}$$

The solution is depicted in Figure 7.10.

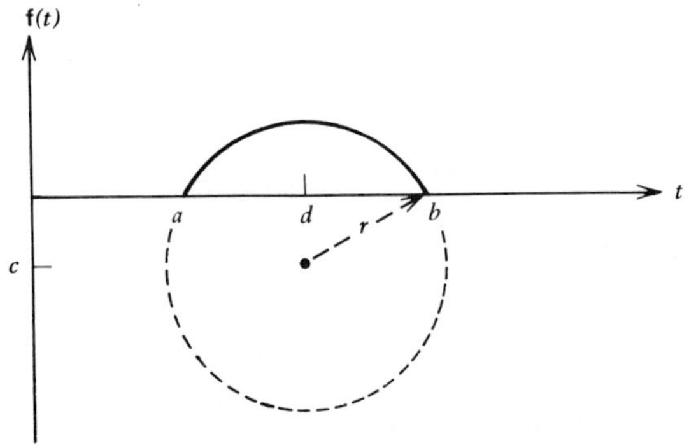

Figure 7.10. The solution to the isoperimetric problem of Figure 7.9.

Optimal Control Via the Adjoint State Equation

In an earlier section we determined the control function $\mathbf{u}$ which would drive the differential system $\ddot{\phi}+\dot{\phi}=\mathbf{u}$ (an electric motor) from the state $\phi(0)=\dot{\phi}(0)=0$ to the state $\phi(1)=1$, $\dot{\phi}(1)=0$ while minimizing $\int_0^1 \mathbf{u}^2(t)\,dt$. In that section we began the solution process by finding and inverting the state equation [see (7.34) and (7.35)]. In this section we attack the problem by working directly with the differential form of the state equation. We define $\mathbf{f}_1 \triangleq \phi$ and $\mathbf{f}_2 \triangleq \dot{\phi}$, and convert the differential equation to the state variable form

$$\begin{aligned} \mathbf{G}_1(\mathbf{f}_1,\mathbf{f}_2) &\triangleq \dot{\mathbf{f}}_1-\mathbf{f}_2=\boldsymbol{\theta}, \qquad \mathbf{f}_1(0)=0, \qquad \mathbf{f}_1(1)=1 \\ \mathbf{G}_2(\mathbf{f}_2,\mathbf{u}) &\triangleq \dot{\mathbf{f}}_2+\mathbf{f}_2-\mathbf{u}=\boldsymbol{\theta}, \qquad \mathbf{f}_2(0)=0, \qquad \mathbf{f}_2(1)=0 \end{aligned} \tag{7.49}$$

Each of the differential equations in (7.49) represents a continuum of constraints; that is, each differential equation is a function of t. The problem consists in finding the set of functions $\mathbf{f}_1$, $\mathbf{f}_2$, and $\mathbf{u}$ in $\mathcal{C}^2(0,1)$ which satisfies the constraints (7.49) and which minimizes

$$\mathbf{F}(\mathbf{u}) \triangleq \tfrac{1}{2}\int_0^1 \mathbf{u}^2(t)\,dt \tag{7.50}$$

We combine the constraints $\mathbf{G}_1$ and $\mathbf{G}_2$ with the objective function (7.50) to obtain the Lagrangian as in (7.40):

$$\begin{aligned} \mathcal{L}(\mathbf{f}_1,\mathbf{f}_2,\mathbf{u},\boldsymbol{\lambda}_1,\boldsymbol{\lambda}_2) &= \mathbf{F}(\mathbf{u}) - \int_0^1 \boldsymbol{\lambda}_1(t)\mathbf{G}_1(\mathbf{f}_1,\mathbf{f}_2)\,dt - \int_0^1 \boldsymbol{\lambda}_2(t)\mathbf{G}_2(\mathbf{f}_2,\mathbf{u})\,dt \\ &= \int_0^1 \left[\tfrac{1}{2}\mathbf{u}^2 - \boldsymbol{\lambda}_1(\dot{\mathbf{f}}_1-\mathbf{f}_2) - \boldsymbol{\lambda}_2(\dot{\mathbf{f}}_2+\mathbf{f}_2-\mathbf{u})\right]dt \end{aligned} \tag{7.51}$$

Notice that there is a Lagrange multiplier *function* corresponding to each constraint function.

We observed in (7.41) that the adjoint equation for a constrained minimization problem is based upon the differential of the Lagrangian with respect to the primal variables; in this instance, the primal variables are $\mathbf{f}_1$, $\mathbf{f}_2$, and $\mathbf{u}$. The differential is

$$\mathbf{d}\mathcal{L}(\mathbf{f}_1,\mathbf{f}_2,\mathbf{u};\,\mathbf{h}_1,\mathbf{h}_2,\mathbf{h}_u)=\frac{d}{d\beta}\int_0^1\Big[\tfrac{1}{2}(\mathbf{u}+\beta\mathbf{h}_u)^2-\boldsymbol{\lambda}_1(\dot{\mathbf{f}}_1+\beta\dot{\mathbf{h}}_1-\mathbf{f}_2-\beta\mathbf{h}_2)$$

$$-\boldsymbol{\lambda}_2(\dot{\mathbf{f}}_2+\beta\dot{\mathbf{h}}_2+\mathbf{f}_2+\beta\mathbf{h}_2-\mathbf{u}-\beta\mathbf{h}_u)\Big]\,dt\Big|_{\beta=0}$$

$$=\int_0^1(\mathbf{u}\mathbf{h}_u-\boldsymbol{\lambda}_1\dot{\mathbf{h}}_1+\boldsymbol{\lambda}_1\mathbf{h}_2-\boldsymbol{\lambda}_2\dot{\mathbf{h}}_2-\boldsymbol{\lambda}_2\mathbf{h}_2+\boldsymbol{\lambda}_2\mathbf{h}_u)\,dt$$

$$=\int_0^1(\mathbf{u}\mathbf{h}_u+\boldsymbol{\lambda}_1\mathbf{h}_2-\boldsymbol{\lambda}_2\mathbf{h}_2+\boldsymbol{\lambda}_2\mathbf{h}_u)\,dt$$

$$+\int_0^1(\dot{\boldsymbol{\lambda}}_1\mathbf{h}_1+\dot{\boldsymbol{\lambda}}_2\mathbf{h}_2)\,dt-\boldsymbol{\lambda}_1\mathbf{h}_1\big|_0^1-\boldsymbol{\lambda}_2\mathbf{h}_2\big|_0^1$$

$$=\int_0^1\big[(\dot{\boldsymbol{\lambda}}_1)\mathbf{h}_1+(\dot{\boldsymbol{\lambda}}_2+\boldsymbol{\lambda}_1-\boldsymbol{\lambda}_2)\mathbf{h}_2+(\mathbf{u}+\boldsymbol{\lambda}_2)\mathbf{h}_u\big]\,dt$$

$$-\boldsymbol{\lambda}_1(1)\mathbf{h}_1(1)+\boldsymbol{\lambda}_1(0)\mathbf{h}_1(0)-\boldsymbol{\lambda}_2(1)\mathbf{h}_2(1)+\boldsymbol{\lambda}_2(0)\mathbf{h}_2(0) \qquad (7.52)$$

In order for $(\mathbf{f}_1,\mathbf{f}_2,\mathbf{u})$ to minimize (7.50), the differential (7.52) must equal zero for all excursions $(\mathbf{h}_1,\mathbf{h}_2,\mathbf{h}_u)$ which are consistent with the boundary conditions in (7.49). As we discovered in connection with the Euler–Lagrange equation (7.25), the fact that $\mathbf{f}_1(0)$, $\mathbf{f}_1(1)$, $\mathbf{f}_2(0)$, and $\mathbf{f}_2(1)$ are fixed requires $\mathbf{h}_1$ and $\mathbf{h}_2$ to satisfy $\mathbf{h}_1(0)=\mathbf{h}_1(1)=\mathbf{h}_2(0)=\mathbf{h}_2(1)=0$; consequently, the boundary terms in (7.52) are zero. In order for the differential (7.52) to be zero for all $\mathbf{h}_1$, $\mathbf{h}_2$, and $\mathbf{h}_u$ in $\mathcal{C}^2(0,1)$ we must have

$$\begin{aligned}&\dot{\boldsymbol{\lambda}}_1=0\\&\dot{\boldsymbol{\lambda}}_2+\boldsymbol{\lambda}_1-\boldsymbol{\lambda}_2=0\\&\mathbf{u}=-\boldsymbol{\lambda}_2\end{aligned} \qquad (7.53)$$

The set of differential equations (7.53) constitutes the adjoint equation(s) for this problem; we also refer to (7.53) as the Euler–Lagrange equations for the problem. Note that there are no boundary conditions associated with the adjoint equations because the primal equations (7.49) have two sets of boundary conditions. See P&C 7.15 for an example in which there are boundary conditions associated with the adjoint equations.

Exercise 1. The primal equations (7.49) are a special case of the more general form

$$\mathbf{G}(\mathbf{x},\mathbf{u}) = \dot{\mathbf{x}} - \mathbf{A}\mathbf{x} - \mathbf{B}\mathbf{u} = \boldsymbol{\theta}$$

$$\mathbf{x}(0), \mathbf{x}(t_f) \text{ given}$$

where $\mathbf{x}(t)$ is $n \times 1$ and $\mathbf{u}(t)$ is $m \times 1$. Using this notation, we express the Lagrangian (7.51) as

$$\mathcal{L}(\mathbf{x},\mathbf{u},\boldsymbol{\lambda}) = \mathbf{F}(\mathbf{u}) - \int_0^{t_f} \boldsymbol{\lambda}^{\mathrm{T}} \mathbf{G}(\mathbf{x},\mathbf{u})\, dt$$

Show that in order for (7.50) to be minimized, the Lagrange multiplier vector must satisfy the following adjoint state equations:

$$\dot{\boldsymbol{\lambda}} + \mathbf{A}^{\mathrm{T}}\boldsymbol{\lambda} = \boldsymbol{\theta}, \qquad \mathbf{u} = -\mathbf{B}^{\mathrm{T}}\boldsymbol{\lambda}$$

The primal equations (7.49) and the adjoint equations (7.53) together constitute a two-point boundary value problem whose solution is the minimizing set of functions, $(\mathbf{f}_1, \mathbf{f}_2, \mathbf{u})$. These equations can be solved numerically by an iterative scheme. On the other hand, using the notation of Exercise 1, we can express the equations jointly in the form

$$\begin{pmatrix} \dot{\mathbf{x}} \\ \dot{\boldsymbol{\lambda}} \end{pmatrix} = \begin{pmatrix} \mathbf{A} & -\mathbf{B}\mathbf{B}^{\mathrm{T}} \\ \boldsymbol{\Theta} & -\mathbf{A}^{\mathrm{T}} \end{pmatrix} \begin{pmatrix} \mathbf{x} \\ \boldsymbol{\lambda} \end{pmatrix} \triangleq \mathbf{Q} \begin{pmatrix} \mathbf{x} \\ \boldsymbol{\lambda} \end{pmatrix}$$

This simple state equation can be solved together with the specified values of $\mathbf{x}(0)$ and $\mathbf{x}(1)$ for $\exp(\mathbf{Q}t)$, $\boldsymbol{\lambda}(0)$, $\mathbf{x}(t)$, $\boldsymbol{\lambda}(t)$, and $\mathbf{u}(t)$. See P&C 4.33. For this particular problem, however, we obtained the optimal control function $\mathbf{u}$ in a previous attack on the problem. From (7.37),

$$\mathbf{u}(t) = \frac{e+1}{3-e} - \frac{2}{3-e} e^{t}$$

This control function and the corresponding trajectories, $\mathbf{f}_1 = \phi$ and $\mathbf{f}_2 = \dot{\phi}$, are displayed in Figure 6.5. From $\mathbf{u}$ and (7.53) it is apparent that $\boldsymbol{\lambda}_2(t) = -\mathbf{u}(t)$ and $\boldsymbol{\lambda}_1(t) = -(e+1)/(3-e)$.

7.3 Optimization with Inequality Constraints

The preceding section on optimization with equality constraints was introduced by means of a gravity problem. From physical considerations we know that a ball dropped on the surfaces shown in Figure 7.5 would come to rest at a point lying on both constraining surfaces. Therefore, we treated the problem mathematically as a minimization of the height (potential energy) of the ball subject to a pair of equality constraints. The formulation of the problem would be more realistic if we were to express the constraints as inequalities, and let the minimization process determine whether or not each inequality is, in fact, an equality. That is, the problem should be expressed as follows: minimize $\mathbf{F}(\mathbf{x}) \triangleq w\xi_3$ subject to $\mathbf{G}_1(\mathbf{x}) \triangleq \xi_3 - \xi_1^2 - \xi_2^2 \geqslant 0$ and $\mathbf{G}_2(\mathbf{x}) \triangleq \xi_2 - 1 \geqslant 0$. In this section we explore problems with inequality constraints. We solve these problems by rephrasing them as problems with equality constraints and drawing on the previous necessary conditions, (7.28) or (7.32).

Kuhn–Tucker Conditions—Finite-Dimensional Case

We first derive the necessary conditions for a problem that involves a finite number of variables and constraints. Afterward we extend the results to infinite-dimensional problems. Let $\mathbf{x} = (\xi_1 \cdots \xi_n)^{\mathrm{T}}$, a vector in $\mathfrak{M}^{n\times 1}$. Define $\mathbf{F}$: $\mathfrak{M}^{n\times 1} \to \mathfrak{R}$ and $\mathbf{G}_i$: $\mathfrak{M}^{n\times 1} \to \mathfrak{R}$, $i = 1,\ldots,m$. We seek to

$$\text{minimize } \mathbf{F}(\mathbf{x}) \text{ subject to } \mathbf{G}_i(\mathbf{x}) \geqslant 0,\ i = 1,\ldots,m \tag{7.54}$$

Definition. Problem (7.54) is commonly referred to as a **mathematical programming** problem. A vector $\mathbf{x}$ that satisfies all the constraints in the problem is called a **feasible point**. The set of all feasible points is the **feasible region** for the problem.

By introducing the new variable s_i, we express the inequality $\mathbf{G}_i(\mathbf{x}) \geqslant 0$ as the equality $\mathbf{G}_i(\mathbf{x}) - s_i^2 = 0$. We call s_i^2 a **surplus variable**; for each $\mathbf{x}$, s_i^2 is positive and equal to the amount by which $\mathbf{G}_i(\mathbf{x})$ exceeds zero.* (If we had, instead, the inequality constraint $\mathbf{G}_i(\mathbf{x}) \leqslant 0$, we would *add* s_i^2 and call it a **slack variable**.) By introducing one surplus variable for each inequality constraint, we can convert the inequalities of (7.54) to equalities. Define

$$\hat{\mathbf{x}} \triangleq (\xi_1 \cdots \xi_n\ s_1 \cdots s_m)^{\mathrm{T}}, \qquad \hat{\mathbf{F}}(\hat{\mathbf{x}}) \triangleq \mathbf{F}(\mathbf{x}), \qquad \text{and } \hat{\mathbf{G}}_i(\hat{\mathbf{x}}) \triangleq \mathbf{G}_i(\mathbf{x}) - s_i^2,$$

$i = 1,\ldots,m$. Then (7.54) becomes

$$\text{minimize } \hat{\mathbf{F}}(\hat{\mathbf{x}}) \text{ subject to } \hat{\mathbf{G}}_i(\hat{\mathbf{x}}) = 0, \qquad i = 1,\ldots,m \tag{7.55}$$

*We use the squared variable s_i^2 in order to guarantee that the surplus variable is positive. Were we to use, instead, the variable s_i as the surplus variable, we would have to introduce the inequality constraint $s_i \geqslant 0$.

According to (7.38)–(7.39), the necessary conditions for a solution to (7.55) can be found by setting to zero the partial derivatives of the Lagrangian function associated with (7.55). The Lagrangian is

$$\begin{aligned}\mathcal{L}(\hat{\mathbf{x}},\boldsymbol{\lambda}) &= \hat{\mathbf{F}}(\hat{\mathbf{x}}) - \sum_{i=1}^{m} \lambda_i \hat{\mathbf{G}}_i(\hat{\mathbf{x}}) \\ &= \mathbf{F}(\mathbf{x}) - \sum_{i=1}^{m} \lambda_i\left(\mathbf{G}_i(\mathbf{x}) - s_i^2\right)\end{aligned} \tag{7.56}$$

From the partial derivatives of (7.56) we find

$$\begin{aligned}&\frac{\partial \mathcal{L}}{\partial \xi_j} = 0 \quad \Rightarrow \quad \frac{\partial \mathbf{F}(\mathbf{x})}{\partial \xi_j} - \sum_{i=1}^{m} \lambda_i \frac{\partial \mathbf{G}_i(\mathbf{x})}{\partial \xi_j} = 0 \quad \text{for } j=1,\dots,n \\ &\frac{\partial \mathcal{L}}{\partial s_i} = 0 \quad \Rightarrow \quad -2\lambda_i s_i = 0 \quad \text{for } i=1,\dots,m \\ &\frac{\partial \mathcal{L}}{\partial \lambda_i} = 0 \quad \Rightarrow \quad \mathbf{G}_i(\mathbf{x}) - s_i^2 = 0 \quad \text{for } i=1,\dots,m\end{aligned} \tag{7.57}$$

The first set of equations in (7.57) can be written as $\nabla\mathbf{F}(\mathbf{x}) = \Sigma_i \lambda_i \nabla\mathbf{G}_i(\mathbf{x})$; this vector equation is the same as the Lagrange multiplier (or adjoint) equation for the equality constraint problem. The last set of equations in (7.57) constitutes the given (primal) constraints, $\mathbf{G}_i(\mathbf{x}) \geqslant 0$, $i=1,\dots,m$. The remaining equations in (7.57) are known as **complementary slackness** conditions; they require that $s_i = 0$ or $\lambda_i = 0$ (or possibly both). If s_i is zero, the constraint $\mathbf{G}_i(\mathbf{x}) \geqslant 0$ is an equality—a **binding constraint**. If $\lambda_i = 0$, the ith constraint is not binding; it drops out of the Lagrange multiplier equation, and can be ignored.* Since $s_i = 0$ if and only if $\mathbf{G}_i(\mathbf{x}) = 0$, the complementary slackness condition is often written in the more meaningful form, $\lambda_i \mathbf{G}_i(\mathbf{x}) = 0$.

Each inequality $\mathbf{G}_i(\mathbf{x}) \geqslant 0$ constrains in only one direction; that is, feasible points lie only to one side of the surface $\mathbf{G}_i(\mathbf{x}) = 0$. The one-sided nature of the inequality constraints implies a one-sided nature for the corresponding Lagrange multipliers. Suppose $\mathbf{x}$ and $\{\lambda_i\}$ are solutions to (7.57) which yield a minimum of $\mathbf{F}(\mathbf{x})$. Denote the constrained minimum value of $\mathbf{F}$ by $\mathbf{F}_{\text{min}}$. Recall from (7.43) and (7.44) that the Lagrange multipliers possess information about the sensitivity of $\mathbf{F}_{\text{min}}$ to slight

*A zero value for λ_i implies only that the ith constraint can be ignored in the neighborhood of the constrained minimum. The ith constraint may make infeasible some distant point $\mathbf{x}$ for which $\mathbf{F}(\mathbf{x})$ is lower than the constrained minimum value.

changes in the equality constraints. Suppose that the ith constraint is binding (i.e., $\mathbf{G}_i(\mathbf{x})=0$). Let us change the ith constraint to $\mathbf{G}_i(\mathbf{x}) \geqslant v_i$, where v_i is a small positive number. Since this modification reduces the feasible region, we say the constraint has been **tightened**. The tightening must result in an increase in $\mathbf{F}_{\min}$ (or at best, no change in $\mathbf{F}_{\min}$); otherwise, the ith constraint could not have been binding. Since a positive change v_i in the value of the constraint results in a nonnegative change in $\mathbf{F}_{\min}$,

$$\lambda_i = \frac{\partial \mathbf{F}_{\min}}{\partial v_i} \geqslant 0$$

We should note that the non-negativity of the Lagrange multipliers is dependent upon the sign convention which we have used for the Lagrangian. If we were to reverse our sign convention, the Lagrange multipliers would be nonpositive. [See the comments following (7.32) and (7.40).]

We rephrase the necessary conditions (7.57) together with the succeeding comments as follows. If (7.54) has a finite solution $\mathbf{x}$, if the linearized versions of $\mathbf{F}$ and the constraints $\mathbf{G}_i$ adequately represent these functions near $\mathbf{x}$, and if the set of constraints is not degenerate at the minimum,† there necessarily exists a unique set of Lagrange multipliers λ_i which together with $\mathbf{x}$ satisfy the following conditions:

$$\begin{aligned} \nabla\mathbf{F}(\mathbf{x}) &= \sum_{i=1}^{m} \lambda_i \nabla\mathbf{G}_i(\mathbf{x}) \\ \mathbf{G}_i(\mathbf{x}) &\geqslant 0 \qquad i=1,\ldots,m \\ \lambda_i(\mathbf{x}) &\geqslant 0 \qquad i=1,\ldots,m \\ \lambda_i\mathbf{G}_i(\mathbf{x}) &= 0 \qquad i=1,\ldots,m \end{aligned} \tag{7.58}$$

The necessary conditions (7.58) are known as the **Kuhn–Tucker conditions.*** The inequalities and the complementary slackness conditions can be expressed jointly in the following way: either $\lambda_i=0$ and $\mathbf{G}_i(\mathbf{x}) \geqslant 0$ (the ith constraint is not binding) or $\mathbf{G}_i(\mathbf{x})=0$ and $\lambda_i \geqslant 0$ (the ith constraint is binding). We note that it is possible to have the "razor's edge" condition, $\lambda_i = \mathbf{G}_i(\mathbf{x})=0$, where there is no surplus in constraint i, but the constraint is not binding.

†See the section below on "Constraint Qualifications" for an example of degeneracy in the constraint set. This degeneracy is also discussed in detail in Hadley [7.7].

*These necessary conditions were first pointed out, in a somewhat different form, by Kuhn and Tucker [7.10].

Example 1. The Kuhn–Tucker Conditions for a Gravity Problem. We determine the rest position of a ball of weight w which is dropped onto the terrain of Figure 7.5. Mathematically, we pick $\mathbf{x}=(\xi_1\ \xi_2\ \xi_3)^{\mathrm{T}}$ to minimize $\mathbf{F}(\mathbf{x}) \triangleq w\xi_3$ subject to $\mathbf{G}_1(\mathbf{x}) \triangleq \xi_3-\xi_1^2-\xi_2^2 \geqslant 0$ and $\mathbf{G}_2(\mathbf{x}) \triangleq \xi_2-1 \geqslant 0$. The necessary conditions (7.58) are

$$\nabla\mathbf{F}(\mathbf{x})=\begin{pmatrix}0\\0\\w\end{pmatrix}=\lambda_1\begin{pmatrix}-2\xi_1\\-2\xi_2\\1\end{pmatrix}+\lambda_2\begin{pmatrix}0\\1\\0\end{pmatrix}$$

$$\begin{array}{rr} \mathbf{G}_1(\mathbf{x})=\xi_3-\xi_1^2-\xi_2^2\geqslant 0 & \mathbf{G}_2(\mathbf{x})=\xi_2-1\geqslant 0\\ \lambda_1\geqslant 0 & \lambda_2\geqslant 0\\ \lambda_1=0\ \text{ or }\ \mathbf{G}_1(\mathbf{x})=0 & \lambda_2=0\ \text{ or }\ \mathbf{G}_2(\mathbf{x})=0 \end{array}$$

From Figure 7.5 it is clear that $\mathbf{G}_1(\mathbf{x})=\mathbf{G}_2(\mathbf{x})=0$. Therefore, the necessary conditions reduce to (7.27). The solution is $\mathbf{x}=(0\ 1\ 1)^{\mathrm{T}}$, $\lambda_1=w$, and $\lambda_2=2w$. The Lagrange multipliers are positive as expected. Let us change the second constraint to $\mathbf{G}_2(\mathbf{x}) \triangleq \xi_2 \geqslant 0$; this second constraint is barely binding. Yet we still have $\mathbf{G}_1(\mathbf{x})=\mathbf{G}_2(\mathbf{x})=0$. The new minimum point is $\mathbf{x}=(0\ 0\ 0)^{\mathrm{T}}$. The Lagrange multipliers are not unique; one solution is $\lambda_1=w$, and $\lambda_2=0$. What would be the solution if the second constraint were changed to $\mathbf{G}_2(\mathbf{x}) \triangleq \xi_2+1 \geqslant 0$?

Kuhn–Tucker Conditions for Hilbert Spaces

The minimization problem (7.54) can be stated in the more general terminology of abstract Hilbert spaces. Define $\mathbf{G}$: $\mathfrak{M}^{n\times 1}\to\mathfrak{M}^{m\times 1}$ by

$$\mathbf{G}(\mathbf{x}) \triangleq \begin{pmatrix}\mathbf{G}_1(\mathbf{x})\\ \vdots \\ \mathbf{G}_m(\mathbf{x})\end{pmatrix}$$

If we let the vector inequality $\mathbf{G}(\mathbf{x})\geqslant \boldsymbol{\theta}$ mean $\mathbf{G}_i(\mathbf{x})\geqslant 0, i=1,\ldots,m$, we can express (7.54) as

$$\text{minimize } \mathbf{F}(\mathbf{x}) \text{ subject to } \mathbf{G}(\mathbf{x})\geqslant\boldsymbol{\theta}$$

The Kuhn–Tucker conditions (7.58) also can be given in Hilbert space notation. Denote $\boldsymbol{\lambda} \triangleq (\lambda_1\cdots\lambda_m)^{\mathrm{T}}$. As we found in (7.32), the Lagrange multiplier equation of (7.58) can be expressed as $\nabla\mathbf{F}(\mathbf{x})=\mathbf{G}'(\mathbf{x})^*\boldsymbol{\lambda}$. Let the vector inequality $\boldsymbol{\lambda}\geqslant\boldsymbol{\theta}$ mean $\lambda_i\geqslant 0$, $i=1,\ldots,m$. Since $\lambda_i\geqslant 0$ and $\mathbf{G}_i(\mathbf{x})\geqslant 0$, it follows that $\lambda_i\mathbf{G}_i(\mathbf{x})=0$ for $i=1,\ldots,m$ if and only if $\sum_{i=1}^m\lambda_i\mathbf{G}_i(\mathbf{x})=0$. Consequently, using the standard inner product for $\mathfrak{M}^{m\times 1}$, we can phrase the complementary slackness conditions as $\langle\mathbf{G}(\mathbf{x}),\boldsymbol{\lambda}\rangle=\boldsymbol{\lambda}^{\mathrm{T}}\mathbf{G}(\mathbf{x})=0$.

The use of Hilbert space terminology for (7.54) and (7.58) naturally leads us to suspect that the Kuhn–Tucker necessary conditions apply for general transformations **F** and **G**. Let **F**: $\mathcal{V}\to\mathcal{R}$ and **G**: $\mathcal{V}\to\mathcal{W}$ be nonlinear transformations defined on Hilbert spaces $\mathcal{V}$ and $\mathcal{W}$. We seek the solution to the problem

$$\text{minimize } \mathbf{F}(\mathbf{x}) \text{ with } \mathbf{G}(\mathbf{x}) \geqslant \boldsymbol{\theta} \tag{7.59}$$

In order to attack (7.59), we must specify the meaning of the vector inequality $\mathbf{G}(\mathbf{x})\geqslant\boldsymbol{\theta}$. If $\mathcal{W}$ is a function space, say, $\mathcal{L}_2(0,1)$, the natural way to extend the element-by-element inequality used for $\mathcal{M}^{m\times 1}$ above is to let $\mathbf{f}\geqslant\boldsymbol{\theta}$ mean $\mathbf{f}(t)\geqslant 0$ for $0\leqslant t\leqslant 1$. We could be more general and define inequalities in terms of "positive convex cones."* However, this generalization is not necessary for our purposes. We use a vector inequality only where it refers to a clear extension of the m-dimensional element-by-element inequality. Problem (7.59) is a Hilbert space extension of (7.54). Therefore we refer to (7.59) as a mathematical programming problem.

Let (7.59) have a solution at a point **x** of $\mathcal{V}$. If **F** and **G** are Fréchet differentiable in the neighborhood of the minimum **x**, and if **G** is not degenerate at **x**, it can be shown that there is a Lagrange multiplier vector $\boldsymbol{\lambda}$ in $\mathcal{W}$ which, together with **x**, satisfies the **Kuhn–Tucker conditions**:†

$$\begin{aligned} \nabla\mathbf{F}(\mathbf{x}) &= \mathbf{G}'(\mathbf{x})^*\boldsymbol{\lambda} \\ \mathbf{G}(\mathbf{x}) &\geqslant \boldsymbol{\theta} \\ \boldsymbol{\lambda} &\geqslant \boldsymbol{\theta} \\ \langle\mathbf{G}(\mathbf{x}),\boldsymbol{\lambda}\rangle &= 0 \end{aligned} \tag{7.60}$$

See the section below on "Constraint Qualifications" for a discussion of degeneracy of the constraint function **G**.

Example 2. Kuhn–Tucker Conditions for Hilbert Spaces. We seek to minimize $\mathbf{F}(\mathbf{x}) \triangleq \frac{1}{2}\|\mathbf{x}\|_{\mathcal{V}}^2$ subject to $\mathbf{G}(\mathbf{x}) \triangleq \mathbf{T}\mathbf{x}-\mathbf{y}\geqslant\boldsymbol{\theta}$, where **T**: $\mathcal{V}\to\mathcal{W}$ is a bounded linear transformation between Hilbert spaces $\mathcal{V}$ and $\mathcal{W}$. This problem is almost identical to the quadratic least-effort problem of Section 6.4; it differs only in the fact that equality is not required in the constraint equation. For this problem, the Kuhn–Tucker conditions are as follows:

$$\begin{aligned} &\mathbf{x}=\mathbf{T}^*\boldsymbol{\lambda}, && \mathbf{T}\mathbf{x}\geqslant\mathbf{y} \\ &\boldsymbol{\lambda}\geqslant\boldsymbol{\theta}, && \langle\mathbf{T}\mathbf{x}-\mathbf{y},\boldsymbol{\lambda}\rangle=0 \end{aligned}$$

*Luenberger [7.11].

†See Luenberger [7.11, p. 249] for a proof of this statement.

These necessary conditions should be compared with the solution (6.46) to the least-effort problem.

To further clarify these results, we compute the solution to the necessary conditions for a specific example. Let $\mathcal{V} = \mathcal{W} = \mathfrak{M}^{2\times 1}$, and

$$\mathbf{G}(\mathbf{x}) = \begin{pmatrix} \xi_1 - 2 \\ \xi_1 + \xi_2 - 1 \end{pmatrix} = \begin{pmatrix} 1 & 0 \\ 1 & 1 \end{pmatrix}\mathbf{x} - \begin{pmatrix} 2 \\ 1 \end{pmatrix} = \mathbf{A}\mathbf{x} - \mathbf{y} \geqslant \boldsymbol{\theta}$$

The feasible region is shown in Figure 7.11. Assume the standard inner product; that is, we minimize the distance to the origin, $\mathbf{F}(\mathbf{x}) = \frac{1}{2}(\xi_1^2 + \xi_2^2)$. The necessary conditions become

$$\begin{aligned}
&\mathbf{x} = \mathbf{A}^{\mathrm{T}}\boldsymbol{\lambda} && \Rightarrow \quad \xi_1 = \lambda_1 + \lambda_2 \quad \text{and} \quad \xi_2 = \lambda_2 \\
&\mathbf{A}\mathbf{x} \geqslant \mathbf{y} && \Rightarrow \quad \xi_1 \geqslant 2 \quad \text{and} \quad \xi_1 + \xi_2 \geqslant 1 \\
&\boldsymbol{\lambda} \geqslant \boldsymbol{\theta} && \Rightarrow \quad \lambda_1 \geqslant 0 \quad \text{and} \quad \lambda_2 \geqslant 0 \\
&\boldsymbol{\lambda}^{\mathrm{T}}(\mathbf{A}\mathbf{x} - \mathbf{y}) = 0 && \Rightarrow \quad \lambda_1(\xi_1 - 2) = 0 \quad \text{and} \quad \lambda_2(\xi_1 + \xi_2 - 1) = 0
\end{aligned}$$

At the solution, either $\lambda_1 = 0$ or $\xi_1 = 2$; either $\lambda_2 = 0$ or $\xi_1 + \xi_2 = 1$. From Figure 7.11 it is clear that the solution is $\xi_1 = 2$, $\xi_2 = 0$, $\lambda_2 = 0$, and $\lambda_1 = 2$. The minimum distance to the origin has zero sensitivity to constraint 2; only constraint 1 is binding.

A Procedure for Minimization with Inequality Constraints

Although the Kuhn–Tucker conditions clarify the mathematical structure of a constrained minimization problem, they do not provide a straightforward procedure for computation of the optimum solution. If we knew which constraints were binding, we could obtain the constrained minima by applying the Lagrange multiplier method (7.28) using only the binding

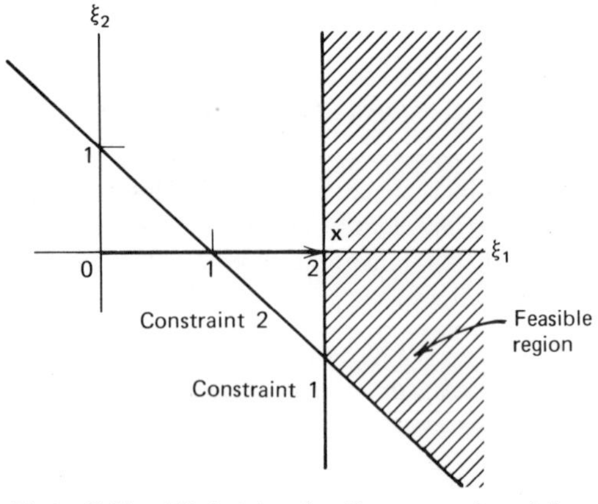

Figure 7.11. Minimizing the distance to the origin.

constraints. In Examples 1 and 2 we were able to determine which constraints were binding by using physical insight and the geometrical pictures of Figures 7.5 and 7.11. We have no way of telling, in general, which constraints are binding and which are not. Usually some sort of trial and error is required. We illustrate with an example the possible relationships between the constraints and the global minimum. Then we describe a conceptual scheme for finding the global minimum.

Consider the one-dimensional function

$$\mathbf{F}(t) \triangleq -t-\sin \pi t \quad \text{defined on } [0,3] \tag{7.61}$$

This function is shown in Figure 7.12. By setting $\mathbf{F}'(t)=0$, we find local minima of $\mathbf{F}(t)$ at $t=(1/\pi)\cos^{-1}(-1/\pi)\approx 0.6$ and $t=2+(1/\pi)\cos^{-1}(-1/\pi)\approx 2.6$. The global minimum is clearly at $t=2.6$. If we add an inequality constraint to the minimization problem, one of three situations can occur:

1. The constraint can be inactive (not binding), wherefore no change in $\mathbf{F}_{\min}$ occurs.
2. The constraint can be binding and cause an increase in $\mathbf{F}_{\min}$ (or no change in $\mathbf{F}_{\min}$ if it is *barely* binding).
3. The constraint can eliminate the global minimum from the feasible region, thereby increasing $\mathbf{F}_{\min}$, yet not be binding.

These three situations are illustrated by the constraints $G_a(t)=3-t\geqslant 0$, $G_b(t)=2-t\geqslant 0$, and $G_c(t)=1-t\geqslant 0$, respectively. The global minimum in

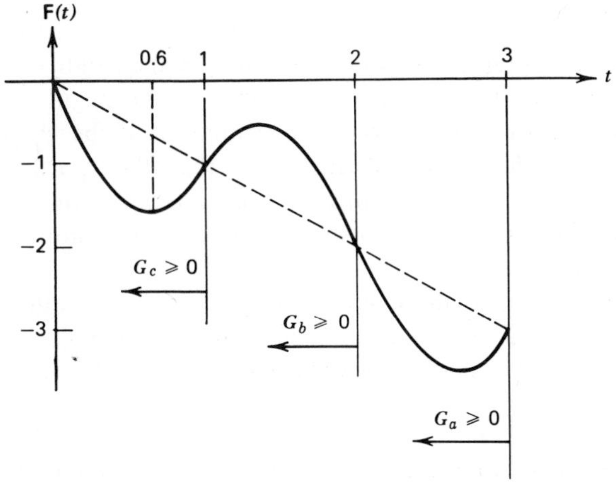

Figure 7.12. A constrained minimization.

the presence of the constraint $\mathbf{G}_b(t) \geqslant 0$ is $t=2$, a point on the constraint $\mathbf{G}_b(t)=0$. The global minimum in the presence of the constraint $\mathbf{G}_c(t) \geqslant 0$ is $t \approx 0.6$, a point which does not lie on the constraint $\mathbf{G}_c(t)=0$.

The fact that the above constraint situations can occur, yet be hidden from our view, should serve to humble us. However, we must not become discouraged. We now describe a conceptual procedure that will find the extremum under *any* conditions. The procedure is usually too long and messy to be practical unless specific properties of the objective function and constraint functions can be used to simplify the steps. Yet the procedure gives us hope. Much of the optimization literature is concerned with characterizing specific classes of problems and tailoring optimization techniques to fit those classes.

It is apparent from Figure 7.12 that introducing an additional constraint can raise $\mathbf{F}_{\min}$, but cannot lower it. We now describe a general procedure that can be used to minimize a function $\mathbf{F}(\mathbf{x})$ subject to inequality constraints $\mathbf{G}_i(\mathbf{x}) \geqslant 0$, $i=1,\ldots,m$. In the procedure we begin with the unconstrained function and introduce constraints one at a time.

Conceptual Procedure for Computing Minima (7.62)

1. Find the *global* minimum of the unconstrained function $\mathbf{F}(\mathbf{x})$. If the unconstrained minimum satisfies all the constraints, it is the solution to the constrained minimization problem, and we are done.
2. If the unconstrained minimum is not a feasible point, then it violates at least one constraint. We now seek a *global* minimum subject to only one of the inequalities (not necessarily one which is violated above). This minimum is either a local minimum of the unconstrained function $\mathbf{F}(\mathbf{x})$ which is already obtained in step 1, or it lies *on* the single constraint. If it lies on the constraint, we find the point by means of the necessary conditions (7.28) for equality constrained minima. If the global minimum subject to the single constraint satisfies all the other constraints, we are done. If not, we repeat the minimization for each other constraint, one at a time, until we find the overall constrained minimum, or until we have tried all the constraints.
3. If the solution to the overall constrained minimization problem is not found in step 2, we seek a *global* minimum of $\mathbf{F}(\mathbf{x})$ subject to two constraints at a time, three at a time, etc., until the overall minimum is found.

The above procedure must find the constrained minimum, for eventually all combinations of binding constraints are tried. The procedure is merely a systematic way of trying the constraints; it often eliminates many (or most) of the trials. Of course, at each trial, determining the global minimum requires the solution of the set of nonlinear equations (7.28); and the number of trials can be large. Furthermore, the equations to be solved may yield critical points other than the global minimum which is desired. Thus for an arbitrary set of functions $\mathbf{F}$ and $\{\mathbf{G}_i\}$, the task can be formidable. Fortunately, some practical problems have a structure which allows simple computer computation at each step. Furthermore, the nature of the problem structure sometimes dictates that only certain combinations of constraints can be active, and allows an efficient computer scheme for evaluating these combinations. Typical of these types of problems are "linear programming" problems (Section 7.4).

Example 3. A Systematic Determination of a Constrained Minimum. We apply the above procedure to find a constrained minimum for the problem illustrated in Figure 7.12: minimize $\mathbf{F}(t) \triangleq -t-\sin \pi t$ on $[0,3]$ subject to $\mathbf{G}_a(t) \triangleq 3-t \geqslant 0$, $\mathbf{G}_b(t) \triangleq 2-t \geqslant 0$, and $\mathbf{G}_c(t) \triangleq 1-t \geqslant 0$. The steps are as follows:

1. Unconstrained minimization of $\mathbf{F}(t)$:
   A local minimum at $t \approx 0.6$ (feasible);
   A global minimum at $t \approx 2.6$ (not feasible).
2. Minimization of $\mathbf{F}(t)$ with a single constraint:
   a. $\mathbf{G}_a(t) \geqslant 0$: A local minimum at $t \approx 0.6$ (feasible);
   A global minimum at $t \approx 2.6$ (not feasible).
   b. $\mathbf{G}_b(t) \geqslant 0$: A local minimum at $t \approx 0.6$ (feasible);
   A global minimum at $t \approx 2$ (not feasible).
   c. $\mathbf{G}_c(t) \geqslant 0$: A global minimum at $t \approx 0.6$ (feasible).

If we had started step 2 with $\mathbf{G}_c(t) \geqslant 0$, we would have eliminated several trials. The solution is clearly $t \approx 0.6$.

Constraint Qualifications

All the previous discussions of necessary conditions for minimization of functionals subject to equality or inequality constraints have assumed that the objective function and constraint functions were smooth (Fréchet differentiable) in the neighborhood of the optimum. Furthermore, the various tangent planes had to be aligned in a nondegenerate manner at the optimum. To clarify this latter point, let us look at two examples in which this condition is not satisfied. In the first example, the Lagrange multiplier equation has no solution; in the second, the solution is not unique.

Let $\mathbf{x}=(\xi_1\ \xi_2)^{\mathrm{T}}$ in $\mathfrak{M}^{2\times 1}$. We minimize $\mathbf{F}(\mathbf{x})=\xi_2$ subject to $\mathbf{G}_1(\mathbf{x})=\xi_2^3-\xi_1 \geqslant 0$ and $\mathbf{G}_2(\mathbf{x})=\xi_2^3+\xi_1 \geqslant 0$. See Figure 7.13*a*. It is apparent from the figure that the minimum occurs at the bottom of the cusp, $\mathbf{x}=\boldsymbol{\theta}$. At the minimum point

$$\nabla \mathbf{G}_1(\boldsymbol{\theta})=\left.\begin{pmatrix}-1\\ 3\xi_2^2\end{pmatrix}\right|_{\mathbf{x}=\boldsymbol{\theta}}=\begin{pmatrix}-1\\ 0\end{pmatrix} \quad \text{and} \quad \nabla \mathbf{G}_2(\boldsymbol{\theta})=\left.\begin{pmatrix}1\\ 3\xi_2^2\end{pmatrix}\right|_{\mathbf{x}=\boldsymbol{\theta}}=\begin{pmatrix}1\\ 0\end{pmatrix}$$

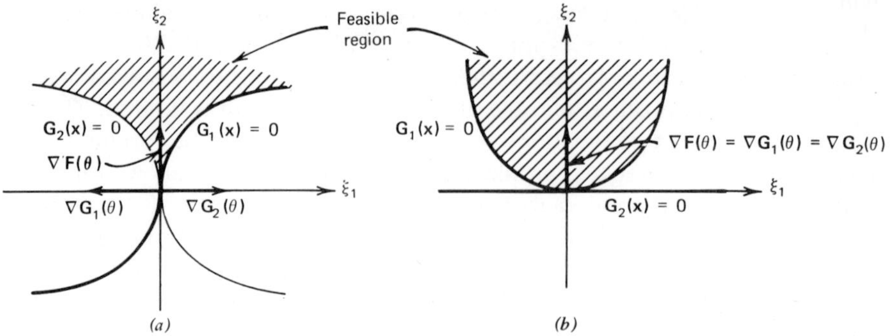

Figure 7.13. Degenerate constraints.

Then according to the Lagrange multiplier equation,

$$\nabla \mathbf{F}(\boldsymbol{\theta}) = \begin{pmatrix} 0 \\ 1 \end{pmatrix} = \lambda_1 \begin{pmatrix} -1 \\ 0 \end{pmatrix} + \lambda_2 \begin{pmatrix} 1 \\ 0 \end{pmatrix} \tag{7.63}$$

The Lagrange multiplier equation has no solution in the usual mathematical sense. Of course, if we view the problem in terms of a physical model, we can obtain a meaningful solution. In effect, we are searching for the rest position of a ball, subject to gravity, which is to be supported by the surfaces shown in Figure 7.13*a*. Because the walls become vertical at the minimum point, the forces which must be exerted by the walls in order to support the ball approach infinity. Consequently, $\lambda_1 = \lambda_2 = \infty$ is, in a limited sense, a solution to the Lagrange multiplier equation. We see also that the sensitivity of $\mathbf{F}_{\min}$ to horizontal motions of the constraints is infinite.

For the second example, minimize $\mathbf{F}(\mathbf{x}) \triangleq \xi_2$ subject to $\mathbf{G}_1(\mathbf{x}) \triangleq \xi_2 - \xi_1^2 \geqslant 0$ and $\mathbf{G}_2(\mathbf{x}) \triangleq \xi_2 \geqslant 0$. See Figure 7.13*b*. The constrained minimum is the origin, $\mathbf{x} = \boldsymbol{\theta}$. The Lagrange multiplier equation is

$$\nabla \mathbf{F}(\boldsymbol{\theta}) = \begin{pmatrix} 0 \\ 1 \end{pmatrix} = \lambda_1 \begin{pmatrix} 0 \\ 1 \end{pmatrix} + \lambda_2 \begin{pmatrix} 0 \\ 1 \end{pmatrix}$$

Thus $\lambda_2 = 1 - \lambda_1$, and the solution to the Lagrange multiplier equation is not unique.

We found in Section 7.2 that for problems with equality constraints, linear dependence of the gradients of the constraints at the constrained minimum results in nonexistence or nonuniqueness of the Lagrange multipliers. We acknowledge such degeneracy in the Lagrange multiplier equation by saying the constraints are *degenerate*. Similar difficulties occur with respect to binding inequality constraints. These difficulties do not necessarily preclude a meaningful solution to the constrained minimization problem. Rather, degeneracy in the constraints merely eliminates or reduces the usefulness of the Lagrange multiplier equation in solving for the minimum, and makes it impossible to determine the Lagrange multipliers. Even *near-degeneracy* in the constraints makes it difficult to use the Lagrange multiplier equation to obtain a solution or to compute accurately the Lagrange multipliers. See the discussion following (7.92).

There are several tests (or constraint qualifications) which we can apply to equality and inequality constraints to assure ourselves that the appropriate necessary conditions, (7.28), (7.32), (7.58), or (7.60), are satisfied by some vector $\boldsymbol{\lambda}$ at the minimum. The simplest (for a finite number of constraints) is a statement that the gradients of the binding constraints exist and are linearly independent at the minimum. This simple qualification also guarantees uniqueness of the Lagrange multiplier vector $\boldsymbol{\lambda}$. Other constraint qualifications are designed to avoid the cusp-like condition demonstrated in Figure 7.13*a* which prevents the existence of Lagrange multipliers without necessarily avoiding the nonuniqueness difficulty demonstrated in Figure 7.13*b*. The Kuhn-Tucker constraint qualification is an example (see Hadley [7.7, p. 194]. The regularity condition discussed by Luenberger [7.11, p. 249] is another. Aoki [7.1, pp. 185–186] discusses Slater's constraint qualification. Geffrion [7.4] also describes a general constraint qualification.

7.4 Mathematical Programming

There are many constrained optimization problems that do not appear, outwardly, to fit the structure of the mathematical programming problem (7.54) or its Hilbert space generalization (7.59). For example, we may wish to maximize a profit rather than minimize a cost; we may wish to use inequalities which are the reverse of those shown in (7.54) and (7.59). Any problem in which we wish to minimize or maximize a function subject to a combination of equality and inequality constraints has come to be called a mathematical programming problem.* In this section we attack this whole class of constrained optimization problems. Our approach is to convert each type of problem to one of the form (7.59), then convert the associated necessary conditions (7.60) back into the terminology of the original problem. We explore a number of specific types of mathematical programming problems which arise frequently in practical optimization. We include applications of the necessary conditions in infinite-dimensional spaces.

Suppose we modify the basic mathematical programming problem (7.59)

*The term "mathematical programming" does not appear to be well chosen, but it has become standard.

by reversing the inequality:

$$\text{minimize } \mathbf{F}(\mathbf{x}) \text{ subject to } \mathbf{G}(\mathbf{x}) \leqslant \boldsymbol{\theta} \tag{7.64}$$

Problem (7.64) can be converted to a problem of the form (7.59) by redefining the constraint transformation to be $\hat{\mathbf{G}}(\mathbf{x}) \stackrel{\Delta}{=} -\mathbf{G}(\mathbf{x})$. That is, $-\mathbf{G}(\mathbf{x}) \geqslant \boldsymbol{\theta}$. Consequently, the Kuhn–Tucker necessary conditions appropriate to (7.64) differ from (7.60) only in the sign on **G**. The necessary conditions associated with the *reversed-inequality* problem (7.64) are

$$\begin{gathered} \nabla \mathbf{F}(\mathbf{x}) = -\mathbf{G}'(\mathbf{x})^* \boldsymbol{\lambda} \\ \mathbf{G}(\mathbf{x}) \leqslant \boldsymbol{\theta}, \qquad \boldsymbol{\lambda} \geqslant \boldsymbol{\theta} \\ \langle \mathbf{G}(\mathbf{x}), \boldsymbol{\lambda} \rangle = 0 \end{gathered} \tag{7.65}$$

Another variation in (7.59) which can be handled by a simple sign change is

$$\text{maximize } \mathbf{F}(\mathbf{x}) \text{ subject to } \mathbf{G}(\mathbf{x}) \geqslant \boldsymbol{\theta} \tag{7.66}$$

Since maximization of $\mathbf{F}(\mathbf{x})$ is equivalent to minimization of $\hat{\mathbf{F}}(\mathbf{x}) \stackrel{\Delta}{=} -\mathbf{F}(\mathbf{x})$, the necessary conditions associated with the maximization problem (7.66) differ from (7.60) only in the sign on **F**. The necessary conditions associated with the *maximization* problem (7.66) are as follows:

$$\begin{gathered} -\nabla \mathbf{F}(\mathbf{x}) = \mathbf{G}'(\mathbf{x})^* \boldsymbol{\lambda} \\ \mathbf{G}(\mathbf{x}) \geqslant \boldsymbol{\theta}, \qquad \boldsymbol{\lambda} \geqslant \boldsymbol{\theta} \\ \langle \mathbf{G}(\mathbf{x}), \boldsymbol{\lambda} \rangle = 0 \end{gathered} \tag{7.67}$$

Exercise 1. Determine the necessary conditions associated with the problem: maximize $\mathbf{F}(\mathbf{x})$ subject to $\mathbf{G}(\mathbf{x}) \leqslant \boldsymbol{\theta}$.

Non-Negativity Constraints

One type of constraint that appears often enough to warrant special treatment is the non-negativity constraint, $\mathbf{x} \geqslant \boldsymbol{\theta}$; that is, the constraint that no element of **x** can be negative. We treat a specific form of problem which includes non-negativity constraints:

$$\text{minimize } \mathbf{F}(\mathbf{x}) \text{ subject to } \mathbf{G}(\mathbf{x}) \geqslant \boldsymbol{\theta} \text{ and } \mathbf{x} \geqslant \boldsymbol{\theta} \tag{7.68}$$

where **x** is in a Hilbert space $\mathcal{V}$ and $\mathbf{G}(\mathbf{x})$ is in a Hilbert space $\mathcal{W}$. We convert this problem to one of the form (7.59) by use of the cartesian product space $\mathcal{V}\times\mathcal{W}$. (The line of thought in this discussion is simplified if we think of $\mathcal{V}$, $\mathcal{W}$, and $\mathcal{V}\times\mathcal{W}$ as $\mathfrak{M}^{n\times 1}$, $\mathfrak{M}^{m\times 1}$, and $\mathfrak{M}^{(n\times m)\times 1}$, respectively).

Denote by $(\mathbf{x},\mathbf{y})$ a vector in $\mathcal{V}\times\mathcal{W}$, where $\mathbf{x}$ is in $\mathcal{V}$ and $\mathbf{y}$ is in $\mathcal{W}$. Define the inner product on $\mathcal{V}\times\mathcal{W}$ by* $\langle(\mathbf{x},\mathbf{y}),(\boldsymbol{\mu},\boldsymbol{\lambda})\rangle_{\mathcal{V}\times\mathcal{W}} \triangleq \langle\mathbf{x},\boldsymbol{\mu}\rangle_{\mathcal{V}} + \langle\mathbf{y},\boldsymbol{\lambda}\rangle_{\mathcal{W}}$. Define $\hat{\mathbf{G}}: \mathcal{V}\to\mathcal{V}\times\mathcal{W}$ by

$$\hat{\mathbf{G}}(\mathbf{x}) \triangleq (\mathbf{x},\mathbf{G}(\mathbf{x})) \tag{7.69}$$

We now seek to minimize $\mathbf{F}(\mathbf{x})$ subject to $\hat{\mathbf{G}}(\mathbf{x})\geqslant\boldsymbol{\theta}$, a problem in the form of (7.59). The necessary conditions (7.60) become

$$\nabla\mathbf{F}(\mathbf{x})=\hat{\mathbf{G}}'(\mathbf{x})^*\hat{\boldsymbol{\lambda}},\qquad \hat{\mathbf{G}}(\mathbf{x})\geqslant\boldsymbol{\theta},\qquad \hat{\boldsymbol{\lambda}}\geqslant\boldsymbol{\theta},\qquad \langle\hat{\mathbf{G}}(\mathbf{x}),\hat{\boldsymbol{\lambda}}\rangle_{\mathcal{V}\times\mathcal{W}}=0$$

for some Lagrange multiplier vector $\hat{\boldsymbol{\lambda}}$ in $\mathcal{V}\times\mathcal{W}$. We will express these necessary conditions in terms of the transformations $\mathbf{F}$ and $\mathbf{G}$ of the original problem, (7.68). The differential of $\hat{\mathbf{G}}$ is

$$\begin{aligned} d\hat{\mathbf{G}}(\mathbf{x},\mathbf{h}) = \hat{\mathbf{G}}'(\mathbf{x})\mathbf{h} &= \frac{d}{d\beta}\hat{\mathbf{G}}(\mathbf{x}+\beta\mathbf{h})\Big|_{\beta=0} \\ &= \left(\frac{d}{d\beta}(\mathbf{x}+\beta\mathbf{h})\Big|_{\beta=0}, \frac{d}{d\beta}\mathbf{G}(\mathbf{x}+\beta\mathbf{h})\Big|_{\beta=0}\right) \\ &= (\mathbf{h},\mathbf{G}'(\mathbf{x})\mathbf{h}) \end{aligned}$$

Let the Lagrange multiplier vector associated with the constraint function (7.69) be denoted by $\hat{\boldsymbol{\lambda}}=(\boldsymbol{\mu},\boldsymbol{\lambda})$; in effect, we let $\boldsymbol{\lambda}$ be the Lagrange multiplier vector associated with the constraint $\mathbf{G}(\mathbf{x})\geqslant\boldsymbol{\theta}$ and let $\boldsymbol{\mu}$ be the Lagrange multiplier vector associated with the non-negativity constraint, $\mathbf{x}\geqslant\boldsymbol{\theta}$. Then, to determine $\hat{\mathbf{G}}'(\mathbf{x})^*\hat{\boldsymbol{\lambda}}$ we follow the procedure (7.33):

$$\begin{aligned} \langle\hat{\mathbf{G}}'(\mathbf{x})\mathbf{h},\hat{\boldsymbol{\lambda}}\rangle_{\mathcal{V}\times\mathcal{W}} &= \langle(\mathbf{h},\mathbf{G}'(\mathbf{x})\mathbf{h}),(\boldsymbol{\mu},\boldsymbol{\lambda})\rangle_{\mathcal{V}\times\mathcal{W}} \\ &= \langle\mathbf{h},\boldsymbol{\mu}\rangle_{\mathcal{V}} + \langle\mathbf{G}'(\mathbf{x})\mathbf{h},\boldsymbol{\lambda}\rangle_{\mathcal{W}} \\ &= \langle\mathbf{h},\boldsymbol{\mu}\rangle_{\mathcal{V}} + \langle\mathbf{h},\mathbf{G}'(\mathbf{x})^*\boldsymbol{\lambda}\rangle_{\mathcal{V}} \\ &= \langle\mathbf{h},\boldsymbol{\mu}+\mathbf{G}'(\mathbf{x})^*\boldsymbol{\lambda}\rangle_{\mathcal{V}} \\ &\triangleq \langle\mathbf{h},\hat{\mathbf{G}}'(\mathbf{x})^*\hat{\boldsymbol{\lambda}}\rangle_{\mathcal{V}} \end{aligned}$$

*P&C 5.3*a*.

Consequently, $\hat{\mathbf{G}}'(\mathbf{x})^*\hat{\boldsymbol{\lambda}} = \boldsymbol{\mu} + \mathbf{G}'(\mathbf{x})^*\boldsymbol{\lambda}$. The complementary slackness condition is

$$\langle \hat{\mathbf{G}}(\mathbf{x}), \hat{\boldsymbol{\lambda}} \rangle_{\mathcal{V}\times\mathcal{W}} = \langle (\mathbf{x}, \mathbf{G}(\mathbf{x})), (\boldsymbol{\mu}, \boldsymbol{\lambda}) \rangle_{\mathcal{V}\times\mathcal{W}}$$
$$= \langle \mathbf{x}, \boldsymbol{\mu} \rangle_{\mathcal{V}} + \langle \mathbf{G}(\mathbf{x}), \boldsymbol{\lambda} \rangle_{\mathcal{W}} = 0$$

But by the sign restrictions on $\hat{\mathbf{G}}(\mathbf{x})$ and $\hat{\boldsymbol{\lambda}}$, each of the arguments in this inner product expression is non-negative. As a result, $\langle \mathbf{x}, \boldsymbol{\mu} \rangle_{\mathcal{V}} = 0$ and $\langle \mathbf{G}(\mathbf{x}), \boldsymbol{\lambda} \rangle_{\mathcal{W}} = 0$. In summary, the necessary conditions associated with (7.68), a problem with *non-negativity* constraints, are

$$\nabla \mathbf{F}(\mathbf{x}) = \mathbf{G}'(\mathbf{x})^*\boldsymbol{\lambda} + \boldsymbol{\mu}$$
$$\mathbf{G}(\mathbf{x}) \geqslant \boldsymbol{\theta}, \qquad \boldsymbol{\lambda} \geqslant \boldsymbol{\theta}, \qquad \langle \mathbf{G}(\mathbf{x}), \boldsymbol{\lambda} \rangle_{\mathcal{W}} = 0 \tag{7.70}$$
$$\mathbf{x} \geqslant \boldsymbol{\theta}, \qquad \boldsymbol{\mu} \geqslant \boldsymbol{\theta}, \qquad \langle \mathbf{x}, \boldsymbol{\mu} \rangle_{\mathcal{V}} = 0$$

We observe that the Lagrange multiplier vector $\boldsymbol{\mu}$ appears in a particularly simple manner in the Lagrange multiplier equation. Since $\boldsymbol{\mu}$ displays the sensitivity of $\mathbf{F}_{\min}$ to changes in the non-negativity constraints, if some of the elements of the minimizing vector $\mathbf{x}$ are strictly positive, the corresponding elements of $\boldsymbol{\mu}$ will be zero.

Example 1. Minimum Cost Feeding of a Dairy Herd. A dairyman wishes to pick a combination of two available feeds which will minimize his feed cost while guaranteeing an adequate diet for his cows. Let ξ_1 and ξ_2 represent the number of pounds of corn silage and soybean meal, respectively, that will be fed per cow per day. We must pick ξ_1 and ξ_2 to guarantee that each cow receives at least 2 lb of protein and 7.5 lb of carbohydrate per day. Assume corn silage is 5% protein and 20% carbohydrate, and costs 2¢/lb. Soybean meal is 20% protein and 60% carbohydrate, and costs 8¢/lb. (These numbers are adjusted for mathematical convenience.) Consequently, we must determine $\mathbf{x} = (\xi_1 \ \ \xi_2)^{\mathrm{T}}$ to minimize $\mathbf{F}(\mathbf{x}) \triangleq 0.02\xi_1 + 0.08\xi_2$ subject to

$$0.05\xi_1 + 0.2\xi_2 \geqslant 2 \text{ lb protein}$$
$$0.2\xi_1 + 0.6\xi_2 \geqslant 7.5 \text{ lb carbohydrate}$$
$$\xi_1 \geqslant 0, \qquad \xi_2 \geqslant 0$$

Because the objective function and constraint functions are linear, this is called a *linear programming* problem. The non-negativity conditions express the obvious requirement that cows cannot be fed negative amounts of feed. Define $\mathbf{G}$: $\mathfrak{M}^{2\times 1} \to \mathfrak{M}^{2\times 1}$ by

$$\mathbf{G}(\mathbf{x}) \triangleq \begin{pmatrix} 0.05 & 0.2 \\ 0.2 & 0.6 \end{pmatrix}\mathbf{x} - \begin{pmatrix} 2 \\ 7.5 \end{pmatrix} \triangleq \mathbf{A}\mathbf{x} - \mathbf{y}$$

Then the problem is identical to (7.68). If we use the standard inner product, the transformation $\mathbf{G}'(\mathbf{x})^*$ consists in multiplication by $\mathbf{A}^{\mathrm{T}}$. Thus the necessary conditions (7.70) become

$$\nabla\mathbf{F}(\mathbf{x})=\begin{pmatrix}0.02\\0.08\end{pmatrix}=\begin{pmatrix}0.05 & 0.2\\0.2 & 0.6\end{pmatrix}\begin{pmatrix}\lambda_1\\\lambda_2\end{pmatrix}+\begin{pmatrix}\mu_1\\\mu_2\end{pmatrix}$$

with $\mathbf{x}\geqslant\boldsymbol{\theta}$, $\boldsymbol{\mu}\geqslant\boldsymbol{\theta}$, and either $\mu_i=0$ or $\xi_i=0$, $i=1, 2$; also, $\mathbf{Ax}\geqslant\mathbf{y}$, $\boldsymbol{\lambda}\geqslant\boldsymbol{\theta}$, and either $\lambda_i=0$ or the ith inequality constraint is binding, $i=1,2$.

The necessary conditions do not indicate which of the constraints are binding. Rather than use the procedure (7.62), we determine graphically which constraints are equalities. The constraints and lines of constant cost are shown in Figure 7.14. It is clear that the minimum occurs for any mix of feeds on the line between $\xi_1=40$ lb, $\xi_2=0$ and $\xi_1=30$ lb, $\xi_2=2.5$ lb. Then $\mathbf{F}_{\min}=\$0.80$ per cow per day. Since $\xi_1>0$, $\mu_1=0$; if we pick $\xi_2<2.5$, the second inequality constraint (carbohydrate) is not binding, and we must have $\lambda_2=0$. From the Lagrange multiplier equation we find $\lambda_1=0.4$ and $\mu_2=0$.

Notice that the units on μ_i and λ_i are dollars per pound; the sensitivity variables are clearly prices. Recall from (7.44) that $\mathbf{F}_{\min}$ increases by an amount $\lambda_1 v_1$ if the first constraint (on protein) is increased by an amount v_1. The units on constraint 1 are pounds of protein. Thus λ_1 is the price (in dollars per pound) of an increase in the protein requirement. In the economics literature, the variables λ_i and μ_i are referred to as *shadow prices*. The shadow price λ_2 is zero for this problem because the carbohydrate constraint is not binding; it costs nothing to increase the carbohydrate requirement by a small amount.

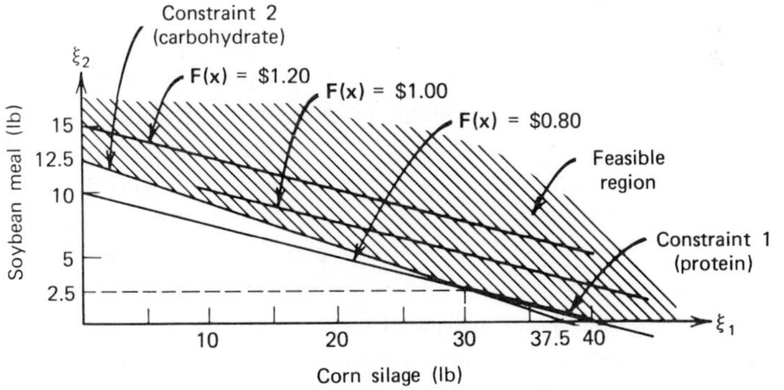

Figure 7.14. A linear programming problem.

Exercise 2. Find the solution $(\mathbf{x},\boldsymbol{\mu},\boldsymbol{\lambda})$ to the dairy feed problem of Example 1 if the price of corn silage rises to (*a*) 2.5¢/lb; (*b*) 3¢/lb.

Linear and Nonlinear Programming

Because the individual inequalities in Example 1 are linear, the feasible region is a polygon. Because **F** is linear, the curves of constant cost are straight lines and the global minimum must lie at a vertex of the polygon (or possible along an edge of the polygon). Problems in which both **F** and **G** are linear are known as **linear programming** problems. In any linear programming problem, for any trial selection of binding constraints, the equations that constitute the necessary conditions are linear. An efficient algorithm, known as the *simplex method*, has been developed for systematically trying certain vertices of the polygon until the minimizing solution has been found.* The method will handle a large (but finite) number of variables and constraints.

Exercise 3. A general form for finite-dimensional linear programming problems is as follows:

$$\text{minimize } \mathbf{F}(\mathbf{x}) \stackrel{\Delta}{=} \mathbf{z}^T\mathbf{x} \text{ subject to } \mathbf{A}\mathbf{x} \geqslant \mathbf{y},\ \mathbf{x} \geqslant \boldsymbol{\theta} \tag{7.71}$$

where $\mathbf{A}$ is $m \times n$, $\mathbf{z}$ and $\mathbf{x}$ are $n \times 1$, and $\mathbf{y}$ is $m \times 1$. Show that the Kuhn–Tucker necessary conditions associated with (7.71) are

$$\begin{gathered} \mathbf{z} = \mathbf{A}^T\boldsymbol{\lambda} + \boldsymbol{\mu} \\ \mathbf{A}\mathbf{x} \geqslant \mathbf{y}, \qquad \boldsymbol{\lambda} \geqslant \boldsymbol{\theta}, \qquad \boldsymbol{\lambda}^T(\mathbf{A}\mathbf{x} - \mathbf{y}) = 0 \\ \mathbf{x} \geqslant \boldsymbol{\theta}, \qquad \boldsymbol{\mu} \geqslant \boldsymbol{\theta}, \qquad \boldsymbol{\mu}^T\mathbf{x} = 0 \end{gathered} \tag{7.72}$$

A mathematical programming problem in which either **F** or **G** is nonlinear is referred to as a **nonlinear programming** problem. A nonlinear programming problem that has a linear constraint transformation **G** and a quadratic objective function **F** deserves special mention. It is called a **quadratic programming** problem. (Note that the least-square problems of Chapter 6 are of this type.) Because derivatives of quadratic functions are linear, the equations which constitute the necessary conditions are linear, and thus easy to solve. The only significant difficulty in solving a quadratic programming problem is in determining which constraints are binding. Since the feasible region is a polygon, the problem has some characteristics in common with linear programming. See P & C 7.18. A variation of the simplex method is used to solve quadratic programming problems (Hadley [7.7]).

*See Hadley [7.6].

Equality and Inequality Constraints

One last constrained minimization problem for which the approach to minimization may not be obvious is

$$\text{minimize } \mathbf{F}(\mathbf{x}) \text{ subject to } \mathbf{G}(\mathbf{x}) \geqslant \boldsymbol{\theta} \text{ and } \mathbf{H}(\mathbf{x}) = \boldsymbol{\theta} \tag{7.73}$$

where $\mathbf{F}$: $\mathcal{V} \to \mathcal{R}$, $\mathbf{G}$: $\mathcal{V} \to \mathcal{W}$, and $\mathbf{H}$: $\mathcal{V} \to \mathcal{H}$, and $\mathcal{V}$, $\mathcal{W}$, and $\mathcal{H}$ are Hilbert spaces. (Think in terms of finite-dimensional spaces: $\mathcal{V} \sim \mathcal{M}^{n\times 1}$, $\mathcal{W} \sim \mathcal{M}^{m\times 1}$, and $\mathcal{H} \sim \mathcal{M}^{p\times 1}$.) Two avenues are open to us. One approach is to convert the inequality constraint $\mathbf{G}(\mathbf{x}) \geqslant \boldsymbol{\theta}$ to an equality by introducing surplus variables in a manner similar to our approach to Section 7.2. A more straightforward approach to (7.73) is to convert the equality constraint $\mathbf{H}(\mathbf{x}) = \boldsymbol{\theta}$ to the pair of inequalities $\mathbf{H}(\mathbf{x}) \geqslant \boldsymbol{\theta}$ and $-\mathbf{H}(\mathbf{x}) \geqslant \boldsymbol{\theta}$. We combine the inequalities in the form

$$\hat{\mathbf{G}}(\mathbf{x}) \triangleq (\mathbf{G}(\mathbf{x}), \mathbf{H}(\mathbf{x}), -\mathbf{H}(\mathbf{x})) \geqslant \boldsymbol{\theta}$$

where $\hat{\mathbf{G}}(\mathbf{x})$ is in $\mathcal{W} \times \mathcal{H} \times \mathcal{H}$, and minimize $\mathbf{F}(\mathbf{x})$ subject to $\hat{\mathbf{G}}(\mathbf{x}) \geqslant \boldsymbol{\theta}$. In order to apply (7.60), we introduce the Lagrange multiplier vector $\hat{\boldsymbol{\lambda}} \triangleq (\boldsymbol{\lambda}, \boldsymbol{\omega}, \boldsymbol{\rho})$, where $\boldsymbol{\lambda}$, $\boldsymbol{\omega}$, and $\boldsymbol{\rho}$ are vectors in $\mathcal{W}$, $\mathcal{H}$, and $\mathcal{H}$, respectively. The differential of $\hat{\mathbf{G}}$ is

$$\begin{aligned} \mathbf{d}\hat{\mathbf{G}}(\mathbf{x}, \mathbf{h}) &= (\mathbf{dG}(\mathbf{x}, \mathbf{h}), \mathbf{dH}(\mathbf{x}, \mathbf{h}), -\mathbf{dH}(\mathbf{x}, \mathbf{h})) \\ &= (\mathbf{G}'(\mathbf{x})\mathbf{h}, \mathbf{H}'(\mathbf{x})\mathbf{h}, -\mathbf{H}'(\mathbf{x})\mathbf{h}) \\ &\triangleq \hat{\mathbf{G}}'(\mathbf{x})\mathbf{h} \end{aligned}$$

To find $\hat{\mathbf{G}}'(\mathbf{x})^*$, we use the standard inner product for the cartesian product space $\mathcal{W} \times \mathcal{H} \times \mathcal{H}$ (P&C 5.3*a*):

$$\begin{aligned} \langle \hat{\mathbf{G}}'(\mathbf{x})\mathbf{h}, \hat{\boldsymbol{\lambda}} \rangle_{\mathcal{W}\times\mathcal{H}\times\mathcal{H}} &= \langle (\mathbf{G}'(\mathbf{x})\mathbf{h}, \mathbf{H}'(\mathbf{x})\mathbf{h}, -\mathbf{H}'(\mathbf{x})\mathbf{h}), (\boldsymbol{\lambda}, \boldsymbol{\omega}, \boldsymbol{\rho}) \rangle_{\mathcal{W}\times\mathcal{H}\times\mathcal{H}} \\ &= \langle \mathbf{G}'(\mathbf{x})\mathbf{h}, \boldsymbol{\lambda} \rangle_{\mathcal{W}} + \langle \mathbf{H}'(\mathbf{x})\mathbf{h}, \boldsymbol{\omega} \rangle_{\mathcal{H}} - \langle \mathbf{H}'(\mathbf{x})\mathbf{h}, \boldsymbol{\rho} \rangle_{\mathcal{H}} \\ &= \langle \mathbf{h}, \mathbf{G}'(\mathbf{x})^*\boldsymbol{\lambda} \rangle_{\mathcal{V}} + \langle \mathbf{h}, \mathbf{H}'(\mathbf{x})^*\boldsymbol{\omega} \rangle_{\mathcal{V}} - \langle \mathbf{h}, \mathbf{H}'(\mathbf{x})^*\boldsymbol{\rho} \rangle_{\mathcal{V}} \\ &= \langle \mathbf{h}, \mathbf{G}'(\mathbf{x})^*\boldsymbol{\lambda} + \mathbf{H}'(\mathbf{x})^*(\boldsymbol{\omega} - \boldsymbol{\rho}) \rangle_{\mathcal{V}} \\ &\triangleq \langle \mathbf{h}, \hat{\mathbf{G}}'(\mathbf{x})^*\hat{\boldsymbol{\lambda}} \rangle_{\mathcal{V}} \end{aligned}$$

According to (7.60), the adjoint equation is

$$\nabla \mathbf{F}(\mathbf{x}) = \mathbf{G}'(\mathbf{x})^*\boldsymbol{\lambda} + \mathbf{H}'(\mathbf{x})^*(\boldsymbol{\omega} - \boldsymbol{\rho})$$

where the Lagrange multiplier vectors are constrained to be non-negative. Since $\boldsymbol{\omega}$ and $\boldsymbol{\rho}$ appear as a difference, we can replace them with the single Lagrange multiplier vector $\boldsymbol{\nu} \triangleq \boldsymbol{\omega}-\boldsymbol{\rho}$ in $\mathcal{H}$, where $\boldsymbol{\nu}$ is *not* constrained in sign. The complementary slackness condition is

$$\langle \hat{\mathbf{G}}(\mathbf{x}), \hat{\boldsymbol{\lambda}} \rangle_{\mathcal{W}\times\mathcal{H}\times\mathcal{H}} = \langle \mathbf{G}(\mathbf{x}), \boldsymbol{\lambda} \rangle_{\mathcal{W}} + \langle \mathbf{H}(\mathbf{x}), \boldsymbol{\omega} \rangle_{\mathcal{H}} + \langle -\mathbf{H}(\mathbf{x}), \boldsymbol{\rho} \rangle_{\mathcal{H}} = 0$$

Since each of the arguments in these inner products is non-negative, $\langle \mathbf{G}(\mathbf{x}), \boldsymbol{\lambda} \rangle_{\mathcal{W}} = 0$, $\langle \mathbf{H}(\mathbf{x}), \boldsymbol{\omega} \rangle_{\mathcal{H}} = 0$, and $\langle -\mathbf{H}(\mathbf{x}), \boldsymbol{\rho} \rangle_{\mathcal{H}} = 0$; the latter two complementary slackness conditions are superfluous since $\mathbf{H}(\mathbf{x}) = \boldsymbol{\theta}$. In summary, the necessary conditions associated with the *equality–inequality* problem (7.73) are*

$$\begin{aligned} &\nabla \mathbf{F}(\mathbf{x}) = \mathbf{G}'(\mathbf{x})^*\boldsymbol{\lambda} + \mathbf{H}'(\mathbf{x})^*\boldsymbol{\nu} \\ &\mathbf{G}(\mathbf{x}) \geqslant \boldsymbol{\theta}, \qquad \boldsymbol{\lambda} \geqslant \boldsymbol{\theta}, \qquad \langle \mathbf{G}(\mathbf{x}), \boldsymbol{\lambda} \rangle_{\mathcal{W}} = 0 \\ &\mathbf{H}(\mathbf{x}) = \boldsymbol{\theta}, \quad \boldsymbol{\nu} \text{ unconstrained} \end{aligned} \tag{7.74}$$

Example 2. Equality and Inequality Constraints. Let $\mathbf{x} = (\xi_1\ \xi_2)^{\mathrm{T}}$ in $\mathfrak{M}^{2\times 1}$. Define $\mathbf{F}(\mathbf{x}) \triangleq \xi_1 + \xi_2$, $\mathbf{G}(\mathbf{x}) \triangleq 4\xi_1 + \xi_2 - 4 \geqslant 0$, and $\mathbf{H}(\mathbf{x}) \triangleq \xi_1^2 + \xi_2^2 - 2 = 0$. We seek to minimize $\mathbf{F}(\mathbf{x})$ subject to $\mathbf{G}(\mathbf{x}) \geqslant 0$ and $\mathbf{H}(\mathbf{x}) = 0$. According to (7.74), $\nabla\mathbf{F}(\mathbf{x}) = \lambda \nabla \mathbf{G}(\mathbf{x}) + \nu \nabla \mathbf{H}(\mathbf{x})$, or

$$\begin{pmatrix} 1 \\ 1 \end{pmatrix} = \lambda \begin{pmatrix} 4 \\ 1 \end{pmatrix} + \nu \begin{pmatrix} 2\xi_1 \\ 2\xi_2 \end{pmatrix}$$

Also $\xi_1^2 + \xi_2^2 = 2$, ν is unconstrained, $4\xi_1 + \xi_2 \geqslant 4$, $\lambda \geqslant 0$, and either $\lambda = 0$ or $4\xi_1 + \xi_2 = 4$. The problem is illustrated in Figure 7.15. The feasible region is the arc of the circle which lies to the right of the line $\mathbf{G}(\mathbf{x}) = 0$. The minimizing vector is clearly at the intersection of the circle and the line: $\mathbf{x} \approx (1.36\ \ -1.46)^{\mathrm{T}}$. From the Lagrange multiplier equation we determine the sensitivity variables $\lambda \approx 0.39$ and $\nu \approx -0.205$. Note that $\lambda > 0$ as predicted. But $\nu < 0$; an increase in $\mathbf{H}(\mathbf{x})$ will decrease $\mathbf{F}_{\min}$.

Summary

By means of the necessary conditions derived in (7.65), (7.67), (7.70), and (7.74), we can state the Kuhn–Tucker necessary conditions for any mathematical programming problem. (Of course, we assume that a constrained optimum exists, that the transformations involved in the problem are smooth near the optimum, and that the set of constraints is not

*Hadley [7.7] presents a proof of the Kuhn–Tucker conditions (7.74) for equality–inequality problems in which $\mathcal{V}$, $\mathcal{W}$, and $\mathcal{H}$ are finite-dimensional. A generalization of (7.74) is developed for infinite-dimensional problems by Varaiya [7.14].

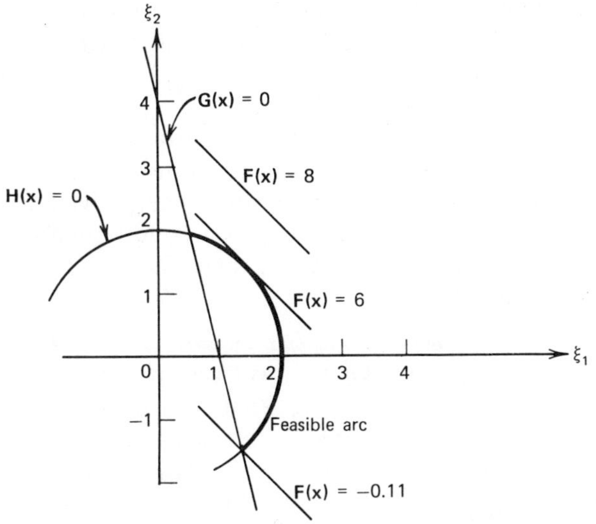

Figure 7.15. Equality and inequality constraints.

degenerate at the optimum.) Suppose, for example, that we wish to

$$\text{maximize } \mathbf{F}(\mathbf{x}) \text{ subject to}$$
$$\mathbf{G}_1(\mathbf{x}) \leqslant \boldsymbol{\theta}, \qquad \mathbf{G}_2(\mathbf{x}) \geqslant \boldsymbol{\theta}, \qquad \mathbf{H}(\mathbf{x}) = \boldsymbol{\theta}, \qquad \mathbf{x} \geqslant \boldsymbol{\theta} \tag{7.75}$$

where $\mathbf{G}_1$: $\mathcal{V} \rightarrow \mathcal{W}$, $\mathbf{G}_2$: $\mathcal{V} \rightarrow \mathcal{Z}$, and $\mathbf{H}$: $\mathcal{V} \rightarrow \mathcal{H}$, and $\mathcal{V}$, $\mathcal{W}$, $\mathcal{Z}$, and $\mathcal{H}$ are Hilbert spaces. Then the necessary conditions for a maximum $\mathbf{x}$ are

$$\begin{aligned}
&-\nabla \mathbf{F}(\mathbf{x}) = -\mathbf{G}_1{}'(\mathbf{x})^*\boldsymbol{\lambda} + \mathbf{G}_2{}'(\mathbf{x})^*\boldsymbol{\omega} + \mathbf{H}'(\mathbf{x})^*\boldsymbol{\nu} + \boldsymbol{\mu} \\
&\mathbf{G}_1(\mathbf{x}) \leqslant \boldsymbol{\theta}, \qquad \boldsymbol{\lambda} \geqslant \boldsymbol{\theta}, \qquad \langle \mathbf{G}_1(\mathbf{x}), \boldsymbol{\lambda} \rangle_{\mathcal{W}} = 0 \\
&\mathbf{G}_2(\mathbf{x}) \geqslant \boldsymbol{\theta}, \qquad \boldsymbol{\omega} \geqslant \boldsymbol{\theta}, \qquad \langle \mathbf{G}_2(\mathbf{x}), \boldsymbol{\omega} \rangle_{\mathcal{Z}} = 0 \\
&\mathbf{H}(\mathbf{x}) = \boldsymbol{\theta}, \qquad \boldsymbol{\nu} \text{ unconstrained} \\
&\mathbf{x} \geqslant \boldsymbol{\theta}, \qquad \boldsymbol{\mu} \geqslant \boldsymbol{\theta}, \qquad \langle \mathbf{x}, \boldsymbol{\mu} \rangle_{\mathcal{V}} = 0
\end{aligned} \tag{7.76}$$

for some set of Lagrange multiplier vectors $\boldsymbol{\lambda}$, $\boldsymbol{\omega}$, $\boldsymbol{\nu}$, and $\boldsymbol{\mu}$.

Exercise 4. Determine the necessary conditions for the problem: minimize $\mathbf{F}(\mathbf{x})$ subject to $\mathbf{G}_1(\mathbf{x}) \leqslant \boldsymbol{\theta}$, $\mathbf{G}_2(\mathbf{x}) \geqslant \boldsymbol{\theta}$, $\mathbf{H}(\mathbf{x}) = \boldsymbol{\theta}$, and $\mathbf{x} \leqslant \boldsymbol{\theta}$.

Of what use are the Lagrange multipliers which seem to permeate all constrained optimization? They have not been introduced arbitrarily. Rather, they are a fundamental part of the constrained optimization problem. As we have noted previously, they are the sensitivities of the optimum value of the objective function to the *specified values* of the constraints. These informative sensitivities exist, whether we compute them or not. As a practical matter, we need to compute them in order to determine a critical point. If some of the constraints are inequalities, we must use some scheme for guessing which constraints are binding. If we pick an incorrect trial set of binding constraints and solve for a feasible $\mathbf{x}$, the fact that we have not found a constrained critical point shows up as a negative sign on one or more of the Lagrange multipliers associated with the *inequality* constraints.

Example 3. Lagrange Multipliers Pinpoint Incorrect Binding Constraints. Let $\mathbf{x}=(\xi_1\ \xi_2)^T$ in $\mathfrak{M}^{2\times1}$. Suppose we seek the maximum and minimum of the function $\mathbf{F}(\mathbf{x}) \stackrel{\Delta}{=} \xi_1+\xi_2$ with $\mathbf{G}(\mathbf{x}) \stackrel{\Delta}{=} 2-\xi_1-2\xi_2 \geqslant 0$, $\mathbf{x} \geqslant \boldsymbol{\theta}$. The problem is illustrated in Figure 7.16. We first seek the constrained minimum. We try only the various combinations of two binding constraints. The Lagrange multiplier equation corresponding to minimization is $\nabla\mathbf{F}(\mathbf{x})=\lambda\nabla\mathbf{G}(\mathbf{x})+\boldsymbol{\mu}$. The other necessary conditions are $\mathbf{x}\geqslant\boldsymbol{\theta}$, $\boldsymbol{\mu}\geqslant\boldsymbol{\theta}$, either $\xi_i=0$ or $\mu_i=0$; $\xi_1+2\xi_2\leqslant 2$, $\lambda\geqslant 0$, either $\lambda=0$ or $\xi_1+2\xi_2=2$.

1. Pick $\xi_1=0$, $\xi_1+2\xi_2=2$, and $\mu_2=0$. It follows that $\xi_2=1$. From the Lagrange multiplier equation

$$\begin{pmatrix}1\\1\end{pmatrix}=\lambda\begin{pmatrix}-1\\-2\end{pmatrix}+\begin{pmatrix}\mu_1\\0\end{pmatrix}$$

we find $\lambda=-\frac{1}{2}\not\geqslant 0$ and $\mu_1=\frac{1}{2}>0$.

2. Pick $\xi_2=0$, $\xi_1+2\xi_2=2$, and $\mu_1=0$. Then $\xi_1=2$, $\lambda=-1\not\geqslant 0$, and $\mu_2=-1\not\geqslant 0$.
3. Pick $\xi_1=0$, $\xi_2=0$, and $\lambda=0$. Then, $\mu_1=1>0$ and $\mu_2=1>0$. The vector $\mathbf{x}=(0\ 0)^T$ is the constrained minimum.

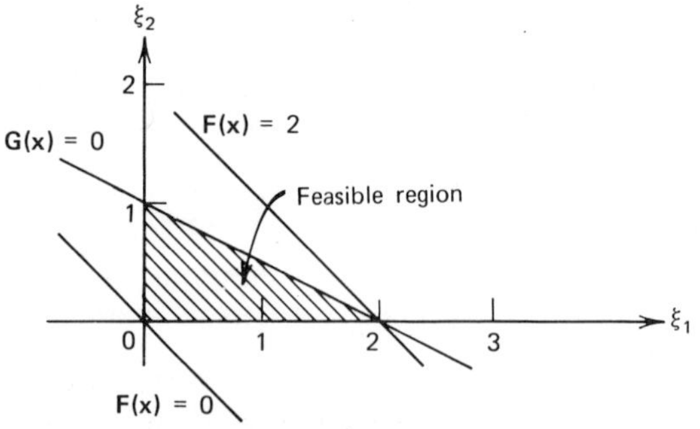

Figure 7.16. Finding constrained extrema.

The Lagrange multiplier equation corresponding to maximization is $-\nabla\mathbf{F}(\mathbf{x}) = \lambda\nabla\mathbf{G}(\mathbf{x}) + \boldsymbol{\mu}$. The other necessary conditions remain as stated above. Trying the same three pairs of binding constraints, we find:

4. For $\xi_1 = 0$, $\xi_1 + 2\xi_2 = 2$, and $\mu_2 = 0$, we have $\xi_2 = 1$, $\lambda = \frac{1}{2} > 0$, and $\mu_1 = -\frac{1}{2} \not\geq 0$.
5. For $\xi_2 = 0$, $\xi_1 + 2\xi_2 = 2$, and $\mu_1 = 0$, we have $\xi_1 = 2$, $\lambda = 1 > 0$, and $\mu_2 = 1 > 0$. The vector $\mathbf{x} = (2\ \ 0)^{\mathrm{T}}$ is the constrained maximum.
6. For $\xi_1 = 0$, $\xi_2 = 0$, and $\lambda = 0$, we have $\mu_1 = -1 \not\geq 0$ and $\mu_2 = -1 \not\geq 0$.

We note that introducing minus signs for certain of the terms in the Lagrange multiplier equation is equivalent to specifying the signs on the corresponding Lagrange multipliers. Since the sign convention for these multipliers is arbitrary, we could, if we wished, change the definition of the sign of certain of the Lagrange multipliers so as to give positive signs for all terms in the Lagrange multiplier equation. The result would be to change the non-negativity requirements for those multipliers to nonpositivity requirements, and to change the sign on λ_i in the sensitivity equation (7.43) for those particular multipliers. Rather than become embroiled in the question of signs on the Lagrange multipliers, we prefer to stick to the assumed sign convention throughout; Lagrange multipliers associated with inequality constraints will always be non-negative.

There is no practical, broadly applicable technique for distinguishing between the various types of constrained critical points for arbitrarily defined objective functions and constraints. Some authors discuss conditions on the objective function and constraints which are sufficient to guarantee that there is only a global maximum or minimum and no other critical points.* Linear programming and quadratic programming problems satisfy such sufficient conditions. However, for many problems we must use some ad hoc scheme for determining whether a given solution to the Kuhn–Tucker necessary conditions is a local or global maximum or minimum or an inflection point. In some instances the physical nature of the problem can be of help in making this determination.

Minimum Energy Control with Inequality Constraints

Throughout Chapters 6–7 we have explored optimal control problems via a specific example—the control of a loaded dc motor. The purpose for repeatedly using the same basic example is to provide a means for comparing various optimization criteria and optimization techniques. The criteria and techniques apply to the control of any system which can be described by differential equations or difference equations. In each prob-

*For example, Hadley [7.7] shows that if the feasible region is "convex" and the objective function is "concave," then any local maximum must be a global maximum.

lem we have selected an armature voltage function **u** which would drive the motor from one rest position to another in a fixed time. We now attack a problem with inequality constraints.

The differential system which describes the motor under consideration is (6.32):

$$\ddot{\phi}(t)+\dot{\phi}(t)=\mathbf{u}(t) \quad \text{with } \phi(0)=\dot{\phi}(0)=0 \tag{7.77}$$

where $\mathbf{u}(t)$ is the armature voltage, and $\phi(t)$ is the angular position of the motor shaft. We now seek a non-negative armature voltage, $\mathbf{u}(t)\geqslant 0$, which will minimize the consumed energy, $\mathbf{F}(\mathbf{u}) \triangleq \frac{1}{2}\int_0^1 \mathbf{u}^2(t)\,dt$, while driving the motor to a state which satisfies $\phi(1)\geqslant 1$, $\dot{\phi}(1)\geqslant 0$. Intuitively, we expect the optimal **u** to be such that the constraint $\dot{\phi}(1)\geqslant 0$ is not binding, because the constraint $\mathbf{u}(t)\geqslant 0$ does not allow the motor to be reversed (brought to a halt). Furthermore, we expect $\phi(1)\geqslant 0$ to be a binding constraint, since extra energy must be consumed in driving the motor beyond its required final position.

As we did beneath (6.32), we relate the required final state and the control input by inverting the state equation corresponding to (7.77). The result is

$$\begin{aligned}\phi(1)&=\int_0^1(1-e^{-1}e^{s})\mathbf{u}(s)\,ds\geqslant 1\\ \dot{\phi}(1)&=\int_0^1 e^{-1}e^{s}\mathbf{u}(s)\,ds\geqslant 0\end{aligned} \tag{7.78}$$

The problem becomes

$$\begin{aligned}&\text{minimize } \mathbf{F}(\mathbf{u})=\tfrac{1}{2}\int_0^1 \mathbf{u}^2(t)\,dt \text{ subject to}\\ &\mathbf{G}_1(\mathbf{u}) \triangleq \int_0^1(1-e^{-1}e^{s})\mathbf{u}(s)\,ds-1\geqslant 0\\ &\mathbf{G}_2(\mathbf{u}) \triangleq \int_0^1 e^{-1}e^{s}\mathbf{u}(s)\,ds\geqslant 0\\ &\mathbf{u}(t)\geqslant 0\end{aligned} \tag{7.79}$$

where $\mathbf{F}$, $\mathbf{G}_1$, and $\mathbf{G}_2$ are transformations from $\mathcal{L}_2(0,1)$ into $\mathcal{R}$. Since the constraints are linear and the objective function is quadratic, (7.79) is an infinite-dimensional example of a quadratic programming problem. The optimal control $\mathbf{u}$ must satisfy the necessary conditions (7.70):

$$\begin{aligned}
&\nabla \mathbf{F}(\mathbf{u}) = \lambda_1 \nabla \mathbf{G}_1(\mathbf{u}) + \lambda_2 \nabla \mathbf{G}_2(\mathbf{u}) + \boldsymbol{\mu} \\
&\mathbf{G}_1(\mathbf{u}) \geqslant 0, \qquad \lambda_1 \geqslant 0, \qquad \lambda_1 \mathbf{G}_1(\mathbf{u}) = 0 \\
&\mathbf{G}_2(\mathbf{u}) \geqslant 0, \qquad \lambda_2 \geqslant 0, \qquad \lambda_2 \mathbf{G}_2(\mathbf{u}) = 0 \\
&\mathbf{u}(t) \geqslant 0, \qquad \boldsymbol{\mu}(t) \geqslant 0, \qquad \int_0^1 \mathbf{u}(t)\,\boldsymbol{\mu}(t)\,dt = 0
\end{aligned} \tag{7.80}$$

Each of the gradients in (7.80) is a function in $\mathcal{L}_2(0,1)$. It can easily be determined that

$$\begin{aligned}
\mathbf{dF}(\mathbf{u},\mathbf{h}) &= \int_0^1 \mathbf{u}(s)\mathbf{h}(s)\,ds \triangleq \int_0^1 (\nabla \mathbf{F})(s)\mathbf{h}(s)\,ds \\
\mathbf{dG}_1(\mathbf{u},\mathbf{h}) &= \int_0^1 (1 - e^{-1}e^{s})\mathbf{h}(s)\,ds \triangleq \int_0^1 (\nabla \mathbf{G}_1)(s)\mathbf{h}(s)\,ds \\
\mathbf{dG}_2(\mathbf{u},\mathbf{h}) &= \int_0^1 e^{-1}e^{s}\mathbf{h}(s)\,ds \triangleq \int_0^1 (\nabla \mathbf{G}_2)(s)\mathbf{h}(s)\,ds
\end{aligned} \tag{7.81}$$

We now must try various combinations of binding constraints in order to find one which satisfies the necessary conditions (7.80).

1. First trial (no constraints): $\lambda_1 = \lambda_2 = \boldsymbol{\mu}(t) = 0$. According to the necessary conditions, it follows that $\mathbf{u}(t) = 0 \geqslant 0$, $\mathbf{G}_1(\mathbf{u}) = -1 \ngeqslant 0$, and $\mathbf{G}_2(\mathbf{u}) = 0 \geqslant 0$. This is not the correct choice; $\mathbf{F}(\mathbf{u})$ is minimized, but $\mathbf{G}_1(\mathbf{u}) \ngeqslant 0$.
2. Second trial (two constraints binding): $\mathbf{G}_1(\mathbf{u}) = \mathbf{G}_2(\mathbf{u}) = \boldsymbol{\mu}(t) = 0$. This set of binding constraints corresponds precisely to the minimum energy control problem of (7.34)–(7.37). We found in that analysis that $\lambda_1 = (e+1)/(3-e) \geqslant 0$, $\lambda_2 = -(e-1)/(3-e) \ngeqslant 0$, and $\mathbf{u}(t) = (e+1-2e^{t})/(3-e)$. Figure 6.5 shows that $\mathbf{u}(t) \ngeqslant 0$ during the latter part of the interval $[0,1]$. That this is not the correct choice of binding constraints is clear. The motor cannot be brought to a halt ($\mathbf{G}_2(\mathbf{u}) = 0$) without reversing the armature voltage.
3. Third trial (one constraint binding): $\mathbf{G}_1(\mathbf{u}) = \lambda_2 = \boldsymbol{\mu}(t) = 0$. The Lagrange multiplier equation becomes $\mathbf{u}(t) = \lambda_1(1 - e^{-1}e^{t})$. Substituting this

form for **u** into the binding constraint, we obtain

$$\mathbf{G}_1(\mathbf{u}) = \lambda_1 \int_0^1 (1 - e^{-1}e^s)^2 ds - 1 = \lambda_1 \left(\frac{e^2 - 4e + 1}{-2e^2} \right) - 1 = 0$$

or $\lambda_1 = (-2e^2)/(e^2 - 4e + 1) \approx 5.95 > 0$; also,

$$\mathbf{G}_2(\mathbf{u}) = \lambda_1 \int_0^1 e^{-1}e^s (1 - e^{-1}e^s)\, ds = \frac{1}{2}\left(\frac{e-1}{e} \right)^2 \lambda_1 > 0$$

Then $\mathbf{u}(t) \approx 5.95(1 - e^{-1}e^t) \geqslant 0$. We have found the constrained minimum. (Because **F** is quadratic and the constraints are linear, there are no other constrained critical points to be concerned about.) The minimizing control function **u** is shown, together with the resulting trajectories, in Figure 7.17. Observe that the control is barely against the constraint $\mathbf{u}(t) \geqslant 0$ at $t = 1$. Otherwise, the energy could be lowered still further.

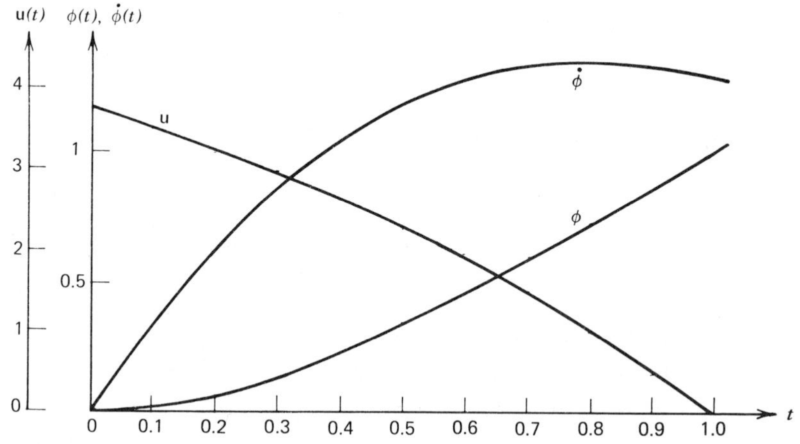

Figure 7.17. Control function and state trajectories for minimum energy.

Minimum Energy Control with Constraints Over a Continuum

Although a continuum of non-negativity constraints were placed on the control in problem (7.79) of the preceding section, these constraints did not play a direct role in the computation; they were not binding except at one of the end points. Therefore we did not have to determine the fraction of the interval over which the non-negativity constraints were binding. In this section we again apply a continuum of constraints; in this instance, however, we are forced to break the continuum of constraints into binding and nonbinding intervals. This infinite-dimensional problem demonstrates that it is sometimes necessary to express the Lagrange multiplier equation in terms of Frechét derivatives of operators rather than in terms of gradients. The problem also illustrates the difficulty in solving nonlinear equations and shows the desirability of using a physical understanding of a problem to aid in choosing binding constraints.

Let the system to be controlled again be described by (7.77). We pick the control function **u** which consumes minimum energy in driving the system precisely to the state $\phi(1)=1$, $\dot{\phi}(1)=0$ while subject to the continuum of constraints $|\mathbf{u}(t)| \leqslant c$, $0 \leqslant t \leqslant 1$, where c is a constant. The final state is again related to the control **u** as in (7.78), but with equalities rather than inequalities. We add the second equation of (7.78) to the first to obtain $\int_0^1 \mathbf{u}(s)\,ds = 1$. Multiplying the second equation by e we find $\int_0^1 e^s\mathbf{u}(s)\,ds = 0$. We define

$$\begin{aligned}
\mathbf{F}(\mathbf{u}) &\triangleq \frac{1}{2}\int_0^1 \mathbf{u}^2(t)\,dt \\
\mathbf{H}_1(\mathbf{u}) &\triangleq \int_0^1 \mathbf{u}(s)\,ds - 1 = 0 \\
\mathbf{H}_2(\mathbf{u}) &\triangleq \int_0^1 e^s\mathbf{u}(s)\,ds = 0 \qquad (7.82)\\
\mathbf{G}_1(\mathbf{u}) &\triangleq c\mathbf{1} - \mathbf{u} \geqslant \boldsymbol{\theta} \\
\mathbf{G}_2(\mathbf{u}) &\triangleq c\mathbf{1} + \mathbf{u} \geqslant \boldsymbol{\theta}
\end{aligned}$$

where **F**, $\mathbf{H}_1$, and $\mathbf{H}_2$ are transformations from $\mathcal{L}_2(0,1)$ into $\mathcal{R}$, but $\mathbf{G}_1$ and $\mathbf{G}_2$ are transformations from $\mathcal{L}_2(0,1)$ into $\mathcal{L}_2(0,1)$.* Then we can express the problem as

$$\begin{gathered}
\text{minimize } \mathbf{F}(\mathbf{u}) \text{ subject to} \\
\mathbf{G}_1(\mathbf{u}) \geqslant \boldsymbol{\theta}, \quad \mathbf{G}_2(\mathbf{u}) \geqslant \boldsymbol{\theta}, \quad \mathbf{H}_1(\mathbf{u}) = 0, \quad \mathbf{H}_2(\mathbf{u}) = 0
\end{gathered} \qquad (7.83)$$

The necessary conditions are those of (7.74):

$$\begin{gathered}
\nabla\mathbf{F}(\mathbf{u}) = \mathbf{G}_1'(\mathbf{u})^*\boldsymbol{\lambda}_1 + \mathbf{G}_2'(\mathbf{u})^*\boldsymbol{\lambda}_2 + \nu_1\nabla\mathbf{H}_1(\mathbf{u}) + \nu_2\nabla\mathbf{H}_2(\mathbf{u}) \\
\mathbf{G}_i(\mathbf{u}) \geqslant \boldsymbol{\theta}, \quad \boldsymbol{\lambda}_i \geqslant \boldsymbol{\theta}, \quad \int_0^1 \boldsymbol{\lambda}_i(s)[\mathbf{G}_i(\mathbf{u})](s)\,ds = 0, \quad i=1,2 \qquad (7.84) \\
\mathbf{H}_i(\mathbf{u}) = 0, \quad \nu_i \text{ unconstrained}, \quad i=1,2
\end{gathered}$$

*The function **1** is defined by $\mathbf{1}(t)=1$ for $0 \leqslant t \leqslant 1$.

Note that ν_1 and ν_2 are scalars, whereas $\boldsymbol{\lambda}_1$ and $\boldsymbol{\lambda}_2$ are functions in $\mathcal{L}_2(0,1)$. In order to express (7.84) explicitly for the transformations defined in (7.82), we must find $\nabla \mathbf{H}_1(\mathbf{u})$, $\nabla \mathbf{H}_2(\mathbf{u})$, $\mathbf{G}_1'(\mathbf{u})^*$, and $\mathbf{G}_2'(\mathbf{u})^*$. First we find

$$\mathbf{dH}_1(\mathbf{u},\mathbf{h}) = \frac{d}{d\beta}\left[\int_0^1 (\mathbf{u}(s)+\beta\mathbf{h}(s))\,ds - 1\right]_{\beta=0}$$

$$= \int_0^1 \mathbf{h}(s)\,ds \stackrel{\Delta}{=} \int_0^1 (\nabla\mathbf{H}_1)(s)\mathbf{h}(s)\,ds$$

or $[\nabla\mathbf{H}_1(\mathbf{u})](t) = 1$ for $0 \leqslant t \leqslant 1$. Similarly, $[\nabla\mathbf{H}_2(\mathbf{u})](t) = e^t$. Next we use (7.9) to determine

$$\mathbf{dG}_1(\mathbf{u},\mathbf{h}) = \lim_{\beta\to 0}\frac{(c\mathbf{1}-\mathbf{u}-\beta\mathbf{h})-(c\mathbf{1}-\mathbf{u})}{\beta} = -\mathbf{h} \stackrel{\Delta}{=} \mathbf{G}_1'(\mathbf{u})\mathbf{h}$$

Consequently, following the procedure (7.33),

$$\langle \mathbf{G}_1'(\mathbf{u})\mathbf{h}, \boldsymbol{\lambda}_1\rangle = \int_0^1 \boldsymbol{\lambda}_1(s)(-\mathbf{h}(s))\,ds$$

$$= \int_0^1 (-\boldsymbol{\lambda}_1(s))\mathbf{h}(s)\,ds = \langle \mathbf{h}, \mathbf{G}_1'(\mathbf{u})^*\boldsymbol{\lambda}_1\rangle$$

or $\mathbf{G}_1'(\mathbf{u})^*\boldsymbol{\lambda}_1 = -\boldsymbol{\lambda}_1$, a function in $\mathcal{L}_2(0,1)$. Similarly, $\mathbf{G}_2'(\mathbf{u})^*\boldsymbol{\lambda}_2 = \boldsymbol{\lambda}_2$. We can now express the necessary conditions (7.84) in the explicit form:

$$\begin{aligned}
&\mathbf{u}(t) = -\boldsymbol{\lambda}_1(t) + \boldsymbol{\lambda}_2(t) + \nu_1 + \nu_2 e^t,\\
&\mathbf{u}(t) \leqslant c, \quad \boldsymbol{\lambda}_1(t) \geqslant 0, \qquad \text{either } \boldsymbol{\lambda}_1(t) = 0 \text{ or } \mathbf{u}(t) = c\\
&\mathbf{u}(t) \geqslant -c, \quad \boldsymbol{\lambda}_2(t) \geqslant 0, \qquad \text{either } \boldsymbol{\lambda}_2(t) = 0 \text{ or } \mathbf{u}(t) = -c\\
&\int_0^1 \mathbf{u}(s)\,ds = 1, \qquad \nu_1 \text{ unconstrained}\\
&\int_0^1 e^s\mathbf{u}(s)\,ds = 0, \qquad \nu_2 \text{ unconstrained}
\end{aligned} \tag{7.85}$$

where the first three lines apply for each t in $[0,1]$.

In order to solve for the constrained minimum, we must select certain of the constraints in (7.82) to be binding. This task is complicated by the fact that we have in $\mathbf{G}_1$ and $\mathbf{G}_2$ a continuum of constraints. If c is very large, we should expect that the inequality constraints are not binding. The solution for this case was determined in trial 2 of the preceding section. The unconstrained control is shown in Figure 7.19. It is clear from the figure that the minimum energy control is unconstrained for $c \geqslant (e-1)/(3-e) \approx 6.1$. If we reduce c below 6.1, we suspect that the control will be against the constraints for a period of time near $t=0$ and $t=1$, but not against these constraints in between. That is, we expect the control $\mathbf{u}$ to be of the form shown in Figure 7.18. We will assume this form of control and solve for the times a and b and the shape of the control in the interval $[a,b]$.

According to (7.85), with the assumed form of control, $\lambda_1(t)=0$ for $a<t\leqslant 1$ and $\lambda_2(t)=0$ for $0\leqslant t<b$. Then the Lagrange multiplier equation of (7.85) is

$$\begin{aligned} c &= -\lambda_1(t)+\nu_1+\nu_2 e^t, && 0\leqslant t\leqslant a \\ \mathbf{u}(t) &= \nu_1+\nu_2 e^t, && a<t<b \\ -c &= \lambda_2(t)+\nu_1+\nu_2 e^t, && b\leqslant t\leqslant 1 \end{aligned} \tag{7.86}$$

The first equality constraint of (7.85) becomes

$$\begin{aligned} \int_0^1 \mathbf{u}(s)\,ds &= \int_0^a (c)\,ds + \int_a^b (\nu_1+\nu_2 e^s)\,ds + \int_b^1 (-c)\,ds \\ &= ca+\nu_1(b-a)+\nu_2(e^b-e^a)+cb-c=1 \end{aligned}$$

The second equality constraint is

$$\begin{aligned} \int_0^1 e^s\mathbf{u}(s)\,ds &= \int_0^a (c)e^s\,ds + \int_a^b e^s(\nu_1+\nu_2 e^s)\,ds + \int_b^1 (-c)e^s\,ds \\ &= ce^a-c+\nu_1(e^b-e^a)+\frac{\nu_2}{2}(e^{2b}-e^{2a})+ce^b-ce=0 \end{aligned}$$

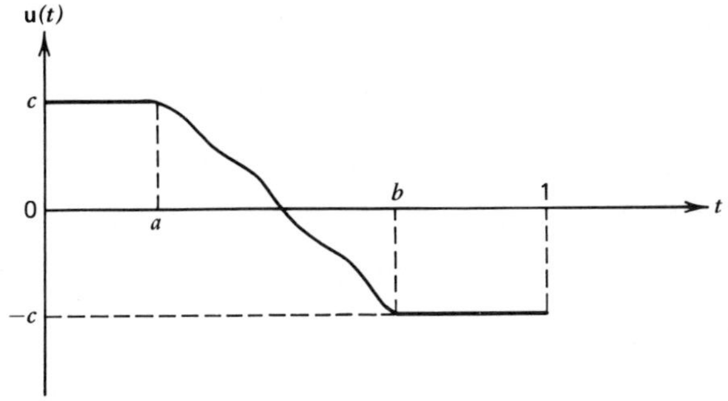

Figure 7.18. Assumed form of control under the constraint $|\mathbf{u}(t)| \leqslant c$.

These two equations can be solved for ν_1 and ν_2, a messy manipulation:

$$\nu_1 = \frac{(e^{2b}-e^{2a})(1-c(a+b-1))-2c(e^b-e^a)(1+e-e^a-e^b)}{(b-a)(e^{2b}-e^{2a})-2(e^b-e^a)^2}$$
$$\nu_2 = \frac{2c(b-a)(1+e-e^a-e^b)-2(e^b-e^a)(1-c(a+b-1))}{(b-a)(e^{2b}-e^{2a})-2(e^b-e^a)^2} \tag{7.87}$$

At the points a and b, $\mathbf{u}(t)$ satisfies

$$\mathbf{u}(a) = \nu_1 + \nu_2 e^a = c, \qquad \mathbf{u}(b) = \nu_1 + \nu_2 e^b = -c \tag{7.88}$$

Substituting (7.87) into (7.88), again a messy manipulation, we obtain a pair of equations in a and b:

$$(a-b+2)e^{2b} + (a-b-2)e^{2a} + 2(a-b)e^{a+b}$$
$$+(1+e)(b-a-2)e^b + (1+e)(b-a+2)e^a = 0 \tag{7.89}$$
$$[(a+b-3)c-1](e^b-e^a)^2 + c(e+1)(b-a)(e^b-e^a) = 0$$

[As a check, we note that the combination $a=0$, $b=1$, $c=(e-1)/(3-e)$ satisfies (7.89).] We used Newton's method, a form of trial and error (Section 8.1), to obtain the following numerical solution to (7.89) for $c=5$:

$$a \approx 0.2028 \qquad b \approx 0.9136$$

Substitution of these values of a and b into (7.87) yields the Lagrange multipliers

$$\nu_1 \approx 14.66 \qquad \nu_2 \approx -7.88 \tag{7.90}$$

Consequently, in the interval $[a,b]$ the optimal control function is

$$\mathbf{u}(t) \approx 14.66 - 7.88e^t \tag{7.91}$$

From (7.86) and our original assumptions concerning the binding constraints, we find

$$\begin{aligned} \boldsymbol{\lambda}_1(t) &= \nu_1 + \nu_2 e^t - c \approx 9.66 - 7.88e^t \text{ on } [0,a] \\ &= 0 \qquad \text{on } [a,1] \\ \boldsymbol{\lambda}_2(t) &= 0 \qquad \text{on } [0,b] \\ &= -\nu_1 - \nu_2 e^t - c \approx 7.88e^t - 9.66 \text{ on } [b,1] \end{aligned} \tag{7.92}$$

The Lagrange multiplier functions (7.92) are plotted in Figure 7.19. Since the Lagrange multipliers $\boldsymbol{\lambda}_1$ and $\boldsymbol{\lambda}_2$ are non-negative, the choice of binding constraints which we made results in satisfaction of the full set of necessary conditions. Note that our solution is unique. Because the objective function is "cup shaped" and the constraints are linear (hyperplanes), it is reasonable to assume that this solution is a global constrained minimum. The inequality constraints are most serious (in terms of their effect on $\mathbf{F}_{\min}$) at the beginning and end of the interval [0, 1], where $\boldsymbol{\lambda}_1(t)$ and $\boldsymbol{\lambda}_2(t)$, respectively, have their maximum values.

If we were to make c very small, the control would not be able to expend enough energy to drive the motor to the required final state. Therefore, there must be some lowest value of c for which the problem (7.83) makes sense. It is intuitively clear that this lowest value of c is precisely the value for which $a=b$ in Figure 7.18; the value for which $\mathbf{u}$ is against one or the other of the constraints at all times. If c is at this lowest value, we have only two numbers to determine, a and c, in order to specify the control.

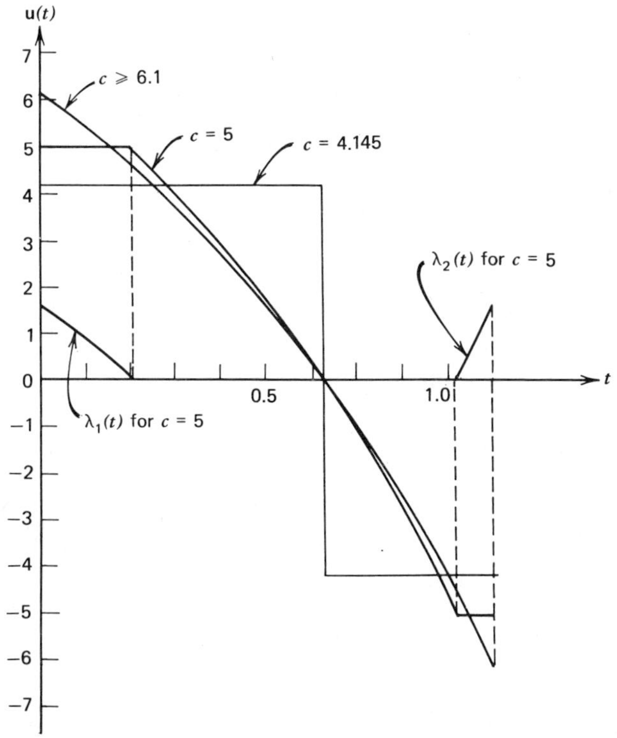

Figure 7.19. Minimum energy control for $|\mathbf{u}(t)| \leqslant c$.

The first equality constraint of (7.85) requires

$$\int_0^1 \mathbf{u}(s)\,ds = \int_0^a (c)\,ds + \int_a^1 (-c)\,ds = ca - c(1-a) = 1$$

or $a=(c+1)/2c$. The second constraint requires

$$\int_0^1 e^s\mathbf{u}(s)\,ds = \int_0^a e^s(c)\,ds + \int_a^1 e^s(-c)\,ds = c(e^a-1) - c(e-e^a) = 0$$

Consequently, $a=\ln((1+e)/2)\approx 0.62$ and $c=1/(2a-1)\approx 4.145$. The control for this limiting case is shown in Figure 7.19. The corresponding constraints are degenerate; the Lagrange multiplier equation (7.86) cannot be solved for $\boldsymbol{\lambda}_1$ and $\boldsymbol{\lambda}_2$ [the equations (7.87) which specify ν_1 and ν_2 are indeterminate]. It should be apparent from Figure 7.19 that the Lagrange multiplier functions, $\boldsymbol{\lambda}_1$ and $\boldsymbol{\lambda}_2$, become very large as $c \rightarrow 4.145$ from above.

Portfolio Selection

Suppose an individual has an opportunity to invest a fixed amount of money. He considers investment in a pair of stocks. The history of stock 1 shows a mean (average) return of 20% and a return variance of 5%. Stock 2 has a mean return of 50% and a variance of 15%. The variance of each stock is a measure of the risk associated with that stock. The covariance between the two stocks is -5%. (Thus when the return on one of the stocks rises, the return on the other tends to fall.)* Let ξ_1 be the fraction of his money the individual invests in stock 1, and ξ_2 the fraction he invests in stock 2. We refer to the "investment" vector $\mathbf{x}=(\xi_1\ \xi_2)^{\mathrm{T}}$ as a *portfolio*. A *physically realizable* portfolio must satisfy $\xi_1 \geqslant 0$, $\xi_2 \geqslant 0$, and $\xi_1+\xi_2=1$. The expected *return* associated with the portfolio $\mathbf{x}$ is given in terms of the individual mean returns by $\mathbf{F}(\mathbf{x}) \triangleq 0.2\xi_1 + 0.5\xi_2$. The *risk* associated with the portfolio can be expressed in terms of the individual variances and covariances by $\mathbf{G}(\mathbf{x}) \triangleq 0.05\xi_1^2 - 0.1\xi_1\xi_2 + 0.15\xi_2^2$. The goal of the investor is to choose his portfolio in such a way as to make the return high and the risk low.

Figure 7.20 illustrates this portfolio selection problem in graphic form. The physically realizable portfolios lie on a straight line between the points $(1\ 0)^{\mathrm{T}}$ and $(0\ 1)^{\mathrm{T}}$ in the $\mathbf{x}$ plane. From the curves of constant return and the curves of constant risk it is apparent that a physically realizable portfolio results in a return between 0.2 and 0.5 and a risk between 0.0167 and 0.15. It would be unwise to choose a portfolio for which the return is less than

*See P&C 5.37 for definitions of mean, variance, and covariance.

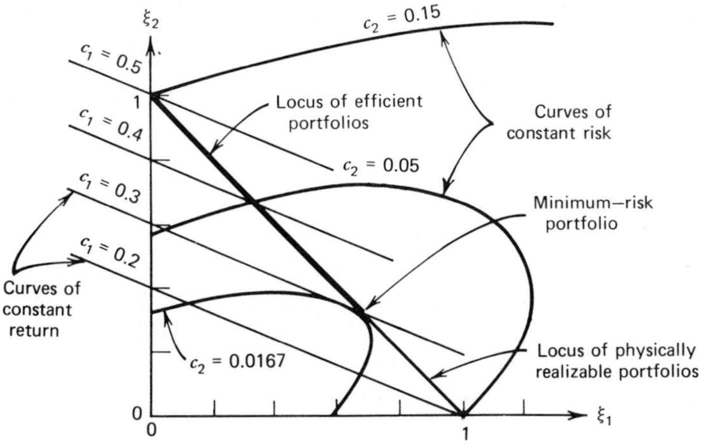

Figure 7.20. A two-dimensional space of portfolios.

0.3, because there would be another portfolio with a higher return and a lower risk. Thus we are led to refer to a physically realizable portfolio as an *efficient portfolio* if there is no other physically realizable portfolio with a higher return at the same risk or a lower risk at the same return.† In Figure 7.20 the *locus of efficient portfolios* is the line joining the points $(\frac{2}{3}\ \frac{1}{3})^T$ and $(0\ 1)^T$.

We can characterize the locus of efficient portfolios in the terminology of constrained optimization. We do so for an n-dimensional portfolio selection problem. Represent the mean returns of n stocks by the vector $\mathbf{y}$; that is, the ith element of $\mathbf{y}$ is the mean return of the ith stock. Denote the covariance matrix for the n stocks by $\mathbf{A}$. That is, the variance of the ith stock is $(\mathbf{A})_{ii}$, and $(\mathbf{A})_{ij}$ is the covariance between stock i and stock j. We assume $\mathbf{A}$ is positive definite. Let a portfolio selection be denoted by $\mathbf{x} = (\xi_1\ \xi_2 \cdots \xi_n)^T$. The expected return on the portfolio $\mathbf{x}$ is $\mathbf{F}(\mathbf{x}) \triangleq \mathbf{y}^T\mathbf{x}$. The associated risk is $\mathbf{G}(\mathbf{x}) \triangleq \mathbf{x}^T\mathbf{A}\mathbf{x}$. If $\mathbf{x}$ is an efficient portfolio with return c_1 and risk c_2, then (by definition of an efficient portfolio) $\mathbf{x}$ must be a solution to the following pair of optimization problems:

$$\text{maximize } \mathbf{y}^T\mathbf{x} \text{ subject to } \mathbf{x}^T\mathbf{A}\mathbf{x} \leqslant c_2, \quad \mathbf{1}^T\mathbf{x} = 1, \quad \mathbf{x} \geqslant \boldsymbol{\theta} \tag{7.93}$$

$$\text{minimize } \mathbf{x}^T\mathbf{A}\mathbf{x} \text{ subject to } \mathbf{y}^T\mathbf{x} \geqslant c_1, \quad \mathbf{1}^T\mathbf{x} = 1, \quad \mathbf{x} \geqslant \boldsymbol{\theta} \tag{7.94}$$

where $\mathbf{1}$ denotes the vector $(1\ 1 \cdots 1)^T$. The locus of efficient portfolios

†Intrilligator [7.9, p. 68].

consists in the set of solutions to this pair of optimization problems as a function of c_1 and c_2.

Because $\mathbf{A}$ is positive definite, the constant risk surfaces in the portfolio space are elliptical objects centered at the origin. See Figure 7.20, for example. Consequently, the physically realizable portfolio of minimum risk is always within the strict interior of the positive quadrant $\mathbf{x} \geqslant \boldsymbol{\theta}$. As a result, the further the physically realizable portfolio $\mathbf{x}$ is from the portfolio of minimum risk, the higher is the risk.

If the mean returns of the individual stocks are all identical, then all physically realizable portfolios have the same return c_1, and the maximization problem (7.93) is trivial. Furthermore, corresponding to the single value of return c_1, the minimization problem (7.94) has a unique solution—the physically realizable portfolio of minimum risk. Thus for this special relationship among the mean returns, the minimum risk portfolio is the only efficient portfolio.

Suppose the mean returns of the individual stocks are not all identical, and let $\mathbf{x}$ be an efficient portfolio with return c_1 and risk c_2 [a solution to (7.93) and (7.94)]. Because risk rises as $\mathbf{x}$ departs from the portfolio of minimum risk, the constraint on risk must be binding in (7.93); in other words, $\mathbf{x}^{\mathrm{T}}\mathbf{A}\mathbf{x} = c_2$. Since the return is a linear function of $\mathbf{x}$, there is always some direction leading away from the portfolio of minimum risk which will increase the return. Consequently the constraint on return must be binding in (7.94); that is, $\mathbf{y}^{\mathrm{T}}\mathbf{x} = c_1$. We now show (under the previous assumption that the elements of $\mathbf{y}$ are not identical), that the solutions of (7.93) are identical to those of (7.94). Either optimization problem is sufficient to describe the locus of efficient portfolio's.

The Kuhn–Tucker necessary conditions corresponding to (7.93) follow from (7.75) and (7.76):

$$\begin{aligned}
&-\mathbf{y} = \lambda(-2\mathbf{A}\mathbf{x}) + \nu\mathbf{1} + \boldsymbol{\mu} \\
&\mathbf{x}^{\mathrm{T}}\mathbf{A}\mathbf{x} \leqslant c_2 \quad \text{and} \quad \lambda \geqslant 0 \quad (\text{either } \mathbf{x}^{\mathrm{T}}\mathbf{A}\mathbf{x} = c_2 \text{ or } \lambda = 0) \\
&\mathbf{x} \geqslant \boldsymbol{\theta} \quad \text{and} \quad \boldsymbol{\mu} \geqslant \boldsymbol{\theta} \quad (\text{either } \xi_i = 0 \text{ or } \mu_i = 0, \quad i = 1,2) \\
&\mathbf{1}^{\mathrm{T}}\mathbf{x} = 1 \quad (\nu \text{ unconstrained in sign})
\end{aligned} \tag{7.95}$$

The necessary conditions corresponding to (7.94) are

$$\begin{aligned}
&2\mathbf{A}\mathbf{x} = \hat{\lambda}\mathbf{y} + \hat{\nu}\mathbf{1} + \hat{\boldsymbol{\mu}} \\
&\mathbf{y}^{\mathrm{T}}\mathbf{x} \geqslant c_1 \quad \text{and} \quad \hat{\lambda} \geqslant 0 \quad (\text{either } \mathbf{y}^{\mathrm{T}}\mathbf{x} = c_1 \text{ or } \hat{\lambda} = 0) \\
&\mathbf{x} \geqslant \boldsymbol{\theta} \quad \text{and} \quad \hat{\boldsymbol{\mu}} \geqslant \boldsymbol{\theta} \quad (\text{either } \xi_i = 0 \text{ or } \hat{\mu}_i = 0, \quad i = 1,2) \\
&\mathbf{1}^{\mathrm{T}}\mathbf{x} = 1 \quad (\hat{\nu} \text{ unconstrained in sign})
\end{aligned} \tag{7.96}$$

The number c_1 is, by definition, the maximum value of $\mathbf{y}^{\mathrm{T}}\mathbf{x}$ which is sought in (7.93). Thus because the risk constraint is binding, the equations $\mathbf{y}^T\mathbf{x}=c_1$ and $\mathbf{x}^T\mathbf{A}\mathbf{x}=c_2$ both must be satisfied at a solution to (7.93). By a similar argument, the same pair of equations must be satisfied at a solution to (7.94). The Lagrange multiplier equation of (7.96) can be rewritten as

$$-\mathbf{y}=\frac{1}{\hat{\lambda}}(-2\mathbf{A}\mathbf{x})+\frac{\hat{\nu}}{\hat{\lambda}}\mathbf{1}+\frac{\hat{\boldsymbol{\mu}}}{\hat{\lambda}} \tag{7.97}$$

If we make the substitutions $\lambda=1/\hat{\lambda}$, $\nu=\hat{\nu}/\hat{\lambda}$, and $\boldsymbol{\mu}=\hat{\boldsymbol{\mu}}/\hat{\lambda}$ in (7.97), and note that the return constraint equation $\mathbf{y}^T\mathbf{x}=c_1$ and the risk constraint equation $\mathbf{x}^T\mathbf{A}\mathbf{x}=c_2$ are both implied by either set of necessary conditions, we see that the conditions (7.96) are equivalent to (7.95).

We conclude our discussion of the portfolio selection problem by solving the necessary conditions (7.96) for the specific two-dimensional example with which we introduced the subject. In this example we have

$$\mathbf{y}=\begin{pmatrix}0.2\\0.5\end{pmatrix} \quad \text{and} \quad \mathbf{A}=\begin{pmatrix}0.05 & -0.05\\ -0.05 & 0.15\end{pmatrix}$$

Assume $\mu_1=\mu_2=0$ (the solution is strictly interior to the positive quadrant, $\mathbf{x}\geqslant\boldsymbol{\theta}$), and $0.2\xi_1+0.5\xi_2=c_1$ (the constraint on return is binding). Then the pair of equations $0.2\xi_1+0.5\xi_2=c_1$ and $\xi_1+\xi_2=1$ requires that

$$\xi_1=\frac{5-10c_1}{3} \quad \text{and} \quad \xi_2=\frac{10c_1-2}{3} \tag{7.98}$$

Substituting these values into the Lagrange multiplier equation of (7.96), we obtain

$$2\begin{pmatrix}0.05 & -0.05\\ -0.05 & 0.15\end{pmatrix}\begin{pmatrix}(5-10c_1)/3\\ (10c_1-2)/3\end{pmatrix}=\hat{\lambda}\begin{pmatrix}0.2\\0.5\end{pmatrix}+\hat{\nu}\begin{pmatrix}1\\1\end{pmatrix}+\begin{pmatrix}0\\0\end{pmatrix}$$

The solution to this vector equation is

$$\hat{\lambda}=-2+\frac{20c_1}{3}, \qquad \hat{\nu}=-2c_1+\frac{19}{30}$$

According to the necessary conditions (7.96), $\hat{\lambda}\geqslant 0$ for an efficient portfolio. Consequently, we must require $-2+20c_1/3\geqslant 0$, or $c_1\geqslant 0.3$. In order that the solutions be physically realizable portfolios, we must have $\xi_1\geqslant 0$ and $\xi_2\geqslant 0$; the conditions

$$\frac{5-10c_1}{3}\geqslant 0 \quad \text{and} \quad \frac{10c_1-2}{3}\geqslant 0$$

require that $0.2 \leqslant c_1 \leqslant 0.5$. Thus the locus of efficient portfolios is given by (7.98) for $0.3 \leqslant c_1 \leqslant 0.5$ (Figure 7.20). The particular efficient portfolio which should be picked by the investor depends upon his personal opinion of the relative importance of risk and return.

7.5 Problems and Comments

*7.1 A *normed linear space* is a vector space on which a norm has been defined. A normed linear space which is *complete* is called a *Banach space*.

(a) A **norm** on a vector space $\mathcal{V}$ is a real-valued function $\|\mathbf{x}\|$ of the vector $\mathbf{x}$ which obeys the following rules for each $\mathbf{x}$ and $\mathbf{y}$ in $\mathcal{V}$:

1. $\|\mathbf{x}\| \geqslant 0$ with equality if and only if $\mathbf{x} = \boldsymbol{\theta}$
2. $\|a\mathbf{x}\| = |a|\,\|\mathbf{x}\|$
3. $\|\mathbf{x}+\mathbf{y}\| \leqslant \|\mathbf{x}\| + \|\mathbf{y}\|$

(b) The **unit ball** in a normed space $\mathcal{V}$ is the set of vectors which satisfy $\|\mathbf{x}\| \leqslant 1$. The unit ball and the norm are equivalent. It can be shown that for any real vector space $\mathcal{V}$ there is a one-to-one correspondence between the set of all norms on $\mathcal{V}$ and the set of all bounded convex bodies in $\mathcal{V}$ which are symmetric about the origin. (The convex bodies must include their boundaries and contain $\boldsymbol{\theta}$ as an interior point. We use the terms body, convex, bounded, and interior in an intuitive sense. See [7.8].)

(c) The minimax (or Chebyshev) norm on $\mathcal{R}^2$ is defined by

$$\|\mathbf{x}\|_\infty \triangleq \max\{|\xi_1|, |\xi_2|\}$$

where $\mathbf{x} = (\xi_1, \xi_2)$. Show that $\|\mathbf{x}\|_\infty$ satisfies the rules given in (*a*). Draw the unit ball for $\|\mathbf{x}\|_\infty$ in the 2-dimensional plane.

(d) Define another norm on $\mathcal{R}^2$ by drawing an appropriate unit ball in the 2-dimensional plane. If possible, also describe this norm mathematically.

(e) Show that the following is a norm on $\mathcal{C}(a,b)$:

$$\|\mathbf{f}\|_\infty \triangleq \max_{a \leqslant t \leqslant b} |\mathbf{f}(t)|$$

*7.2 Define $\mathbf{G}$: $\mathcal{V} \to \mathcal{W}$, where $\mathcal{V}$ and $\mathcal{W}$ are normed spaces. Then $\mathbf{dG}(\mathbf{x},\mathbf{h}) = \mathbf{G}'(\mathbf{x})\mathbf{h}$. Let $\mathcal{L}(\mathcal{V}, \mathcal{W})$ denote the space of linear transformations from $\mathcal{V}$ into $\mathcal{W}$ (P&C 2.6). We call $\mathbf{G}'$: $\mathcal{V} \to \mathcal{L}(\mathcal{V}, \mathcal{W})$ the **Fréchet derivative** of $\mathbf{G}$. Generally, $\mathbf{G}'$ is a nonlinear function of $\mathbf{x}$.

(*a*) The *chain rule* extends to Fréchet derivatives (see [7.11]). Let $\mathcal{Z}$ be a normed space, and define $\mathbf{H}$: $\mathcal{W} \to \mathcal{Z}$. Then $\mathbf{HG}$: $\mathcal{V} \to \mathcal{Z}$, and $(\mathbf{HG})'(\mathbf{x}) = \mathbf{H}'(\mathbf{G}(\mathbf{x}))\mathbf{G}'(\mathbf{x})$. Verify this result for $\mathcal{V} = \mathcal{W} = \mathfrak{M}^{2\times 1}$, $\mathcal{Z} = \mathcal{R}$, $\mathbf{x} = (\xi_1 \;\; \xi_2)^{\mathrm{T}}$, $\mathbf{y} = (\eta_1 \;\; \eta_2)^{\mathrm{T}}$, $\mathbf{G}(\mathbf{x}) \triangleq (\xi_1^2 \;\; 2\xi_1\xi_2)^{\mathrm{T}}$, and $\mathbf{H}(\mathbf{y}) \triangleq \eta_1^3 + \eta_2^3$.

(*b*) We can define *higher-order Fréchet differentials and Fréchet derivatives* (see [7.12]). For example, if $\mathbf{G}'$ is a sufficiently smooth function of $\mathbf{x}$, then we know

$$\mathbf{dG}'(\mathbf{x},\mathbf{k}) = \mathbf{G}''(\mathbf{x})\mathbf{k} = \lim_{\alpha \to 0} \frac{\mathbf{G}'(\mathbf{x}+\alpha\mathbf{k}) - \mathbf{G}'(\mathbf{x})}{\alpha}$$

where $\mathbf{G}''(\mathbf{x})$: $\mathcal{V} \to \mathcal{L}(\mathcal{V}, \mathcal{W})$ is linear. We call $\mathbf{G}''$: $\mathcal{V} \to \mathcal{L}(\mathcal{V}, \mathcal{L}(\mathcal{V}, \mathcal{W}))$ the *second Fréchet derivative* of $\mathbf{G}$. Note that $\mathbf{dG}'(\mathbf{x},\mathbf{k})$ is a linear transformation from $\mathcal{V}$ into $\mathcal{W}$; that is, if $\mathbf{h}$ is in $\mathcal{V}$, $\mathbf{dG}'(\mathbf{x},\mathbf{k})\mathbf{h}$ is in $\mathcal{W}$. We define the *second Fréchet differential* of $\mathbf{G}$ by

$$\mathbf{d}^2\mathbf{G}(\mathbf{x},\mathbf{h},\mathbf{k}) \triangleq \mathbf{dG}'(\mathbf{x},\mathbf{k})\mathbf{h} = \mathbf{G}''(\mathbf{x})\mathbf{k}\mathbf{h}$$

If $\mathbf{G}$ is a functional ($\mathcal{W} = \mathcal{R}$), then

$$\begin{aligned}\mathbf{d}^2\mathbf{G}(\mathbf{x},\mathbf{h},\mathbf{k}) &= \frac{\partial}{\partial\alpha}\mathbf{G}'(\mathbf{x}+\alpha\mathbf{k})\Big|_{\alpha=0}\mathbf{h} \\ &= \frac{\partial^2}{\partial\alpha\partial\beta}\mathbf{G}(\mathbf{x}+\alpha\mathbf{k}+\beta\mathbf{h})\Big|_{\alpha=\beta=0}\end{aligned}$$

Find the first and second Fréchet differentials and derivatives for $\mathcal{V} = \mathfrak{M}^{2\times 1}$, $\mathcal{W} = \mathcal{R}$, $\mathbf{x} = (\xi_1 \;\; \xi_2)^{\mathrm{T}}$, and $\mathbf{G}(\mathbf{x}) \triangleq \xi_1^3 + \xi_1\xi_2^3$.

(*c*) *Taylor's formula with remainder* extends to transformations (see [7.3]). The three-term Taylor expansion is

$$\mathbf{G}(\mathbf{x}+\mathbf{h}) = \mathbf{G}(\mathbf{x}) + \mathbf{dG}(\mathbf{x},\mathbf{h}) + \frac{1}{2!}\mathbf{d}^2\mathbf{G}(\hat{\mathbf{x}},\mathbf{h},\mathbf{h})$$

where $\hat{\mathbf{x}}$ is on the line between $\mathbf{x}$ and $\mathbf{x}+\mathbf{h}$; that is, $\hat{\mathbf{x}} = c\mathbf{x} + (1-c)(\mathbf{x}+\mathbf{h})$ for some c in $[0,1]$. Find the three-term Taylor expansion for the functional $\mathbf{G}(\mathbf{x}) = \xi_1^3 + \xi_1\xi_2^3$ used in (*b*).

7.3 For each of the following definitions of the functional **F**, find the Fréchet differential **dF(f, h)** and the gradient ∇**F(f)** relative to the inner product defined on the space.

(*a*) Let $\mathcal{V}$ be the space of twice-differentiable functions on [0, 1] with zero end points. Assume the standard inner product on $\mathcal{V}$. Define **F**: $\mathcal{V} \to \mathcal{R}$ by $\mathbf{F}(\mathbf{f}) \triangleq \int_0^1 [\mathbf{f}(s) + \dot{\mathbf{f}}(s)]^2 \, ds$.

(*b*) Let $\mathcal{L}_2(\omega; 0, 1)$ denote the Hilbert space of functions which have finite norm under the inner product $\langle \mathbf{f}, \mathbf{g} \rangle_\omega = \int_0^1 \omega(s) \mathbf{f}(s) \mathbf{g}(s) \, ds$, where ω is an appropriate weight function. Define **F**: $\mathcal{L}_2(\omega; 0, 1) \to \mathcal{R}$ by $\mathbf{F}(\mathbf{f}) \triangleq \|\mathbf{f}\|_\omega^2$.

7.4 The definition of the Fréchet differential depends upon norms, but not on inner products. Define **F**: $\mathcal{C}(-1, 1) \to \mathcal{R}$ by $\mathbf{F}(\mathbf{f}) = \mathbf{f}^2(0)$, where the norm on $\mathcal{C}(-1, 1)$ is defined by $\|\mathbf{f}\|_\infty = \max_t |\mathbf{f}(t)|$.

(*a*) Find the Fréchet differential, **dF(f, h)**.

(*b*) Describe **F′(f)**, the Fréchet derivative of **F** at **f**.

(*c*) Show that the tangent function, **F(f) + dF(f, h)**, approximates **F(f + h)** in the manner required for a Fréchet differential.

7.5 Let Ω be the rectangle $0 \leqslant s \leqslant a$, $0 \leqslant t \leqslant b$ in $\mathcal{R}^2$. Define **F**: $\mathcal{L}_2(\Omega) \to \mathcal{R}$ by

$$\mathbf{F}(\mathbf{f}) = \tfrac{1}{2} \int_0^a \int_0^b e^{s-t} \mathbf{f}^2(s, t) \, dt \, ds$$

(*a*) Find the differential **dF(f, h)**.

(*b*) Choose an inner product for $\mathcal{L}_2(\Omega)$. Find ∇**F(f)** relative to this inner product.

(*c*) Show that **dF(f, h)** is a Fréchet differential. [Use the norm from the inner product defined in (b).]

7.6 The differential system $\dot{\mathbf{x}} = \mathbf{G}(\mathbf{x}, \mathbf{u})$, $\mathbf{x}(0)$ given, describes a dynamic system; $\mathbf{x}(t)$ is in $\mathfrak{M}^{n \times 1}$; $\mathbf{u}(t)$ is in $\mathfrak{M}^{m \times 1}$; **G** is a nonlinear transformation between two function spaces. Derive a *linear* differential system which describes the dynamic behavior of the system for small excursions of the vector functions **u** and **x** about the nominal control $\mathbf{u}_0$ and the *corresponding* nominal trajectory $\mathbf{x}_0$.

7.7 Let **A** be an $m \times n$ matrix of rank n with $m > n$. We wish to solve the equation $\mathbf{Ax} = \mathbf{y}$ in a least-square sense. Define **F**: $\mathfrak{M}^{n \times 1} \to \mathcal{R}$ by $\mathbf{F}(\mathbf{x}) \triangleq \frac{1}{2} \|\mathbf{Ax} - \mathbf{y}\|_R^2$, where the norm is derived from the inner product $\langle \mathbf{y}, \mathbf{z} \rangle_R \triangleq \mathbf{z}^{\mathrm{T}} \mathbf{R} \mathbf{y}$ and **R** is an $m \times m$ symmetric positive-definite matrix.

(*a*) Find **dF(x, h)**.

(*b*) State the necessary conditions which must be satisfied by the minimum error solution **x**.

*7.8 *Nonlinear least squares*: let **G** be a nonlinear transformation from $\mathfrak{M}^{n\times 1}$ into $\mathfrak{M}^{m\times 1}$. We desire to fit the nonlinear model $\mathbf{G}(\mathbf{x})$ to a set of observations **y** by proper choice of the elements of **x**. We pick **x** to minimize $\|\mathbf{G}(\mathbf{x})-\mathbf{y}\|^2$, where the norm is derived from some inner product on $\mathfrak{M}^{m\times 1}$.

(*a*) Determine the necessary conditions on **x** in order that it be the minimizing vector.

(*b*) Show that if **G** is linear and the inner product is the standard inner product, then the necessary conditions reduce to the normal equations.

7.9 Suppose a man has a sum of money, c, which he can spend or invest as he wishes over his lifetime. Let $\mathbf{f}(t)$ represent his total capital at time t. Then **f** obeys $\dot{\mathbf{f}}(t)=a\mathbf{f}(t)-\mathbf{g}(t)$, where a is the interest rate on his investments and $\mathbf{g}(t)$ is his rate of expenditure. Obviously, he begins with $\mathbf{f}(0)=c$. Assume $\mathbf{f}(t_f)=0$, where t_f is his remaining life expectancy. The man would like to maximize his enjoyment over his lifetime. He expresses his enjoyment at time t by the utility function $\mathbf{u}(t)=\sqrt{\mathbf{g}(t)}$; his enjoyment increases with increased spending. Thus he selects a spending program g which will maximize $\int_0^{t_f} e^{-bt}\mathbf{u}(t)\,dt$, where the factor e^{-bt} accounts for the fact that future enjoyment is of less value today.

(*a*) Express the integral to be maximized in terms of **f**; that is, eliminate **u** and **g**. Determine the Euler–Lagrange equation associated with this unconstrained minimization problem.

(*b*) Solve the Euler–Lagrange equation together with the boundary conditions to determine the man's investment and spending plan (**f** and **g**).

7.10 We wish to find a least-norm solution to the equations $\mathbf{Ax}=\mathbf{y}$, where **A** is an $m\times n$ matrix of rank m with $m<n$. Use the Lagrange multiplier method to find the vector **x** which minimizes $\|\mathbf{x}\|_Q^2$ subject to $\mathbf{Ax}=\mathbf{y}$, where the norm is defined by the inner product $\langle\mathbf{x},\mathbf{z}\rangle_Q \triangleq \mathbf{z}^T\mathbf{Qx}$ and **Q** is an $n\times n$ symmetric positive-definite matrix.

7.11 Let **Q** be a symmetric $n\times n$ matrix with eigenvalues $\lambda_1>\lambda_2>\cdots>\lambda_n$ and corresponding eigenvectors $\mathbf{x}_1,\ldots,\mathbf{x}_n$. Find the points **x** in $\mathfrak{M}^{n\times 1}$ which are the global maximum and minimum of the function $\mathbf{x}^T\mathbf{Qx}$ subject to the constraint $\mathbf{x}^T\mathbf{x}=1$. Also find the corresponding Lagrange multipliers and values of $\mathbf{x}^T\mathbf{Qx}$.

7.12 Let $\mathbf{x}_1,\ldots,\mathbf{x}_n$ be random variables with zero mean ($\mathbf{E}(\mathbf{x}_i)=0$). The correlation matrix for this set of random variables is

$$\mathbf{R} \triangleq \begin{pmatrix} \sigma_{11}^2 & \cdots & \sigma_{1n}^2 \\ \vdots & & \\ \sigma_{n1}^2 & & \sigma_{nn}^2 \end{pmatrix}$$

where $\sigma_{ij}^2 \triangleq \mathbf{E}(\mathbf{x}_i\mathbf{x}_j)$. We call the quantity $\sigma^2 \triangleq \mathrm{trace}(\mathbf{R})=\Sigma_{i=1}^n \sigma_{ii}^2$ the *total system variance*. Define a zero mean random variable $\mathbf{y}=\Sigma_{i=1}^n c_i\mathbf{x}_i$, where $\Sigma_{i=1}^n c_i^2=1$. Denote the variance of $\mathbf{y}$ by σ_y^2; that is, $\sigma_y^2=\mathbf{E}(\mathbf{y}^2)$. Pick the coefficients $\{c_i\}$ so that $\mathbf{y}$ best explains the total system variance in the sense that the quantity σ_y^2/σ^2 is maximized. Also determine the maximum value of σ_y^2/σ^2 and the Lagrange multiplier associated with the constraint on the set $\{c_i\}$. (Hint: express σ_y^2 in terms of the matrix $\mathbf{R}$.)

7.13 Define $\mathbf{F}(\mathbf{x}) \triangleq \frac{1}{2}\|\mathbf{y}-\mathbf{A}\mathbf{x}\|^2$, where $\mathbf{A}$ is an $m\times n$ matrix of rank n, $m>n$, and the norm is defined in terms of the standard inner product on $\mathfrak{M}^{m\times 1}$. We wish to maximize $\mathbf{F}(\mathbf{x})$ subject to the constraint $\mathbf{B}\mathbf{x}=\mathbf{z}$, where $\mathbf{B}$ is a $p\times n$ matrix of rank p, $p<n$.

(*a*) Determine the necessary conditions which must be satisfied at the constrained maximum. Solve the necessary conditions.

(*b*) Compute the solution for

$$\mathbf{A}=\begin{pmatrix} 1 & 1 \\ 0 & 1 \\ 1 & 1 \end{pmatrix}, \qquad \mathbf{y}=\begin{pmatrix} 1 \\ 0 \\ 0 \end{pmatrix}, \qquad \mathbf{B}=(1\ \ 1), \qquad \text{and } \mathbf{z}=(1)$$

7.14 *Linear, quadratic-loss control*: a certain control system is described by the state equation $\dot{\mathbf{x}}=\mathbf{A}\mathbf{x}+\mathbf{B}\mathbf{u}$, where $\mathbf{x}(t)$ is an $n\times 1$ matrix and $\mathbf{u}(t)$ is an $m\times 1$ matrix. Let $\mathbf{x}(0)=\mathbf{x}_0$. We wish to determine the control $\mathbf{u}$ which will minimize $\mathbf{F}(\mathbf{x},\mathbf{u})=\frac{1}{2}\int_0^{t_f}\{\mathbf{u}^{\mathrm{T}}(t)\mathbf{Q}\mathbf{u}(t)+[\mathbf{x}(t)-\mathbf{x}_d(t)]^{\mathrm{T}}\mathbf{R}[\mathbf{x}(t)-\mathbf{x}_d(t)]\}\,dt$. The vector function $\mathbf{x}_d$ is the desired state trajectory of the system. (The problem is sometimes called "flying a wire.") Assume the matrices $\mathbf{Q}$ and $\mathbf{R}$ are symmetric and positive definite.

(*a*) Determine the Euler–Lagrange equations (or adjoint equations) for this problem by equating to zero the differential of the Lagrangian, as illustrated in the derivation of (7.53).

(*b*) Note that $\mathbf{u}(t)$ is a linear transformation of the Lagrange multiplier $\boldsymbol{\lambda}(t)$. In order to express the control $\mathbf{u}(t)$ in feedback form as a linear transformation of the state $\mathbf{x}(t)$, we seek an

$n\times n$ matrix function $\mathbf{P}(t)$ such that $\boldsymbol{\lambda}(t)=\mathbf{P}(t)\mathbf{x}(t)$. For the special case where $\mathbf{x}_d=\boldsymbol{\theta}$, determine the matrix differential equation which $\mathbf{P}(t)$ must satisfy. This equation is known as the *Ricatti differential equation*.

7.15 *Optimal control*: a certain control system is described by the differential system $\dot{\mathbf{x}}=\mathbf{G}(\mathbf{x},\mathbf{u})$, $\mathbf{x}(0)=\mathbf{x}_0$, where $\mathbf{x}(t)$ and $\mathbf{u}(t)$ are $n\times 1$ and $m\times 1$ matrices, respectively. We wish to determine the control $\mathbf{u}$ which will minimize $\int_0^{t_f}\mathbf{F}(\mathbf{x},\mathbf{u})\,dt$. Find the Euler–Lagrange equations for this problem by equating to zero the differential of the Lagrangian,

$$\mathcal{L}(\mathbf{x},\boldsymbol{\lambda},\mathbf{u})=\int_0^{t_f}\left[\mathbf{F}(\mathbf{x},\mathbf{u})-\boldsymbol{\lambda}^{\mathrm{T}}(\dot{\mathbf{x}}-\mathbf{G}(\mathbf{x},\mathbf{u}))\right]dt$$

7.16 Let the point $\mathbf{y}=(\eta_1,\eta_2)$ in $\mathfrak{R}^2$ lie on the circle $\eta_1^2+\eta_2^2=1$. Let the point $\mathbf{z}=(\zeta_1,\zeta_2)$ in $\mathfrak{R}^2$ lie on the line $\zeta_1+\zeta_2=3$. We wish to pick $\mathbf{y}$ and $\mathbf{z}$ so that $\|\mathbf{y}-\mathbf{z}\|^2$ is minimum (using the standard inner product for $\mathfrak{R}^2$).

(*a*) Use the Lagrange multiplier technique to determine the set of necessary conditions which specify the optimum $\mathbf{y}$ and $\mathbf{z}$.

(*b*) Solve graphically or analytically for the optimum $\mathbf{y}$ and $\mathbf{z}$. Then determine the minimum value of $\|\mathbf{y}-\mathbf{z}\|^2$ and the sensitivites of this minimum value to perturbations in the positions of the two constraints.

7.17 Determine graphically the vector $\mathbf{x}=(\xi_1\ \xi_2)^{\mathrm{T}}$ in $\mathfrak{M}^{2\times 1}$ which will maximize $\mathbf{F}(\mathbf{x})=\xi_1-\xi_2$ subject to the conditions $\mathbf{G}_1(\mathbf{x})=-\xi_1+2\xi_2\leqslant 2$, $\mathbf{G}_2(\mathbf{x})=\xi_1+2\xi_2\leqslant 4$, $\mathbf{G}_3(\mathbf{x})=\xi_1\leqslant 2$, $\xi_1\geqslant 0$, and $\xi_2\geqslant 0$. Also determine the Lagrange multipliers.

7.18 Determine graphically the vector $\mathbf{x}=(\xi_1\ \xi_2)^{\mathrm{T}}$ which maximizes $\mathbf{F}(\mathbf{x})=\xi_1^2+\xi_2^2$ subject to the conditions $\mathbf{G}_1(\mathbf{x})=-\xi_1+2\xi_2\leqslant 2$, $\mathbf{G}_2(\mathbf{x})=\xi_1+2\xi_2\leqslant 4$, $\mathbf{G}_3(\mathbf{x})=\xi_1\leqslant 2$, $\xi_1\geqslant 0$, and $\xi_2\geqslant 0$. Also determine the Lagrange multipliers.

7.19 Let $\mathbf{x}=(\xi_1,\xi_2)$. Define $\mathbf{F}$: $\mathfrak{R}^2\to\mathfrak{R}$ and $\mathbf{G}$: $\mathfrak{R}^2\to\mathfrak{R}$ by $\mathbf{F}(\mathbf{x})\triangleq\xi_1^2+10\xi_2^2$, $\mathbf{G}(\mathbf{x})\triangleq 2-\xi_1-\xi_2$. We wish to minimize $\mathbf{F}(\mathbf{x})$ subject to $\mathbf{G}(\mathbf{x})\geqslant 0$ and $\mathbf{x}\geqslant\boldsymbol{\theta}$.

(*a*) Find the Kuhn–Tucker necessary conditions.

(*b*) Solve the necessary conditions for the minimum point and the Lagrange multipliers.

(*c*) Solve the necessary conditions under the assumption that only the constraint $\mathbf{G}(\mathbf{x})=0$ is binding. Interpret the resulting solution.

*7.20 *Duality in linear programming*: let $\mathbf{x}$ and $\mathbf{z}$ be $n \times 1$ matrices, let $\boldsymbol{\lambda}$ and $\mathbf{y}$ be $m \times 1$ matrices, and let $\mathbf{A}$ be an $m \times n$ matrix. Find the Kuhn–Tucker necessary conditions associated with the following two problems:

(*a*) Minimize $\mathbf{z}^{\mathrm{T}}\mathbf{x}$ subject to $\mathbf{A}\mathbf{x} \geqslant \mathbf{y}$, $\mathbf{x} \geqslant \boldsymbol{\theta}$;

(*b*) Maximize $\mathbf{y}^{\mathrm{T}}\boldsymbol{\lambda}$ subject to $\mathbf{A}^{\mathrm{T}}\boldsymbol{\lambda} \leqslant \mathbf{z}$, $\boldsymbol{\lambda} \geqslant \boldsymbol{\theta}$.

(*c*) Show that (*a*) and (*b*) are dual problems in the sense that the maximizing point $\boldsymbol{\lambda}$ in (*b*) is the Lagrange multiplier for the matrix constraint in (*a*) and the minimizing point $\mathbf{x}$ in (*a*) is the Lagrange multiplier for the matrix constraint in (*b*). [Hint: denote the Lagrange multiplier of (*a*) by $\boldsymbol{\lambda}$ and the Lagrange multiplier of (*b*) by $\mathbf{x}$.] What roles are played by the surplus variable vector associated with the matrix constraint of (*a*) and the slack variable vector associated with the matrix constraint of (*b*)? How are the optimal values of the two objective functions in (*a*) and (*b*) related?

(*d*) Verify the results of (*c*) for

$$\mathbf{A} = \begin{pmatrix} -1 & 1 & 1 \\ 2 & 2 & 0 \end{pmatrix}, \quad \mathbf{z} = \begin{pmatrix} 2 \\ 4 \\ 2 \end{pmatrix}, \quad \text{and } \mathbf{y} = \begin{pmatrix} 1 \\ -1 \end{pmatrix}$$

7.21 Define $\mathbf{F}(\xi_1, \xi_2) \triangleq \xi_1 + \xi_2$ and $\mathbf{G}(\xi_1, \xi_2) \triangleq \xi_1^2 + 2\xi_2^2 - 1$. We wish to find all extrema of $\mathbf{F}(\xi_1, \xi_2)$ subject to the constraints $\mathbf{G}(\xi_1, \xi_2) \leqslant 0$, $\xi_1 \geqslant 0$, and $\xi_2 \geqslant 0$.

(*a*) Find the Kuhn–Tucker necessary conditions associated with a constrained minimum of $\mathbf{F}$.

(*b*) Solve for the constrained maxima and minima and for the corresponding sensitivity variables.

7.6 References

[7.1] Aoki, M., *Introduction to Optimization Techniques*, Macmillan, New York, 1971.

[7.2] Apostol, Tom M., *Mathematical Analysis*, Addison-Wesley, Reading, Mass., 1957.

*[7.3] Collatz, Lothar, *Functional Analysis and Numerical Mathematics*, Academic Press, New York, 1966.

[7.4] Geffrion, A. M., "Duality in Nonlinear Programming: A Simplified Applications-Oriented Approach," *SIAM Rev.*, **13**, 1 (January 1971), 1–37.

*[7.5] Gelfand, I. M. and S. V. Fomin, *Calculus of Variations*, transl. by R. A. Silverman, Prentice-Hall, Englewood Cliffs, N.J., 1963.

[7.6] Hadley, G., *Linear Programming*, Addison-Wesley, Reading, Mass., 1962.

*[7.7] Hadley, G., *Nonlinear and Dynamic Programming*, Addison-Wesley, Reading, Mass., 1964.

[7.8] Householder, A. S., "The Approximate Solution of Matrix Problems," *J. Assoc. Comp. Mach.*, **5**, 1958, 205–243.

*[7.9] Intrilligator, M. D., *Mathematical Optimization and Economic Theory*, Prentice-Hall, Englewood Cliffs, N.J. 1971.

[7.10] Kuhn, H. W. and A. W. Tucker, "Nonlinear Programming," *Proceedings of the Second Berkeley Symposium on Mathematical Statistics and Probability*, University of California Press, Berkeley, Calif., 1951, 481–492.

*[7.11] Luenberger, David G., *Optimization by Vector Space Methods*, Wiley, New York, 1969.

[7.12] Lusternik, L. A. and V. J. Sobolev, *Elements of Functional Analysis*, Gordon and Breach Publishers, New York (for Hindustan Publishing Corp., Delhi, India), 1961.

[7.13] Pearson, J. B., Jr. and R. Sridhar, "A Discrete Optimal Control Problem," *IEEE Trans. Autom. Control*, **AC-11**, 2 (April 1966), 171–174.

[7.14] Pervozvanskiy, A. A., "Relationship Between the Theorems of Mathematical Programming and the Maximum Principle," *Eng. Cybern.*, **5**, 1 (January–February 1967), 6–11.

[7.15] Porter, William A., *Modern Foundations of Systems Engineering*, The Macmillan Co., New York, 1966.

[7.16] Varaiya, P. P., "Nonlinear Programming in Banach Space," *SIAM J. Appl. Math.*, **15**, 2 (March 1967), 284–293.

8

Computing the Optimum: Iteration in a Hilbert Space

Chapters 6–8 deal with techniques for optimization. As discussed in the introduction to Chapter 7, these techniques can be used to pick optimum model parameters. They can also be used to determine how best to operate a system. The approach to optimization is different in each of these chapters. Chapter 6 treats ideal optimization problems—problems with quadratic objective functions and linear equality constraints—by means of orthogonal projection techniques. In Chapter 7 we remove most of the restrictions on the objective functions and the constraint relations. The primary restriction in Chapter 7 is the requirement that the objective functions and constraint functions be differentiable in the regions of interest in order that we be able to use linear approximations. The optimization approach of Chapter 7 is *indirect*. That is, we first find necessary conditions for an extremum; then we solve these necessary conditions in order to *obtain* an extremum. If the objective function or the constraint functions are complicated, the necessary conditions will probably be complicated nonlinear equations which are difficult to solve.

This chapter is concerned with ways to *compute* extrema. In contrast to Chapter 7, this chapter dwells on *direct* methods of optimization, methods that avoid the necessary conditions altogether. In a direct method, an extremum is computed by working directly with the objective function and the constraint functions. Once again linearization is our primary tool, and thus certain differentiability assumptions are required of the objective functions and constraint functions*

Section 8.1 provides a transition between the indirect approach of Chapter 7 and the direct approach of Sections 8.2–8.6. Chapter 7 is incomplete in that it does not treat techniques for solving the nonlinear equations which usually constitute the necessary conditions for an extremum. Section 8.1 explores Newton's method for solving a nonlinear

* Except for simulation, the only tool that appears to have *broad* applicability in the analysis and optimization of nonlinear systems is linearization.

operator equation. Newton's method is essentially a process of repetitive linearization; the nonlinear equation is replaced by a sequence of linear equations. By viewing the linearizations of the nonlinear equation as operations on an objective function, we reinterpret Newton's method as a simple direct method of minimization.

Direct methods of optimization depend very much on the nature of the constraints. Section 8.2 treats the most straightforward direct method for unconstrained minimization—steepest descent. More sophisticated (and more efficient) descent techniques are explored in Section 8.3. In Section 8.4 we use "gradient projections" to extend the descent techniques of the previous two sections to problems which have *linear* equality and inequality constraints. The "penalty function" approach for handling *nonlinear* equality and inequality constraints is treated in Section 8.5. The final section provides a brief summary of the most useful direct methods of optimization.

Direct methods for computing extrema are generally more versatile than indirect methods. Yet the necessary conditions derived in Chapter 7 still perform a useful function. For optimization in the presence of constraints the necessary conditions pinpoint the role of the Lagrange multipliers (or sensitivity variables) associated with the constraints. These multipliers form an important part of the answer to the optimization problem. In Sections 8.4 and 8.5 we determine how to compute these Lagrange multipliers during the process of "direct" optimization.

Any optimization problem of significant size requires the use of a computer. The field of computer optimization algorithms is undergoing rapid change. Many optimization techniques have been published; relatively few of these have been published as specific computer algorithms; fewer yet have been tried extensively in numerical computation. Even for linear programming problems, the most straightforward of inequality constrained optimization problems, the library programs available at most computer centers still require cautious use if the problem is very large. The range of potential optimization problems is too broad to be covered by a few general optimization algorithms. An algorithm that is suitable for a small, highly nonlinear problem is probably not appropriate for a large nearly linear problem. In order that an algorithm be efficient, it must be designed to take full advantage of the specialized nature of a restricted class of problems. Rather than attempt to cover the broad subject of algorithm design, we dwell in this chapter on principles that are fundamental to many practical optimization techniques in order to provide insight into these techniques and to place the various techniques in perspective. We work primarily with smooth (Fréchet differentiable) functions in a Hilbert space setting. In the derivations of algorithms, we focus on finite-dimensional spaces. However, we demonstrate the applicability of the concepts to functions which are defined in infinite-dimensional function spaces.*

8.1 Solving Nonlinear Equations by Newton's Method

In the indirect approach to optimization discussed in Chapter 7, the extrema can be obtained only by solving a set of necessary conditions. If the optimization pertains to a static system (one that does not evolve with time), these necessary conditions usually take the form of a set of nonlinear algebraic equations. If the optimization involves a dynamic system, the necessary conditions usually take the form of a nonlinear differential equation and an associated set of boundary conditions. In either case, we can view the necessary conditions as a nonlinear vector equation. Nonlinear vector equations are usually very difficult to solve. [Attempt, for example, to solve the pair of algebraic equations (7.89) that arises in connection with the optimal control problem of Section 7.4.] We now explore a scalar solution technique which generalizes easily to vector equations. Suppose we wish to solve the scalar equation $g(t)=0$. We can approximate g in the neighborhood of a particular value of t by the tangent function, $g(t)\approx g(t_0)+g'(t_0)h$, where $t=t_0+h$. If t_0 is at all close to the zero of $g(t)$, then the zero of the tangent function should be even closer to the zero of $g(t)$. Consequently, given an estimate t_0 of the zero of $g(t)$, we can use the zero of the tangent function, denoted t_1, as an improved estimate. Mathematically, $t_1=t_0+h_1$, where h_1 satisfies $g(t_0)+g'(t_0)h_1=0$; that is,

$$t_1=t_0-\frac{g(t_0)}{g'(t_0)} \tag{8.1}$$

It is apparent from Figure 8.1 that the estimate t_1 of the solution to $g(t)=0$ still needs improvement. We obtain this improvement by repeating the linearization:

$$t_{k+1}=t_k-\frac{g(t_k)}{g'(t_k)}, \qquad k=0,1,2,\ldots \tag{8.2}$$

In the example pictured in Figure 8.1, t_k approaches the zero of $g(t)$ as $k\to\infty$, and the approach is rapid. The process of successive linearization

* See Zoutendijk [8.36] and Powell [8.25] for surveys of computational methods. Luenberger [8.20] explores many of the concepts of this chapter in the setting of normed spaces.

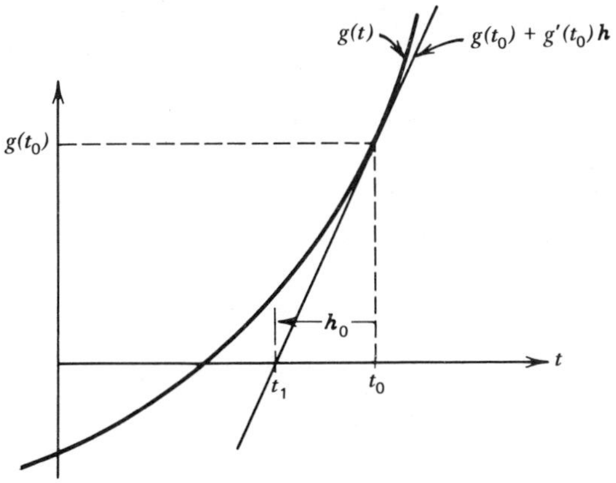

Figure 8.1. Approximating a root by linearization.

summarized by (8.2) is commonly known as *Newton's method* for obtaining the solutions to a nonlinear equation.

Example 1. Solving an Equation by Newton's Method. Suppose $g(t)=t^2-t$. This function has zeros at $t=0,1$. For this function, (8.2) becomes

$$t_{k+1}=t_k-\frac{t_k^2-t_k}{2t_k-1}$$

Letting $t_0=2$, we obtain the following sequence of approximations to a zero of $g(t)$:

$$\begin{aligned}
t_0&=2.0000\\
t_1&=1.3333\ldots\\
t_2&=1.0666\ldots\\
t_3&=1.00392\ldots\\
t_4&=1.0000152\ldots\\
&\downarrow\\
&1.00\ldots
\end{aligned}$$

Exercise 1. Repeat the iteration in Example 1 using $t_0=-1$ and $t_0=\frac{1}{2}$.

There is no conceptual difficulty in extending Newton's method to a general nonlinear equation,

$$\mathbf{G}(\mathbf{x})=\boldsymbol{\theta} \tag{8.3}$$

where **G** is a Fréchet differentiable nonlinear operator on a Hilbert space

$\mathcal{V}$. Assume we have an initial estimate $\mathbf{x}_0$ of a solution to (8.3). We linearize $\mathbf{G}$ about $\mathbf{x}_0$:

$$\mathbf{G}(\mathbf{x}) \approx \mathbf{G}(\mathbf{x}_0) + \mathbf{G}'(\mathbf{x}_0)\mathbf{h} = \boldsymbol{\theta}$$

where $\mathbf{x} = \mathbf{x}_0 + \mathbf{h}$. We solve this *linear* equation for $\mathbf{h}$, obtaining $\mathbf{h}_1 = -\mathbf{G}'(\mathbf{x}_0)^{-1}\mathbf{G}(\mathbf{x}_0)$. Then our improved estimate of the solution to (8.3) is

$$\mathbf{x}_1 = \mathbf{x}_0 + \mathbf{h}_1 = \mathbf{x}_0 - \mathbf{G}'(\mathbf{x}_0)^{-1}\mathbf{G}(\mathbf{x}_0)$$

Assuming that $\mathbf{x}_1$ is closer to the solution of (8.3) than is $\mathbf{x}_0$, we repeat the operation:

$$\mathbf{x}_{k+1} = \mathbf{x}_k - \mathbf{G}'(\mathbf{x}_k)^{-1}\mathbf{G}(\mathbf{x}_k), \qquad k = 0, 1, 2, \ldots \tag{8.4}$$

The iteration (8.4) constitutes **Newton's method** for finding the solutions of (8.3). In certain contexts the method is also referred to as the Newton–Raphson method, the Newton–Kantorovich method, or quasilinearization.

Example 2. Simultaneous Nonlinear Equations. Let $\mathbf{x} = (\xi_1 \ \xi_2)^{\mathrm{T}}$ in $\mathfrak{M}^{2\times 1}$. We use Newton's method to solve

$$\mathbf{G}\begin{pmatrix}\xi_1\\ \xi_2\end{pmatrix} \triangleq \begin{pmatrix}\xi_1\xi_2^{\,2}\\ \xi_1^{\,2}\xi_2\end{pmatrix} = \begin{pmatrix}0\\ 0\end{pmatrix}$$

(This set of equations constitutes the necessary conditions for a minimum of $\mathbf{F}(\mathbf{x}) = \xi_1^{\,2}\xi_2^{\,2}/2$.) According to (7.17), the operator $\mathbf{G}'(\mathbf{x})$ for this problem consists in multiplication by the Jacobian matrix:

$$\frac{\partial \mathbf{G}}{\partial \mathbf{x}} = \begin{pmatrix}\xi_2^{\,2} & 2\xi_1\xi_2\\ 2\xi_1\xi_2 & \xi_1^{\,2}\end{pmatrix}$$

Then the iteration (8.4) becomes

$$\mathbf{x}_{k+1} = \begin{pmatrix}\xi_1\\ \xi_2\end{pmatrix}_{k+1} = \begin{pmatrix}\xi_1\\ \xi_2\end{pmatrix}_k - \frac{1}{3}\begin{pmatrix}-1/\xi_2^{\,2} & 2/\xi_1\xi_2\\ 2/\xi_1\xi_2 & -1/\xi_1^{\,2}\end{pmatrix}_k \begin{pmatrix}\xi_1\xi_2^{\,2}\\ \xi_1^{\,2}\xi_2\end{pmatrix}_k$$

$$= \begin{pmatrix}\xi_1\\ \xi_2\end{pmatrix}_k - \frac{1}{3}\begin{pmatrix}\xi_1\\ \xi_2\end{pmatrix}_k = \frac{2}{3}\begin{pmatrix}\xi_1\\ \xi_2\end{pmatrix}_k = \frac{2}{3}\mathbf{x}_k$$

Thus $\mathbf{x}_{k+1} = (\frac{2}{3})\mathbf{x}_k = (\frac{2}{3})^k\mathbf{x}_0 \to \boldsymbol{\theta}$ for any initial estimate $\mathbf{x}_0$. Clearly, $\mathbf{x}_\infty = (0 \quad 0)^{\mathrm{T}}$ is the correct solution to the vector equation.

Exercise 2. Let $\mathbf{G}(\mathbf{x}) = \mathbf{A}\mathbf{x} - \mathbf{y} = \boldsymbol{\theta}$, where $\mathbf{A}$ is an invertible $n \times n$ matrix.

Show that Newton's method (8.4) for this linear equation converges to the exact solution in one step.

Quasilinearization—Newton's Method for Differential Systems.

Newton's method as applied to a differential system is known as **quasilinearization**. A nonlinear differential equation with boundary conditions at a single point is easy to solve, in principle, by "marching" forward or backward in small increments of the independent variable from the point where the boundary conditions are known (P&C 3.1, 3.2). This approach is not possible if the boundary conditions occur at more than one point. Let the differential system to be solved be

$$\mathbf{f}'' = \mathbf{F}(\mathbf{f}, \mathbf{u}), \quad \mathbf{f}(0) = \mathbf{f}(1) = 0 \tag{8.5}$$

where $\mathbf{F}$ is a nonlinear function of $\mathbf{f}$, and $\mathbf{u}$ is a specific input function. We rephrase the differential system in terms of an operator $\mathbf{G}$,

$$\mathbf{G}(\mathbf{f}) \stackrel{\Delta}{=} \mathbf{f}'' - \mathbf{F}(\mathbf{f}, \mathbf{u}) = \boldsymbol{\theta}$$

where $\mathbf{G}$ acts only on the subspace of functions in $\mathcal{C}^2(0,1)$ which satisfy the boundary conditions. Note that $\mathbf{G}(\mathbf{f})$ is a function. According to (7.9), the differential of $\mathbf{G}$ can be obtained from

$$\begin{aligned}[\mathbf{dG}(\mathbf{f}, \mathbf{h})](\mathbf{t}) &= \left[\lim_{\beta \to 0} \frac{\mathbf{G}(\mathbf{f} + \beta \mathbf{h}) - \mathbf{G}(\mathbf{f})}{\beta}\right](t) \\ &= \lim_{\beta \to 0} \frac{\mathbf{G}(\mathbf{f}(t) + \beta \mathbf{h}(t)) - \mathbf{G}(\mathbf{f}(t))}{\beta} \\ &= \frac{d}{d\beta} \mathbf{G}(\mathbf{f}(t) + \beta \mathbf{h}(t))\big|_{\beta=0} \end{aligned} \tag{8.6}$$

[The result (8.6) is a function space extension of (7.15). It assumes that the limit converges in a pointwise sense.] The differential of $\mathbf{G}$ is

$$\begin{aligned}[\mathbf{dG}(\mathbf{f}, \mathbf{h})](t) &= \frac{d}{d\beta}\left[\mathbf{f}''(t) + \beta \mathbf{h}''(t) - \mathbf{F}(\mathbf{f}(t) + \beta \mathbf{h}(t), \mathbf{u}(t))\right]\big|_{\beta=0} \\ &= \mathbf{h}''(t) - \frac{\partial \mathbf{F}(\mathbf{f}(t), \mathbf{u}(t))}{\partial \mathbf{f}(t)} \mathbf{h}(t) \end{aligned}$$

According to Newton's method, if $\mathbf{f}_0$ is an initial approximation to the solution of the differential system, then the exact solution should be the limit $\mathbf{f}_\infty$ of the sequence of approximations $\{\mathbf{f}_k, k=0,1,2,\ldots\}$, where $\mathbf{f}_{k+1} = \mathbf{f}_k + \mathbf{h}_k$ and $\mathbf{h}_k$ is the solution to the *linear* equation

$$\mathbf{G}(\mathbf{f}_k) + \mathbf{dG}(\mathbf{f}_k, \mathbf{h}_k) = \boldsymbol{\theta}$$

Since $\mathbf{h}_k = \mathbf{f}_{k+1} - \mathbf{f}_k$, we can also express this successive approximation process directly in terms of the sequence of approximate solutions $\mathbf{f}_k$:

$$\begin{aligned}
&[\mathbf{G}(\mathbf{f}_k)](t) + [\mathbf{dG}(\mathbf{f}_k, \mathbf{f}_{k+1} - \mathbf{f}_k)](t) \\
&\qquad = \mathbf{f}_k''(t) - \mathbf{F}(\mathbf{f}_k(t), \mathbf{u}(t)) \\
&\qquad\quad + [\mathbf{f}_{k+1}''(t) - \mathbf{f}_k''(t)] - \frac{\partial \mathbf{F}(\mathbf{f}_k(t), \mathbf{u}(t))}{\partial \mathbf{f}(t)} [\mathbf{f}_{k+1}(t) - \mathbf{f}_k(t)] \\
&\qquad = \mathbf{f}_{k+1}''(t) - \mathbf{F}(\mathbf{f}_k(t), \mathbf{u}(t)) - \frac{\partial \mathbf{F}(\mathbf{f}_k(t), \mathbf{u}(t))}{\partial \mathbf{f}(t)} [\mathbf{f}_{k+1}(t) - \mathbf{f}_k(t)] \\
&\qquad = 0
\end{aligned}$$

or

$$\mathbf{f}_{k+1}'' - \mathbf{F}_{\mathbf{f}}(\mathbf{f}_k, \mathbf{u})\mathbf{f}_{k+1} = \mathbf{F}(\mathbf{f}_k, \mathbf{u}) - \mathbf{F}_{\mathbf{f}}(\mathbf{f}_k, \mathbf{u})\mathbf{f}_k \tag{8.7}$$

The quantities on the right side of this differential equation are known at the kth iteration. For each value of k, the boundary conditions are those inherent in the subspace used to define $\mathbf{G}$:

$$\mathbf{f}_{k+1}(0) = \mathbf{f}_{k+1}(1) = 0 \tag{8.8}$$

We have converted the problem of solving the nonlinear differential system (8.5) to the problem of solving a sequence of *linear* differential systems (8.7). Of course, the solution of a linear variable-coefficient two-point boundary value problem is still not a trivial task; but it is tractable.* Quasilinearization applies to differential systems of any order and to arbitrary boundary conditions. See Bellman and Kalaba [8.3] or Sage [8.31,

* One practical numerical approach to solving the linear two-point boundary value problem (8.7)–(8.8) is to approximate derivatives by finite differences, thereby converting the differential equation to a set of linear algebraic equations as we did in (3.2). Another approach is to replace the final boundary condition $\mathbf{f}_{k+1}(1)=0$ by the initial condition $\mathbf{f}'_{k+1}(0)=c$, and solve this modified differential system by numerical integration techniques. By trying several values for c, we quickly find one which results in $\mathbf{f}_{k+1}(1)=0$ (see Acton [8.1, Chap. 6]. A third alternative is to apply to (8.7)–(8.8) the analytical techniques of inversion or eigenfunction expansion (Chapters 3 and 5).

p. 439] for numerical applications of quasilinearization to the nonlinear equations which arise in problems of optimal control.

Example 3. Newton's Method for a Nonlinear Differential System. Let $\mathbf{f}''(t) = \mathbf{f}^2(t) + \mathbf{u}(t)$, with $\mathbf{f}(0) = \mathbf{f}(1) = 0$. Let $\mathbf{u}(t) = -\pi^2 \sin \pi t - \sin^2 \pi t$, for which the exact solution is $\mathbf{f}(t) = \sin \pi t$. Then the function $\mathbf{F}$ of (8.5) is $\mathbf{F}(\mathbf{f}, \mathbf{u}) = \mathbf{f}^2 + \mathbf{u}$; consequently, $\mathbf{F}_{\mathbf{f}}(\mathbf{f}, \mathbf{u}) = 2\mathbf{f}$. Newton's method (8.7)–(8.8) requires

$$\mathbf{f}''_{k+1}(t) - 2\mathbf{f}_k(t)\mathbf{f}_{k+1}(t) = \mathbf{u}(t) - \mathbf{f}_k^2(t), \qquad \mathbf{f}_{k+1}(0) = \mathbf{f}_{k+1}(1) = 0$$

We pick as the initial estimate of the solution the function $\mathbf{f}_0(t) = t - 0.5$.* At each iteration we must solve a linear variable-coefficient two-point boundary value problem. The sequence of solutions was obtained numerically by means of a finite-difference approximation. The results of the first two iterations, $\mathbf{f}_1$ and $\mathbf{f}_2$, are compared with the exact solution $\mathbf{f}$ in Figure 8.2.

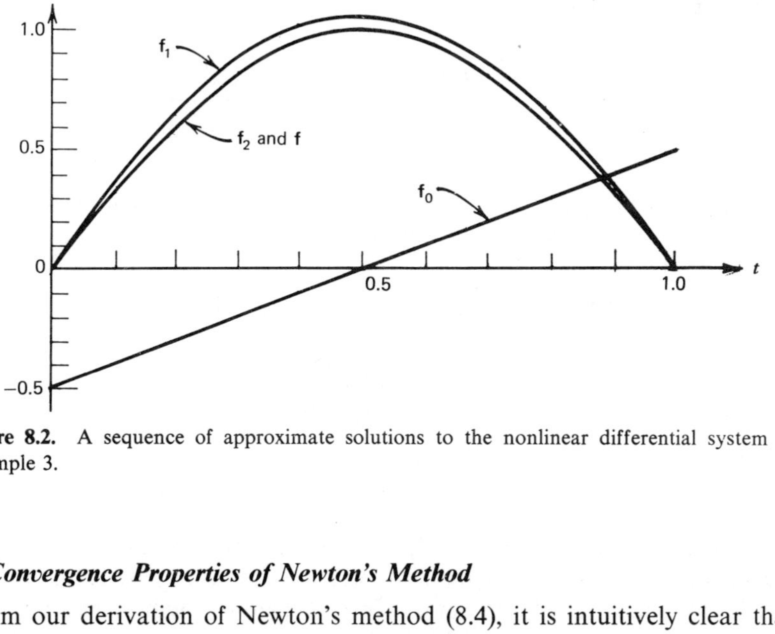

Figure 8.2. A sequence of approximate solutions to the nonlinear differential system of Example 3.

Convergence Properties of Newton's Method

From our derivation of Newton's method (8.4), it is intuitively clear that the method will produce a zero of the transformation $\mathbf{G}$ if $\mathbf{G}$ is sufficiently smooth near the zero, if $\mathbf{G}'(\mathbf{x})$ is not singular near the zero, and if the initial estimate $\mathbf{x}_0$ is sufficiently close to the zero. In some instances, the method converges to a zero of $\mathbf{G}$ for any initial estimate (see Example 2). It is

* Note that $\mathbf{f}_0$ has nonzero end points. We could as well modify $\mathbf{f}_0(t)$ near $t=0$ and $t=1$ so that $\mathbf{f}_0$ did have zero end points; $\mathbf{f}_1$ would be affected very little by the change.

useful to know how rapidly Newton's method converges to a solution. We return to the scalar case to explore the rate of convergence.

Let t_∞ denote a zero of a smooth scalar function $g(t)$. Assume $g'(t)$ is nonzero near t_∞. Then, according to (8.2), Newton's method can be expressed as $t_{k+1}=f(t_k), k=0,1,2,\ldots,$ where

$$f(t) \triangleq t-\frac{g(t)}{g'(t)}$$

Since $t_\infty=f(t_\infty)$, the error in the $(k+1)$st approximation to the solution can be expressed as

$$|t_{k+1}-t_\infty|=|f(t_k)-f(t_\infty)|$$

We expand f in a three-term Taylor series about t_∞: $f(t_k)=f(t_\infty)+f'(t_\infty)(t_k-t_\infty)+[f''(s)/2](t_k-t_\infty)^2$, where s is some number between t_k and t_∞. The first derivative of f is

$$f'(t)=1-\frac{(g'(t))^2-g(t)g''(t)}{(g'(t))^2}=\frac{g(t)g''(t)}{(g'(t))^2} \tag{8.9}$$

Since $g(t_\infty)=0$ and $g'(t_\infty)\neq 0$, we have $f'(t_\infty)=0$. Consequently, the error expression for Newton's method becomes

$$\begin{aligned}|t_{k+1}-t_\infty| &= \left|\frac{f''(s)}{2}(t_k-t_\infty)^2\right| \\ &= \left|\frac{f''(s)}{2}\right||t_k-t_\infty|^2 \\ &\leqslant c|t_k-t_\infty|^2 \end{aligned} \tag{8.10}$$

where c is a bound on $|f''(t)/2|$ for t in the neighborhood of t_∞.* A sequence $\{t_k\}$ which approaches its limit t_∞ as fast as (8.10) (but no faster) is said to exhibit **quadratic convergence**. Newton's method usually converges quadratically for t_0 sufficiently close to t_∞. In words, quadratic convergence implies that the error in the $(k+1)$st iterate is smaller by the factor $c|t_k-t_\infty|$ than the error in the kth iterate; the convergence becomes more and more rapid as $t_k \to t_\infty$.

* The existence of a bound on $f''(t)$ requires that $g'''(t)$ be bounded; it follows that $g''(t)$ is also bounded.

Example 4. Quadratic Convergence. Suppose that $|t_k - t_\infty| = 0.1$ and $c = 1$ in (8.10). Then $|t_{k+1} - t_\infty| \leqslant 0.01$, $|t_{k+2} - t_\infty| \leqslant 10^{-4}$, etc. The number of correct digits doubles at each iteration. In Example 1, $c = 2$; although the number of correct digits does not quite double at each iteration in this case, the increasing rate of convergence of Newton's method is still exhibited.

It can be shown that the application of Newton's method to *vector* equations also can be expected to produce quadratic convergence. That is, if the operator $\mathbf{G}$ of (8.3) is Fréchet differentiable near the zero $\mathbf{x}_\infty$, if $\mathbf{G}'(\mathbf{x})$ is invertible near $\mathbf{x}_\infty$, and if $\mathbf{x}_0$ is sufficiently close to $\mathbf{x}_\infty$, then for $k = 0, 1, 2, \ldots$ the errors in the iterates satisfy

$$\|\mathbf{x}_{k+1} - \mathbf{x}_\infty\| \leqslant c \|\mathbf{x}_k - \mathbf{x}_\infty\|^2 \tag{8.11}$$

for some finite number c.†

Clustered Zeros

Example 2 demonstrates that the convergence of Newton's method is not *always* quadratic. The rate of convergence in that example does not increase as $k \to \infty$. In order to understand the reason for this slow convergence, we again examine a scalar case, $g(t) = 0$. According to the derivation of (8.10), convergence is at least quadratic if the derivative (8.9) is zero at the root t_∞. If t_∞ is a simple zero of g, then $g(t_\infty) = 0, g'(t_\infty) \neq 0$, and we can usually feel confident that (8.9) is zero. Suppose, however, that t_∞ is a *double zero* of g. Then both $g(t_\infty) = 0$ and $g'(t_\infty) = 0$, and we cannot conclude that (8.9) is zero. Rather, using a two-term Taylor series, we conclude only that

$$\begin{aligned} |t_{k+1} - t_\infty| &= |f'(s)(t_k - t_\infty)| \\ &= |f'(s)||t_k - t_\infty| \\ &\leqslant c|t_k - t_\infty| \end{aligned} \tag{8.12}$$

where s is some point between t_k and t_∞, and c is a bound on $|f'(s)|$. A sequence $\{t_k\}$ which obeys (8.12), but not (8.10), is said to exhibit **linear convergence**.*

Example 5. Linear Convergence. Suppose that $|t_k - t_\infty| = 1$ and $c = 0.1$ in (8.12). Then $|t_{k+1} - t_\infty| \leqslant 0.1$, $|t_{k+2} - t_\infty| \leqslant 0.01$, etc. The number of correct digits increases

† Equation (8.11) also requires boundedness of $\mathbf{G}''(\mathbf{x})$ and $\mathbf{G}'''(\mathbf{x})$, higher Fréchet derivatives of $\mathbf{G}$. See Luenberger [8.20, p. 28].

* See Collatz [8.5, pp. 290–292] for a general treatment of order of convergence in normed spaces.

by one at each iteration. If c were greater than 0.1, the number of correct digits would increase somewhat more slowly than one per iteration; yet the rate of increase in the number of correct digits would still be constant. If c were nearly 1, the convergence would be very slow.

An analysis of the scalar case by Forsythe [8.12] shows that we should expect Newton's method to exhibit linear convergence whenever two or more zeros of g are close together relative to the position of t_0; that is, two or more zeros of g appear to be in approximately the same position as viewed from t_0. Only when the iteration gets close enough to the zeros to distinguish them clearly does the convergence become quadratic. If the sequence $\{t_k\}$ approaches a true multiple zero of g, the convergence remains linear (and thus, relatively slow) for all k. Forsythe also shows that the slow convergence which results from close zeros or multiple zeros of g can be compensated for, and sometimes quadratic convergence restored, by "overrelaxation"—increasing the size of the step in Newton's method:

$$t_{k+1} = t_k - \frac{\omega g(t_k)}{g'(t_k)}, \qquad \omega \geqslant 1 \tag{8.13}$$

In most practical situations the overrelaxation factor ω must be adjusted empirically in order to enhance the rate of convergence of the iteration (P&C 8.4).

Example 6. Slow Convergence in Newton's Method. Let $g(t) = t^2 - t = 0$, an equation with roots at $t_\infty = 0, 1$. From the point $t_0 = 100$, the pair of zeros appears to be a double zero; Newton's method should converge linearly until t_k approaches the neighborhood of the origin (say, $-1 \leqslant t_k \leqslant 2$). This conclusion is verified by the first few iterations:

$$t_1 \approx 50.25, \qquad t_2 \approx 25.39, \qquad t_3 \approx 12.96$$

On the other hand, we can conclude from Example 1 that after $t_k \approx 2$, the convergence is approximately quadratic. In P&C 8.4 we improve the rate of convergence for this problem by using the overrelaxed Newton's method (8.13).

A vector sequence $\{\mathbf{x}_k\}$ is said to exhibit *linear convergence* if it obeys the analogue of (8.12),

$$\|\mathbf{x}_{k+1} - \mathbf{x}_\infty\| \leqslant c\|\mathbf{x}_k - \mathbf{x}_\infty\|, \qquad k = 0, 1, 2, \ldots \tag{8.14}$$

but does not obey (8.11).* We should expect the practical application of Newton's method in the vector case (8.3)–(8.4) to be affected by multiple

* Linear convergence is sometimes referred to as geometric convergence; the magnitudes of the iterates form a geometric sequence.

zeros or clustered zeros in the same way the scalar case is affected; the convergence will be relatively slow (linear rather than quadratic) throughout most, if not at all, of the iteration. Again, overrelaxation may improve the rate of convergence.†

Modifications of Newton's Method

Newton's method (8.4) for solving the equation $\mathbf{G}(\mathbf{x})=\boldsymbol{\theta}$ has several disadvantages. First, the Fréchet derivative $\mathbf{G}'(\mathbf{x})$ must be determined analytically. (In some instances we can approximate $\mathbf{G}'(\mathbf{x}_k)$ at each iteration by using finite differences.) Second, the method is expensive, requiring the evaluation of $\mathbf{G}(\mathbf{x}_k)$ and $\mathbf{G}'(\mathbf{x}_k)$, and the inversion of $\mathbf{G}'(\mathbf{x}_k)$ at each iteration.* Third, it is difficult to find a starting vector $\mathbf{x}_0$ which is sufficiently close to a solution $\mathbf{x}_s$ to obtain convergence to $\mathbf{x}_s$. It is natural, therefore, to seek other methods for solving the nonlinear equation. If $\mathbf{x}_k$ is near $\mathbf{x}_s$, and if $\mathbf{G}'(\mathbf{x})$ is not ill-conditioned (sensitive to slight changes in $\mathbf{x}$), then $\mathbf{G}'(\mathbf{x})^{-1}$ changes very little over several iterations. Therefore, one useful modification of Newton's method is to avoid the repeated computation of $\mathbf{G}'(\mathbf{x}_k)^{-1}$:

$$\mathbf{x}_{k+1}=\mathbf{x}_k-\mathbf{G}'(\mathbf{x}_0)^{-1}\mathbf{G}(\mathbf{x}_k), \qquad k=0,1,2,\ldots \tag{8.15}$$

The operator $\mathbf{G}'(\mathbf{x}_0)^{-1}$ can be updated periodically if necessary. A further simplification of Newton's method consists in replacing $\mathbf{G}'(\mathbf{x}_k)^{-1}$ by a single scalar multiplier α:

$$\mathbf{x}_{k+1}=\mathbf{x}_k-\alpha\mathbf{G}(\mathbf{x}_k), \qquad k=0,1,2,\ldots \tag{8.16}$$

The scalar α can be changed at each iteration if desired. Both these modifications considerably reduce the amount of computation required at each iteration. Balanced against the computational saving per iteration, however, is a reduction in the rate of convergence (an increase in the number of iterations required to achieve a given accuracy). The primary advantage of Newton's method, as compared with other methods, is its relatively rapid rate of convergence once a good approximate solution $\mathbf{x}_0$ has been obtained. Most other methods achieve only linear convergence.

† See Acton [8.1] for a discussion of other techniques which are sometimes helpful in obtaining the roots of sets of nonlinear equations. Rall [8.27] discusses the convergence of Newton's method to multiple roots of operator equations. Collatz [8.5] discusses Newton's method and its various modifications in the context of metric spaces.

* In a practical algorithm it would be more efficient to avoid inverting $\mathbf{G}'(\mathbf{x}_k)$; rather, we would merely solve $\mathbf{G}(\mathbf{x}_k)+\mathbf{G}'(\mathbf{x}_k)\mathbf{h}_k=\boldsymbol{\theta}$ for $\mathbf{h}_k$, and let $\mathbf{x}_{k+1}=\mathbf{x}_k+\mathbf{h}_k$ (see P&C 1.3). The problem of noninvertibility of $\mathbf{G}'(\mathbf{x}_k)$ is addressed in P&C 8.5.

The relative efficiency of various methods for solving a particular class of nonlinear equations is best determined by experimentation.

Unfortunately, modifications like (8.15) and (8.16) do not address the problem of finding a suitable starting vector. In fact, the likelihood of achieving convergence to a solution $\mathbf{x}_s$ from a given starting vector $\mathbf{x}_0$ is decreased by the modifications. Beginning in the next section, we introduce a "descent" approach to the solution of nonlinear equations; the success of this approach is much less dependent upon the choice of a good initial approximation $\mathbf{x}_0$ than is Newton's method, but the rate of convergence is lower. A practical approach to the solution of nonlinear equations is to try Newton's method using a few different starting vectors. If the iterations do not converge to the desired solutions, we turn to a slower method which does not require a good initial estimate; we switch over to Newton's method after the slower method has produced a vector $\mathbf{x}_0$ in the neighborhood of a solution $\mathbf{x}_s$.

Descent Interpretation of Newton's Method

The determination of extrema by solving necessary conditions constitutes an indirect approach to optimization. In order to provide a bridge between this indirect approach and the direct optimization approaches of the succeeding sections, we now reinterpret Newton's method for solving equations and view it as a technique for direct minimization of objective functions. Suppose we wish to solve an equation $\mathbf{G}(\mathbf{x}) = \boldsymbol{\theta}$ ($\mathbf{G}: \mathcal{V} \to \mathcal{V}$). If this equation constitutes the necessary condition for an extremum of some objective function $\mathbf{F}$, then $\mathbf{G}(\mathbf{x}) = \nabla\mathbf{F}(\mathbf{x}) = \boldsymbol{\theta}$. Furthermore, it is possible that there is a function $\mathbf{F}$ such that $\mathbf{G} = \nabla\mathbf{F}$ even if the equation does not arise out of an optimization problem.* On the other hand, if there is no functional $\mathbf{F}$ for which $\nabla\mathbf{F} = \mathbf{G}$, we can always invent a functional which has minima at the solutions to $\mathbf{G}(\mathbf{x}) = \boldsymbol{\theta}$. For example, define $\mathbf{F}(\mathbf{x}) \triangleq \frac{1}{2}\|\mathbf{G}(\mathbf{x})\|^2$. Then $\mathbf{F}(\mathbf{x}) \geqslant 0$, with equality occurring at the solutions to the equation; the solutions to the equation are minima of $\mathbf{F}(\mathbf{x})$. For this particular definition of $\mathbf{F}$ the differential of $\mathbf{F}$ is

$$\begin{aligned} \mathbf{dF}(\mathbf{x}, \mathbf{h}) &= \frac{d}{d\beta} \tfrac{1}{2}\langle \mathbf{G}(\mathbf{x} + \beta\mathbf{h}), \mathbf{G}(\mathbf{x} + \beta\mathbf{h})\rangle|_{\beta=0} \\ &= \langle \mathbf{G}'(\mathbf{x})\mathbf{h}, \mathbf{G}(\mathbf{x})\rangle = \langle \mathbf{h}, \mathbf{G}'(\mathbf{x})^*\mathbf{G}(\mathbf{x})\rangle \end{aligned}$$

The critical points of $\mathbf{F}(\mathbf{x})$ occur at the solutions to $\nabla\mathbf{F}(\mathbf{x}) = \mathbf{G}'(\mathbf{x})^*\mathbf{G}(\mathbf{x}) = \boldsymbol{\theta}$.

* It can be shown that if $\mathbf{G}'(\mathbf{x})$ is continuous in $\mathbf{x}$, there is a functional $\mathbf{F}$ such that $\mathbf{G} = \nabla\mathbf{F}$ if and only if $\mathbf{G}'(\mathbf{x})$ is self-adjoint (Rothe [8.30]).

If $\mathbf{G}'(\mathbf{x})^*$ is nonsingular (i.e., if $\mathbf{G}'(\mathbf{x})$ is onto), then the minima of $\mathbf{F}(\mathbf{x})$ and the solutions to $\mathbf{G}(\mathbf{x})=\boldsymbol{\theta}$ are identical. It is evident that the problem of solving an equation and the problem of minimizing a functional can be thought of as essentially equivalent.

Suppose the functional $\mathbf{F}$ and the operator $\mathbf{G}$ are related by $\mathbf{G}=\nabla\mathbf{F}$. Suppose also that the zeros of $\mathbf{G}(\mathbf{x})$ are identical to minima of $\mathbf{F}(\mathbf{x})$. Then the application of Newton's method to the equation $\mathbf{G}(\mathbf{x})=\nabla\mathbf{F}(\mathbf{x})=\boldsymbol{\theta}$ can produce the minima of $\mathbf{F}(\mathbf{x})$ and the zeros of $\mathbf{G}(\mathbf{x})$ simultaneously. We now explore Newton's method as an iterative procedure on the functional $\mathbf{F}$ rather than on the operator $\mathbf{G}$. According to (8.4), the iteration is

$$\mathbf{x}_{k+1}=\mathbf{x}_k-(\nabla\mathbf{F})'(\mathbf{x}_k)^{-1}\nabla\mathbf{F}(\mathbf{x}_k), \qquad k=0,1,2,\ldots \tag{8.17}$$

What type of operator is $(\nabla\mathbf{F})'(\mathbf{x})$? Consider the analogous scalar problem, minimization of $f(t)$; the gradient of f is $g(t)=f'(t)$. Newton's iteration becomes

$$t_{k+1}=t_k-f''(t_k)^{-1}f'(t_k), \qquad k=0,1,2,\ldots \tag{8.18}$$

Clearly, the operator $(\nabla\mathbf{F})'(\mathbf{x})$ involves second derivatives of $\mathbf{F}$.

Suppose $\mathbf{F}$: $\mathfrak{M}^{n\times 1}\rightarrow\mathfrak{R}$; then $\mathbf{G}$ is an operator on $\mathfrak{M}^{n\times 1}$. Let $\mathbf{x}=(\xi_1\cdots\xi_n)^{\mathrm{T}}$, and denote the functional $\partial\mathbf{F}/\partial\xi_i$ by $\mathbf{F}_i$. Then

$$\mathbf{G}(\mathbf{x})=\nabla\mathbf{F}(\mathbf{x})=(\mathbf{F}_1(\mathbf{x})\cdots\mathbf{F}_n(\mathbf{x}))^{\mathrm{T}}$$

Let $\mathbf{h}=(h_1\cdots h_n)^{\mathrm{T}}$. Using (7.15) and (7.11), we find

$$\mathbf{dG}(\mathbf{x},\mathbf{h})=\begin{pmatrix}\mathbf{dF}_1(\mathbf{x},\mathbf{h})\\ \vdots\\ \mathbf{dF}_n(\mathbf{x},\mathbf{h})\end{pmatrix}=\begin{pmatrix}\dfrac{\partial\mathbf{F}_1(\mathbf{x})}{\partial\xi_1}h_1+\cdots+\dfrac{\partial\mathbf{F}_1(\mathbf{x})}{\partial\xi_n}h_n\\ \vdots \qquad\qquad \vdots\\ \dfrac{\partial\mathbf{F}_n(\mathbf{x})}{\partial\xi_1}h_1+\cdots+\dfrac{\partial\mathbf{F}_n(\mathbf{x})}{\partial\xi_n}h_n\end{pmatrix}$$

$$=\begin{pmatrix}\dfrac{\partial^2\mathbf{F}(\mathbf{x})}{\partial\xi_1^2} & \cdots & \dfrac{\partial^2\mathbf{F}(\mathbf{x})}{\partial\xi_n\partial\xi_1}\\ \vdots & & \vdots\\ \dfrac{\partial^2\mathbf{F}(\mathbf{x})}{\partial\xi_1\partial\xi_n} & \cdots & \dfrac{\partial^2\mathbf{F}(\mathbf{x})}{\partial\xi_n^2}\end{pmatrix}\mathbf{h}\overset{\Delta}{=}\frac{\partial^2\mathbf{F}(\mathbf{x})}{\partial\mathbf{x}^2}\mathbf{h} \tag{8.19}$$

The matrix of second partial derivatives, which we have denoted by $\partial^2\mathbf{F}(\mathbf{x})/\partial\mathbf{x}^2$, is commonly known as the **Hessian matrix** of the functional $\mathbf{F}$.* In this instance, where $\mathbf{x}$ is an n-dimensional column vector, the operator $(\nabla\mathbf{F})'(\mathbf{x})$ consists in multiplication by the Hessian matrix. In P&C 7.2 we define a generalization of $(\nabla\mathbf{F})'$, called the second Fréchet derivative of $\mathbf{F}$, which we denote $\mathbf{F}''$. Thus $\mathbf{F}''(\mathbf{x})\mathbf{h}=[\partial^2\mathbf{F}(\mathbf{x})/\partial\mathbf{x}^2]\mathbf{h}$. In terms of the second Fréchet derivative notation, the Newton iteration (8.17) applies to functionals in an arbitrary Hilbert space $\mathcal{V}$:

$$\mathbf{x}_{k+1}=\mathbf{x}_k-\mathbf{F}''(\mathbf{x}_k)^{-1}\nabla\mathbf{F}(\mathbf{x}_k), \qquad k=0,1,2,\ldots \tag{8.20}$$

We have assumed that the process of solving $\mathbf{G}(x)=\boldsymbol{\theta}$ corresponds to the determination of a minimum of $\mathbf{F}(\mathbf{x})$. We are inclined to conclude that as the iteration (8.20) proceeds, the sequence of estimates $\{\mathbf{x}_k\}$ must produce a descending set of values $\{\mathbf{F}(\mathbf{x}_k)\}$. However, we must remember that Newton's method will produce a minimum of $\mathbf{F}(\mathbf{x})$ (a zero of $\mathbf{G}(\mathbf{x})$) only if $\mathbf{x}_0$ is sufficiently close to that minimum. Furthermore, even if the sequence converges to a minimum, the descent of $\{\mathbf{F}(\mathbf{x}_k)\}$ to the minimum need not be a monotonic descent.

Example 7. Solution of Equations by Descent. Let $\mathbf{y}=(\xi_1\ \xi_2)^{\mathrm{T}}$. Suppose we wish to minimize $\hat{\mathbf{F}}(\mathbf{y}) \triangleq \frac{1}{2}(\xi_1^2+\xi_2^2)$ subject to $\hat{\mathbf{G}}(\mathbf{y}) \triangleq 1-\xi_1-\xi_2=0$. According to (7.28), at the minimum the Lagrange multiplier equation must be satisfied:

$$\nabla\hat{\mathbf{F}}(\mathbf{y})-\lambda\nabla\hat{\mathbf{G}}(\mathbf{y})=\begin{pmatrix}\xi_1+\lambda\\ \xi_2+\lambda\end{pmatrix}=\begin{pmatrix}0\\0\end{pmatrix}$$

Define $\mathbf{x} \triangleq (\xi_1\ \xi_2\ \lambda)^{\mathrm{T}}$,

$$\mathbf{G}(\mathbf{x}) \triangleq \begin{pmatrix}\nabla\hat{\mathbf{F}}(\mathbf{y})-\lambda\nabla\hat{\mathbf{G}}(\mathbf{y})\\ \hat{\mathbf{G}}(\mathbf{y})\end{pmatrix},$$

and

$$\begin{aligned}\mathbf{F}(\mathbf{x}) &\triangleq \tfrac{1}{2}\mathbf{G}^{\mathrm{T}}(\mathbf{x})\mathbf{G}(\mathbf{x})\\ &=\tfrac{1}{2}(\xi_1+\lambda)^2+\tfrac{1}{2}(\xi_2+\lambda)^2+\tfrac{1}{2}(1-\xi_1-\xi_2)^2\end{aligned}$$

We solve the necessary conditions by applying the Newton iteration (8.20) to $\mathbf{F}(\mathbf{x})$. The gradient of $\mathbf{F}$ is

$$\nabla\mathbf{F}(\mathbf{x})=\begin{pmatrix}(\xi_1+\lambda)-(1-\xi_1-\xi_2)\\ (\xi_2+\lambda)-(1-\xi_1-\xi_2)\\ (\xi_1+\lambda)+(\xi_2+\lambda)\end{pmatrix}$$

* The Hessian matrix of $\mathbf{F}$ is also the Jacobian matrix of $\nabla\mathbf{F}$.

The Hessian matrix of **F** is

$$\frac{\partial^2 \mathbf{F}(\mathbf{x})}{\partial \mathbf{x}^2} = \begin{pmatrix} 2 & 1 & 1 \\ 1 & 2 & 1 \\ 1 & 1 & 2 \end{pmatrix}$$

Let $\mathbf{x}_0 = (0\ \ 0\ \ 0)^{\mathrm{T}}$. Then according to (8.20),

$$\mathbf{x}_1 = \mathbf{x}_0 - \left(\frac{\partial^2 \mathbf{F}(\mathbf{x}_0)}{\partial \mathbf{x}^2} \right)^{-1} \nabla \mathbf{F}(\mathbf{x}_0)$$

$$= \begin{pmatrix} 0 \\ 0 \\ 0 \end{pmatrix} - \tfrac{1}{4} \begin{pmatrix} 3 & -1 & -1 \\ -1 & 3 & -1 \\ -1 & -1 & 3 \end{pmatrix} \begin{pmatrix} -1 \\ -1 \\ 0 \end{pmatrix} = \begin{pmatrix} \frac{1}{2} \\ \frac{1}{2} \\ -\frac{1}{2} \end{pmatrix} = \begin{pmatrix} \xi_1 \\ \xi_2 \\ \lambda \end{pmatrix}$$

Because $\mathbf{F}(\mathbf{x})$ is quadratic, $\nabla \mathbf{F}(\mathbf{x})$ is linear, and the iteration has converged in one step.

In the next section we explore in detail the concept of an iterative descent of a functional $\mathbf{F}(\mathbf{x})$. By guaranteeing that every iteration produces a decrease in $\mathbf{F}(\mathbf{x})$, we usually are able to obtain convergence to a minimum, even from poor initial estimates $\mathbf{x}_0$. The price we pay for an increased likelihood of convergence is a reduction in the *rate* of convergence. The quadratic convergence rate which is usually associated with Newton's method becomes the ideal for which we strive. After we have fully explored descent techniques for minimizing unconstrained functionals, we turn to techniques that permit us to deal explicitly with linear and nonlinear constraints.

8.2 Steepest Descent

Throughout the remaining sections of this chapter we seek a sequence of vectors $\{\mathbf{x}_k\}$ which converges to an extremum of a functional $\mathbf{F}(\mathbf{x})$. By convention, we focus on minimization problems. (Of course, the methods we discuss apply equally well to maximization problems.) The techniques that we use involve a descent of the objective functional $\mathbf{F}(\mathbf{x})$ in a direction suggested by the gradient of **F**. Thus the techniques are usually referred to as *descent methods* or *gradient methods*. The determination of the gradient of **F** is equivalent to linearization of **F**. Thus each of the gradient techniques that we discuss consists in a succession of linearizations. The details of a descent method depend upon the nature of the constraints. In Sections 8.2 and 8.3, the functional to be minimized is unconstrained. A gradient projection method for handling minimization with linear constraints is covered in Section 8.4. Section 8.5 explores the penalty function approach for dealing with nonlinear constraints.

Throughout this section we assume that the objective function $\mathbf{F}: \mathcal{V} \to \mathcal{R}$ is unconstrained. In any iterative minimization, the $(k+1)$st iterate is related to the kth iterate by the equation $\mathbf{x}_{k+1} = \mathbf{x}_k + \mathbf{h}_k$. We refer to the vector $\mathbf{h}_k$ as the kth *step*. We call $\{\mathbf{x}_k\}$ a *descending sequence* if $\mathbf{F}(\mathbf{x}_{k+1}) < \mathbf{F}(\mathbf{x}_k)$ for each k. Suppose the sequence of scalars $\{\mathbf{F}(\mathbf{x}_k)\}$ is a descending sequence which is bounded from below by $\mathbf{F}(\mathbf{x}_m)$, where $\mathbf{x}_m$ is the global minimum of $\mathbf{F}(\mathbf{x})$. Then the sequence $\{\mathbf{F}(\mathbf{x}_k)\}$ approaches a limit. Of course, even if the limit of $\{\mathbf{F}(\mathbf{x}_k)\}$ is $\mathbf{F}(\mathbf{x}_m)$, it is possible that the sequence $\{\mathbf{x}_k\}$ does not approach $\mathbf{x}_m$. Yet if we compute at each iteration a descent step [a step vector $\mathbf{h}_k$ for which $\mathbf{F}(\mathbf{x}_{k+1}) < \mathbf{F}(\mathbf{x}_k)$], it seems likely that the sequence $\{\mathbf{x}_k\}$ will converge to at least a local minimum of $\mathbf{F}(\mathbf{x})$. In point of fact, this descent strategy is usually successful even if the initial vector $\mathbf{x}_0$ is far from a minimum.* However, as we shall demonstrate, the rate of convergence can be extremely low.

Directions of Descent

It is obvious that many step directions will produce a descent in $\mathbf{F}(\mathbf{x})$. In order to guarantee descent we need to clearly distinguish these descent directions from the directions of ascent. One direction that surely produces a descent is the direction of the negative gradient, $-\nabla\mathbf{F}(\mathbf{x})$. Thus $\mathbf{h}_k = -\alpha k \nabla\mathbf{F}(\mathbf{x}_k)$ is a descent step if the scalar α_k is sufficiently small and positive. We usually think of the step direction $-\nabla\mathbf{F}(\mathbf{x})$ as the **direction of steepest descent** (P&C 8.9). We should remember, however, that the direction of $\nabla\mathbf{F}(\mathbf{x})$ depends upon the inner product which defines $\nabla\mathbf{F}(\mathbf{x})$.†

What other step directions produce descent? If $\nabla\mathbf{F}(\mathbf{x}_k) \neq \boldsymbol{\theta}$ and $\mathbf{y}$ is an arbitrary vector in $\mathcal{V}$, then we can find a linear operator $\mathbf{H}_k$ for which $\mathbf{H}_k\nabla\mathbf{F}(\mathbf{x}_k) = \mathbf{y}$. Consequently, we can express any step $\mathbf{h}_k$ in the form $\mathbf{h}_k = -\alpha_k\mathbf{H}_k\nabla\mathbf{F}(\mathbf{x}_k)$, where $\mathbf{H}_k$ is a linear operator on $\mathcal{V}$. Thus *any* iteration in the space $\mathcal{V}$ can be described by

$$\mathbf{x}_{k+1} = \mathbf{x}_k - \alpha_k\mathbf{H}_k\nabla\mathbf{F}(\mathbf{x}_k), \qquad k = 0,1,2,\ldots \tag{8.21}$$

The iteration produces a descent in $\mathbf{F}(\mathbf{x})$ if $\mathbf{F}(\mathbf{x}_{k+1}) < \mathbf{F}(\mathbf{x}_k)$. But

$$\begin{aligned}\mathbf{F}(\mathbf{x}_{k+1}) - \mathbf{F}(\mathbf{x}_k) &= \mathbf{dF}(\mathbf{x}_k, \mathbf{h}_k) + \epsilon \\ &= \langle \mathbf{h}_k, \nabla\mathbf{F}(\mathbf{x}_k)\rangle + \epsilon\end{aligned}$$

* See Wolfe [8.34] for a theoretical discussion of the conditions for convergence of descent methods.

† See (7.12).

where ϵ can be made as small as desired by using a sufficiently small step size, $\|\mathbf{h}_k\|$. Thus the step $\mathbf{h}_k$ used in (8.21) is a descent step if and only if the step size is sufficiently small and

$$\langle \mathbf{h}_k, \nabla \mathbf{F}(\mathbf{x}_k)\rangle < 0 \tag{8.22}$$

The boundary between the ascending and descending directions is the hyperplane $\mathbf{dF}(\mathbf{x}_k, \mathbf{h}_k) = \langle \mathbf{h}_k, \nabla \mathbf{F}(\mathbf{x}_k)\rangle = 0$. The condition (8.22) states that the descent directions are those which lie on the same side of this hyperplane as does the vector $-\nabla \mathbf{F}(\mathbf{x}_k)$. We divide the inequality (8.22) by the negative number $-\alpha_k$ to express the descent condition as a condition on $\mathbf{H}_k$:

$$\langle \mathbf{H}_k \nabla \mathbf{F}(\mathbf{x}_k), \nabla \mathbf{F}(\mathbf{x}_k)\rangle > 0 \tag{8.23}$$

Thus $-\mathbf{H}_k \nabla \mathbf{F}(\mathbf{x}_k)$ points in a descent direction if and only if (8.23) is satisfied. By P&C 5.28, if $\mathbf{H}_k$ is self-adjoint and positive definite, then (8.23) is satisfied for all vectors $\nabla \mathbf{F}(\mathbf{x}_k)$; that is, for any $\mathbf{F}$, $-\mathbf{H}_k \nabla \mathbf{F}(\mathbf{x}_k)$ points in a descent direction. We will assume that the "step direction" operators $\{\mathbf{H}_k\}$ are self-adjoint and positive definite, thereby guaranteeing that the iteration (8.21) is a descent.

We noted in P&C 5.30 that if $\mathbf{T}$ is a self-adjoint positive-definite linear operator on $\mathcal{V}$, then $\langle \mathbf{x}, \mathbf{y}\rangle_{\mathrm{T}} \triangleq \langle \mathbf{x}, \mathbf{T}\mathbf{y}\rangle$ defines a new inner product on $\mathcal{V}$. Using the $\mathbf{T}$ inner product, we can express the differential of $\mathbf{F}$ as

$$\mathbf{dF}(\mathbf{x}, \mathbf{h}) = \langle \mathbf{h}, \nabla_{\mathrm{T}} \mathbf{F}(\mathbf{x})\rangle_{\mathrm{T}} = \langle \mathbf{h}, \mathbf{T}\nabla_{\mathrm{T}} \mathbf{F}(\mathbf{x})\rangle = \langle \mathbf{h}, \nabla \mathbf{F}(\mathbf{x})\rangle$$

The two gradients are related by $\nabla \mathbf{F} = \mathbf{T}\nabla_{\mathrm{T}} \mathbf{F}$. Since we have assumed the kth step direction operator $\mathbf{H}_k$ is self-adjoint and positive definite, $\mathbf{H}_k^{-1}$ is also self-adjoint and positive definite; consequently, relative to the inner product for which $\mathbf{T} = \mathbf{H}_k^{-1}$, we have $\nabla \mathbf{F} = \mathbf{H}_k^{-1} \nabla_{\mathrm{H}^{-1}} \mathbf{F}$ or $\nabla_{\mathrm{H}^{-1}} \mathbf{F} = \mathbf{H}_k \nabla \mathbf{F}$. Then

$$\mathbf{h}_k = -\alpha_k \mathbf{H}_k \nabla \mathbf{F}(\mathbf{x}_k) = -\alpha_k \nabla_{\mathrm{H}^{-1}} \mathbf{F}(\mathbf{x}_k) \tag{8.24}$$

Thus *any descent step is a steepest descent step* relative to *some* inner product on $\mathcal{V}$. We will refer to the usual steepest descent, where $\mathbf{H}_k = \mathbf{I}$ and the gradient is defined in terms of a standard inner product on $\mathcal{V}$, as the **standard steepest descent**.

Step Length

We are interested in computing a sequence of vectors $\{\mathbf{x}_k\}$ as described in (8.21) in order to produce a descent in the values of $\mathbf{F}(\mathbf{x}_k)$. Suppose we pick the step directions of standard steepest descent: $\mathbf{x}_{k+1}=\mathbf{x}_k-\alpha_k\nabla\mathbf{F}(\mathbf{x}_k)$. If we were to use very small values for the multipliers $\{\alpha_k\}$, the sequence $\{\mathbf{x}_k\}$ would closely follow the true path of steepest descent. However, the use of very small step sizes would result in a very slow crawl along the descent path and would require an extremely large number of evaluations of $\nabla\mathbf{F}(\mathbf{x}_k)$. We cannot afford the amount of computation associated with the small-step approach. It is much more efficient to use large step sizes even though the resulting sequence $\{\mathbf{x}_k\}$ can deviate drastically from the path of steepest descent. After all, the steepest descent path is not necessarily the most direct path to the minimum. In order to keep the total number of iterations low, we pick the step sizes which provide the greatest amount of descent at each step.

This "best step" concept applies as well to the more general descent iteration (8.21); we pick α_k to minimize $\mathbf{F}(\mathbf{x}_{k+1})$. This choice of α_k constitutes a one-dimensional minimization of $\mathbf{F}(\mathbf{x})$ along the step path $-\mathbf{H}_k\nabla\mathbf{F}(\mathbf{x}_k)$ from the point $\mathbf{x}_k$. The minimum occurs for that step $\mathbf{h}_k=-\alpha_k\mathbf{H}_k\nabla\mathbf{F}(\mathbf{x}_k)$ which is orthogonal to $-\nabla\mathbf{F}(\mathbf{x}_{k+1})$, as depicted in Figure 8.3. We show this mathematically. Define $f(\alpha_k)\triangleq\mathbf{F}(\mathbf{x}_k-\alpha_k\mathbf{H}_k\nabla\mathbf{F}(\mathbf{x}_k))=\mathbf{F}(\mathbf{x}_{k+1})$; then

$$\begin{aligned}\frac{d}{d\alpha_k}\mathbf{F}(\mathbf{x}_{k+1})&=\frac{d}{d\alpha_k}f(\alpha_k)=\frac{d}{d\beta}f(\alpha_k+\beta)|_{\beta=0}\\&=\frac{d}{d\beta}\mathbf{F}(\mathbf{x}_{k+1}-\beta\mathbf{H}_k\nabla\mathbf{F}(\mathbf{x}_k))|_{\beta=0}\\&=\mathbf{dF}(\mathbf{x}_{k+1},-\mathbf{H}_k\nabla\mathbf{F}(\mathbf{x}_k))=\langle-\mathbf{H}_k\nabla\mathbf{F}(\mathbf{x}_k),\nabla\mathbf{F}(\mathbf{x}_{k+1})\rangle=0\end{aligned}\tag{8.25}$$

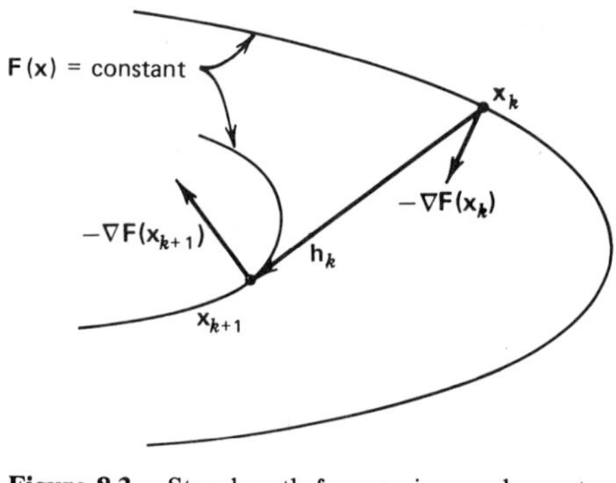

Figure 8.3. Step length for maximum descent.

The smallest positive root of the equation

$$\frac{d}{d\alpha_k}\mathbf{F}(\mathbf{x}_k - \alpha_k \mathbf{H}_k \nabla \mathbf{F}(\mathbf{x}_k)) = 0 \tag{8.26}$$

is probably the best value for α_k. (Of course, if the surface $\mathbf{F}(\mathbf{x})$ has a number of hills and valleys, the best choice for α_k may be a larger root of the equation which results in a step to another valley.)

The determination of the best step size is sometimes called the "one-dimensional minimization," the "univariate search," or the "line search." One obvious approach to the one-dimensional minimization is to solve the one-dimensional equation (8.26) by Newton's method. An alternative approach is to evaluate $\mathbf{F}(\mathbf{x})$ at three points along the step direction, fit a quadratic function through the points, and determine the point which minimizes the quadratic. This minimizing point is taken as $\mathbf{x}_{k+1}$. A third approach is a systematic trial and error scheme in which $\mathbf{F}(\mathbf{x})$ is evaluated at a sequence of points along the step direction. Each point in the sequence is chosen in such a way that it significantly reduces the interval in which the best value of $\mathbf{x}_{k+1}$ is known to lie. The golden section search is one such trial and error technique.* A relatively rough computation of the best step length will usually yield a near minimum in the total number of iterations of (8.21) required to compute a minimum of $\mathbf{F}(\mathbf{x})$ to a given accuracy. The computation required for accurate determination of the best step length is seldom justified by a significant further reduction in the number of iterations unless $\mathbf{x}_k$ is very close to the minimum point.

We now carry out the determination of the best step length for a specific form of functional. Define the quadratic functional $\mathbf{F}$: $\mathfrak{M}^{n\times 1} \to \mathfrak{R}$ by

$$\mathbf{F}(\mathbf{x}) \triangleq \tfrac{1}{2}\mathbf{x}^{\mathrm{T}}\mathbf{A}\mathbf{x} - \mathbf{y}^{\mathrm{T}}\mathbf{x} + d \tag{8.27}$$

Assume that the $n \times n$ matrix $\mathbf{A}$ is symmetric and positive definite. Then $\mathbf{F}(\mathbf{x})$ has a unique minimum at the point $\mathbf{x}_m$ which satisfies $\mathbf{dF}(\mathbf{x},\mathbf{h}) = (\mathbf{A}\mathbf{x} - \mathbf{y})\mathbf{h}^{\mathrm{T}} = 0$. Thus $\mathbf{x}_m$ is the solution to the linear equation $\mathbf{A}\mathbf{x} = \mathbf{y}$. We apply the standard steepest descent in $\mathfrak{M}^{n\times 1}$; that is, we assume the standard inner product, and use $\mathbf{H}_k = \mathbf{I}$. The gradient of $\mathbf{F}$ is

$$\nabla \mathbf{F}(\mathbf{x}) = \mathbf{A}\mathbf{x} - \mathbf{y} \tag{8.28}$$

At the kth iteration the descent step is $\mathbf{h}_k = -\alpha_k \nabla \mathbf{F}(\mathbf{x}_k)$. We determine the best step length by means of (8.25). According to the last line of

* See Pierre [8.24] for a detailed discussion of one-dimensional minimization techniques.

(8.25), $\langle \nabla \mathbf{F}(\mathbf{x}_{k+1}), \nabla \mathbf{F}(\mathbf{x}_k)\rangle = 0$, or

$$\begin{aligned}\nabla \mathbf{F}(\mathbf{x}_k)^{\mathrm{T}}\nabla \mathbf{F}(\mathbf{x}_{k+1}) &= \nabla \mathbf{F}(\mathbf{x}_k)^{\mathrm{T}}\left[\mathbf{A}(\mathbf{x}_k - \alpha_k \nabla \mathbf{F}(\mathbf{x}_k)) - \mathbf{y}\right] \\ &= \nabla \mathbf{F}(\mathbf{x}_k)^{\mathrm{T}}\left[\nabla \mathbf{F}(\mathbf{x}_k) - \alpha_k \mathbf{A}\nabla \mathbf{F}(\mathbf{x}_k)\right] \\ &= \nabla \mathbf{F}(\mathbf{x}_k)^{\mathrm{T}}\nabla \mathbf{F}(\mathbf{x}_k) - \alpha_k \nabla \mathbf{F}(\mathbf{x}_k)^{\mathrm{T}}\mathbf{A}\nabla \mathbf{F}(\mathbf{x}_k) = 0\end{aligned}$$

Thus for the quadratic function (8.27) and standard steepest descent, the best step length at the kth step is

$$\alpha_k = \frac{\nabla \mathbf{F}(\mathbf{x}_k)^{\mathrm{T}}\nabla \mathbf{F}(\mathbf{x}_k)}{\nabla \mathbf{F}(\mathbf{x}_k)^{\mathrm{T}}\mathbf{A}\nabla \mathbf{F}(\mathbf{x}_k)} \tag{8.29}$$

Example 1. Standard Steepest Descent for a Quadratic Functional. Let $\mathbf{x} = (\xi_1\ \xi_2)^{\mathrm{T}}$ in $\mathfrak{M}^{2\times 1}$. Define $\mathbf{F}(\mathbf{x}) \stackrel{\Delta}{=} \frac{1}{2}(a\xi_1^2 + b\xi_2^2)$, where $0 < a < b$. Corresponding to (8.27) and (8.28), we have

$$\mathbf{A} = \begin{pmatrix} a & 0 \\ 0 & b \end{pmatrix}, \qquad \nabla \mathbf{F}(\mathbf{x}) = \begin{pmatrix} a\xi_1 \\ b\xi_2 \end{pmatrix}, \qquad \mathbf{y} = \begin{pmatrix} 0 \\ 0 \end{pmatrix}$$

The standard steepest descent iteration is

$$\begin{aligned}\mathbf{x}_{k+1} &= \mathbf{x}_k - \alpha_k \nabla \mathbf{F}(\mathbf{x}_k) \\ &= \begin{pmatrix} \xi_1 \\ \xi_2 \end{pmatrix} - \left(\frac{a^2\xi_1^2 + b^2\xi_2^2}{a^3\xi_1^2 + b^3\xi_2^2}\right)_k \begin{pmatrix} a\xi_1 \\ b\xi_2 \end{pmatrix}_k \\ &= \left(\frac{1}{a^3\xi_1^2 + b^3\xi_2^2}\right)_k \begin{pmatrix} (b-a)b^2\xi_2^2 & 0 \\ 0 & -(b-a)a^2\xi_1^2 \end{pmatrix}_k \begin{pmatrix} \xi_1 \\ \xi_2 \end{pmatrix}_k\end{aligned}$$

Since the minimizing vector is $\mathbf{x}_m = \boldsymbol{\theta}$, the error at the kth iteration is just $\mathbf{x}_k$. If the initial vector, $\mathbf{x}_0 = (\xi_1\ \xi_2)_0^{\mathrm{T}}$, is chosen such that $a^2\xi_1^2 = b^2\xi_2^2$, then for $k = 0, 1, 2, \ldots$,

$$\mathbf{x}_{k+1} = \begin{pmatrix} (b-a)/(b+a) & 0 \\ 0 & -(b-a)/(b+a) \end{pmatrix}\mathbf{x}_k$$

Thus for this initial vector the error is reduced by the factor $(b-a)/(b+a)$ at each iteration. This steepest descent is illustrated in Figure 8.4 for $a = 1$, $b = 10$, $\mathbf{x}_0 = (5\ \ 0.5)^{\mathrm{T}}$, and $(b-a)/(b+a) \approx 0.8182$.

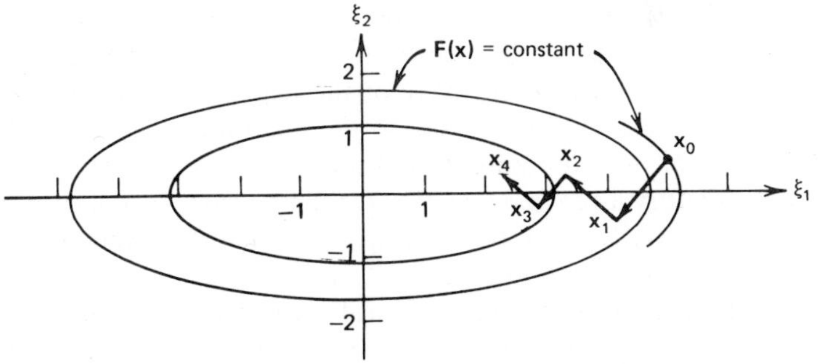

Figure 8.4. Standard steepest descent for a quadratic functional.

Convergence Rate

Example 1 provides a good illustration of the convergence characteristics of descent methods. A functional is necessarily "cuplike" in the immediate vicinity of a minimum. Consequently, in this vicinity the functional is typically near quadratic.* Thus during the latter portion of the minimization of an arbitrary functional, the iteration should behave as it would for a quadratic functional. We now examine the rate of convergence of steepest descent methods for quadratic functionals.

The level surfaces of the quadratic functional (8.27) are concentric ellipsoids in n dimensions. The n axes of the ellipsoids are each collinear with one of the eigenvectors of the matrix $\mathbf{A}$. The lengths of the axes of each ellipsoid are inversely proportional to the square roots of the corresponding eigenvalues. Thus in Example 1 and Figure 8.4 the axes of the two-dimensional ellipsoids lie along the eigenvectors, $\mathbf{y}_1 = (1 \;\; 0)^{\mathrm{T}}$ and $\mathbf{y}_2 = (0 \;\; 1)^{\mathrm{T}}$, of the matrix $\mathbf{A}$. The lengths of these axes are proportional, respectively, to

$$\frac{1}{\sqrt{\lambda_1}} = \frac{1}{\sqrt{a}} = 1 \qquad \text{and} \qquad \frac{1}{\sqrt{\lambda_2}} = \frac{1}{\sqrt{b}} = \frac{1}{\sqrt{10}}$$

The direction of the gradient $\nabla\mathbf{F}(\mathbf{x})$ depends not only on the point $\mathbf{x}$ and on the inner product, but also on the relative lengths of the axes of the ellipsoids. If the ellipsoids are long and thin (i.e., if the eigenvalues of $\mathbf{A}$ differ significantly), then, for some starting vectors $\mathbf{x}_0$, each of the

* It is possible that the quadratic term is missing from the Taylor series expansion about the minimum, as in the function $\mathbf{F}(\xi_1, \xi_2) \triangleq \xi_1^4 + \xi_2^4$. Then the functional is not near quadratic in the neighborhood of the minimum.

gradients $\{\nabla \mathbf{F}(\mathbf{x}_k)\}$ in the standard descent iteration is poorly aimed. In this situation the iterates $\{\mathbf{x}_k\}$ "hemstitch," as illustrated in Figure 8.4, and convergence is slow. A good measure of the hemstitching tendency of the standard steepest descent, as applied to the quadratic functional (8.27), is the ratio λ_L/λ_s of the largest and smallest eigenvalues of $\mathbf{A}$. (In Section 6.6 we call this quantity a *condition number* for the matrix.) If $\mathbf{A}$ is even slightly ill conditioned, say, $\lambda_L/\lambda_S = 10$, we can expect the iterations to converge slowly for most starting vectors.

In Example 1, where $\mathbf{x}_0$ has been carefully selected, the error vector $\mathbf{x}_k$ is reduced in size at each iteration by precisely the factor $(b-a)/(b+a)$. In P&C 8.10 we find that this factor is a bound on the convergence rate for *any* starting vector; that is,

$$\|\mathbf{x}_{k+1}\| \leqslant \left(\frac{b-a}{b+a}\right)\|\mathbf{x}_k\|$$

for any $\mathbf{x}_0$. It is apparent from the vector $\mathbf{x}_0$ used in Example 1 that the bound cannot be tightened. Clearly the steepest descent of Example 1 does not exhibit the quadratic convergence which is characteristic of Newton's method. Rather, the convergence is linear. Furthermore, if b/a is large, the rate of the linear convergence is low; that is $(b-a)/(b+a) \approx 1$. We note, however, that if $\mathbf{x}_0$ is nearly on one of the axes of the ellipsoids, the rate of the linear convergence is higher than that indicated by the bound $(b-a)/(b+a)$.

Exercise 1. Show that if $\mathbf{x}_0$ is on one of the axes of the ellipsoids in Example 1, then convergence occurs in one step for any inner product.

Exercise 2. Show that if $b=a$ in Example 1, then convergence occurs in one step for any $\mathbf{x}_0$ and any inner product.

Define the general quadratic functional $\mathbf{F}$ on the real Hilbert space $\mathcal{V}$ by

$$\mathbf{F}(\mathbf{x}) \triangleq \tfrac{1}{2}\langle \mathbf{x}, \mathbf{T}\mathbf{x}\rangle - \langle \mathbf{x}, \mathbf{y}\rangle + d \tag{8.30}$$

where $\mathbf{T}$ is a self-adjoint positive-definite bounded linear operator on $\mathcal{V}$. The minimum of $\mathbf{F}(\mathbf{x})$ is the unique vector $\mathbf{x}_m = \mathbf{T}^{-1}\mathbf{y}$. The standard steepest descent minimization of $\mathbf{F}(\mathbf{x})$, by analogy with (8.28) and (8.29), is

$$\begin{aligned} \mathbf{x}_{k+1} &= \mathbf{x}_k - \alpha_k \nabla \mathbf{F}(\mathbf{x}_k) \\ \nabla \mathbf{F}(\mathbf{x}_k) &= \mathbf{T}\mathbf{x} - \mathbf{y} \\ \alpha_k &= \frac{\langle \nabla \mathbf{F}(\mathbf{x}_k), \nabla \mathbf{F}(\mathbf{x}_k)\rangle}{\langle \nabla \mathbf{F}(\mathbf{x}_k), \mathbf{T}\nabla \mathbf{F}(\mathbf{x}_k)\rangle} \end{aligned} \tag{8.31}$$

Kantorovich [8.17, Chapter 15] has shown that the error in the descent (8.31) obeys

$$\|\mathbf{x}_{k+1}-\mathbf{x}_m\| \leqslant \left(\frac{\lambda_L-\lambda_S}{\lambda_L+\lambda_S}\right)\|\mathbf{x}_k-\mathbf{x}_m\| \tag{8.32}$$

for $k=1,2,\ldots$, where λ_L and λ_S are the largest and smallest eigenvalues of $\mathbf{T}$. The convergence estimate (8.32) is a generalization of the estimate derived for the two-dimensional problem of Example 1. That example demonstrates that the bound $(\lambda_L-\lambda_S)/(\lambda_L+\lambda_S)$ gives a precise estimate of the rate of convergence for at least some starting vectors $\mathbf{x}_0$. Thus we conclude that if $\mathbf{T}$ is slightly ill-conditioned (say, $\lambda_L/\lambda_S=10$), we should expect the standard steepest descent to be slow in converging to the minimum of (8.30). We should also expect similar convergence properties in the latter stages of the standard steepest descent for nonquadratic functionals—if the valley is long and thin near the minimum, convergence will be slow.

Scaling and the Newton Descent

An obvious approach to improving the convergence of a steepest descent iteration is to rescale the variables in the problem so that the valley near the minimum of interest is nearly circular. We explore this scaling via the finite-dimensional case. For the quadratic functional (8.27), the rescaling should produce a quadratic form whose matrix has all its eigenvalues equal. Then all gradients will point directly toward the minimum (for the standard inner product), the convergence estimate of (8.32) will be $(\lambda_L-\lambda_S)/(\lambda_L+\lambda_S)=0$, and convergence will occur in one step. We can describe an arbitrary rescaling of the elements of the $n\times 1$ vector $\mathbf{x}$ by $\mathbf{z}=\mathbf{Q}\mathbf{x}$, where $\mathbf{Q}$ is a symmetric positive definite $n\times n$ matrix. Using the substitution $\mathbf{x}=\mathbf{Q}^{-1}\mathbf{z}$, (8.27) becomes

$$\hat{\mathbf{F}}(\mathbf{z}) \stackrel{\Delta}{=} \mathbf{F}(\mathbf{x})=\tfrac{1}{2}\mathbf{z}^{\mathrm{T}}\mathbf{Q}^{-1}\mathbf{A}\mathbf{Q}^{-1}\mathbf{z}-\mathbf{y}^{\mathrm{T}}\mathbf{Q}^{-1}\mathbf{z}+d$$

We select $\mathbf{Q}$ such that all the eigenvalues of $\mathbf{Q}^{-1}\mathbf{A}\mathbf{Q}^{-1}$ are equal. An obvious choice is $\mathbf{Q}=\sqrt{\mathbf{A}}$, which produces $\mathbf{Q}^{-1}\mathbf{A}\mathbf{Q}^{-1}=\mathbf{I}$ and

$$\hat{\mathbf{F}}(\mathbf{z})=\tfrac{1}{2}\mathbf{z}^{\mathrm{T}}\mathbf{z}-\left(\sqrt{\mathbf{A}}^{\,-1}\mathbf{y}\right)^{\mathrm{T}}\mathbf{z}+d$$

The standard steepest descent method will produce the minimum of this rescaled functional in one step. However, the rescaling requires that we determine $\sqrt{\mathbf{A}}$. To do so we must first compute the eigenvalues of $\mathbf{A}$ (Section 4.6).

A more suitable way of improving the convergence of the steepest descent method is to leave $\mathbf{F}(\mathbf{x})$ untouched, but to change (rescale) the inner product. This rescaling modifies the directions of the gradients so that they point directly toward the minimum of the functional (8.27). Let $\langle\cdot,\cdot\rangle$ denote the standard inner product on $\mathfrak{M}^{n\times 1}$. Recall that for any symmetric positive-definite matrix $\mathbf{Q}$, $\langle\mathbf{x},\mathbf{y}\rangle_{\mathbf{Q}} \triangleq \langle\mathbf{Qx},\mathbf{y}\rangle$ defines a new inner product on $\mathfrak{M}^{n\times 1}$. Since $\mathbf{A}$ is symmetric and positive definite, we can define

$$\langle\mathbf{x},\mathbf{y}\rangle_{\mathbf{A}} \triangleq \langle\mathbf{Ax},\mathbf{y}\rangle = \mathbf{y}^{\mathrm{T}}\mathbf{Ax} \tag{8.33}$$

In terms of this new inner product, (8.27) becomes

$$\mathbf{F}(\mathbf{x}) = \tfrac{1}{2}\langle\mathbf{x},\mathbf{x}\rangle_{\mathbf{A}} - \langle\mathbf{x},\mathbf{A}^{-1}\mathbf{y}\rangle_{\mathbf{A}} + d$$

According to the discussion preceding (8.24), $\nabla\mathbf{F} = \mathbf{A}\nabla_{\mathbf{A}}\mathbf{F}$. Consequently, the gradient and the first steepest descent step relative to the new inner product are*

$$\begin{aligned}\nabla_{\mathbf{A}}\mathbf{F}(\mathbf{x}) &= \mathbf{A}^{-1}\nabla\mathbf{F}(\mathbf{x}) = \mathbf{x} - \mathbf{A}^{-1}\mathbf{y}\\ \mathbf{x}_1 &= \mathbf{x}_0 - \alpha_k\nabla_{\mathbf{A}}\mathbf{F}(\mathbf{x}_0)\end{aligned} \tag{8.34}$$

Using (8.25) to compute the best step length in a manner identical to the derivation of (8.29), we obtain

$$\begin{aligned}\langle\nabla\mathbf{F}(\mathbf{x}_1), \nabla_{\mathbf{A}}\mathbf{F}(\mathbf{x}_0)\rangle &= \nabla_{\mathbf{A}}\mathbf{F}(\mathbf{x}_0)^{\mathrm{T}}\nabla\mathbf{F}(\mathbf{x}_1)\\ &= \nabla_{\mathbf{A}}\mathbf{F}(\mathbf{x}_0)^{\mathrm{T}}\left[\mathbf{A}(\mathbf{x}_0 - \alpha_0\nabla_{\mathbf{A}}\mathbf{F}(\mathbf{x}_0)) - \mathbf{y}\right]\\ &= \nabla_{\mathbf{A}}\mathbf{F}(\mathbf{x}_0)^{\mathrm{T}}\left[\nabla\mathbf{F}(\mathbf{x}_0) - \alpha_0\nabla\mathbf{F}(\mathbf{x}_0)\right] = 0\end{aligned}$$

and $\alpha_0 = 1$. Consequently,

$$\mathbf{x}_1 = \mathbf{x}_0 - \nabla_{\mathbf{A}}\mathbf{F}(\mathbf{x}_0) = \mathbf{A}^{-1}\mathbf{y} = \mathbf{x}_m$$

and convergence occurs in one step. As we discovered previously, choosing an inner product amounts to choosing a particular step direction operator in the descent iteration (8.21). We have chosen the inner product which produces the gradients $\nabla_{\mathbf{A}}\mathbf{F}(\mathbf{x}_0) = \mathbf{A}^{-1}\nabla\mathbf{F}(\mathbf{x}_0)$, where $\nabla\mathbf{F}$ is the gradient

* The choice of $\mathbf{Q} = \mathbf{A}$ in the previous change of variables, $\mathbf{z} = \mathbf{Qx}$, produces this same descent direction; that is, $\nabla\hat{\mathbf{F}}(\mathbf{z}) = \nabla_{\mathbf{A}}\mathbf{F}(\mathbf{x})$ if $\mathbf{z} = \mathbf{x}$.

relative to the standard inner product. Thus the step-direction operator $\mathbf{H}_k$ consisting in multiplication by $\mathbf{A}^{-1}$ results in one-step convergence to the minimum of a quadratic.

We are not likely to actually apply steepest descent to a quadratic functional, because the solution can be obtained by direct solution of the linear equation $\nabla \mathbf{F}(\mathbf{x}) = \boldsymbol{\theta}$. However, a nonquadratic functional is typically near quadratic in the neighborhood of a minimum. Once a steepest descent iteration has reached the vicinity of the minimum, we should be able to obtain approximate one-step convergence to the minimum by proper choice of the step-direction operator $\mathbf{H}_k$. Assume we can approximate the nonquadratic functional $\mathbf{F}(\mathbf{x})$ near its minimum by the three-term Taylor series

$$\mathbf{F}(\mathbf{x}_{k+1}) = \mathbf{F}(\mathbf{x}_k + \mathbf{h}_k) \approx \mathbf{F}(\mathbf{x}_k) + \mathbf{h}_k^{\mathrm{T}} \nabla \mathbf{F}(\mathbf{x}_k) + \tfrac{1}{2}\mathbf{h}_k^{\mathrm{T}} \frac{\partial^2 \mathbf{F}(\mathbf{x}_k)}{\partial \mathbf{x}^2} \mathbf{h}_k \stackrel{\Delta}{=} \hat{\mathbf{F}}(\mathbf{h}_k) \tag{8.35}$$

Note that the Hessian matrix of $\mathbf{F}$ is the equivalent of the matrix $\mathbf{A}$ in (8.27). We pick the step $\mathbf{h}_k$ to minimize the quadratic functional $\hat{\mathbf{F}}(\mathbf{h}_k)$. The minimum occurs for

$$\nabla \hat{\mathbf{F}}(\mathbf{h}_k) = \nabla \mathbf{F}(\mathbf{x}_k) + \frac{\partial^2 \mathbf{F}(\mathbf{x}_k)}{\partial \mathbf{x}^2} \mathbf{h}_k = \boldsymbol{\theta}$$

or

$$\mathbf{h}_k = -\left(\frac{\partial^2 \mathbf{F}(\mathbf{x}_k)}{\partial \mathbf{x}^2} \right)^{-1} \nabla \mathbf{F}(\mathbf{x}_k) \tag{8.36}$$

Thus near a minimum, the best step-direction operator consists in multiplication by the inverse Hessian matrix of $\mathbf{F}(\mathbf{x})$. As the iterates approach the minimum point $\mathbf{x}_m$, the quadratic approximation (8.35) becomes progressively more accurate. Consequently, the step (8.36) should provide a progressively increasing reduction in the error $\|\mathbf{x}_k - \mathbf{x}_m\|$; that is, convergence should be faster than linear. In point of fact, the step (8.36) is identical to the descent version of Newton's method, (8.20). Consequently, convergence must be quadratic for $\mathbf{x}_k$ near $\mathbf{x}_m$.

Since $\mathbf{F}$ is not purely quadratic near $\mathbf{x}_m$, we can usually obtain a greater amount of descent from each step by searching for a minimum along the step direction rather than just using the "nominal" Newton step (8.36). That is, we use a one-dimensional minimization technique to pick the best value of α_k in the iteration*:

$$\mathbf{x}_{k+1}=\mathbf{x}_k-\alpha_k\left(\frac{\partial^2\mathbf{F}(\mathbf{x}_k)}{\partial\mathbf{x}^2}\right)^{-1}\nabla\mathbf{F}(\mathbf{x}_k) \tag{8.37}$$

We call (8.37) the **best-step Newton descent** (or sometimes just the "Newton descent"). We could as well call (8.37) the "best" steepest descent; it is, in fact, steepest descent relative to the inner product $\langle\mathbf{x},\mathbf{y}\rangle_{\mathrm{F}}\triangleq\langle\mathbf{x},(\partial^2\mathbf{F}(\mathbf{x}_k)/\partial\mathbf{x}^2)\mathbf{y}\rangle$. The best-step Newton descent provides the fastest convergence of all practical minimization techniques *once* $\mathbf{x}_k$ *has approached* $\mathbf{x}_m$. If $\mathbf{x}_k$ is not sufficiently close to $\mathbf{x}_m$, however, the inverse Hessian may not be positive definite, and the Newton step direction (8.36) may not even be a descent direction; that is, the iteration may not converge. Thus until $\mathbf{x}_k$ is definitely close enough to $\mathbf{x}_m$ to ensure positive definiteness of the Hessian matrix, we should use a slower technique (such as standard steepest descent) which has a good likelihood of finding the vicinity of the minimum. We should also note that rapid convergence does not necessarily imply computational efficiency. In point of fact, the amount of computation per iteration can be high for the Newton descent (P&C 8.11).

Example 2. Best-Step Newton Descent. Let $\mathbf{x}=(\xi_1\ \xi_2)^{\mathrm{T}}$ in $\mathfrak{M}^{2\times1}$. Define the nonquadratic functional $\mathbf{F}$ on $\mathfrak{M}^{2\times1}$ by $\mathbf{F}(\mathbf{x})\triangleq\frac{1}{2}(\xi_1^2+10\xi_2^2)+\xi_2^4$. This functional has a minimum at $\mathbf{x}=(0\ 0)^{\mathrm{T}}$. We seek the minimum by means of the best-step Newton descent (8.37). The gradient and inverse Hessian are

$$\nabla\mathbf{F}(\mathbf{x})=\begin{pmatrix}\xi_1\\10\xi_2+4\xi_2^3\end{pmatrix},\qquad\left(\frac{\partial^2\mathbf{F}(\mathbf{x})}{\partial\mathbf{x}^2}\right)^{-1}=\begin{pmatrix}1&0\\0&1/(10+12\xi_2^2)\end{pmatrix}$$

The inverse Hessian is positive definite for all $\mathbf{x}$, and the Newton iteration (8.37) is a descent. The iteration is

$$\mathbf{x}_{k+1}=\mathbf{x}_k-\alpha_k\left(\frac{\partial^2\mathbf{F}(\mathbf{x}_k)}{\partial\mathbf{x}^2}\right)^{-1}\nabla\mathbf{F}(\mathbf{x}_k)$$

$$=\begin{pmatrix}\xi_1\\\xi_2\end{pmatrix}_k-\alpha_k\begin{pmatrix}\xi_1\\\xi_2(10+4\xi_2^2)/(10+12\xi_2^2)\end{pmatrix}_k$$

* The use of the adjustable step length α_k is equivalent to the use of an adjustable overrelaxation factor in (8.13).

Let $\mathbf{x}_0=(10\ \ 1)^{\mathrm{T}}$. From this initial point, the Newton step direction is

$$-\left(\frac{\partial^2\mathbf{F}(\mathbf{x}_0)}{\partial\mathbf{x}^2}\right)^{-1}\nabla\mathbf{F}(\mathbf{x}_0)=\begin{pmatrix}-10\\-0.636\end{pmatrix}$$

From Figure 8.5 it is clear that a step in this direction will move relatively close to the origin. The step length is adjusted to minimize $\mathbf{F}(\mathbf{x}_1)$; according to the figure, the point $\mathbf{x}_1=(-0.2\ \ 0.35)^{\mathrm{T}}$ yields an approximate minimum of $\mathbf{F}(\mathbf{x})$ along the step path. The point $\mathbf{x}_1$ lies well within the unit circle. Consequently, the fourth-order term is negligible, the functional is essentially quadratic, and the next iterate, $\mathbf{x}_2$, should be a very good approximation to the minimum. The new Newton step direction is

$$-\left(\frac{\partial^2\mathbf{F}(\mathbf{x}_1)}{\partial\mathbf{x}^2}\right)^{-1}\nabla\mathbf{F}(\mathbf{x}_1)=\begin{pmatrix}0.2\\-0.321\end{pmatrix}$$

The point $\mathbf{x}_2$ is indistinguishable from the origin in the scale used for Figure 8.5. The first two steps of the standard steepest descent from the same starting point are also shown in the figure for comparison. The standard descent performs a slow hemstitch much like the one in Figure 8.4.

The best-step Newton descent (8.37) was derived for a functional $\mathbf{F}$ which acts on the n-dimensional space $\mathfrak{M}^{n\times 1}$. However, it can be applied to any smooth functional $\mathbf{F}$ which acts on a real Hilbert space $\mathcal{V}$. The Hilbert space version of the best-step Newton descent, analogous to (8.37), is

$$\mathbf{x}_{k+1}=\mathbf{x}_k-\alpha_k\mathbf{F}''(\mathbf{x}_k)^{-1}\nabla\mathbf{F}(\mathbf{x}_k) \tag{8.38}$$

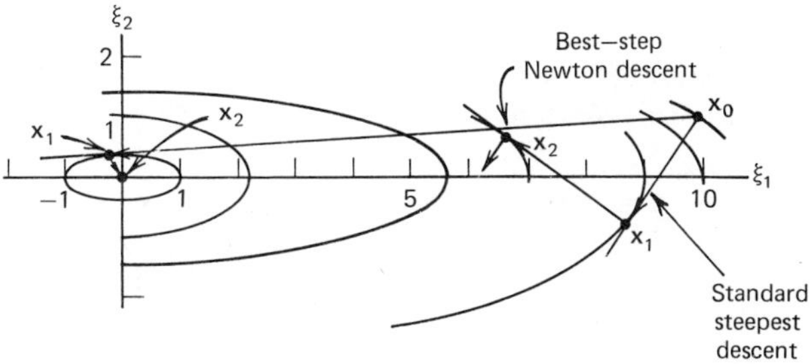

Figure 8.5. A comparison of best-step Newton descent and standard steepest descent.

where $\mathbf{F}''(\mathbf{x})$ is the second Fréchet derivative of $\mathbf{F}$ at $\mathbf{x}$, and $\nabla\mathbf{F}$ is the gradient of $\mathbf{F}$ relative to the inner product on $\mathcal{V}$.*

Example 3. Scaling in a Function Space. Define $\mathbf{F}$: $\mathcal{L}_2(0,1)\to\mathcal{R}$ by $\mathbf{F}(\mathbf{u}) \triangleq \frac{1}{2}\int_0^1 \omega(t)\mathbf{u}^2(t)\,dt$, where $0<\omega(t)<\infty$ on $[0,1]$. Then $\nabla\mathbf{F}(\mathbf{u})=\omega\mathbf{u}$, and the minimum of $\mathbf{F}(\mathbf{u})$ occurs at $\mathbf{u}=\boldsymbol{\theta}$. We explore the minimization of $\mathbf{F}(\mathbf{u})$ by standard steepest descent and by the best-step Newton descent. The steepest descent iteration is

$$\mathbf{u}_{k+1}=\mathbf{u}_k-\alpha_k\nabla\mathbf{F}(\mathbf{u}_k)$$

The best choice for α_k satisfies

$$\begin{aligned}\langle\nabla\mathbf{F}(\mathbf{u}_{k+1}),\nabla\mathbf{F}(\mathbf{u}_k)\rangle&=\langle\omega\mathbf{u}_{k+1},\nabla\mathbf{F}(\mathbf{u}_k)\rangle\\&=\langle\omega\mathbf{u}_k-\alpha_k\omega\nabla\mathbf{F}(\mathbf{u}_k),\nabla\mathbf{F}(\mathbf{u}_k)\rangle\\&=\langle\nabla\mathbf{F}(\mathbf{u}_k)-\alpha_k\omega\nabla\mathbf{F}(\mathbf{u}_k),\nabla\mathbf{F}(\mathbf{u}_k)\rangle=0\end{aligned}$$

or

$$\alpha_k=\frac{\langle\nabla\mathbf{F}(\mathbf{u}_k),\nabla\mathbf{F}(\mathbf{u}_k)\rangle}{\langle\nabla\mathbf{F}(\mathbf{u}_k),\omega\nabla\mathbf{F}(\mathbf{u}_k)\rangle}=\frac{\int_0^1\omega^2(t)\mathbf{u}_k^2(t)\,dt}{\int_0^1\omega^3(t)\mathbf{u}_k^2(t)\,dt}$$

Let $\omega(t)=e^t$ and $\mathbf{u}_0(t)=e^{-t}$. Then $\alpha_0=1/(e-1)$, and the first steepest descent step produces

$$\begin{aligned}\mathbf{u}_1(t)&=e^{-t}-\alpha_0e^te^{-t}\\&=e^{-t}-\frac{1}{e-1}\end{aligned}$$

The function $\mathbf{u}_1$ is slightly more like the minimizing function than is $\mathbf{u}_0$.

In order to apply the Newton descent, we must first find $\mathbf{F}''(\mathbf{u})$. We do so by defining $\mathbf{G}(\mathbf{u})\triangleq\Delta\mathbf{F}(\mathbf{u})=\omega\mathbf{u}$ and finding $\mathbf{dG}(\mathbf{u},\mathbf{h})$:

$$\mathbf{dG}(\mathbf{u},\mathbf{h})=\lim_{\beta\to0}\frac{\omega(\mathbf{u}+\beta\mathbf{h})-\omega\mathbf{u}}{\beta}=\omega\mathbf{h}=\mathbf{F}''(\mathbf{u})\mathbf{h}$$

(See P&C 7.2 for a general definition of second Fréchet derivatives.) According to (8.38), the first Newton descent step is

$$\begin{aligned}\mathbf{u}_1&=\mathbf{u}_0-\alpha_0\mathbf{F}''(\mathbf{u}_0)^{-1}\nabla\mathbf{F}(\mathbf{u}_0)\\&=\mathbf{u}_0-\alpha_0\frac{1}{\omega}(\omega\mathbf{u}_0)\\&=\mathbf{u}_0-\alpha_0\mathbf{u}_0\end{aligned}$$

* See P&C 7.2 for a discussion of the second Fréchet derivative. For $\mathcal{V}=\mathfrak{M}^{n\times1}$, $\mathbf{F}''(\mathbf{x})$ can be described in terms of the Hessian matrix, $\mathbf{F}''(\mathbf{x})\mathbf{z}=(\partial^2\mathbf{F}(\mathbf{x})/\partial\mathbf{x}^2)\mathbf{z}$.

Clearly, the best step length is $\alpha_0 = 1$ which produce the minimizing vector $\mathbf{u}_1 = \boldsymbol{\theta}$ in one step, regardless of the choice of ω or of $\mathbf{u}_0$.

Ill-Conditioned Objective Functions

Let $\mathbf{F}(\mathbf{x})$ be the quadratic functional of (8.27). If $\mathbf{A}$ is *very* ill-conditioned, say, $\lambda_L/\lambda_s = 10^6$, then the ellipsoids $\mathbf{F}(\mathbf{x}) = \text{constant}$ (discussed previously in connection with convergence rates) are extremely long and thin, and the scaling discussed in the preceding section is mandatory. Otherwise, according to (8.32), the convergence would be interminably slow. However, as indicated in (8.34), scaling requires the manipulation of the ill-conditioned matrix $\mathbf{A}$. Even with a carefully designed algorithm and double precision arithmetic, the computed value of the vector $\mathbf{A}^{-1}\mathbf{y}$ of (8.34) will be inaccurate. The error will consist primarily of a vector in the near null-space of $\mathbf{A}$ (the eigenspace of $\mathbf{A}$ corresponding to the smallest eigenvalue, λ_s; see Sections 2.4 and 4.2). But the eigenvectors for the smallest eigenvalue are oriented along the longest axis of the ellipsoids. Consequently the scaled descent (8.34) tends to step *along* the valley even though, theoretically, it should step directly to the minimum. The same difficulty occurs if we apply the best-step Newton descent (8.37) to a *nonquadratic* functional for which the minimum lies at the bottom of an extremely long, thin valley. The Hessian matrix for such a functional is very ill conditioned near the minimum, and the computed value of the step in (8.37) tends to be aligned with the long dimension of the valley rather than directed toward the minimum. There is nothing that we can do about extreme ill conditioning except to be very accurate in scaling and computing gradients.

8.3 Other Descent Methods

In the preceding section we explored two extremes of steepest descent minimization: standard steepest descent and the "perfectly scaled" best-step Newton descent. Our examination showed that the Newton descent is ideal from the standpoint of speed of convergence. No other method can surpass it without the use of third or higher derivatives. However, the Newton descent is applicable only to the final portion of the descent iteration, where the functional is positive definite. Even in this terminal stage of the minimization, the Newton descent has a drawback. If the functional $\mathbf{F}(\mathbf{x})$ is complicated, the second derivatives are expensive to evaluate; consequently, the low number of iterations necessary to achieve satisfactory convergence are offset by the large amount of computation required for each iteration. In contrast to the Newton descent, standard steepest descent is slow, yet relatively sure of convergence to a minimum.

Thus it is a natural choice for the early stages of the minimization. Because only first derivatives (the gradient) must be evaluated, there is relatively little computation for each iteration; but the number of iterations required to reach the neighborhood of the minimum is likely to be high.

A great variety of descent techniques have been devised in an attempt to achieve a good compromise between the amount of computation required per iteration and the total number of iterations required for convergence. Many of these techniques work well for some restricted class of problems. The best choice from among the available techniques depends upon the degree of nonlinearity of the functional and the difficulty in evaluating its first and second derivatives.* In this section we explore several of the techniques which have found real success, and attempt to understand the principles which have led to that success. We first approximate the Newton descent by an approach which we refer to as "quasi-Newton"; we develop the DFP descent method using this quasi-Newton philosophy. Then we examine the "conjugate-direction" approach, an approach based on n-dimensional Fourier series. Conjugate-direction methods produce the minimum of an n-dimensional quadratic function in no more than n steps. Again the DFP descent is an example. Other conjugate-direction methods which we explore are "conjugate-gradient descent" and "accelerated steepest descent."

Quasi-Newton Methods

We continue to use the basic descent iteration (8.21), $\mathbf{x}_{k+1}=\mathbf{x}_k-\alpha_k\mathbf{H}_k\boldsymbol{\nabla}\mathbf{F}(\mathbf{x}_k)$, where the step-direction operator $\mathbf{H}_k$ is self-adjoint and positive definite, and α_k is selected for maximum descent at each iteration. Because the best-step Newton descent (8.37)–(8.38) is so efficient in the neighborhood of a minimum, it is logical to seek descent methods which act like the Newton descent once the iterates approach the minimum. A descent method for which $\mathbf{H}_k \to \mathbf{F}''(\mathbf{x}_k)^{-1}$ as $k\to\infty$ is called a **quasi-Newton method**. Clearly, any quasi-Newton method exhibits quadratic convergence in the final stages of the iteration and produces, in the limit, the second derivative information $\mathbf{F}''(\mathbf{x})^{-1}$ at the minimizing point $\mathbf{x}_m$.

The least complicated quasi-Newton method consists in using the standard steepest descent until $\mathbf{x}_k$ appears to be approaching a limit, then switching to Newton's method. Mathematically, we can express this method by the following sequence of step-direction operators:

$$\begin{aligned}\mathbf{H}_k &= \mathbf{I}, & k&=0,1,\ldots,N\\ &= \mathbf{F}''(\mathbf{x}_k)^{-1}, & k&=N+1,N+2,\ldots\end{aligned} \tag{8.39}$$

*See Powell [8.25] for a survey of these methods for finite-dimensional spaces.

where the number N must be determined by the user. Another quasi-Newton iteration combines the standard steepest descent and Newton descent concepts in a more sophisticated way*:

$$\mathbf{H}_k = (\beta_k \mathbf{I} + \mathbf{F}''(\mathbf{x}_k))^{-1} \tag{8.40}$$

The positive weight β_k is reduced in some fashion from a large number to zero. Positive definiteness of $\mathbf{H}_k$ [descent of $\mathbf{F}(\mathbf{x}_k)$] can be assured for any $\mathbf{x}_k$ by keeping β_k slightly larger than the magnitude of the most negative eigenvalue of $\mathbf{F}''(\mathbf{x}_k)$.

For most functionals, the descent methods (8.39) and (8.40) will serve the purpose of ensuring convergence to a minimum from poor starting vectors $\mathbf{x}_0$. However, the high computational costs associated with the slow convergence of standard steepest descent and with the evaluation of second derivatives in the Newton descent have not been reduced. A simple attack on the computational problem would be to periodically rescale the variables during the initial iterations in order to reduce the hemstitching. During the final iterations, where $\mathbf{x}_k$ and the second derivatives of $\mathbf{F}$ are changing little, a further computational saving can be made by updating $\mathbf{F}''(\mathbf{x}_k)^{-1}$ only occasionally. Of course, all these techniques require judgment on the part of the user.

One of the most sophisticated and most successful quasi-Newton methods is due to Davidon [8.7] and Fletcher and Powell [8.10].† The step-direction operators used in the Davidon–Fletcher–Powell (DFP) method require only values and first derivatives of the objective function $\mathbf{F}(\mathbf{x})$. However, the influence of the second derivatives is included by changing $\mathbf{H}_k$ at each iteration in accordance with the change in the gradient (first derivatives) which occurs during the most recent step. Beginning with a positive-definite $\mathbf{H}_0$ (usually $\mathbf{H}_0 = \mathbf{I}$), corrections are made to $\mathbf{H}_k$ at each iteration which maintain its positive definiteness, yet make it more like $\mathbf{F}''(\mathbf{x}_k)^{-1}$. In the limit, we obtain not only the minimum, $\mathbf{x}_m$, but also the second-derivative information $\mathbf{F}''(\mathbf{x}_m)^{-1}$.‡ Indicative of the efficiency of the DFP method is the fact that it converges to the minimum of any n-dimensional *quadratic* functional in no more than n steps. Thus in terms of speed of convergence it is much better than steepest descent. Yet the computational effort per iteration is usually considerably less than that required for the Newton step.

*Goldfeld, Quandt, and Trotter [8.14].

†See Broyden [8.4] for an extensive discussion and comparison of quasi-Newton methods.

‡Because the sequence of positive-definite operators $\{\mathbf{H}_k\}$ defines a sequence of inner products, as indicated in P&C 5.30, the DFP method is sometimes called a "variable metric" method. In point of fact, any descent method can be interpreted via (8.24) as a variable-metric steepest descent.

The DFP Descent

We develop the DFP method for the n-dimensional quadratic functional $\mathbf{F}$: $\mathfrak{M}^{n\times 1}\to\mathfrak{R}$ given in (8.27). (We later apply it to nonquadratic functionals.) For this development we let the symbol $\mathbf{H}_k$ of the descent iteration denote an $n\times n$ *step-direction matrix* rather than a step-direction operator:

$$\mathbf{x}_{k+1}=\mathbf{x}_k-\alpha_k\mathbf{H}_k\nabla\mathbf{F}(\mathbf{x}_k) \tag{8.41}$$

We use the standard inner product to define gradients for $\mathfrak{M}^{n\times 1}$; thus $\nabla\mathbf{F}(\mathbf{x})=\mathbf{Ax}-\mathbf{y}$. We denote the kth step by $\mathbf{h}_k=-\alpha_k\mathbf{H}_k\nabla\mathbf{F}(\mathbf{x}_k)$. At each step we pick the step length (or α_k) for maximum descent. Thus the steps satisfy $\langle\mathbf{h}_k,\nabla\mathbf{F}(\mathbf{x}_{k+1})\rangle=0$, as indicated in (8.25) and Figure 8.3. The inverse Hessian matrix of $\mathbf{F}$ is $\mathbf{A}^{-1}$. We seek a sequence of matrices $\{\mathbf{H}_k\}$ for which $\mathbf{H}_k\to\mathbf{A}^{-1}$ as the iteration proceeds. Expressed another way, we wish to have $\mathbf{H}_k\mathbf{A}\to\mathbf{I}$. We obtain this property by forcing a successively increasing number of eigenvalues of $\mathbf{H}_k\mathbf{A}$ to the value $\lambda=1$ as k increases.

We start with an initial point $\mathbf{x}_0$ and an initial symmetric, positive-definite step-direction matrix, $\mathbf{H}_0=\mathbf{I}$. Thus the initial step is $\mathbf{h}_0=-\alpha_0\nabla\mathbf{F}(\mathbf{x}_0)$, and $\mathbf{x}_1=\mathbf{x}_0+\mathbf{h}_0$. We pick the second step direction matrix $\mathbf{H}_1$ such that $\mathbf{h}_0$ is an eigenvector of $\mathbf{H}_1\mathbf{A}$ for the eigenvalue $\lambda=1$. We pick $\mathbf{H}_2$ such that the first two step vectors, $\mathbf{h}_0$ and $\mathbf{h}_1$, are eigenvectors of $\mathbf{H}_2\mathbf{A}$ for the eigenvalue $\lambda=1$. Continuing this process for a total of n steps, we obtain a matrix $\mathbf{H}_n$ such that $\mathbf{H}_n\mathbf{A}$ has n eigenvectors, $\mathbf{h}_0,\mathbf{h}_1,\ldots,\mathbf{h}_{n-1}$, for the eigenvalue $\lambda=1$. The eigenvectors $\{\mathbf{h}_i\}$ turn out to be $\mathbf{A}$-orthogonal, and consequently linearly independent. Therefore, in n steps we produce $\mathbf{H}_n=\mathbf{A}^{-1}$.

We view the determination of the matrix $\mathbf{H}_{k+1}$ as the addition of a correction matrix to $\mathbf{H}_k$:

$$\mathbf{H}_{k+1}=\mathbf{H}_k+\mathbf{C}_k \qquad k=0,1,\ldots \tag{8.42}$$

We will show that the correction matrix $\mathbf{C}_k$ does not destroy the symmetry or positive definiteness of $\mathbf{H}_k$, yet makes it more like $\mathbf{A}^{-1}$. The first correction matrix $\mathbf{C}_0$ must satisfy

$$\mathbf{H}_1\mathbf{Ah}_0=\mathbf{H}_0\mathbf{Ah}_0+\mathbf{C}_0\mathbf{Ah}_0=\mathbf{Ah}_0+\mathbf{C}_0\mathbf{Ah}_0=\mathbf{h}_0 \tag{8.43}$$

or $\mathbf{C}_0\mathbf{Ah}_0=\mathbf{h}_0-\mathbf{Ah}_0$. Define

$$\mathbf{z}_k\stackrel{\Delta}{=}\mathbf{Ah}_k \qquad k=0,1,\ldots \tag{8.44}$$

Then the requirement on $\mathbf{C}_0$ becomes

$$\mathbf{C}_0\mathbf{z}_0=\mathbf{h}_0-\mathbf{z}_0 \tag{8.45}$$

Consider the matrices

$$\frac{\mathbf{h}_0\mathbf{h}_0^T}{\mathbf{h}_0^T\mathbf{z}_0} \quad \text{and} \quad \frac{\mathbf{z}_0\mathbf{z}_0^T}{\mathbf{z}_0^T\mathbf{z}_0}$$

Both matrices are symmetric. The first converts $\mathbf{z}_0$ into $\mathbf{h}_0$; the second acts like the identity on $\mathbf{z}_0$. Thus the choice

$$\mathbf{C}_0 = \frac{\mathbf{h}_0\mathbf{h}_0^T}{\mathbf{h}_0^T\mathbf{z}_0} - \frac{\mathbf{z}_0\mathbf{z}_0^T}{\mathbf{z}_0^T\mathbf{z}_0} \tag{8.46}$$

satisfies (8.45). By employing the symmetry of $\mathbf{H}_1$, the eigenvalue condition (8.43), and the step-length condition (8.25), we find

$$\begin{aligned}\langle \mathbf{h}_1, \mathbf{A}\mathbf{h}_0 \rangle &= -\alpha_1 \langle \mathbf{H}_1 \nabla \mathbf{F}(\mathbf{x}_1), \mathbf{A}\mathbf{h}_0 \rangle = -\alpha_1 \langle \nabla \mathbf{F}(\mathbf{x}_1), \mathbf{H}_1 \mathbf{A}\mathbf{h}_0 \rangle \\ &= -\alpha_1 \langle \nabla \mathbf{F}(\mathbf{x}_1), \mathbf{h}_0 \rangle = 0\end{aligned}$$

Thus the first two steps are **A**-orthogonal. Note that this **A**-orthogonality is a consequence of the symmetry of $\mathbf{H}_1$ and the eigenvector relationship (8.43), rather than the specific choice of $\mathbf{C}_0$.

The second correction matrix $\mathbf{C}_1$ must satisfy the pair of conditions

$$\begin{aligned}\mathbf{H}_2\mathbf{A}\mathbf{h}_1 &= \mathbf{H}_2\mathbf{z}_1 = \mathbf{H}_1\mathbf{z}_1 + \mathbf{C}_1\mathbf{z}_1 = \mathbf{h}_1 \\ \mathbf{H}_2\mathbf{A}\mathbf{h}_0 &= \mathbf{H}_2\mathbf{z}_0 = \mathbf{H}_1\mathbf{z}_0 + \mathbf{C}_1\mathbf{z}_0 = \mathbf{h}_0\end{aligned} \tag{8.47}$$

or

$$\mathbf{C}_1\mathbf{z}_1 = \mathbf{h}_1 - \mathbf{H}_1\mathbf{z}_1 \quad \text{and} \quad \mathbf{C}_1\mathbf{z}_0 = \mathbf{h}_0 - \mathbf{H}_1\mathbf{z}_0 = \boldsymbol{\theta} \tag{8.48}$$

where the final equality is a consequence of (8.43) and (8.44). A comparison of the first condition of (8.48) with (8.45)–(8.46) suggests

$$\mathbf{C}_1 = \frac{\mathbf{h}_1\mathbf{h}_1^T}{\mathbf{h}_1^T\mathbf{z}_1} - \frac{(\mathbf{H}_1\mathbf{z}_1)(\mathbf{H}_1\mathbf{z}_1)^T}{(\mathbf{H}_1\mathbf{z}_1)^T\mathbf{z}_1} \tag{8.49}$$

By inserting (8.49) into the second condition of (8.48), we find

$$\mathbf{C}_1\mathbf{z}_0 = \mathbf{h}_1 \frac{\mathbf{h}_1^T\mathbf{z}_0}{\mathbf{h}_1^T\mathbf{z}_1} - \mathbf{H}_1\mathbf{z}_1 \frac{(\mathbf{H}_1\mathbf{z}_1)^T\mathbf{z}_0}{(\mathbf{H}_1\mathbf{z}_1)^T\mathbf{z}_1}$$

but

$$\mathbf{h}_1^T\mathbf{z}_0 = \langle \mathbf{h}_1, \mathbf{z}_0 \rangle = \langle \mathbf{h}_1, \mathbf{A}\mathbf{h}_0 \rangle = 0$$

and

$$(\mathbf{H}_1\mathbf{z}_1)^{\mathrm{T}}\mathbf{z}_0=\langle \mathbf{z}_0,\mathbf{H}_1\mathbf{z}_1\rangle=\langle \mathbf{H}_1\mathbf{z}_0,\mathbf{z}_1\rangle=\langle \mathbf{H}_1\mathbf{A}\mathbf{h}_0,\mathbf{A}\mathbf{h}_1\rangle=\langle \mathbf{h}_0,\mathbf{A}\mathbf{h}_1\rangle=0$$

Thus $\mathbf{C}_1\mathbf{z}_0=\boldsymbol{\theta}$, and (8.49) satisfies both of the eigenvalue conditions (8.47). The **A**-orthogonality of $\mathbf{h}_2$ and $\mathbf{h}_1$ can be verified by the same approach which we used to show that $\langle \mathbf{h}_1,\mathbf{A}\mathbf{h}_0\rangle=0$. Note that $\mathbf{x}_2=\mathbf{x}_1+\mathbf{h}_1$, and consequently, for the quadratic functional **F** of (8.27),

$$\nabla\mathbf{F}(\mathbf{x}_2)=\mathbf{A}\mathbf{x}_2-\mathbf{y}=\mathbf{A}\mathbf{x}_1-\mathbf{y}+\mathbf{A}\mathbf{h}_1=\nabla\mathbf{F}(\mathbf{x}_1)+\mathbf{A}\mathbf{h}_1$$

We use this fact together with the symmetry of $\mathbf{H}_2$, (8.47), the step-length condition $\langle \nabla\mathbf{F}(\mathbf{x}_1),\mathbf{h}_0\rangle=0$, and the **A**-orthogonality condition $\langle \mathbf{A}\mathbf{h}_1,\mathbf{h}_0\rangle=0$ to show that

$$\begin{aligned}\langle \mathbf{h}_2,\mathbf{A}\mathbf{h}_0\rangle&=-\alpha_2\langle \mathbf{H}_2\nabla\mathbf{F}(\mathbf{x}_2),\mathbf{A}\mathbf{h}_0\rangle\\&=-\alpha_2\langle \nabla\mathbf{F}(\mathbf{x}_2),\mathbf{H}_2\mathbf{A}\mathbf{h}_0\rangle\\&=-\alpha_2\langle \nabla\mathbf{F}(\mathbf{x}_1)+\mathbf{A}\mathbf{h}_1,\mathbf{h}_0\rangle\\&=0\end{aligned}$$

Thus the set $\{\mathbf{h}_0,\mathbf{h}_1,\mathbf{h}_2\}$ is **A**-orthogonal.

Exercise 1. Show by induction that for the n-dimensional quadratic functional (8.27) treated above, the conditions

$$(\mathbf{H}_{k+1}\mathbf{A})\mathbf{h}_i=\mathbf{h}_i,\qquad i=0,1,\ldots,k$$

can be satisfied by the symmetric step-direction matrices

$$\mathbf{H}_{k+1}=\mathbf{H}_k+\frac{\mathbf{h}_k\mathbf{h}_k^{\mathrm{T}}}{\mathbf{h}_k^{\mathrm{T}}\mathbf{z}_k}-\frac{(\mathbf{H}_k\mathbf{z}_k)(\mathbf{H}_k\mathbf{z}_k)^{\mathrm{T}}}{(\mathbf{H}_k\mathbf{z}_k)^{\mathrm{T}}\mathbf{z}_k},\qquad \mathbf{H}_0=\mathbf{I}\tag{8.50}$$

for $k=0,1,\ldots,n-1$. Show also that the resulting step vectors $\{\mathbf{h}_0,\ldots,\mathbf{h}_{n-1}\}$ are **A**-orthogonal. Note that $\mathbf{h}_k=-\alpha_k\mathbf{H}_k\nabla\mathbf{F}(\mathbf{x}_k)$, $\mathbf{z}_k=\mathbf{A}\mathbf{h}_k$, and $\nabla\mathbf{F}(\mathbf{x}_{k+1})=\nabla\mathbf{F}(\mathbf{x}_k+\mathbf{h}_k)=\nabla\mathbf{F}(\mathbf{x}_k)+\mathbf{A}\mathbf{h}_k$. Since the step vectors $\{\mathbf{h}_0,\ldots,\mathbf{h}_{n-1}\}$ are **A**-orthogonal, they are necessarily independent. Because each is an eigenvector of $\mathbf{H}_n\mathbf{A}$ for the eigenvalue $\lambda=1$, $\mathbf{H}_n\mathbf{A}=\mathbf{I}$ and $\mathbf{H}_n=\mathbf{A}^{-1}$. Thus the Newton step (8.38) is used in the nth iteration, and convergence to the minimum is complete after n steps.

The **DFP descent method** proceeds as follows. We begin with $\mathbf{x}_0$ $(k=0)$ and compute $\mathbf{h}_0=-\alpha_0\nabla\mathbf{F}(\mathbf{x}_0)$, where α_0 is chosen to provide maximum descent in the value of $\mathbf{F}(\mathbf{x}_1)$.* Thus the first iteration produces $\mathbf{x}_1 = \mathbf{x}_0 + \mathbf{h}_0$. At the end of the kth iteration we have $\mathbf{x}_k = \mathbf{x}_{k-1} + \mathbf{h}_{k-1}$ plus the quantities necessary to compute $\mathbf{h}_{k-1} = -\alpha_{k-1}\mathbf{H}_{k-1}\nabla\mathbf{F}(\mathbf{x}_{k-1})$. In the $(k+1)$st iteration we first find $\nabla\mathbf{F}(\mathbf{x}_k)$. Then since $\nabla\mathbf{F}(\mathbf{x}) = \mathbf{A}\mathbf{x} - \mathbf{y}$ for the quadratic functional (8.27),

$$\mathbf{z}_{k-1} = \mathbf{A}\mathbf{h}_{k-1} = \mathbf{A}(\mathbf{x}_k - \mathbf{x}_{k-1}) = \nabla\mathbf{F}(\mathbf{x}_k) - \nabla\mathbf{F}(\mathbf{x}_{k-1})$$

Because first derivatives are generally easier to compute than second derivatives, we compute $\mathbf{z}_{k-1}$ from the gradients, $\nabla\mathbf{F}(\mathbf{x}_k) - \nabla\mathbf{F}(\mathbf{x}_{k-1})$, rather than from $\mathbf{h}_{k-1}$ and the Hessian matrix $\mathbf{A}$. Next we find $\mathbf{H}_k$ by means of (8.50). Finally, we find $\mathbf{h}_k = -\alpha_k\mathbf{H}_k\nabla\mathbf{F}(\mathbf{x}_k)$, where the step size (or α_k) is selected to provide maximum descent in the value of $\mathbf{F}(\mathbf{x}_{k+1})$. Thus we obtain $\mathbf{x}_{k+1} = \mathbf{x}_k + \mathbf{h}_k$. For an n-dimensional quadratic functional $\mathbf{F}$, the DFP iteration converges to the minimum in n steps.

Example 1. Minimization of a Quadratic Functional by the DFP Descent. Let $\mathbf{x} = (\xi_1\ \xi_2)^{\mathrm{T}}$ and $\mathbf{F}(\mathbf{x}) = \frac{1}{2}(\xi_1^2 + 10\xi_2^2)$. Then

$$\nabla\mathbf{F}(\mathbf{x}) = \begin{pmatrix} \xi_1 \\ 10\xi_2 \end{pmatrix} \quad \text{and} \quad \mathbf{A} = \begin{pmatrix} 1 & 0 \\ 0 & 10 \end{pmatrix}$$

where $\mathbf{A}$ is the Hessian matrix of $\mathbf{F}$. Let $\mathbf{x}_0 = (10\ 1)^{\mathrm{T}}$. Then $\nabla\mathbf{F}(\mathbf{x}_0) = (10\ 10)^{\mathrm{T}}$. In the DFP descent, $\mathbf{H}_0 = \mathbf{I}$ and the first iteration is $\mathbf{x}_1 = \mathbf{x}_0 - \alpha_0\nabla\mathbf{F}(\mathbf{x}_0)$. By solving $d\mathbf{F}(\mathbf{x}_1)/d\alpha_0 = 0$ we find $\alpha_0 = 2/11$. Thus

$$\mathbf{x}_1 = \tfrac{9}{11}\begin{pmatrix} 10 \\ -1 \end{pmatrix}, \qquad \mathbf{h}_0 = -\tfrac{20}{11}\begin{pmatrix} 1 \\ 1 \end{pmatrix}, \qquad \nabla\mathbf{F}(\mathbf{x}_1) = \tfrac{90}{11}\begin{pmatrix} 1 \\ -1 \end{pmatrix}, \qquad \mathbf{z}_0 = -\tfrac{20}{11}\begin{pmatrix} 1 \\ 10 \end{pmatrix}$$

Since $\mathbf{H}_0 = \mathbf{I}$, $\mathbf{H}_0\mathbf{z}_0 = \mathbf{z}_0$. The new step direction is determined by (8.50) with $k = 0$:

$$\mathbf{H}_1 = \begin{pmatrix} 1 & 0 \\ 0 & 1 \end{pmatrix} + \tfrac{1}{11}\begin{pmatrix} 1 & 1 \\ 1 & 1 \end{pmatrix} - \tfrac{1}{101}\begin{pmatrix} 1 & 10 \\ 10 & 100 \end{pmatrix}$$
$$= \tfrac{1}{1111}\begin{pmatrix} 1201 & -9 \\ -9 & 112 \end{pmatrix}$$

We easily verify that $\mathbf{h}_0$ is an eigenvector of $\mathbf{H}_1\mathbf{A}$ for $\lambda = 1$. The second DFP iteration is

$$\mathbf{x}_2 = \mathbf{x}_1 - \alpha_1\mathbf{H}_1\nabla\mathbf{F}(\mathbf{x}_1)$$
$$= \tfrac{9}{11}\begin{pmatrix} 10 \\ -1 \end{pmatrix} - \tfrac{90}{101}\alpha_1\begin{pmatrix} 10 \\ -1 \end{pmatrix}$$

*The DFP method does not necessarily require $\mathbf{H}_0 = \mathbf{I}$. Any symmetric positive-definite matrix will do. A choice for which $\mathbf{H}_0 \approx (\partial^2\mathbf{F}(\mathbf{x})/\partial\mathbf{x}^2)^{-1}$ is best.

By solving $(d/d\alpha_1)\mathbf{F}(\mathbf{x}_2)=0$, we find $\alpha_1=101/110$. Then

$$\mathbf{x}_2=\begin{pmatrix}0\\0\end{pmatrix},\quad \mathbf{h}_1=\tfrac{9}{11}\begin{pmatrix}-10\\1\end{pmatrix},\quad \nabla\mathbf{F}(\mathbf{x}_2)=\begin{pmatrix}0\\0\end{pmatrix},\quad \mathbf{z}_1=\tfrac{90}{11}\begin{pmatrix}-1\\1\end{pmatrix}$$

We have found the minimum of the two-dimensional quadratic function in two steps. Thus the method is slower than that of the Newton descent (which converges in one step) but considerably faster than standard steepest descent (Figure 8.4). The $\mathbf{A}$-orthogonality of $\mathbf{h}_0$ and $\mathbf{h}_1$ is easily verified. The inverse Hessian matrix can be obtained from (8.50) with $k=1$:

$$\mathbf{H}_2=\tfrac{1}{1111}\begin{pmatrix}1201&-9\\-9&112\end{pmatrix}+\tfrac{1}{110}\begin{pmatrix}100&-10\\-10&1\end{pmatrix}-\tfrac{1}{101}\begin{pmatrix}100&-10\\-10&1\end{pmatrix}$$
$$=\begin{pmatrix}1&0\\0&0.1\end{pmatrix}=\mathbf{A}^{-1}$$

Note that we have obtained the inverse Hessian without computing second derivatives.

We have dealt extensively with quadratic functionals throughout this chapter, primarily in order to provide motivation for the design of various descent techniques and to compare the relative merits of the techniques. However, the actual minimization of a quadratic functional $\mathbf{F}(\mathbf{x})$ is usually best carried out by solving the linear necessary condition, $\nabla\mathbf{F}(\mathbf{x})=\boldsymbol{\theta}$. Thus, although the DFP descent works extremely well for quadratic functionals, its real usefulness is for nonquadratic functionals.

We now show that even if the functional $\mathbf{F}(\mathbf{x})$ is not quadratic, the DFP method is a descent method. We do so by showing that if $\mathbf{H}_k$ is positive definite (a descent-direction matrix), then $\mathbf{H}_{k+1}$ is also positive definite. Let $\langle\mathbf{x},\mathbf{y}\rangle\stackrel{\Delta}{=}\mathbf{y}^{\mathrm{T}}\mathbf{x}$. Then $\langle\mathbf{x},\mathbf{y}\rangle_{\mathrm{Q}}\stackrel{\Delta}{=}\mathbf{y}^{\mathrm{T}}\mathbf{Q}\mathbf{x}$ defines an inner product on $\mathfrak{M}^{n\times 1}$ for any real symmetric positive-definite $n\times n$ matrix $\mathbf{Q}$. If $\mathbf{H}_k$ is positive definite, $\mathbf{Q}=\mathbf{H}_k$ defines an inner product. From (8.50) and (8.44) we find, for any $\mathbf{x}$ in $\mathfrak{M}^{n\times 1}$, that

$$\mathbf{x}^{\mathrm{T}}\mathbf{H}_{k+1}\mathbf{x}=\mathbf{x}^{\mathrm{T}}\mathbf{H}_k\mathbf{x}+\frac{\mathbf{x}^{\mathrm{T}}\mathbf{h}_k\mathbf{h}_k^{\mathrm{T}}\mathbf{x}}{\mathbf{h}_k^{\mathrm{T}}\mathbf{z}_k}-\frac{\mathbf{x}^{\mathrm{T}}\mathbf{H}_k\mathbf{z}_k(\mathbf{H}_k\mathbf{z}_k)^{\mathrm{T}}\mathbf{x}}{(\mathbf{H}_k\mathbf{z}_k)^{\mathrm{T}}\mathbf{z}_k}$$

$$=\|\mathbf{x}\|_{\mathrm{H}}^2+\frac{\langle\mathbf{x},\mathbf{h}_k\rangle^2}{\langle\mathbf{h}_k,\mathbf{z}_k\rangle}-\frac{\langle\mathbf{x},\mathbf{z}_k\rangle_{\mathrm{H}}^2}{\|\mathbf{z}_k\|_{\mathrm{H}}^2}$$

$$=\frac{\langle\mathbf{x},\mathbf{h}_k\rangle^2}{\langle\mathbf{h}_k,\mathbf{A}\mathbf{h}_k\rangle}+\frac{\|\mathbf{x}\|_{\mathrm{H}}^2\|\mathbf{z}_k\|_{\mathrm{H}}^2-\langle\mathbf{x},\mathbf{z}_k\rangle_{\mathrm{H}}^2}{\|\mathbf{z}_k\|_{\mathrm{H}}^2}$$

The first term of the last line is $(\langle \mathbf{x},\mathbf{h}_k\rangle/\|\mathbf{h}_k\|_{\mathbf{A}})^2$, which is non-negative, and becomes zero only if $\langle \mathbf{x},\mathbf{h}_k\rangle=0$. The second term of the last line must also be non-negative, by the Cauchy–Schwartz inequality (P&C 5.4), and becomes zero only if $\mathbf{x}$ is collinear with $\mathbf{z}_k=\mathbf{A}\mathbf{h}_k$. These two terms cannot be zero simultaneously unless $\mathbf{A}\mathbf{h}_k$ is orthogonal to $\mathbf{h}_k$. This orthogonality cannot occur because $\mathbf{A}$ is positive definite. Thus $\mathbf{x}^{\mathrm{T}}\mathbf{H}_{k+1}\mathbf{x}>0$ and $\mathbf{H}_{k+1}$ defines a descent step.

Of course, the set of descent steps is not $\mathbf{A}$-orthogonal if $\mathbf{F}(\mathbf{x})$ is not quadratic. Nor is $\mathbf{H}_n$ the inverse Hessian. However, the technique still exhibits the properties of most practical interest. Convergence is rapid—it is more like the Newton descent than it is like standard steepest descent. Furthermore, as $\mathbf{x}_k\to\mathbf{x}_m$, $\mathbf{H}_k\to(\partial^2\mathbf{F}(\mathbf{x}_m)/\partial\mathbf{x}^2)^{-1}$ and convergence becomes quadratic.*

By means of a slight change in notation, the DFP descent can be applied to functionals $\mathbf{F}$ defined on any separable Hilbert space $\mathcal{V}$. Let $\mathbf{x}$ and $\mathbf{y}$ be vectors in $\mathfrak{M}^{n\times 1}$. Represent the matrix $\mathbf{x}\mathbf{y}^{\mathrm{T}}$ by the **outer product** (or dyad) notation.

$$\mathbf{x}\rangle\langle\mathbf{y} \triangleq \mathbf{x}\mathbf{y}^{\mathrm{T}}$$

Then we can express the matrix correction process (8.50) of the DFP descent in the form

$$\mathbf{H}_{k+1}=\mathbf{H}_k+\frac{\mathbf{h}_k\rangle\langle\mathbf{h}_k}{\langle\mathbf{h}_k,\mathbf{z}_k\rangle}-\frac{\mathbf{H}_k\mathbf{z}_k\rangle\langle\mathbf{H}_k\mathbf{z}_k}{\langle\mathbf{z}_k,\mathbf{H}_k\mathbf{z}_k\rangle} \tag{8.51}$$

We can apply the DFP descent to $\mathbf{F}$: $\mathcal{V}\to\mathfrak{R}$ if we can assign an appropriate meaning to the symbols and operations in (8.51). A simple function space example is treated in P&C 8.15. Tokumaru, Adachi, and Goto [8.33] apply the DFP method to control problems in a function space.

Conjugate-Direction Methods

One of the significant characteristics of the DFP method of the preceding section is that it converges to the minimum of a *finite-dimensional quadratic* function in a finite number of steps. This "n-step" convergence is a consequence of the $\mathbf{A}$-orthogonality of the step directions. In this section we explore the philosophy behind a number of descent techniques which exhibit n-step convergence.

*Pearson [8.23] notes that some improvement in the rate of convergence is obtained if the iteration is occasionally restarted with $\mathbf{H}_k=\mathbf{I}$.

Once again we focus on the finite-dimensional quadratic functional $\mathbf{F}$: $\mathfrak{M}^{n\times 1}\to\mathfrak{R}$ defined by

$$\mathbf{F}(\mathbf{x})=\tfrac{1}{2}\mathbf{x}^{\mathrm{T}}\mathbf{A}\mathbf{x}-\mathbf{y}^{\mathrm{T}}\mathbf{x}+d \tag{8.52}$$

where $\mathbf{A}$ is symmetric and positive definite. The gradient of $\mathbf{F}$ relative to the standard inner product is $\nabla\mathbf{F}(\mathbf{x})=\mathbf{A}\mathbf{x}-\mathbf{y}$, and the minimizing point is $\mathbf{x}_m=\mathbf{A}^{-1}\mathbf{y}$. Corresponding to a specific starting vector $\mathbf{x}_0$ is a specific initial error vector, $\mathbf{x}_m-\mathbf{x}_0$, which must be driven to zero during the course of the iteration. Suppose the sequence of steps is $\mathbf{A}$-orthogonal (as in the DFP method). Then there is a unique set of step lengths, determined by the Fourier coefficients of the error vector, such that the sequence of steps drives the error to zero in precisely n steps. That is, $\mathbf{x}_m-\mathbf{x}_0$ can be expanded in an n-term Fourier series.

Let us be specific. Suppose $\{\mathbf{v}_0,\ldots,\mathbf{v}_{n-1}\}$ is an $\mathbf{A}$-orthogonal set of vectors. The Fourier series expansion of the initial error vector in terms of the set $\{\mathbf{v}_k\}$ is

$$\begin{aligned}\mathbf{x}_m-\mathbf{x}_0&=\sum_{k=0}^{n-1}\frac{\langle\mathbf{x}_m-\mathbf{x}_0,\mathbf{v}_k\rangle_{\mathbf{A}}}{\langle\mathbf{v}_k,\mathbf{v}_k\rangle_{\mathbf{A}}}\mathbf{v}_k\\&=\sum_{k=0}^{n-1}\frac{\langle\mathbf{v}_k,\mathbf{A}(\mathbf{x}_m-\mathbf{x}_0)\rangle}{\langle\mathbf{v}_k,\mathbf{A}\mathbf{v}_k\rangle}\mathbf{v}_k\end{aligned}$$

where $\langle\cdot,\cdot\rangle$ denotes the standard inner product for $\mathfrak{M}^{n\times 1}$. This expansion can also be written in the form

$$\mathbf{x}_m=\mathbf{x}_0+\sum_{k=0}^{n-1}\frac{\langle\mathbf{v}_k,\mathbf{y}-\mathbf{A}\mathbf{x}_0\rangle}{\langle\mathbf{v}_k,\mathbf{A}\mathbf{v}_k\rangle}\mathbf{v}_k \tag{8.53}$$

Since $\mathbf{y}-\mathbf{A}\mathbf{x}_0$ is a known quantity [it is, in fact, $-\nabla\mathbf{F}(\mathbf{x}_0)$], the Fourier coefficients can be computed. Define

$$\mathbf{h}_k\triangleq(\langle\mathbf{v}_k,\mathbf{y}-\mathbf{A}\mathbf{x}_0\rangle/\langle\mathbf{v}_k,\mathbf{A}\mathbf{v}_k\rangle)\mathbf{v}_k$$

and

$$\mathbf{x}_{k+1}\triangleq\mathbf{x}_0+\mathbf{h}_0+\cdots+\mathbf{h}_k$$

for $k=0,1,\ldots,n-1$. Then by (8.53), $\mathbf{x}_n=\mathbf{x}_m$, and we can compute $\mathbf{x}_m$ by the n-step iteration

$$\mathbf{x}_{k+1}=\mathbf{x}_k+\frac{\langle \mathbf{v}_k,\mathbf{y}-\mathbf{A}\mathbf{x}_0\rangle}{\langle \mathbf{v}_k,\mathbf{A}\mathbf{v}_k\rangle}\mathbf{v}_k, \qquad k=0,1,\dots,n-1 \tag{8.54}$$

According to the definitions of $\mathbf{x}_{k+1}$ and $\mathbf{h}_k$, the vector $\mathbf{x}_{k+1}-\mathbf{x}_0$ is in $\operatorname{span}\{\mathbf{v}_0,\mathbf{v}_1,\dots,\mathbf{v}_k\}$. On the other hand, the vector $\mathbf{v}_{k+1}$ is $\mathbf{A}$-orthogonal to $\operatorname{span}\{\mathbf{v}_0,\mathbf{v}_1,\dots,\mathbf{v}_k\}$. Therefore, $\mathbf{v}_{k+1}$ and $\mathbf{x}_{k+1}-\mathbf{x}_0$ are $\mathbf{A}$-orthogonal. We express this fact mathematically with $k+1$ replaced by k: $\langle \mathbf{v}_k,\mathbf{A}(\mathbf{x}_k-\mathbf{x}_0)\rangle=0$ or $\langle \mathbf{v}_k,\mathbf{A}\mathbf{x}_0\rangle=\langle \mathbf{v}_k,\mathbf{A}\mathbf{x}_k\rangle$. We use this identity to express the iteration (8.54) in terms of the most recent iterate $\mathbf{x}_k$ rather than the initial guess $\mathbf{x}_0$:

$$\begin{aligned}\mathbf{x}_{k+1}&=\mathbf{x}_k+\frac{\langle \mathbf{v}_k,\mathbf{y}-\mathbf{A}\mathbf{x}_k\rangle}{\langle \mathbf{v}_k,\mathbf{A}\mathbf{v}_k\rangle}\mathbf{v}_k\\ &=\mathbf{x}_k-\frac{\langle \mathbf{v}_k,\nabla\mathbf{F}(\mathbf{x}_k)\rangle}{\langle \mathbf{v}_k,\mathbf{A}\mathbf{v}_k\rangle}\mathbf{v}_k, \qquad k=0,1,\dots,n-1\end{aligned} \tag{8.55}$$

Because the step directions $\{\mathbf{v}_k\}$ are $\mathbf{A}$-orthogonal, we refer to any iteration of the form (8.55) as a **conjugate-direction method** for minimizing $\mathbf{F}(\mathbf{x})$. Any such conjugate-direction method will produce the minimum of the finite-dimensional quadratic function (8.52) in precisely n steps.

We now show that conjugate-direction methods (as applied to quadratic functions) are descent methods. The kth error vector is $\mathbf{x}_m-\mathbf{x}_k=\mathbf{h}_k+\cdots+\mathbf{h}_{n-1}$. By the Pythagorean theorem (P&C 5.4) and the $\mathbf{A}$-orthogonality of $\{\mathbf{h}_k\}$,

$$\|\mathbf{x}_m-\mathbf{x}_k\|_{\mathrm{A}}^2=\|\mathbf{h}_k\|_{\mathrm{A}}^2+\cdots+\|\mathbf{h}_{n-1}\|_{\mathrm{A}}^2$$

Consequently, $\|\mathbf{x}_m-\mathbf{x}_{k+1}\|_{\mathrm{A}}^2\leqslant\|\mathbf{x}_m-\mathbf{x}_k\|_{\mathrm{A}}^2$, and the error vector decreases monotonically. The locus of solutions $\mathbf{x}$ to the equation $\|\mathbf{x}_m-\mathbf{x}\|_{\mathrm{A}}^2=\text{constant}$ is an ellipsoid in $\mathfrak{M}^{n\times 1}$. But

$$\begin{aligned}\|\mathbf{x}_m-\mathbf{x}\|_{\mathrm{A}}^2&=\langle \mathbf{x}_m-\mathbf{x},\mathbf{x}_m-\mathbf{x}\rangle_{\mathrm{A}}\\ &=\langle \mathbf{x},\mathbf{x}\rangle_{\mathrm{A}}-2\langle \mathbf{x}_m,\mathbf{x}\rangle_{\mathrm{A}}+\langle \mathbf{x}_m,\mathbf{x}_m\rangle_{\mathrm{A}}\\ &=\mathbf{x}^{\mathrm{T}}\mathbf{A}\mathbf{x}-2\mathbf{x}^{\mathrm{T}}\mathbf{A}\mathbf{x}_m+\|\mathbf{x}_m\|_{\mathrm{A}}^2\\ &=2\mathbf{F}(\mathbf{x})+\|\mathbf{x}_m\|_{\mathrm{A}}^2-2d\end{aligned}$$

Therefore, the locus of constant $\|\mathbf{x}_m-\mathbf{x}\|_{\mathrm{A}}^2$ is also a locus of constant $\mathbf{F}(\mathbf{x})$, and the monotonic decrease in $\|\mathbf{x}_m - \mathbf{x}\|_{\mathbf{A}}^2$ implies a monotonic descent in $\mathbf{F}(\mathbf{x})$. We will show further that the Fourier coefficients produce maximum descent at each step. Suppose $\mathbf{v}_k$ specifies a descent direction from $\mathbf{x}_k$. We let $\mathbf{x}_{k+1} = \mathbf{x}_k + \alpha_k \mathbf{v}_k$, and pick α_k to minimize $\mathbf{F}(\mathbf{x}_{k+1})$. According to (8.25),

$$\begin{aligned}\langle \nabla \mathbf{F}(\mathbf{x}_{k+1}), \mathbf{v}_k \rangle &= \langle \mathbf{A}(\mathbf{x}_{k+1}) - \mathbf{y}, \mathbf{v}_k \rangle \\ &= \langle \mathbf{A}\mathbf{x}_k + \alpha_k \mathbf{A}\mathbf{v}_k - \mathbf{y}, \mathbf{v}_k \rangle \\ &= \langle \mathbf{A}\mathbf{x}_k - \mathbf{y}, \mathbf{v}_k \rangle + \alpha_k \langle \mathbf{v}_k, \mathbf{A}\mathbf{v}_k \rangle = 0\end{aligned}$$

or $\alpha_k = \langle \mathbf{v}_k, \mathbf{y} - \mathbf{A}\mathbf{x}_k \rangle / \langle \mathbf{v}_k, \mathbf{A}\mathbf{v}_k \rangle$. But this is just the Fourier coefficient used in (8.55). If $\mathbf{v}_k$ were an ascent direction, the sign of the numerator of α_k would be negative, and the step would still be a descent.

In summary, a conjugate-direction method produces the minimum of an n-dimensional quadratic function by means of a sequence of n $\mathbf{A}$-orthogonal descent steps, each step sized for maximum descent. Conjugate-direction methods apply to infinite-dimensional quadratic functionals as well. If we replace the finite-dimensional functional (8.52) by its infinite-dimensional counterpart (8.30),

$$\mathbf{F}(\mathbf{x}) = \tfrac{1}{2}\langle \mathbf{x}, \mathbf{T}\mathbf{x} \rangle - \langle \mathbf{x}, \mathbf{y} \rangle + d$$

the only changes necessary in (8.55) are a replacement of the matrix $\mathbf{A}$ by the linear operator $\mathbf{T}$ and the extension of the iteration to an infinite number of steps. The iteration generates a $\mathbf{T}$-orthogonal basis $\{\mathbf{v}_k\}$ for the infinite-dimensional space. The error vector $\mathbf{x}_m - \mathbf{x}_k$ decreases in a manner determined by the properties of Fourier coefficients for infinite-dimensional spaces.

Various conjugate-direction methods differ in the manner in which the $\mathbf{A}$-orthogonal step directions are obtained. A brute force approach is to determine the Hessian matrix $\mathbf{A}$ of the quadratic function by taking derivatives of $\mathbf{F}(\mathbf{x})$, then obtain an $\mathbf{A}$-orthogonal set $\{\mathbf{v}_k\}$ by applying the Gram–Schmidt procedure to some basis for $\mathfrak{M}^{n\times 1}$. The genius of the DFP method of the preceding section is that it avoids the direct determination of the Hessian matrix. Rather, it uses the initial estimate $\mathbf{x}_0$ and the values and gradients of $\mathbf{F}(\mathbf{x})$ at each step to generate an $\mathbf{A}$-orthogonal set of steps $\{\mathbf{h}_k\}$ and, in the limit, $\mathbf{A}^{-1}$. We verified in the previous section that the DFP method is indeed a conjugate-direction method; that is, it is a descent method, the step directions are $\mathbf{A}$-orthogonal, and it produces the minimum of an n-dimensional quadratic function in n steps. In the next section we describe another conjugate-direction technique which is simpler to carry out than is the DFP method.

Because quadratic functions are relatively easy to minimize, our primary interest is in the minimization of nonquadratic functions. If $\mathbf{F}(\mathbf{x})$ is nearly quadratic, we still expect to be able to minimize $\mathbf{F}(\mathbf{x})$ by applying a conjugate-direction technique, using the Hessian matrix of $\mathbf{F}$ to specify a quadratic approximation to $\mathbf{F}$. Of course, we do not obtain pure conjugacy of the steps or n-step convergence. Yet we expect convergence to a minimum of $\mathbf{F}(\mathbf{x})$ to be faster than for steepest descent. In the preceding section we found that each step in the DFP method is guaranteed to be a descent step even for a nonquadratic $\mathbf{F}$. Thus the DFP descent is an appropriate method even for the initial stages of the minimization of a nonquadratic function. Like the DFP descent, the conjugate-direction method of the next two sections produces a descent toward the minimum at each step even for a nonquadratic $\mathbf{F}$.

Minimization by Conjugate Gradients

Fletcher and Reeves [8.9] have developed an efficient conjugate-direction method, known as the *conjugate-gradient* (or CG) method, which is suitable for minimization of nonquadratic functionals, both finite and infinite-dimensional. We first introduce the technique as it pertains to the n-step minimization of the n-dimensional quadratic functional (8.52) with Hessian matrix $\mathbf{A}$. Later we apply the results to nonquadratic and infinite-dimensional functions. The conjugate-gradient method is based upon the $\mathbf{A}$-orthogonalization of the sequence of negative gradients at the interates, $\{-\nabla\mathbf{F}(\mathbf{x}_k)\}$; that is, the initial step direction is $\mathbf{v}_0=-\nabla\mathbf{F}(\mathbf{x}_0)$, and the kth step direction is

$$\mathbf{v}_k=-\nabla\mathbf{F}(\mathbf{x}_k)-\sum_{i=0}^{k-1}\frac{\langle-\nabla\mathbf{F}(\mathbf{x}_k),\mathbf{A}\mathbf{v}_i\rangle}{\langle\mathbf{v}_i,\mathbf{A}\mathbf{v}_i\rangle}\mathbf{v}_i \tag{8.56}$$

where $\langle\cdot,\cdot\rangle$ is the standard inner product for $\mathfrak{M}^{n\times 1}$. It is apparent that the set $\{\mathbf{v}_0,\ldots,\mathbf{v}_{n-1}\}$ is an $\mathbf{A}$-orthogonal set if the vectors $\{-\nabla\mathbf{F}(\mathbf{x}_0),\ldots,-\nabla\mathbf{F}(\mathbf{x}_{n-1})\}$ are independent. We show for the quadratic function (8.52) that the set $\{-\nabla\mathbf{F}(\mathbf{x}_k)\}$ is, in fact, orthogonal with respect to the standard inner product. We also show that only the $(k-1)$st Fourier coefficient is nonzero in the orthogonalization equation (8.56); this property considerably reduces the required computations. Finally, we show that the remaining Fourier coefficient needed in the determination of $\mathbf{v}_k$ can be computed from recent gradient information alone—the Hessian matrix $\mathbf{A}$ need not be computed.

Our approach for showing that the negative gradients $\{-\nabla\mathbf{F}(\mathbf{x}_k)\}$ are orthogonal (independent) is similar to the approach of Beckman [8.2]. In any conjugate-direction method, the iteration (8.55) generates a sequence of iterates $\{\mathbf{x}_k\}$ and a corresponding sequence of negative gradients $\{-\nabla\mathbf{F}(\mathbf{x}_k)\}$. In the conjugate-gradient method, these negative gradients are $\mathbf{A}$-orthogonalized as they are produced, yielding the sequence of step directions $\{\mathbf{v}_k\}$ used in (8.55). We show that for a quadratic function the sequence of negative gradients $\{-\nabla\mathbf{F}(\mathbf{x}_0),\ldots,-\nabla\mathbf{F}(\mathbf{x}_{n-1})\}$ produced by this conjugate-gradient iteration is identical to the sequence obtained by carrying out a Gram–Schmidt orthogonalization (in the standard inner product) of the vectors $\{\mathbf{v}_0,\mathbf{A}\mathbf{v}_0,\mathbf{A}\mathbf{v}_1,\ldots,\mathbf{A}\mathbf{v}_{n-2}\}$. Consequently, the conjugate-gradient algorithm produces, in effect, a pair of parallel Gram-Schmidt orthogonalizations. Each orthogonalization provides the sequence of independent vectors needed for the other.

The vectors of an orthogonal set each can be scaled without changing the orthogonality of the set. Therefore, we can express in the following form the Gram–Schmidt orthogonalization of a set $\{\mathbf{w}_i\}$ to produce a set $\{\mathbf{z}_i\}$*:

$$\mathbf{z}_{k+1} = -\frac{\langle \mathbf{z}_k,\mathbf{z}_k\rangle}{\langle \mathbf{w}_{k+1},\mathbf{z}_k\rangle}\left[\mathbf{w}_{k+1}-\frac{\langle \mathbf{w}_{k+1},\mathbf{z}_0\rangle}{\langle \mathbf{z}_0,\mathbf{z}_0\rangle}\mathbf{z}_0-\cdots-\frac{\langle \mathbf{w}_{k+1},\mathbf{z}_k\rangle}{\langle \mathbf{z}_k,\mathbf{z}_k\rangle}\mathbf{z}_k\right]$$
$$= \mathbf{z}_k-\frac{\langle \mathbf{z}_k,\mathbf{z}_k\rangle}{\langle \mathbf{w}_{k+1},\mathbf{z}_k\rangle}\left[\mathbf{w}_{k+1}-\frac{\langle \mathbf{w}_{k+1},\mathbf{z}_0\rangle}{\langle \mathbf{z}_0,\mathbf{z}_0\rangle}\mathbf{z}_0-\cdots-\frac{\langle \mathbf{w}_{k+1},\mathbf{z}_{k-1}\rangle}{\langle \mathbf{z}_{k-1},\mathbf{z}_{k-1}\rangle}\mathbf{z}_{k-1}\right] \tag{8.57}$$

We apply (8.57) to the set $\{\mathbf{w}_0,\ldots,\mathbf{w}_{n-1}\} \triangleq \{\mathbf{v}_0,\mathbf{A}\mathbf{v}_0,\ldots,\mathbf{A}\mathbf{v}_{n-2}\}$, where $\mathbf{v}_0=-\nabla\mathbf{F}(\mathbf{x}_0)$. Then $\mathbf{z}_0=\mathbf{v}_0=-\nabla\mathbf{F}(\mathbf{x}_0)$, and

$$\mathbf{z}_1=\mathbf{z}_0-\frac{\langle \mathbf{z}_0,\mathbf{z}_0\rangle}{\langle \mathbf{w}_1,\mathbf{z}_0\rangle}\mathbf{w}_1=-\nabla\mathbf{F}(\mathbf{x}_0)+\frac{\langle \mathbf{v}_0,\nabla\mathbf{F}(\mathbf{x}_0)\rangle}{\langle \mathbf{v}_0,\mathbf{A}\mathbf{v}_0\rangle}\mathbf{A}\mathbf{v}_0. \tag{8.58}$$

Since $\nabla\mathbf{F}(\mathbf{x})=\mathbf{A}\mathbf{x}-\mathbf{y}$ for the quadratic function (8.52), the iteration (8.55) yields the following gradients

$$-\nabla\mathbf{F}(\mathbf{x}_{k+1})=-\mathbf{A}\mathbf{x}_{k+1}+\mathbf{y}=-\mathbf{A}\mathbf{x}_k+\frac{\langle \mathbf{v}_k,\nabla\mathbf{F}(\mathbf{x}_k)\rangle}{\langle \mathbf{v}_k,\mathbf{A}\mathbf{v}_k\rangle}\mathbf{A}\mathbf{v}_k+\mathbf{y}$$
$$=-\nabla\mathbf{F}(\mathbf{x}_k)+\frac{\langle \mathbf{v}_k,\nabla\mathbf{F}(\mathbf{x}_k)\rangle}{\langle \mathbf{v}_k,\mathbf{A}\mathbf{v}_k\rangle}\mathbf{A}\mathbf{v}_k \tag{8.59}$$

Comparing (8.58) to (8.59) with $k=0$, we find that $\mathbf{z}_1=-\nabla\mathbf{F}(\mathbf{x}_1)$.

*Compare with (5.18).

We use induction to show that continuation of the orthogonalization produces the rest of the sequence of negative gradients. Assume that orthogonalization of $\{\mathbf{v}_0, \mathbf{A}\mathbf{v}_0, \ldots, \mathbf{A}\mathbf{v}_{k-1}\}$ has produced $\{\mathbf{z}_0, \ldots, \mathbf{z}_k\} = \{-\nabla\mathbf{F}(\mathbf{x}_0), \ldots, -\nabla\mathbf{F}(\mathbf{x}_k)\}$. Then, by (8.57),

$$\mathbf{z}_{k+1} = -\nabla\mathbf{F}(\mathbf{x}_k) - \frac{\langle \mathbf{z}_k, \mathbf{z}_k\rangle}{\langle \mathbf{A}\mathbf{v}_k, \mathbf{z}_k\rangle} \times \left[\mathbf{A}\mathbf{v}_k + \frac{\langle \mathbf{A}\mathbf{v}_k, \nabla\mathbf{F}(\mathbf{x}_0)\rangle}{\langle \mathbf{z}_0, \mathbf{z}_0\rangle}\mathbf{z}_0 + \cdots + \frac{\langle \mathbf{A}\mathbf{v}_k, \nabla\mathbf{F}(\mathbf{x}_{k-1})\rangle}{\langle \mathbf{z}_{k-1}, \mathbf{z}_{k-1}\rangle}\mathbf{z}_{k-1}\right]$$

Since $\mathbf{v}_k$ is obtained as the last step in the **A**-orthogonalization of the set $\{-\nabla\mathbf{F}(\mathbf{x}_0), \ldots, -\nabla\mathbf{F}(\mathbf{x}_k)\}$, $\mathbf{v}_k$ is **A**-orthogonal to span $\{\nabla\mathbf{F}(\mathbf{x}_0), \ldots, \nabla\mathbf{F}(\mathbf{x}_{k-1})\}$. Consequently, $\langle \mathbf{A}\mathbf{v}_k, \nabla\mathbf{F}(\mathbf{x}_i)\rangle = 0$ for $i = 0, \ldots, k-1$ and

$$\mathbf{z}_{k+1} = -\nabla\mathbf{F}(\mathbf{x}_k) + \frac{\langle \nabla\mathbf{F}(\mathbf{x}_k), \nabla\mathbf{F}(\mathbf{x}_k)\rangle}{\langle \mathbf{A}\mathbf{v}_k, \nabla\mathbf{F}(\mathbf{x}_k)\rangle}\mathbf{A}\mathbf{v}_k \tag{8.60}$$

We now show that (8.59) and (8.60) are equal. Since $\nabla\mathbf{F}(\mathbf{x}_k)$ is orthogonal to span$\{\mathbf{v}_0, \mathbf{A}\mathbf{v}_0, \ldots, \mathbf{A}\mathbf{v}_{k-2}\}$, (8.56) simplifies to

$$\mathbf{v}_k = -\nabla\mathbf{F}(\mathbf{x}_k) + \frac{\langle \nabla\mathbf{F}(\mathbf{x}_k), \mathbf{A}\mathbf{v}_{k-1}\rangle}{\langle \mathbf{v}_{k-1}, \mathbf{A}\mathbf{v}_{k-1}\rangle}\mathbf{v}_{k-1} \tag{8.61}$$

The step lengths in any descent method are such that the gradient at the kth step is orthogonal to the previous step direction (Figure 8.3), $\langle \nabla\mathbf{F}(\mathbf{x}_k), \mathbf{v}_{k-1}\rangle = 0$. Using this fact together with (8.61), we find that

$$\langle \mathbf{v}_k, \nabla\mathbf{F}(\mathbf{x}_k)\rangle = -\langle \nabla\mathbf{F}(\mathbf{x}_k), \nabla\mathbf{F}(\mathbf{x}_k)\rangle + \frac{\langle \nabla\mathbf{F}(\mathbf{x}_k), \mathbf{A}\mathbf{v}_{k-1}\rangle}{\langle \mathbf{v}_{k-1}, \mathbf{A}\mathbf{v}_{k-1}\rangle}\underbrace{\langle \mathbf{v}_{k-1}, \nabla\mathbf{F}(\mathbf{x}_k)\rangle}_{0}$$

Thus the numerators of the third terms of (8.59) and (8.60) differ only in sign. We use (8.61) together with the **A**-orthogonality of the set $\{\mathbf{v}_k\}$ to determine that

$$\langle \mathbf{v}_k, \mathbf{A}\mathbf{v}_k\rangle = \langle \mathbf{A}\mathbf{v}_k, \mathbf{v}_k\rangle = -\langle \mathbf{A}\mathbf{v}_k, \nabla\mathbf{F}(\mathbf{x}_k)\rangle + \frac{\langle \nabla\mathbf{F}(\mathbf{x}_k), \mathbf{A}\mathbf{v}_{k-1}\rangle}{\langle \mathbf{v}_{k-1}, \mathbf{A}\mathbf{v}_{k-1}\rangle}\underbrace{\langle \mathbf{A}\mathbf{v}_k, \mathbf{v}_{k-1}\rangle}_{0}$$

Consequently, the denominators of the third terms of (8.59) and (8.60) differ only in sign. It follows that the right-hand sides of (8.59) and (8.60) are equal, and therefore, $\mathbf{z}_{k+1} = -\nabla\mathbf{F}(\mathbf{x}_{k+1})$.

We have shown that the conjugate-gradient method contains two parallel orthogonalization processes, the **A**-orthogonalization of the vectors $\{-\nabla\mathbf{F}(\mathbf{x}_0),\ldots,-\nabla\mathbf{F}(\mathbf{x}_{n-1})\}$ to produce $\{\mathbf{v}_0,\ldots,\mathbf{v}_{n-1}\}$ and the standard orthogonalization of the set $\{\mathbf{v}_0,\mathbf{A}\mathbf{v}_0,\ldots,\mathbf{A}\mathbf{v}_{n-2}\}$ to produce $\{-\nabla\mathbf{F}(\mathbf{x}_0),\ldots,-\nabla\mathbf{F}(\mathbf{x}_{n-1})\}$. Consequently, we know that the sequence of negative gradients forms an orthogonal set; our motivation for referring to the method as the conjugate-gradient method is obvious.

In the process of showing the orthogonality of the gradients, we showed in (8.61) that only one Fourier coefficient must be computed in order to obtain $\mathbf{v}_k$ from $-\nabla\mathbf{F}(\mathbf{x}_k)$. We now show that this coefficient can be computed without knowledge of the Hessian matrix **A**. Replacing k by $k-1$ in (8.59), we find

$$\frac{\mathbf{A}\mathbf{v}_{k-1}}{\langle \mathbf{v}_{k-1},\mathbf{A}\mathbf{v}_{k-1}\rangle}=\frac{\nabla\mathbf{F}(\mathbf{x}_{k-1})-\nabla\mathbf{F}(\mathbf{x}_k)}{\langle \mathbf{v}_{k-1},\nabla\mathbf{F}(\mathbf{x}_{k-1})\rangle}$$

We substitute this identity into the Fourier coefficient of (8.61), then apply the identity $\langle \mathbf{v}_{k-1},\ \nabla\mathbf{F}(\mathbf{x}_{k-1})\rangle=-\|\nabla\mathbf{F}(\mathbf{x}_{k-1})\|^2$ (derived previously) to obtain

$$\frac{\langle \nabla\mathbf{F}(\mathbf{x}_k),\mathbf{A}\mathbf{v}_{k-1}\rangle}{\langle \mathbf{v}_{k-1},\mathbf{A}\mathbf{v}_{k-1}\rangle}=\frac{\cancelto{0}{\langle\nabla\mathbf{F}(\mathbf{x}_k),\nabla\mathbf{F}(\mathbf{x}_{k-1})\rangle}-\langle \nabla\mathbf{F}(\mathbf{x}_k),\nabla\mathbf{F}(\mathbf{x}_k)\rangle}{\langle \mathbf{v}_{k-1},\nabla\mathbf{F}(\mathbf{x}_{k-1})\rangle}$$

$$=\frac{-\|\nabla\mathbf{F}(\mathbf{x}_k)\|^2}{-\|\nabla\mathbf{F}(\mathbf{x}_{k-1})\|^2}$$

and

$$\mathbf{v}_k=-\nabla\mathbf{F}(\mathbf{x}_k)+\frac{\|\nabla\mathbf{F}(\mathbf{x}_k)\|^2}{\|\nabla\mathbf{F}(\mathbf{x}_{k-1})\|^2}\mathbf{v}_{k-1}\tag{8.62}$$

Thus $\mathbf{v}_k$ can be computed from the gradients alone.

The Conjugate-Gradient Algorithm

The conjugate-gradient descent consists in successively computing $\mathbf{x}_{k+1}$ from $\mathbf{v}_k$ using (8.55), computing $\nabla\mathbf{F}(\mathbf{x}_{k+1})$, then computing $\mathbf{v}_{k+1}$ from $\nabla\mathbf{F}(\mathbf{x}_{k+1})$ using (8.62). The first step is always a steepest descent step; that is, $\mathbf{v}_0=-\nabla\mathbf{F}(\mathbf{x}_0)$. We determined in the preceding section that the step length given in (8.55) is also the step length which yields a minimum of $\mathbf{F}(\mathbf{x})$ along the direction $\mathbf{v}_k$ from the point $\mathbf{x}_k$. Since the step length as expressed in (8.55) depends upon knowledge of the Hessian matrix **A**, we ordinarily determine the step length by means of a one-dimensional minimization. In summary, the **conjugate-gradient (CG) descent** algorithm is

$$\mathbf{v}_0 = -\nabla \mathbf{F}(\mathbf{x}_0)$$

$$\mathbf{x}_{k+1} = \mathbf{x}_k + \alpha_k \mathbf{v}_k \tag{8.63}$$

$$\mathbf{v}_{k+1} = -\nabla \mathbf{F}(\mathbf{x}_{k+1}) + \frac{\|\nabla \mathbf{F}(\mathbf{x}_{k+1})\|^2}{\|\nabla \mathbf{F}(\mathbf{x}_k)\|^2}\mathbf{v}_k$$

for $k = 0, 1, 2, \ldots$ where α_k is chosen by some one-dimensional minimization technique to minimize $\mathbf{F}(\mathbf{x}_{k+1})$. A computer algorithm corresponding to (8.63) is given in [8.9].

Example 2. Minimization of a Quadratic Function by CG Descent. Let $\mathbf{x} = (\xi_1\ \xi_2)^T$ and $\mathbf{F}(\mathbf{x}) = \frac{1}{2}(\xi_1^2 + 10\xi_2^2)$. Then $\nabla \mathbf{F}(\mathbf{x}) = (\xi_1\ 10\xi_2)^T$. Let $\mathbf{x}_0 = (10\ \ 1)^T$. According to (8.63), $\mathbf{v}_0 = -\nabla \mathbf{F}(\mathbf{x}_0) = (-10\ \ -10)^T$ and $\|\nabla \mathbf{F}(\mathbf{x}_0)\|^2 = 200$. The first CG iteration requires minimization of $\mathbf{F}(\mathbf{x}_1)$, where

$$\mathbf{x}_1 = \mathbf{x}_0 + \alpha_0 \mathbf{v}_0 = \begin{pmatrix} 10 - 10\,\alpha_0 \\ 1 - 10\,\alpha_0 \end{pmatrix}$$

By solving $(d/d\alpha_0)\ \mathbf{F}(\mathbf{x}_1) = 0$ we find $\alpha_0 = 2/11$. Then

$$\mathbf{x}_1 = \tfrac{9}{11}\begin{pmatrix} 10 \\ -1 \end{pmatrix}, \qquad \nabla \mathbf{F}(\mathbf{x}_1) = \tfrac{90}{11}\begin{pmatrix} 1 \\ -1 \end{pmatrix}, \qquad \|\nabla \mathbf{F}(\mathbf{x}_1)\|^2 = \tfrac{16{,}200}{121},$$

and

$$\mathbf{v}_1 = -\nabla \mathbf{F}(\mathbf{x}_1) - \frac{\|\nabla \mathbf{F}(\mathbf{x}_1)\|^2}{\|\nabla \mathbf{F}(\mathbf{x}_0)\|^2}\nabla \mathbf{F}(\mathbf{x}_0) = \tfrac{180}{121}\begin{pmatrix} -10 \\ 1 \end{pmatrix}$$

The second CG iteration requires minimization of $\mathbf{F}(\mathbf{x}_2)$, where

$$\mathbf{x}_2 = \mathbf{x}_1 + \alpha_1 \mathbf{v}_1 = \tfrac{1}{121}\begin{pmatrix} 990 - 1800\,\alpha_1 \\ -99 + 180\,\alpha_1 \end{pmatrix}$$

By solving $(d/d\alpha_1)\ \mathbf{F}(\mathbf{x}_2) = 0$ we find $\alpha_1 = 11/20$. Then $\mathbf{x}_2 = (0\ \ 0)^T$, the minimum of $\mathbf{F}(\mathbf{x})$.

A comparison of Example 2 with Example 1 shows that the DFP algorithm and the CG algorithm produce the same sequence of iterates for the particular function treated in the examples. It can be shown that the two methods *always* produce the same sequence of iterates if the DFP descent is begun with a standard steepest descent step. The DFP algorithm is the less efficient of the two methods in that it requires the computation and storage of the step-direction matrices (8.50). However, the CG method yields only the minimizing point, not the curvature information contained in the inverse Hessian matrix.

The CG descent extends to infinite-dimensional problems more easily than does the DFP descent. To minimize the infinite-dimensional quadratic functional

$$\mathbf{F}(\mathbf{x}) = \tfrac{1}{2}\langle \mathbf{x}, \mathbf{T}\mathbf{x}\rangle - \langle \mathbf{x}, \mathbf{y}\rangle + d$$

where $\mathbf{T}$ is a bounded self-adjoint positive-definite operator on a Hilbert space, we simply carry out the steps described by (8.63). [The derivation of (8.63) applies unchanged except for an occasional replacement of the matrix $\mathbf{A}$ by the operator $\mathbf{T}$.] Of course, the iteration does not terminate in a finite number of steps for an infinite-dimensional $\mathbf{F}(\mathbf{x})$. Daniel [8.6] shows that for an infinite-dimensional quadratic function, the final stages of the CG descent converge to the minimum point $\mathbf{x}_m$ according to

$$\|\mathbf{x}_{k+1} - \mathbf{x}_m\| \leqslant \frac{\sqrt{\lambda_{\mathrm{L}}} - \sqrt{\lambda_{\mathrm{S}}}}{\sqrt{\lambda_{\mathrm{L}}} + \sqrt{\lambda_{\mathrm{S}}}} \|\mathbf{x}_k - \mathbf{x}_m\| \tag{8.64}$$

where λ_{L} and λ_{S} are the largest and smallest eigenvalues of $\mathbf{T}$. Comparison of this convergence ratio with the ratio (8.32) for standard steepest descent shows that the asymptotic convergence rate of the conjugate-gradient method is much higher than that for steepest descent. However, both methods can be improved by scaling the functional $\mathbf{F}$ to reduce the condition number of $\mathbf{T}$, $\lambda_{\mathrm{L}}/\lambda_{\mathrm{S}}$.

Example 3. An Infinite-Dimensional CG Descent. Define $\mathbf{F}$: $\mathcal{L}_2(0,1) \to \mathcal{R}$ by $\mathbf{F}(\mathbf{u}) \triangleq \frac{1}{2}\int_0^1 e^t u^2(t)\,dt$. Then $\nabla\mathbf{F}(\mathbf{u}) = e^t\mathbf{u}$ and the minimum of $\mathbf{F}(\mathbf{u})$ occurs at $\mathbf{u} = \boldsymbol{\theta}$.* Let $\mathbf{u}_0(t) = e^{-t}$. Then $\mathbf{v}_0 = -\nabla\mathbf{F}(\mathbf{u}_0) = -\mathbf{1}$, where $\mathbf{1}$ is the unit function. The first step in the conjugate-gradient descent is

$$\mathbf{u}_1 = \mathbf{u}_0 + \alpha_0\mathbf{v}_0 \quad \text{or} \quad \mathbf{u}_1(t) = e^{-t} - \alpha_0$$

We choose α_0 to minimize $\mathbf{F}(\mathbf{u}_1)$:

$$\frac{d}{d\alpha_0}\mathbf{F}(\mathbf{u}_1) = \frac{d}{d\alpha_0}\frac{1}{2}\int_0^1 e^t\left(e^{-t} - \alpha_0\right)^2 dt$$

$$= -\int_0^1 \left(1 - \alpha_0 e^t\right) dt$$

$$= -1 + \alpha_0(e-1) = 0$$

*In this example we do not distinguish carefully between the exponential function and its value at t.

Thus $\alpha_0 = 1/(e-1)$, $\mathbf{u}_1(t) = e^{-t} - 1/(e-1)$, and $\nabla\mathbf{F}(\mathbf{u}_1) = 1 - e^t/(e-1)$. Furthermore,

$$\|\nabla\mathbf{F}(\mathbf{u}_0)\|^2 = \int_0^1 (1)^2\,dt = 1$$

$$\|\nabla\mathbf{F}(\mathbf{u}_1)\|^2 = \int_0^1 \left(1 - e^t/(e-1)\right)^2 dt = \frac{3-e}{2(e-1)}$$

Consequently,

$$\begin{aligned}\mathbf{v}_1(t) &= -\left(1 - e^t/(e-1)\right) - (3-e)/2(e-1)\\ &= \frac{e^t}{e-1} - \frac{e+1}{2(e-1)}\end{aligned}$$

These first two CG steps serve to demonstrate the nature of the CG descent method in a function space. The integrations required to carry out the iteration are usually performed numerically by computer. The second CG step direction, $\mathbf{v}_1$, differs only slightly from the standard steepest-descent step direction, $-\nabla\mathbf{F}(\mathbf{u}_1) = (e^t/(e-1)) - 1$. The improved convergence rate of the CG descent does not become apparent until a number of steps are taken.

Although a quadratic **F** was assumed in the derivation of the conjugate-gradient method, the algorithm (8.63) can be carried out for any smooth **F**. If **F** is n-dimensional and nearly quadratic, we should expect the procedure to yield a reasonable approximation to the minimum of **F** in n steps. To obtain a closer approximation to the minimum, we need only repeat the iteration, starting again with a steepest descent step. It is not immediately evident what the consequences will be if we apply (8.63) to a very nonquadratic **F**. Will the iteration be a descent? We show that it is. At the kth step of the iteration (8.63), we have

$$\begin{aligned}\langle \mathbf{v}_k, \nabla\mathbf{F}(\mathbf{x}_k)\rangle &= \left\langle -\nabla\mathbf{F}(\mathbf{x}_k) + \frac{\|\nabla\mathbf{F}(\mathbf{x}_k)\|^2}{\|\nabla\mathbf{F}(\mathbf{x}_{k-1})\|^2}\mathbf{v}_{k-1}, \nabla\mathbf{F}(\mathbf{x}_k)\right\rangle\\ &= -\|\nabla\mathbf{F}(\mathbf{x}_k)\|^2 + \frac{\|\nabla\mathbf{F}(\mathbf{x}_k)\|^2}{\|\nabla\mathbf{F}(\mathbf{x}_{k-1})\|^2}\underbrace{\langle \mathbf{v}_{k-1}, \nabla\mathbf{F}(\mathbf{x}_k)\rangle}_{0}\end{aligned}$$

The second term is zero because the step length α_{k-1} is chosen to minimize $\mathbf{F}(\mathbf{x})$ along the step path $\mathbf{v}_{k-1}$ as shown in Figure 8.3. Therefore, $\langle \mathbf{v}_k, \nabla\mathbf{F}(\mathbf{x}_k)\rangle < 0$ for $\nabla\mathbf{F}(\mathbf{x}_k) \neq \boldsymbol{\theta}$, and by (8.22), $\mathbf{v}_k$ is a descent direction. We conclude that it is appropriate to use the CG descent (8.63) to minimize any smooth functional. Of course, we should not expect the gradients to be orthogonal for a nonquadratic functional. Nor should we expect the iteration to converge in n steps if $\mathbf{F}(\mathbf{x})$ is n-dimensional. In point of fact, Daniel [8.6] shows that the convergence of the conjugate-gradient method to the minimum of an n-dimensional function is $(1/n)$-quadratic in the sense that n CG steps achieve the effect of one step of a quadratically convergent descent (such as the Newton descent).

Experience in applying the CG descent to nonquadratic functions shows that the method converges quickly and somewhat more efficiently than the DFP method for well-behaved functions. However, for functions that have an ill-conditioned Hessian matrix, the DFP method seems to be the more successful of the two [8.23]. The success of the CG method depends upon the first step being a standard steepest descent step. After n steps, the successive step directions $\mathbf{v}_k$ become nearly parallel, and the identity of the initial step is lost. Consequently, Fletcher and Reeves recommend that the algorithm (8.63) be restarted with a steepest descent step after each $n+1$ iterations. We treat a nonquadratic example in P&C 8.13.

Accelerated Steepest Descent

The last minimization technique that we describe is an intuitively simple scheme for accelerating the convergence of standard steepest descent. Observation of a steepest descent iteration for any two-dimensional quadratic function shows that a straight line can be passed through all the even (or odd) iterates. This line also passes through the minimum of the quadratic (see Figure 8.4). If after the first two steepest descent steps we take the third step in the direction of the line from $\mathbf{x}_0$ through $\mathbf{x}_2$, the third step will terminate at the minimum. Forsythe and Motzkin [8.13] suggest the occasional use of such an "acceleration step" as a means for increasing the rate of convergence of the steepest descent method for n-dimensional nonquadratic functions. Shah, Buehler, and Kempthorne [8.32] incorporate this acceleration concept into an algorithm which converges to the minimum of any n-dimensional quadratic function in no more than $(2n-1)$ steps. Their algorithm is as follows. For notational convenience they begin with $\mathbf{x}_1 = \mathbf{x}_0$. Then

$$\begin{aligned}
\mathbf{x}_2 &= \mathbf{x}_0 - \alpha_0 \nabla \mathbf{F}(\mathbf{x}_0) \\
\mathbf{x}_{k+1} &= \mathbf{x}_k + \alpha_k \mathbf{v}_k \qquad k = 2,3,4,\ldots \\
\mathbf{v}_k &= -\nabla \mathbf{F}(\mathbf{x}_k), \qquad k = 2,4,6,\ldots \\
&= (\mathbf{x}_k - \mathbf{x}_{k-2}), \qquad k = 3,5,7,\ldots
\end{aligned} \tag{8.65}$$

At each step the scalar α_k is chosen to minimize $\mathbf{F}(\mathbf{x}_{k+1})$. We refer to (8.65) as **accelerated steepest descent** (in [8.32] the algorithm is called (PARTAN)). The algorithm is illustrated abstractly in Figure 8.6.

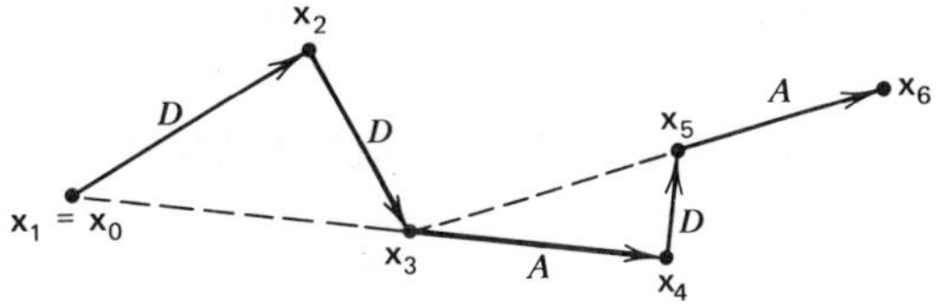

Figure 8.6. Accelerated steepest descent: the steepest descent steps are labeled D; the acceleration steps are labeled A.

Shah, Buehler, and Kempthorne [8.32] show that for an n-dimensional quadratic function the sequence of iterates $x_0, x_2, x_4, \ldots$ is identical to the full sequence of iterates generated by the conjugate-gradient descent, (8.63). Between each pair of even-numbered iterates of (8.65) (except the first) are a steepest descent step and an acceleration step. Since the conjugate-gradient descent takes no more than n steps to reach the minimum of the n-dimensional quadratic, the accelerated steepest descent takes no more than $(2n-1)$ steps.

Exercise 2. Verify for the two-dimensional quadratic $\mathbf{F}$ and the starting vector $\mathbf{x}_0$ of Example 2 that accelerated steepest descent generates the conjugate-gradient iterates.

For a nonquadratic function we merely continue the iteration (8.65) until changes from iteration to iteration are considered insignificant. The first two steepest descent steps for a nonquadratic function are shown in Figure 8.5 (Section 8.2). Graphical determination of the first acceleration step is easy. For this example, the initial acceleration step compares favorably with the first Newton descent step. Experience has shown that accelerated steepest descent is an effective minimization technique even for functions with narrow curved valleys. (A numerical example is contained in [8.32].) The iteration (8.65) also can be used as it stands for minimization of infinite-dimensional functionals.

8.4 The Gradient Projection Method

Sections 8.2 and 8.3 explore a variety of descent techniques for minimizing functionals without constraints. We now turn to constrained minimization problems. In this section we examine a concept that is well suited for minimization with linear constraints—gradient projection, introduced by Rosen [8.28]. Nonlinear constraints are handled more easily by the penalty function approach which we discuss in Section 8.5. Problems which contain both linear and nonlinear constraints can be treated by a combination of the two approaches.

Let $\mathbf{x}=(\xi_1\ \xi_2)^{\mathrm{T}}$. Suppose we wish to

$$
\begin{aligned}
&\text{minimize} \quad && \mathbf{F}(\mathbf{x}) \stackrel{\Delta}{=} \tfrac{1}{2}(\xi_1^2+10\xi_2^2) \\
&\text{with} && \mathbf{T}_1(\mathbf{x}) \stackrel{\Delta}{=} \xi_1+\xi_2 \geqslant 2 \\
& && \mathbf{T}_2(\mathbf{x}) \stackrel{\Delta}{=} \xi_1 \geqslant 0 \\
& && \mathbf{T}_3(\mathbf{x}) \stackrel{\Delta}{=} \xi_2 \geqslant 0
\end{aligned}
\tag{8.66}
$$

See Figure 8.7. Were it not for the constraints, we could minimize $\mathbf{F}(\mathbf{x})$ by any of the descent methods described in Sections 8.2 and 8.3. In the gradient projection method we begin with a point $\mathbf{x}_0$ in the feasible region and proceed according to one of the descent methods until a step strikes the boundary of the feasible region. Rather than allow the iterates to leave the feasible region, we terminate the step at the boundary. We allow the succeeding steps to descend only within the confines of the feasible region. In order to simplify the presentation of the gradient projection method, we use steepest descent as the underlying unconstrained descent method throughout Section 8.4.

Because the constraint functions $\mathbf{T}_1$, $\mathbf{T}_2$, and $\mathbf{T}_3$ of (8.66) are linear, the constraint surfaces and their intersections are hyperplanes. Consequently the "locally best" step direction *in the presence of the constraints* is the orthogonal projection of the negative gradient onto the constraining hyperplane (P&C 8.9). We use this projected negative gradient as the next step direction. We repeat the iteration until there is no longer a feasible direction of descent. Because we guarantee feasibility and descent at each step, we expect convergence to a minimum. The procedure may require an

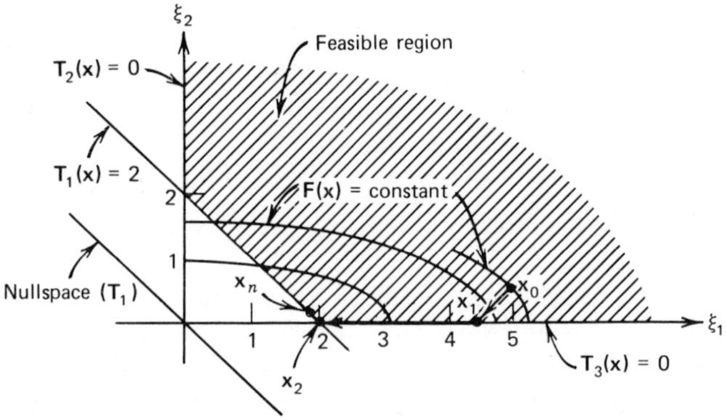

Figure 8.7. Minimization by the gradient projection method.

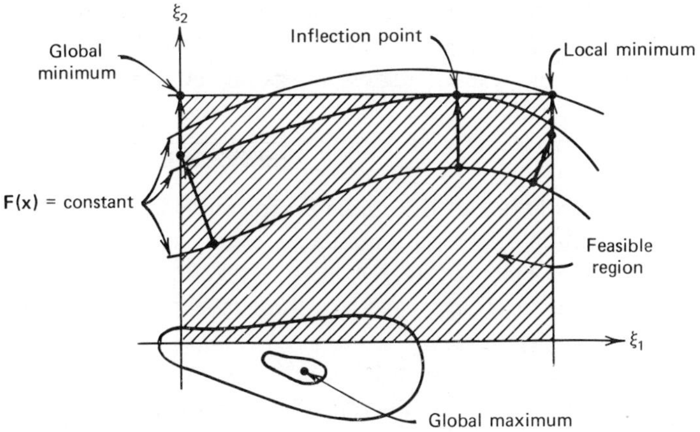

Figure 8.8. Possible outcomes of the gradient projection method.

infinite sequence of steps to achieve full convergence, as is the case in the absence of constraints. The constrained descent may find a global constrained minimum $\mathbf{x}_m$ as in Figure 8.7; it seldom fails to produce at least a local constrained minimum. However, as demonstrated in Figure 8.8, a constrained horizontal inflection point is also a possible outcome.

Although the gradient projection method uses the constraints directly, rather than including their effects by means of the Lagrange multiplier equation (7.28), the Lagrange multipliers are still inherent in the minimization problem. In point of fact, we will find that the method requires the computation of Lagrange multipliers at certain points in the iteration in order to make decisions concerning the succeeding steps. As we introduce the details of the gradient projection method, we focus primarily on finite-dimensional examples. However, the method works as well for minimization of infinite-dimensional functionals subject to linear constraints.

Projecting Onto the Constraining Boundary

Suppose we wish to solve the following minimization problem:

$$\begin{aligned} &\text{minimize} \quad \mathbf{F}(\mathbf{x}) \\ &\text{subject to} \quad \mathbf{T}_i(\mathbf{x}) \geqslant c_i, \quad i=1,\ldots,m \end{aligned} \tag{8.67}$$

where $\mathbf{F}$ and $\mathbf{T}_i$ are functionals acting on a Hilbert space $\mathcal{V}$ with inner product $\langle \cdot, \cdot \rangle$, and $\mathbf{T}_i$ is linear. Each of the equations $\mathbf{T}_i(\mathbf{x}) = c_i$ defines a hyperplane of co-dimension one in $\mathcal{V}$. Parallel to the ith hyperplane is a subspace, nullspace($\mathbf{T}_i$), also of co-dimension one. The feasible region, defined by the set $\{\mathbf{T}_i(\mathbf{x}) \geqslant c_i\}$, forms a polyhedron in $\mathcal{V}$ (Figure 8.7). A boundary point of the feasible region is a point for which at least one of the constraint inequalities is an equality. Thus the boundary consists in portions of the intersections of various of the individual constraint hyperplanes.

Let $\mathbf{x}_k$ be the kth iterate of the gradient projection sequence. Suppose $\mathbf{x}_k$ lies in the intersection of a certain number p of the constraint hyperplanes. For notational convenience we *renumber the constraints* so that the p constraints defining the intersection are the first p constraints in the set. Define the subspace

$$\mathcal{W}_p \triangleq \text{nullspace}(\mathbf{T}_1) \cap \cdots \cap \text{nullspace}(\mathbf{T}_p) \tag{8.68}$$

Then the constraint intersection in which $\mathbf{x}_k$ lies is the hyperplane $\mathbf{x}_k + \mathcal{W}_p$. Suppose it is possible to descend from $\mathbf{x}_k$ without leaving the constraint hyperplane $\mathbf{x}_k + \mathcal{W}_p$; that is, suppose there are vectors in $\mathcal{W}_p$ which specify directions of descent from $\mathbf{x}_k$. (This situation occurs at $\mathbf{x}_1$ in Figure 8.7; $\mathcal{W}_p$ is the ξ_1 axis, and $p = 1$.) According to P&C 8.9 the greatest rate of decrease in $\mathbf{F}(\mathbf{x})$, subject to the condition that the step vector lie in a subspace $\mathcal{W}_p$, is obtained by stepping in the direction of the orthogonal projection of $-\nabla\mathbf{F}(\mathbf{x}_k)$ on $\mathcal{W}_p$; that is, the projection on $\mathcal{W}_p$ along $\mathcal{W}_p^{\perp}$. If we denote the orthogonal projector on $\mathcal{W}_p$ by $\mathbf{P}_{\mathcal{W}}$ and the projector on $\mathcal{W}_p^{\perp}$ by $\mathbf{P}_{\mathcal{W}^{\perp}}$, then the step direction vector is

$$\begin{aligned}\mathbf{P}_{\mathcal{W}}(-\nabla\mathbf{F}(\mathbf{x}_k)) &= (\mathbf{I} - \mathbf{P}_{\mathcal{W}^{\perp}})(-\nabla\mathbf{F}(\mathbf{x}_k))\\ &= -\nabla\mathbf{F}(\mathbf{x}_k) + \mathbf{P}_{\mathcal{W}^{\perp}}(\nabla\mathbf{F}(\mathbf{x}_k))\end{aligned} \tag{8.69}$$

Because $\mathbf{T}_i$ is linear, $\nabla\mathbf{T}_i(\mathbf{x})$ does not depend on $\mathbf{x}$, and $\mathbf{T}_i\mathbf{x} = \langle \mathbf{x}, \nabla\mathbf{T}_i \rangle$. As a result, $\nabla\mathbf{T}_i$ is orthogonal to nullspace($\mathbf{T}_i$), and

$$\mathcal{W}_p^{\perp} = \text{span}\{\nabla\mathbf{T}_1, \ldots, \nabla\mathbf{T}_p\} \tag{8.70}$$

Consequently, we can perform the orthogonal projection of $-\nabla\mathbf{F}(\mathbf{x}_k)$ on $\mathcal{W}_p$ by applying the Gram–Schmidt procedure to the set $\{\nabla\mathbf{T}_1, \ldots, \nabla\mathbf{T}_p, -\nabla\mathbf{F}(\mathbf{x}_k)\}$; that is, we orthogonalize the set $\{\nabla\mathbf{T}_i\}$, project $-\nabla\mathbf{F}(\mathbf{x}_k)$ onto the orthogonalized set to obtain $\mathbf{P}_{\mathcal{W}^{\perp}}(-\nabla\mathbf{F}(\mathbf{x}_k))$, then subtract the result from $-\nabla\mathbf{F}(\mathbf{x}_k)$ to get $\mathbf{P}_{\mathcal{W}}(-\nabla\mathbf{F}(\mathbf{x}_k))$.

Example 1. Orthogonal Projection on $\mathcal{W}_p$. Let $\mathbf{F}$ and $\mathbf{T}_i$ of (8.67) be as defined in (8.66), where $\mathcal{V} = \mathfrak{M}^{2\times 1}$. Assume the standard inner product. Then $\nabla\mathbf{F}(\mathbf{x}) = (\xi_1 \; 10\xi_2)^{\mathrm{T}}$. According to Figure 8.7, where $\mathbf{x}_0 = (5 \; \frac{1}{2})^{\mathrm{T}}$, the first steepest descent step produces $\mathbf{x}_1 = (4.5 \; 0)^{\mathrm{T}}$, a point on the single constraint, $\mathbf{T}_3(\mathbf{x}) = 0$. From the point $\mathbf{x}_1$

it is possible to reduce $\mathbf{F}(\mathbf{x})$ while moving along the constraint. We show this fact mathematically. Note that $p=1$ and $\nabla\mathbf{T}_3=(0\ 1)^{\mathrm{T}}$. (We do not bother to renumber the constraints.) Orthogonalization of the set $\{\nabla\mathbf{T}_3\}$ is trivial; we merely define $\mathbf{z}_1=\nabla\mathbf{T}_3=(0\ 1)^{\mathrm{T}}$. The negative gradient is $-\nabla\mathbf{F}(\mathbf{x}_1)=(-4.5\ 0)^{\mathrm{T}}$. The orthogonal projection of $-\nabla\mathbf{F}(\mathbf{x}_1)$ on the set $\{\mathbf{z}_1\}$ is

$$\mathbf{P}_{\mathcal{W}^\perp}(-\nabla\mathbf{F}(\mathbf{x}_1))=\frac{\langle-\nabla\mathbf{F}(\mathbf{x}_1),\mathbf{z}_1\rangle}{\|\mathbf{z}_1\|^2}\mathbf{z}_1=\tfrac{0}{1}\begin{pmatrix}0\\1\end{pmatrix}=\boldsymbol{\theta}$$

Consequently, $\mathbf{P}_{\mathcal{W}}(-\nabla\mathbf{F}(\mathbf{x}_1))=-\nabla\mathbf{F}(\mathbf{x}_1)=(-4.5\ 0)^{\mathrm{T}}$.

A Pseudoinverse Approach to Projection

In many presentations of the gradient projection method, $\mathbf{P}_{\mathcal{W}}$ and $\mathbf{P}_{\mathcal{W}^\perp}$ are expressed in term of pseudoinverses. The relationship between pseudoinverse computations and Gram–Schmidt orthogonalization is explored thoroughly in Chapter 6. We will review this relationship for finite-dimensional problems. Let $\mathcal{V}=\mathfrak{M}^{n\times 1}$. We can use the gradients of the p contraints which pass through $\mathbf{x}_k$ to define the $n\times p$ matrix

$$\mathbf{A}_p=\left(\nabla\mathbf{T}_1\,\vdots\,\cdots\,\vdots\,\nabla\mathbf{T}_p\right) \tag{8.71}$$

Then $\mathcal{W}_p^\perp$ consists in linear combinations of the columns of $\mathbf{A}_p$, or $\mathcal{W}_p^\perp=\text{range}(\mathbf{A}_p)$. In order to compute (8.69), we seek $\mathbf{P}_{\mathcal{W}^\perp}$, the orthogonal projector on $\mathcal{W}_p^\perp$, and its complement, $\mathbf{P}_{\mathcal{W}}$. Let $\mathbf{A}_p^\dagger$ denote the pseudoinverse of $\mathbf{A}_p$ relative to the inner product on $\mathfrak{M}^{n\times 1}$; let $\mathbf{y}$ be an arbitrary vector in $\mathfrak{M}^{n\times 1}$. According to the definition of the pseudoinverse (Section 6.5), the vector $\mathbf{x}=\mathbf{A}_p^\dagger\mathbf{y}$ minimizes $\|\mathbf{y}-\mathbf{A}_p\mathbf{x}\|$, and $\mathbf{A}_p\mathbf{x}$ is the orthogonal projection of $\mathbf{y}$ on range$(\mathbf{A}_p)$. Therefore,

$$\mathbf{P}_{\mathcal{W}^\perp}\mathbf{y}=\mathbf{A}_p\mathbf{x}=\mathbf{A}_p\mathbf{A}_p^\dagger\mathbf{y} \tag{8.72}$$

Thus $\mathbf{P}_{\mathcal{W}^\perp}$ consists in multiplication by the matrix $\mathbf{A}_p\mathbf{A}_p^\dagger$, and $\mathbf{P}_{\mathcal{W}}$ consists in multiplication by $\mathbf{I}-\mathbf{A}_p\mathbf{A}_p^\dagger$. Of course, to actually compute $\mathbf{A}_p^\dagger$ we may revert to the projection technique discussed at the end of Section 6.6.

Exercise 1. Define an operator $\mathbf{U}$: $\mathcal{V}\to\mathcal{R}^p$ which is a Hilbert space analogue of multiplication by $\mathbf{A}_p$. Show that $\mathcal{W}_p^\perp=\text{range}(\mathbf{U})$, $\mathbf{P}_{\mathcal{W}^\perp}=\mathbf{U}\mathbf{U}^\dagger$, and $\mathbf{P}_{\mathcal{W}}=\mathbf{I}-\mathbf{U}\mathbf{U}^\dagger$.

Example 2. Projection on $\mathcal{W}_p$ by Pseudoinverse Computations. Consider the two-dimensional problem of Example 1 and Figure 8.7, where the single constraint $\mathbf{T}_3\mathbf{x}=\xi_2=0$ is active at the iterate $\mathbf{x}_1=(4.5\ 0)^{\mathrm{T}}$. The matrix (8.71) is

$$\mathbf{A}_1=(\nabla\mathbf{T}_3)=\begin{pmatrix}0\\1\end{pmatrix}$$

Since the inner product is standard, the pseudoinverse is given by (6.39):

$$\mathbf{A}_1{}^{\dagger}=(\mathbf{A}_1^{\mathrm{T}}\mathbf{A}_1)^{-1}\mathbf{A}_1^{\mathrm{T}}=(0\ 1)$$

Then, according to (8.72), we have

$$\mathbf{P}_{\mathscr{U}^{\perp}}\mathbf{y}=\mathbf{A}_1\mathbf{A}_1{}^{\dagger}\mathbf{y}=\begin{pmatrix}0\\1\end{pmatrix}(0\ 1)\mathbf{y}=\begin{pmatrix}0&0\\0&1\end{pmatrix}\mathbf{y}$$

$$\mathbf{P}_{\mathscr{U}}\mathbf{y}=(\mathbf{I}-\mathbf{A}_1\mathbf{A}_1{}^{\dagger})\mathbf{y}=\begin{pmatrix}1&0\\0&0\end{pmatrix}\mathbf{y}$$

and the best constrained descent direction from $\mathbf{x}_1$ is

$$\mathbf{P}_{\mathscr{U}}(-\nabla\mathbf{F}(\mathbf{x}_1))=\begin{pmatrix}1&0\\0&0\end{pmatrix}\begin{pmatrix}-4.5\\0\end{pmatrix}=\begin{pmatrix}-4.5\\0\end{pmatrix}$$

Step Length

Using the best constrained descent direction (8.69), we perform the descent step

$$\mathbf{x}_{k+1}=\mathbf{x}_k-\alpha_k\mathbf{P}_{\mathscr{U}}(\nabla\mathbf{F}(\mathbf{x}_k)) \tag{8.73}$$

We must determine the value of α_k which minimizes $\mathbf{F}(\mathbf{x}_{k+1})$ subject to the constraints. We can find the value of α_k which yields the *unconstrained* minimum of $\mathbf{F}(\mathbf{x}_{k+1})$ by means of a one-dimensional minimization as we do for unconstrained steepest descent [see (8.26)]. On the other hand, the value of α_k which places $\mathbf{x}_{k+1}$ on the jth constraint surface satisfies

$$\begin{aligned}\mathbf{T}_j(\mathbf{x}_{k+1})&=\langle\mathbf{x}_{k+1},\nabla\mathbf{T}_j\rangle\\&=\langle\mathbf{x}_k,\nabla\mathbf{T}_j\rangle-\alpha_k\langle\mathbf{P}_{\mathscr{U}}(\nabla\mathbf{F}(\mathbf{x}_k)),\nabla\mathbf{T}_j\rangle\\&=c_j\end{aligned}$$

or

$$\alpha_k=\frac{\mathbf{T}_j(\mathbf{x}_k)-c_i}{\langle\mathbf{P}_{\mathscr{U}}(\nabla\mathbf{F}(\mathbf{x}_k)),\nabla\mathbf{T}_j\rangle} \tag{8.74}$$

To determine the correct value of step length* α_k, we compute the value of α_k as given by (8.74) for each constraint which does not pass through $\mathbf{x}_k$; we compute the value of α_k which minimizes $\mathbf{F}(\mathbf{x}_{k+1})$ in the absence of

*We use the term "step length" rather loosely; α_k is proportional to, but not equal to the length of the step vector, $\|-\alpha_k\mathbf{P}_{\mathscr{U}}(\nabla\mathbf{F}(\mathbf{x}_k))\|$.

constraints; finally, we accept as the correct step length the smallest of the non-negative values (a negative value corresponds to an ascent step).

Example 3. Determination of Step Length. Once again, we turn to the two-dimensional problem of (8.66) and Figure 8.7. From Example 1, $\mathbf{x}_1 = (4.5 \; 0)^T$ and the projected gradient is $\mathbf{P}_{\mathcal{U}}(\nabla\mathbf{F}(\mathbf{x}_1)) = (4.5 \; 0)^T$. Therefore, (8.73) becomes

$$\mathbf{x}_2 = \mathbf{x}_1 - \alpha_1 \mathbf{P}_{\mathcal{U}}(\nabla\mathbf{F}(\mathbf{x}_1))$$

$$= \begin{pmatrix} 4.5 \\ 0 \end{pmatrix} - \alpha_1 \begin{pmatrix} 4.5 \\ 0 \end{pmatrix}$$

From Figure 8.7 it is clear that the unconstrained minimum of $\mathbf{F}(\mathbf{x}_2)$ occurs for $\alpha_1 = 1$ (or $\mathbf{x}_2 = \boldsymbol{\theta}$); this fact can be verified mathematically by solving $d\mathbf{F}(\mathbf{x}_2)$ $d\alpha_1 = 0$. Comparison of (8.66) with (8.67) shows that $\mathbf{T}_1(\mathbf{x}_1) = 4.5$ and $c_1 = 2$. According to (8.74), the value of α_1 which places $\mathbf{x}_2$ on constraint 1 is

$$\alpha_1 = \frac{\mathbf{T}_1(\mathbf{x}_1) - c_1}{\langle \mathbf{P}_{\mathcal{U}}(\nabla\mathbf{F}(\mathbf{x}_1)), \nabla\mathbf{T}_1 \rangle} = \frac{4.5 - 2}{\begin{pmatrix} 4.5 \\ 0 \end{pmatrix}^T \begin{pmatrix} 1 \\ 1 \end{pmatrix}} = \frac{5}{9}$$

Similarly, $\mathbf{x}_2$ will lie on constraint 2 if

$$\alpha_1 = \frac{4.5 - 0}{\begin{pmatrix} 4.5 \\ 0 \end{pmatrix}^T \begin{pmatrix} 1 \\ 0 \end{pmatrix}} = 1$$

The smallest of these prospective step lengths is $\alpha_1 = \frac{5}{9}$. Thus the size of the step is restricted by constraint 1, and $\mathbf{x}_2 = (2 \; 0)^T$; this result is verified by Figure 8.7.

If we are confident that the one-dimensional function $\mathbf{F}(\mathbf{x}_{k+1}(\alpha_k))$ is concave (has a single minimum as a function of α_k) in the feasible region, we may be able to avoid the one-dimensional unconstrained minimization required to find α_k. First determine the constrained values of α_k given in (8.74); the smallest (non-negative) of these determines the *largest feasible step*. Let $\hat{\mathbf{x}}_{k+1}$ denote the new iterate produced by this step, and check the inner product

$$\langle \nabla\mathbf{F}(\hat{\mathbf{x}}_{k+1}), \mathbf{P}_{\mathcal{U}}(\nabla\mathbf{F}(\mathbf{x}_k)) \rangle \tag{8.75}$$

If (8.75) is zero, then, according to (8.25), $\hat{\mathbf{x}}_{k+1}$ is also the one-dimensional unconstrained minimum of $\mathbf{F}(\mathbf{x}_{k+1}(\alpha_k))$. If (8.75) is positive, the one-dimensional unconstrained minimum is beyond the point $\hat{\mathbf{x}}_{k+1}$ along the step direction $\mathbf{P}_{\mathcal{U}}(-\nabla\mathbf{F}(\mathbf{x}_k)$. In either case, we let $\mathbf{x}_{k+1} = \hat{\mathbf{x}}_{k+1}$. On the other hand, if (8.75) is negative, the one-dimensional unconstrained minimum of $\mathbf{F}(\mathbf{x}_{k+1}(\alpha_k))$ lies between $\mathbf{x}_k$ and $\hat{\mathbf{x}}_{k+1}$. Thus if (8.75) is negative, we must perform a one-dimensional search in order to determine $\mathbf{x}_{k+1}$. In Example 3, the largest feasible step produces $\hat{\mathbf{x}}_2=(2\ 0)^{\mathrm{T}}$, corresponding to the smallest constrained step length, $\alpha_1=5/9$. The inner product (8.75) is

$$\langle \nabla\mathbf{F}(\mathbf{x}_2), \mathbf{P}_{\mathcal{W}}(\nabla\mathbf{F}(\mathbf{x}_1))\rangle = \begin{pmatrix}2\\0\end{pmatrix}^{\mathrm{T}}\begin{pmatrix}4.5\\0\end{pmatrix} = 9 > 0$$

Therefore, $\mathbf{x}_2=\hat{\mathbf{x}}_2=(2\ 0)^{\mathrm{T}}$, as we found previously.

The Lagrange Multipliers and Optimality

To this point we have determined how to project the negative gradient onto a constraint hyperplane in order to determine a feasible descent direction which is best in a local sense. We have also determined how to find the largest feasible step length in a given feasible descent direction. We must now determine how to recognize a constrained minimum of $\mathbf{F}(\mathbf{x})$. Assume the iterate $\mathbf{x}_k$ lies at the intersection of p constraints; we renumber these constraints so that they are the first p constraints in the set; that is, $\mathbf{T}_i(\mathbf{x}_k)=c_i$, $i=1,\ldots,p$. The remaining constraints are satisfied as inequalities, $\mathbf{T}_i(\mathbf{x}_k)>c_i$, $i=p+1,\ldots,m$. Assume the constraints are independent; that is, $\{\nabla\mathbf{T}_i, i=1,\ldots,p\}$ is an independent set. According to the Kuhn–Tucker conditions (7.58), if $\mathbf{x}_k$ is a constrained minimum then

$$\begin{aligned} \nabla\mathbf{F}(\mathbf{x}_k) &= \sum_{i=1}^{p} \lambda_i \nabla\mathbf{T}_i \\ \lambda_i \geqslant 0, &\qquad i=1,\ldots,p \end{aligned} \tag{8.76}$$

The key to recognition of a constrained minimum is an investigation of the conditions (8.76). By (8.70), if $\nabla\mathbf{F}(\mathbf{x}_k)=\sum_{i=1}^{p}\lambda_i\nabla\mathbf{T}_i$, then $\nabla\mathbf{F}(\mathbf{x}_k)$ is in $\mathcal{W}_p^{\perp}$, and the projected negative gradient is $\mathbf{P}_{\mathcal{W}}(-\nabla\mathbf{F}(\mathbf{x}_k))=\boldsymbol{\theta}$. Thus a straightforward determination of the projection of the negative gradient on the hyperplane of constraints which pass through the point $\mathbf{x}_k$ will disclose whether or not the Lagrange multiplier equation is satisfied at $\mathbf{x}_k$; it is satisfied if and only if the projected gradient is $\boldsymbol{\theta}$.

Suppose the Lagrange multiplier equation is satisfied at $\mathbf{x}_k$, but one of the multipliers is negative. Then according to (8.76), $\mathbf{x}_k$ is not a constrained minimum. More specifically, the constraint corresponding to the negative multiplier is not binding. If we treat it as binding, we allow it to prevent a feasible descent step. Thus even though $\mathbf{x}_k$ lies on a nonbinding constraint, we should ignore that constraint in the determination of $\mathcal{W}_p$ and the projected gradient. An example of this situation occurs at the iterate $\mathbf{x}_2=(2\ 0)^{\mathrm{T}}$ in the problem of (8.66) and Figure 8.7. The point $\mathbf{x}_2$ lies on the pair of constraints $\mathbf{T}_1(\mathbf{x})=2$ and $\mathbf{T}_3(\mathbf{x})=0$, and $p=2$. (Note that we have avoided renumbering the constraints in order to keep consistency with Figure 8.7.) According to (8.70), we define

$$\mathcal{W}_2^{\perp}=\text{span}\{\nabla\mathbf{T}_1,\nabla\mathbf{T}_3\}=\text{span}\left\{\begin{pmatrix}1\\1\end{pmatrix},\begin{pmatrix}0\\1\end{pmatrix}\right\}$$

Since $\mathcal{W}_2^{\perp}$ is the whole two-dimensional space, $\mathcal{W}_2=\{\boldsymbol{\theta}\}$, and the projected negative gradient is $\mathbf{P}_{\mathcal{W}}(\nabla\mathbf{F}(\mathbf{x}_2))=\boldsymbol{\theta}$. The Lagrange multiplier equation at $\mathbf{x}_2$ is

$$\nabla\mathbf{F}(\mathbf{x}_2)=\begin{pmatrix}2\\0\end{pmatrix}=\lambda_1\begin{pmatrix}1\\1\end{pmatrix}+\lambda_3\begin{pmatrix}0\\1\end{pmatrix}=\lambda_1\nabla\mathbf{T}_1+\lambda_3\nabla\mathbf{T}_3$$

The solution is $\lambda_1=2$ and $\lambda_3=-2$. The negative multiplier associated with constraint 3 implies that that constraint (as an equality) prevents further feasible descent. This implication is verified by Figure 8.7. Therefore, we should ignore constraint 3 in the determination of the descent step from the point $\mathbf{x}_2$.

It is possible that a single nonbinding constraint will result in more than one negative Lagrange multiplier. On the other hand, several nonbinding constraints may result in only a single negative Lagrange multiplier. Thus we must be careful in our use of negative Lagrange multipliers as indicators of the nonbinding constraints. However, if treating all the constraints which pass through $\mathbf{x}_k$ as binding constraints prevents a feasible descent step [i.e., $\mathbf{P}_{\mathcal{W}}(-\nabla\mathbf{F}(\mathbf{x}_k))=\boldsymbol{\theta}$], and if at least one of the Lagrange multipliers is negative, then in order to obtain the freedom to take a descent step it is sufficient to ignore any single constraint for which the associated Lagrange multiplier is negative. It is perhaps most appropriate to ignore the constraint associated with the most negative Lagrange multiplier; it is that constraint which most hampers the descent.

Example 4. Interpreting the Lagrange Multipliers. Consider a minimization in $\mathcal{M}^{2\times 1}$ with the standard inner product. Let $\mathbf{x}\triangleq(\xi_1\ \xi_2)^{\mathrm{T}}$ denote a vector in $\mathcal{M}^{2\times 1}$. Suppose $\mathbf{x}_k=\boldsymbol{\theta}$ and $-\nabla\mathbf{F}(\boldsymbol{\theta})=(0\ 1)^{\mathrm{T}}$. Suppose further that the following constraints pass through $\mathbf{x}_k=\boldsymbol{\theta}$:

$$\mathbf{T}_1(\mathbf{x})\triangleq\xi_1+\xi_2\geqslant 0\quad\text{and}\quad\mathbf{T}_2(\mathbf{x})\triangleq-5\xi_1-\xi_2\geqslant 0$$

The Lagrange multiplier equation (assuming both constraints are binding) is

$$\nabla\mathbf{F}(\boldsymbol{\theta})=\begin{pmatrix}0\\-1\end{pmatrix}=\lambda_1\begin{pmatrix}1\\1\end{pmatrix}+\lambda_2\begin{pmatrix}-5\\-1\end{pmatrix}=\lambda_1\nabla\mathbf{T}_1+\lambda_2\nabla\mathbf{T}_2$$

The solution is $\lambda_1 = -5/4$ and $\lambda_2 = -1/4$. According to the above paragraph, we can obtain a nonzero projected gradient if we ignore either one of the constraints and treat the other as binding. The correctness of this conclusion is verified by Figure 8.9*a*.

Suppose, on the other hand, that the constraints which pass through $\mathbf{x}_k = \boldsymbol{\theta}$ are $T_1(\mathbf{x}) \stackrel{\Delta}{=} \xi_1 + \xi_2 \geqslant 0$ and $T_2(\mathbf{x}) \stackrel{\Delta}{=} 5\xi_1 + \xi_2 \geqslant 0$. Then the Lagrange multiplier equation and its solution (assuming both constraints are binding) are

$$\begin{pmatrix} 0 \\ -1 \end{pmatrix} = \lambda_1 \begin{pmatrix} 1 \\ 1 \end{pmatrix} + \lambda_2 \begin{pmatrix} 5 \\ 1 \end{pmatrix}$$

$$\lambda_1 = -\tfrac{5}{4}, \qquad \lambda_2 = \tfrac{1}{4}$$

Since λ_1 is negative, we can obtain a feasible descent step by dropping constraint 1. This conclusion is verified by Figure 8.9*b*.

In summary, the projected gradient and the Lagrange multipliers are the key to the progress of the gradient projection method. We must determine the Lagrange multipliers at each iterate for which the projected gradient is zero. If the multipliers are all positive, the iterate is the constrained minimum. If any of the multipliers is negative, one of the constraints passing through the iterate must be ignored in order to proceed further. Rather than being an extraneous calculation, the Lagrange multiplier determination is closely related to the projection process. Consider a problem defined on $\mathcal{V} = \mathfrak{M}^{n \times 1}$. As in (8.71), we define the matrix $\mathbf{A}_p$ in terms of the gradients of the p constraint surfaces which intersect at $\mathbf{x}_k$. We

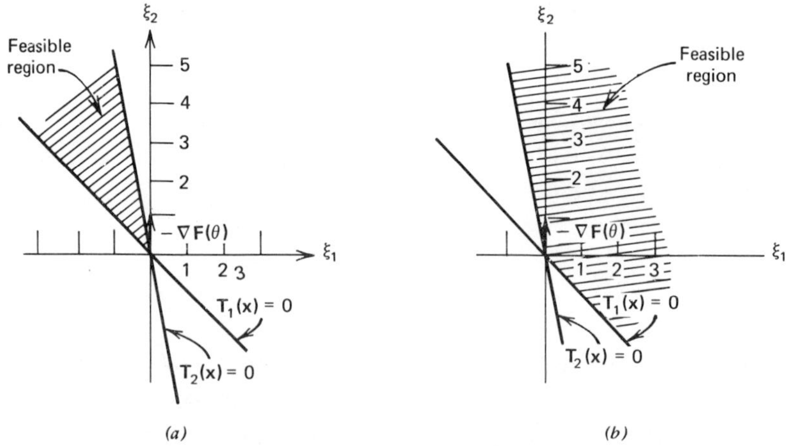

Figure 8.9. Constraint configurations for Example 4.

also define the vector

$$\boldsymbol{\lambda}_k \stackrel{\Delta}{=} \mathbf{A}_p^{\dagger} \nabla \mathbf{F}(\mathbf{x}_k) \tag{8.77}$$

Then by (8.69) and (8.72),

$$\mathbf{P}_{\mathscr{US}}(-\nabla \mathbf{F}(\mathbf{x}_k)) = -\nabla \mathbf{F}(\mathbf{x}_k) + \mathbf{A}_p \boldsymbol{\lambda}_k \tag{8.78}$$

At each iterate where $\mathbf{P}_{\mathscr{US}}(-\nabla \mathbf{F}(\mathbf{x}_k)) = \boldsymbol{\theta}$, we have $\nabla \mathbf{F}(\mathbf{x}_k) = \mathbf{A}_p \boldsymbol{\lambda}_k$, and the vector $\boldsymbol{\lambda}_k$ is precisely the required Lagrange multiplier vector. Note that the vector $\boldsymbol{\lambda}_k$ is *defined* for each iterate, including those iterates where the Lagrange multiplier equation is not satisfied. At the end of the iteration the vectors $\mathbf{x}_k$ and $\boldsymbol{\lambda}_k$ become the constrained minimum and the associated Lagrange multiplier vector, respectively.

Crossing the Feasible Region

Except at an iterate $\mathbf{x}_k$ where the intersecting constraints (treated as equalities) satisfy the Lagrange multiplier equation, it is difficult to recognize the presence of constraints which could be ignored. However, there is one situation where this recognition is not difficult. If the negative gradient $-\nabla \mathbf{F}(\mathbf{x}_k)$ is directed into the feasible region, *all* constraint hyperplanes which pass through $\mathbf{x}_k$ should be ignored—the step should cut directly across the feasible region. If such a step were not permitted, there would be no way for the method to find a minimum which lies in the interior of the feasible region. Suppose $-\nabla \mathbf{F}(\mathbf{x}_k)$ is directed into the feasible region. Then for small α_k the vector $\mathbf{x}_{k+1} = \mathbf{x}_k - \alpha_k \nabla \mathbf{F}(\mathbf{x}_k)$ is feasible with respect to each of the p constraint hyperplanes which pass through $\mathbf{x}_k$. Suppose $\mathbf{T}_j$ passes through $\mathbf{x}_k$; that is, $\mathbf{T}_j \mathbf{x}_k = \langle \mathbf{x}_k, \nabla \mathbf{T}_j \rangle = c_j$. Then strict feasibility of $\mathbf{x}_{k+1}$ requires

$$\begin{aligned} \mathbf{T}_j \mathbf{x}_{k+1} &= \langle \mathbf{x}_{k+1}, \nabla \mathbf{T}_j \rangle \\ &= \underbrace{\langle \mathbf{x}_k, \nabla \mathbf{T}_j \rangle}_{c_j} - \alpha_k \langle \nabla \mathbf{F}(\mathbf{x}_k), \nabla \mathbf{T}_j \rangle > c_j \end{aligned}$$

or

$$\langle -\nabla \mathbf{F}(\mathbf{x}_k), \nabla \mathbf{T}_j \rangle > 0, \qquad j = 1, \ldots, p \tag{8.79}$$

We can use (8.79) to check for the opportunity to permit the gradient projection descent to leave the constraint boundaries and cut across the feasible region.

Example 5. Cutting Across the Feasible Region. Assume the kth iterate of a two-dimensional minimization produces the configuration shown in Figure 8.9*b*, where $\mathbf{x}_k=\boldsymbol{\theta}$, $-\nabla\mathbf{F}(\mathbf{x}_k)=(0\quad 1)^{\mathrm{T}}$, and $\mathbf{x}_k$ lies on the two constraints, $\mathbf{T}_1(\mathbf{x}) \triangleq \xi_1+\xi_2 \geqslant 0$ and $\mathbf{T}_2(\mathbf{x}) \triangleq 5\xi_1+\xi_2 \geqslant 0$. According to the test (8.79),

$$\langle -\nabla\mathbf{F}(\mathbf{x}_k), \nabla\mathbf{T}_1\rangle = \begin{pmatrix}0\\1\end{pmatrix}^{\mathrm{T}}\begin{pmatrix}1\\1\end{pmatrix}=1>0$$

$$\langle -\nabla\mathbf{F}(\mathbf{x}_k), \nabla\mathbf{T}_2\rangle = \begin{pmatrix}0\\1\end{pmatrix}^{\mathrm{T}}\begin{pmatrix}5\\1\end{pmatrix}=1>0$$

Clearly, $-\nabla\mathbf{F}(\mathbf{x}_k)$ is feasible relative to both constraints, and both should be ignored.

A Gradient Projection Algorithm

In order to summarize the concepts of the previous sections, we now state an algorithm for the gradient projection method which applies for both finite- and infinite-dimensional problems*:

0. Find a feasible starting vector $\mathbf{x}_0$; determine the hyperplane $\mathbf{x}_0+\mathcal{W}_p$ which is defined by the constraints which pass through $\mathbf{x}_0$, as in (8.68).
1. Compute $-\nabla\mathbf{F}(\mathbf{x}_k)$; if this descent direction is feasible, as determined by (8.79), let $p=0$, let $\mathcal{W}_p=\mathcal{V}$, and go directly to step 5 (using the negative gradient rather than a projected gradient).
2. Compute $\mathbf{P}_{\mathcal{W}}(-\nabla\mathbf{F}(\mathbf{x}_k)$ using (8.69); if the projected gradient is nonzero, go directly to step 5.
3. $\mathbf{P}_{\mathcal{W}}(-\nabla\mathbf{F}(\mathbf{x}_k))=\boldsymbol{\theta}$; compute the Lagrange multipliers using (8.76); if none of the multipliers is negative, $\mathbf{x}_k$ is the constrained minimum, and the iteration terminates.
4. At least one of the Lagrange multipliers is negative; redefine the hyperplane $\mathbf{x}_k+\mathcal{W}_p$ by deleting from the definition (8.68) the constraint with the most negative multiplier; replace p by $p-1$; return to 2.†
5. $\mathbf{P}_{\mathcal{W}}(-\nabla\mathbf{F}(\mathbf{x}_k))\neq\boldsymbol{\theta}$; compute the best step length α_k by means of (8.74), (8.75), and one-dimensional minimization; compute the next iterate

*We assume that the constraints which intersect at the iterate $\mathbf{x}_k$ are independent; that is, the set $\{\nabla\mathbf{T}_i, i=1,\ldots,p\}$ is independent. The gradient projection algorithm of Rosen [8.28] allows for dependent constraints; by means of an auxiliary linear programming problem, the dependent set is replaced by an appropriate independent set. This auxiliary linear program is also discussed by Kunzi et al. [8.18].

†In certain circumstances this simple criterion for determining which constraint to delete can lead to hemstitching of the iterates. More sophisticated criteria for deleting constraints, designed to avoid hemstitching, are discussed by Fletcher [8.11]. Rosen's algorithm [8.28] uses one of these criteria.

$\mathbf{x}_{k+1}$ using (8.73); the step vector lies in $\mathcal{W}_p$; add to the definition (8.68) of $\mathcal{W}_p$ any constraints which pass through $\mathbf{x}_{k+1}$ but not through $\mathbf{x}_k$, and increase p accordingly; replace k by $k+1$; return to 1.

Notice that at each iteration of the algorithm the constraining hyperplane $\mathbf{x}_k + \mathcal{W}_p$ may need to be modified. The step may encounter additional constraints at $\mathbf{x}_{k+1}$, resulting in a decrease in the dimension of $\mathcal{W}_p$; the step may depart from one or more constraints at $\mathbf{x}_{k+1}$, resulting in an increase in the dimension of $\mathcal{W}_p$.

Example 6. Minimization by Gradient Projection. Let $\mathbf{x} = (\xi_1 \ \xi_2)^{\mathrm{T}}$ in $\mathfrak{M}^{2\times 1}$ and $\mathbf{F}(\mathbf{x}) = \xi_1^2 + \frac{1}{2}(\xi_2 - 3)^2$. We minimize $\mathbf{F}(\mathbf{x})$ subject to the constraints

$$\mathbf{T}_1(\mathbf{x}) = 4\xi_1 + \xi_2 \geqslant 4$$

$$\mathbf{T}_2(\mathbf{x}) = 3\xi_1 + 2\xi_2 \geqslant 6$$

$$\mathbf{T}_3(\mathbf{x}) = \xi_2 \geqslant 0$$

We follow the steps of the gradient projection algorithm, using the standard inner product:

(0) Let the initial feasible point be $\mathbf{x}_0 = (3 \ 0)^{\mathrm{T}}$. The only constraint which passes through $\mathbf{x}_0$ is $\mathbf{T}_3$. Thus $\mathcal{W}_p^{\perp} = \mathcal{W}_1^{\perp} = \mathrm{span}\{\nabla \mathbf{T}_3\}$.

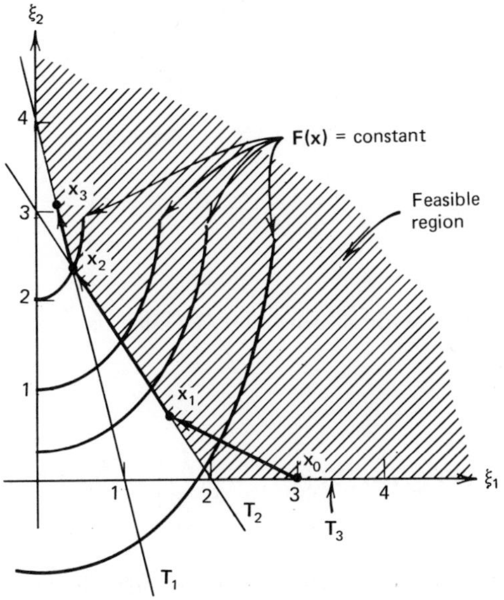

Figure 8.10. Gradient projection minimization of Example 6.

(1) The gradient of $\mathbf{F}$ is $\nabla\mathbf{F}(\mathbf{x})=(2\xi_1\ \ \xi_2-3)^{\mathrm{T}}$. Therefore, $\nabla\mathbf{F}(\mathbf{x}_0)=(6\ \ -3)^{\mathrm{T}}$. According to (8.79), $\langle -\nabla\mathbf{F}(\mathbf{x}_0), \nabla\mathbf{T}_3\rangle = 3 > 0$. Consequently, the negative gradient is feasible, and we drop the constraint $\mathbf{T}_3$ (Figure 8.10); $\mathcal{W}_1$ is replaced by $\mathcal{W}_0 = \mathfrak{M}^{2\times 1}$ (or $\mathcal{W}_0^{\perp}=\{\boldsymbol{\theta}\}$).

(5) According to (8.74), the intersection with $\mathbf{T}_1$ yields $\alpha_0=\frac{8}{21}$; the intersection with $\mathbf{T}_2$ yields $\alpha_0=\frac{1}{4}$. The smaller of these step lengths (corresponding to $\mathbf{T}_2$) determines the largest feasible step, and $\hat{\mathbf{x}}_1=\mathbf{x}_0-\frac{1}{4}\nabla\mathbf{F}(\mathbf{x}_0)=(\frac{3}{2}\ \ \frac{3}{4})^{\mathrm{T}}$. By (8.75), $\langle\nabla\mathbf{F}(\hat{\mathbf{x}}_1),\nabla\mathbf{F}(\mathbf{x}_0)\rangle=18+\frac{27}{4}>0$, so that $\mathbf{x}_1=\hat{\mathbf{x}}_1$. The iterate $\mathbf{x}_1$ lies on the constraint $\mathbf{T}_2$. Therefore, we replace $\mathcal{W}_0^{\perp}$ by $\mathcal{W}_1^{\perp}=\operatorname{span}\{\nabla\mathbf{T}_2\}$.

(1) $-\nabla\mathbf{F}(\mathbf{x}_1)=(-3\ \ \frac{9}{4})^{\mathrm{T}}$. This negative gradient is not feasible.

(2) By (8.69),

$$\mathbf{P}_{\mathcal{W}}(-\nabla\mathbf{F}(\mathbf{x}_1)) = -\nabla\mathbf{F}(\mathbf{x}_1) - \frac{\langle -\nabla\mathbf{F}(\mathbf{x}_1),\nabla\mathbf{T}_2\rangle}{\|\nabla\mathbf{T}_2\|^2}\nabla\mathbf{T}_2$$

$$= \begin{pmatrix} -3 \\ \frac{9}{4} \end{pmatrix} - \frac{-9/2}{13}\begin{pmatrix} 3 \\ 2 \end{pmatrix} = \frac{51}{52}\begin{pmatrix} -2 \\ 3 \end{pmatrix}$$

(5) The best step yields $\mathbf{x}_2=(\frac{2}{5}\ \ \frac{12}{5})^{\mathrm{T}}$ at the intersection with $\mathbf{T}_1$. Because the iterate $\mathbf{x}_2$ lies on the additional constraint $\mathbf{T}_1$, we replace $\mathcal{W}_1^{\perp}$ by $\mathcal{W}_2^{\perp}=\operatorname{span}\{\nabla\mathbf{T}_2,\nabla\mathbf{T}_1\}$.

(1) $-\nabla\mathbf{F}(\mathbf{x}_2)=(-\frac{4}{5}\ \ \frac{3}{5})^{\mathrm{T}}$. This negative gradient is not feasible.

(2) We compute the projected gradient by means of (8.69) and (8.72); $\mathbf{A}_2=(\nabla\mathbf{T}_1 \ \vdots\ \nabla\mathbf{T}_2)=\begin{pmatrix} 4 & 3 \\ 1 & 2 \end{pmatrix}$, and $\mathbf{A}_2^{\dagger}=\mathbf{A}_2^{-1}=\frac{1}{5}\begin{pmatrix} 2 & -3 \\ -1 & 4 \end{pmatrix}$. Consequently,

$$\mathbf{P}_{\mathcal{W}}(-\nabla\mathbf{F}(\mathbf{x}_2))=(\mathbf{I}-\mathbf{A}_2\mathbf{A}_2^{\dagger})(-\nabla\mathbf{F}(\mathbf{x}_2))=\boldsymbol{\theta}.$$

(3) From (8.76), $\nabla\mathbf{F}(\mathbf{x}_2)=\lambda_1\nabla\mathbf{T}_1+\lambda_2\nabla\mathbf{T}_2$, or by (8.77),

$$\begin{pmatrix} \lambda_1 \\ \lambda_2 \end{pmatrix} = \mathbf{A}_2^{\dagger}\nabla\mathbf{F}(\mathbf{x}_2) = \begin{pmatrix} \frac{17}{25} \\ -\frac{16}{25} \end{pmatrix}$$

(4) Since $\lambda_2<0$, we drop constraint $\mathbf{T}_2$; that is, we replace $\mathcal{W}_2^{\perp}$ by $\mathcal{W}_1^{\perp}=\operatorname{span}\{\nabla\mathbf{T}_1\}$.

(2) By (8.69),

$$\mathbf{P}_{\mathcal{W}}(-\nabla\mathbf{F}(\mathbf{x}_2)) = -\nabla\mathbf{F}(\mathbf{x}_2) - \frac{\langle -\nabla\mathbf{F}(\mathbf{x}_2),\nabla\mathbf{T}_1\rangle}{\|\nabla\mathbf{T}_1\|^2}\nabla\mathbf{T}_1 = \frac{16}{85}\begin{pmatrix} -1 \\ 4 \end{pmatrix}$$

(5) By a one-dimensional minimization, we find the best step length, $\alpha_2=\frac{17}{18}$, and $\mathbf{x}_3=(\frac{2}{9}\ \ \frac{28}{9})^{\mathrm{T}}$. No additional constraints pass through $\mathbf{x}_3$, so $\mathcal{W}_1$ is unchanged.

(1) $-\nabla\mathbf{F}(\mathbf{x}_3)=(-\frac{4}{9}\ \ -\frac{1}{9})^{\mathrm{T}}$.

(2) $$\mathbf{P}_{\mathcal{U}}(-\nabla\mathbf{F}(\mathbf{x}_3))=-\nabla\mathbf{F}(\mathbf{x}_3)-\frac{\langle-\nabla\mathbf{F}(\mathbf{x}_3),\nabla\mathbf{T}_1\rangle}{\|\nabla\mathbf{T}_1\|^2}\nabla\mathbf{T}_1=\boldsymbol{\theta}.$$

(3) The Lagrange multiplier equation is

$$\nabla\mathbf{F}(\mathbf{x}_3)=\begin{pmatrix}\frac{4}{9}\\ \frac{1}{9}\end{pmatrix}=\lambda\begin{pmatrix}4\\1\end{pmatrix}$$

or $\lambda=\frac{1}{9}>0$. Thus $\mathbf{x}_3$ is the constrained minimum.

Example 7. Gradient Projection in a Function Space. Suppose we wish to minimize $\mathbf{F}(\mathbf{u})=\frac{1}{2}\int_0^1\mathbf{u}^2(t)\,dt$ subject to

$$\mathbf{T}_1\mathbf{u}=\int_0^1(1-e^{-1}e^s)\mathbf{u}(s)\,ds=\phi(1)\geqslant 1$$

$$\mathbf{T}_2\mathbf{u}=\int_0^1 e^{-1}e^s\mathbf{u}(s)\,ds=\dot{\phi}(1)\geqslant 0$$

This is essentially the problem (7.79), wherein we seek to determine the control voltage $\mathbf{u}$ which will drive a motor shaft to an angle $\phi(1)\geqslant 1$ and velocity $\dot{\phi}(1)\geqslant 0$ while consuming minimum energy. We solve this problem by the gradient projection method. Let us assume the standard function space inner product. Then $\nabla\mathbf{F}(\mathbf{u})=\mathbf{u}$, $[\nabla\mathbf{T}_1(\mathbf{u})](t)=1-e^{-1}e^t$, and $[\nabla\mathbf{T}_2(\mathbf{u})](t)=e^{-1}e^t$. We begin by selecting a simple trial solution, $\mathbf{u}_0(t)=2e$. Then

$$\mathbf{T}_1\mathbf{u}_0=\int_0^1 2e(1-e^{-1}e^s)\,ds=2\geqslant 1$$

$$\mathbf{T}_2\mathbf{u}_0=\int_0^1 2e(e^{-1}e^s)\,ds=2(e-1)\geqslant 0$$

The initial solution is feasible and unconstrained (see Figure 8.11). Thus the initial step is in the direction $-\nabla\mathbf{F}(\mathbf{u}_0)=-\mathbf{u}_0$, and $\mathbf{u}_1=\mathbf{u}_0-\alpha_0\mathbf{u}_0$. According to (8.74), the

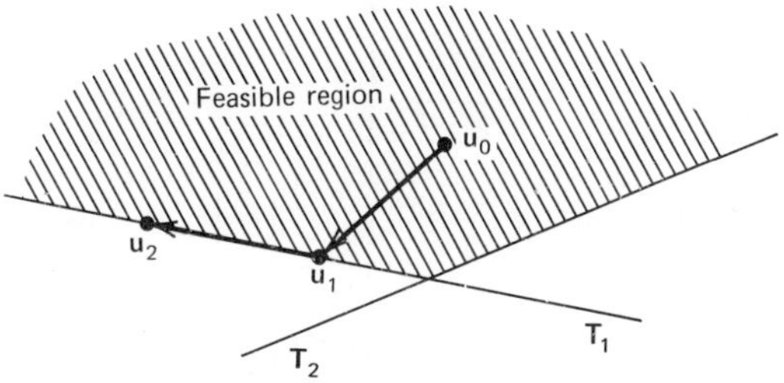

Figure 8.11. Abstract representation of Example 7.

intersection with $\mathbf{T}_1$ occurs for

$$\alpha_0 = \frac{\int_0^1 (1-e^{-1}e^s)\mathbf{u}_0(s)\,ds - 1}{\int_0^1 \mathbf{u}_0(s)(1-e^{-1}e^s)\,ds} = \tfrac{1}{2}$$

The intersection with $\mathbf{T}_2$ occurs for $\alpha_0 = 1$. Thus the largest feasible step intersects $\mathbf{T}_1$, and

$$\hat{\mathbf{u}}_1 = \mathbf{u}_0 - \tfrac{1}{2}\mathbf{u}_0 = \tfrac{1}{2}\mathbf{u}_0$$

or $\hat{\mathbf{u}}_1(t) = e$. According to (8.75),

$$\langle \nabla \mathbf{F}(\hat{\mathbf{u}}_1), \nabla \mathbf{F}(\mathbf{u}_0) \rangle = \int_0^1 (e)(2e)\,dt = 2e^2 > 0$$

Therefore, $\mathbf{u}_1(t) = \hat{\mathbf{u}}_1(t) = e$.

The new gradient is $\nabla \mathbf{F}(\mathbf{u}_1) = \mathbf{u}_1$ (or $[\nabla \mathbf{F}(\mathbf{u}_1)](t) = e$). We project the negative gradient on $\mathbf{T}_1$ as indicated in (8.69):

$$\mathbf{P}_{\mathcal{U}}(-\nabla \mathbf{F}(\mathbf{u}_1)) = -\nabla \mathbf{F}(\mathbf{u}_1) + \frac{\langle \nabla \mathbf{F}(\mathbf{u}_1), \nabla \mathbf{T}_1 \rangle}{\|\nabla \mathbf{T}_1\|^2} \nabla \mathbf{T}_1$$

$$= -\mathbf{u}_1 + \frac{\int_0^1 (e)(1-e^{-1}e^s)\,ds}{\int_0^1 (1-e^{-1}e^s)^2\,ds} \nabla \mathbf{T}_1$$

$$= -\mathbf{u}_1 - \left(\frac{2e^2}{e^2 - 4e + 1}\right) \nabla \mathbf{T}_1$$

or

$$[\mathbf{P}_{\mathcal{U}}(-\nabla \mathbf{F}(\mathbf{u}_1))](t) = -e - \left(\frac{2e^2}{e^2-4e+1}\right)(1 - e^{-1}e^t)$$

$$= \frac{2ee^t - e(e-1)^2}{e^2 - 4e + 1}$$

Then $\mathbf{u}_2 = \mathbf{u}_1 + \alpha_1 \mathbf{P}_{\mathcal{U}}(-\nabla \mathbf{F}(\mathbf{u}_1))$. No positive α_1 yields an intersection with $\mathbf{T}_2$. Setting $(d/d\alpha_1)\mathbf{F}(\mathbf{u}_2(\alpha_1)) = 0$ shows that

$$\alpha_1 = -\frac{\langle \mathbf{u}_1, \mathbf{P}_{\mathcal{U}}(-\nabla \mathbf{F}(\mathbf{u}_1)) \rangle}{\|\mathbf{P}_{\mathcal{U}}(-\nabla \mathbf{F}(\mathbf{u}_1))\|^2}$$

A messy calculation finds $\alpha_1 = 1$. It follows that

$$\mathbf{u}_2(t) = \left(-\frac{2e^2}{e^2 - 4e + 1} \right)(1 - e^{-1}e^t)$$

The Lagrange multiplier equation, $\nabla \mathbf{F}(\mathbf{u}_2) = \lambda \nabla \mathbf{T}_1$, shows that the coefficient in the expression for $\mathbf{u}_2$ is the Lagrange multiplier; its value is $\lambda \approx 5.95 > 0$. We have found the constrained minimum. The minimization process is illustrated abstractly in Figure 8.11.

Additional Comments

Several additional comments are worth mentioning. First, the gradient projection method still applies if linear *equality constraints* are present. The only changes required are that the equality constraints are always included in the definition of $\mathfrak{W}_p$ and $\mathfrak{W}_p^{\perp}$, and the signs of those Lagrange multipliers which correspond to equality constraints are not required to be non-negative at the constrained minimum. Second, if an initial feasible point is not easily determined from the physical or mathematical nature of the problem, some scheme for finding such a point is required. One such scheme is considered in P&C 8.18. Other techniques are described in [8.28] and [8.16]. Third, a technique for reducing the amount of computation required for a finite-dimensional gradient projection iteration is discussed in [8.18].

The gradient projection method is suitable for solving linear programming problems. For certain of these problems it is more efficient than the well-known simplex method of linear programming [8.15]. The amount of computation required for a gradient projection iteration is greater than that for a simplex iteration. The efficiency of the gradient projection method arises out of its ability to cut directly across the feasible region. If both methods are applied to the same problem with the same initial boundary point $\mathbf{x}_0$, and if the gradient projection method is not permitted to leave the boundary and cut across the feasible region, both methods produce precisely the same sequence of iterates [8.28].

The final stages of the gradient projection descent may consist of an infinite sequence of steepest descent steps occurring within a specific constraint hyperplane (P&C 8.17). Luenberger [8.21] derives an estimate of the asymptotic rate of convergence of such infinite sequences. His estimate is expressed in a form similar to the steepest descent estimate, (8.32). A number of other methods exist for descending a functional $\mathbf{F}(\mathbf{x})$ while maintaining feasibility. These "methods of feasible directions" are discussed by Zoutendijk [8.37] and Hadley [8.16].

8.5 Penalty Functions

The preceding discussion of the gradient projection method has demonstrated that the presence of constraints, even linear constraints, considerably complicates the process of computing maxima and minima of functionals. To this point, the only technique we have considered for handling optimization problems with nonlinear constraints is the use of Newton's method in solving the nonlinear necessary conditions (Section 8.1). If inequality constraints are involved, this approach requires considerable trial and error, both to determine which constraints are binding and to determine an acceptable starting solution for Newton's method.

In this section we introduce a general "penalty function" approach for performing minimization with nonlinear constraints. This approach treats the rigid boundary conditions as limiting cases of "constraining forces" on the iterates. The actual computations are just the computations required for one of the techniques for unconstrained minimization (Sections 8.2–8.3). Before we carry out the unconstrained minimization, we attach to the objective function $\mathbf{F}(\mathbf{x})$ some function of the constraints in order to penalize or avoid nonfeasible solutions. These penalized objective functions are reminiscent of the Lagrangian function, (7.38). In point of fact, for some problems the Lagrangian is a suitable penalized objective function.

The Penalty Function Concept

Assume we wish to solve the problem

$$\begin{aligned} &\text{minimize} \quad \mathbf{F}(\mathbf{x}) \\ &\text{with} \qquad\quad \mathbf{G}_i(\mathbf{x}) \geqslant 0, \quad i=1,\ldots,m \end{aligned} \tag{8.80}$$

where $\mathbf{F}$ and $\mathbf{G}_i$ are functionals acting on a Hilbert space $\mathcal{V}$. Assume further that (8.80) has solutions and that $\mathbf{F}$ and $\mathbf{G}_i$ are Fréchet differentiable.* We define the **penalized objective function** $\psi_c(\mathbf{x})$ by

$$\psi_c(\mathbf{x}) \triangleq \mathbf{F}(\mathbf{x}) + \sum_{i=1}^{m} c_i \boldsymbol{\Phi}_i(\mathbf{G}_i(\mathbf{x})) \tag{8.81}$$

where $\boldsymbol{\Phi}_i$ is a **penalty function**, and c_i is a **penalty constant**. The penalty function $\boldsymbol{\Phi}_i(s)$ is a continuous scalar valued function of s. There are two

*Continuity of $\mathbf{F}$ and $\mathbf{G}_i$ is sufficient to guarantee most of the results of this section. Differentiability enables us to use gradient techniques.

basic penalty function approaches, interior and exterior. In the **interior approach** we begin with a feasible estimate of the solution, and pick an interior penalty function $\mathbf{\Phi}_i$ which severely penalizes too near an approach to the feasible boundary specified by $\mathbf{G}_i(\mathbf{x})=0$; that is, $\mathbf{\Phi}_i$ is designed so that $\mathbf{\Phi}_i(s)\to\infty$ as $s\to 0$ from positive values. In the **exterior approach** we begin with any initial estimate of the solution, and use an exterior penalty function $\mathbf{\Phi}_i$ which penalizes vectors $\mathbf{x}$ which lie outside the feasible region; specifically, we select $\mathbf{\Phi}_i$ such that $\mathbf{\Phi}_i(s)>0$ for $s<0$ (infeasible x) and $\mathbf{\Phi}_i(s)=0$ for $s\geqslant 0$ (feasible $\mathbf{x}$).

The Penalty Function Algorithm (8.82)

0. Define a penalty function for each constraint; pick an initial iterate $\mathbf{x}_0$ ($\mathbf{x}_0$ must be feasible with respect to those constraints for which interior penalty functions are used).
1. Pick the penalty constants $\{c_i\}$ to provide a suitable balance between the function value $\mathbf{F}(\mathbf{x}_0)$ and the value of the penalty terms, $\sum_{i=1}^{m} c_i\mathbf{\Phi}_i(\mathbf{G}_i(\mathbf{x}_0))$.
2. Determine the minimum, $\mathbf{x}_{j+1}$, of the unconstrained penalized objective function (8.81) by using the present iterate $\mathbf{x}_j$ and one of the descent techniques of Sections 8.2–8.3.
3. Pick new penalty constants $\{c_i\}$ in order to modify (or rebalance) the magnitudes of the penalty terms (the magnitudes of interior penalties should be reduced to allow a closer approach to the boundary of the feasible region; the magnitudes of exterior penalties should be increased to force a closer approach to the boundary); replace $j+1$ by j; return to 2.

Under the assumptions inherent in (8.80), the sequence of unconstrained minima $\{\mathbf{x}_j\}$ generated by the algorithm (8.82) approaches a constrained minimum of (8.80). The iteration is terminated after the convergence is considered adequate. We either accept the final value of $\mathbf{x}_j$ as an adequate approximation to a point of constrained minimum, or extrapolate numerically from the sequence $\{\mathbf{x}_j\}$ to obtain a better estimate of the minimum point. Fiacco and McCormick, in their comprehensive treatment of penalty functions [8.8], refer to (8.82) as the "sequential unconstrained minimization technique (SUMT)".*

The Interior Penalty Approach

An interior penalty function $\Phi(s)$ must approach infinity as s approaches zero from positive values. One function which exhibits this characteristic is

*Lootsma [8.19] also presents a clear discussion of the penalty function approach to constrained minimization.

the logarithmic penalty function,

$$\mathbf{\Phi}(s) \triangleq -\ln(s) \tag{8.83}$$

Another is the **inverse penalty function,**

$$\mathbf{\Phi}(s) \triangleq \frac{1}{s} \tag{8.84}$$

These penalty functions are illustrated in Figure 8.12.

Consider performing the unconstrained minimization required in step 2 of (8.82) by a descent method (say, the DFP method). The descent begins in the feasible region. Because of the penalty terms, $\psi_c(\mathbf{x}) \to \infty$ as $\mathbf{x}$ approaches the constraint boundary. Thus the minimum of $\psi_c(\mathbf{x})$ cannot lie very close to the boundary. However, by decreasing the penalty constants $\{c_i\}$ drastically, we can move the minimum of $\psi_c(\mathbf{x})$ much closer to the boundary. It is clear that interior penalty functions force the minimum $\mathbf{x}_j$ of the penalized objective function (8.81) to be in the region of strict feasibility ($\mathbf{G}_i(\mathbf{x})>0$, $i=1,\ldots,m$). As a consequence, *interior penalty methods cannot be extended to handle equality constraints*; an equality constraint defines a feasible region which has no interior points. Fiacco and McCormick [8.8] show that for interior penalty functions, finite-dimensional $\mathcal{V}$, and small penalty constants $\{c_i\}$, the penalized objective function (8.81) has unconstrained minima. They also show that as the penalty constants are reduced to zero, the penalty terms $c_i\mathbf{\Phi}_i(\mathbf{G}_i(\mathbf{x}_j))$ approach zero, even though $\mathbf{\Phi}_i(\mathbf{G}_i(\mathbf{x}_j))\to\infty$ for the sequence of minima $\mathbf{x}_j$

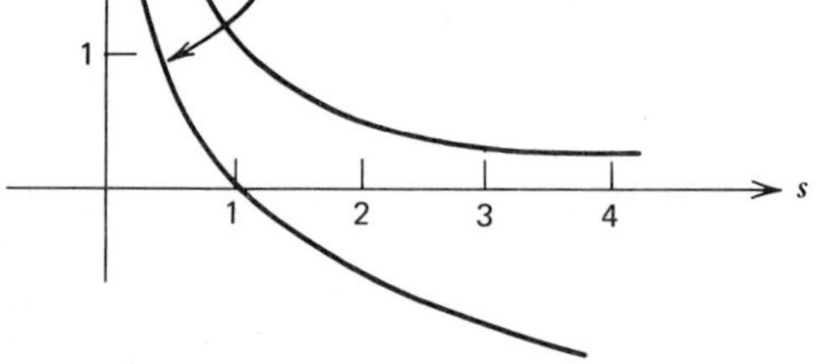

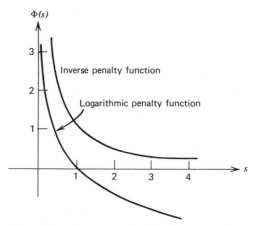

Figure 8.12. Interior penalty functions.

generated by the iteration (8.82). They show, further, that the sequence of values $\psi_c(\mathbf{x}_j)$ approaches a local minimum value of $\mathbf{F}(\mathbf{x})$ monotonically from above as the penalty constants approach zero. We would ordinarily expect the sequence $\{\mathbf{x}_j\}$ to approach simultaneously the corresponding solution to (8.80).

Example 1. Minimization by Means of Logarithmic Penalty Functions. Let $\mathbf{x}=(\xi_1 \ \xi_2)^{\mathrm{T}}$. We will minimize $\mathbf{F}(\mathbf{x}) \triangleq \xi_1+2\xi_2$ subject to $\mathbf{G}_1(\mathbf{x}) \triangleq \xi_1-\xi_2^2 \geqslant 0$ and $\mathbf{G}_2(\mathbf{x}) \triangleq \xi_2-1 \geqslant 0$. It is clear from Figure 8.13 that the constrained minimum lies at $\mathbf{x}=(1 \ \ 1)^{\mathrm{T}}$. We incorporate the constraints into the objective function by means of logarithmic penalty functions (8.83). For simplicity, we use the same penalty constant for each constraint. The penalized objective function (8.81) is

$$\psi_c(\mathbf{x})=\xi_1+2\xi_2-c\ln(\xi_1-\xi_2^2)-c\ln(\xi_2-1)$$

If we were minimizing numerically, we would find the sequence of minima $\{\mathbf{x}_j\}$ of $\psi_c(\mathbf{x})$ corresponding to a decreasing sequence of penalty constants. We would use the minimum $\mathbf{x}_j$ as the initial iterate in determining $\mathbf{x}_{j+1}$. Because of the simple nature of the functions, we can determine analytically the minimum of $\psi_c(\mathbf{x})$ as a function of c. Consequently, we will not need to concern ourselves with an initial feasible vector $\mathbf{x}_0$. We merely solve

$$\nabla\psi_c(\mathbf{x})=\begin{pmatrix} 1-\dfrac{c}{\xi_1-\xi_2^2} \\ 2+\dfrac{2c\xi_2}{\xi_1-\xi_2^2}-\dfrac{c}{\xi_2-1} \end{pmatrix}=\begin{pmatrix} 0 \\ 0 \end{pmatrix}$$

The result is

$$\mathbf{x}=\begin{pmatrix} \xi_1 \\ \xi_2 \end{pmatrix}=\begin{pmatrix} \sqrt{1+c/2} \\ 1+3c/2 \end{pmatrix}$$

We conclude by noting the minima of $\psi_c(\mathbf{x})$ corresponding to a decreasing sequence of penalty constants (see Figure 8.13):

c	ξ_1	ξ_2	$\psi_c(\mathbf{x})$
1	2.5	1.225	6.44
0.5	1.75	1.18	5.48
0.1	1.15	1.025	3.8
0	1	1	3

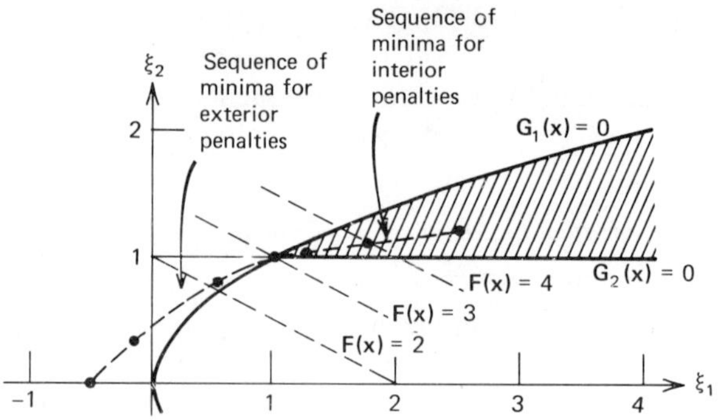

Figure 8.13. Minimization by penalty functions.

The Exterior Penalty Approach

An exterior penalty function $\mathbf{\Phi}(s)$ must be positive for $s<0$ and zero for $s \geqslant 0$. One function which has these properties is

$$\begin{aligned} \mathbf{\Phi}(s) &\triangleq -s, \qquad s \leqslant 0 \\ &\triangleq 0, \qquad s \geqslant 0 \end{aligned} \tag{8.85}$$

A more common exterior function is the **quadratic loss penalty function.**

$$\begin{aligned} \mathbf{\Phi}(s) &\triangleq s^2, \qquad s \leqslant 0 \\ &\triangleq 0, \qquad s \geqslant 0 \end{aligned} \tag{8.86}$$

These two penalty functions are illustrated in Figure 8.14. Note that the quadratic loss function is differentiable at $s=0$, whereas (8.85) is not.

Suppose we use a descent method to achieve the unconstrained minimization required in step 2 of (8.82). The unconstrained minimum of $\mathbf{F}(\mathbf{x})$ probably lies outside the feasible region. (Otherwise, no constraints are binding and the problem simplifies.) Because of the penalty terms, however, the minimum of $\psi_c(\mathbf{x})$ cannot be very far outside of the feasible region; for as $\mathbf{x}$ strays outside the feasible boundary, $\mathbf{G}_i(\mathbf{x})$ increases, and thus $c_i\mathbf{\Phi}_i(\mathbf{G}_i(\mathbf{x}))$ and $\psi_c(\mathbf{x})$ increase. Furthermore, by increasing the penalty constant c_i, we can increase the emphasis on the penalty term, and thereby force the minimum of $\psi_c(\mathbf{x})$ closer to the feasible region. Fiacco and McCormick [8.8] show that for exterior penalty functions, finite-dimensional $\mathcal{V}$, and large penalty constants, the penalized objective func-

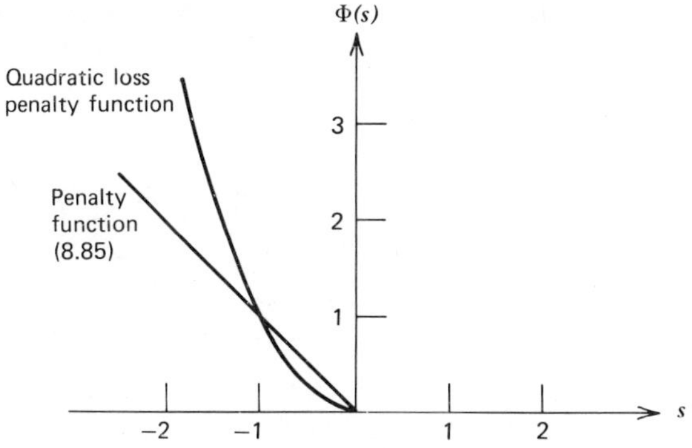

Figure 8.14. Exterior penalty functions.

tion has unconstrained minima. They also show that as the penalty constants approach infinity, the values of the penalty terms $c_i \boldsymbol{\Phi}_i(\mathbf{G}_i(\mathbf{x}_j))$ approach zero. Furthermore, the sequence of values $\psi_c(\mathbf{x}_j)$ approaches a local minimum value of $\mathbf{F}(\mathbf{x})$ monotonically from below as the penalty constants approach infinity. We would normally expect the sequence $\{\mathbf{x}_j\}$ to approach simultaneously the corresponding point of local minimum.

Example 2. Minimization by Means of Quadratic Loss Penalty Functions. We treat the same problem which is solved via interior penalty functions in Example 1. We minimize $\mathbf{F}(\mathbf{x}) \stackrel{\Delta}{=} \xi_1 + 2\xi_2$ subject to $\mathbf{G}_1(\mathbf{x}) \stackrel{\Delta}{=} \xi_1 - \xi_2^2 \geqslant 0$ and $\mathbf{G}_2(\mathbf{x}) \stackrel{\Delta}{=} \xi_2 - 1 \geqslant 0$. (See Figure 8.13.) We attach the constraints to the objective function by means of quadratic loss penalty functions, (8.86), using the same penalty constant for each constraint:

$$\psi_c(\mathbf{x}) = \xi_1 + 2\xi_2 + c\boldsymbol{\Phi}(\xi_1 - \xi_2^2) + c\boldsymbol{\Phi}(\xi_2 - 1)$$

Once again, the simple nature of the functions permits us to perform the unconstrained minimizations analytically. Assume that at the minimum of $\psi_c(\mathbf{x})$, both constraints are violated. Then $\boldsymbol{\Phi}(\mathbf{G}_i(\mathbf{x})) = \mathbf{G}_i^2(\mathbf{x})$ and

$$\nabla\psi_c(\mathbf{x}) = \begin{pmatrix} 1 + 2c(\xi_1 - \xi_2^2) \\ 2 - 4c\xi_2(\xi_1 - \xi_2^2) + 2c(\xi_2 - 1) \end{pmatrix} = \begin{pmatrix} 0 \\ 0 \end{pmatrix}$$

The solution is

$$\mathbf{x} = \begin{pmatrix} \xi_1 \\ \xi_2 \end{pmatrix} = \begin{pmatrix} \dfrac{(c-1)^2}{(c+1)^2} - \dfrac{1}{2c} \\ \dfrac{c-1}{c+1} \end{pmatrix}$$

This solution does, in fact, violate both constraints as assumed. We conclude by displaying the minima of $\psi_c(\mathbf{x})$ corresponding to an increasing sequence of penalty constants (see Figure 8.13):

c	ξ_1	ξ_2	$\psi_c(x)$
1	−0.5	0	0.75
2	−0.139	0.333	1.54
10	0.619	0.818	2.58
∞	1	1	3

One significant difference between the interior and exterior penalty methods is the fact that the latter can handle equality constraints. Suppose we attach to the problem (8.80) the single constraint $\mathbf{H}(\mathbf{x})=0$. We can replace the equality constraint by the pair of constraints

$$\mathbf{H}(\mathbf{x}) \geqslant 0 \quad \text{and} \quad -\mathbf{H}(\mathbf{x}) \geqslant 0$$

We attach these constraints to $\psi_c(\mathbf{x})$ by means of quadratic loss penalty functions. Because $\mathbf{H}(\mathbf{x})$ is positive if and only if $-\mathbf{H}(\mathbf{x})$ is negative, the added penalty terms become

$$c\boldsymbol{\Phi}(\mathbf{H}(\mathbf{x}))+c\boldsymbol{\Phi}(-\mathbf{H}(\mathbf{x}))=c\mathbf{H}^2(\mathbf{x})$$

Thus the **quadratic loss penalty function for equality constraints** is the simple quadratic function

$$\boldsymbol{\Phi}(s) \triangleq s^2 \tag{8.87}$$

Lagrange Multipliers

Inherent in the constrained optimization problem (8.80) is a set of Lagrange multipliers (or sensitivities of the minimum value of $\mathbf{F}(\mathbf{x})$ to perturbations in the constraints). Therefore, we should expect to be able to determine these Lagrange multipliers at the termination of the penalty function algorithm, if only by solving for $\{\lambda_i\}$ via the Lagrange multiplier equation of (7.58). In point of fact, as the algorithm (8.82) progresses, the sequence of values $\boldsymbol{\Phi}_i(\mathbf{G}_i(\mathbf{x}_j))$ of the ith penalty function provides a sequence of estimates of that Lagrange multiplier λ_i which corresponds to $\mathbf{G}_i$. We can determine the relationship between the penalty functions and the Lagrange multipliers from the expression for $\nabla\psi_c(\mathbf{x})$. The Fréchet differential of (8.81) is

$$\begin{aligned}
\mathbf{d}\psi_c(\mathbf{x},\mathbf{h}) &= \frac{d}{d\beta}\left[\psi_c(\mathbf{x}+\beta\mathbf{h})\right]_{\beta=0} \\
&= \frac{d}{d\beta}\mathbf{F}(\mathbf{x}+\beta\mathbf{h})_{\beta=0} + \sum_{i=1}^{m} c_i \frac{\partial \mathbf{\Phi}_i(\mathbf{G}_i(\mathbf{x}))}{\partial \mathbf{G}_i(\mathbf{x})} \frac{d}{d\beta}\mathbf{G}_i(\mathbf{x}+\beta\mathbf{h})_{\beta=0} \\
&= \mathbf{dF}(\mathbf{x},\mathbf{h}) + \sum_{i=1}^{m} c_i \frac{\partial \mathbf{\Phi}_i}{\partial \mathbf{G}_i}\mathbf{dG}_i(\mathbf{x},\mathbf{h}) \\
&= \langle \mathbf{h}, \nabla \mathbf{F}(\mathbf{x})\rangle + \sum_{i=1}^{m} c_i \frac{\partial \mathbf{\Phi}_i}{\partial \mathbf{G}_1}\langle \mathbf{h}, \nabla \mathbf{G}_i(\mathbf{x})\rangle
\end{aligned}$$

where all gradients are defined in terms of some inner product $\langle\cdot,\cdot\rangle$ on $\mathcal{V}$. Consequently,

$$\nabla\psi_c(\mathbf{x}) = \nabla\mathbf{F}(\mathbf{x}) + \sum_{i=1}^{m} c_i \frac{\partial \mathbf{\Phi}_i(\mathbf{G}_i(\mathbf{x}))}{\partial \mathbf{G}_i(\mathbf{x})}\nabla\mathbf{G}_i(\mathbf{x}) \tag{8.88}$$

The jth iteration of (8.82) produces $\mathbf{x}_j$, a point of unconstrained minimum of $\psi_c(\mathbf{x})$. Therefore, $\nabla\psi_c(\mathbf{x}_j) = \boldsymbol{\theta}$. But, according to (8.88), the equation $\nabla\psi_c(\mathbf{x}_j) = \boldsymbol{\theta}$ is of the same form as the Lagrange multiplier equation (7.58):

$$\nabla\mathbf{F}(\mathbf{x}_j) = \sum_{i=1}^{m} \lambda_i \nabla\mathbf{G}_i(\mathbf{x}_j)$$

Consequently, the scalars

$$\lambda_i(\mathbf{x}_j) \triangleq -c_i \left.\frac{\partial \mathbf{\Phi}_i(s)}{\partial s}\right|_{s=\mathbf{G}_i(\mathbf{x}_j)}, \qquad i = 1,\ldots,m \tag{8.89}$$

are approximations to the Lagrange multipliers of the original problem (8.80) in the same sense that $\mathbf{x}_j$ is an approximation to a constrained minimum. As $j\to\infty$, $\mathbf{x}_j$ approaches a point of constrained minimum of $\mathbf{F}(\mathbf{x})$ and $\psi_c(\mathbf{x}_j)$ approaches the corresponding constrained minimum value of $\mathbf{F}(\mathbf{x})$. Consequently, in the limit, $\Delta\psi_c(\mathbf{x}_j) = \boldsymbol{\theta}$ becomes the Lagrange multiplier equation associated with (8.80), and the approximations (8.89) become the exact Lagrange multipliers of (8.80). It is appropriate to represent the multipliers of (8.89) by the vector $\boldsymbol{\lambda}_j = (\lambda_1(\mathbf{x}_j)\cdots\lambda_m(\mathbf{x}_j))^{\mathrm{T}}$. Then we can think of the jth iterate produced by the penalty function algorithm as the pair $(\mathbf{x}_j, \boldsymbol{\lambda}_j)$, where $\mathbf{x}_j$ is an approximate constrained minimum and $\boldsymbol{\lambda}_j$ is an approximation to the Lagrange multiplier vector associated with that constrained minimum.

We easily determine specific forms for (8.89) for each of the penalty functions (8.83), (8.84), and (8.86). For the *logarithmic penalty function* (8.83),

$$\lambda_i(\mathbf{x}_j) = \frac{c_i}{\mathbf{G}_i(\mathbf{x}_j)} \tag{8.90}$$

For the *inverse penalty function* (8.84),

$$\lambda_i(\mathbf{x}_j) = \frac{c_i}{\mathbf{G}_i^2(\mathbf{x}_j)} \tag{8.91}$$

For the *quadratic loss penalty function* (8.86),

$$\begin{aligned} \lambda_i(\mathbf{x}_j) &= -2c_i\mathbf{G}_i(\mathbf{x}_j), \qquad & \mathbf{G}_i(\mathbf{x}_j) \leqslant 0 \\ &= 0, & \mathbf{G}_i(\mathbf{x}_j) \geqslant 0 \end{aligned} \tag{8.92}$$

In practice we can evaluate $\boldsymbol{\lambda}_j$ at the termination of the algorithm (8.82), and accept that vector as an adequate approximation to the true $\boldsymbol{\lambda}$. On the other hand, we can evaluate $\boldsymbol{\lambda}_j$ at each iteration and extrapolate numerically from this sequence to estimate the true $\boldsymbol{\lambda}$.

Example 3. Determination of Lagrange Multipliers. Let $\mathbf{F}$, $\mathbf{G}_1$, and $\mathbf{G}_2$ be as defined in Examples 1 and 2 (Figure 8.13). The minimum of $\mathbf{F}(\mathbf{x})$ subject to $\mathbf{G}_1(\mathbf{x}) \geqslant 0$ and $\mathbf{G}_2(\mathbf{x}) \geqslant 0$ is easily found to be $\mathbf{x} = (1 \;\; 1)^{\mathrm{T}}$ with $\boldsymbol{\lambda} = (1 \;\; 4)^{\mathrm{T}}$. Using (8.90) together with the sequence of interior iterates $\{\mathbf{x}_j\}$ computed in Example 1, we obtain a corresponding sequence of iterates $\{\boldsymbol{\lambda}_j\}$:

c	$\mathbf{G}_1(\mathbf{x}_j)$	$\mathbf{G}_2(\mathbf{x}_j)$	$\lambda_1(\mathbf{x}_j)$	$\lambda_2(\mathbf{x}_j)$
1	1	0.225	1	4.45
0.5	0.36	0.18	1.39	2.78
0.1	0.1	0.025	1.0	4.0
0	0	0	1	4

The limiting values of $\lambda_i(\mathbf{x}_j)$ are numerically indeterminate. They have been evaluated by applying L'Hospital's rule to

$$\lambda_i(\mathbf{x}_j) = \lim_{c \to 0} \frac{c}{\mathbf{G}_i(\mathbf{x}_j(c))}$$

By using (8.92), we also obtain the Lagrange multiplier iterates corresponding to the exterior sequence $\{\mathbf{x}_j\}$ computed in Example 2:

c	$\mathbf{G}_1(\mathbf{x}_j)$	$\mathbf{G}_2(\mathbf{x}_j)$	$\lambda_1(\mathbf{x}_j)$	$\lambda_2(\mathbf{x}_j)$
1	−0.5	−1	1	2
2	−0.25	−0.667	1	2.66
10	−0.05	−0.182	1	3.64
∞	0	0	1	4

The limiting values of $\lambda_i(\mathbf{x}_j)$ were obtained by evaluating

$$\lambda_i(\mathbf{x}_j) = \lim_{c\to\infty}\left[-2c\mathbf{G}_i(\mathbf{x}_j(c))\right]$$

Additional Comments

The interior penalty method provides at each iteration a feasible approximation to a local constrained minimum. Thus any one of the iterates may be an *acceptable* solution to the problem. Associated with this feature is the requirement that we begin with an initial feasible vector, $\mathbf{x}_0$. If physical considerations do not suggest a choice for $\mathbf{x}_0$, we must develop an algorithm which will generate a feasible vector (P&C 8.18). Unfortunately, the interior approach cannot handle equality constraints. A more significant drawback of the interior penalty method is that $\psi_c(\mathbf{x})$ includes all the constraints at every iteration. Consequently, the computation of values and gradients of $\psi_c(\mathbf{x})$ during the unconstrained descent phase of the iteration can be very costly.

The exterior penalty method is less strict than the interior method in the sense that it merely prevents straying very far into the infeasible region. As a result, the method is suitable for handling equality constraints, and does not require an initial feasible vector. Another advantage of the exterior approach is that the function $\psi_c(\mathbf{x})$ includes at each iteration only those constraints which are binding at that iteration (the exterior penalty terms are zero in the feasible region). As a result, the computation of values and gradients of $\psi_c(\mathbf{x})$ during each unconstrained descent is considerably simplified. Unavoidably associated with this computational advantage is restricted differentiability of $\psi_c(\mathbf{x})$ at the boundary of the feasible region; the quadratic loss penalty function has only first derivatives at $s=0$. In contrast, because of the extreme smoothness of typical interior penalty functions, if interior penalty functions are used, the penalty terms in $\psi_c(\mathbf{x})$ are as differentiable as the functions $\mathbf{F}$ and $\mathbf{G}_i$.

In both the interior and exterior approaches, the penalty constants $\{c_i\}$ are selected at each iteration to provide a suitable balance between the influence of $\mathbf{F}(\mathbf{x})$ and the influence of the penalty terms. For interior penalties, if the gradient of the penalty term $c_i\mathbf{\Phi}_i(\mathbf{G}_i(\mathbf{x}))$ is very small compared to the gradient of $\mathbf{F}(\mathbf{x})$ (i.e., if c_i is too small), then the unconstrained minimum of $\psi_c(\mathbf{x})$ is very much closer to the boundary of the feasible region than is $\mathbf{x}_j$ (Figure 8.15). Consequently, small numerical errors can cause a descent step to cross out of the feasible region. On the other hand, if c_i is too large, the minimization of $\psi_c(\mathbf{x})$ almost ignores $\mathbf{F}(\mathbf{x})$. For exterior penalties, if the gradient of the penalty term $c_i\mathbf{\Phi}_i(\mathbf{G}_i(\mathbf{x}_j))$ is very large compared to $\nabla\mathbf{F}(\mathbf{x}_j)$ (i.e., if c_i is too large), then the gradients computed in the descent minimization of $\psi_c(\mathbf{x})$ tend to be unneccessarily large and inaccurate (Figure 8.16); consequently, the descent minimization of $\psi_c(\mathbf{x})$ tends to be inefficient. If c_i is too small, the minimization of $\psi_c(\mathbf{x})$ almost ignores the constraint $\mathbf{G}_i(\mathbf{x}) \geqslant 0$. The sequence of approximations (8.89) to the Lagrange multipliers can be used as an indication of the relative importance of the various constraints. A careful choice of c_i is relatively unimportant if $\lambda_i(\mathbf{x}_j)$ remains small compared to the other Lagrange multipliers.

As illustrated in Figure 8.13, the sequence of unconstrained minima usually follows a very smooth path. The convergence of the sequence will usually be accelerated if we occasionally estimate a new point on the curve by extrapolation from the preceeding minima rather than by carrying out the next numerical minimization in the sequence. See Fiacco and McCormick [8.8].

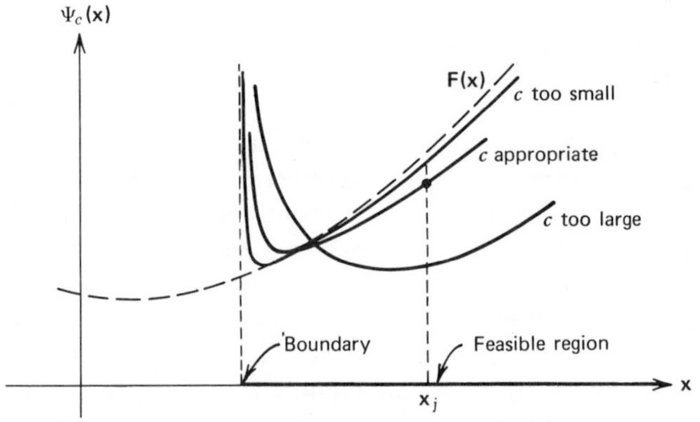

Figure 8.15. Choosing c for an interior penalty function.

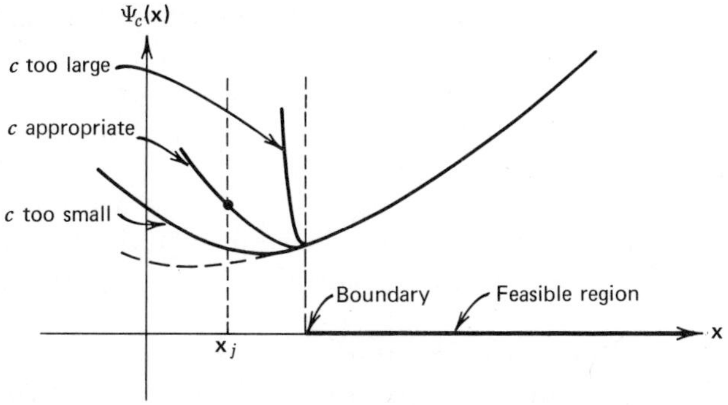

Figure 8.16. Choosing c for an exterior penalty function.

For both interior and exterior penalty functions, as the iterates $\{\mathbf{x}_j\}$ approach the constrained minimum, the function $\psi_c(\mathbf{x})$ develops a steep narrow valley which is close to the boundary of the feasible region and which follows the shape of that boundary. Therefore, the Hessian matrix of $\psi_c(\mathbf{x})$ becomes ill-conditioned near the minimum of $\psi_c(\mathbf{x})$ as the algorithm (8.82) progresses. Experience seems to indicate that the numerical difficulty which results from this ill conditioning is more pronounced for exterior penalties than for interior penalties. The computational difficulty which results from this ill conditioning is serious enough that considerable effort is being spent in seeking a good alternative to the use of penalty functions for treating nonlinear constraints. Some of the promising alternatives are discussed by Powell [8.26] and by Rosen and Kreuser [8.29]. These alternative approaches bear some of the features of the approaches we have discussed previously: gradient projection, the use of penalties, and the Lagrangian function of Section 7.2.

The penalty function concepts of this section apply unchanged for functionals $\mathbf{F}$ and $\mathbf{G}_i$ defined on an infinite-dimensional function space. For instance, the problem of Example 7, Section 8.4 can be solved in this manner. However, even simple function space examples require extensive use of computer integration routines.

8.6 Summary

Throughout the preceding four sections we have treated a variety of descent techniques for minimization.* The technique that is most

*These techniques can be used to maximize an objective function $\mathbf{F}(\mathbf{x})$ by instead minimizing $-\mathbf{F}(\mathbf{x})$.

appropriate for a particular problem depends strongly on the nature of the problem. We can often tailor the descent technique to take advantage of the specific properties of the minimization problem, and thereby improve the efficiency of the minimization (reduce the amount of computation required). For example, the penalty function method of solving constrained minimization problems does not take into account the special nature of linear constraint functions. Consequently, the penalty function approach is less efficient than the gradient projection approach for linear constraints.

Generally speaking, for unconstrained minimization the DFP descent and the conjugate gradient descent are among the most efficient techniques available. For problems in which the constraints are linear, a gradient projection method is usually best. (If the objective function is also linear, a linear programming algorithm may be more efficient.) Problems in which the constraints are nonlinear usually call for the use of penalty functions. (New techniques are presently under development to overcome the ill-conditioning problems which arise from the use of penalty functions.) An appropriate approach for a minimization problem with a variety of constraints is to attach the nonlinear constraints to the objective function by means of penalty functions (using exterior penalties for the nonlinear equality constraints), then use a gradient projection method to minimize the penalized objective function subject to the linear constraints. Specific computer implementations of any one of these techniques vary in the way gradients are obtained (e.g., analytically or by finite-difference approximation) and in the way the one-dimensional minimization is performed.

Numerous variations of the algorithms of Sections 8.2–8.5 can be found in textbooks on optimization, in journals on computing, and in computer libraries. Such algorithms are often untried, and should be used with caution. An untried algorithm may fail to converge for certain problems; it may converge to an incorrect solution; or it may converge so slowly that its use is impractical. An algorithm cannot be considered to be reliable until it has been programmed for the computer and used successfully for a variety of optimization problems.

8.7 Problems and Comments

8.1 Use Newton's method, together with a calculator or computer, to solve the following equation to six significant digits: $e^p+2p=0$.

8.2 Use Newton's method, together with a digital computer, to solve the following pair of equations to four significant digits:

$$(a-b+2)e^{2b}+(a-b-2)e^{2a}+2(a-b)e^{a+b}$$
$$+(1+e)(b-a-2)e^{b}+(1+e)(b-a+2)e^{a}=0$$

$$\left[5(a+b-3)-1\right](e^{b}-e^{a})^{2}+5(e+1)(b-a)(e^{b}-e^{a})=0$$

(These are equations (7.89); they arise out of an optimal control problem.)

8.3 Minimization of the following objective function corresponds to adjusting the capacitances c_1 and c_2 in the interstage coupling of a tuned amplifier in such a manner as to maximize the power transfer to the load:

$$F(c_1,c_2)=(11-c_1-c_2)^2+(1+c_1+10c_2-c_1c_2)^2$$

Starting from the point $(c_1,c_2)=(14,3)$, and using a computer or calculator, determine the minimizing values of c_1 and c_2 to three significant digits by means of the following:

(*a*) Standard steepest descent;

(*b*) Best-step Newton descent.

Use any convenient scheme for performing the one-dimensional minimizations.

(*c*) What happens if the Newton descent is started from the point $(c_1,c_2)=(18,3)$?

8.4 The function $g(t)=t^2-t$ has roots at $t=0,1$. Newton's method converges slowly toward these roots if the initial estimate t_0 is far from the roots. The convergence can be accelerated by means of an overrelaxation factor ω, as in (8.13). Show that in order to maximize the rate of convergence, we need $\omega\approx 2$ while t_k is far from the roots and $\omega\approx 1$ while t_k is close to the roots.

8.5 *Degenerate nonlinear equations*: we wish to solve the nonlinear equation $\mathbf{G}(\mathbf{x})=\boldsymbol{\theta}$, where $\mathbf{G}:\mathcal{V}\to\mathcal{W}$, and $\mathcal{V}$ and $\mathcal{W}$ are Hilbert spaces. Newton's method works only if $\mathbf{G}'(\mathbf{x}_k)$ is invertible at each iteration. If $\mathbf{G}'(\mathbf{x}_k)$ becomes degenerate, or if $\dim(\mathcal{V})\neq\dim(\mathcal{W})$, then we can carry out a "least-square" Newton iteration based on the pseudoinverse:

$$\mathbf{x}_{k+1}=\mathbf{x}_k-\mathbf{G}'(\mathbf{x}_k)^{\dagger}\mathbf{G}(\mathbf{x}_k), \qquad k=0,1,2,\ldots$$

(*a*) Assume $\mathbf{G}'(\mathbf{x})\mathbf{G}'(\mathbf{x})^*$ is invertible. Then the least-square Newton iteration can be expressed as

$$\mathbf{x}_{k+1}=\mathbf{x}_k-\mathbf{G}'(\mathbf{x}_k)^*(\mathbf{G}'(\mathbf{x}_k)\mathbf{G}'(\mathbf{x}_k)^*)^{-1}\mathbf{G}(\mathbf{x}_k)$$

Apply this iteration to the equation $\mathbf{G}(\xi_1,\xi_2)=\xi_1\xi_2=0$. In what sense does it solve the equation?

(*b*) Assume $\mathbf{G}'(\mathbf{x})^*\mathbf{G}'(\mathbf{x})$ is invertible. Then the least-square Newton iteration can be expressed as

$$\mathbf{x}_{k+1}=\mathbf{x}_k-(\mathbf{G}'(\mathbf{x}_k)^*\mathbf{G}'(\mathbf{x}_k))^{-1}\mathbf{G}'(\mathbf{x}_k)^*\mathbf{G}(\mathbf{x}_k)$$

(This iteration is the Gauss–Newton method for obtaining least-square solutions to nonlinear equations.) Apply this iteration to the equation

$$\mathbf{G}(\xi)=\begin{pmatrix}\xi^2\\ \xi^2-4\end{pmatrix}=\begin{pmatrix}0\\0\end{pmatrix}$$

8.6 Determine how to apply the quasilinearization technique of (8.5)–(8.8) to a differential system with *nonhomogeneous* boundary conditions.

*8.7 *Contraction operators*: let $\mathcal{V}$ be an inner product space. Let the closed set $\mathcal{S}$ in $\mathcal{V}$ be invariant under the operator $\mathbf{G}$; that is, assume $\mathbf{G}(\mathcal{S})$ is in $\mathcal{S}$. We say $\mathbf{G}$ is a **contraction operator** on $\mathcal{S}$ if for all $\mathbf{x}$ and $\mathbf{y}$ in $\mathcal{S}$,

$$\|\mathbf{G}(\mathbf{x})-\mathbf{G}(\mathbf{y})\|\leqslant c\|\mathbf{x}-\mathbf{y}\|$$

for some constant $c<1$. A solution to the equation $\mathbf{G}(\mathbf{x})=\mathbf{x}$ is called a *fixed point* of $\mathbf{G}$.

(*a*) Show that if $\mathbf{G}$ is a contraction operator on $\mathcal{S}$, the successive approximation $\mathbf{x}_{k+1}=\mathbf{G}(\mathbf{x}_k)$, $k=0,1,\ldots,$ converges to a fixed point of $\mathbf{G}$ for any starting vector $\mathbf{x}_0$ in $\mathcal{S}$. (Hint: show that the sequence $\|\mathbf{x}_{k+1}-\mathbf{x}_k\|\to 0$ as $k\to\infty$.)

(*b*) Show that if $\mathbf{G}$ is a contraction operator on $\mathcal{S}$, the fixed point in $\mathcal{S}$ is unique; that is, repeated application of $\mathbf{G}$ contracts the set $\mathcal{S}$ into a single point.

(*c*) Show that if $\|\mathbf{G}'(\mathbf{x})\|<1$ for all $\mathbf{x}$ in a convex set $\mathcal{S}$, and if $\mathcal{S}$ is invariant under $\mathbf{G}$, then $\mathbf{G}$ is a contraction operator on $\mathcal{S}$. (Hint: use a two-term Taylor expansion of $\mathbf{G}(\mathbf{x})$.)

8.8 *Solving equations by iteration*: we wish to solve the matrix equation

$\mathbf{Ax}=\mathbf{y}$, where $\mathbf{A}$ is $n\times n$ and $\mathbf{x}$ is $n\times 1$. If n is large and $\mathbf{A}$ is sparse, Gaussian elimination is more difficult and costly than iterative methods of solution. Assume the equations are normalized so that the diagonal elements of $\mathbf{A}$ are $a_{ii}=1, i=1,\ldots,n$.

(*a*) Pick an initial estimate of the solution, $\mathbf{x}_0=((\xi_1)_0\cdots(\xi_n)_0)^{\mathrm{T}}$. We obtain an improved iterate $\mathbf{x}_{k+1}$ from $\mathbf{x}_k$ in the following manner: we solve the ith equation for a new value of the ith variable $(\xi_i)_{k+1}$ in terms of the previous values of the other variables; we carry out this operation for $i=1,\ldots,n$. This iterative process is known as the **Jacobi method** (or the method of simultaneous displacements). The iteration can be expressed in the general matrix form $\mathbf{x}_{k+1}=\mathbf{Bx}_k+\mathbf{y}$. Express $\mathbf{B}$ in terms of $\mathbf{A}$. Under what conditions does the iteration converge to a solution of the equation $\mathbf{Ax}=\mathbf{y}$? That is, under what conditions is the operator $\mathbf{G}(\mathbf{x})\triangleq\mathbf{Bx}+\mathbf{y}$ a contraction operator (P&C 8.7)?

(*b*) The rate of convergence of the Jacobi method can be improved by using updated information as soon as it is available. That is, we solve the ith equation for $(\xi_i)_{k+1}$ in terms of $(\xi_1)_{k+1},\ldots,(\xi_{i-1})_{k+1},(\xi_{i+1})_k,\ldots,(\xi_n)_k$. This iterative process is known as the **Gauss–Seidel method** (or the method of successive displacements). Show that the Gauss–Seidel method can be expressed as the matrix iteration $\mathbf{x}_{k+1}=-(\mathbf{I}+\mathbf{L})^{-1}\mathbf{Ux}_k+(\mathbf{I}+\mathbf{L})^{-1}\mathbf{y}$, where $\mathbf{L}$ and $\mathbf{U}$ are the lower and upper triangular parts of $\mathbf{A}$. The iteration converges if and only if the eigenvalues of $-(\mathbf{I}+\mathbf{L})^{-1}\mathbf{U}$ are less than one in magnitude.

(*c*) The Gauss–Seidel method is sometimes suitable for solving nonlinear equations. Use the method to solve the equation

$$\mathbf{G}\begin{pmatrix}\xi_1\\ \xi_2\end{pmatrix}=\begin{pmatrix}\xi_1\xi_2^2\\ \xi_1^2+\xi_2\end{pmatrix}=\begin{pmatrix}1\\ 2\end{pmatrix}$$

using the starting vector $\mathbf{x}_0=(10\ \ 10)^{\mathrm{T}}$. Note that $\mathbf{x}=(1\ \ 1)^{\mathrm{T}}$ is a solution to these equations. Why does the iteration not converge to this solution?

*8.9 *Direction of steepest descent*: define a functional $\mathbf{F}$: $\mathcal{V}\rightarrow\mathcal{R}$ where $\mathcal{V}$ is a Hilbert space with inner product $\langle\cdot,\cdot\rangle$. Let $\mathbf{h}$ be a vector in $\mathcal{V}$. The rate of decrease of $\mathbf{F}(\mathbf{x})$ in the direction $\mathbf{h}$ is defined to be

$$-\lim_{\|\mathbf{h}\|\rightarrow 0}\frac{\mathbf{F}(\mathbf{x}+\mathbf{h})-\mathbf{F}(\mathbf{x})}{\|\mathbf{h}\|}$$

(*a*) Show that the rate of decrease of $\mathbf{F}(\mathbf{x})$ in the direction $\mathbf{h}$ is maximum for $\mathbf{h}$ aligned with $-\nabla\mathbf{F}(\mathbf{x})$.

(*b*) Assume $\mathbf{h}$ is constrained to lie in a subspace $\mathcal{W}$ of $\mathcal{V}$. Show that the rate of decrease of $\mathbf{F}(\mathbf{x})$ in the direction $\mathbf{h}$ is maximum for $\mathbf{h}$ aligned with the orthogonal projection of $-\nabla\mathbf{F}(\mathbf{x})$ on $\mathcal{W}$. Note that the definitions of the rate of decrease, the gradient, and the orthogonal projection all depend upon the inner product.

8.10 Let $\mathbf{x}=(\xi_1\ \xi_2)^{\mathrm{T}}$ denote a vector in $\mathfrak{M}^{2\times 1}$, and define $\mathbf{F}(\mathbf{x})=\frac{1}{2}(a\xi_1^2+b\xi_2^2)$, where $0<a\leqslant b$. In Example 1, Section 8.2 we apply the standard steepest descent iteration to $\mathbf{F}$, and show that for a starting vector $\mathbf{x}_0$ such that $a^2\xi_1^2=b^2\xi_2^2$ we have $\|\mathbf{x}_{k+2}\|=[(b-a)^2/(b+a)^2]\|\mathbf{x}_k\|$. Show that for *any* starting vector, the iterates are reduced by *at least* the factor $(b-a)^2/(b+a)^2$ with each two iterations.

8.11 *Computational efficiency of descent methods*: the number of multiplications (and divisions) required to carry out a computation is usually used as a measure of the cost of that computation.

(*a*) The number of multiplications involved in computing a single value of a function depends strongly upon the nature of the function and the desired accuracy. Function evaluations often require an iterative computation. Determine the number of multiplications required to evaluate $f(0.5)=e^{0.5}$ to four significant figures using the Taylor series expansion, $e^t=1+t+t^2/2!+t^3/3!+\cdots$.

(*b*) Determine the number of function evaluations and multiplications required to carry out one steepest descent step for an n-dimensional scalar-valued function, $\mathbf{F}(\mathbf{x})$. Assume that 20 evaluations of $\mathbf{F}(\mathbf{x}_{k+1})$ are required to determine the step length α_k.

(*c*) Repeat (*b*) for one step of the best-step Newton descent. Assume a matrix inversion requires $n^3/3$ multiplications.

(*d*) Repeat (*b*) for one step of the DFP descent.

(*e*) Repeat (*b*) for one step of the conjugate-gradient descent.

(*f*) Compare (*b*)–(*e*) by expressing each result in terms of multiplications per iteration. Assume each n-dimensional function evaluation requires $10n$ multiplications.

8.12 Let $\mathbf{x}=(\xi_1\ \xi_2\ \xi_3)^{\mathrm{T}}$ denote a vector in $\mathfrak{M}^{3\times 1}$. Beginning with $\mathbf{x}_0=(1\ 1\ 1)^{\mathrm{T}}$, minimize $\mathbf{F}(\mathbf{x})=\frac{1}{2}(\xi_1^2+4\xi_2^2+9\xi_3^2)$ by means of

(*a*) DFP descent (with $\mathbf{H}_0=\mathbf{I}$);

(*b*) Conjugate-gradient descent;

(*c*) Accelerated steepest descent.

8.13 Let $\mathbf{x}=(\xi_1\ \xi_2)^{\mathrm{T}}$ denote a vector in $\mathfrak{M}^{2\times 1}$. Beginning with $\mathbf{x}_0=(10$

$1)^T$, determine the first two iterates ($\mathbf{x}_1$ and $\mathbf{x}_2$) in the descent of the function $\mathbf{F}(\mathbf{x}) = \frac{1}{2}(\xi_1^2 + 10\xi_2^2) + \xi_2^4$, using

(*a*) DFP descent (with $\mathbf{H}_0 = \mathbf{I}$);

(*b*) Conjugate-gradient descent;

(*c*) Accelerated steepest descent.

(*d*) Compare the results of (*a*)–(*c*) with the steepest descent and Newton descent iterates of Figure 8.5 (Example 2, Section 8.2).

8.14 Let $\mathbf{F}(\mathbf{x}) = \frac{1}{2}\mathbf{x}^T\mathbf{A}\mathbf{x} - \mathbf{y}^T\mathbf{x}$, where $\mathbf{x}$ and $\mathbf{y}$ are $n \times 1$ matrices and $\mathbf{A}$ is a symmetric positive-definite $n \times n$ matrix. Let $\mathbf{Q}$ be another symmetric positive-definite $n \times n$ matrix. We wish to minimize $\mathbf{F}(\mathbf{x})$ by steepest descent using the gradients defined by the inner product $\langle \mathbf{x}, \mathbf{y} \rangle \triangleq \mathbf{y}^T\mathbf{Q}\mathbf{x}$. That is, we wish to use the iteration $\mathbf{x}_{k+1} = \mathbf{x}_k - \alpha_k \nabla\mathbf{F}(\mathbf{x}_k)$.

(*a*) Find $\nabla\mathbf{F}(\mathbf{x})$.

(*b*) Assume α_k is selected to maximize the descent at the kth iteration. Find α_k explicitly in terms of $\mathbf{x}_k$ and $\nabla\mathbf{F}(\mathbf{x}_k)$.

8.15 Define $\mathbf{F}$: $\mathcal{L}_2(0,1) \to \mathcal{R}$ by $\mathbf{F}(\mathbf{u}) \triangleq \frac{1}{2}\int_0^1 e^t \mathbf{u}^2(t)\,dt$. We wish to minimize $\mathbf{F}(\mathbf{u})$ by means of the DFP descent (with $\mathbf{H}_0 = \mathbf{I}$). Let $\mathbf{u}_0(t) = e^{-t}$. The first iterate, $\mathbf{u}_1(t) = e^{-t} - 1/(e-1)$, is computed in Example 3, Section 8.2. Find the *direction* of the second DFP descent step; that is, find $-\mathbf{H}_1\nabla\mathbf{F}(\mathbf{u}_1)$.

8.16 Let $\mathbf{x} = (\xi_1\ \xi_2\ \xi_3)^T$ denote a vector in the standard inner product space $\mathfrak{M}^{3\times 1}$. Use the gradient projection method to minimize $\mathbf{F}(\mathbf{x}) \triangleq \frac{1}{2}(\xi_1^2 + \xi_2^2 + \xi_3^2)$ subject to $\mathbf{T}_1(\mathbf{x}) \triangleq \xi_2 + \xi_3 = 10$ and $\mathbf{T}_2(\mathbf{x}) \triangleq \xi_3 \geq 2$. Let the initial iterate be $\mathbf{x}_0 = (10\ 0\ 10)^T$.

8.17 Let $\mathbf{x} = (\xi_1\ \xi_2\ \xi_3)^T$ denote a vector in the standard inner product space $\mathfrak{M}^{3\times 1}$. We wish to minimize $\mathbf{F}(\mathbf{x}) \triangleq \frac{1}{2}(\xi_1^2 + 4\xi_2^2 + 9\xi_3^2)$ subject to $\mathbf{T}_1(\mathbf{x}) \triangleq \xi_2 + \xi_3 \geq 10$ and $\mathbf{T}_2(\mathbf{x}) \triangleq \xi_3 \geq -5$. Determine the first two steps of the gradient projection iteration for $\mathbf{x}_0 = (10\ 10\ 10)^T$.

8.18 *Finding a feasible point*: some techniques for constrained minimization require an initial iterate which satisfies all the constraints which are imposed upon the variables. Suppose the constraints are $\mathbf{G}_i(\mathbf{x}) \geq 0$, $i = 1, \ldots, m$, where $\mathbf{G}_i$ is a functional acting on a Hilbert space $\mathcal{V}$. The penalty function concept can be used to find a point $\mathbf{x}$ which is feasible. Let $\mathbf{x}_0$ be chosen arbitrarily; $\mathbf{x}_0$ may be feasible with respect to some of the constraints and infeasible with respect to the others. Define the functional

$$\mathbf{F}(\mathbf{x}) \triangleq -\sum_{\substack{i \\ \text{infeasible}}} \mathbf{G}_i(\mathbf{x}) - c \sum_{\substack{i \\ \text{feasible}}} \ln(\mathbf{G}_i(\mathbf{x}))$$

Apply an unconstrained descent technique to $\mathbf{F}(\mathbf{x})$. The descending sequence of iterates will become feasible with respect to one or more of the "infeasible" constraints. When this happens, remove the corresponding constraint from the infeasible sum, and include it in the feasible sum. Continuation of this process will eventually produce an iterate which is feasible with respect to all the constraints. (Note that we do not need to let c approach zero.) Use this procedure to find a point $\mathbf{x}=(\xi_1\ \xi_2)^{\mathrm{T}}$ in $\mathfrak{M}^{2\times 1}$ which is feasible with respect to the constraints $\mathbf{G}_1(\mathbf{x})=\xi_1-\xi_2^2\geqslant 0$ and $\mathbf{G}_2(\mathbf{x})=\xi_2-1\geqslant 0$. Begin with the vector $\mathbf{x}_0=(0\ 1)^{\mathrm{T}}$.

8.19 Let $\mathbf{x}=(\xi_1,\xi_2)$ denote a vector in $\mathfrak{R}^2$. Apply the penalty function approach to determine the minimum of $\mathbf{F}(\mathbf{x}) \triangleq \xi_1+2\xi_2$ subject to $\mathbf{G}_1(x) \triangleq \xi_1-\xi_2^2\geqslant 0$ and $\mathbf{G}_2(x) \triangleq \xi_2-1=0$; use the logarithmic penalty function for $\mathbf{G}_1$ and the quadratic loss penalty function for $\mathbf{G}_2$. Solve analytically for the minimum $\mathbf{x}(c_1,c_2)$ of the penalized objective function (as a function of the two penalty constants). Solve also for the corresponding approximation to the Lagrange multiplier vector, $\boldsymbol{\lambda}(c_1,c_2)$. Compute the values of $\mathbf{x}(c_1,c_2)$ and $\boldsymbol{\lambda}(c_1,c_2)$ for a sequence of penalty constants which produces convergence toward the exact solution.

8.20 Let $\mathbf{x}=(\xi_1,\xi_2)$ denote a vector in $\mathfrak{R}^2$. We can minimize $\mathbf{F}(\mathbf{x}) \triangleq \xi_1+2\xi_2$ subject to $\mathbf{G}_1(\mathbf{x}) \triangleq \xi_1-\xi_2^2\geqslant 0$ and $\mathbf{G}_2(\mathbf{x}) \triangleq \xi_2-1\geqslant 0$ by the following approach. Attach the nonlinear constraint $\mathbf{G}_1$ to the objective function $\mathbf{F}$ by means of a logarithmic penalty function. Then apply the gradient projection method to the penalized objective function, subject to the linear constraint $\mathbf{G}_2$. Repeat the minimization for a decreasing sequence of penalty constants.

(*a*) Begin with $\mathbf{x}_0=(6,2)$, and compute the constrained minimum of the penalized objective function for the penalty constant $c=1$.

(*b*) Compute the approximate Lagrange multiplier vector which corresponds to $c=1$.

8.8 References

*[8.1] Acton, Forman S., *Numerical Methods that Work*, Harper & Row, New York, 1970.

[8.2] Beckman, F. S., "The Solution of Linear Equations by the Conjugate Gradient Method," *Mathematical Methods for Digital Computers*, A. Ralston and H. S. Wilf (Eds.), Wiley, New York, 1960.

[8.3] Bellman, Richard E. and Robert E. Kalaba, *Quasilinearization and Nonlinear Boundary-Value Problems*, American Elsevier, New York, 1965.

[8.4] Broyden, C. G., "Quasi-Newton Methods," Chapter 6 of *Numerical Methods for Unconstrained Optimization*, W. Murray (Ed.), Academic Press, New York, 1972.

*[8.5] Collatz, Lothar, *Functional Analysis and Numerical Mathematics*, Academic Press, New York, 1966.

[8.6] Daniel, James W., "Convergence of the Conjugate Gradient Method with Computationally Convenient Modifications," *Numer. Math.*, **10** (1967), 125–131.

[8.7] Davidon, W. C., "Variable Metric Method for Minimization," *A.E.C. Res. Dev. Rep. ANL*-5990 (rev.), 1959.

[8.8] Fiacco, Anthony V. and Garth P. McCormick, *Nonlinear Programming: Sequential Unconstrained Minimization Techniques*, Wiley, New York, 1968.

[8.9] Fletcher, R. and C. M. Reeves, "Function Minimization by Conjugate Gradients," *Computer J.*, **7** (1964), 149–154.

[8.10] Fletcher, R. and M. J. D. Powell, "A Rapidly Converging Descent Method for Minimization," *Computer J.*, **6** (1963), 163–168.

[8.11] Fletcher, R., "Minimizing General Functions Subject to Linear Constraints," Chapter 20 of *Numerical Methods for Non-Linear Optimization*, F. A. Lootsma (Ed.), Academic Press, New York, 1972.

[8.12] Forsythe, George E., "Singularity and Near Singularity in Numerical Analysis," *Am. Math. Mon.*, **65** (1958), 229–240.

[8.13] Forsythe, G. E. and T. S. Motzkin, "Asymptotic Properties of the Optimum Gradient Method," *Bull. Am. Math. Soc.*, **57**, (1951), 183.

[8.14] Goldfeld, S. M., R. E. Quandt, and H. F. Trotter, "Maximization by Quadratic Hill-Climbing," *Econometrica*, **34** (1966), 541–551.

[8.15] Hadley, G., *Linear Programming*, Addison-Wesley, Reading, Mass., 1962.

*[8.16] Hadley, G., *Nonlinear and Dynamic Programming*, Addison-Wesley, Reading, Mass., 1964.

[8.17] Kantorovich, L. V. and G. P. Akilov, *Functional Analysis on Normed Spaces*, Macmillan, New York, 1964.

[8.18] Kunzi, H. P., W. Krelle, and W. Oettli, *Nonlinear Programming*, Blaisdell, Waltham, Mass., 1966.

[8.19] Lootsma, F. A. (Ed.), *Numerical Methods for Non-Linear Optimization*, Academic Press, New York, 1972.

*[8.20] Luenberger, D. G., *Optimization by Vector Space Methods*, Wiley, New York, 1969.

[8.21] Luenberger, David G., "The Gradient Projection Method Along Geodesics," Management Sci., **18**, 11 (July 1972), 620–631.

[8.22] Lusternik, L. A. and V. J. Sobolev, *Elements of Functional Analysis*, Gordon and Breach, New York (for Hindustan Publishing Corp., Delhi, India), 1961.

[8.23] Pearson, J. D., "Variable Metric Methods of Minimization," *Computer J.*, **12** (1969), 171–178.

*[8.24] Pierre, Donald A., *Optimization Theory with Applications*, Wiley, New York, 1969.

[8.25] Powell, M. J. D., "A Survey of Numerical Methods for Unconstrained Optimization," *SIAM Rev.*, **12**, 1 (1970), 79–97.

[8.26] Powell, M. J. D., "Problems Related to Unconstrained Optimization," Chapter 3 of *Numerical Methods for Non-Linear Optimization*, F. A. Lootsma (Ed.), Academic Press, New York, 1972.

[8.27] Rall, L. B., "Convergence of the Newton Process to Multiple Solutions," *Numer. Math.*, **9** (1966), 23–37.

[8.28] Rosen, J. B., "The Gradient Projection Method for Nonlinear Programming, Part I, Linear Constraints," *J. Soc. Indust. Appl. Math.*, **8** (1960), 181–217.

[8.29] Rosen, J. B. and J. Kreuser, "A Gradient Projection Algorithm for Non-Linear Constraints," Chapter 21 of *Numerical Methods for Non-Linear Optimization*, F. A. Lootsma (Ed.), Academic Press, New York, 1972.

[8.30] Rothe, E. H., "Gradient Mappings," *Bull. Am. Math. Soc.*, **57** (1953), 5–19.

[8.31] Sage, Andrew P., *Optimum Systems Control*, Prentice-Hall, Englewood Cliffs, N. J., 1968.

[8.32] Shah, B. V., R. J. Buehler, and O. Kempthorne, "Some Algorithms for Minimizing a Function of Several Variables," *J. Soc. Indust. Appl. Math.*, **12**, 1 (March 1964), 74–92.

[8.33] Tokumaru, H., N. Adachi, and K. Goto, "Davidon's Method for Minimization Problems in Hilbert Space with an Application to Control Problems," *SIAM J. Control*, **8** (May 1970), 163–178.

[8.34] Wolfe, Philip, "Convergence Conditions for Ascent Methods," SIAM Review, Vol. 11, No. 2, April 1969, 226–235.

[8.35] Zoutendijk, G. "Nonlinear Programming: A Numerical Survey," SIAM J. Control, Vol. 4, No. 1, 1966, 194–210.

[8.36] Zoutendijk, G., "Computational Methods in Nonlinear Programming," *Integer and Nonlinear Programming*, (J. Abadie, ed.), North-Holland Publishing Co., Amsterdam 37–86.

[8.37] Zoutendijk, G., *Methods of Feasible Directions. A Study in Linear and Non-Linear Programming*, Elsevier Publishing Co., New York, 1960.

Appendix 1. Matrices and Determinants

Matrices

An $m \times n$ matrix is a rectangular array of numbers with m rows and n columns. We often represent a matrix by a single symbol; for example,

$$\mathbf{A} \triangleq \begin{pmatrix} a_{11} & a_{12} & \cdots & a_{1n} \\ a_{21} & a_{22} & & \vdots \\ \vdots & \vdots & & \vdots \\ a_{m1} & a_{m2} & \cdots & a_{mn} \end{pmatrix}$$

We sometimes refer to the single element in row i and column j of $\mathbf{A}$ by the symbol a_{ij}. We also refer to the set of all such elements as $\{a_{ij}\}$. Matrices are used as models for various physical and mathematical objects. (See Section 1.5, where matrices are used to represent algebraic equations.) However, in this appendix we merely define rules for manipulating matrices.

Two matrices are **equal** if and only if they have precisely the same elements; thus $\mathbf{A}=\mathbf{B}$ means $\mathbf{A}$ and $\mathbf{B}$ have the same number of rows and columns, and that $a_{ij}=b_{ij}$ for all rows i and columns j. A matrix all of whose elements are zero is called the **zero matrix** and denoted $\mathbf{\Theta}$. We define the **sum** (or difference) of two matrices $\mathbf{A}$ and $\mathbf{B}$ as the matrix obtained by adding (or subtracting) corresponding elements of $\mathbf{A}$ and $\mathbf{B}$. Then $\mathbf{A}+\mathbf{B}$ has in its ith row and jth column the element $a_{ij}+b_{ij}$. Addition of $\mathbf{A}$ and $\mathbf{B}$ is defined only if $\mathbf{A}$ and $\mathbf{B}$ have the same number of rows and columns. If c is a scalar, multiplication of $\mathbf{A}$ by c means multiplication of each element of $\mathbf{A}$; the i,j element of $c\mathbf{A}$ is ca_{ij}.

Two matrices $\mathbf{A}$ and $\mathbf{B}$ are **conformable in the order AB** if the number of columns in $\mathbf{A}$ equals the number of rows in $\mathbf{B}$. If $\mathbf{A}$ and $\mathbf{B}$ are conformable in the order **AB**, we can define the matrix product **AB** by $\mathbf{C} \triangleq \mathbf{AB}$, where

$$c_{ij} = \sum_k a_{ik} b_{kj}$$

Example 1. Matrix Multiplication. Let

$$\mathbf{A} = \begin{pmatrix} 1 & 2 & 3 \\ 2 & 4 & 1 \end{pmatrix} \quad \text{and} \quad \mathbf{B} = \begin{pmatrix} 2 & 3 & -1 \\ 5 & 3 & 0 \\ 4 & 1 & 1 \end{pmatrix}$$

Then

$$\mathbf{AB} = \begin{pmatrix} 1 & 2 & 3 \\ 2 & 4 & 1 \end{pmatrix} \begin{pmatrix} 2 & 3 & -1 \\ 5 & 3 & 0 \\ 4 & 1 & 1 \end{pmatrix} = \begin{pmatrix} 24 & 12 & 2 \\ 28 & 19 & -1 \end{pmatrix} = \mathbf{C}$$

Thus the i,j element of **AB** is obtained by multiplying the elements in column j of **B** by the elements in row i of **A** and adding the individual products. In Example 1, row i of **B** and column j of **A** do not contain the same number of elements; they cannot be multiplied element by element. Because **A** and **B** are not conformable in the order **BA**, the product **BA** is not defined.

Exercise 1. Verify by simple examples the following properties of matrices:

1. Addition of matrices is commutative and associative:

$$\mathbf{A} + \mathbf{B} = \mathbf{B} + \mathbf{A}$$
$$\mathbf{A} + (\mathbf{B} + \mathbf{C}) = (\mathbf{A} + \mathbf{B}) + \mathbf{C}$$

2. Multiplication of matrices is distributive over addition:

$$\mathbf{A}(\mathbf{B} + \mathbf{C}) = \mathbf{AB} + \mathbf{AC}$$

where **B** and **C** have the same number of rows and columns, and **A** is conformable to **B** and **C**.

3. Multiplication of matrices is associative but generally not commutative (even if **A** and **B** are conformable in either order):

$$\mathbf{A}(\mathbf{BC}) = (\mathbf{AB})\mathbf{C}$$
$$\mathbf{AB} \neq \mathbf{BA} \text{ except in special cases}$$

The **transpose** of a matrix **A**, denoted by $\mathbf{A}^T$, is the matrix obtained by interchanging the rows and columns of **A**. The i,j element of $\mathbf{A}^T$ is a_{ji}. The **main diagonal** of a matrix **A** consists in the elements $a_{11}, a_{22}, \ldots, a_{pp}$, where p

is the smaller of m and n. Thus in Example 1, the main diagonal of **C** consists in the elements 24 and 19. Clearly, $\mathbf{A}^T$ is just **A**, flipped about its main diagonal. If $\mathbf{A}^T = \mathbf{A}$, we say **A** is **symmetric**. If $\mathbf{A}^T = -\mathbf{A}$, we call **A** **skew symmetric**.

Exercise 2. Show that if **A** is conformable to **B**, then

$$(\mathbf{AB})^T = \mathbf{B}^T\mathbf{A}^T$$

An $n \times n$ matrix is called a **square matrix** of **order n**. A square matrix in which the elements not on the main diagonal are zero is called a **diagonal matrix**. A diagonal matrix with all its diagonal elements equal to 1 is a **unit (or identity) matrix**, denoted by the symbol **I**.

Exercise 3. Verify that if **I** is conformable to **A**, then $\mathbf{IA} = \mathbf{A}$.

A matrix consisting of a single column is called a **column matrix**. A matrix consisting of a single row is a **row matrix**.

Determinants

The determinant of a square matrix **A**, denoted by det(**A**) or $|\mathbf{A}|$ is a scalar-valued function of the elements of **A**. For a 2×2 matrix, the determinant is defined by

$$\begin{vmatrix} a_{11} & a_{12} \\ a_{21} & a_{22} \end{vmatrix} \stackrel{\Delta}{=} a_{11}a_{22} - a_{12}a_{21} \tag{A1.1}$$

The solution to the pair of linear equations

$$\begin{aligned} a_{11}\xi_1 + a_{12}\xi_2 &= \eta_1 \\ a_{21}\xi_1 + a_{22}\xi_2 &= \eta_2 \end{aligned} \tag{A1.2}$$

can be expressed simply in terms of determinants:

$$\begin{aligned} \xi_1 &= \frac{\eta_1 a_{22} - \eta_2 a_{12}}{a_{11}a_{22} - a_{21}a_{12}} = \frac{\begin{vmatrix} \eta_1 & a_{12} \\ \eta_2 & a_{22} \end{vmatrix}}{\begin{vmatrix} a_{11} & a_{12} \\ a_{21} & a_{22} \end{vmatrix}} \\ \xi_2 &= \frac{\begin{vmatrix} a_{11} & \eta_1 \\ a_{21} & \eta_2 \end{vmatrix}}{\begin{vmatrix} a_{11} & a_{12} \\ a_{21} & a_{22} \end{vmatrix}} \end{aligned} \tag{A1.3}$$

In point of fact, this relationship between linear equations and determinants is the motivation behind the definition of determinants. Both (A1.1) and (A1.3) extend to nth-order matrices. Let

$$\begin{aligned} a_{11}\xi_1 + a_{12}\xi_2 + \cdots + a_{1n}\xi_n &= \eta_1 \\ a_{21}\xi_1 + a_{22}\xi_2 + \cdots + a_{2n}\xi_n &= \eta_2 \\ \vdots \qquad\qquad & \quad \vdots \\ a_{n1}\xi_1 + a_{n2}\xi_2 + \cdots + a_{nn}\xi_n &= \eta_n \end{aligned} \tag{A1.4}$$

and define

$$\mathbf{A} \triangleq \begin{pmatrix} a_{11} & \cdots & a_{1n} \\ \vdots & & \vdots \\ a_{n1} & \cdots & a_{nn} \end{pmatrix}$$

Were we write the solution for ξ_i of (A1.4) as a ratio of determinants in the manner of (A1.3), we would find that $|\mathbf{A}|$ must consist in a sum of all possible products of elements from $\mathbf{A}$ that can be formed using only one element from each row and each column of $\mathbf{A}$. Fortunately, $|\mathbf{A}|$ can be expressed in terms of the determinants of its smaller submatrices. For a third-order matrix,

$$\begin{vmatrix} a_{11} & a_{12} & a_{13} \\ a_{21} & a_{22} & a_{23} \\ a_{31} & a_{32} & a_{33} \end{vmatrix} = a_{11}\begin{vmatrix} a_{22} & a_{23} \\ a_{32} & a_{33} \end{vmatrix} - a_{21}\begin{vmatrix} a_{12} & a_{13} \\ a_{32} & a_{33} \end{vmatrix} + a_{31}\begin{vmatrix} a_{12} & a_{13} \\ a_{22} & a_{23} \end{vmatrix}$$

This third-order determinant is written as an expansion based on elements from the first column of the matrix; it could also be written as an expansion based on any other row or column of the matrix. We must attach to each element of the row or column on which the expansion is based, a sign and a second-order determinant; for the element a_{ij}, the sign is $(-1)^{i+j}$; the determinant which we attach is $|\mathbf{A}(i|j)|$, where $\mathbf{A}(i|j)$ is the matrix $\mathbf{A}$ with row i and column j removed.

The above expansion of $|\mathbf{A}|$ in terms of lower-order determinants is known as the **Laplace expansion** of $|\mathbf{A}|$. We will use it as our definition of $|\mathbf{A}|$. Thus for the nth-order matrix $\mathbf{A}$, we define the **determinant** by

$$|\mathbf{A}| \triangleq \sum_{i \text{ or } j} (-1)^{i+j} a_{ij} |\mathbf{A}(i|j)| \tag{A1.5}$$

where the sum is taken either over all i (for some column j) or over all j

(for some row i). The determinant $|\mathbf{A}(i|j)|$ is of order $n-1$; it is called the **minor** of the element a_{ij}. If the sign is included, $(-1)^{i+j}|\mathbf{A}(i|j)|$, it is called the **co-factor** of the element a_{ij}.

If $|\mathbf{A}|\neq 0$, the solution to (A1.4) can be expressed as a ratio of determinants in the manner of (A1.3),

$$\xi_i = \frac{|\mathbf{A}(i)|}{|\mathbf{A}|} \tag{A1.6}$$

where $\mathbf{A}(i)$ is the matrix $\mathbf{A}$ with its ith column replaced by the column of numbers $\{\eta_i\}$ in the right hand side of (A1.4). Equation (A1.6) is known as **Cramer's formula.**

We can compute $|\mathbf{A}|$ by repeated use of the Laplace expansion (A1.5). Each determinant is expanded in terms of lower-order determinants until we have expanded even the second-order determinants as in (A1.1). The amount of computation involved in this process is extremely large. (There are $n!$ terms, each of which is a product of n elements from $\mathbf{A}$.) We reduce the computation drastically by using the following properties of determinants to simplify $\mathbf{A}$ before computing $|\mathbf{A}|$.

1. The value of $|\mathbf{A}|$ is not changed if we add to one row of $\mathbf{A}$ some multiple of another row of $\mathbf{A}$.
2. The sign of $|\mathbf{A}|$ is reversed if we interchange any two rows of $\mathbf{A}$.
3. If we multiply one row of $\mathbf{A}$ by c, then $|\mathbf{A}|$ is multiplied by c.

Exercise 4. Verify properties 1, 2, and 3 for a third-order determinant.

Example 2. Simplification of Determinants. Employing property 1, we use multiples of the rows enclosed in dotted lines to introduce zeros below the main diagonal. Using property 3, we factor a constant out of each row in order to make the diagonal elements equal to one. Evaluation of the determinant of the resulting triangular matrix $\mathbf{B}$ (which has ones on the main diagonal) is trivial; the expansion of $|\mathbf{B}|$ in terms of its first column using (A1.5) yields $|\mathbf{B}|=1$.

$$\begin{vmatrix} 1 & 2 & 2 \\ 2 & 3 & 5 \\ 3 & 2 & 5 \end{vmatrix} = \begin{vmatrix} 1 & 2 & 2 \\ 0 & -1 & 1 \\ 0 & -4 & -1 \end{vmatrix} = (-1)\begin{vmatrix} 1 & 2 & 2 \\ 0 & 1 & -1 \\ 0 & -4 & -1 \end{vmatrix}$$

$$= (-1)\begin{vmatrix} 1 & 2 & 2 \\ 0 & 1 & -1 \\ 0 & 0 & -5 \end{vmatrix} = (-1)(-5)\begin{vmatrix} 1 & 2 & 2 \\ 0 & 1 & -1 \\ 0 & 0 & 1 \end{vmatrix} = 5$$

Note that we also could have introduced zeros above the main diagonal, obtaining the identity matrix $\mathbf{I}$ rather than the triangular matrix $\mathbf{B}$. Then $|\mathbf{I}|=1$.

There are two additional properties of determinants that are useful on occasion:

4. $|\mathbf{A}|=|\mathbf{A}^{\mathrm{T}}|$. This property is clearly true for second-order **A** [see (A1.1)]. For third-order **A** we see by (A1.5) that the expansion of $|\mathbf{A}|$ along column 1 is precisely the expansion of $|\mathbf{A}^{\mathrm{T}}|$ along row 1. Proof for nth-order **A** is by induction.

5. $|\mathbf{AB}|=|\mathbf{A}|\cdot|\mathbf{B}|$, where **A** and **B** are both $n\times n$ matrices. To understand this property, we write

$$|\mathbf{AB}|=\begin{vmatrix} \mathbf{A}^{(1)}\mathbf{B}_{(1)} & \cdots & \mathbf{A}^{(1)}\mathbf{B}_{(n)} \\ \vdots & & \vdots \\ \mathbf{A}^{(n)}\mathbf{B}_{(1)} & \cdots & \mathbf{A}^{(n)}\mathbf{B}_{(n)} \end{vmatrix} \tag{A1.7}$$

where $\mathbf{A}^{(i)}$ is row i of **A** and $\mathbf{B}_{(j)}$ is column j of **B**. Applying property 1 of determinants to (A1.7), we subtract from row 1 a multiple, c, of row n:

$$|\mathbf{AB}|=\begin{vmatrix} \left[\mathbf{A}^{(1)}-c\mathbf{A}^{(n)}\right]\mathbf{B}_{(1)} & \cdots & \left[\mathbf{A}^{(1)}-c\mathbf{A}^{(n)}\right]\mathbf{B}_{(n)} \\ \vdots & & \vdots \\ \mathbf{A}^{(n)}\mathbf{B}_{(1)} & \cdots & \mathbf{A}^{(n)}\mathbf{B}_{(n)} \end{vmatrix}$$

Clearly, we can use properties 1, 2, and 3 to simplify **A** as in Example 2, even though **A** has been multiplied on the right by **B**. Using these properties, we reduce **A** to the identity matrix **I**; the factors which we take outside the determinant bars constitute $|\mathbf{A}|$. (Some of these factors can be zero.) Therefore,

$$\begin{aligned} |\mathbf{AB}| &= |\mathbf{A}|\cdot\begin{vmatrix} \mathbf{I}_{(1)}\mathbf{B}^{(1)} & \cdots & \mathbf{I}_{(1)}\mathbf{B}^{(n)} \\ \vdots & & \vdots \\ \mathbf{I}_{(n)}\mathbf{B}^{(1)} & \cdots & \mathbf{I}_{(n)}\mathbf{B}^{(n)} \end{vmatrix} \\ &= |\mathbf{A}|\cdot|\mathbf{IB}| \\ &= |\mathbf{A}|\cdot|\mathbf{B}| \end{aligned}$$

Determinants are discussed extensively in Hohn [A.1].

Inverses

If **A** is square and there is a matrix **B** that satisfies $\mathbf{AB}=\mathbf{BA}=\mathbf{I}$, we denote **B** by $\mathbf{A}^{-1}$ and call it the **inverse of A**.

Example 3. An Inverse Matrix. If $\mathbf{A}=\begin{pmatrix}1 & 1\\ 2 & 3\end{pmatrix}$, then $\mathbf{A}^{-1}=\begin{pmatrix}3 & -1\\ -2 & 1\end{pmatrix}$.

The classical method of computing $\mathbf{A}^{-1}$ involves determinants and cofactors of **A**. However, the method is very inefficient. (An efficient method for computing $\mathbf{A}^{-1}$ is described in Section 1.5.) Still, the determinant of **A** does supply useful information about $\mathbf{A}^{-1}$. If $|\mathbf{A}|=0$, the matrix **A** is in some sense degenerate. When $|\mathbf{A}|=0$, Cramer's formula (A1.6) will not yield a solution to (A1.4); in point of fact, the equations represented by **A** are not independent. The inverse of **A** does not exist under these circumstances. We call a matrix **A** **singular** if $|\mathbf{A}|=0$.

Exercise 5. A matrix is singular if an only if its rows are dependent; that is, if and only if the simplification of the matrix using the determinant properties 1, 2, and 3 generates a zero row in the matrix. Evaluate the determinant of the following singular matrix.

$$\begin{pmatrix}2 & 1 & 2\\ 1 & 3 & 2\\ 4 & 7 & 6\end{pmatrix}$$

Exercise 6. Let **A** and **B** be nonsingular square $n\times n$ matrices. Use Exercise 2 and the definition of $\mathbf{A}^{-1}$ to show that

(*a*) $(\mathbf{AB})^{-1}=\mathbf{B}^{-1}\mathbf{A}^{-1}$
(*b*) $(\mathbf{A}^{\mathrm{T}})^{-1}=(\mathbf{A}^{-1})^{\mathrm{T}}$

References

* [A1.1] Hohn, Franz E., *Elementary Matrix Algebra*, 2nd ed., Macmillan, New York, 1964.

Appendix 2. Delta Functions and Linear System Equations

The motivation behind this appendix is the linear time-invariant dynamic system (or its model, the linear constant-coefficient differential equation with initial conditions). The complex exponential functions are well suited to the analysis of these systems. These functions are in some sense characteristic of constant-coefficient differential equations; for example, if we let the input u to the differential equation

$$\frac{df(t)}{dt}+af(t)=u(t) \tag{A2.1}$$

be the exponential $e^{\mu t}$, the output f is $e^{\mu t}/(\mu+a)$, an exponential of the same form. This "characteristic" property of the complex exponentials is the motivation behind Fourier series and Laplace transforms. In Fourier series, we resolve a periodic input into a sum of sines and cosines (complex exponentials), then solve the equation for each component of the input. In Laplace transformation, we resolve a more general input into a continuum of functions $\{e^{st}\}$ (where s is a complex variable taking values on a line parallel to the axis of imaginaries in the complex s plane). For each exponential input the output is easily determined. Thus the simple system (A2.1) acts like multiplication by $1/(s+a)$ on the input e^{st}. We call the function $1/(s+a)$ the *transfer function* for the system. The transfer function is essentially equivalent to the differential equation (with zero initial conditions) as a model for the system. Analysis of a system by decomposing inputs into complex exponentials (e.g., sines and cosines of various frequencies) constitutes **frequency-domain analysis**. See Schwartz and Friedland [A2.3] for a discussion of Fourier series and the Laplace transform.

There is another set of functions that are in a sense characteristic of linear time-invariant systems. Abruptly changing inputs, such as fast-rising steps or sharp pulses, yield system responses which adequately describe the system. Because these responses are experimentally measurable functions of time, we refer to their use as **time-domain analysis** of the system. We first discuss the mathematical properties of these abruptly changing functions, then describe the sense in which the responses to these functions characterize the system.

Delta Functions

Figure A2.1*a* shows a sequence of "step" functions which are progressively more rapidly rising as n increases. Figure A2.1*b* shows a sequence of progressively sharper pulses; the nth pulse is the derivative of the nth step. It is computationally more convenient to work with the idealizations of

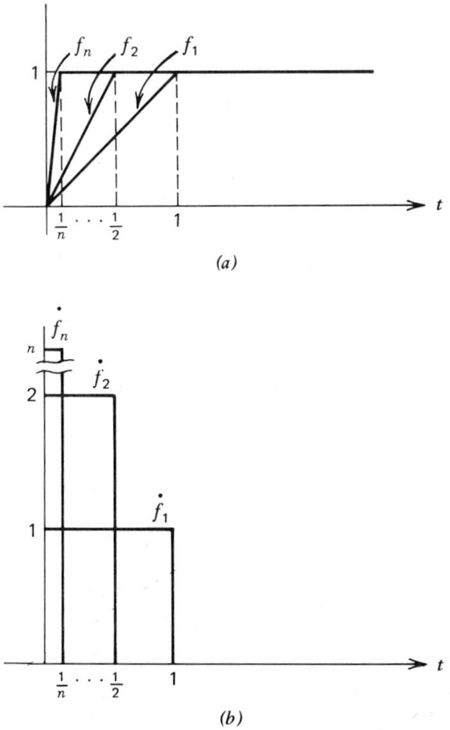

Figure A2.1. Approximations to step and impulse functions.

these functions—their limits as $n \to \infty$. Unfortunately, neither the step nor the pulse is a true mathematical function in the limit. Their values are undefined at $t=0$. We take a practical approach to this difficulty. In the physical world, variables are continuous with time. We cannot, without infinite expenditure of energy, move things abruptly. Even the sharp-cornered functions of Figure A2.1*a* are idealizations. We think only in terms of smooth functions which are arbitrarily close, in an obvious sense, to the idealized step or pulse. The response of a physical system to the idealized inputs can be approached arbitrarily closely by using smooth inputs. For instance, a bat hitting a baseball is sufficiently like an ideal pulse (or impulse) if the energy of the bat is transferred to the ball before the ball moves appreciably; the time constants of the system must be large compared to the duration of the impulse. The ideal impulse represents an instantaneous insertion of energy into the system.

Although a sequence of smooth functions is a suitable way of viewing a step or an impulse, the sequence is not unique. Very different sequences can approximate the same limiting function. The sequences shown in Figure A2.2 approach the same step and impulse as do those in Figure A2.1. Fortunately, the sequences of Figures A2.1 and A2.2 do have a significant property in common. In both cases the integral of the impulse is 1 (i.e., the step size is 1). This property is the key to the mathematical usage of the idealized functions. Equal area impulses insert equal amounts of energy into the system.

We formally define the **unit impulse** (or **Dirac delta function**) at $t=t_0$ by

$$\begin{aligned} \delta(t-t_0) &\triangleq 0, && t \neq t_0 \\ &\triangleq \text{undefined}, && t=t_0 \end{aligned} \qquad \text{(A2.2)}$$

$$\int_{-\infty}^{\infty} f(t)\delta(t-t_0)\,dt = f(t_0)$$

for every continuous function f. The integral property in (A2.2) is commonly known as the **sifting property** of delta functions. Applied to the function $f(t)=1$, it requires $\delta(t-t_0)$ to have unit area. A unit impulse is often represented graphically by an arrow as shown in Figure A2.3.

Delta functions have been rigorously defined by Schwartz [A2.2] as part of his theory of "distributions" (or **generalized functions**). Friedman [A2.1] and Zadeh and DeSoer [A2.4] present simplified versions of the theory. In essence, the theory of distributions shows that we can manipulate delta functions as if they were ordinary functions as long as we do not work with their values. We take values only of their integrals as in (A2.2). In point of fact, the first property of (A2.2) is not a necessary part of the definition of

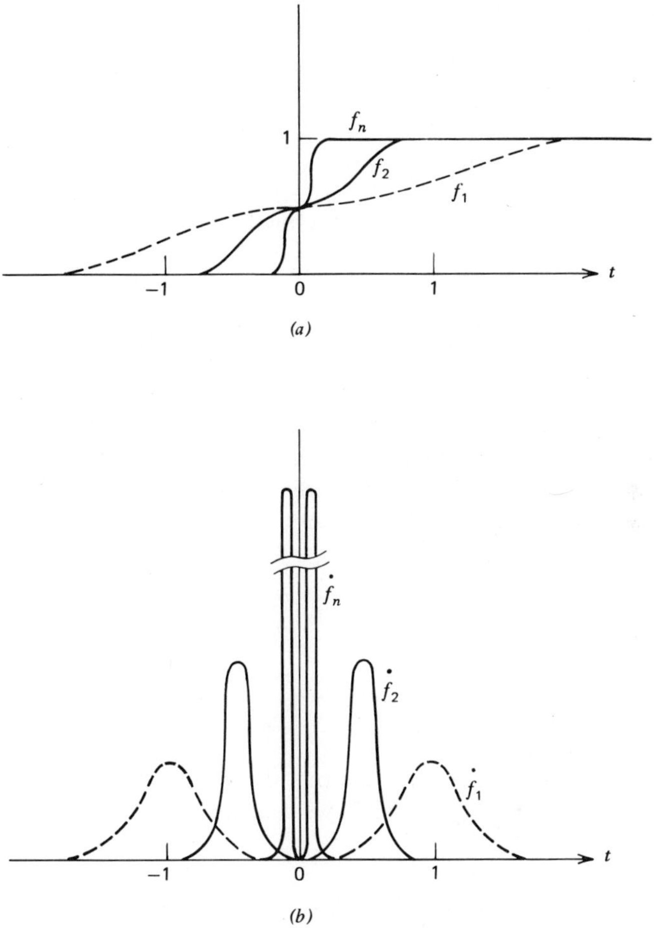

Figure A2.2. Smooth approximations to step and impulse functions.

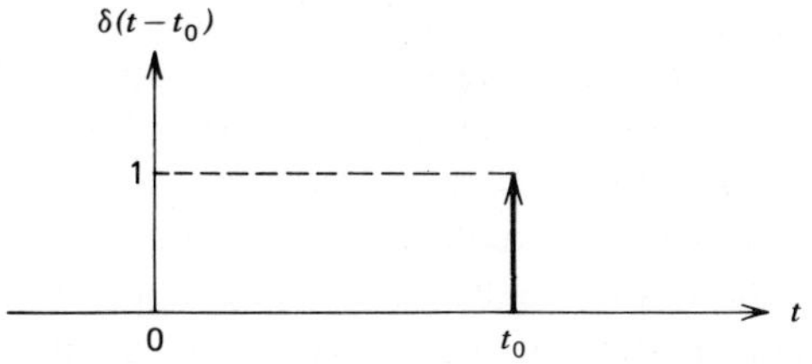

Figure A2.3. Graphic representation of a unit impulse.

$\delta(t-t_0)$; it follows from the sifting property. We retain the first property because it is useful.

The derivatives and integrals of delta functions also exist. We denote the first integral or unit step at t_0 by $\delta^{(-1)}(t-t_0)$. It is defined by

$$\begin{aligned} \delta^{(-1)}(t-t_0) &\triangleq 0, & t<t_0 \\ &\triangleq \text{undefined}, & t=t_0 \\ &\triangleq 1, & t>t_0 \end{aligned} \tag{A2.3}$$

$$\int_{-\infty}^{\infty} f(t)\delta^{(-1)}(t-t_0)\,dt = \int_{t_0}^{\infty} f(t)\,dt$$

The sifting property of (A2.3) can be derived directly from that of the unit impulse by integration by parts. The impulse is clearly the derivative of the step. We also use integration by parts of the sifting integral in (A2.2) to show, for a function f which is continuously differentiable, that

$$\begin{aligned} f^{(1)}(t_0) &= \int_{-\infty}^{\infty} \frac{df}{dt}(t)\delta(t-t_0)\,dt \\ &= \underbrace{f(t)\delta(t-t_0)|_{-\infty}^{\infty}}_{0} - \int_{-\infty}^{\infty} f(t)\frac{d\delta}{dt}(t-t_0)\,dt \end{aligned}$$

This is the sifting property for the first derivative $\delta^{(1)}(t-t_0)$. In fact, from the definition of $\delta(t-t_0)$ in (A2.2) we can show the following properties of the ***n*th derivative** of $\delta(t-t_0)$:

$$\begin{aligned} \delta^{(n)}(t-t_0) &= 0, & t\neq t_0 \\ &= \text{undefined}, & t=t_0 \end{aligned} \tag{A2.4}$$

$$\int_{-\infty}^{\infty} f(t)\delta^{(n)}(t-t_0)\,dt = (-1)^n f^{(n)}(t)$$

where f is any function with a continuous nth derivative at t_0.

In a sense, any function that we encounter in applications is as differentiable as we need it to be. We merely must allow for delta functions and their derivatives in the result. If we differentiate the set of all continuous functions, we get a set of functions some of which have discontinuities (steps). In applications we usually do not encounter functions with more than a few discontinuities. If we differentiate again, we get a set of functions some of which contain discontinuities, some of which contain impulses, and some of which contain both. Again, we usually encounter only a few steps or impulses.

Example 1. Derivative of a Discontinuous Function. Let $f(t)=1+\frac{1}{2}t-\delta^{(-1)}(t-2)$. Then the derivative of f is $f'(t)=\frac{1}{2}-\delta(t-2)$. See Figure A2.4.

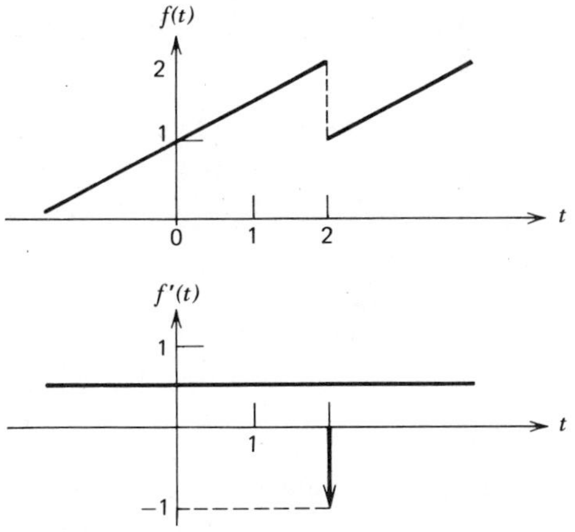

Figure A2.4. Derivative of a discontinuous function.

Time-Domain Analysis

The justification given earlier for introducing delta functions was that the response of a linear system for a delta function input is characteristic of the system; furthermore, it is a useful description of the system. We justify that earlier statement by an intuitive derivation of the solution of a linear system in terms of its impulse response.

We define the **impulse response** k of a linear system as the output of the system with no initial stored energy (zero initial conditions) and with a unit impulse input $\delta(t)$. For the system of (A2.1) with $f(0)=0$, the response to the impulse $\delta(t-t_0)$ is

$$\begin{aligned} k(t,t_0)&=0, && t<t_0 \\ &=\exp\left[-a(t-t_0)\right], && t>t_0 \end{aligned} \tag{A2.5}$$

This impulse response can be verified by substituting k into (A2.1):

$$\frac{dk}{dt}(t,t_0)+ak(t,t_0)=\delta(t-t_0),\ \ k(0,t_0)=0 \tag{A2.6}$$

Differentiation of k produces the unit impulse at $t=t_0$.

We wish to express the response f of the system to an input u in terms of the impulse response k. We begin by decomposing u into a sequence of finite-width pulses (see Figure A2.5*a*). We replace each pulse by an impulse of approximately the same area (see Figure A2.5*b*). This approximate input $\hat{u}$ can be expressed as

$$\hat{u}(t) = \sum_{j=-\infty}^{\infty} \left[u(j\Delta t)\Delta t\right]\delta(t - j\Delta t)$$

Because the system is linear, we can apply the principle of superposition —the response to $\hat{u}(t)$ is the sum of the responses to the individual impulses. The response to a unit impulse at time $j\Delta t$ is $k(t,j\Delta t)$. Therefore,

$$\hat{f}(t) = \sum_{j=-\infty}^{\infty} u(j\Delta t)k(t,j\Delta t)\Delta t$$

If we let Δt be very small, then $\hat{u}$ is essentially u; that is, the finite pulses are very sharp and have essentially the correct area. In that case, $\hat{f}$ is essentially f. We define new variables $\tau \stackrel{\Delta}{=} j\Delta t$ and $d\tau \stackrel{\Delta}{=} \Delta t$. Then, in the limit as $\Delta t \to 0$, we write the superposition sum as the **superposition integral**:

$$f(t) = \int_{-\infty}^{\infty} u(\tau)k(t,\tau)\,d\tau \qquad \text{(A2.7)}$$

Example 2. Solution of (A2.1) using the Impulse Response. The impulse response for (A2.1) with $f(0)=0$ is given in (A2.5). Let $u(t)=\delta^{(-1)}(t)$, a unit step. Then by

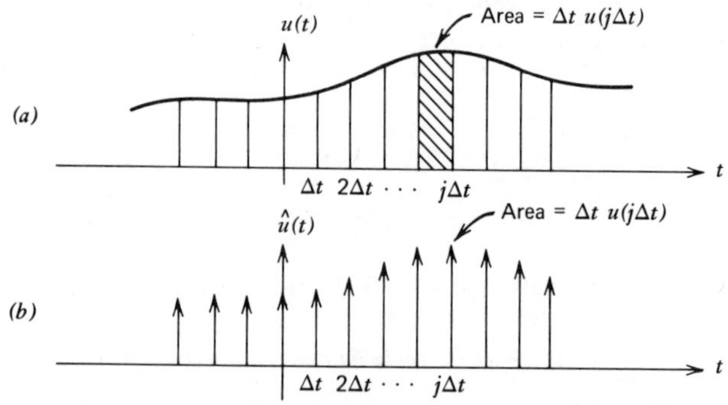

Figure A2.5. Impulse approximation of an input function.

the superposition integral (A2.7), for $t>0$,

$$f(t)=\int_{-\infty}^{0}(0)k(t,\tau)\,d\tau+\int_{0}^{\infty}(1)k(t,\tau)\,d\tau$$

$$=\int_{0}^{t}e^{-a(t-\tau)}\,d\tau$$

$$=\frac{1}{a}(1-e^{-at})$$

This solution can be verified by substituting it into (A2.1).

Exercise 1. Use the superposition integral to find the solution to (A2.1) for the input $u(t)=e^{bt}\delta^{(-1)}(t)$, for an arbitrary scalar b.

In Example 2, the impulse response can be expressed as $k(t,\tau)=k(t-\tau,0)$. This property of the system in the example is a basic characteristic of linear **time-invariant** systems: the shape of the impulse response does not depend on the time τ of application of the impulse, but only on the time $(t-\tau)$ since the application of the impulse. For linear time-invariant systems we denote the impulse response by $g(t-\tau)\stackrel{\Delta}{=}k(t,\tau)$. Then (A2.7) becomes

$$f(t)=\int_{-\infty}^{\infty}u(\tau)g(t-\tau)\,d\tau \tag{A2.8}$$

In this form, f is known as the **convolution** of the two functions u and g.

Exercise 2. Verify for the system of Example 2 the following general property of convolution:

$$\int_{-\infty}^{\infty}u(\tau)g(t-\tau)\,d\tau=\int_{-\infty}^{\infty}u(t-\tau)g(\tau)\,d\tau$$

Frequency-Domain Analysis

We have explored the manner in which the impulse response of a linear dynamic system represents that system. As a model for the system, the superposition integral is equivalent to the differential equation with zero initial conditions. If the linear system is also time-invariant, Laplace transforms provide a third model for the system with zero initial conditions. As noted at the beginning of this appendix, the Laplace transform model of the system is the transfer function. We now indicate briefly the relationship between the impulse response and the transfer function.

The one-sided Laplace transform F of a function f is defined by*

$$F(s) \triangleq \int_0^\infty f(t)e^{-st}\,dt$$

The **transfer function** of a linear time-invariant system is defined as the Laplace transform of the output with zero initial conditions divided by the Laplace transform of the input. We state without proof that the same transfer function applies to all inputs; it characterizes the system. The Laplace transform of the unit impulse $\delta(t)$ is

$$\int_0^\infty \delta(t)e^{-st}\,dt = e^0 = 1$$

(In order to avoid "splitting" the delta function, we either integrate over $[0^-,\infty]$ or think in terms of the pulse sequence of Figure A2.1b, which is entirely contained within $[0,\infty]$.) When the system input is $\delta(t)$, the system output is $g(t)$, the impulse response. Therefore, the transfer function is the Laplace transform $G(s)$ of the impulse response. Then if U and F are the transforms of the input and output, respectively, the definition of the transfer function requires

$$F(s) = G(s)U(s) \tag{A2.9}$$

for any linear time-invariant system. Equation (A2.9) is the frequency domain equivalent of the convolution in (A2.8).

Example 3. Solution of (A2.1) Using the Transfer Function. Let the input to (A2.1) be $u(t) = \delta^{(-1)}(t)$, a unit step at the origin. Its Laplace transform is

$$\begin{aligned} U(s) &= \int_0^\infty \delta^{(-1)}(t)e^{-st}\,dt \\ &= \int_0^\infty e^{-st}\,dt \\ &= \frac{1}{s} \end{aligned} \tag{A2.10}$$

(The integral converges only if the real part of s is positive.) The impulse response g, from (A2.5) is

$$\begin{aligned} g(t) \triangleq k(t,0) &= 0, \qquad t<0 \\ &= e^{-at}, \qquad t>0 \end{aligned}$$

* See Schwartz and Friedland [A2.3].

Its Laplace transform G is the transfer function of the system:

$$G(s) = \int_0^\infty e^{-at} e^{-st}\,dt$$

$$= \left. \frac{e^{-(s+a)t}}{-(s+a)} \right|_0^\infty$$

$$= \frac{1}{s+a}$$

(The integral converges only if the real part of $s+a$ is positive.) By (A2.9),

$$F(s) = \left(\frac{1}{s+a}\right)\left(\frac{1}{s}\right)$$

$$= \frac{1/a}{s} - \frac{1/a}{s+a}$$

Since F is a combination of the transforms of $\delta^{(-1)}$ and g, we know

$$f(t) = \frac{1}{a}\delta^{(-1)}(t) - \frac{1}{a}g(t)$$

$$= 0, \qquad t < 0$$

$$= \frac{1 - e^{-at}}{a}, \qquad t > 0$$

Compare this solution to that of Example 2.

The Discrete Delta Function

We complete our discussion of delta functions by extending the definition to discrete cases. The **discrete** (or **Kronecker**) **delta function** δ_{ij} is defined by

$$\begin{aligned} \delta_{ij} &= 0, \qquad i \neq j \\ &= 1, \qquad i = j \end{aligned} \tag{A2.11}$$

for integer values of i and j. The function δ_{i0} is the analogue of $\delta(t)$; that is, δ_{i0} is nonzero only for $i=0$. The function δ_{ij} for fixed j is the analogue of $\delta(t-t_0)$. It satisfies the sifting property

$$\sum_i \delta_{ij}\xi_i = \xi_j \tag{A2.12}$$

where $(\ldots, \xi_{-1}, \xi_0, \xi_1, \xi_2, \ldots)$ is a finite or infinite sequence of real numbers.

The Kronecker delta is useful in the analysis of **discrete** linear time-invariant systems such as those encountered in digital computer simulation of continuous systems. One discrete approximation to (A2.1) is

$$\xi_{i+1} - \xi_i + a\xi_i = \eta_i$$

or

$$\begin{aligned} \xi_{i+1} &= (1-a)\xi_i + \eta_i \\ &= b\xi_i + \eta_i \qquad i = 0, 1, 2, \ldots \end{aligned} \tag{A2.13}$$

where $b \overset{\Delta}{=} 1 - a$ and $\xi_0 = 0$. We could generate discrete analogues of the impulse response and the convolution integral of (A2.8). We could also derive a discrete transform (called the z-transform) which yields transfer function characteristics analogous to (A2.9). See Schwartz and Friedland [A2.3] for a discussion of these topics.

Example 4. A Discrete Impulse Response. We apply to the system of (A2.13) the impulse input δ_{i0}. By repeated use of (A2.13) we find

$$\xi_1 = b\xi_0 + \delta_{00} = b(0) + 1 = 1$$

$$\xi_2 = b\xi_1 + \delta_{10} = b(1) + 0 = b$$

$$\xi_3 = b\xi_2 + \delta_{20} = b(b) + 0 = b^2$$

$$\vdots$$

Thus the impulse response k of the discrete system is

$$k_i = b^i, \qquad i = 0, 1, 2, \ldots$$

Exercise 3. Derive the discrete analogue of (A2.8). Use it, along with the above impulse response, to obtain the output sequence for (A2.13) corresponding to the input sequence:

$$\eta_i = 1, \qquad i = 0, 1, 2, \ldots$$

To check the result, generate the output directly by repetitive use of (A2.13). Sketch the solution for a specific value of a, and compare with the solution to (A2.1).

References

[A2.1] Friedman, Bernard, *Principles and Techniques of Applied Mathematics*, Wiley, New York, 1956.

[A2.2] Schwartz, L., *Theorie des Distributions*, Vols. 1 and 2, Hermann & Cie, Paris, 1951, 1957.

*[A2.3] Schwartz, Ralph J. and Bernard Friedland, *Linear Systems*, McGraw-Hill, New York, 1965.

*[A2.4] Zadeh, Lofti A. and Charles A. Desoer, *Linear System Theory*, McGraw-Hill, New York, 1963.

Appendix 3. Decomposition Theorems

The purpose of this appendix is to prove the theorems used in the derivation of the Jordan canonical form of a matrix. These theorems are illustrated by the examples in that derivation (Section 4.4).

Theorem 1. Let $\mathbf{U}$ be a linear operator on an n-dimensional vector space $\mathcal{V}$. Let $\mathcal{N}_g(\mathbf{U})$ and $\mathcal{R}_g(\mathbf{U})$ be the generalized nullspace and generalized range of $\mathbf{U}$, respectively. Let q be the index of annihilation of $\mathbf{U}$ (see Section 4.4). Then

(*a*) $\mathcal{V} = \mathcal{N}_g(\mathbf{U}) \oplus \mathcal{R}_g(\mathbf{U})$.

(*b*) Both $\mathcal{N}_g(\mathbf{U})$ and $\mathcal{R}_g(\mathbf{U})$ are invariant under $\mathbf{U}$.

(*c*) $\mathbf{U}$ is invertible on $\mathcal{R}_g(\mathbf{U})$ and nilpotent on $\mathcal{N}_g(\mathbf{U})$.

(A **nilpotent operator** is one some power of which is zero.)

Proof

(*a*) Assume $\mathbf{x}$ is in $\mathcal{R}_g(\mathbf{U})$; that is, $\mathbf{U}^q\mathbf{y} = \mathbf{x}$ for some $\mathbf{y}$ in $\mathcal{V}$. Assume $\mathbf{x}$ is also in $\mathcal{N}_g(\mathbf{U})$; that is, $\mathbf{U}^q\mathbf{x} = \boldsymbol{\theta}$. Substituting the first equation into the second, we find $\mathbf{U}^{2q}\mathbf{y} = \boldsymbol{\theta}$, and $\mathbf{y}$ must be in $\mathcal{N}_g(\mathbf{U})$. By definition of q, $\mathbf{U}^q\mathbf{y} = \boldsymbol{\theta}$. Thus $\mathbf{x}$ must be $\boldsymbol{\theta}$, and $\mathcal{N}_g(\mathbf{U})$ and $\mathcal{R}_g(\mathbf{U})$ are disjoint. The direct sum follows from (2.43):

$$\begin{aligned}\dim(\mathcal{V}) &= \operatorname{rank}(\mathbf{U}^q) + \operatorname{nullity}(\mathbf{U}^q)\\ &= \dim(\mathcal{R}_g(\mathbf{U})) + \dim(\mathcal{N}_g(\mathbf{U}))\end{aligned}$$

(*b*) If $\mathbf{U}^q\mathbf{x} = \boldsymbol{\theta}$ then $\mathbf{U}^{q+1}\mathbf{x} = \boldsymbol{\theta}$. Therefore, if $\mathbf{x}$ is in $\mathcal{N}_g(\mathbf{U})$ so is $\mathbf{U}\mathbf{x}$; $\mathcal{N}_g(\mathbf{U})$ is invariant under $\mathbf{U}$. Since range $(\mathbf{U}^{k+1})$ is in range$(\mathbf{U}^k)$ for $k = 0, 1, 2, \ldots$, $\mathcal{R}_g(\mathbf{U})$ is invariant under $\mathbf{U}$.

(*c*) By definition of q, $\mathbf{U}$ annihilates no nonzero vectors in $\mathcal{R}_g(\mathbf{U})$. Therefore, $\mathbf{U}$ is nonsingular (or invertible) on $\mathcal{R}_g(\mathbf{U})$. By definition of $\mathcal{N}_g(\mathbf{U})$, $\mathbf{U}^q\mathbf{x} = \boldsymbol{\theta}$ for any $\mathbf{x}$ in $\mathcal{N}_g(\mathbf{U})$. Therefore, $\mathbf{U}$ is nilpotent on $\mathcal{N}_g(\mathbf{U})$.

It follows from parts (*a*) and (*b*) of Theorem 1 that **U** decomposes into smaller operators

$$\mathbf{U}_1 : \mathcal{R}_g(\mathbf{U}) \to \mathcal{R}_g(\mathbf{U}) \qquad \mathbf{U}_2 : \mathcal{N}_g(\mathbf{U}) \to \mathcal{N}_g(\mathbf{U})$$

If $\mathfrak{X}_{\mathcal{R}}$ and $\mathfrak{X}_{\mathcal{N}}$ are bases for $\mathcal{R}_g(\mathbf{U})$ and $\mathcal{N}_g(\mathbf{U})$, respectively, then, by part (*a*) of the theorem, $\mathfrak{X} \triangleq \{\mathfrak{X}_{\mathcal{R}}, \mathfrak{X}_{\mathcal{N}}\}$ is a basis for $\mathcal{V}$. It follows from part (*b*) that

$$[\mathbf{U}]_{\mathfrak{X}\mathfrak{X}} = \begin{pmatrix} [\mathbf{U}_1]_{\mathfrak{X}_{\mathcal{R}}\mathfrak{X}_{\mathcal{R}}} & 0 \\ 0 & [\mathbf{U}_2]_{\mathfrak{X}_{\mathcal{N}}\mathfrak{X}_{\mathcal{N}}} \end{pmatrix} \tag{A3.1}$$

From part (*c*) of the theorem we recognize that $\mathbf{U}_1$ is invertible and $\mathbf{U}_2$ is nilpotent ($\mathbf{U}_2^q = \boldsymbol{\theta}$). Thus $[\mathbf{U}_1]_{\mathfrak{X}_{\mathcal{R}}\mathfrak{X}_{\mathcal{R}}}$ is an invertible matrix, whereas the matrix $[\mathbf{U}_2]_{\mathfrak{X}_{\mathcal{N}}\mathfrak{X}_{\mathcal{N}}}$ is nilpotent. See Exercise 1 of Section 4.4 and P&C 4.3.

Theorem 2. Let **A** be an $n \times n$ matrix with the characteristic polynomial

$$c(\lambda) = \det(\lambda\mathbf{I} - \mathbf{A}) = (\lambda - \lambda_1)^{m_1} \cdots (\lambda - \lambda_p)^{m_p}$$

Define $\mathcal{W}_i \triangleq \mathcal{N}_g(\mathbf{A} - \lambda_i\mathbf{I})$, $i = 1, 2, \ldots, p$. Then

$$\dim(\mathcal{W}_i) = m_i$$

Proof. Define the operator **T** on $\mathfrak{M}^{n\times 1}$ by $\mathbf{Tx} \triangleq \mathbf{Ax}$. Define the operator **U** on $\mathfrak{M}^{n\times 1}$ by

$$\mathbf{Ux} \triangleq \mathbf{Tx} - \lambda_i\mathbf{x} = (\mathbf{A} - \lambda_i\mathbf{I})\mathbf{x}$$

where λ_i is the *i*th eigenvalue of **A**. Let $\mathfrak{X}_{\mathcal{N}}$ be some basis for $\mathcal{W}_i$, the generalized nullspace of $\mathbf{A} - \lambda_i\mathbf{I}$. Let $\mathfrak{X}_{\mathcal{R}}$ be some basis for $\mathcal{R}_g(\mathbf{U})$, the generalized range of $(\mathbf{A} - \lambda_i\mathbf{I})$. We use $\mathfrak{X} = \{\mathfrak{X}_{\mathcal{R}}, \mathfrak{X}_{\mathcal{N}}\}$ as a basis for $\mathfrak{M}^{n\times 1}$. Then $[\mathbf{U}]_{\mathfrak{X}\mathfrak{X}}$ is given by (A3.1). The generalized eigenspace $\mathcal{W}_i$ is invariant under multiplication by $(\mathbf{A} - \lambda_i\mathbf{I})$, by Theorem 1. Since $\lambda_i\mathbf{I}$ is just a scalar operation, $\mathcal{W}_i$ is also invariant under multiplication by **A**. Therefore, $[\mathbf{T}]_{\mathfrak{X}\mathfrak{X}}$ is also in the form of (A3.1):

$$[\mathbf{T}]_{\mathfrak{X}\mathfrak{X}} = \begin{pmatrix} [\mathbf{T}_1]_{\mathfrak{X}_{\mathcal{R}}\mathfrak{X}_{\mathcal{R}}} & 0 \\ 0 & [\mathbf{T}_2]_{\mathfrak{X}_{\mathcal{N}}\mathfrak{X}_{\mathcal{N}}} \end{pmatrix} \tag{A3.2}$$

where $\mathbf{T}_1$ and $\mathbf{T}_2$ describe the effect of **T** on $\mathcal{R}_g(\mathbf{U})$ and $\mathcal{N}_g(\mathbf{U})$, respectively. The matrices **A** and $[\mathbf{T}]_{\mathfrak{X}\mathfrak{X}}$ have the same eigenvalues. Therefore,

$c(\lambda)$ can be evaluated by applying to $\det(\lambda\mathbf{I}-[\mathbf{T}]_{\mathfrak{X}\mathfrak{X}})$ the rules for simplification of determinants (Example 2, Appendix 1):

$$\begin{aligned} c(\lambda) &= \det(\lambda\mathbf{I}-[\mathbf{T}]_{\mathfrak{X}\mathfrak{X}}) \\ &= \begin{vmatrix} \lambda\mathbf{I}-[\mathbf{T}_1]_{\mathfrak{X}_{\mathfrak{R}}\mathfrak{X}_{\mathfrak{R}}} & 0 \\ 0 & \lambda\mathbf{I}-[\mathbf{T}_2]_{\mathfrak{X}_{\mathfrak{N}}\mathfrak{X}_{\mathfrak{N}}} \end{vmatrix} \\ &= \det(\lambda\mathbf{I}-[\mathbf{T}_1]_{\mathfrak{X}_{\mathfrak{R}}\mathfrak{X}_{\mathfrak{R}}})\det(\lambda\mathbf{I}-[\mathbf{T}_2]_{\mathfrak{X}_{\mathfrak{N}}\mathfrak{X}_{\mathfrak{N}}}) \\ &= (\lambda-\lambda_1)^{m_1}\cdots(\lambda-\lambda_p)^{m_p} \end{aligned} \tag{A3.3}$$

Observe that λ_i is not an eigenvalue of $\mathbf{T}_1$, since $\mathbf{U}_1=\mathbf{T}_1-\lambda_i\mathbf{I}$, and $\mathbf{U}_1$ is invertible. We now show that λ_i is the only eigenvalue of $\mathbf{T}_2$. Let λ_j be a *second* eigenvalue of $\mathbf{T}_2$ with eigenvector $\mathbf{z}$; that is, $\mathbf{T}_2\mathbf{z}=\lambda_j\mathbf{z}$. But $\mathbf{T}_2$ acts only on $\mathfrak{N}_g(\mathbf{A}-\lambda_i\mathbf{I})$. Therefore, $\mathbf{z}$ must be a generalized eigenvector of some rank, say, r, for the *first* eigenvalue λ_i; we attach the subscript r to $\mathbf{z}$, and observe that $\mathbf{T}_2\mathbf{z}_r=\lambda_i\mathbf{z}_r+\mathbf{z}_{r-1}$, where $\mathbf{z}_{r-1}$ is a generalized eigenvector of rank $r-1$ for λ_i. Thus

$$\lambda_j\mathbf{z}_r=\lambda_i\mathbf{z}_r+\mathbf{z}_{r-1}$$

Operating on this equation with $(\mathbf{T}_2-\lambda_i\mathbf{I})^{r-1}$ yields

$$(\lambda_i-\lambda_j)(\mathbf{T}_2-\lambda_i\mathbf{I})^{r-1}\mathbf{z}_r+(\mathbf{T}_2-\lambda_i\mathbf{I})^{r-1}\mathbf{z}_{r-1}=\boldsymbol{\theta}$$

The second term is zero since $\mathbf{z}_{r-1}$ is of rank $r-1$. The vector $(\mathbf{T}_2-\lambda_i\mathbf{I})^{r-1}\mathbf{z}_r$ is a nonzero eigenvector of $\mathbf{T}_2$. Therefore $(\lambda_i-\lambda_j)=0$; λ_i is the only eigenvalue of $\mathbf{T}_2$.

Because λ_i is not an eigenvalue of $\mathbf{T}_1$ but is the only eigenvalue of $\mathbf{T}_2$, it follows from (A3.3) that

$$\det(\lambda\mathbf{I}-[\mathbf{T}_2]_{\mathfrak{X}_{\mathfrak{N}}\mathfrak{X}_{\mathfrak{N}}})=(\lambda-\lambda_i)^{m_i} \tag{A3.4}$$

The matrix $[\mathbf{T}_2]_{\mathfrak{X}_{\mathfrak{N}}\mathfrak{X}_{\mathfrak{N}}}$ is an $m_i\times m_i$ matrix. The space $\mathfrak{W}_i$ on which $\mathbf{T}_2$ operates has dimension m_i.

Theorem 3. Let $\mathbf{A}$ be an $n\times n$ matrix with the characteristic polynomial

$$c(\lambda)=(\lambda-\lambda_1)^{m_1}\cdots(\lambda-\lambda_p)^{m_p}$$

Define $\mathfrak{W}_i\stackrel{\Delta}{=}\mathfrak{N}_g(\mathbf{A}-\lambda_i\mathbf{I})$, $i=1,2,\ldots,p$. Then

$$\mathfrak{M}^{n\times 1}=\mathfrak{W}_1\oplus\cdots\oplus\mathfrak{W}_p$$

Proof. Let $\mathbf{x}_j$ denote a vector from $\mathcal{W}_j$. We know that $(\mathbf{A}-\lambda_j\mathbf{I})$ is nilpotent on $\mathcal{W}_j$ and is invertible on $\mathcal{W}_i$ $(i\neq j)$. Therefore, $(\mathbf{A}-\lambda_j\mathbf{I})^{q_j}$ takes $\mathbf{x}_j$ to zero, and takes any nonzero vector in $\mathcal{W}_i$ to another nonzero vector in $\mathcal{W}_i$. Any vector $\mathbf{x}$ in $\Sigma_{j\neq i}\mathcal{W}_j$ can be written as

$$\mathbf{x}=\sum_{j\neq i} c_j\mathbf{x}_j, \qquad \mathbf{x}_j \text{ in } \mathcal{W}_j$$

Multiplying this equation successively by $(\mathbf{A}-\lambda_j\mathbf{I})^{q_j}$ for each $j\neq i$, we successively annihilate each of the components $c_j\mathbf{x}_j$ of $\mathbf{x}$. Therefore,

$$\prod_{j\neq i}(\mathbf{A}-\lambda_j\mathbf{I})^{q_j}\mathbf{x}=\boldsymbol{\theta}$$

Since each of the multiplications is invertible on $\mathcal{W}_i$, we conclude that $\mathbf{x}$ is not in $\mathcal{W}_i$, and

$$\mathcal{W}_i\cap\left(\sum_{j\neq i}\mathcal{W}_j\right)=\boldsymbol{\theta}$$

The additional fact that

$$\begin{aligned}\dim(\mathcal{V})&=m_1+m_2+\cdots+m_p\\&=\dim(\mathcal{W}_1)+\cdots+\dim(\mathcal{W}_p)\end{aligned}$$

guarantees that the $\mathcal{W}_i$ sum directly to yield $\mathcal{V}$.

Answers to Selected Problems

Chapter 1

1.1 (*a*)

$$\left(\begin{array}{cccc:ccc} 1 & 0 & \frac{1}{3} & 1 & 0 & \frac{5}{6} & -\frac{1}{6} \\ 0 & 1 & \frac{1}{3} & 1 & 0 & -\frac{2}{3} & \frac{1}{3} \\ 0 & 0 & 0 & 0 & 1 & \frac{1}{2} & -\frac{1}{2} \end{array}\right)$$

(*b*) Range (**A**) = vectors of the form $((\eta_3-\eta_2)/2 \quad \eta_2 \quad \eta_3)^T$

(*c*) Nullspace (**A**) = vectors fo the form

$$\left(-\tfrac{1}{3}\xi_3-\xi_4 \quad -\tfrac{1}{3}\xi_3-\xi_4 \quad \xi_3 \quad \xi_4\right)^T$$

1.2 (d) If $-2\eta_1+\eta_3=0$, then

$$\mathbf{x}=\begin{pmatrix} \eta_1-\frac{1}{3}\eta_2+\frac{2}{3}\eta_4 \\ -\eta_1+\frac{5}{6}\eta_2-\frac{2}{3}\eta_4 \\ -\frac{1}{6}\eta_2+\frac{1}{3}\eta_4 \end{pmatrix}$$

Otherwise it is inconsistent

1.3 (*a*) $(4n^3+n-3)/3$

(*b*) $(3n^3+n)/2$

(*c*) $(n^3+2n-3)/3$

1.4 (*a*) (2)

$$\begin{pmatrix} 1 & 0 & 0 & 1 & 0 \\ 0 & 1 & 0 & 0 & 0 \\ 0 & 0 & 1 & 0 & 0 \\ 0 & 0 & 0 & 1 & 0 \\ 0 & 0 & 0 & 0 & 1 \end{pmatrix}$$

Chapter 2

2.3 (*a*) Linearly dependent

2.5 Dimension $= mn$

2.7 (*a*)

$$\left(\begin{array}{cccc:ccc} 1 & 0 & \frac{1}{3} & 1 & 0 & \frac{5}{6} & -\frac{1}{6} \\ 0 & 1 & \frac{1}{3} & 1 & 0 & -\frac{2}{3} & \frac{1}{3} \\ 0 & 0 & 0 & 0 & 1 & \frac{1}{2} & -\frac{1}{2} \end{array}\right)$$

(*b*)

$$\text{Basis} = \left\{ \begin{pmatrix} -\frac{1}{2} \\ 1 \\ 0 \end{pmatrix}, \begin{pmatrix} \frac{1}{2} \\ 0 \\ 4 \end{pmatrix} \right\}$$

2.8 (*c*) If $\eta_3 = 2\eta_1$, then $\mathbf{y}$ is in span$\{\mathbf{y}_i\}$, and

$$\mathbf{y} = (\eta_1 - \tfrac{1}{3}\eta_2 + \tfrac{2}{3}\eta_4)\mathbf{y}_1 + (-\eta_1 + \tfrac{5}{6}\eta_2 - \tfrac{2}{3}\eta_4)\mathbf{y}_2 + (-\tfrac{1}{6}\eta_2 + \tfrac{1}{3}\eta_4)\mathbf{y}_3$$

2.9 One basis for span$\{\mathbf{f}_1, \mathbf{f}_2, \mathbf{f}_3, \mathbf{f}_4\}$ is $\{\mathbf{g}_1, \mathbf{g}_2\}$, where $\mathbf{g}_1(t) = 1 + t + 2t^2$ and $\mathbf{g}_2(t) = 2t + t^2 + t^3$. Include $\mathbf{g}_3(t) = 1$ and $\mathbf{g}_4(t) = t$ to make a basis for $\mathcal{P}^4$.

2.10 (*a*) $[\mathbf{x}]_{\mathcal{X}} = (3 \ \ -2 \ \ -1)^{\mathrm{T}}$

2.11 $[\mathbf{f}]_{\mathcal{F}} = (1 \ \ -\frac{1}{2} \ \ -\frac{1}{2})^{\mathrm{T}}$

2.16 (*a*) $\mathbf{f}(12) = \$229.52$
(*b*) $\mathbf{f}(12) = \$229.86$
(*c*) $\mathbf{f}(12) = \$218.95$

2.17 (*a*) Dim (nullspace$(\mathbf{D}^2)$) $= 2$

2.18 (*a*) $\mathbf{A} = (\frac{1}{2} \ \ 1 \ \ 1 \ \ 1 \ \ 1 \ \ \frac{1}{2})$
(*b*)

$$\mathbf{A} = \begin{pmatrix} 0 & 0 & 0 & 0 & 0 & 0 \\ \frac{1}{2} & \frac{1}{2} & 0 & 0 & 0 & 0 \\ \frac{1}{2} & 1 & \frac{1}{2} & 0 & 0 & 0 \\ \frac{1}{2} & 1 & 1 & \frac{1}{2} & 0 & 0 \\ \frac{1}{2} & 1 & 1 & 1 & \frac{1}{2} & 0 \\ \frac{1}{2} & 1 & 1 & 1 & 1 & \frac{1}{2} \end{pmatrix}$$

2.19 (*a*) One basis for range(**A**) is $\{(1\ 0\ 2)^T, (0\ 1\ 1)^T\}$; one basis for nullspace(**A**) is $\{(\frac{4}{3}\ -\frac{8}{3}\ 0\ 1)^T, (-2\ 1\ 1\ 0)^T\}$

2.23 (*a*)

$$\begin{aligned}\omega_{\mathbf{y}}(y) &= \tfrac{1}{3} \text{ for } y=2\\ &= \tfrac{2}{3} \text{ for } y=0\end{aligned}$$

(*b*) $\mathbf{E}(\mathbf{y}) = \frac{2}{3}$

(*c*) $[\mathbf{y}]_{\mathcal{X}} = (2\ 2\ 0\ 0\ 0\ 0)^T$; $[\mathbf{E}]_{\mathcal{X}\mathcal{E}} = (\frac{1}{6}\ \frac{1}{6}\ \frac{1}{6}\ \frac{1}{6}\ \frac{1}{6}\ \frac{1}{6})$

(*d*) $\mathbf{E}(\mathbf{y}^2) = \frac{4}{3}$

2.25 (*b*) A basis is $\{\mathbf{x}_1, \mathbf{x}_2\}$, where $\mathbf{x}_1(t) = \begin{pmatrix} 1 \\ 0 \end{pmatrix}$ and $\mathbf{x}_2(t) = \begin{pmatrix} 1-e^{-t} \\ e^{-t} \end{pmatrix}$

2.26 A basis is $\{(0\ 1\ 0)^T, (0\ 0\ 1)^T\}$

2.27 $[\mathbf{T}]_{\mathcal{X}\mathcal{X}} = \begin{pmatrix} 1 & 1 \\ 2 & -2 \end{pmatrix}$

2.28 (*b*) $[\mathbf{T}]_{\mathcal{X}\mathcal{Y}} = \begin{pmatrix} 4 & 1 & 2 \\ -3 & 1 & -1 \end{pmatrix}$

2.31

$$\mathbf{S}_{\mathcal{X}\mathcal{Y}} = \frac{1}{3}\begin{pmatrix} 1 & 1 & 2 \\ -2 & 1 & 2 \\ 2 & 2 & 1 \end{pmatrix}$$

2.33 (*a*)

$$[\mathbf{T}]_{\mathcal{X}\mathcal{Y}} = \frac{1}{2}\begin{pmatrix} 1 & -1 \\ 1 & 1 \\ 2 & 4 \end{pmatrix}$$

(*b*)

$$\mathbf{S}_{\mathcal{X}\mathcal{Y}} = \frac{1}{3}\begin{pmatrix} 1 & -3 \\ 1 & 3 \end{pmatrix}, \quad \mathbf{S}_{\mathcal{Y}\mathcal{X}} = \begin{pmatrix} 1 & 1 & -1 \\ 0 & -2 & 2 \\ -2 & 2 & 0 \end{pmatrix}$$

(*c*)

$$[\mathbf{T}]_{\mathcal{Z}\mathcal{X}} = \begin{pmatrix} 1 & -1 \\ 0 & 3 \\ -1 & 1 \end{pmatrix}$$

Chapter 3

3.1 $\mathbf{f}(1)=10$; $\mathbf{f}(8)\approx 4.75$

3.3 (*a*) First iteration: for $n=0,1,\ldots,5$,

$$\mathbf{f}(n)=0,\ -0.5,\ -0.75,\ -0.88,\ -0.94,0$$

3.5

$$\boldsymbol{\phi}(t)=\int_0^t(1-e^{-(t-s)})\mathbf{u}(s)\,ds+\boldsymbol{\phi}(0)+\frac{1-e^{-t}}{e-1}\left[\int_0^1 e^s\mathbf{u}(s)\,ds-e\alpha_3\right]$$

Solutions exist if and only if

$$\int_0^1 \mathbf{u}(s)\,ds=\alpha_3-\alpha_1 \qquad \text{and} \qquad \alpha_1+\frac{\alpha_2}{2}+\frac{\alpha_3}{2}=0$$

3.8 (*a*) A basis for nullspace(**T**) is $\{\mathbf{f}_1,\mathbf{f}_2\}$, where

$$\mathbf{f}_1(k)=(\tfrac{1}{2})^k \text{ and } \mathbf{f}_2(k)=1, \qquad k=0,1,2,\ldots$$

3.11

$$k(t,s)=\begin{cases}\dfrac{(e^t-e^{2t})(e^{-2s}-e^{-b}e^{-s})}{(1-e^{-b})} & \text{for} \quad 0\leqslant t\leqslant s\\[2ex] \dfrac{(e^t-e^{-b}e^{2t})(e^{-2s}-e^{-s})}{(1-e^{-b})} & \text{for} \quad s\leqslant t\leqslant b\end{cases}$$

$$\rho_1(t)=\frac{e^be^t-e^{2t}}{e^b-1}, \qquad \rho_2(t)=\frac{e^{-b}}{e^b-1}(e^{2t}-e^t)$$

3.12 (*b*)

$$\mathbf{f}(t)=\int_0^t e^{-(t-s)}\sin(t-s)\mathbf{u}(s)\,ds+\alpha_1e^{-t}(\sin t+\cos t)$$
$$+\alpha_2e^{-t}\sin t$$

3.14 (*a*) $\mathbf{f}_c(t)=c_1t$

(*b*)

$$k(t,s)=\begin{cases} 0 & \text{for} \quad t_1<t<s \\ t/s^2 & \text{for} \quad t>s \end{cases}$$

(*c*) $\rho_1(t)=t$

(*d*) $\mathbf{f}(t)=\mathrm{t}\displaystyle\int_{t_1}^{t}\frac{\mathbf{u}(s)}{s^2}\,\mathrm{ds}+\alpha t$

3.17 (*c*)

$$\mathbf{f}(i)=\sum_{j=0}^{i-2}\left[2-\left(\tfrac{1}{2}\right)^{i-j-2}\right]\mathbf{u}(i+1)+\alpha_1\left[-1+2\left(\tfrac{1}{2}\right)^{i}\right]$$

$$+(\alpha_2+\alpha_1)\left[2-2\left(\tfrac{1}{2}\right)^{i}\right] \qquad \text{for} \quad i=0,1,2,\ldots$$

3.20 (*b*)

$$e^{\mathbf{A}t}=\begin{pmatrix} 2e^{-t}-e^{-2t} & e^{-t}+e^{-2t} \\ e^{-2t}-e^{-t} & 2e^{-2t}-e^{-t} \end{pmatrix}$$

3.21 (*a*) For $\mathbf{x}(k\tau)\overset{\Delta}{=}(\mathbf{f}(k\tau)\;\mathbf{f}((k+1)\tau)\cdots\mathbf{f}((k+n-1)\tau))^{\mathrm{T}}$, $\mathbf{x}((k+1)\tau)=\mathbf{A}\mathbf{x}(k\tau)+\mathbf{B}\mathbf{u}(k\tau)$, $\mathbf{x}(0)$ given, where

$$\mathbf{A}=\begin{pmatrix} 0 & 1 & 0 & \ldots & 0 \\ 0 & 0 & 1 & \ldots & 0 \\ \vdots & \vdots & & & \vdots \\ -a_1 & -a_2 & & \ldots & -a_n \end{pmatrix} \text{ and } \mathbf{B}=\begin{pmatrix} 0 \\ \vdots \\ 0 \\ 1 \end{pmatrix}$$

(*b*)

$$\mathbf{x}(k\tau)=\mathbf{A}^k\mathbf{x}(0)+\sum_{j=0}^{k-1}\mathbf{A}^{k-j-1}\mathbf{B}\mathbf{u}(j\tau)$$

Chapter 4

4.1 (*b*)

$$[\mathbf{P}_1]_{\mathcal{E}\mathcal{E}}=\begin{pmatrix} 0 & 0 & 1 \\ 0 & 0 & 0 \\ 0 & 0 & 1 \end{pmatrix}$$

4.3 (*a*)

$$[\mathbf{T}_a]_{\mathscr{X}_a\mathscr{X}_a} = \begin{pmatrix} 1 & -1 \\ 1 & 0 \end{pmatrix}, \ [\mathbf{T}_b]_{\mathscr{X}_b\mathscr{X}_b} = (-1)$$

(*b*)

$$[\mathbf{T}]_{\mathscr{X}\mathscr{X}} = \left(\begin{array}{c|c} [\mathbf{T}_a]_{\mathscr{X}_a\mathscr{X}_a} & \boldsymbol{\Theta} \\ \hline \boldsymbol{\Theta} & [\mathbf{T}_b]_{\mathscr{X}_b\mathscr{X}_b} \end{array}\right)$$

(*c*)

$$[\mathbf{T}]_{\mathscr{X}\mathscr{X}} = \left(\begin{array}{c|c} [\mathbf{T}_1]_{\mathscr{X}_1\mathscr{X}_1} & \boldsymbol{\Theta} \\ \hline \boldsymbol{\Theta} & [\mathbf{T}_2]_{\mathscr{X}_2\mathscr{X}_2} \end{array}\right) = \left(\begin{array}{cc|c} 1 & -1 & 0 \\ 1 & 0 & 0 \\ \hline 0 & 0 & -1 \end{array}\right)$$

(*d*)

$$[\mathbf{T}]_{\mathscr{X}\mathscr{X}} = \left(\begin{array}{cc|c} 1 & 1 & 0 \\ 1 & 0 & 0 \\ \hline 0 & 0 & 2 \end{array}\right)$$

4.5 (*a*) Approximately n^4 multiplications by the trace iteration; approximately $4n^3/3$ multiplications by Krylov's method

4.7 (*a*)

$$\mathbf{A} = \begin{pmatrix} 0 & \frac{2}{3} & \frac{2}{3} \\ \frac{1}{2} & 0 & \frac{1}{3} \\ \frac{1}{2} & \frac{1}{3} & 0 \end{pmatrix}$$

(*b*) $\lambda = 1, -\frac{2}{3}, -\frac{1}{3}$

(*c*)

$$\mathbf{S} = \begin{pmatrix} 0.4 & -2 & 0 \\ 0.3 & 1 & 1 \\ 0.3 & 1 & -1 \end{pmatrix}$$

(*d*)

$$\lim_{n\to\infty} \mathbf{x}_n = \begin{pmatrix} 0.4 \\ 0.3 \\ 0.3 \end{pmatrix}$$

4.12 (*a*) $\lambda_k = -k^2, \mathbf{f}_k(t) = \sin kt, k = 1, 2, 3, \ldots$

(*b*) Any $\lambda \neq 0, \mathbf{f}_\lambda(t) = e^{\sqrt{\lambda} t} - e^{-\sqrt{\lambda}(t-\pi)}$

(*c*) No eigenvalues

(*d*) $\lambda_k = -4k^2, \mathbf{f}_k(t) = b_1 \cos 2kt + b_2 \sin 2kt$ (two independent eigenvectors for each eigenvalue)

4.17 (*b*) The exact dominant eigendata are $\lambda_1 = -\dfrac{1}{\pi^2}, \mathbf{f}_1(\mathrm{t}) = \sin \pi t$

4.18 (*b*)

$$\boldsymbol{\Lambda} = \begin{pmatrix} 4 & 1 & 0 & & & \\ 0 & 4 & 1 & & & \\ 0 & 0 & 4 & & & \\ & & & 4 & 1 & \\ & & & 0 & 4 & \\ & & & & & 2 \end{pmatrix}$$

4.21 (*a*)

$$f(\mathbf{A}) = f(1)\begin{pmatrix} 1 & -1 & 0 \\ 0 & 0 & 0 \\ 0 & 0 & 0 \end{pmatrix} + f(2)\begin{pmatrix} 0 & 1 & -1 \\ 0 & 1 & -1 \\ 0 & 0 & 0 \end{pmatrix} + f(3)\begin{pmatrix} 0 & 0 & 1 \\ 0 & 0 & 1 \\ 0 & 0 & 1 \end{pmatrix}$$

(*b*)

$$f(\mathbf{A}) = f(1)\begin{pmatrix} 1 & 0 & -1 \\ 0 & 1 & -1 \\ 0 & 0 & 0 \end{pmatrix} + f'(1)\begin{pmatrix} 0 & 1 & -1 \\ 0 & 0 & 0 \\ 0 & 0 & 0 \end{pmatrix} + f(2)\begin{pmatrix} 0 & 0 & 1 \\ 0 & 0 & 1 \\ 0 & 0 & 1 \end{pmatrix}$$

4.25 (*a*)

$$(s\mathbf{I} - \mathbf{A})^{-1} = \begin{pmatrix} \dfrac{s-3}{(s-2)^2} & \dfrac{1}{(s-2)^2} & 0 \\ \dfrac{1}{(s-2)^2} & \dfrac{s-1}{(s-2)^2} & 0 \\ \dfrac{-1}{(s-2)^2} & \dfrac{1}{(s-2)^2} & \dfrac{1}{s-2} \end{pmatrix}$$

(*b*)

$$(s\mathbf{I} - \mathbf{A})^{-1} = \frac{1}{s-2}\begin{pmatrix} 1 & 0 & 0 \\ 0 & 1 & 0 \\ 0 & 0 & 1 \end{pmatrix} + \frac{1}{(s-2)^2}\begin{pmatrix} -1 & 1 & 0 \\ -1 & 1 & 0 \\ -1 & 1 & 0 \end{pmatrix}$$

4.27 (*b*)

$$e^{\mathbf{A}t}=\begin{pmatrix} e^{t} & e^{2t}-e^{t} & (e^{3t}+2e^{2t}-3e^{t})/2 \\ 0 & e^{2t} & e^{2t}-e^{3t} \\ 0 & 0 & 3e^{2t}+e^{3t} \end{pmatrix}$$

4.33 (*b*) $\lambda=0,0,1,-1$
(*c*) $e^{\mathbf{Q}t}=\mathbf{E}_{10}^{\mathbf{Q}}+t\mathbf{E}_{11}^{\mathbf{Q}}+e^{-t}\mathbf{E}_{20}^{\mathbf{Q}}+e^{t}\mathbf{E}_{30}^{\mathbf{Q}},\ \ \mathbf{y}(t)=e^{\mathbf{Q}t}\mathbf{y}(0)$

Chapter 5

5.5 (*b*)

$$\mathbf{Q}_{\mathfrak{X}}=\begin{pmatrix} 1 & 0 \\ 0 & 3 \end{pmatrix}$$

5.7 (*b*) $\langle\mathbf{x},\mathbf{y}\rangle=\mathbf{y}^{\mathrm{T}}\mathbf{Q}_{\mathcal{E}}\,\mathbf{x}$, where

$$\mathbf{Q}_{\mathcal{E}}=\tfrac{1}{4}\begin{pmatrix} 3 & -1 & -1 \\ -1 & 3 & -1 \\ -1 & -1 & 3 \end{pmatrix}$$

5.8 $\langle\mathbf{x},\mathbf{y}\rangle=\mathbf{y}^{\mathrm{T}}\mathbf{Q}\,\mathbf{x}$, where

$$\mathbf{Q}=\begin{pmatrix} 6 & -3 & -4 \\ -3 & 2 & 2 \\ -4 & 2 & 3 \end{pmatrix}$$

5.14 $\{(1\ \ 1\ \ 0)^{\mathrm{T}}/\sqrt{2}\,,\ (1\ \ -1\ \ 1)^{\mathrm{T}}/\sqrt{3}\,\}$
5.17 $\mathcal{W}^{\perp}=\text{span}\ \{\mathbf{g}_2(t)=t,\mathbf{g}_3(t)=t^2-\frac{3}{5}\}$
5.20 $(\mathbf{T}^*\mathbf{g})(t)=b(t)\int_t^1\mathbf{g}(s)\,ds$
5.21 $(\mathbf{T}^*\mathbf{y})(s)=\mathbf{Q}(s)^{\mathrm{T}}\mathbf{y}$
5.22 (*a*) $\mathbf{T}^*\mathbf{y}=\mathbf{Q}^{-1}\mathbf{A}^{\mathrm{T}}\mathbf{R}^{\mathrm{T}}\mathbf{y}$
(*b*) $\mathbf{Q}=\mathbf{R}, \mathbf{A}$ square, $\mathbf{QA}=(\mathbf{QA})^{\mathrm{T}}$
5.25 $\mathbf{L}^*\mathbf{g}=\mathbf{g}''+\mathbf{g}',\mathbf{g}'(b)+2\mathbf{g}(b)=\mathbf{g}'(a)+2\mathbf{g}(a)=0$
5.29 (*a*) $\|\mathbf{A}\|=\sqrt{15}$
(*c*) $\|\mathbf{A}\|_E=\sqrt{15}$
(*d*) $\|\mathbf{T}^{-1}\|=[\frac{1}{4}+(\pi/b)^2]^{-1}$
5.30 (*c*) $\mathbf{Ux}\triangleq\sqrt{\mathbf{Q}}\,\mathbf{x}$, where

$$\sqrt{\mathbf{Q}}=(\sqrt{2}/2)\begin{pmatrix} 5 & 1 \\ 1 & 5 \end{pmatrix}$$

5.31 (*c*) $\mathbf{E}_{i0}^{\mathbf{A}}=\Sigma \mathbf{x}_j \mathbf{y}_j^{\mathrm{T}}$, where the sum is taken over all the dyads that correspond to the eigenvalue associated with the ith constituent matrix

5.34 (*b*)

$$\mathbf{f}(t)=\sum_{k=1}^{\infty}\frac{2b^2/r_k^2}{b+\cos^2 r_k}\int_0^b \mathbf{u}(s)\sin\left(\frac{r_k s}{b}\right)ds\,\sin\left(\frac{r_k t}{b}\right)$$

where r_k is the kth root of $\tan r_k=-r_k/b$

5.35 (*c*)

$$\mathbf{i}_1(t)=\frac{\omega}{2\pi}\sum_{k=-\infty}^{\infty}(R+jk\omega L)^{-1}\int_0^{2\pi/\omega}\mathbf{e}(s)e^{-jk\omega s}\,ds\,e^{jk\omega t}$$

5.36 (*b*)

$$\mathbf{f}(s,t)=c-\frac{4ab}{\pi^2}\sum_{k=0}^{\infty}\sum_{m=0}^{\infty}\frac{\int_a^b \mathbf{u}(\sigma,\tau)\cos(m\pi\sigma/a)\cos(k\pi\tau/b)\,d\sigma\,d\tau}{m^2b^2+k^2a^2}$$
$$\times\cos\left(\frac{m\pi s}{a}\right)\cos\left(\frac{k\pi t}{b}\right)$$

where c is an arbitrary constant, and in the summation k and m are not zero simultaneously

5.38 (*b*) $\mathbf{y}_1=(1\ \ -1)^{\mathrm{T}}/\sqrt{2}$, $\mathbf{y}_2=(1\ \ 1)^{\mathrm{T}}/\sqrt{2}$

Chapter 6

6.3 (*b*) $b_0=27.683$, $b_1=0.11532$, $b_2=-0.3452$

6.4 $\dfrac{\text{trace}(\mathbf{A})}{n}\mathbf{I}$

6.6 $\xi_1=28-10e$, $\xi_2=18-6e$

6.7 $\xi_0=\frac{1}{2}$, $\xi_1=\frac{3}{4}$, $\xi_2=-\frac{1}{4}$

6.8 (*a*) $\mathbf{w}_j=\mathbf{T}\mathbf{v}_j$, $j=1,\ldots,n$
(*b*) $\mathbf{f}(t)\approx\mathbf{f}_a(t)=-t-\frac{7}{30}$

6.11 (*b*) $\int_0^1\mathbf{u}(s)\,(1-s)\,\mathrm{d}s=\mathbf{f}(1)$, $\int_0^1\mathbf{u}(s)\,\mathrm{d}s=\dot{\mathbf{f}}(1)$
(*c*) $\mathbf{u}_s(t)=6-12t$

6.12 $\mathbf{u}_s(t) = \dfrac{e(e-1)e^t - e}{e^2 - 3e + 1} \approx 19.9e^{-t} - 11.6$

6.14 (*a*) $\mathbf{A}^\dagger = (\mathbf{A}^T\mathbf{R}\mathbf{A})^{-1}\mathbf{A}^T\mathbf{R}$

6.16 (*a*) $\mathbf{A}^\dagger = \mathbf{Q}^{-1}\mathbf{A}^T(\mathbf{A}\mathbf{Q}^{-1}\mathbf{A}^T)^{-1}$

6.17 (*a*) $(\mathbf{T}^*\mathbf{z})(s) = \mathbf{Q}^{-1}\mathbf{B}^T e^{\mathbf{A}^T(t_f - s)}\mathbf{z}$

(*b*) $\mathbf{C} = \int_0^{t_f} e^{\mathbf{A}(t_f - s)}\mathbf{B}\mathbf{Q}^{-1}\mathbf{B}^T e^{\mathbf{A}^T(t_f - s)}\,ds$

(*d*)

$$\mathbf{u}(t) = (\mathbf{T}^\dagger\mathbf{y})(t) = \begin{pmatrix} 6-12t & 6-12t & 0 & 0 \\ 0 & 0 & 6-12t & 6-12t \end{pmatrix}\mathbf{y}$$

6.19 (*a*)

$$\mathbf{A}^\dagger = \tfrac{1}{6}\begin{pmatrix} 2 & -1 & 5 & -3 \\ -2 & 2 & -6 & 4 \\ 2 & -1 & 5 & -3 \end{pmatrix}$$

6.21 (*b*) $\mathbf{A}^\dagger$ = transpose of answer to Problem 6.19a

6.25

$$\mathbf{A}^\dagger = \tfrac{1}{27}\begin{pmatrix} 10 & -8 & 2 \\ -7 & 11 & 4 \\ -2 & 7 & 5 \end{pmatrix}$$

Chapter 7

7.2 (*b*)

$$d^2\mathbf{G}(\mathbf{x},\mathbf{h},\mathbf{k}) = (k_1 k_2)\begin{pmatrix} 6\xi_1 & 3\xi_2^2 \\ 3\xi_2^2 & 6\xi_1\xi_2 \end{pmatrix}\begin{pmatrix} h_1 \\ h_2 \end{pmatrix}$$

7.3 (*b*) $\nabla F(\mathbf{f}) = 2\mathbf{f}$

7.5 (*b*) $[\nabla F(\mathbf{f})](s,t) = e^{s-t}\mathbf{f}(s,t)$, assuming the standard inner product

7.6 $\dot{\mathbf{y}} = \dfrac{\partial \mathbf{G}}{\partial \mathbf{x}}(\mathbf{x}_0, \mathbf{u}_0)\mathbf{y} + \dfrac{\partial \mathbf{G}}{\partial \mathbf{u}}(\mathbf{x}_0, \mathbf{u}_0)\mathbf{v}$, $\mathbf{y}(0) = \boldsymbol{\theta}$, where $\mathbf{y} \triangleq \mathbf{x} - \mathbf{x}_0$ and $\mathbf{v} \triangleq \mathbf{u} - \mathbf{u}_0$

7.8 (*a*) $\mathbf{G}'(\mathbf{x})^*\mathbf{G}(\mathbf{x}) = \mathbf{G}'(\mathbf{x})^*\mathbf{y}$

7.9 (*a*) $\ddot{\mathbf{f}} + (2b - 3a)\dot{\mathbf{f}} + 2a(a-b)\mathbf{f} = \boldsymbol{\theta}$

(*b*) $\mathbf{g}(t) = \dfrac{c(2b-a)}{1 - e^{(a-2b)t_f}}\, e^{2(a-b)t}$

7.12 The maximum is attained if the vector of coefficients $\mathbf{c}$ is an eigenvector corresponding to the largest eigenvalue of the correlation matrix $\mathbf{R}$; the Lagrange multiplier equals the largest eigenvalue

7.13 (*a*) $\mathbf{A}^T\mathbf{A}\mathbf{x}-\mathbf{A}^T\mathbf{y}=\mathbf{B}^T\boldsymbol{\lambda},\ \mathbf{B}\mathbf{x}=\mathbf{z}$

7.14 (*a*) $\mathbf{u}=-\mathbf{Q}^{-1}\mathbf{B}^T\boldsymbol{\lambda},\ \dot{\boldsymbol{\lambda}}=-\mathbf{A}^T\boldsymbol{\lambda}-\mathbf{R}(\mathbf{x}-\mathbf{x}_d),\ \boldsymbol{\lambda}(t_f)=\boldsymbol{\theta}$

(*b*) $\dot{\mathbf{P}}+\mathbf{P}\mathbf{A}+\mathbf{A}^T\mathbf{P}-\mathbf{P}\mathbf{B}\mathbf{Q}^{-1}\mathbf{B}^T\mathbf{P}+\mathbf{R}=\boldsymbol{\Theta},\ \mathbf{P}(t_f)=\boldsymbol{\Theta}$

7.15

$$-\dot{\boldsymbol{\lambda}}=\left(\frac{\partial \mathbf{G}}{\partial \mathbf{x}}\right)^T\boldsymbol{\lambda}+\left(\frac{\partial \mathbf{F}}{\partial \mathbf{x}}\right)^T,\ \boldsymbol{\lambda}(t_f)=\boldsymbol{\theta}$$

$$\boldsymbol{\theta}=\left(\frac{\partial \mathbf{G}}{\partial \mathbf{u}}\right)^T\boldsymbol{\lambda}+\left(\frac{\partial \mathbf{F}}{\partial \mathbf{u}}\right)^T$$

7.18 $\mathbf{x}=(2\ \ 1)^T,\ \boldsymbol{\lambda}=(0\ \ 1\ \ 3)^T,\ \boldsymbol{\mu}=(0\ \ 0)^T$

7.19 (*b*) $\xi_1=0,\ \xi_2=2,\ \lambda=40$ (constraint $\mathbf{G}$), $\mu_1=40,\ \mu_2=0$

(*c*) $\xi_1=\frac{20}{11},\ \xi_2=\frac{2}{11},\ \lambda=\frac{40}{11},\ \mu_1=\mu_i=0$ (a saddle point, not a minimum)

Chapter 8

8.1 $p_5=-0.351733711$ (for $p_0=1$)

8.2 $a\approx 0.2028,\ b\approx 0.9136$

8.11 (*a*) Eight multiplications

(*f*) $10n^2+221n$ (steepest descent)

$\frac{16}{3}n^3+15n^2+221n$ (Newton)

$13n^2+225n$ (DFP)

$10n^2+223n$ (conjugate gradient)

8.14 (*a*) $\nabla\mathbf{F}(\mathbf{x})=\mathbf{Q}^{-1}(\mathbf{A}\mathbf{x}-\mathbf{y})$

(*b*) $\alpha_k=\|\nabla\mathbf{F}(\mathbf{x}_k)\|^2_{\mathbf{Q}^{-1}}/\|\nabla\mathbf{F}(\mathbf{x}_k)\|^2_{\mathbf{Q}^{-1}\mathbf{A}\mathbf{Q}^{-1}}$

8.15 $[\mathbf{H}_1\nabla\mathbf{F}(\mathbf{u}_1)](t)=1-2e^t/(e+1)$

8.16 At the minimum, $\mathbf{x}=(0\ \ 8\ \ 2)^T$

8.19

$$\mathbf{x}=\left(c_1+\left(\frac{c_2-1}{c_2+1}\right)^2,\left(\frac{c_2-1}{c_2+1}\right)^2\right)$$

$$\boldsymbol{\lambda}=(1,4c_2/(c_2+1))$$

8.20 (*a*) The constrained minimum of the penalized objective function is $\mathbf{x}=(2,1)$

(*b*) Corresponding to (*a*) is $\boldsymbol{\lambda}=(1,4)$

Index